THEORY AND ANALYSIS
OF ELASTIC PLATES

Theory and Analysis of Elastic Plates

J. N. Reddy

University Distinguished Professor
Department of Mechanical Engineering
Texas A&M University, College Station
Texas, USA 77843–3123

USA	Publishing Office:	TAYLOR & FRANCIS 325 Chestnut Street Philadelphia, PA 19106 Tel: (215) 625-8900 Fax: (215) 625-2940
	Distribution Center:	TAYLOR & FRANCIS 47 Runway Road Suite G Levittown, PA 19057 Tel: (215) 269-0400 Fax: (215) 269-0363
UK		TAYLOR & FRANCIS LTD. 1 Gunpowder Square London EC4A 3DE Tel: 171 583 0490 Fax: 171 583 0581

THEORY AND ANALYSIS OF ELASTIC PLATES

Copyright © 1999 Taylor & Francis. All rights reserved. Printed in the United States of America. Except as permitted under the United States Copyright Act of 1976, no part of this publication may be reproduced or distributed in any form or by any means, or stored in a database or retrieval system, without the prior written permission of the publisher.

1 2 3 4 5 6 7 8 9 0

Edited by Edward A. Cilurso and James Reed. Printed by EdwardsBrothers, Ann Arbor, MI, 1998.

A CIP catalog record for this book is available from the British Library.
∞ The paper in this publication meets the requirements of the ANSI Standard Z39.48-1984 (Permanence of Paper)

Library of Congress Cataloging-in-Publication Data
Reddy, J. N. (Junuthula Naraimha), 1945-
    Theory and analysis of elastic plates / J.N. Reddy
       p.   cm.
  Includes bibliographical references and index.
  ISBN 1-56032-705-7 (cloth : alk. paper)
  1. Elastic plates and shells. I. Title
QA935 .R347 1999   98-48726
624.1'776--dc21   CIP

ISBN 1-56032-705-7 (case)

"Whence all creation had its origin,
he, whether he fashioned it or whether he did not,
he, who surveys it all from highest heaven,
he knows—or maybe even he does not know."

Rigveda

Contents

Preface xv

1 Basic Equations of Elasticity 1
   1.1 Introduction 1
   1.2 Vectors, Tensors, and Matrices 2
      1.2.1 Preliminary Comments 2
      1.2.2 Components of Vectors and Tensors 2
      1.2.3 Summation Convention 4
      1.2.4 The Del Operator 6
      1.2.5 Matrices and Cramer's Rule 13
   1.3 Transformation of Vector and Tensor Components 16
      1.3.1 Vector Transformations 16
      1.3.2 Tensor Transformations 18
   1.4 Equations of an Elastic Body 21
      1.4.1 Introduction 21
      1.4.2 Kinematics 21
      1.4.3 Compatibility Equations 26
      1.4.4 Stress and Its Measures 28
      1.4.5 Equations of Motion 31
      1.4.6 Constitutive Equations 33
   1.5 Transformation of Stresses, Strains, and Stiffnesses 38
      1.5.1 Introduction 38
      1.5.2 Transformation of Stress Components 40
      1.5.3 Transformation of Strain Components 41
      1.5.4 Transformation of Material Stiffnesses 41
   Exercises 43
   References for Additional Reading 46

2 Energy Principles and Variational Methods 47

2.1 Virtual Work 47
2.1.1 Introduction 47
2.1.2 Virtual Displacements and Forces 48
2.1.3 External and Internal Virtual Work 51
2.1.4 The Variational Operator 56
2.1.5 Functionals 57
2.1.6 Fundamental Lemma of Variational Calculus 58
2.1.7 Euler–Lagrange Equations 58

2.2 Energy Principles 61
2.2.1 Introduction 61
2.2.2 The Principle of Virtual Displacements 62
2.2.3 Hamilton's Principle 65
2.2.4 The Principle of Minimum Total Potential Energy ... 70

2.3 Variational Methods 72
2.3.1 Introduction 72
2.3.2 The Rayleigh–Ritz Method 73
2.3.3 The Galerkin Method 92

2.4 Summary 97

Exercises 97

References for Additional Reading 105

3 The Classical Theory of Plates 107

3.1 Introduction 107
3.2 Assumptions of the Theory 108
3.3 Displacement Field and Strains 110
3.4 Equations of Motion 113
3.5 Boundary and Initial Conditions 120
3.6 Plate Elastic Stiffnesses 126
3.7 Characterization of Orthotropic Materials 133
3.8 Equations of Motion in Terms of Displacements 137

Exercises 141

References for Additional Reading 142

4 Analysis of Plate Strips 143

4.1 Introduction 143
4.2 Governing Equations 143
4.3 Bending Analysis 145
    4.3.1 Without Axial Force 145
    4.3.2 With Axial Force 148
    4.3.3 Plate Strip on an Elastic Foundation 153

4.4 Buckling Under In-plane Compressive Load 155
    4.4.1 Introduction 155
    4.4.2 Simply Supported Plate Strips 157
    4.4.3 Clamped Plate Strips 158
    4.4.4 Other Boundary Conditions 159

4.5 Free Vibration 161
    4.5.1 General Formulation 161
    4.5.2 Simply Supported Plate Strips 164
    4.5.3 Clamped Plate Strips 165

4.6 Transient Analysis 167
    4.6.1 Preliminary Comments 167
    4.6.2 The Navier Solution 168
    4.6.3 The Rayleigh–Ritz Solution 169
    4.6.4 Transient Response 171

Exercises 174

References for Additional Reading 177

5 Analysis of Circular Plates 179

5.1 Introduction 179
5.2 Governing Equations 179
    5.2.1 Transformation of Equations from Rectangular
          Coordinates to Polar Coordinates 179
    5.2.2 Direct Derivation of Equations
          in Polar Coordinates 184

5.3 Axisymmetric Bending 192
    5.3.1 Governing Equations 192
    5.3.2 Analytical Solutions 194
    5.3.3 Rayleigh–Ritz Formulation 198
    5.3.4 Simply Supported Circular Plate Under
          Distributed Load 200

5.3.5 Simply Supported Circular Plate Under
Central Point Load 205
5.3.6 Annular Plate with Simply Supported Outer Edge . . 207
5.3.7 Clamped Circular Plate Under Distributed Load 212
5.3.8 Clamped Circular Plate Under Central Point Load . . 214
5.3.9 Annular Plates with Clamped Outer Edges 215
5.3.10 Circular Plates on Elastic Foundation.............. 220
5.3.11 Bending of Circular Plates Under Thermal Loads . . 223

5.4 Asymmetrical Bending..................................... 225
 5.4.1 General Solution 225
 5.4.2 General Solution of Circular Plates Under
  Linearly Varying Asymmetric Loading 226
 5.4.3 Clamped Plate Under Asymmetric Loading 228
 5.4.4 Simply Supported Plate Under
  Asymmetric Loading 229
 5.4.5 Circular Plates Under Non-central Point Load 230
 5.4.6 Rayleigh–Ritz Formulation 233

5.5 Free Vibration 238
 5.5.1 Introduction... 238
 5.5.2 General Analytical Solution.......................... 239
 5.5.3 Clamped Circular Plates............................. 240
 5.5.4 Simply Supported Circular Plates.................... 243
 5.5.5 Rayleigh–Ritz Solutions 244

Exercises... 247

References for Additional Reading 251

6 Bending of Simply Supported Rectangular Plates 253

6.1 Introduction... 253
 6.1.1 Governing Equations 253
 6.1.2 Boundary Conditions................................ 254

6.2 The Navier Solutions 256
 6.2.1 Solution Procedure 256
 6.2.2 Calculation of Bending Moments, Shear Forces,
  and Stresses.. 260
 6.2.3 Sinusoidally Loaded Plates 265
 6.2.4 Plates with Distributed and Point Loads............. 269
 6.2.5 Plates with Thermal Loads.......................... 275

6.3 Lévy's Solutions 278
    6.3.1 Solution Procedure 278
    6.3.2 Analytical Solution 281
    6.3.3 Plates Under Distributed Transverse Loads 285
    6.3.4 Plates with Distributed Edge Moments 291
    6.3.5 An Alternate Form of the Lévy Solution 294
    6.3.6 The Rayleigh–Ritz Solutions 300

Exercises 305

References for Additional Reading 308

7 Bending of Rectangular Plates with Various Boundary Conditions 309

7.1 Introduction 309

7.2 Lévy Solutions 309
    7.2.1 Basic Equations 309
    7.2.2 Plates with Edges $x = 0, a$ Clamped (CCSS) 313
    7.2.3 Plates with Edge $x = 0$ Clamped and Edge $x = a$ Simply Supported (CSSS) 320
    7.2.4 Plates with Edge $x = 0$ Clamped and Edge $x = a$ Free (CFSS) 324
    7.2.5 Plates with Edge $x = 0$ Simply Supported and Edge $x = a$ Free (SFSS) 326
    7.2.6 Solution by the Method of Superposition 330

7.3 Approximate Solutions by the Rayleigh–Ritz Method 334
    7.3.1 Analysis of the Lévy Plates 334
    7.3.2 Formulation for General Plates 339
    7.3.3 Clamped Plates (CCCC) 344

Exercises 346

References for Additional Reading 352

8 General Buckling of Rectangular Plates 353

8.1 Buckling of Simply Supported Plates Under Compressive Loads 353
    8.1.1 Governing Equations 353
    8.1.2 The Navier Solution 354
    8.1.3 Biaxial Compression of a Plate 355
    8.1.4 Biaxial Loading of a Plate 357

8.1.5 Uniaxial Compression of a Rectangular Plate 358

8.2 Buckling of Plates Simply Supported Along Two Opposite Sides and Compressed in the Direction Perpendicular to Those Sides 364
    8.2.1 The Lévy Solution...................................364
    8.2.2 Buckling of SSSF Plates............................366
    8.2.3 Buckling of SSCF Plates............................367
    8.2.4 Buckling of SSCC Plates 371

8.3 Buckling of Rectangular Plates Using the Rayleigh–Ritz Method 374
    8.3.1 Analysis of the Lévy Plates........................374
    8.3.2 General Formulation................................376
    8.3.3 Buckling of a Simply Supported Plate Under Combined Bending and Compression................379
    8.3.4 Buckling of a Simply Supported Plate Under In-plane Shear......................................382
    8.3.5 Buckling of Clamped Plates Under In-plane Shear...385

Exercises...388

References for Additional Reading 390

9 Dynamic Analysis of Rectangular Plates 391

9.1 Introduction..391
    9.1.1 Governing Equations 391
    9.1.2 Natural Vibration 391
    9.1.3 Transient Analysis..................................392

9.2 Natural Vibrations of Simply Supported Plates............392

9.3 Natural Vibration of Plates with Two Parallel Sides Simply Supported 395
    9.3.1 The Lévy Solution..................................395
    9.3.2 Analytical Solution 396
    9.3.3 Vibration of SSSF Plates............................397
    9.3.4 Vibration of SSCF Plates...........................399
    9.3.5 Vibration of SSCC Plates...........................402
    9.3.6 Vibration of SSCS Plates...........................404
    9.3.7 Vibration of SSFF Plates...........................406
    9.3.8 The Rayleigh–Ritz Solutions.......................408

9.4 Natural Vibration of Plates with General
  Boundary Conditions................................... 410
  9.4.1 The Rayleigh–Ritz Solution........................ 410
  9.4.2 Simply Supported Plates (SSSS).................... 411
  9.4.3 Clamped Plates (CCCC) 413
  9.4.4 CCCS Plates...................................... 415
  9.4.5 CSCS Plates 415
  9.4.6 CFCF, CCFF, and CFFF Plates.................... 416

9.5 Transient Analysis....................................... 419
  9.5.1 Spatial Variation of the Solution................... 419
  9.5.2 Time Integration 421

Exercises.. 422

References for Additional Reading........................... 423

10 Analysis of Rectangular Plates Using Shear Deformation Plate Theories................................ 425

10.1 First-Order Shear Deformation Plate Theory............. 425
  10.1.1 Preliminary Comments............................ 425
  10.1.2 Kinematics...................................... 426
  10.1.3 Equations of Motion............................. 428
  10.1.4 Plate Constitutive Equations 432
  10.1.5 Equations of Motion in Terms of Displacements.... 433

10.2 The Navier Solutions of FSDT......................... 434
  10.2.1 General Solution 434
  10.2.2 Bending Analysis................................ 436
  10.2.3 Buckling Analysis 440
  10.2.4 Natural Vibration 443

10.3 Third-Order Plate Theories............................ 446
  10.3.1 General Comments 446
  10.3.2 Displacement Field 446
  10.3.3 Strains and Stresses 448
  10.3.4 Equations of Motion............................. 449

10.4 The Navier Solutions of TSDT......................... 454
  10.4.1 Preliminary Comments........................... 454
  10.4.2 General Solution 455
  10.4.3 Bending Analysis................................ 457
  10.4.4 Buckling Analysis 461
  10.4.5 Natural Vibration 462

Exercises 463
References for Additional Reading 466

11 Finite Element Models of Beams and Plates 469

11.1 Introduction 469
11.2 Finite Element Models of Beams 470
    11.2.1 Euler–Bernoulli Beam Elements 470
    11.2.2 Timoshenko Beam Elements 484

11.3 Finite Element Models of CPT 494
    11.3.1 Introduction 494
    11.3.2 General Formulation 494
    11.3.3 Plate Bending Elements 497
    11.3.4 Fully Discretized Finite Element Models 501
    11.3.5 Numerical Results 502

11.4 Finite Element Models of FSDT 507
    11.4.1 Virtual Work Statements 507
    11.4.2 Lagrange Interpolation Functions 508
    11.4.3 Finite Element Model 512
    11.4.4 Numerical Results 515

Exercises 524
References for Additional Reading 531

Subject Index 535

Preface

The objective of this book is to present a complete and up-to-date treatment of classical as well as shear deformation plate theories and their solutions by analytical as well as numerical methods. *Beams* and *plates* are common structural elements of most engineering structures, including aerospace, automotive, and civil engineering structures, and their study, both from theoretical and analysis points of view, is fundamental to the understanding of the behavior of such structures.

There exists a number of books on the theory of plates, and most of them cover the classical, Kirchhoff plate theory in detail and present the Navier solutions of the theory for rectangular plates. Much of the latest developments in shear deformation plate theories and their finite element models have not been compiled in a textbook form. The present book is aimed at filling this void in the literature.

The motivation that led to the writing of the present book has come from many years of the author's research in the development of shear deformation plate theories and their analysis by the finite element method, and also from the fact that there does not exist a book that contains a detailed coverage of shear deformation beam and plate theories, analytical solutions, and finite element models in one volume. The present book fulfills the need for a complete treatment of the classical and shear deformation theories of plates and their solution by analytical and numerical methods.

Some mathematical preliminaries, equations of elasticity, and virtual work principles and variational methods are reviewed in Chapters 1 and 2. A reader who has had a course in elasticity or energy and variational principles of mechanics may skip these chapters and go directly to Chapter 3, where a complete derivation of the equations of motion of the classical plate theory (CPT) is presented. Solutions for cylindrical bending, buckling, natural vibration, and transient response of plate strips are developed in Chapter 4. A detailed treatment of circular plates is undertaken in Chapter 5, and analytical and Rayleigh–Ritz solutions of axisymmetric and asymmetric bending are presented for various boundary conditions and loads. A brief discussion of natural vibrations of circular plates is also included here.

Chapter 6 is dedicated to the bending of rectangular plates with all edges simply supported, and the Navier and Rayleigh–Ritz solutions are presented. Bending of rectangular plates with general boundary conditions are treated in Chapter 7. The Lévy solutions are presented for rectangular plates with two parallel edges simply supported while the other two have arbitrary boundary conditions; the Rayleigh–Ritz solutions are presented for rectangular plates with arbitrary conditions. General buckling of rectangular plates under various boundary conditions is presented in Chapter 8. The Navier, Lévy, and Rayleigh–Ritz solutions are developed here. Chapter 9 is devoted to the dynamic analysis of rectangular plates, where solutions are developed for free vibration and transient response.

The first-order and third-order shear deformation plate theories are treated in Chapter 10. Analytical solutions presented in these chapters are limited to rectangular plates with simply supported boundary conditions on all four edges (the Navier solution). Parametric effects of the material orthotropy and plate aspect ratio on bending deflections and stresses, buckling loads, and vibration frequencies are discussed. Finally, Chapter 11 deals with the linear finite element analysis of beams and plates. Finite element models based on both classical and first-order shear deformation plate theories are developed and numerical results are presented.

The book is suitable as a textbook for a first course on theory of plates in civil, aerospace, mechanical, and mechanics curricula. It can be used as a reference by engineers and scientists working in industry and academia. An introductory course on mechanics of materials and elasticity should prove to be helpful but not necessary because a review of the basics is included in the first two chapters of this book.

The author's research in the area of plates over the years has been supported through research grants from the Air Force Office of Scientific Research (AFOSR), the Army Research Office (ARO), and the Office of Naval Research (ONR). The support is gratefully acknowledged. The author also wishes to express his appreciation to Dr. Filis T. Kokkinos for his help with the illustrations in this book.

J. N. Reddy
College Station, Texas

Chapter One

Basic Equations of Elasticity

1.1 Introduction

The primary objective of this book is to study theories and analytical and numerical solutions of plate structures, i.e., structural elements undergoing bending. The bending theories considered in this book are valid for *thin* and *moderately thick* plates. These 2D plate theories are obtained from the equations of 3D elasticity by integrating them through the thickness.

The governing equations of plates can be derived by either vector mechanics or energy and variational principles. In vector mechanics, forces and moments on a typical element of the plate are summed to obtain the equations of equilibrium or motion. In energy methods, principles of virtual work or their derivatives, such as the principles of minimum total potential energy or maximum complementary energy, are used to obtain the equations. While both methods can give the same equations, the energy methods have the advantage of providing information on the form of the boundary conditions. The energy and variational principles also provide a natural means of determining solutions to the governing equations. Hence, the variational approach is adopted in the present study to derive the governing equations of plates.

In order to study theories of plates, a good understanding of the basic equations of elasticity and the concepts of work and energy is required. A study of these topics in turn requires familiarity with the notions of vectors, tensors, transformations of vector and tensor components, and matrices. Therefore, a brief review of vectors and tensors is presented first, followed by a review of the equations of elasticity. Readers familiar with these topics can skip the remaining portion of this chapter and go directly to Chapter 2, where the principles of virtual work are discussed.

1.2 Vectors, Tensors, and Matrices
1.2.1 Preliminary Comments

All quantities appearing in analytical descriptions of physical laws can be classified into two categories: *scalars* and *nonscalars*. The scalars are given by a single real or complex number. Nonscalar quantities not only need a specified magnitude, but also additional information, such as direction and/or differentiability. Time, temperature, volume, and mass density are examples of scalars. Displacement, temperature gradient, force, moment, and acceleration are examples of nonscalars.

The term *vector* is used to imply a nonscalar that has magnitude and "direction" and obeys the parallelogram law of vector addition and rules of scalar multiplication. A vector in modern mathematical analysis is an abstraction of the elementary notion of a *physical vector*, and it is 'an element from a linear vector space.' While the definition of a vector in abstract analysis does not require it to have a magnitude, in nearly all cases of practical interest it does, in which case the vector is said to belong to a 'normed vector space.' In this book we only need vectors from a special normed vector space – that is, physical vectors.

Not all nonscalar quantities are vectors. Some quantities require the specification of magnitude and two directions. For example, the specification of stress requires not only a force, but also an area upon which the force acts. A stress is a *tensor*. Vector is a tensor of order one, while stress is a second order tensor.

1.2.2 Components of Vectors and Tensors

In the analytical description of a physical phenomenon, a coordinate system in the chosen frame of reference is introduced and various physical quantities involved in the description are expressed in terms of measurements made in that system. The form of the equations thus depends upon the chosen coordinate system and may appear different in another type of coordinate system. The laws of nature, however, should be independent of the choice of a coordinate system, and we may seek to represent the law in a manner independent of a particular coordinate system. A way of doing this is provided by vector and tensor notation. When vector notation is used, a particular coordinate system need not be introduced. Consequently, use of vector notation in formulating natural laws leaves them *invariant* to coordinate transformations.

Often a specific coordinate system is chosen to express governing equations of a problem that facilitates their solution. Then the vector and tensor quantities are expressed in terms of their components in that coordinate system. For example, a vector $\mathbf{A}$ in a three-dimensional space may be expressed in terms of its components a_1, a_2, and a_3 and *basis vectors* $\mathbf{e}_1$, $\mathbf{e}_2$, and $\mathbf{e}_3$ as

$$\mathbf{A} = a_1 \mathbf{e}_1 + a_2 \mathbf{e}_2 + a_3 \mathbf{e}_3 \tag{1.2.1}$$

If the basis vectors of a coordinate system are constants, i.e., with fixed lengths and directions, the coordinate system is called a *Cartesian coordinate system*. The general Cartesian system is oblique. When the Cartesian system is orthogonal, it is called *rectangular Cartesian*. When the basis vectors are of unit length and mutually orthogonal, they are called *orthonormal*. We denote an orthonormal Cartesian basis by

$$(\hat{\mathbf{e}}_1, \hat{\mathbf{e}}_2, \hat{\mathbf{e}}_3) \quad \text{or} \quad (\hat{\mathbf{e}}_x, \hat{\mathbf{e}}_y, \hat{\mathbf{e}}_z) \tag{1.2.2}$$

The Cartesian coordinates are denoted by

$$(x_1, x_2, x_3) \quad \text{or} \quad (x, y, z) \tag{1.2.3}$$

The familiar rectangular Cartesian coordinate system is shown in Figure 1.2.1. We shall always use a right-hand coordinate system.

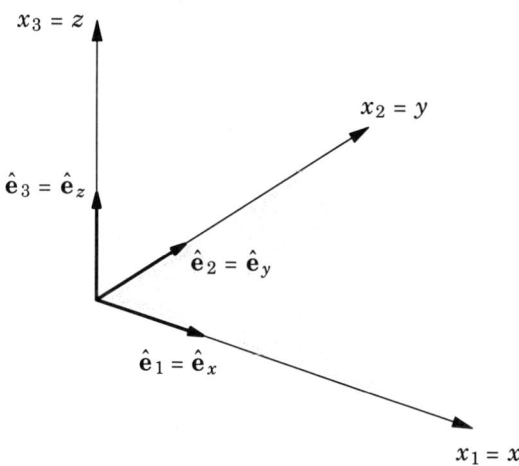

Figure 1.2.1. A rectangular Cartesian coordinate system, $(x_1, x_2, x_3) = (x, y, z)$; $(\hat{\mathbf{e}}_1, \hat{\mathbf{e}}_2, \hat{\mathbf{e}}_3) = (\hat{\mathbf{e}}_x, \hat{\mathbf{e}}_y, \hat{\mathbf{e}}_z)$ are the unit basis vectors.

4   THEORY AND ANALYSIS OF ELASTIC PLATES

A second-order tensor can be expressed in terms of its rectangular Cartesian system as

$$\begin{aligned}\boldsymbol{\Phi} = {}& \varphi_{11}\hat{\mathbf{e}}_1\hat{\mathbf{e}}_1 + \varphi_{12}\hat{\mathbf{e}}_1\hat{\mathbf{e}}_2 + \varphi_{13}\hat{\mathbf{e}}_1\hat{\mathbf{e}}_3 \\ & + \varphi_{21}\hat{\mathbf{e}}_2\hat{\mathbf{e}}_1 + \varphi_{22}\hat{\mathbf{e}}_2\hat{\mathbf{e}}_2 + \varphi_{23}\hat{\mathbf{e}}_2\hat{\mathbf{e}}_3 \\ & + \varphi_{31}\hat{\mathbf{e}}_3\hat{\mathbf{e}}_1 + \varphi_{32}\hat{\mathbf{e}}_3\hat{\mathbf{e}}_2 + \varphi_{33}\hat{\mathbf{e}}_3\hat{\mathbf{e}}_3\end{aligned} \quad (1.2.4)$$

Here we have selected a rectangular Cartesian basis to represent the tensor. The first and second order tensors (i.e., vectors and dyads) will be of greatest utility in the present study.

1.2.3 Summation Convention

It is convenient to abbreviate a summation of terms by understanding that a once repeated index means summation over all values of that index. For example, the component form of vector $\mathbf{A}$

$$\mathbf{A} = a^1\mathbf{e}_1 + a^2\mathbf{e}_2 + a^3\mathbf{e}_3 \quad (1.2.5)$$

where $(\mathbf{e}_1, \mathbf{e}_2, \mathbf{e}_3)$ are basis vectors (not necessarily unit), can be expressed in the form

$$\mathbf{A} = \sum_{j=1}^{3} a^j\mathbf{e}_j = a^j\mathbf{e}_j \quad (1.2.6)$$

The repeated index is a *dummy index* in the sense that any other symbol that is not already used in that expression can be used:

$$\mathbf{A} = a^j\mathbf{e}_j = a^k\mathbf{e}_k = a^m\mathbf{e}_m \quad (1.2.7)$$

The range of summation is always known in the context of the discussion. For example, in the present context the range of j, k, and m is 1 to 3, because we are discussing vectors in a three-dimensional space.

A second-order tensor $\mathbf{P}$ can be expressed in a short form using the summation convention

$$\mathbf{P} = P_{ij}\hat{\mathbf{e}}_i\hat{\mathbf{e}}_j \quad (1.2.8)$$

and an nth-order tensor has the form

$$\boldsymbol{\Phi} = \varphi_{ijk\ell\ldots}\hat{\mathbf{e}}_i\hat{\mathbf{e}}_j\hat{\mathbf{e}}_k\hat{\mathbf{e}}_\ell \cdots \quad (1.2.9)$$

A unit second-order tensor **I** is defined by

$$\mathbf{I} = \delta_{ij}\hat{\mathbf{e}}_i\hat{\mathbf{e}}_j \tag{1.2.10}$$

where δ_{ij}, called the *Kronecker delta*, is defined as

$$\delta_{ij} = \begin{cases} 0, & \text{if } i \neq j \\ 1, & \text{if } i = j \end{cases} \tag{1.2.11}$$

for any values of i and j. Note that $\delta_{ij} = \delta_{ji}$.

In Eqs. (1.2.8)–(1.2.10) we have chosen a rectangular Cartesian basis to represent the tensors. First and second order tensors (i.e., vectors and dyads) will be of greatest utility in the present study.

The *dot product* (or *scalar product*) and *cross product* (or *vector product*) of vectors can be defined in terms of their components with the help of Kronecker delta symbol and

$$\mathcal{E}_{ijk} = \begin{cases} 1, & \text{if } i,j,k \text{ are in cyclic order and } i \neq j \neq k \\ -1, & \text{if } i,j,k \text{ are not in cyclic order and } i \neq j \neq k \\ 0, & \text{if any of } i,j,k \text{ are repeated} \end{cases} \tag{1.2.12}$$

The symbol $\mathcal{E}_{ijk}$ is called the *alternating symbol* (permutation symbol) or *alternating tensor*, since it is a Cartesian component of a third-order tensor.

In an orthonormal basis, the scalar and vector products can be expressed in the index form using the Kronecker delta $\delta_{ij} \equiv \hat{\mathbf{e}}_i \cdot \hat{\mathbf{e}}_j$ and the alternating symbols

$$\mathbf{A} \cdot \mathbf{B} = (A_i\hat{\mathbf{e}}_i) \cdot (B_j\hat{\mathbf{e}}_j) = A_iB_j\delta_{ij} = A_iB_i$$

$$\mathbf{A} \times \mathbf{B} = (A_i\hat{\mathbf{e}}_i) \times (B_j\hat{\mathbf{e}}_j) = A_iB_j\mathcal{E}_{ijk}\hat{\mathbf{e}}_k \tag{1.2.13}$$

Note that

$$\mathcal{E}_{ijk} = \mathcal{E}_{kij} = \mathcal{E}_{jki}; \quad \mathcal{E}_{ijk} = -\mathcal{E}_{jik} = -\mathcal{E}_{ikj} \tag{1.2.14}$$

That is, a cyclic permutation of the indices does not change the sign, while the interchange of any two indices will change the sign. Further, the Kronecker delta and the permutation symbol are related by the identity, known as the $\mathcal{E} - \delta$ *identity*

$$\mathcal{E}_{ijk}\mathcal{E}_{imn} = \delta_{jm}\delta_{kn} - \delta_{jn}\delta_{km} \tag{1.2.15}$$

The permutation symbol and the Kronecker delta prove to be very useful in proving vector identities and simplifying vector equations. The following example illustrates some of the uses of δ_{ij} and $\mathcal{E}_{ijk}$.

Example 1.2.1 _____

We wish to simplify the vector operation $(\mathbf{A} \times \mathbf{B}) \cdot (\mathbf{C} \times \mathbf{D})$ in an alternate vector form and thereby establish a vector identity. We begin with

$$\begin{aligned}
(\mathbf{A} \times \mathbf{B}) \cdot (\mathbf{C} \times \mathbf{D}) &= (A_i \hat{\mathbf{e}}_i) \times (B_j \hat{\mathbf{e}}_j) \cdot (C_m \hat{\mathbf{e}}_m) \times (D_n \hat{\mathbf{e}}_n) \\
&= (A_i B_j \mathcal{E}_{ijk} \hat{\mathbf{e}}_k) \cdot (C_m D_n \mathcal{E}_{mnp} \hat{\mathbf{e}}_p) \\
&= A_i B_j C_m D_n \mathcal{E}_{ijk} \mathcal{E}_{mnp} \delta_{kp} \\
&= A_i B_j C_m D_n \mathcal{E}_{ijk} \mathcal{E}_{mnk} \\
&= A_i B_j C_m D_n (\delta_{im} \delta_{jn} - \delta_{in} \delta_{jm}) \\
&= A_i B_j C_m D_n \delta_{im} \delta_{jn} - A_i B_j C_m D_n \delta_{in} \delta_{jm}
\end{aligned}$$

where we have used the $\mathcal{E} - \delta$ identity. Since $C_m \delta_{im} = C_i$ (or $A_i \delta_{im} = A_m$), and $A_i C_i = \mathbf{A} \cdot \mathbf{C}$, and so on, we can write

$$\begin{aligned}
(\mathbf{A} \times \mathbf{B}) \cdot (\mathbf{C} \times \mathbf{D}) &= A_i B_j C_i D_j - A_i B_j C_j D_i \\
&= A_i C_i B_j D_j - A_i B_j D_j C_i \\
&= (\mathbf{A} \cdot \mathbf{C})(\mathbf{B} \cdot \mathbf{D}) - (\mathbf{A} \cdot \mathbf{D})(\mathbf{B} \cdot \mathbf{C})
\end{aligned}$$

Although the above vector identity is established using an orthonormal basis, it holds in a general coordinate system.

___

1.2.4 The Del Operator

A position vector to an arbitrary point (x, y, z) or (x_1, x_2, x_3) in a body, measured from the origin, is given by

$$\begin{aligned}
\mathbf{r} &= x \hat{\mathbf{e}}_x + y \hat{\mathbf{e}}_y + z \hat{\mathbf{e}}_z \\
&= x_1 \hat{\mathbf{e}}_1 + x_2 \hat{\mathbf{e}}_2 + x_3 \hat{\mathbf{e}}_3
\end{aligned} \quad (1.2.16)$$

or, in summation notation, by

$$\mathbf{r} = x_j \hat{\mathbf{e}}_j \quad (1.2.17)$$

Consider a scalar field ϕ which is a function of the position vector, $\phi = \phi(\mathbf{r})$. The differential change is given by

$$d\phi = \frac{\partial \phi}{\partial x_i} dx_i \qquad (1.2.18)$$

The differentials dx_i are components of $d\mathbf{r}$, that is,

$$d\mathbf{r} = dx_i \hat{\mathbf{e}}_i$$

We would now like to write $d\phi$ in such a way that we account for the *direction* as well as the magnitude of $d\mathbf{r}$. Since $dx_i = d\mathbf{r} \cdot \hat{\mathbf{e}}_i$, we can write

$$d\phi = d\mathbf{r} \cdot \left(\hat{\mathbf{e}}_i \frac{\partial \phi}{\partial x_i} \right)$$

Let us now denote the magnitude of $d\mathbf{r}$ by $ds \equiv |d\mathbf{r}|$. Then $\hat{\mathbf{e}} = d\mathbf{r}/ds$ is a unit vector in the direction of $d\mathbf{r}$, and we have

$$\left(\frac{d\phi}{ds} \right)_{\hat{\mathbf{e}}} = \hat{\mathbf{e}} \cdot \left(\hat{\mathbf{e}}_i \frac{\partial \phi}{\partial x_i} \right) \qquad (1.2.19)$$

The derivative $(d\phi/ds)_{\hat{\mathbf{e}}}$ is called the *directional derivative of ϕ*. We see that *it is the rate of change of ϕ with respect to distance and that it depends on the direction $\hat{\mathbf{e}}$ in which the distance is taken*.

The vector that is scalar multiplied by $\hat{\mathbf{e}}$ can be obtained immediately whenever the scalar field is given. Because the magnitude of this vector is equal to the maximum value of the directional derivative, it is called the *gradient vector* and is denoted by grad ϕ:

$$\text{grad } \phi \equiv \hat{\mathbf{e}}_i \frac{\partial \phi}{\partial x_i} \qquad (1.2.20)$$

We interpret grad ϕ as some operator operating on ϕ, that is, grad $\phi = \nabla \phi$. This operator is denoted, in a general coordinate system (q^1, q^2, q^3) with basis $(\mathbf{e}^1, \mathbf{e}^2, \mathbf{e}^3)$, by

$$\nabla \equiv \mathbf{e}^1 \frac{\partial}{\partial q^1} + \mathbf{e}^2 \frac{\partial}{\partial q^2} + \mathbf{e}^3 \frac{\partial}{\partial q^3} = \mathbf{e}^i \frac{\partial}{\partial q^i} \qquad (1.2.21)$$

and is called the del operator. In rectangular Cartesian system, we have

$$\nabla \equiv \hat{\mathbf{e}}_x \frac{\partial}{\partial x} + \hat{\mathbf{e}}_y \frac{\partial}{\partial y} + \hat{\mathbf{e}}_z \frac{\partial}{\partial z} \qquad (1.2.22)$$

8   THEORY AND ANALYSIS OF ELASTIC PLATES

The del operator is a *vector differential* operator. However, it is important to note that the del operator has some of the properties of a vector, but it does not have them all because it is an operator. For instance $\nabla \cdot \mathbf{A}$ is a scalar, called the *divergence* of $\mathbf{A}$, whereas $\mathbf{A} \cdot \nabla$ is a scalar *differential operator*. Thus the del operator does not commute in this sense. The operation $\nabla \times \mathbf{A}$ is called the *curl* of $\mathbf{A}$.

Example 1.2.2

Here we illustrate the use of index notation and the del operator to rewrite the following vector expressions in alternative forms:

(a) grad $(\mathbf{r} \cdot \mathbf{A})$.

(b) div $(\mathbf{r} \times \mathbf{A})$.

(c) curl $(\mathbf{r} \times \mathbf{A})$.

where $\mathbf{A} = \mathbf{A}(\mathbf{r})$ is a vector and $\mathbf{r}$ is the position vector. For simplicity we use the rectangular Cartesian basis.

(a) Consider

$$\begin{aligned}
\nabla(\mathbf{r} \cdot \mathbf{A}) &= \left(\hat{\mathbf{e}}_i \frac{\partial}{\partial x_i}\right)(x_j \hat{\mathbf{e}}_j \cdot A_k \hat{\mathbf{e}}_k) \\
&= \left(\hat{\mathbf{e}}_i \frac{\partial}{\partial x_i}\right)(x_j A_j) \\
&= \hat{\mathbf{e}}_i \left(\frac{\partial x_j}{\partial x_i} A_j + x_j \frac{\partial A_j}{\partial x_i}\right) \\
&= \hat{\mathbf{e}}_i (\delta_{ij} A_j + x_j A_{j,i}) \\
&= \hat{\mathbf{e}}_i A_i + \hat{\mathbf{e}}_i x_j A_{j,i}
\end{aligned}$$

The first expression is clearly equal to $\mathbf{A}$. A close examination of the second expression on the right side of the equality indicates that the basis vector goes with the differential operator because both have the same subscript i. Also x_j and A_j have the same subscript, implying that there is a dot product operation between them. Hence we can write

$$\hat{\mathbf{e}}_i \frac{\partial A_j}{\partial x_i} x_j = (\nabla \mathbf{A}) \cdot \mathbf{r}$$

Thus we have

$$\nabla(\mathbf{r} \cdot \mathbf{A}) = \mathbf{A} + (\nabla \mathbf{A}) \cdot \mathbf{r}$$

Note that $\nabla \mathbf{A}$ is a second-order tensor with components $A_{j,i}\hat{\mathbf{e}}_i\hat{\mathbf{e}}_j$.

(b) Consider

$$\nabla \cdot (\mathbf{r} \times \mathbf{A}) = \left(\hat{\mathbf{e}}_i \frac{\partial}{\partial x_i}\right) \cdot (x_j\hat{\mathbf{e}}_j \times A_k\hat{\mathbf{e}}_k)$$

$$= \hat{\mathbf{e}}_i \cdot \left(\frac{\partial}{\partial x_i}\right)(x_j A_k \mathcal{E}_{jkm}\hat{\mathbf{e}}_m)$$

$$= \mathcal{E}_{jkm}\left(\frac{\partial x_j}{\partial x_i}A_k + x_j\frac{\partial A_k}{\partial x_i}\right)\hat{\mathbf{e}}_i \cdot \hat{\mathbf{e}}_m$$

$$= \mathcal{E}_{jkm}(\delta_{ij}A_k + x_j A_{k,i})\delta_{im}$$

$$= \mathcal{E}_{jkj}A_k + \mathcal{E}_{jkm}x_j A_{k,m}$$

Clearly the first term is zero on account of the repeated subscript in the permutation symbol. Now consider the second term

$$\mathcal{E}_{jkm}x_j A_{k,m} = -A_{k,m}\mathcal{E}_{mkj}x_j = -(\nabla \times \mathbf{A}) \cdot \mathbf{r}$$

Thus we have

$$\nabla \cdot (\mathbf{r} \times \mathbf{A}) = -(\nabla \times \mathbf{A}) \cdot \mathbf{r}$$

(c) Consider

$$\nabla \times (\mathbf{r} \times \mathbf{A}) = \left(\hat{\mathbf{e}}_i \frac{\partial}{\partial x_i}\right) \times (x_j\hat{\mathbf{e}}_j \times A_k\hat{\mathbf{e}}_k)$$

$$= \hat{\mathbf{e}}_i \times \left(\frac{\partial}{\partial x_i}\right)(x_j A_k \mathcal{E}_{jkm}\hat{\mathbf{e}}_m)$$

$$= \mathcal{E}_{jkm}\left(\frac{\partial x_j}{\partial x_i}A_k + x_j\frac{\partial A_k}{\partial x_i}\right)\hat{\mathbf{e}}_i \times \hat{\mathbf{e}}_m$$

$$= \mathcal{E}_{jkm}(\delta_{ij}A_k + x_j A_{k,i})\mathcal{E}_{imn}\hat{\mathbf{e}}_n$$

$$= (\delta_{jn}\delta_{ki} - \delta_{ji}\delta_{kn})(\delta_{ij}A_k + x_j A_{k,i})\hat{\mathbf{e}}_n$$

$$= (A_n + x_n A_{i,i} - 3A_n - x_i A_{n,i})\hat{\mathbf{e}}_n$$

$$= \mathbf{r}(\nabla \cdot \mathbf{A}) - 2\mathbf{A} - (\mathbf{r} \cdot \nabla)\mathbf{A}$$

$$= \mathbf{r}\,\mathrm{div}\mathbf{A} - 2\mathbf{A} - \mathbf{r} \cdot \mathrm{grad}\mathbf{A}$$

In an orthogonal curvilinear coordinate system, the del operator can be expressed in terms of its orthonormal basis. Two commonly used orthogonal curvilinear coordinate systems are *cylindrical coordinates* and *spherical coordinates,* which are shown in Figure 1.2.2. Here we summarize the relationships between the rectangular Cartesian and the two orthogonal curvilinear coordinate systems and the forms of the del and Laplacian operators.

10   THEORY AND ANALYSIS OF ELASTIC PLATES

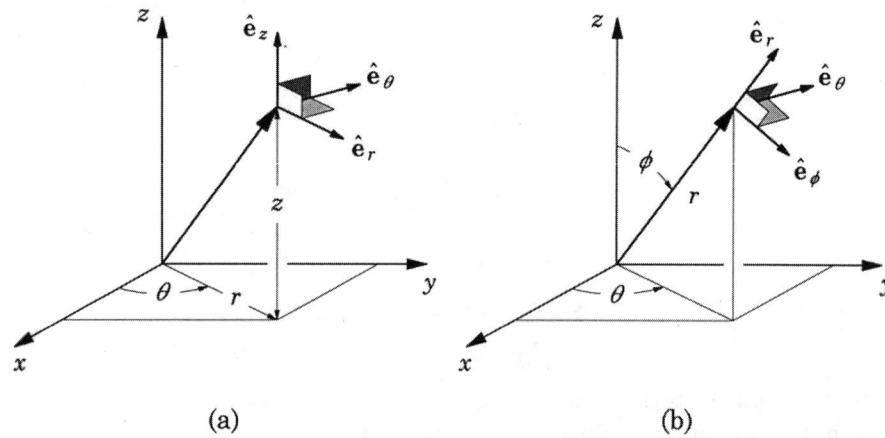

Figure 1.2.2. Orthogonal curvilinear coordinate systems: (a) cylindrical coordinate system; (b) spherical coordinate system.

Cylindrical Coordinates (r, θ, z)

The rectangular Cartesian coordinates (x, y, z) are related to the cylindrical coordinates (r, θ, z) by

$$x = r\cos\theta, \quad y = r\sin\theta, \quad z = z \tag{1.2.23}$$

and the inverse relations are

$$r = \sqrt{x^2 + y^2}, \quad \theta = \tan^{-1}\left(\frac{y}{x}\right), \quad z = z \tag{1.2.24}$$

The basis vectors are

$$\mathbf{e}_r = \hat{\mathbf{e}}_r, \quad \mathbf{e}_\theta = r\hat{\mathbf{e}}_\theta, \quad \mathbf{e}_z = \hat{\mathbf{e}}_z \tag{1.2.25}$$

In terms of the Cartesian basis we have

$$\begin{aligned}\hat{\mathbf{e}}_r &= \cos\theta\,\hat{\mathbf{e}}_x + \sin\theta\,\hat{\mathbf{e}}_y \\ \hat{\mathbf{e}}_\theta &= -\sin\theta\,\hat{\mathbf{e}}_x + \cos\theta\,\hat{\mathbf{e}}_y \\ \hat{\mathbf{e}}_z &= \hat{\mathbf{e}}_z\end{aligned} \tag{1.2.26}$$

The Cartesian basis in terms of the cylindrical orthonormal basis is

$$\begin{aligned}\hat{\mathbf{e}}_x &= \cos\theta\,\hat{\mathbf{e}}_r - \sin\theta\,\hat{\mathbf{e}}_\theta \\ \hat{\mathbf{e}}_y &= \sin\theta\,\hat{\mathbf{e}}_r + \cos\theta\,\hat{\mathbf{e}}_\theta \\ \hat{\mathbf{e}}_z &= \hat{\mathbf{e}}_z\end{aligned} \tag{1.2.27}$$

The position vector is given by

$$\mathbf{r} = r\hat{\mathbf{e}}_r + z\hat{\mathbf{e}}_z \tag{1.2.28}$$

The nonzero derivatives of the basis vectors are

$$\frac{\partial \hat{\mathbf{e}}_r}{\partial \theta} = -\sin\theta\,\hat{\mathbf{e}}_x + \cos\theta\,\hat{\mathbf{e}}_y = \hat{\mathbf{e}}_\theta$$

$$\frac{\partial \hat{\mathbf{e}}_\theta}{\partial \theta} = -\cos\theta\,\hat{\mathbf{e}}_x - \sin\theta\,\hat{\mathbf{e}}_y = -\hat{\mathbf{e}}_r \tag{1.2.29}$$

The del (∇) and Laplacian (∇^2) operators are given by

$$\nabla = \hat{\mathbf{e}}_r \frac{\partial}{\partial r} + \frac{\hat{\mathbf{e}}_\theta}{r} \frac{\partial}{\partial \theta} + \hat{\mathbf{e}}_z \frac{\partial}{\partial z} \tag{1.2.30}$$

$$\nabla^2 = \frac{1}{r}\left[\frac{\partial}{\partial r}\left(r\frac{\partial}{\partial r}\right) + \frac{1}{r}\frac{\partial^2}{\partial \theta^2} + r\frac{\partial^2}{\partial z^2}\right] \tag{1.2.31}$$

Then the gradient and divergence of a vector $\mathbf{u}$ in the cylindrical coordinate system can be obtained as (see Exercise 1.9)

$$\nabla\mathbf{u} = \frac{\partial u_r}{\partial r}\hat{\mathbf{e}}_r\hat{\mathbf{e}}_r + \frac{1}{r}\left(u_r + \frac{\partial u_\theta}{\partial \theta}\right)\hat{\mathbf{e}}_\theta\hat{\mathbf{e}}_\theta + \frac{\partial u_z}{\partial z}\hat{\mathbf{e}}_z\hat{\mathbf{e}}_z$$

$$+ \frac{1}{r}\frac{\partial u_z}{\partial \theta}\hat{\mathbf{e}}_\theta\hat{\mathbf{e}}_z + \frac{\partial u_\theta}{\partial z}\hat{\mathbf{e}}_z\hat{\mathbf{e}}_\theta + \frac{\partial u_r}{\partial z}\hat{\mathbf{e}}_z\hat{\mathbf{e}}_r + \frac{\partial u_z}{\partial r}\hat{\mathbf{e}}_r\hat{\mathbf{e}}_z$$

$$+ \frac{\partial u_\theta}{\partial r}\hat{\mathbf{e}}_r\hat{\mathbf{e}}_\theta + \frac{1}{r}\left(\frac{\partial u_r}{\partial \theta} - u_\theta\right)\hat{\mathbf{e}}_\theta\hat{\mathbf{e}}_r \tag{1.2.32}$$

$$\nabla \cdot \mathbf{u} = \left(\hat{\mathbf{e}}_r \frac{\partial}{\partial r} + \hat{\mathbf{e}}_\theta \frac{1}{r}\frac{\partial}{\partial \theta} + \hat{\mathbf{e}}_z \frac{\partial}{\partial z}\right) \cdot (u_r\hat{\mathbf{e}}_r + u_\theta\hat{\mathbf{e}}_\theta + u_z\hat{\mathbf{e}}_z)$$

$$= \hat{\mathbf{e}}_r \cdot \hat{\mathbf{e}}_r \frac{\partial u_r}{\partial r} + \hat{\mathbf{e}}_\theta \cdot \frac{\partial \hat{\mathbf{e}}_r}{\partial \theta}\frac{u_r}{r} + \hat{\mathbf{e}}_\theta \cdot \hat{\mathbf{e}}_\theta \frac{1}{r}\frac{\partial u_\theta}{\partial \theta} + \hat{\mathbf{e}}_z \cdot \hat{\mathbf{e}}_z \frac{\partial u_z}{\partial z}$$

$$= \frac{\partial u_r}{\partial r} + \frac{u_r}{r} + \frac{1}{r}\frac{\partial u_\theta}{\partial \theta} + \frac{\partial u_z}{\partial z}$$

$$= \frac{1}{r}\frac{\partial (ru_r)}{\partial r} + \frac{1}{r}\frac{\partial u_\theta}{\partial \theta} + \frac{\partial u_z}{\partial z} \tag{1.2.33}$$

Spherical Coordinates (r, ϕ, θ)

The rectangular Cartesian coordinates (x, y, z) are related to the spherical coordinates (r, ϕ, θ) by

$$\begin{aligned} x &= r \sin \phi \cos \theta \\ y &= r \sin \phi \sin \theta \\ z &= r \cos \phi \end{aligned} \quad (1.2.34)$$

and the inverse relations are

$$\begin{aligned} r &= \sqrt{x^2 + y^2 + z^2} \\ \phi &= \cos^{-1} \frac{z}{\sqrt{x^2 + y^2 + z^2}} \\ \theta &= \tan^{-1} \left(\frac{y}{x} \right) \end{aligned} \quad (1.2.35)$$

The basis vectors are

$$\mathbf{e}_r = \hat{\mathbf{e}}_r, \quad \mathbf{e}_\phi = \hat{\mathbf{e}}_\phi, \quad \mathbf{e}_\theta = r \sin \theta \, \hat{\mathbf{e}}_\theta \quad (1.2.36)$$

It is clear that the basis are not unit vectors in general. In terms of the Cartesian basis we have

$$\begin{aligned} \hat{\mathbf{e}}_r &= \sin \phi \cos \theta \, \hat{\mathbf{e}}_x + \sin \phi \sin \theta \, \hat{\mathbf{e}}_y + \cos \phi \, \hat{\mathbf{e}}_z \\ \hat{\mathbf{e}}_\phi &= \cos \phi \cos \theta \, \hat{\mathbf{e}}_x + \cos \phi \sin \theta \, \hat{\mathbf{e}}_y - \sin \phi \, \hat{\mathbf{e}}_z \\ \hat{\mathbf{e}}_\phi &= -\sin \theta \, \hat{\mathbf{e}}_x + \cos \theta \, \hat{\mathbf{e}}_y \end{aligned} \quad (1.2.37)$$

On the other hand, the Cartesian basis in terms of the spherical orthonormal basis is

$$\begin{aligned} \hat{\mathbf{e}}_x &= \sin \phi \cos \theta \hat{\mathbf{e}}_r + \cos \phi \cos \theta \, \hat{\mathbf{e}}_\phi - \sin \theta \, \hat{\mathbf{e}}_\theta \\ \hat{\mathbf{e}}_y &= \sin \phi \sin \theta \, \hat{\mathbf{e}}_r + \cos \phi \sin \theta \, \hat{\mathbf{e}}_\phi + \cos \phi \, \hat{\mathbf{e}}_\theta \\ \hat{\mathbf{e}}_z &= \cos \phi \, \hat{\mathbf{e}}_r - \sin \phi \, \hat{\mathbf{e}}_\phi \end{aligned} \quad (1.2.38)$$

The position vector is given by

$$\mathbf{r} = r \hat{\mathbf{e}}_r \quad (1.2.39)$$

and the nonzero derivatives of the basis vectors are given by

$$\frac{\partial \hat{\mathbf{e}}_r}{\partial \phi} = \hat{\mathbf{e}}_\phi, \qquad \frac{\partial \hat{\mathbf{e}}_r}{\partial \theta} = \sin \phi \, \hat{\mathbf{e}}_\phi, \qquad \frac{\partial \hat{\mathbf{e}}_\phi}{\partial \phi} = -\hat{\mathbf{e}}_r$$

$$\frac{\partial \hat{\mathbf{e}}_\phi}{\partial \theta} = \cos \phi \, \hat{\mathbf{e}}_\theta, \qquad \frac{\partial \hat{\mathbf{e}}_\theta}{\partial \theta} = -\sin \phi \, \hat{\mathbf{e}}_r - \cos \phi \, \hat{\mathbf{e}}_\phi$$

(1.2.40)

The del (∇) and Laplacian (∇^2) operators are given by

$$\nabla = \hat{\mathbf{e}}_r \frac{\partial}{\partial \phi} + \frac{\hat{\mathbf{e}}_\phi}{r} \frac{\partial}{\partial \phi} + \frac{1}{r \sin \phi} \hat{\mathbf{e}}_\theta \frac{\partial}{\partial \theta} \qquad (1.2.41)$$

$$\nabla^2 = \frac{1}{r^2} \frac{\partial}{\partial r}\left(r^2 \frac{\partial}{\partial r}\right) + \frac{1}{r^2 \sin \phi} \frac{\partial}{\partial \phi}\left(\sin \phi \frac{\partial}{\partial \phi}\right) + \frac{1}{r^2 \sin^2 \phi} \frac{\partial^2}{\partial \theta^2} \qquad (1.2.42)$$

1.2.5 Matrices and Cramer's Rule

A second-order tensor has nine components in any coordinate system. Equation (1.2.4) contains the Cartesian components of a tensor $\boldsymbol{\Phi}$. The form of the equation suggests writing down the scalars φ_{ij} (jth component in the ith row) in the rectangular array $[\Phi]$:

$$[\Phi] = \begin{bmatrix} \varphi_{11} & \varphi_{12} & \varphi_{13} \\ \varphi_{21} & \varphi_{22} & \varphi_{23} \\ \varphi_{31} & \varphi_{32} & \varphi_{33} \end{bmatrix}$$

The rectangular array of φs is called a *matrix* when it satisfies certain properties.

Two matrices are equal if and only if they are identical, i.e., all elements of the two arrays are the same. If a matrix has m rows and n columns, we will say that it is m by n ($m \times n$), the number of rows always being listed first. The element in the ith row and jth column of a matrix $[A]$ is generally denoted by a_{ij}, and we will sometimes write $[A] = [a_{ij}]$ to denote this. A square matrix is one that has the same number of rows as columns. The elements of a square matrix for which the row number and the column number are the same (i.e., a_{ij}) are called *diagonal elements* or simply the *diagonal*. A square matrix is said to be a *diagonal matrix* if all of the off-diagonal elements are zero. An *identity matrix*, denoted by $[I]$, is a diagonal matrix whose elements are all 1's. The elements of the identity matrix can be written as $[I] = [\delta_{ij}]$.

Let $[A] = [a_{ij}]$ be an $n \times n$ matrix. We wish to associate with $[A]$ a scalar that in some sense measures the "size" of $[A]$ and indicates whether $[A]$ is singular or nonsingular. The *determinant* of the matrix $[A] = [a_{ij}]$ is defined to be the scalar $\det[A] = |A|$ computed according to the rule

$$\det[A] = |a_{ij}| = \sum_{i=1}^{n}(-1)^{i+1}a_{i1}|A_{i1}| \qquad (1.2.43)$$

where $|A_{ij}|$ is the determinant of the $(n-1)\times(n-1)$ matrix that remains after deleting the ith row and the first column of $[A]$. For 1×1 matrices, the determinant is defined according to $|a_{11}| = a_{11}$. For convenience we define the determinant of a zeroth-order matrix to be unity. In the above definition, special attention is given to the first column of the matrix $[A]$. We call it the expansion of $|A|$ according to the first column of $[A]$. One can expand $|A|$ according to any column or row:

$$|A| = \sum_{i=1}^{n}(-1)^{i+j}a_{ij}|A_{ij}| \qquad (1.2.44)$$

where $|A_{ij}|$ is the determinant of the matrix obtained by deleting the ith row and jth column of matrix $[A]$.

For an $n \times n$ matrix $[A]$, the determinant of the $(n-1) \times (n-1)$ submatrix of $[A]$ obtained by deleting row i and column j of $[A]$ is called *minor* of a_{ij} and is denoted by $M_{ij}(A)$. The quantity $\text{cof}_{ij}(A) \equiv (-1)^{i+j}M_{ij}(A)$ is called the *cofactor* of a_{ij}. The determinant of $[A]$ can be cast in terms of the minor and cofactor of a_{ij} for any value of j:

$$\det[A] = \sum_{i=1}^{n} a_{ij}\,\text{cof}_{ij}(A) \qquad (1.2.45)$$

The *adjunct* (also called *adjoint*) of a matrix $[A]$ is the transpose of the matrix obtained from $[A]$ by replacing each element by its cofactor. The adjunct of $[A]$ is denoted by $\text{Adj}(A)$.

Let $[A]$ be an $n \times n$ square matrix with elements a_{ij}. The cofactor of a_{ij}, denoted here by A_{ij}, is related to the minor M_{ij} of a_{ij} by

$$A_{ij} = (-1)^{i+j}M_{ij} \qquad (1.2.46)$$

By definition, the determinant $|A|$ is given by

$$|A| = \sum_{k=1}^{n} a_{ik} A_{ik} = \sum_{k=1}^{n} a_{kj} A_{kj} \quad \text{(no sum on } i \text{ and } j\text{)} \tag{1.2.47}$$

for any value of i and j $(i, j \leq n)$. Then we have

$$\sum_{k=1}^{n} a_{rk} A_{ik} = |A| \, \delta_{ri}, \quad \sum_{k=1}^{n} a_{ks} A_{kj} = |A| \, \delta_{sj} \tag{1.2.48}$$

Using the above result, one can show that the solution to a set of n linear equations in n unknown quantities

$$\sum_{j=1}^{n} a_{ij} u_j = f_i, \quad i = 1, 2, \ldots, n \tag{1.2.49}$$

in the case when the determinant of the matrix of coefficients is not zero ($|A| \neq 0$) is given by

$$u_j = \frac{1}{|A|} \sum_{i=1}^{n} A_{ij} b_i, \quad j = 1, 2, \ldots, n \tag{1.2.50}$$

The result in Eq. (1.2.50) is known as *Cramer's rule*.

Example 1.2.3

Consider the system of equations

$$\begin{bmatrix} 6 & 3 & -1 \\ 1 & 4 & -3 \\ 2 & -2 & 5 \end{bmatrix} \begin{Bmatrix} u_1 \\ u_2 \\ u_3 \end{Bmatrix} = \begin{Bmatrix} f_1 \\ f_2 \\ f_3 \end{Bmatrix} = \begin{Bmatrix} 7 \\ -5 \\ 8 \end{Bmatrix}$$

Using Cramer's rule, we obtain ($\bar{u}_i = u_i |A|$)

$$\bar{u}_1 = \begin{vmatrix} f_1 & 3 & -1 \\ f_2 & 4 & -3 \\ f_3 & -2 & 5 \end{vmatrix}, \quad \bar{u}_2 = \begin{vmatrix} 6 & f_1 & -1 \\ 1 & f_2 & -3 \\ 2 & f_3 & 5 \end{vmatrix}, \quad \bar{u}_3 = \begin{vmatrix} 6 & 3 & f_1 \\ 1 & 4 & f_2 \\ 2 & -2 & f_3 \end{vmatrix}$$

where $|A|$ is the determinant of the coefficient matrix

$$|A| = \sum_{i=1}^{3} (-1)^{i+1} a_{i1} |A_{i1}|$$

$$= 6 \begin{vmatrix} 4 & -3 \\ -2 & 5 \end{vmatrix} - 3 \begin{vmatrix} 1 & -3 \\ 2 & 5 \end{vmatrix} + (-1) \begin{vmatrix} 1 & 4 \\ 2 & -2 \end{vmatrix}$$

$$= 6(20 - 6) - 3(5 + 6) - (-2 - 8) = 61$$

16   THEORY AND ANALYSIS OF ELASTIC PLATES

Hence the solution becomes

$$u_1 = \frac{1}{61}(14f_1 - 13f_2 - 5f_3) = \frac{123}{61}$$
$$u_2 = \frac{1}{61}(-11f_1 + 32f_2 + 17f_3) = -\frac{101}{61}$$
$$u_3 = \frac{1}{61}(-10f_1 + 18f_2 + 21f_3) = \frac{8}{61}$$

1.3 Transformation of Vector and Tensor Components

1.3.1 Vector Transformations

In structural analysis, one is required to refer all quantities used in the analytical description of a structure to a common coordinate system. Scalars, by definition, are independent of any coordinate system. While vectors and tensors are independent of a particular coordinate system, their components are not. The same vector can have different components in different coordinate systems. Any two sets of components of a vector can be related by writing one set of components in terms of the other. Such relationships are called transformations. To establish the rules of the transformation of vector components, we consider barred $(\bar{x}_1, \bar{x}_2, \bar{x}_3)$ and unbarred (x_1, x_2, x_3) coordinate systems that are related by the equations

$$x_1 = x_1(\bar{x}_1, \bar{x}_2, \bar{x}_3)$$
$$x_2 = x_2(\bar{x}_1, \bar{x}_2, \bar{x}_3)$$
$$x_3 = x_3(\bar{x}_1, \bar{x}_2, \bar{x}_3) \tag{1.3.1}$$

The inverse transformation is

$$\bar{x}_1 = \bar{x}_1(x_1, x_2, x_3)$$
$$\bar{x}_2 = \bar{x}_2(x_1, x_2, x_3)$$
$$\bar{x}_3 = \bar{x}_3(x_1, x_2, x_3) \tag{1.3.2}$$

The bases in the two systems are denoted $(\hat{e}_1, \hat{e}_2, \hat{e}_3)$ and $(\hat{\bar{e}}_1, \hat{\bar{e}}_2, \hat{\bar{e}}_3)$ (see Figure 1.3.1).

BASIC EQUATIONS OF ELASTICITY   17

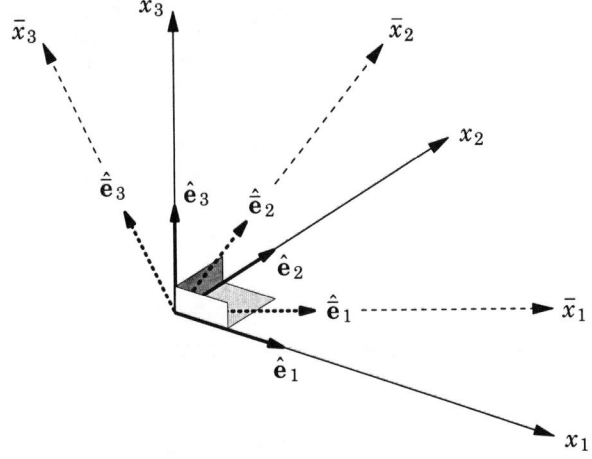

Figure 1.3.1. Unbarred and barred rectangular coordinate systems.

The relations between the bases of the barred and unbarred coordinate systems can be written as

$$\hat{\bar{\mathbf{e}}}_i = (\hat{\bar{\mathbf{e}}}_i \cdot \hat{\mathbf{e}}_k)\, \hat{\mathbf{e}}_k = a_{ik}\hat{\mathbf{e}}_k \qquad (1.3.3a)$$

In matrix notation, we can write Eq. (1.3.3a) as

$$\begin{Bmatrix} \hat{\bar{\mathbf{e}}}_1 \\ \hat{\bar{\mathbf{e}}}_2 \\ \hat{\bar{\mathbf{e}}}_3 \end{Bmatrix} = \begin{bmatrix} a_{11} & a_{12} & a_{13} \\ a_{21} & a_{22} & a_{23} \\ a_{31} & a_{32} & a_{33} \end{bmatrix} \begin{Bmatrix} \hat{\mathbf{e}}_1 \\ \hat{\mathbf{e}}_2 \\ \hat{\mathbf{e}}_3 \end{Bmatrix} \qquad (1.3.3b)$$

where [A] denotes the 3×3 matrix array whose elements are the direction cosines, a_{ij}. We also have the inverse relation

$$\hat{\mathbf{e}}_j = (\hat{\mathbf{e}}_j \cdot \hat{\bar{\mathbf{e}}}_k)\, \hat{\bar{\mathbf{e}}}_k = a_{kj}\hat{\bar{\mathbf{e}}}_k \qquad (1.3.4a)$$

or

$$\begin{Bmatrix} \hat{\mathbf{e}}_1 \\ \hat{\mathbf{e}}_2 \\ \hat{\mathbf{e}}_3 \end{Bmatrix} = \begin{bmatrix} a_{11} & a_{21} & a_{31} \\ a_{12} & a_{22} & a_{32} \\ a_{13} & a_{23} & a_{33} \end{bmatrix} \begin{Bmatrix} \hat{\bar{\mathbf{e}}}_1 \\ \hat{\bar{\mathbf{e}}}_2 \\ \hat{\bar{\mathbf{e}}}_3 \end{Bmatrix} \qquad (1.3.4b)$$

From Eqs. (1.3.3a) and (1.3.4a), we note that

$$a_{ik}a_{jk} = a_{ki}a_{kj} = \delta_{ij} \qquad (1.3.5a)$$

or
$$[A][A]^{\mathrm{T}} = [I] \tag{1.3.5b}$$

In other words, $[A]^{\mathrm{T}}$ is equal to its inverse. Such transformations are called orthogonal transformations and $[A]$ is called an orthogonal matrix.

The transformations (1.3.3) and (1.3.4) between two orthogonal sets of bases also hold for their respective coordinates:

$$\bar{x}_i = a_{ik} x_k\,, \qquad x_j = a_{kj} \bar{x}_k \tag{1.3.6}$$

Analogous to Eq. (1.3.6), the components of a vector **u** in the barred and unbarred coordinate systems are related by the expressions

$$\bar{u}_i = a_{ij} u_j\,, \qquad u_i = a_{ji} \bar{u}_j \tag{1.3.7}$$

1.3.2 Tensor Transformations

A second-order tensor can be expressed in two different coordinate systems using the corresponding bases. In a barred system, we have

$$\boldsymbol{\Phi} = \bar{\varphi}_{mn} \hat{\bar{\mathbf{e}}}_m \hat{\bar{\mathbf{e}}}_n \tag{1.3.8}$$

and in an unbarred system it is expressed as

$$\boldsymbol{\Phi} = \varphi_{ij} \hat{\mathbf{e}}_i \hat{\mathbf{e}}_j \tag{1.3.9}$$

Using Eq. (1.3.4) for $\hat{\mathbf{e}}_i$ and $\hat{\mathbf{e}}_j$ in Eq. (1.3.8), we arrive at the equation

$$\boldsymbol{\Phi} = \varphi_{ij} a_{mi} a_{nj} \hat{\bar{\mathbf{e}}}_m \hat{\bar{\mathbf{e}}}_n \tag{1.3.10}$$

Comparing Eq. (1.3.10) with Eq. (1.3.8), we arrive at the relation

$$\bar{\varphi}_{mn} = \varphi_{ij} a_{mi} a_{nj} \tag{1.3.11a}$$

or, in matrix form, we have

$$[\bar{\varphi}] = [A][\varphi][A]^T \tag{1.3.11b}$$

The inverse relation can be derived using the orthogonality property of $[A]$: $[A]^{-1} = [A]^T$. Premultiply both sides of Eq. (1.3.11b) with

$[A]^{-1} = [A]^T$ and postmultiply both sides of the resulting equation with $([A]^T)^{-1} = [A]$, and obtain the result

$$[\varphi] = [A]^T[\bar{\varphi}][A] \tag{1.3.12}$$

Equations (1.3.11b) and (1.3.12) are useful in transforming second-order tensors (e.g., stresses and strains) from one coordinate system to another coordinate system (see section 1.5).

Example 1.3.1

Suppose that $(\bar{x}_1, \bar{x}_2, \bar{x}_3)$ is obtained from (x_1, x_2, x_3) by rotating the x_1x_2-plane counter-clockwise by an angle θ about the x_3-axis (see Figure 1.3.2). Then the two sets of coordinates are related by

$$\begin{Bmatrix} \bar{x}_1 \\ \bar{x}_2 \\ \bar{x}_3 \end{Bmatrix} = \begin{bmatrix} \cos\theta & \sin\theta & 0 \\ -\sin\theta & \cos\theta & 0 \\ 0 & 0 & 1 \end{bmatrix} \begin{Bmatrix} x_1 \\ x_2 \\ x_3 \end{Bmatrix} \tag{1.3.13a}$$

and the inverse relation is given by

$$\begin{Bmatrix} x_1 \\ x_2 \\ x_3 \end{Bmatrix} = \begin{bmatrix} \cos\theta & -\sin\theta & 0 \\ \sin\theta & \cos\theta & 0 \\ 0 & 0 & 1 \end{bmatrix} \begin{Bmatrix} \bar{x}_1 \\ \bar{x}_2 \\ \bar{x}_3 \end{Bmatrix} \tag{1.3.13b}$$

Clearly, we have $[A]^{-1} = [A]^T$. The components $\bar{\varphi}_{ij}$ and φ_{ij} of a dyad (or a second-order tensor) Φ in the two coordinate systems, $(\bar{x}_1, \bar{x}_2, \bar{x}_3)$ and (x_1, x_2, x_3), respectively, are related by Eqs. (1.3.11b) and (1.3.12). Equation (1.3.11b) in the present case gives the relations

$$\begin{aligned}
\bar{\varphi}_{11} &= \varphi_{11}\cos^2\theta + \varphi_{22}\sin^2\theta + (\varphi_{12} + \varphi_{21})\cos\theta\sin\theta \\
\bar{\varphi}_{12} &= (\varphi_{22} - \varphi_{11})\cos\theta\sin\theta + \varphi_{12}\cos^2\theta - \varphi_{21}\sin^2\theta \\
\bar{\varphi}_{13} &= \varphi_{13}\cos\theta + \varphi_{23}\sin\theta \\
\bar{\varphi}_{21} &= (\varphi_{22} - \varphi_{11})\cos\theta\sin\theta + \varphi_{21}\cos^2\theta - \varphi_{12}\sin^2\theta \\
\bar{\varphi}_{22} &= \varphi_{11}\sin^2\theta + \varphi_{22}\cos^2\theta - (\varphi_{12} + \varphi_{21})\cos\theta\sin\theta \\
\bar{\varphi}_{23} &= \varphi_{23}\cos\theta - \varphi_{13}\sin\theta \\
\bar{\varphi}_{31} &= \varphi_{31}\cos\theta + \varphi_{32}\sin\theta \\
\bar{\varphi}_{32} &= \varphi_{32}\cos\theta - \varphi_{31}\sin\theta \\
\bar{\varphi}_{33} &= \varphi_{33}
\end{aligned} \tag{1.3.14}$$

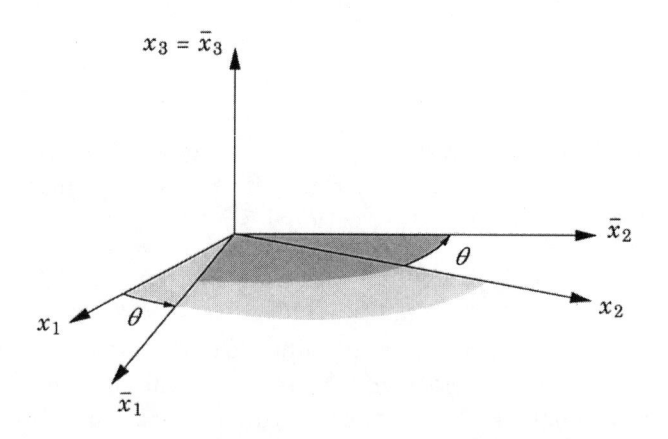

Figure 1.3.2. A special rotational transformation between unbarred and barred rectangular Cartesian coordinate systems.

The inverse relations are provided by Eq. (1.3.12):

$$\varphi_{11} = \bar{\varphi}_{11} \cos^2 \theta + \bar{\varphi}_{22} \sin^2 \theta - (\bar{\varphi}_{12} + \bar{\varphi}_{21}) \cos \theta \sin \theta$$
$$\varphi_{12} = (\bar{\varphi}_{11} - \bar{\varphi}_{22}) \cos \theta \sin \theta + \bar{\varphi}_{12} \cos^2 \theta - \bar{\varphi}_{21} \sin^2 \theta$$
$$\varphi_{13} = \bar{\varphi}_{13} \cos \theta - \bar{\varphi}_{23} \sin \theta$$
$$\varphi_{21} = (\bar{\varphi}_{11} - \bar{\varphi}_{22}) \cos \theta \sin \theta + \bar{\varphi}_{21} \cos^2 \theta - \bar{\varphi}_{12} \sin^2 \theta$$
$$\varphi_{22} = \bar{\varphi}_{11} \sin^2 \theta + \bar{\varphi}_{22} \cos^2 \theta + (\bar{\varphi}_{12} + \bar{\varphi}_{21}) \cos \theta \sin \theta$$
$$\varphi_{23} = \bar{\varphi}_{23} \cos \theta + \bar{\varphi}_{13} \sin \theta$$
$$\varphi_{31} = \bar{\varphi}_{31} \cos \theta - \bar{\varphi}_{32} \sin \theta$$
$$\varphi_{32} = \bar{\varphi}_{32} \cos \theta + \bar{\varphi}_{31} \sin \theta$$
$$\varphi_{33} = \bar{\varphi}_{33} \tag{1.3.15}$$

The transformation law (1.3.11a) is often taken to be the *definition* of a second-order tensor. In other words, $\mathbf{\Phi}$ is a second-order tensor if and only if its components transform according to Eq. (1.3.11a). In general, an nth order tensor transforms according to the formula

$$\bar{\varphi}_{mnpq\ldots} = \varphi_{ijk\ell\ldots} a_{mi} a_{nj} a_{pk} a_{q\ell} \cdots \tag{1.3.16}$$

1.4 Equations of an Elastic Body
1.4.1 Introduction

The objective of this section is to review the basic equations of an elastic body. The equations governing the motion and deformation of a solid body can be classified into four basic categories:

(1) Kinematics (strain-displacement equations)

(2) Kinetics (conservation of momenta)

(3) Thermodynamics (first and second laws of thermodynamics)

(4) Constitutive equations (stress-strain relations).

Kinematics is a study of the geometric changes or deformation in a body without the consideration of forces causing the deformation or the nature of the body. *Kinetics* is the study of the static or dynamic equilibrium of forces acting on a body. The thermodynamic principles are concerned with the conservation of energy and relations among heat, mechanical work, and thermodynamic properties of the body. The constitutive equations describe the constitutive behavior of the body and relate the dependent variables introduced in the kinetic description to those in the kinematic and thermodynamic descriptions. These equations are supplemented by appropriate boundary and initial conditions of the problem.

In the following sections, an overview of the kinematic, kinetic, and constitutive equations of an elastic body is presented. The thermodynamic principles are not reviewed as we will account for thermal effects only through constitutive relations.

1.4.2 Kinematics

The term *deformation* of a body refers to relative displacements and changes in the geometry experienced by the body. In a rectangular Cartesian frame of reference (X_1, X_2, X_3), every particle X in the body corresponds to a position $\mathbf{X} = (X_1, X_2, X_3)$. When the body is deformed, the particle X moves to a new position $\mathbf{x} = (x_1, x_2, x_3)$. If the displacement of every particle in the body is known, we can construct the deformed configuration $\mathcal{C}$ from the reference (or undeformed) configuration $\mathcal{C}^0$ (see Figure 1.4.1). In the Lagrangian description, the displacements are expressed in terms of the *material coordinates* $\mathbf{X}$

$$\mathbf{u}(\mathbf{X},t) = \mathbf{x}(\mathbf{X},t) - \mathbf{X} \qquad (1.4.1)$$

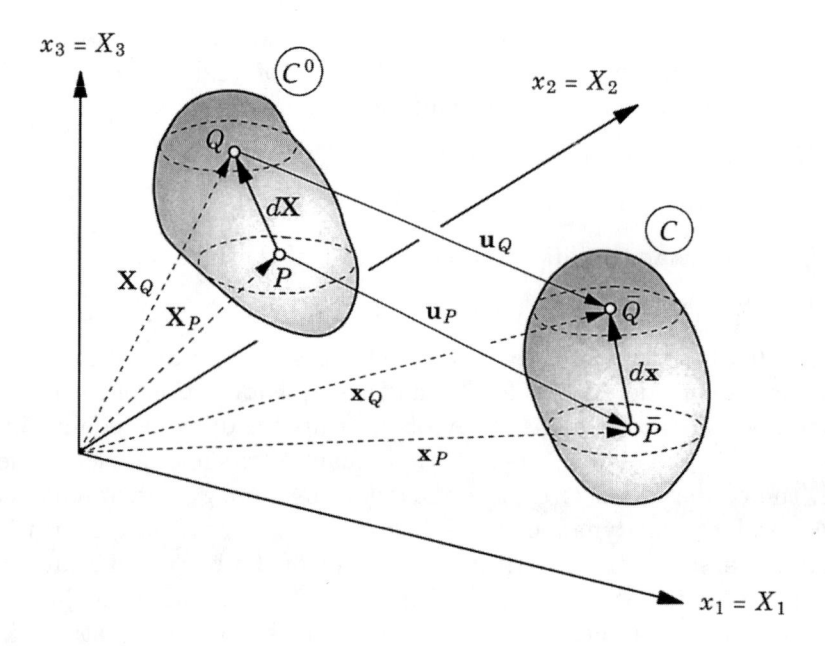

Figure 1.4.1. Undeformed and deformed configurations of a body.

The deformation of a body can be measured in terms of the strain tensor **E**, which is defined such that it gives the change in the square of the length of the material vector $d\mathbf{X}$

$$\begin{aligned}2d\mathbf{X} \cdot \mathbf{E} \cdot d\mathbf{X} &= d\mathbf{x} \cdot d\mathbf{x} - d\mathbf{X} \cdot d\mathbf{X} \\ &= d\mathbf{X} \cdot (\mathbf{I} + \nabla \mathbf{u}) \cdot [d\mathbf{X} \cdot (\mathbf{I} + \nabla \mathbf{u})] - d\mathbf{X} \cdot d\mathbf{X} \\ &= d\mathbf{X} \cdot \left[(\mathbf{I} + \nabla \mathbf{u}) \cdot (\mathbf{I} + \nabla \mathbf{u})^T - \mathbf{I}\right] \cdot d\mathbf{X} \end{aligned} \quad (1.4.2)$$

where ∇ denotes the gradient operator with respect to the material coordinates $\mathbf{X}$ and $\mathbf{E}$ is known as the *Green–Lagrange strain tensor*

$$\mathbf{E} = \frac{1}{2}\left[\nabla \mathbf{u} + (\nabla \mathbf{u})^T + \nabla \mathbf{u} \cdot (\nabla \mathbf{u})^T\right] \quad (1.4.3)$$

Note that the Green–Lagrange strain tensor is symmetric, $\mathbf{E} = \mathbf{E}^T$. The strain components defined in Eq. (1.4.3) are called finite strain components because no assumption concerning the smallness (compared to unity) of the strains is made.

The rectangular Cartesian component form of **E** is

$$E_{jk} = \frac{1}{2}\left(\frac{\partial u_j}{\partial X_k} + \frac{\partial u_k}{\partial X_j} + \frac{\partial u_m}{\partial X_j}\frac{\partial u_m}{\partial X_k}\right) \quad (1.4.4)$$

where summation on repeated subscripts is implied. In explicit form, we have

$$E_{11} = \frac{\partial u_1}{\partial X_1} + \frac{1}{2}\left[\left(\frac{\partial u_1}{\partial X_1}\right)^2 + \left(\frac{\partial u_2}{\partial X_1}\right)^2 + \left(\frac{\partial u_3}{\partial X_1}\right)^2\right]$$

$$E_{12} = \frac{1}{2}\left(\frac{\partial u_1}{\partial X_2} + \frac{\partial u_2}{\partial X_1} + \frac{\partial u_1}{\partial X_1}\frac{\partial u_1}{\partial X_2} + \frac{\partial u_2}{\partial X_1}\frac{\partial u_2}{\partial X_2} + \frac{\partial u_3}{\partial X_1}\frac{\partial u_3}{\partial X_2}\right)$$

$$E_{13} = \frac{1}{2}\left(\frac{\partial u_1}{\partial X_3} + \frac{\partial u_3}{\partial X_1} + \frac{\partial u_1}{\partial X_1}\frac{\partial u_1}{\partial X_3} + \frac{\partial u_2}{\partial X_1}\frac{\partial u_2}{\partial X_3} + \frac{\partial u_3}{\partial X_1}\frac{\partial u_3}{\partial X_3}\right)$$

$$E_{22} = \frac{\partial u_2}{\partial X_2} + \frac{1}{2}\left[\left(\frac{\partial u_1}{\partial X_2}\right)^2 + \left(\frac{\partial u_2}{\partial X_2}\right)^2 + \left(\frac{\partial u_3}{\partial X_2}\right)^2\right]$$

$$E_{23} = \frac{1}{2}\left(\frac{\partial u_2}{\partial X_3} + \frac{\partial u_3}{\partial X_2} + \frac{\partial u_1}{\partial X_2}\frac{\partial u_1}{\partial X_3} + \frac{\partial u_2}{\partial X_2}\frac{\partial u_2}{\partial X_3} + \frac{\partial u_3}{\partial X_2}\frac{\partial u_3}{\partial X_3}\right)$$

$$E_{33} = \frac{\partial u_3}{\partial X_3} + \frac{1}{2}\left[\left(\frac{\partial u_1}{\partial X_3}\right)^2 + \left(\frac{\partial u_2}{\partial X_3}\right)^2 + \left(\frac{\partial u_3}{\partial X_3}\right)^2\right] \quad (1.4.5)$$

The strain components in other coordinate systems can be derived from Eq. (1.4.3) by expressing the tensor **E** and del operator ∇ in that coordinate system. See Eqs. (1.2.30) and (1.2.41) for the definition of ∇ in cylindrical and spherical coordinates systems, respectively. For example, the Green–Lagrange strain tensor components in the cylindrical coordinate system are given by

$$E_{rr} = \frac{\partial u_r}{\partial r} + \frac{1}{2}\left[\left(\frac{\partial u_r}{\partial r}\right)^2 + \left(\frac{\partial u_\theta}{\partial r}\right)^2 + \left(\frac{\partial u_z}{\partial r}\right)^2\right]$$

$$E_{\theta\theta} = \frac{u_r}{r} + \frac{1}{r}\frac{\partial u_\theta}{\partial \theta} + \frac{1}{2}\left[\left(\frac{1}{r}\frac{\partial u_r}{\partial \theta}\right)^2 + \left(\frac{1}{r}\frac{\partial u_\theta}{\partial \theta}\right)^2 + \left(\frac{1}{r}\frac{\partial u_z}{\partial \theta}\right)^2\right]$$

$$+ \frac{1}{r^2}\left(2u_r\frac{\partial u_\theta}{\partial \theta} - 2u_\theta\frac{\partial u_r}{\partial \theta} + u_\theta^2 + u_r^2\right)$$

$$E_{zz} = \frac{\partial u_z}{\partial z} + \frac{1}{2}\left[\left(\frac{\partial u_r}{\partial z}\right)^2 + \left(\frac{\partial u_\theta}{\partial z}\right)^2 + \left(\frac{\partial u_z}{\partial z}\right)^2\right]$$

$$E_{r\theta} = \frac{1}{2}\left(\frac{1}{r}\frac{\partial u_r}{\partial \theta} + \frac{\partial u_\theta}{\partial r} - \frac{u_\theta}{r}\right) + \frac{1}{2r}\left(\frac{\partial u_r}{\partial r}\frac{\partial u_r}{\partial \theta} + \frac{\partial u_\theta}{\partial \theta}\frac{\partial u_\theta}{\partial \theta}\right.$$
$$\left. + \frac{\partial u_z}{\partial r}\frac{\partial u_z}{\partial \theta} + u_r\frac{\partial u_\theta}{\partial r} - u_\theta\frac{\partial u_r}{\partial r}\right)$$

$$E_{z\theta} = \frac{1}{2}\left(\frac{\partial u_\theta}{\partial z} + \frac{1}{r}\frac{\partial u_z}{\partial \theta}\right) + \frac{1}{2r}\left(\frac{\partial u_r}{\partial \theta}\frac{\partial u_r}{\partial z} + \frac{\partial u_\theta}{\partial \theta}\frac{\partial u_\theta}{\partial z} + \frac{\partial u_z}{\partial \theta}\frac{\partial u_z}{\partial z}\right.$$
$$\left. - u_\theta\frac{\partial u_r}{\partial z} + u_r\frac{\partial u_\theta}{\partial z}\right)$$

$$E_{rz} = \frac{1}{2}\left(\frac{\partial u_r}{\partial z} + \frac{\partial u_z}{\partial r}\right) + \frac{1}{2}\left(\frac{\partial u_r}{\partial r}\frac{\partial u_r}{\partial z} + \frac{\partial u_\theta}{\partial r}\frac{\partial u_\theta}{\partial z} + \frac{\partial u_z}{\partial r}\frac{\partial u_z}{\partial z}\right) \quad (1.4.6)$$

If the displacement gradients are so small, $|u_{i,j}| \ll 1$, that their squares and products are negligible compared to $|u_{i,j}|$ and the difference between $\frac{\partial u_i}{\partial x_j}$ and $\frac{\partial u_i}{\partial X_j}$ vanishes, then the Green–Lagrange strain tensor reduces to the *infinitesimal strain tensor*:

$$\varepsilon = \frac{1}{2}\left[\nabla \mathbf{u} + (\nabla \mathbf{u})^T\right] \quad (1.4.7)$$

or in Cartesian component form

$$\varepsilon_{ij} = \frac{1}{2}\left(\frac{\partial u_i}{\partial x_j} + \frac{\partial u_j}{\partial x_i}\right) \equiv \frac{1}{2}(u_{i,j} + u_{j,i}) \quad (1.4.8)$$

Suppose that we use the notation $x_1 = x$, $x_2 = y$, and $x_3 = z$, and let $(u_1, u_2, u_3) = (u, v, w)$ denote the displacements along (x, y, z). Then the infinitesimal strain components in Eq. (1.4.8) become

$$\varepsilon_{xx} = \frac{\partial u}{\partial x}, \quad \varepsilon_{yy} = \frac{\partial v}{\partial y}, \quad \varepsilon_{zz} = \frac{\partial w}{\partial z} \quad (1.4.9)$$

$$2\varepsilon_{xy} = \frac{\partial u}{\partial y} + \frac{\partial v}{\partial x}$$
$$2\varepsilon_{xz} = \frac{\partial u}{\partial z} + \frac{\partial w}{\partial x} \quad (1.4.10)$$
$$2\varepsilon_{yz} = \frac{\partial v}{\partial z} + \frac{\partial w}{\partial y}$$

Example 1.4.1

The displacement field of a beam under the Euler–Bernoulli kinematic hypothesis (i.e., a transverse normal line before deformation remains straight, inextensible, and normal after deformation) can be expressed as

$$u(x,y,z) = u_0(x) - z\frac{dw_0}{dx}, \quad v = 0, \quad w(x,y,z) = w_0(x) \quad (1.4.11)$$

Then the only nonzero nonlinear strains are given by

$$\begin{aligned}
\varepsilon_{xx} &= \frac{du}{dx} + \frac{1}{2}\left[\left(\frac{du}{dx}\right)^2 + \left(\frac{dw}{dx}\right)^2\right] \\
&= \frac{du_0}{dx} + \frac{1}{2}\left[\left(\frac{du_0}{dx}\right)^2 + \left(\frac{dw_0}{dx}\right)^2\right] \\
&\quad - z\left(\frac{d^2w_0}{dx^2} + \frac{du_0}{dx}\frac{d^2w_0}{dx^2}\right) + \frac{z^2}{2}\left(\frac{d^2w_0}{dx^2}\right)^2 \\
\varepsilon_{zz} &= \frac{dw}{dz} + \frac{1}{2}\left[\left(\frac{du}{dz}\right)^2 + \left(\frac{dw}{dz}\right)^2\right] = \frac{1}{2}\left(\frac{dw_0}{dx}\right)^2 \\
2\varepsilon_{xz} &= \frac{du}{dz} + \frac{dw}{dx} + \frac{du}{dx}\frac{du}{dz} + \frac{dw}{dx}\frac{dw}{dz} \\
&= -\frac{du_0}{dx}\frac{dw_0}{dx} + z\frac{d^2w_0}{dx^2}\frac{dw_0}{dx} \quad (1.4.12)
\end{aligned}$$

If we assume that the following terms are negligible compared to du_0/dx,

$$\left(\frac{du_0}{dx}\right)^2, \quad h\frac{du_0}{dx}\frac{d^2w_0}{dx^2}, \quad h^2\left(\frac{d^2w_0}{dx^2}\right)^2, \quad \frac{du_0}{dx}\frac{dw_0}{dx}, \quad h\frac{dw_0}{dx}\frac{d^2w_0}{dx^2} \quad (1.4.13)$$

where h denotes the height of the beam, then the nonlinear strains in Eq. (1.4.12) can be simplified to

$$\varepsilon_{xx} = \frac{du_0}{dx} + \frac{1}{2}\left(\frac{dw_0}{dx}\right)^2 - z\frac{d^2w_0}{dx^2}, \quad \varepsilon_{zz} = \frac{1}{2}\left(\frac{dw_0}{dx}\right)^2, \quad 2\varepsilon_{xz} = 0 \quad (1.4.14)$$

1.4.3 Compatibility Equations

By definition, the components of the strain tensor can be computed from a differentiable displacement field using Eq. (1.4.3). However, if the *six* components of strain tensor are given and if we are required to find the *three* displacement components, the strains given should be such that a unique solution to the six differential equations relating the strains and displacements exists. For example, consider the relations

$$\frac{\partial u}{\partial x} = x^2 + 2y, \quad \frac{\partial u}{\partial y} = 3x + y^2$$

The two equations cannot be solved for u because they are inconsistent. The inconsistency can be verified by computing the second derivative $\partial^2 u/\partial x \partial y$ from the two relations. The first equation gives 2 and the second equation gives 3, which are unequal.

The existence of a unique solution to the six partial differential equations defined by the strain-displacement relations is guaranteed if the infinitesimal strain components satisfy certain integrability conditions, known in elasticity as the *compatibility conditions*. In Cartesian component form, these are

$$\mathcal{E}_{ikm}\mathcal{E}_{j\ell n}\,\varepsilon_{ij,k\ell} = 0 \tag{1.4.15}$$

or in vector form

$$\nabla \times (\nabla \times \varepsilon)^{\mathrm{T}} = \mathbf{0} \quad \text{(zero dyad)} \tag{1.4.16}$$

Note that m and n are the free subscripts. In the three-dimensional case, Eq. (1.4.15) gives 81 relations among the strain components. However, due to the symmetry of $\varepsilon_{ij,k\ell}$ with respect to i and j and k and ℓ, there are only six independent relations. The sufficiency of the statements (1.4.15) and (1.4.16) can be verified.

Equation (1.4.15) can be written as

$$\frac{\partial^2 \varepsilon_{ij}}{\partial x_m \partial x_n} + \frac{\partial^2 \varepsilon_{mn}}{\partial x_i \partial x_j} - \frac{\partial^2 \varepsilon_{im}}{\partial x_j \partial x_n} - \frac{\partial^2 \varepsilon_{jn}}{\partial x_i \partial x_m} = 0 \tag{1.4.17}$$

for any $i, j, m, n = 1, 2, 3$. For the two-dimensional case, Eq. (1.4.17) reduces to the following single compatibility equation

$$\frac{\partial^2 \varepsilon_{11}}{\partial x_2^2} + \frac{\partial^2 \varepsilon_{22}}{\partial x_1^2} - 2\frac{\partial^2 \varepsilon_{12}}{\partial x_1 \partial x_2} = 0 \tag{1.4.18}$$

BASIC EQUATIONS OF ELASTICITY   27

It should be noted that the strain compatibility equations are satisfied automatically when the strains are computed from a displacement field. Thus, one needs to verify the compatibility conditions only when the strains are computed from stresses that are in equilibrium.

Example 1.4.2

(a) Consider the following two-dimensional strain field:

$$\varepsilon_{11} = c_1 x_1 \left(x_1^2 + x_2^2\right), \quad \varepsilon_{22} = \frac{1}{3} c_2 x_1^3, \quad \varepsilon_{12} = c_3 x_1^2 x_2$$

where c_1, c_2, and c_3 are constants. We wish to check if the strain field is compatible. Using Eq. (1.4.18) we obtain

$$\frac{\partial^2 \varepsilon_{11}}{\partial x_2^2} + \frac{\partial^2 \varepsilon_{22}}{\partial x_1^2} - 2\frac{\partial^2 \varepsilon_{12}}{\partial x_1 \partial x_2} = 2c_1 x_1 + 2c_2 x_1 - 4c_3 x_1$$

Thus the strain field is in general not compatible, unless $c_1 = c_2 = c_3$.

(b) Given the strain tensor $\mathbf{E} = E_{rr}\hat{\mathbf{e}}_r\hat{\mathbf{e}}_r + E_{\theta\theta}\hat{\mathbf{e}}_\theta\hat{\mathbf{e}}_\theta$ in an axisymmetric body (i.e., E_{rr} and $E_{\theta\theta}$ are functions of r and z only), we wish to determine the compatibility conditions on E_{rr} and $E_{\theta\theta}$.

Using the vector form of compatibility conditions, Eq. (1.4.16), we obtain (see Exercise 1.11)

$$\mathbf{F} \equiv \nabla \times \mathbf{E} = \left(\frac{\partial E_{\theta\theta}}{\partial r} + \frac{E_{\theta\theta} - E_{rr}}{r}\right)\hat{\mathbf{e}}_z\hat{\mathbf{e}}_\theta + \frac{\partial E_{rr}}{\partial z}\hat{\mathbf{e}}_\theta\hat{\mathbf{e}}_r - \frac{\partial E_{\theta\theta}}{\partial z}\hat{\mathbf{e}}_r\hat{\mathbf{e}}_\theta$$

and

$$\nabla \times (\mathbf{F})^{\mathrm{T}} = \frac{\partial}{\partial r}\left(\frac{\partial E_{\theta\theta}}{\partial r} + \frac{E_{\theta\theta} - E_{rr}}{r}\right)(\hat{\mathbf{e}}_r \times \hat{\mathbf{e}}_\theta)\hat{\mathbf{e}}_z$$
$$- \frac{\partial^2 E_{\theta\theta}}{\partial r \partial z}(\hat{\mathbf{e}}_r \times \hat{\mathbf{e}}_\theta)\hat{\mathbf{e}}_r + \frac{1}{r}\left(\frac{\partial E_{\theta\theta}}{\partial r} + \frac{E_{\theta\theta} - E_{rr}}{r}\right)\left(\hat{\mathbf{e}}_\theta \times \frac{\partial\hat{\mathbf{e}}_\theta}{\partial \theta}\right)\hat{\mathbf{e}}_z$$
$$+ \frac{1}{r}\frac{\partial E_{rr}}{\partial z}\left(\hat{\mathbf{e}}_\theta \times \frac{\partial\hat{\mathbf{e}}_r}{\partial \theta}\right)\hat{\mathbf{e}}_\theta + \frac{1}{r}\frac{\partial E_{rr}}{\partial z}(\hat{\mathbf{e}}_\theta \times \hat{\mathbf{e}}_r)\frac{\partial\hat{\mathbf{e}}_\theta}{\partial \theta}$$
$$- \frac{1}{r}\frac{\partial E_{\theta\theta}}{\partial z}\left(\hat{\mathbf{e}}_\theta \times \frac{\partial\hat{\mathbf{e}}_\theta}{\partial \theta}\right)\hat{\mathbf{e}}_r + \frac{\partial}{\partial z}\left(\frac{\partial E_{\theta\theta}}{\partial r} + \frac{E_{\theta\theta} - E_{rr}}{r}\right)(\hat{\mathbf{e}}_z \times \hat{\mathbf{e}}_\theta)\hat{\mathbf{e}}_z$$
$$+ \frac{\partial^2 E_{rr}}{\partial z^2}(\hat{\mathbf{e}}_z \times \hat{\mathbf{e}}_r)\hat{\mathbf{e}}_\theta - \frac{\partial^2 E_{\theta\theta}}{\partial z^2}(\hat{\mathbf{e}}_z \times \hat{\mathbf{e}}_r)\hat{\mathbf{e}}_r = 0$$

Noting that
$$\frac{\partial \hat{e}_r}{\partial \theta} = \hat{e}_\theta, \quad \frac{\partial \hat{e}_\theta}{\partial \theta} = -\hat{e}_r$$
$$\hat{e}_\theta \times \hat{e}_r = \hat{e}_z, \quad \hat{e}_\theta \times \hat{e}_z = \hat{e}_r, \quad \hat{e}_z \times \hat{e}_r = \hat{e}_\theta \tag{1.4.19}$$

and that a tensor is zero only when all its components are zero, we obtain

$$\hat{e}_z \hat{e}_z: \quad \frac{\partial}{\partial r}\left(\frac{\partial E_{\theta\theta}}{\partial r} + \frac{E_{\theta\theta} - E_{rr}}{r}\right) + \frac{1}{r}\left(\frac{\partial E_{\theta\theta}}{\partial r} + \frac{E_{\theta\theta} - E_{rr}}{r}\right) = 0$$

$$\hat{e}_z \hat{e}_r: \quad -\frac{\partial^2 E_{\theta\theta}}{\partial r \partial z} + \frac{1}{r}\frac{\partial}{\partial z}(E_{rr} - E_{\theta\theta}) = 0$$

$$\hat{e}_r \hat{e}_z: \quad -\frac{\partial}{\partial z}\left(\frac{\partial E_{\theta\theta}}{\partial r} + \frac{E_{\theta\theta} - E_{rr}}{r}\right) = 0$$

$$\hat{e}_\theta \hat{e}_\theta: \quad \frac{\partial^2 E_{rr}}{\partial z^2} = 0, \qquad \hat{e}_\theta \hat{e}_r: \quad -\frac{\partial^2 E_{\theta\theta}}{\partial z^2} = 0$$

1.4.4 Stress and Its Measures

Stress at a point is a measure of force per unit area. The force per unit area acting on an elemental area ds of the deformed body is called the *stress vector* acting on the element. The concept also applies to the surface created by slicing the deformed body with a plane. The stress vector on the area element Δs is the vector $\Delta \mathbf{f}/\Delta s$, which has the same direction as $\Delta \mathbf{f}$ but with a magnitude equal to $|\Delta \mathbf{f}|/\Delta s$. The stress vector at a point P on Δs is defined by

$$\mathbf{T} \equiv \lim_{\Delta s \to 0} \frac{\Delta \mathbf{f}}{\Delta s} \tag{1.4.20}$$

Equation (1.4.20) implies that the stress vector at a point depends on the direction and magnitude of the force vector and the surface area on which it acts. The surface area in turn depends on the orientation of the plane used to slice the body. For example, consider a constant cross-section member. If it is sliced with a plane perpendicular to its axis, the resulting surface area A_0 is the same as the area of cross section of the member. However, if it is sliced at the same point with a plane oriented at an angle $0° < \theta < 90°$ to the vertical axis, the surface area A obtained will be different: $A = A_0/\cos\theta$. Thus the stress vector at a point depends on the force vector as well as the orientation of the surface on which it acts. For this reason, the stress vector acting on a plane with normal $\hat{n}$ is denoted by $\mathbf{T}^{(\hat{n})}$.

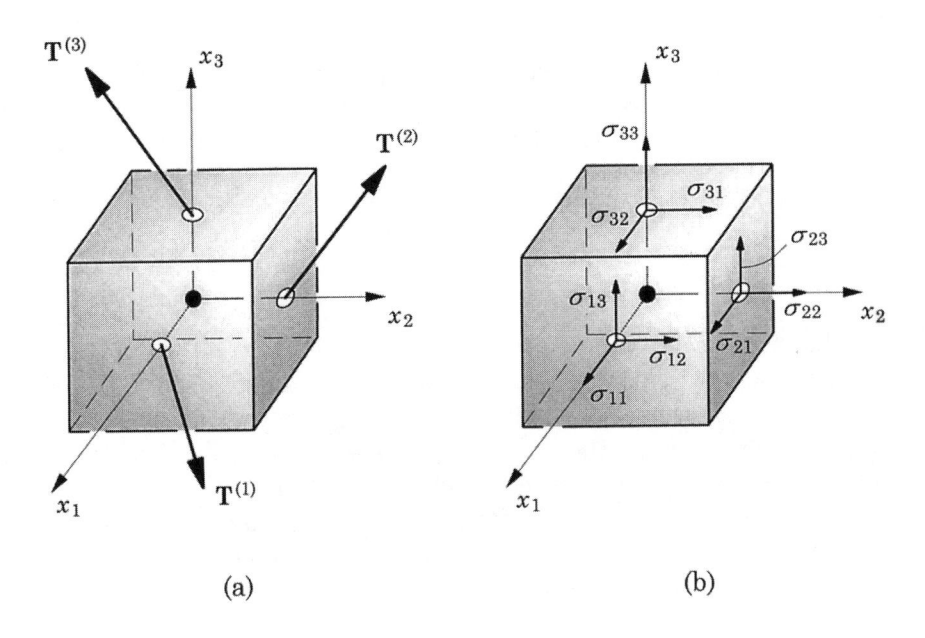

Figure 1.4.2. Stress vectors (a) and components of stress tensor (b) on coordinate planes.

The state of stress at a point inside a body can be expressed in terms of stress vectors on three mutually perpendicular planes, say planes perpendicular to the rectangular coordinate axes. Let $\mathbf{T}^{(i)}$ denote the stress vector at a point P on a plane perpendicular to the x_i-axis (see Figure 1.4.2a). Each vector $\mathbf{T}^{(i)}$, $(i = 1, 2, 3)$ can be resolved into components along the coordinate lines

$$\mathbf{T}^{(i)} = \sigma_{ij} \hat{\mathbf{e}}_j \quad (i = 1, 2, 3) \tag{1.4.21}$$

where the symbol σ_{ij} denotes the component of the stress vector $\mathbf{T}^{(i)}$ at the point P along the x_j-direction, and $\hat{\mathbf{e}}_j$ is the unit base vector along the x_j-direction. There are a total of nine stress components σ_{ij}, $(i, j = 1, 2, 3)$ at the point P. It can be shown that the quantities σ_{ij} transform like the components of a second-order tensor called the stress tensor, σ. The stress components σ_{ij} are shown on three perpendicular planes in Figure 1.4.2b. Note that the cube that is used to indicate the stress components has no dimensions; it is a point cube so that all nine components are acting at the point P.

Using the linear momentum principle or Newton's second law of motion, it can be shown that the stress vector $\mathbf{T}^{(\hat{n})}$ on a plane with unit normal $\hat{\mathbf{n}}$ is related to the stress tensor σ by the relation

$$\mathbf{T}^{(\hat{n})} = \hat{\mathbf{n}} \cdot \sigma = \sigma^T \cdot \hat{\mathbf{n}} \quad \text{or} \quad \{T^{(\hat{n})}\} = [\sigma]^T \{n\} \qquad (1.4.22)$$

Equation (1.4.22) is known as the *Cauchy stress formula* and σ is called the *Cauchy stress tensor*.

In the above discussion it is understood that stress at a point in a deformed body is measured as the force per unit area in the deformed body. The area element Δs in the deformed body corresponds to an area element ΔS in the reference configuration. The stress vector $\mathbf{T}^{(\hat{n})}$ is *measured* with respect to the area element in the deformed body. Stress vectors can be *referred* to either the reference coordinate system (X_1, X_2, X_3) or the spatial coordinate system (x_1, x_2, x_3). The stress that is measured and referred to the deformed configuration $\mathcal{C}$ is called the *Cauchy stress* or *true stress*, σ. The components σ_{ij} can be interpreted as the jth component of the force per unit area in the current configuration $\mathcal{C}$ acting on a surface segment whose outward normal at $\mathbf{x}$ is in the ith direction.

A suitable strain measure to use with the Cauchy stress tensor σ would be the infinitesimal strain tensor ε of Eq. (1.4.7). Since we defined strain tensor $\mathbf{E}$ as a function of the material point $\mathbf{X}$ in a reference state, the stress must also be expressed as a function of the material point. The stress measure that is used in nonlinear analysis of solid bodies is the second Piola–Kirchhoff stress tensor, which is measured in the deformed body but referred to the material coordinates, X_i. The strain measure to use with the second Piola–Kirchhoff stress is the Green–Lagrange strain tensor. For small deformation problems, the difference between the two measures of stress disappears. In the present study, we consider only small strain problems and therefore will not distinguish between various measures of stress, although it should be understood that σ denotes the second Piola–Kirchhoff stress tensor in the rest of the study. The principle of conservation of moment of momentum, in the absence of any distributed moments, leads to the symmetry of the stress tensor, $\sigma = \sigma^T$ ($\sigma_{ij} = \sigma_{ji}$). Thus, there are only six independent components of the Cauchy and second Piola–Kirchhoff stress tensors.

1.4.5 Equations of Motion

Consider a given mass of a material body $\mathcal{B}$ at any instant occupying a volume Ω bounded by a surface Γ. Suppose that the body is acted upon by external forces $\mathbf{T}$ (per unit surface area) and $\mathbf{f}$ (per unit mass). The principle of conservation of linear momentum states that the rate of change of the total linear momentum of a given continuous medium equals the vector sum of all the external forces acting on the body $\mathcal{B}$, which initially occupied a configuration $\mathcal{C}^0$, provided Newton's Third Law of action and reaction governs the internal forces. The principle leads to the following result:

$$\nabla \cdot \sigma + \rho_0 \mathbf{f} = \rho_0 \frac{\partial^2 \mathbf{u}}{\partial t^2} \tag{1.4.23}$$

where ∇ is the gradient operator with respect to the coordinates $\mathbf{X} \approx \mathbf{x}$ and ρ_0 is the material density in the undeformed body. In writing Eq. (1.4.23), all variables are assumed to be functions of $\mathbf{X}$ and time t. For the static case, Eq. (1.4.23) reduces to the equilibrium equations

$$\nabla \cdot \sigma + \rho_0 \mathbf{f} = 0 \tag{1.4.24}$$

The equations of motion or equilibrium are supplemented by the boundary conditions on the displacement vector and/or stress vector:

$$\mathbf{u} = \hat{\mathbf{u}}, \quad \sigma \cdot \hat{\mathbf{n}} = \hat{\mathbf{T}} \tag{1.4.25}$$

where $\hat{\mathbf{u}}$ and $\hat{\mathbf{T}}$ are specified displacement and stress vectors, respectively.

In rectangular Cartesian coordinates (x, y, z), the equations of motion (1.4.23) can be expressed as

$$\begin{aligned}
\frac{\partial \sigma_{xx}}{\partial x} + \frac{\partial \sigma_{xy}}{\partial y} + \frac{\partial \sigma_{xz}}{\partial z} + \rho_0 f_x &= \rho_0 \frac{\partial^2 u}{\partial t^2} \\
\frac{\partial \sigma_{xy}}{\partial x} + \frac{\partial \sigma_{yy}}{\partial y} + \frac{\partial \sigma_{yz}}{\partial z} + \rho_0 f_y &= \rho_0 \frac{\partial^2 v}{\partial t^2} \\
\frac{\partial \sigma_{xz}}{\partial x} + \frac{\partial \sigma_{yz}}{\partial y} + \frac{\partial \sigma_{zz}}{\partial z} + \rho_0 f_z &= \rho_0 \frac{\partial^2 w}{\partial t^2}
\end{aligned} \tag{1.4.26}$$

For coordinate systems other than the rectangular Cartesian systems, the equations of motion can be derived from the vector form (1.4.23) by expressing σ, $\mathbf{f}$, $\mathbf{u}$, and ∇ in the chosen coordinate system. For example, the equations of motion (1.4.23) in the cylindrical coordinate system (r, θ, z) are given by

$$\frac{\partial \sigma_{rr}}{\partial r} + \frac{1}{r}\frac{\partial \sigma_{r\theta}}{\partial \theta} + \frac{\partial \sigma_{rz}}{\partial z} + \frac{1}{r}(\sigma_{rr} - \sigma_{\theta\theta}) + \rho_0 f_r = \rho_0 \frac{\partial^2 u_r}{\partial t^2}$$

$$\frac{\partial \sigma_{r\theta}}{\partial r} + \frac{1}{r}\frac{\partial \sigma_{\theta\theta}}{\partial \theta} + \frac{\partial \sigma_{\theta z}}{\partial z} + \frac{2\sigma_{r\theta}}{r} + \rho_0 f_\theta = \rho_0 \frac{\partial^2 u_\theta}{\partial t^2} \quad (1.4.27)$$

$$\frac{\partial \sigma_{rz}}{\partial r} + \frac{1}{r}\frac{\partial \sigma_{\theta z}}{\partial \theta} + \frac{\partial \sigma_{zz}}{\partial z} + \frac{\sigma_{rz}}{r} + \rho_0 f_z = \rho_0 \frac{\partial^2 u_z}{\partial t^2}$$

where u_r, u_θ, and u_z are, for example, the components of the displacement vector $\mathbf{u}$ in the cylindrical coordinate system (see Exercise 1.15).

Example 1.4.3

Here we illustrate the derivation of the equations of equilibrium of an elastic body using Newton's laws. Consider the static equilibrium of an infinitesimal parallelepiped of dimensions dx_1, dx_2, and dx_3 and with its sides parallel to the coordinate planes, as shown in Figure 1.4.3. The stresses acting on typical planes are shown in the figure. Each stress component on a surface perpendicular to the negative x_k–axis is assumed to increase from its value σ_{ij} to

$$\sigma_{ij} + \frac{\partial \sigma_{ij}}{\partial x_k} dx_k$$

on the plane perpendicular to the positive x_k–axis. For example, the forces acting on the plane perpendicular to the negative x_1–axis are $\sigma_{11}dx_2dx_3$, $\sigma_{12}dx_2dx_3$, and $\sigma_{13}dx_2dx_3$, all in the opposite directions to the positive coordinates.

Summing the forces along the x_1–axis, we obtain

$$-\sigma_{11}dx_2dx_3 - \sigma_{21}dx_1dx_3 - \sigma_{31}dx_1dx_2 + \left(\sigma_{11} + \frac{\partial \sigma_{11}}{\partial x_1}dx_1\right)dx_2dx_3$$
$$+ \left(\sigma_{21} + \frac{\partial \sigma_{21}}{\partial x_2}dx_2\right)dx_1dx_3 + \left(\sigma_{31} + \frac{\partial \sigma_{31}}{\partial x_3}dx_3\right)dx_1dx_2$$
$$+ \rho_0 f_1 dx_1 dx_2 dx_3 = 0$$

BASIC EQUATIONS OF ELASTICITY 33

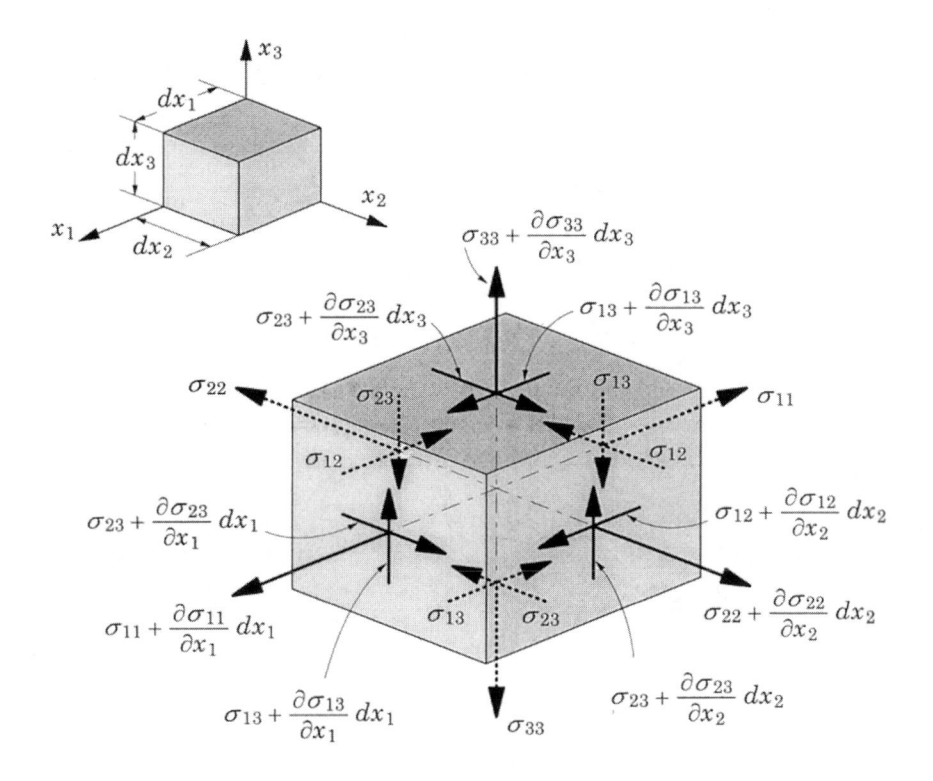

Figure 1.4.3. Equilibrating stress components on a parallelepiped.

or
$$\frac{\partial \sigma_{11}}{\partial x_1} + \frac{\partial \sigma_{21}}{\partial x_2} + \frac{\partial \sigma_{31}}{\partial x_3} + \rho_0 f_1 = 0$$

Similarly, summation of forces in the x_2- and x_3-directions yield

$$\frac{\partial \sigma_{12}}{\partial x_1} + \frac{\partial \sigma_{22}}{\partial x_2} + \frac{\partial \sigma_{32}}{\partial x_3} + \rho_0 f_2 = 0, \quad \frac{\partial \sigma_{13}}{\partial x_1} + \frac{\partial \sigma_{23}}{\partial x_2} + \frac{\partial \sigma_{33}}{\partial x_3} + \rho_0 f_3 = 0$$

1.4.6 Constitutive Equations

The kinematic relations and the mechanical and thermodynamic principles are applicable to any continuum irrespective of its physical constitution. Here we consider equations characterizing the individual material and its reaction to applied loads. These equations are called the *constitutive equations*. The formulation of the constitutive equations for a given material is guided by certain rules (i.e., constitutive axioms). We will not discuss them here but will review the linear constitutive relations for solids undergoing small deformations.

A material body is said to be *homogeneous* if the material properties are the same throughout the body. In a *heterogeneous* body, the material properties are a function of position. An *anisotropic* body is one that has different values of a material property in different directions at a point, i.e., material properties are direction-dependent. An *isotropic* body is one for which every material property in all directions at a point is the same. An isotropic or anisotropic material can be nonhomogeneous or homogeneous.

A material body is said to be *ideally elastic* when, under isothermal conditions, the body recovers its original form completely upon removal of the forces causing deformation, and there is a one to one relationship between the state of stress and the state of strain. The constitutive equations described here do not include creep at constant stress and stress relaxation at constant strain. Thus, the material coefficients that specify the constitutive relationship between the stress and strain components are assumed to be constant during the deformation. This does not automatically imply that we neglect temperature effects on deformation. We account for the thermal expansion of the material, which can produce strains or stresses as large as those produced by the applied mechanical forces. The dependence of the constitutive properties on temperature and strains can be accounted for if required. A more detailed discussion of the thermoelastic constitutive relations will be presented in the chapter on mechanical behavior of a lamina. Here we review the basic constitutive equations of linear elasticity (i.e., generalized Hooke's law) for small displacements.

The generalized Hooke's law relates the six components of stress to the six components of strain as

$$\begin{Bmatrix} \sigma_1 \\ \sigma_2 \\ \sigma_3 \\ \sigma_4 \\ \sigma_5 \\ \sigma_6 \end{Bmatrix} = \begin{bmatrix} C_{11} & C_{12} & C_{13} & C_{14} & C_{15} & C_{16} \\ C_{21} & C_{22} & C_{23} & C_{24} & C_{25} & C_{26} \\ C_{31} & C_{32} & C_{33} & C_{34} & C_{35} & C_{36} \\ C_{41} & C_{42} & C_{43} & C_{44} & C_{45} & C_{46} \\ C_{51} & C_{52} & C_{53} & C_{54} & C_{55} & C_{56} \\ C_{61} & C_{62} & C_{63} & C_{64} & C_{65} & C_{66} \end{bmatrix} \begin{Bmatrix} \varepsilon_1 \\ \varepsilon_2 \\ \varepsilon_3 \\ \varepsilon_4 \\ \varepsilon_5 \\ \varepsilon_6 \end{Bmatrix} \quad (1.4.28)$$

BASIC EQUATIONS OF ELASTICITY 35

where C_{ij} are the elastic coefficients and

$$\sigma_1 = \sigma_{11}, \ \sigma_2 = \sigma_{22}, \ \sigma_3 = \sigma_{33}, \ \sigma_4 = \sigma_{23}, \ \sigma_5 = \sigma_{13}, \ \sigma_6 = \sigma_{12}$$
$$\varepsilon_1 = \varepsilon_{11}, \ \varepsilon_2 = \varepsilon_{22}, \ \varepsilon_3 = \varepsilon_{33}, \ \varepsilon_4 = 2\varepsilon_{23}, \ \varepsilon_5 = 2\varepsilon_{13}, \ \varepsilon_6 = 2\varepsilon_{12}$$
(1.4.29)

The single subscript notation (1.4.29) used for stresses and strains is called the *contracted notation* or the *Voigt-Kelvin notation*. The two-subscript components C_{ij} are obtained from $C_{ijk\ell}$ by the following change of subscripts:

$$11 \to 1 \quad 22 \to 2 \quad 33 \to 3 \quad 23 \to 4 \quad 13 \to 5 \quad 12 \to 6$$

The resulting C_{ij} are also symmetric ($C_{ij} = C_{ji}$) by virtue of the assumption that there exists a potential function $U_0 = U_0(\varepsilon_{ij})$, called the *strain energy density function*, whose derivative with respect to a strain component determines the corresponding stress component

$$\sigma_{ij} = \frac{\partial U_0}{\partial \varepsilon_{ij}} \tag{1.4.30}$$

Such materials are termed *hyperelastic* materials. For hyperelastic materials, there are only 21 independent coefficients of the matrix $[C]$.

When three mutually orthogonal planes of material symmetry exist, the number of elastic coefficients is reduced to 9, and such materials are called *orthotropic*. The stress-strain relations for an orthotropic material take the form

$$\begin{Bmatrix} \sigma_1 \\ \sigma_2 \\ \sigma_3 \\ \sigma_4 \\ \sigma_5 \\ \sigma_6 \end{Bmatrix} = \begin{bmatrix} C_{11} & C_{12} & C_{13} & 0 & 0 & 0 \\ C_{12} & C_{22} & C_{23} & 0 & 0 & 0 \\ C_{13} & C_{23} & C_{33} & 0 & 0 & 0 \\ 0 & 0 & 0 & C_{44} & 0 & 0 \\ 0 & 0 & 0 & 0 & C_{55} & 0 \\ 0 & 0 & 0 & 0 & 0 & C_{66} \end{bmatrix} \begin{Bmatrix} \varepsilon_1 \\ \varepsilon_2 \\ \varepsilon_3 \\ \varepsilon_4 \\ \varepsilon_5 \\ \varepsilon_6 \end{Bmatrix} \tag{1.4.31}$$

The inverse relations, strain-stress relations, are given by

$$\begin{Bmatrix} \varepsilon_1 \\ \varepsilon_2 \\ \varepsilon_3 \\ \varepsilon_4 \\ \varepsilon_5 \\ \varepsilon_6 \end{Bmatrix} = \begin{bmatrix} S_{11} & S_{12} & S_{13} & 0 & 0 & 0 \\ S_{12} & S_{22} & S_{23} & 0 & 0 & 0 \\ S_{13} & S_{23} & S_{33} & 0 & 0 & 0 \\ 0 & 0 & 0 & S_{44} & 0 & 0 \\ 0 & 0 & 0 & 0 & S_{55} & 0 \\ 0 & 0 & 0 & 0 & 0 & S_{66} \end{bmatrix} \begin{Bmatrix} \sigma_1 \\ \sigma_2 \\ \sigma_3 \\ \sigma_4 \\ \sigma_5 \\ \sigma_6 \end{Bmatrix}$$

$$= \begin{bmatrix} \frac{1}{E_1} & -\frac{\nu_{21}}{E_2} & -\frac{\nu_{31}}{E_3} & 0 & 0 & 0 \\ -\frac{\nu_{12}}{E_1} & \frac{1}{E_2} & -\frac{\nu_{32}}{E_3} & 0 & 0 & 0 \\ -\frac{\nu_{13}}{E_1} & -\frac{\nu_{23}}{E_2} & \frac{1}{E_3} & 0 & 0 & 0 \\ 0 & 0 & 0 & \frac{1}{G_{23}} & 0 & 0 \\ 0 & 0 & 0 & 0 & \frac{1}{G_{13}} & 0 \\ 0 & 0 & 0 & 0 & 0 & \frac{1}{G_{12}} \end{bmatrix} \begin{Bmatrix} \sigma_1 \\ \sigma_2 \\ \sigma_3 \\ \sigma_4 \\ \sigma_5 \\ \sigma_6 \end{Bmatrix} \quad (1.4.32)$$

where S_{ij} denote the compliance coefficients, $[S] = [C]^{-1}$, E_1, E_2, E_3 are Young's moduli in 1, 2, and 3 material directions, respectively, ν_{ij} is Poisson's ratio, defined as the ratio of transverse strain in the jth direction to the axial strain in the ith direction when stressed in the i-direction, and G_{23}, G_{13}, G_{12} = shear moduli in the 2-3, 1-3, and 1-2 planes, respectively. Since the compliance matrix $[S]$ is the inverse of the stiffness matrix $[C]$ and the inverse of a symmetric matrix is symmetric, it follows that the compliance matrix $[S]$ is also a symmetric matrix. This in turn implies, in view of Eq. (1.4.31), that the following reciprocal relations hold:

$$\frac{\nu_{21}}{E_2} = \frac{\nu_{12}}{E_1}; \quad \frac{\nu_{31}}{E_3} = \frac{\nu_{13}}{E_1}; \quad \frac{\nu_{32}}{E_3} = \frac{\nu_{23}}{E_2} \quad (1.4.33a)$$

or, in short,

$$\frac{\nu_{ij}}{E_i} = \frac{\nu_{ji}}{E_j} \quad \text{(no sum on } i, j) \quad (1.4.33b)$$

for $i, j = 1, 2, 3$. Thus, there are only nine independent material coefficients

$$E_1,\ E_2,\ E_3,\ G_{23},\ G_{13},\ G_{12},\ \nu_{12},\ \nu_{13},\ \nu_{23}$$

for an orthotropic material.

When there exist no preferred directions in the material (i.e., the material has infinite number of planes of material symmetry), the number of independent elastic coefficients reduces to 2. Such materials are called *isotropic*. For isotropic materials we have $E_1 = E_2 = E_3 = E$, $G_{12} = G_{13} = G_{23} \equiv G$, and $\nu_{12} = \nu_{23} = \nu_{13} \equiv \nu$. Of the three constants (E, ν, G), only two are independent and the third one is related to the other two by the relation

$$G = \frac{E}{2(1+\nu)} \quad (1.4.34)$$

BASIC EQUATIONS OF ELASTICITY   37

A *plane stress state* is defined to be one in which all transverse stresses are negligible. The strain-stress relations of an orthotropic body in plane stress of state can be written as

$$\begin{Bmatrix} \varepsilon_1 \\ \varepsilon_2 \\ \varepsilon_6 \end{Bmatrix} = \begin{bmatrix} S_{11} & S_{12} & 0 \\ S_{12} & S_{22} & 0 \\ 0 & 0 & S_{66} \end{bmatrix} \begin{Bmatrix} \sigma_1 \\ \sigma_2 \\ \sigma_6 \end{Bmatrix}$$

$$= \begin{bmatrix} \frac{1}{E_1} & -\frac{\nu_{21}}{E_2} & 0 \\ -\frac{\nu_{12}}{E_1} & \frac{1}{E_2} & 0 \\ 0 & 0 & \frac{1}{G_{12}} \end{bmatrix} \begin{Bmatrix} \sigma_1 \\ \sigma_2 \\ \sigma_6 \end{Bmatrix} \quad (1.4.35)$$

The strain-stress relations (1.4.35) are inverted to obtain the stress-strain relations

$$\begin{Bmatrix} \sigma_1 \\ \sigma_2 \\ \sigma_6 \end{Bmatrix} = \begin{bmatrix} Q_{11} & Q_{12} & 0 \\ Q_{12} & Q_{22} & 0 \\ 0 & 0 & Q_{66} \end{bmatrix} \begin{Bmatrix} \varepsilon_1 \\ \varepsilon_2 \\ \varepsilon_6 \end{Bmatrix} \quad (1.4.36)$$

where the Q_{ij}, called the *plane stress-reduced stiffnesses*, are given by

$$Q_{11} = \frac{S_{22}}{S_{11}S_{22} - S_{12}^2} = \frac{E_1}{1 - \nu_{12}\nu_{21}}$$

$$Q_{12} = \frac{S_{12}}{S_{11}S_{22} - S_{12}^2} = \frac{\nu_{12}E_2}{1 - \nu_{12}\nu_{21}}$$

$$Q_{22} = \frac{S_{11}}{S_{11}S_{22} - S_{12}^2} = \frac{E_2}{1 - \nu_{12}\nu_{21}}$$

$$Q_{66} = \frac{1}{S_{66}} = G_{12} \quad (1.4.37)$$

Note that the reduced stiffnesses involve four independent material constants, E_1, E_2, ν_{12}, and G_{12}.

The transverse shear stresses are related to the transverse shear strains in an orthotropic material by the relations

$$\begin{Bmatrix} \sigma_4 \\ \sigma_5 \end{Bmatrix} = \begin{bmatrix} Q_{44} & 0 \\ 0 & Q_{55} \end{bmatrix} \begin{Bmatrix} \varepsilon_4 \\ \varepsilon_5 \end{Bmatrix} \quad (1.4.38)$$

where $Q_{44} = C_{44} = G_{23}$ and $Q_{55} = C_{55} = G_{13}$.

When temperature changes occur in the elastic body, we account for the thermal expansion of the material, even though the variation of elastic constants with temperature is neglected. When the strains, geometric changes, and temperature variations are sufficiently small, all governing equations are linear and superposition of mechanical and thermal effects is possible. The linear thermoelastic constitutive equations have the form

$$\sigma_j = C_{ji}[-\alpha_i(T - T_0) + \varepsilon_i] \tag{1.4.39}$$

$$\varepsilon_j = S_{ji}\sigma_i + \alpha_i(T - T_0) \tag{1.4.40}$$

where α_i ($i = 1, 2, 3$) are the linear coefficients of thermal expansion, T denotes temperature, and T_0 is the reference temperature of the undeformed body. In writing Eqs. (1.4.39) and (1.4.40), it is assumed that α_i and C_{ij} are independent of strain and temperature. For an isotropic material we have $\alpha_1 = \alpha_2 = \alpha_3 \equiv \alpha$. The plane stress strain-stress relations for a thermoelastic case are given by

$$\begin{Bmatrix} \sigma_1 \\ \sigma_2 \\ \sigma_6 \end{Bmatrix} = \begin{bmatrix} Q_{11} & Q_{12} & 0 \\ Q_{12} & Q_{22} & 0 \\ 0 & 0 & Q_{66} \end{bmatrix} \begin{Bmatrix} \varepsilon_1 - \alpha_1 \Delta T \\ \varepsilon_2 - \alpha_2 \Delta T \\ \varepsilon_6 \end{Bmatrix} \tag{1.4.41}$$

where $\Delta T = T - T_0$ is the temperature change from the reference temperature T_0 and α_i ($i = 1, 2$) are the coefficients of thermal expansion of an orthotropic material in the x_i−coordinate direction. In addition, Eq. (1.4.38) holds for the thermoelastic case.

Tables 1.4.1 and 1.4.2 contain the values of typical engineering constants for some commonly used metals and polymeric materials. The engineering properties of polymeric materials depend on the processing conditions, which can be different for different manufacturers.

1.5 Transformation of Stresses, Strains, and Stiffnesses

1.5.1 Introduction

In general, the coordinate system used in the solution of a problem does not coincide with the material coordinate system. Therefore, there is a need to establish transformation relations among stresses and strains in one coordinate system to the corresponding quantities in another coordinate system. These relations can be used to transform constitutive equations from the material coordinates to the coordinates used in the solution of a problem.

Table 1.4.1. Values of the engineering constants for several materials*.

Material	E_1	E_2	G_{12}	G_{13}	G_{23}	ν_{12}
Aluminum	10.6	10.6	3.38	3.38	3.38	0.33
Copper	18.0	18.0	6.39	6.39	6.39	0.33
Steel	30.0	30.0	11.24	11.24	11.24	0.29
Gr.–Ep (AS)	20.0	1.3	1.03	1.03	0.90	0.30
Gr.–Ep (T)	19.0	1.5	1.00	0.90	0.90	0.22
Gl.–Ep (1)	7.8	2.6	1.30	1.30	0.50	0.25
Gl.–Ep (2)	5.6	1.2	0.60	0.60	0.50	0.26
Br.–Ep	30.0	3.0	1.00	1.00	0.60	0.30

*Moduli are in msi = million psi; 1 psi = 6.895 kN/m^2.

Table 1.4.2. Values of additional engineering constants for the materials listed in Table 1.4.1*.

Material	E_3	ν_{13}	ν_{23}	α_1	α_2
Aluminum	10.6	0.33	0.33	13.1	13.1
Copper	18.0	0.33	0.33	18.0	18.0
Steel	30.0	0.29	0.29	10.0	10.0
Gr.–Ep (AS)	1.3	0.30	0.49	1.0	30.0
Gr.–Ep (T)	1.5	0.22	0.49	−0.167	15.6
Gl.–Ep (1)	2.6	0.25	0.34	3.5	11.4
Gl.–Ep (2)	1.3	0.26	0.34	4.8	12.3
Br.–Ep	3.0	0.25	0.25	2.5	8.0

* Units of E_3 are msi, and the units of α_1 and α_2 are 10^{-6} in./in./°F.

Gr.–Ep (AS) = graphite–epoxy (AS/3501); Gr.–Ep (T) = graphite–epoxy (T300/934); Gl.–Ep = glass–epoxy; Br.–Ep = boron–epoxy.

Here we consider a special coordinate transformation. Let the principal material coordinates be denoted by (x_1, x_2, x_3) and the plate coordinates be (x, y, z). Further assume that the $x_1 x_2$–plane is parallel to the xy–plane, but rotated counterclockwise about the $z = x_3$–axis by an angle θ (see Figure 1.5.1). Conversely, (x, y, z) is obtained from

40   THEORY AND ANALYSIS OF ELASTIC PLATES

(x_1, x_2, x_3) by rotating the x_1x_2–plane clockwise by an angle θ. Then the coordinates of a material point in the two coordinate systems are related by Eqs. (1.3.13a,b). We wish to write the relationships between the components of stress and strain in the two coordinate systems.

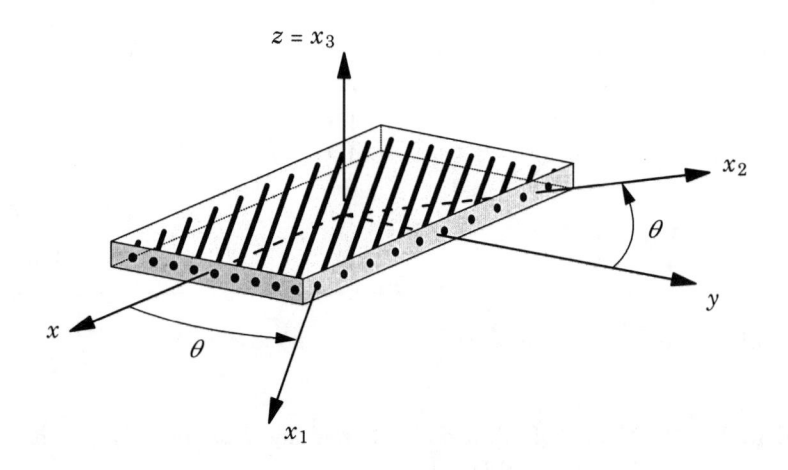

Figure 1.5.1. A plate with global and material coordinate systems.

1.5.2 Transformation of Stress Components

Let $\sigma_{11}, \sigma_{12}, \cdots, \sigma_{33}$ denote the stress components in the material coordinates (x_1, x_2, x_3) and $\sigma_{xx}, \sigma_{xy}, \cdots, \sigma_{zz}$ be the stress components referred to the plate coordinates (x, y, z). In the notation of section 1.2, (x_1, x_2, x_3) is the system with bars, (x, y, z) is the system without bars, and θ is the angle of rotation about the $x_3 = z$ axis in the counterclockwise direction. Since stress is a second-order tensor, we have from Eqs. (1.3.14) and (1.3.15) the following relations among the two sets of components:

$$\begin{aligned}
\sigma_{11} &= \sigma_{xx} \cos^2\theta + \sigma_{yy} \sin^2\theta + 2\sigma_{xy} \cos\theta \sin\theta \\
\sigma_{12} &= (\sigma_{yy} - \sigma_{xx}) \cos\theta \sin\theta + \sigma_{xy}(\cos^2\theta - \sin^2\theta) \\
\sigma_{13} &= \sigma_{xz} \cos\theta + \sigma_{yz} \sin\theta \\
\sigma_{22} &= \sigma_{xx} \sin^2\theta + \sigma_{yy} \cos^2\theta - 2\sigma_{xy} \cos\theta \sin\theta \\
\sigma_{23} &= \sigma_{yz} \cos\theta - \sigma_{xz} \sin\theta \\
\sigma_{33} &= \sigma_{zz}
\end{aligned} \quad (1.5.1)$$

$$\sigma_{xx} = \sigma_{11}\cos^2\theta + \sigma_{22}\sin^2\theta - 2\sigma_{12}\cos\theta\sin\theta$$
$$\sigma_{xy} = (\sigma_{11} - \sigma_{22})\cos\theta\sin\theta + \sigma_{12}(\cos^2\theta - \sin^2\theta)$$
$$\sigma_{xz} = \sigma_{13}\cos\theta - \sigma_{23}\sin\theta$$
$$\sigma_{yy} = \sigma_{11}\sin^2\theta + \sigma_{22}\cos^2\theta + 2\sigma_{12}\cos\theta\sin\theta$$
$$\sigma_{yz} = \sigma_{23}\cos\theta + \sigma_{13}\sin\theta$$
$$\sigma_{zz} = \sigma_{33} \tag{1.5.2}$$

1.5.3 Transformation of Strain Components

Since strains are also tensor quantities, transformation equations derived for stresses are also valid for *tensor* components of strains. We have

$$\varepsilon_{11} = \varepsilon_{xx}\cos^2\theta + \varepsilon_{yy}\sin^2\theta + 2\varepsilon_{xy}\cos\theta\sin\theta$$
$$\varepsilon_{12} = (\varepsilon_{yy} - \varepsilon_{xx})\cos\theta\sin\theta + \varepsilon_{xy}(\cos^2\theta - \sin^2\theta)$$
$$\varepsilon_{13} = \varepsilon_{xz}\cos\theta + \varepsilon_{yz}\sin\theta$$
$$\varepsilon_{22} = \varepsilon_{xx}\sin^2\theta + \varepsilon_{yy}\cos^2\theta - 2\varepsilon_{xy}\cos\theta\sin\theta$$
$$\varepsilon_{23} = \varepsilon_{yz}\cos\theta - \varepsilon_{xz}\sin\theta$$
$$\varepsilon_{33} = \varepsilon_{zz} \tag{1.5.3}$$

$$\varepsilon_{xx} = \varepsilon_{11}\cos^2\theta + \varepsilon_{22}\sin^2\theta - 2\varepsilon_{12}\cos\theta\sin\theta$$
$$\varepsilon_{xy} = (\varepsilon_{11} - \varepsilon_{22})\cos\theta\sin\theta + \varepsilon_{12}(\cos^2\theta - \sin^2\theta)$$
$$\varepsilon_{xz} = \varepsilon_{13}\cos\theta - \varepsilon_{23}\sin\theta$$
$$\varepsilon_{yy} = \varepsilon_{11}\sin^2\theta + \varepsilon_{22}\cos^2\theta + 2\varepsilon_{12}\cos\theta\sin\theta$$
$$\varepsilon_{yz} = \varepsilon_{23}\cos\theta + \varepsilon_{13}\sin\theta$$
$$\varepsilon_{zz} = \varepsilon_{33} \tag{1.5.4}$$

1.5.4 Transformation of Material Stiffnesses

The material stiffnesses $C_{ijk\ell}$ in their original form are the components of a fourth-order tensor. Hence, the tensor transformation law holds again. The fourth-order elasticity tensor components $\bar{C}_{ijk\ell}$ in the problem coordinates can be related to the components C_{mnpq} in the material coordinates by the tensor transformation law

$$\bar{C}_{ijk\ell} = a_{im}a_{jn}a_{kp}a_{\ell q}C_{mnpq} \tag{1.5.5}$$

However, the above equation involves five matrix multiplications with four-subscript material coefficients. Alternatively, the same result can be obtained by using the stress-strain relations in the principal material coordinates and the stress and strain transformation equations in Eqs. (1.5.1)–(1.5.4).

For the plane stress case, the elastic stiffnesses Q_{ij} in the principal material system are related to $\bar{Q}_{ij}$ in the transformed system by the equations

$$\bar{Q}_{11} = Q_{11} \cos^4 \theta + 2(Q_{12} + 2Q_{66}) \sin^2 \theta \cos^2 \theta + Q_{22} \sin^4 \theta$$
$$\bar{Q}_{12} = (Q_{11} + Q_{22} - 4Q_{66}) \sin^2 \theta \cos^2 \theta + Q_{12}(\sin^4 \theta + \cos^4 \theta)$$
$$\bar{Q}_{22} = Q_{11} \sin^4 \theta + 2(Q_{12} + 2Q_{66}) \sin^2 \theta \cos^2 \theta + Q_{22} \cos^4 \theta$$
$$\bar{Q}_{16} = (Q_{11} - Q_{12} - 2Q_{66}) \sin \theta \cos^3 \theta + (Q_{12} - Q_{22} + 2Q_{66}) \sin^3 \theta \cos \theta$$
$$\bar{Q}_{26} = (Q_{11} - Q_{12} - 2Q_{66}) \sin^3 \theta \cos \theta + (Q_{12} - Q_{22} + 2Q_{66}) \sin \theta \cos^3 \theta$$
$$\bar{Q}_{66} = (Q_{11} + Q_{22} - 2Q_{12} - 2Q_{66}) \sin^2 \theta \cos^2 \theta + Q_{66}(\sin^4 \theta + \cos^4 \theta)$$
$$\bar{Q}_{44} = Q_{44} \cos^2 \theta + Q_{55} \sin^2 \theta$$
$$\bar{Q}_{45} = (Q_{55} - Q_{44}) \cos \theta \sin \theta$$
$$\bar{Q}_{55} = Q_{55} \cos^2 \theta + Q_{44} \sin^2 \theta \tag{1.5.6}$$

The thermal coefficients of expansion α_{ij} are the components of a second-order tensor, and therefore they transform like the stress and strain components. For an orthotropic material, the only nonzero components of thermal expansion tensor referred to the principal materials coordinates are α_1, α_2, and α_3. Hence, we can write the transformation relations among α_{ij} and $(\alpha_{xx}, \alpha_{yy}, \alpha_{xy})$ [c.f., Eq. (1.5.4)]

$$\alpha_{xx} = \alpha_1 \cos^2 \theta + \alpha_2 \sin^2 \theta$$
$$\alpha_{yy} = \alpha_1 \sin^2 \theta + \alpha_2 \cos^2 \theta$$
$$\alpha_{xy} = 2(\alpha_1 - \alpha_{22}) \sin \theta \cos \theta$$
$$\alpha_{xz} = 0, \quad \alpha_{yz} = 0, \quad \alpha_{zz} = \alpha_3 \tag{1.5.7}$$

The stress-strain relations in the problem coordinates (x, y, z) for the thermoelastic case can now be expressed as

$$\begin{Bmatrix} \sigma_{xx} \\ \sigma_{yy} \\ \sigma_{yz} \\ \sigma_{xz} \\ \sigma_{xy} \end{Bmatrix} = \begin{bmatrix} \bar{Q}_{11} & \bar{Q}_{12} & 0 & 0 & \bar{Q}_{16} \\ \bar{Q}_{12} & \bar{Q}_{22} & 0 & 0 & \bar{Q}_{26} \\ 0 & 0 & \bar{Q}_{44} & \bar{Q}_{45} & 0 \\ 0 & 0 & \bar{Q}_{45} & \bar{Q}_{55} & 0 \\ \bar{Q}_{16} & \bar{Q}_{26} & 0 & 0 & \bar{Q}_{66} \end{bmatrix} \begin{Bmatrix} \varepsilon_{xx} - \alpha_{xx} \Delta T \\ \varepsilon_{yy} - \alpha_{yy} \Delta T \\ \varepsilon_{yz} \\ \varepsilon_{xz} \\ \varepsilon_{xy} - \alpha_{xy} \Delta T \end{Bmatrix} \tag{1.5.8}$$

Exercises

1.1 Prove the following properties of δ_{ij} and $\mathcal{E}_{ijk}$ (assume $i, j = 1, 2, 3$ when they are dummy indices):

(a) $F_{ij}\delta_{jk} = F_{ik}$.

(b) $\delta_{ij}\delta_{ij} = \delta_{ii} = 3$.

(c) $\mathcal{E}_{ijk}\mathcal{E}_{ijk} = 6$.

(d) $\mathcal{E}_{ijk}F_{ij} = 0$ whenever $F_{ij} = F_{ji}$ (symmetric).

1.2 Let $\mathbf{r}$ denote a position vector. Show that:

(a) grad $(r^n) = nr^{n-2}\mathbf{r}$.

(b) $\nabla^2(r^n) = n(n+1)r^{n-2}$.

(c) div $(\mathbf{r}) = 3$.

(d) curl $(f(r)\mathbf{r}) = \mathbf{0}$, where $f(r)$ is an arbitrary continuous function of r with continuous first derivatives.

1.3 If $[A]$ is a symmetric $n \times n$ matrix, and $[B]$ is any $n \times n$ matrix, show that $[B]^T[A][B]$ is symmetric.

1.4 Establish the $\mathcal{E} - \delta$ identity of Eq. (1.2.15).

1.5 Prove that the determinant of a 3×3 matrix $[A]$ can be expressed in the form

$$|A| = \mathcal{E}_{ijk}a_{1i}a_{2j}a_{3k} \qquad (a)$$

and thus prove

$$|A| = \frac{1}{6}\mathcal{E}_{ijk}\mathcal{E}_{rst}a_{ir}a_{js}a_{kt} \qquad (b)$$

where a_{ij} is the element occupying the ith row and the jth column of the matrix.

1.6 Using Cramer's rule determine the solution to the following equations:

(a) (Ans: $x_1 = \frac{9}{4}$, $x_2 = \frac{7}{2}$, $x_3 = \frac{11}{4}$)

$$2x_1 - x_2 = 1$$
$$-x_1 + 2x_2 - x_3 = 2$$
$$-x_2 + 2x_3 = 2$$

44 THEORY AND ANALYSIS OF ELASTIC PLATES

(b) [Ans: $x_1 = \frac{3}{2}\alpha$, $x_2 = \frac{1}{2h}\alpha$, $x_3 = -\frac{2}{h}\alpha$, $\alpha = (f_0 h^4/24b)$]

$$\frac{2b}{h^3} \begin{bmatrix} 12 & 0 & 3h \\ 0 & 4h^2 & h^2 \\ 3h & h^2 & 2h^2 \end{bmatrix} \begin{Bmatrix} x_1 \\ x_2 \\ x_3 \end{Bmatrix} = \frac{f_0 h}{12} \begin{Bmatrix} 12 \\ 0 \\ h \end{Bmatrix}$$

where b, f_0 and h are constants.

1.7 Let [A] be a 3×3 matrix, [I] be a 3×3 identity matrix, and λ be a scalar. Show that

$$\det[A - \lambda I] = \lambda^3 - I_1 \lambda^2 + I_2 \lambda - I_3$$

where

$$I_1 = a_{ii}, \quad I_2 = \frac{1}{2}(a_{ii}a_{jj} - a_{ij}a_{ij}), \quad I_3 = |A|$$

1.8 Use the definition $\nabla^2 = \nabla \cdot \nabla$ to show that the Laplacian operator in the cylindrical coordinate system is given by

$$\nabla^2 = \frac{1}{r}\left[\frac{\partial}{\partial r}\left(r\frac{\partial}{\partial r}\right) + \frac{1}{r}\frac{\partial^2}{\partial \theta^2} + r\frac{\partial^2}{\partial z^2}\right]$$

1.9 Show that the gradient of a vector **u** in the cylindrical coordinate system is given by

$$\nabla \mathbf{u} = \frac{\partial u_r}{\partial r}\hat{e}_r\hat{e}_r + \frac{1}{r}\left(u_r + \frac{\partial u_\theta}{\partial \theta}\right)\hat{e}_\theta\hat{e}_\theta + \frac{\partial u_z}{\partial z}\hat{e}_z\hat{e}_z$$
$$+ \frac{1}{r}\frac{\partial u_z}{\partial \theta}\hat{e}_\theta\hat{e}_z + \frac{\partial u_\theta}{\partial z}\hat{e}_z\hat{e}_\theta + \frac{\partial u_r}{\partial z}\hat{e}_z\hat{e}_r + \frac{\partial u_z}{\partial r}\hat{e}_r\hat{e}_z$$
$$+ \frac{\partial u_\theta}{\partial r}\hat{e}_r\hat{e}_\theta + \frac{1}{r}\left(\frac{\partial u_r}{\partial \theta} - u_\theta\right)\hat{e}_\theta\hat{e}_r$$

1.10 Show that the curl of a vector **u** in the cylindrical coordinate system is given by

$$\nabla \times \mathbf{u} = \hat{e}_r\left(\frac{1}{r}\frac{\partial u_z}{\partial \theta} - \frac{\partial u_\theta}{\partial z}\right) + \hat{e}_\theta\left(\frac{\partial u_r}{\partial z} - \frac{\partial u_z}{\partial r}\right)$$
$$+ \hat{e}_z\left(\frac{\partial u_\theta}{\partial r} + \frac{u_\theta}{r} - \frac{1}{r}\frac{\partial u_r}{\partial \theta}\right)$$

BASIC EQUATIONS OF ELASTICITY   45

1.11 Show that the curl of a tensor $\mathbf{E} = E_{rr}(r,z)\hat{e}_r\hat{e}_r + E_{\theta\theta}(r,z)\hat{e}_\theta\hat{e}_\theta$ in the cylindrical coordinate system is given by

$$\nabla \times \mathbf{E} = \left(\frac{\partial E_{\theta\theta}}{\partial r} + \frac{E_{\theta\theta} - E_{rr}}{r}\right)\hat{e}_z\hat{e}_\theta + \frac{\partial E_{rr}}{\partial z}\hat{e}_\theta\hat{e}_r - \frac{\partial E_{\theta\theta}}{\partial z}\hat{e}_r\hat{e}_\theta$$

1.12 Find the linear strains associated with the 2-D displacement field

$$u_1 = -c_0 x_1 x_2 + c_1 x_2 + c_2 x_3 + c_4$$
$$u_2 = c_0 \left[\nu\left(x_2^2 - x_1^2\right) + x_1^2\right] + c_4 x_3 + c_5 x_1 + c_6$$
$$u_3 = 0$$

where c_i and ν are constants.

1.13 Consider the displacement vector in polar axisymmetric system

$$\mathbf{u} = u_r \hat{e}_r + u_z \hat{e}_z$$

The strain tensor in the cylindrical coordinate system can be written in the dyadic form as

$$\varepsilon = \varepsilon_{rr}\hat{e}_r\hat{e}_r + \varepsilon_{\theta\theta}\hat{e}_\theta\hat{e}_\theta + \varepsilon_{zz}\hat{e}_z\hat{e}_z + \varepsilon_{r\theta}\left(\hat{e}_r\hat{e}_\theta + \hat{e}_\theta\hat{e}_r\right)$$
$$+ \varepsilon_{rz}\left(\hat{e}_r\hat{e}_z + \hat{e}_z\hat{e}_r\right) + \varepsilon_{\theta z}\left(\hat{e}_\theta\hat{e}_z + \hat{e}_z\hat{e}_\theta\right)$$

Show that the only nonzero linear strains are given by

$$\varepsilon_{rr} = \frac{\partial u_r}{\partial r}, \quad \varepsilon_{\theta\theta} = \frac{u_r}{r}, \quad \varepsilon_{rz} = \frac{1}{2}\left(\frac{\partial u_r}{\partial z} + \frac{\partial u_z}{\partial r}\right), \quad \varepsilon_{zz} = \frac{\partial u_z}{\partial z}$$

1.14 Using the definitions of ∇ and $\mathbf{u}$ in the cylindrical coordinate system, show that the linear strains (1.4.7) in the cylindrical coordinate system are given by

$$\varepsilon_{rr} = \frac{\partial u_r}{\partial r}, \quad \varepsilon_{r\theta} = \frac{1}{2}\left(\frac{1}{r}\frac{\partial u_r}{\partial \theta} + \frac{\partial u_\theta}{\partial r} - \frac{u_\theta}{r}\right), \quad \varepsilon_{rz} = \frac{1}{2}\left(\frac{\partial u_r}{\partial z} + \frac{\partial u_z}{\partial r}\right)$$

$$\varepsilon_{\theta\theta} = \frac{u_r}{r} + \frac{1}{r}\frac{\partial u_\theta}{\partial \theta}, \quad \varepsilon_{z\theta} = \frac{1}{2}\left(\frac{\partial u_\theta}{\partial z} + \frac{1}{r}\frac{\partial u_z}{\partial \theta}\right), \quad \varepsilon_{zz} = \frac{\partial u_z}{\partial z}$$

1.15 Show that the equations of motion in the cylindrical coordinate system are given by

$$\frac{\partial \sigma_{rr}}{\partial r} + \frac{1}{r}\frac{\partial \sigma_{r\theta}}{\partial \theta} + \frac{\partial \sigma_{rz}}{\partial z} + \frac{1}{r}(\sigma_{rr} - \sigma_{\theta\theta}) + \rho_0 f_r = \rho_0 \frac{\partial^2 u_r}{\partial t^2}$$

$$\frac{\partial \sigma_{r\theta}}{\partial r} + \frac{1}{r}\frac{\partial \sigma_{\theta\theta}}{\partial \theta} + \frac{\partial \sigma_{\theta z}}{\partial z} + \frac{2\sigma_{r\theta}}{r} + \rho_0 f_\theta = \rho_0 \frac{\partial^2 u_\theta}{\partial t^2}$$

$$\frac{\partial \sigma_{rz}}{\partial r} + \frac{1}{r}\frac{\partial \sigma_{\theta z}}{\partial \theta} + \frac{\partial \sigma_{zz}}{\partial z} + \frac{\sigma_{rz}}{r} + \rho_0 f_z = \rho_0 \frac{\partial^2 u_z}{\partial t^2}$$

1.16 Derive the transformation relations between the rectangular Cartesian components of stress $(\sigma_{xx}, \sigma_{yy}, \sigma_{xy})$ and the components of stress $(\sigma_{rr}, \sigma_{\theta\theta}, \sigma_{r\theta})$ in the cylindrical coordinate system.

1.17 Derive the transformation relations between the rectangular Cartesian components of strain $(\varepsilon_{xx}, \varepsilon_{yy}, \varepsilon_{xy})$ and the components of strain $(\varepsilon_{rr}, \varepsilon_{\theta\theta}, \varepsilon_{r\theta})$ in the cylindrical coordinate system.

References for Additional Reading

1. Boley, B. A., and Weiner, J. H., *Theory of Thermal Stresses*, John Wiley, New York (1960).

2. Bowen, R. M., and Wang, C. C., *Introduction to Vectors and Tensors*, Vols. I and II, Plenum, New York (1976).

3. Eringen, A. C., *Mechanics of Continua*, John Wiley, New York (1967).

4. Fung, Y. C., *Foundations of Solid Mechanics*, Prentice–Hall, Englewood Cliff, NJ (1965).

5. Jeffreys, H., *Cartesian Tensors*, Cambridge University Press, London (1965).

6. Malvern, L. E., *Introduction to the Mechanics of a Continuous Medium*, Prentice–Hall, Englewood Cliffs, NJ (1969).

7. Nowinski, J. L., *Theory of Thermoelasticity with Applications*, Sijthoff & Noordhoff, Alphen aan den Rijn, The Netherlands (1978).

8. Reddy, J. N., and Rasmussen, M. L., *Advanced Engineering Analysis*, John Wiley, New York (1982); reprinted by Krieger, Melbourne, Florida (1990).

9. Reddy, J. N., *Energy and Variational Methods in Applied Mechanics*, John Wiley, New York (1984).

10. Reddy, J. N., *Mechanics of Laminated Composite Plates: Theory and Analysis*, CRC Press, Boca Raton, Florida (1997).

Chapter Two

Energy Principles and Variational Methods

2.1 Virtual Work
2.1.1 Introduction

The governing equations of a continuum can be derived using the laws (or principles) of physics. For solid mechanics problems the laws of physics may be expressed in several alternative forms. For example, the principle of conservation of linear momentum, which leads to the equations of motion, can be derived using either Newton's Second Law of motion or by using the *principle of virtual displacements*. The former is termed a vector approach and the latter a variational approach.

The use of Newton's laws to determine the governing equations of a structural problem requires the isolation of a typical volume element of the structure with all its applied and reactive forces (i.e., the free-body diagram of the element). Then the vector sum of all static and dynamic forces and moments acting on the element is set to zero to obtain the equations of motion. For simple mechanical systems for which the free-body diagram can be set up, the vector approach provides an easy and direct way of deriving the governing equations. However, for complicated systems the procedure becomes more cumbersome and intractable. In addition, the type of boundary conditions to be used in conjunction with the derived equations is not always clear.

In the variational approach, the total work done or energy stored in the body due to actual forces in moving through *virtual* displacements that are consistent with the geometric constraints of a body is set to zero to obtain the equations of motion. The variational approach yields not only the equations of motion but also the force boundary conditions and indicate the form of the variables involved in the specification of geometric boundary conditions. In addition, the variational statements are useful in obtaining approximate solutions by direct variational methods, such as the Rayleigh–Ritz and finite element methods.

In most of the present study, the principles of virtual work will be used to derive the equations of motion of plates. We begin with the concepts of virtual displacements and forces and work and energy.

2.1.2 Virtual Displacements and Forces

From purely geometrical considerations, a given mechanical system can take many possible configurations consistent with the geometric constraints on the system. Of all the possible configurations, only one corresponds to the actual configuration of the system under the prescribed forces. It is this configuration that satisfies Newton's second law (i.e., equations of equilibrium or motion of the system). The set of configurations that satisfy the geometric constraints but not necessarily Newton's second law is called the *set of admissible configurations*. These configurations are restricted to a neighborhood of the true configuration so that they are obtained from infinitesimal variations of the true configuration. During such variations, the geometric constraints of the system are not violated and all the forces are fixed at their actual values. When a mechanical system experiences such variations in its configuration, it is said to undergo *virtual displacements* from its actual configuration. These displacements need not have any relationship to the actual displacements that might occur due to a change in the applied loads. The displacements are called virtual because they are *imagined* to take place while the actual loads are acting at their fixed values. *The virtual displacements at the boundary points at which the geometric boundary conditions are specified are necessarily zero.*

A boundary condition in which the specified value is zero is called a *homogeneous* boundary condition. For example, $u = \hat{u}$ is a nonhomogeneous boundary condition. If $\hat{u} = 0$, then the boundary condition is said to be homogeneous. The *homogeneous form* of a boundary condition $u = \hat{u}$ is $u = 0$.

For example, a bar fixed at one end and subjected to an axial load at the other end (see Figure 2.1.1) can be imagined to have virtual displacements that are zero at the fixed end, where the displacement is specified (to be zero). One such displacement is $v(x) = cx$, where c is a constant and the origin of the coordinate x is taken at the fixed end. In

ENERGY PRINCIPLES AND VARIATIONAL METHODS 49

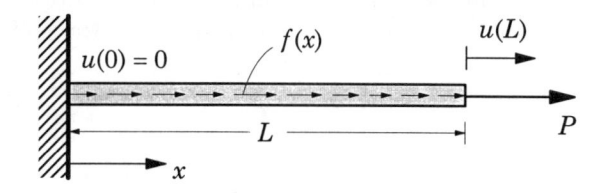

(a) Actual displacements and forces

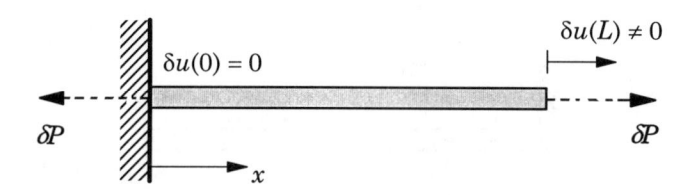

(b) Virtual displacements and forces

Figure 2.1.1. Virtual displacements and forces for a bar fixed at the left end. Actual forces are shown in solid lines and the virtual forces are shown in dashed lines.

fact, polynomials of the form $v(x) = c_1 x + c_2 x^2 + \cdots + c_n x^n$ with at least one of the constants c_k not zero constitute a set of admissible variations. Note that any polynomial with a nonzero constant coefficient is *not* an admissible variation because it is not zero at $x = 0$. For a cantilever beam of length L (i.e., a beam fixed at $x = 0$ and free at $x = L$; see Figure 2.1.2), the set of virtual displacements consists of polynomials of the form $v(x) = c_1 x^2 + c_2 x^3 + \cdots + c_n x^{n+1}$ with at least one of the c's nonzero. Note that v and dv/dx are zero at $x = 0$, consistent with the geometric boundary conditions of the beam.

Analogous to the concept of virtual displacements, one can think of virtual forces on a system. The virtual forces must be a set of forces that are in equilibrium among themselves. These forces can be internal and/or external and need not have any relation to the actual forces of the system. For example, a bar fixed at one end and subjected to an axial load at the other end can be imagined to have a variety of virtual forces. One such set is a force P applied, in addition to any actual forces, at the right and left ends in opposite directions. Similarly, for a cantilever beam of length L, a virtual force system consists of applying a virtual point load F acting upward at the free end and downward at the fixed end, and a moment of $M = F \cdot L$ acting clock-wise at the fixed end. The set of virtual forces $(F, -F, M = FL)$ is in self-equilibrium as can be verified by summing virtual forces and moments. The virtual forces and moments are shown by dotted lines in Figures 2.1.1 and 2.1.2.

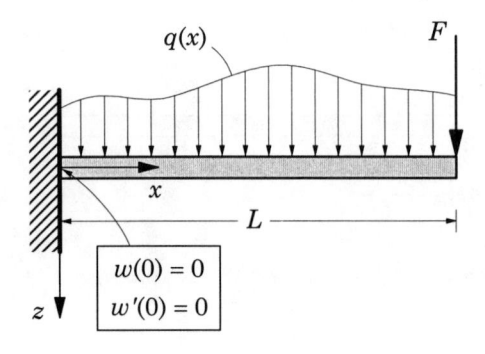

(a) Actual displacements and forces

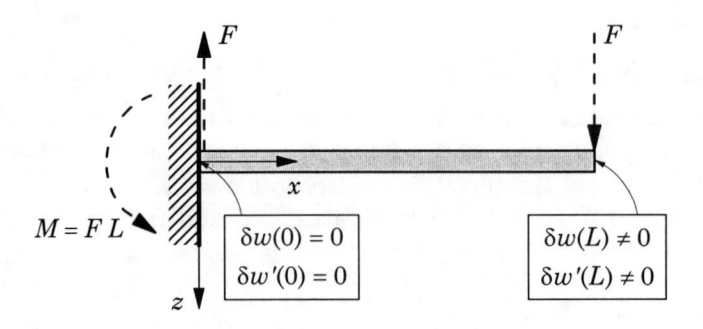

(b) Virtual displacements and forces

Figure 2.1.2. Virtual displacements and forces for a cantilever beam. Actual forces are shown in solid lines and the virtual forces are shown in dashed lines.

2.1.3 External and Internal Virtual Work

Let us first recall the concept of work. When a force acts on a material point and moves through a displacement, the work done by the force is defined by the *projection* of the force in the direction of the displacement times the magnitude of the displacement. The work done by the actual forces through virtual displacements or the work done by virtual forces through actual displacements is called *virtual work*. The virtual work done by actual forces $\mathbf{F}$ in a body Ω_0 in moving through the virtual displacements $\delta\mathbf{u}$ is given by

$$\delta W = \int_{\Omega_0} \mathbf{F} \cdot \delta\mathbf{u} \, dv$$

where dv denotes the volume element in the material body Ω_0. The symbol δ is used to denote a virtual change in the variable it operates on. Similarly, the virtual work done by virtual forces $\delta\mathbf{F}$ in moving through the actual displacement $\mathbf{u}$ is

$$\delta W^* = \int_{\Omega_0} \delta\mathbf{F} \cdot \mathbf{u} \, dv$$

External Virtual Work

The external virtual work done due to virtual displacements $\delta\mathbf{u}$ in a solid body Ω_0 subjected to body forces $\mathbf{f}$ per unit volume and surface tractions $\mathbf{T}$ per unit area of the boundary Γ_σ is given by

$$\delta V = -\left(\int_{\Omega_0} \mathbf{f} \cdot \delta\mathbf{u} \, dv + \int_{\Gamma_\sigma} \mathbf{T} \cdot \delta\mathbf{u} \, ds \right) \qquad (2.1.1)$$

where ds denotes a surface element and Γ_σ denotes the portion of the boundary on which stresses are specified. The negative sign in Eq. (2.1.1) indicates that the work is performed on the body. It is understood that the displacements are specified on the remaining portion $\Gamma_u = \Gamma - \Gamma_\sigma$ of the boundary Γ. Therefore, the virtual displacements are zero on Γ_u, irrespective of whether $\mathbf{u}$ is specified to be zero or not. For example, a bar fixed at one end ($x = 0$) and subjected to an axial load at the other end ($x = L$) can be imagined to have a virtual displacement $\delta u(x)$, provided $\delta u(0) = 0$, because the actual displacement is specified at $x = 0$.

52 THEORY AND ANALYSIS OF ELASTIC PLATES

The external virtual work done due to virtual body forces $\delta\mathbf{f}$ and surface tractions $\delta\mathbf{T}$ on Γ_u of a solid body Ω_0 in moving through the actual displacements $\mathbf{u}$ is given by

$$\delta V^* = -\left(\int_{\Omega_0} \delta\mathbf{f} \cdot \mathbf{u}\, dv + \int_{\Gamma_u} \delta\mathbf{T} \cdot \mathbf{u}\, ds\right) \tag{2.1.2}$$

Internal Virtual Work

The forces applied on a deformable body cause it to deform and the body experiences internal stresses. The relative movement of the material particles in the body can be measured in terms of strains. The forces associated with the internal stress field displace the material particles through displacements corresponding to the strain field in the body, and hence work is done. The work done by these internal forces in moving through displacements of the material particles is called *internal work*. Note that the work done on the body is responsible for the internal work.

The internal virtual work due to the virtual displacement $\delta\mathbf{u}$ can be computed as follows. Suppose that an infinitesimal material element of volume $dv = dx_1 dx_2 dx_3$ of the body experiences small virtual strains $\delta\varepsilon_{ij}$ due to the virtual displacements δu_i where

$$\delta\varepsilon_{ij} = \frac{1}{2}(\delta u_{i,j} + \delta u_{j,i}), \quad u_{i,j} \equiv \frac{\partial u_i}{\partial x_j} \tag{2.1.3}$$

The work done by the force due to actual stress σ_{11}, for example, in moving through the virtual displacement $\delta u_1 = \delta\varepsilon_{11}\, dx_1$ is (see Figure 2.1.3)

$$(\sigma_{11}\, dx_2 dx_3)(\delta\varepsilon_{11}\, dx_1) = \sigma_{11}\delta\varepsilon_{11}\, dx_1 dx_2 dx_3$$

Similarly, the work done by the force due to stress σ_{12} in shearing the body is (see Figure 2.1.4)

$$(\sigma_{12}\, dx_2 dx_3)(2\delta\varepsilon_{12}\, dx_1) = 2\sigma_{12}\delta\varepsilon_{12}\, dx_1 dx_2 dx_3$$

Thus, the total virtual work done by forces due to all the stresses in a volume element in moving through their respective displacements is

$$(\sigma_{11}\delta\varepsilon_{11} + \sigma_{22}\delta\varepsilon_{22} + \sigma_{33}\delta\varepsilon_{33} + 2\sigma_{12}\delta\varepsilon_{12} 2\sigma_{13}\delta\varepsilon_{13} + 2\sigma_{23}\delta\varepsilon_{23})\, dx_1 dx_2 dx_3$$

$$= \sigma_{ij}\, \delta\varepsilon_{ij}\, dv \tag{2.1.4}$$

ENERGY PRINCIPLES AND VARIATIONAL METHODS   53

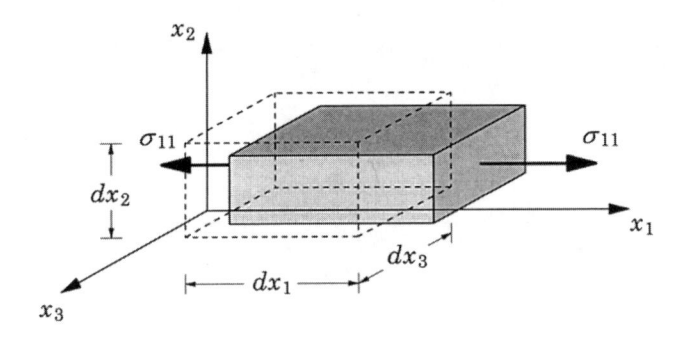

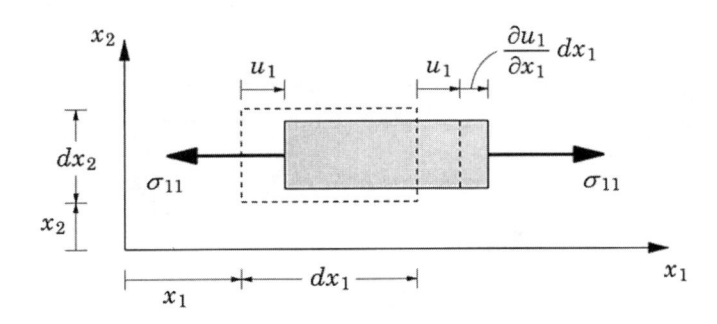

Figure 2.1.3. Virtual work done by normal stress σ_{11}.

The total internal virtual work done, denoted by δU, is obtained by integrating the above expression over the entire volume of the body

$$\delta U = \int_{\Omega_0} \sigma_{ij}\, \delta\varepsilon_{ij}\, dv = \int_{\Omega_0} \sigma : \delta\varepsilon\, dv \qquad (2.1.5)$$

where ':' denotes the double dot product. Equation (2.1.5) is valid for any material body irrespective of its constitutive behavior. The expression in Eq. (2.1.5) is called the *virtual strain energy*. Although Eq. (2.1.5) was derived for infinitesimal strain case, it also holds for the finite strain case.

The internal virtual work done by virtual stresses $\delta\sigma_{ij}$ in moving through the actual strains ε_{ij} is given by

$$\delta U^* = \int_{\Omega_0} \varepsilon_{ij}\, \delta\sigma_{ij}\, dv. = \int_{\Omega_0} \varepsilon : \delta\sigma\, dv \qquad (2.1.6)$$

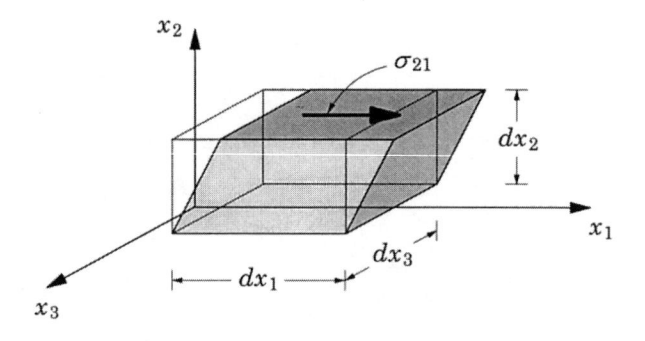

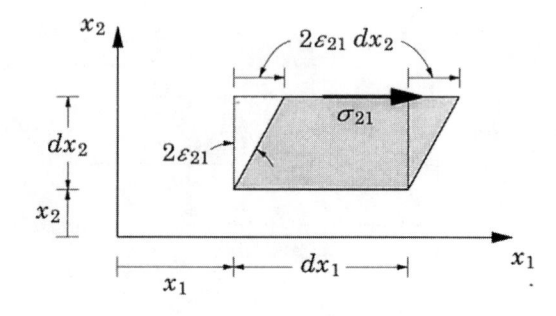

Figure 2.1.4. Virtual work done by shear stress σ_{21}.

The expression in Eq. (2.1.6) is also known as the *virtual complementary strain energy*. The virtual forces ($\delta\mathbf{f}, \delta\mathbf{T}$) and virtual stresses ($\delta\sigma$) should be such that the stress equilibrium equations

$$\frac{\partial \delta\sigma_{ij}}{\partial x_j} + \rho_0\,\delta f_i = 0 \quad \text{or} \quad \nabla \cdot \delta\sigma + \rho_0\,\delta\mathbf{f} = \mathbf{0} \tag{2.1.7}$$

and stress boundary conditions

$$\delta\sigma_{ij}\, n_j = \delta T_i \quad \text{or} \quad \delta\sigma \cdot \hat{\mathbf{n}} = \mathbf{T} = \mathbf{0} \tag{2.1.8}$$

are satisfied.

Next we consider an example of virtual work done by actual forces in moving through virtual displacements.

Example 2.1.1

Consider the bending and stretching of an Euler–Bernoulli beam on an elastic foundation and subjected to an applied transverse load $q(x)$ at the top (see Figure 2.1.5). If we assume that the strains are of the form given in Eq. (1.4.14), the internal and external virtual works are given by

$$\delta W_E = -\int_0^L q(x)\delta w\left(x, \frac{h}{2}\right) dx - \int_0^L F_s \delta w\left(x, -\frac{h}{2}\right) dx$$

$$= -\int_0^L q(x)\delta w_0(x)\, dx + \int_0^L k\delta w_0(x)\, dx \qquad (a)$$

$$\delta W_I = \int_0^L \int_A (\sigma_{xx}\delta\varepsilon_{xx} + \sigma_{zz}\delta\varepsilon_{zz})\, dx dA$$

$$= \int_0^L \int_A \left[\sigma_{xx}\left(\frac{d\delta u_0}{dx} + \frac{dw_0}{dx}\frac{d\delta w_0}{dx} - z\frac{d^2\delta w_0}{dx^2}\right)\right.$$

$$\left. + \sigma_{zz}\left(\frac{dw_0}{dx}\frac{d\delta w_0}{dx}\right)\right] dx dA \qquad (b)$$

where k is the modulus of the elastic foundation, u_0 the axial displacement of a point at $z = 0$, w_0 the transverse deflection, L the length, and A is the cross-sectional area of the beam. The foundation reaction force F_s is replaced with $F_s = -kw_0(x)$ using the linear elastic constitutive equation for the foundation.

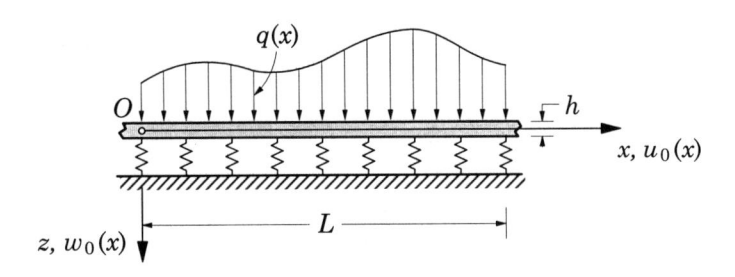

Figure 2.1.5. A beam on elastic foundation.

2.1.4 The Variational Operator

The delta symbol 'δ' used in conjunction with virtual displacements and forces can be interpreted as an operator, called the *variational operator*. It is used to denote a variation (or change) in a given quantity, i.e., δu denotes a variation in u. Thus δ is an operator that produces virtual change or variation δu in a dependent variable u, in much the same way that dx denotes a change in x. In fact, the variational operator δ *is a differential operator with respect to the dependent variables*. Indeed, the laws of variation of sums, products, ratios, powers, and so forth are completely analogous to the corresponding laws of differentiation. For example, if $F_1 = F_1(u)$ and $F_2 = F_2(u)$, we have

(a) $\delta(F_1 \pm F_2) = \delta F_1 \pm \delta F_2$

(b) $\delta(F_1 \, F_2) = \delta F_1 \, F_2 + F_1 \, \delta F_2$

(c) $\delta(\frac{F_1}{F_2}) = \frac{\delta F_1 \, F_2 - F_1 \, \delta F_2}{F_2^2}$

(d) $\delta(F_1)^n = n(F_1)^{n-1} \delta F_1$ \hfill (2.1.9a)

If $G = G(u, v, w)$ is a function of several dependent variables (and possibly their derivatives), the total variation is the sum of partial variations:

$$\delta G = \delta_u G + \delta_v G + \delta_w G \tag{2.1.9b}$$

where, for example, δ_u denotes the partial variation with respect to u.

The variational operator can be interchanged with differential and integral operators (commutativity):

(a) $\delta\left(\frac{du}{dx}\right) = \alpha \frac{dv}{dx} = \frac{d}{dx}(\alpha v) = \frac{d}{dx}(\delta u)$

(b) $\delta\left(\int_\Omega u \, d\Omega\right) = \alpha \int_\Omega v \, d\Omega = \int_\Omega \alpha v \, d\Omega = \int_\Omega \delta u \, d\Omega$ \hfill (2.1.9c)

where Ω denotes a one-, two-, or three-dimensional domain and $d\Omega$ is an element of the domain.

The variational operator proves to be very useful in constructing virtual work statements and deriving governing equations from virtual work principles, as will be shown shortly. The delta operator can be viewed as a differential operator with respect to dependent variables.

2.1.5 Functionals

A formal mathematical definition of a *functional* requires concepts from functional analysis. It is a real number (or scalar) obtained by operating on functions (dependent variables) from a given set of functions (or vector space). Thus, a functional is a real number $I(u)$ obtained by an operator $I(\cdot)$ that maps functions u of a vector space H into a real number $I(u)$ in the set of real numbers R:

$$I : H \to R \qquad (2.1.10)$$

Note that $I(\cdot)$ is an operator and $I(u)$ is a functional. For example, the integral expression

$$I(u) = \int_0^L [au(x) + bu'(x) + cu''(x)] \, dx$$

qualifies as a functional for all integrable and square-integrable functions $u(x)$. Note that the value of the functional $I(u)$ depends on the choice of u.

A functional is said to be *linear* if

$$I(c_1 u + c_2 v) = c_1 I(u) + c_2 I(v) \qquad (2.1.11)$$

for all constants c_1 and c_2 and dependent variables u and v. A *quadratic* functional is one which satisfies the relation

$$I(cu) = c^2 I(u) \qquad (2.1.12)$$

for all constants c and the dependent variable u.

The first variation of a functional of u (and its derivatives) can be calculated as follows. Let $I(u)$ denote the integral defined in the interval (a, b)

$$I(u) = \int_a^b F(x, u, u') \, dx \qquad (2.1.13)$$

where F is a function, in general, of x, u, and $u' \equiv du/dx$. The first variation of the functional $I(u)$ is

$$\delta I(u) = \delta \int_a^b F(x, u, u') \, dx = \int_a^b \delta F \, dx = \int_a^b \left(\frac{\partial F}{\partial u} \delta u + \frac{\partial F}{\partial u'} \delta u' \right) dx \qquad (2.1.14)$$

Thus, the variation of a functional can be readily calculated.

2.1.6 Fundamental Lemma of Variational Calculus

The *fundamental lemma of calculus of variations* is useful in obtaining differential equations from integral statements. The lemma can be stated as follows: for any integrable function G, if the statement

$$\int_a^b G \cdot \eta \, dx = 0 \qquad (2.1.15)$$

holds for any arbitrary continuous function $\eta(x)$, for all x in (a,b), then it follows that $G = 0$ in (a,b). A mathematical proof of the lemma can be found in most books on variational calculus. A simple proof of the lemma follows. Since η is arbitrary, it can be replaced by G. We have

$$\int_a^b G^2 \, dx = 0$$

Since an integral of a positive function is positive, the above statement implies that $G = 0$.

A more general statement of the Fundamental Lemma is as follows: If η is arbitrary in $a < x < b$ and $\eta(a)$ is arbitrary, then the statement

$$\int_a^b G\eta \, dx + B(a)\eta(a) = 0 \qquad (2.1.16)$$

implies that $G = 0$ in $a < x < b$ and $B(a) = 0$ because $\eta(x)$ is independent of $\eta(a)$.

2.1.7 Euler–Lagrange Equations

Consider the problem of finding the extremum (i.e., minimum or maximum) of the functional

$$I(u) = \int_a^b F(x, u, u') \, dx, \quad u(a) = u_a, \quad u(b) = u_b \qquad (2.1.17)$$

The necessary condition for the functional to have a minimum or maximum is (analogous to maxima or minima of functions) that its first variation be zero:

$$\delta I(u) = 0 \qquad (2.1.18)$$

We have

$$0 = \int_a^b \left(\frac{\partial F}{\partial u} \delta u + \frac{\partial F}{\partial u'} \delta u' \right) dx$$

Note that $\delta u' = \delta(du/dx) = d(\delta u)/dx$. We cannot use the Fundamental Lemma in the above equation because it is not in the form of Eq. (2.1.15) or Eq. (2.1.16). To recast the above equation in the form of Eq. (2.1.15), we integrate the second term by parts and obtain

$$\begin{aligned} 0 &= \int_a^b \left(\frac{\partial F}{\partial u} \delta u + \frac{\partial F}{\partial u'} \delta u' \right) dx \\ &= \int_a^b \left(\frac{\partial F}{\partial u} \delta u + \frac{\partial F}{\partial u'} \frac{d\delta u}{dx} \right) dx \\ &= \int_a^b \left[\frac{\partial F}{\partial u} - \frac{d}{dx}\left(\frac{\partial F}{\partial u'}\right) \right] \delta u \, dx + \left[\frac{\partial F}{\partial u'} \delta u \right]_a^b \end{aligned} \quad (2.1.19)$$

Let us first examine the boundary expression:

$$\left[\frac{\partial F}{\partial u'} \right] \cdot \delta u$$

There are two parts to this expression: a varied quantity and its coefficient. The variable u that is subjected to variation is called the *primary variable*. The coefficient of the varied quantity, i.e., the expression next to δu in the boundary term is called a *secondary variable*. The product of the primary variable (or its variation) with the secondary variable often represents the work done (or virtual work done). The specification of the primary variable at a boundary point is termed the *essential boundary condition*, and the specification of the secondary variable is called the *natural boundary condition*. In solid mechanics, these are known as the *geometric* and *force* boundary conditions, respectively. All admissible variations must satisfy the homogeneous form of the essential (or geometric) boundary conditions: $\delta u(a) = 0$ and $\delta u(b) = 0$. Elsewhere, $a < x < b$, δu is arbitrary.

Returning to Eq. (2.1.19), we note that the boundary terms drop out because of the conditions on δu. We have

$$0 = \int_a^b \left[\frac{\partial F}{\partial u} - \frac{d}{dx}\left(\frac{\partial F}{\partial u'}\right) \right] \delta u \, dx$$

which must hold for any δu in (a, b). In view of the fundamental lemma of calculus of variations ($\eta = \delta u$), it follows that [see Eq. (2.1.15)]

$$G \equiv \frac{\partial F}{\partial u} - \frac{d}{dx}\left(\frac{\partial F}{\partial u'}\right) = 0 \quad \text{in} \quad a < x < b \quad (2.1.20)$$

60   THEORY AND ANALYSIS OF ELASTIC PLATES

Thus the necessary condition for $I(u)$ to be an extremum at $u = u(x)$ is that $u(x)$ be the solution of Eq. (2.1.20), which is known as the *Euler–Lagrange equation* associated with the functional $I(u)$, i.e., $\delta I = 0$ is equivalent to Eq. (2.1.20).

If $u(a) = u_a$ and $\delta u(b)$ is arbitrary (i.e., $u(a)$ is specified but u is not specified at $x = b$), then $\delta u(a) = 0$ and we have from Eq. (2.1.19) the result

$$0 = \int_a^b \left[\frac{\partial F}{\partial u} - \frac{d}{dx}\left(\frac{\partial F}{\partial u'}\right) \right] \delta u \, dx + \left(\frac{\partial F}{\partial u'}\right)_{x=b} \delta u(b) \qquad (2.1.21)$$

Since δu is arbitrary in (a, b) and $\delta u(b)$ is arbitrary, the above equation implies, in view of Eq. (2.1.16), that both the integral expression and the boundary term be zero separately:

$$\frac{\partial F}{\partial u} - \frac{d}{dx}\left(\frac{\partial F}{\partial u'}\right) = 0, \quad a < x < b \qquad (2.1.22a)$$

$$\left(\frac{\partial F}{\partial u'}\right) = 0 \text{ at } x = b \qquad (2.1.22b)$$

Both Eq. (2.1.22a) and Eq. (2.1.22b) are called the Euler–Lagrange equations. Note that the boundary conditions that are a part of the Euler–Lagrange equations always belong to the class of natural boundary conditions.

Example 2.1.2

Consider the functional

$$I(u) = \int_\Omega \left[\frac{1}{2}(\nabla u \cdot \nabla u) - fu \right] dv - \int_{\Gamma_2} \hat{t} u \, ds \qquad (a)$$

where u and f are functions of position, ds and dv denote the surface and volume elements of the domain Ω, and Γ_2 denotes a portion of the boundary Γ. We wish to determine the Euler–Lagrange equations associated with the functional when u is specified on the other portion of the boundary

$$u = \hat{u} \text{ on } \Gamma_1, \quad \Gamma_1 + \Gamma_2 = \Gamma$$

The first variation of $I(u)$ is given by

$$\delta I(u) = \int_\Omega \left[\frac{1}{2}(\nabla \delta u \cdot \nabla u + \nabla u \cdot \nabla \delta u) - f \delta u \right] dv - \int_{\Gamma_2} \hat{t} \delta u \, ds$$

$$= \int_\Omega (-\nabla \cdot \nabla u + f) \delta u \, dv + \oint_\Gamma \hat{n} \cdot \nabla u \, \delta u \, ds - \int_{\Gamma_2} \hat{t} \delta u \, ds \qquad (b)$$

where the *divergence theorem* of vector calculus

$$\int_\Omega \nabla F \, dv = \oint_\Gamma \hat{\mathbf{n}} \cdot \nabla F \, ds \tag{2.1.23}$$

is used to relieve δu of the gradient operation:

$$\int_\Omega \nabla u \cdot \nabla \delta u \, dv = \int_\Omega [\nabla \cdot (\nabla u \delta u) - (\nabla \cdot \nabla u) \, \delta u] \, dv$$
$$= \oint_\Gamma (\hat{\mathbf{n}} \cdot \nabla u) \, \delta u \, ds - \int_\Omega (\nabla \cdot \nabla u) \, \delta u \, dv$$

A circle on the boundary integral denotes integration over the closed boundary. Since the total boundary Γ is the direct sum of Γ_1 and Γ_2 and $\delta u = 0$ on Γ_1, the integral on the total boundary of (b) is reduced to the boundary integral over Γ_2 and we obtain ($\nabla \cdot \nabla = \nabla^2$)

$$\delta I(u) = \int_\Omega \left(-\nabla^2 u + f\right) \delta u \, dv + \int_{\Gamma_2} \left(\frac{\partial u}{\partial n} - \hat{t}\right) \delta u \, ds \tag{c}$$

where

$$\frac{\partial u}{\partial n} = \hat{\mathbf{n}} \cdot \nabla u$$

is the *normal derivative* of u. Now, setting $\delta I = 0$ and using the fundamental lemma, we arrive at the Euler–Lagrange equations

$$-\nabla^2 u + f = 0 \quad \text{in } \Omega \tag{d}$$

$$\frac{\partial u}{\partial n} - \hat{t} = 0 \quad \text{on } \Gamma_2 \tag{e}$$

2.2 Energy Principles
2.2.1 Introduction

Recall that the virtual work due to virtual displacements is the work done by actual forces in moving the material particles of the body through virtual displacements that are consistent with the geometric constraints. All applied forces are kept constant during the virtual displacements. Consider a rigid body acted upon by a set of applied forces $\mathbf{F}_1, \mathbf{F}_2, \ldots \mathbf{F}_n$, and suppose that the points of application of these forces are subjected to the virtual displacements $\delta \mathbf{u}_1$, $\delta \mathbf{u}_2$, $\cdots$, $\delta \mathbf{u}_n$, respectively. The virtual displacement $\delta \mathbf{u}_i$ has no relation to $\delta \mathbf{u}_j$ for $i \neq j$. The external virtual work done by the virtual displacements is

$$\delta V = -[\mathbf{F}_1 \cdot \delta \mathbf{u}_1 + \mathbf{F}_2 \cdot \delta \mathbf{u}_2 + \cdots + \mathbf{F}_n \cdot \delta \mathbf{u}_n] = -\mathbf{F}_i \cdot \delta \mathbf{u}_i \quad (2.2.1)$$

where the sum on repeated indices (over the range of 1 to n) is implied. The internal virtual work done δU is zero because a rigid body does not undergo any strains (hence, virtual strains are zero). In addition, the virtual displacements $\delta \mathbf{u}_1$, $\delta \mathbf{u}_2$, $\cdots$, $\delta \mathbf{u}_n$ should all be the same, say $\delta \mathbf{u}$, for a rigid body. Thus, we have

$$\delta V = -\mathbf{F}_i \cdot \delta \mathbf{u}_i = -\left(\sum_{i=1}^n \mathbf{F}_i\right) \cdot \delta \mathbf{u} \quad \text{and} \quad \delta U = 0 \quad (2.2.2)$$

But by Newton's second law, the vector sum of the forces acting on a body in static equilibrium is zero. This implies that the total virtual work, $\delta U + \delta V$, is equal to zero. Thus, for a body in equilibrium the total virtual work done due to virtual displacements is zero. This statement is known as *the principle of virtual displacements*. The principle also holds for continuous, deformable bodies for which δU is not zero.

2.2.2 The Principle of Virtual Displacements

Consider a continuous body $\mathcal{B}$ in equilibrium under the action of body forces $\mathbf{f}$ and surface tractions $\mathbf{T}$. Let the reference configuration be the initial configuration $\mathcal{C}^0$ whose volume is denoted as Ω_0. Suppose that over portion Γ_u of the total boundary Γ of the region Ω_0, the displacements are specified to be $\hat{\mathbf{u}}$, and on portion Γ_σ, the tractions are specified to be $\hat{\mathbf{T}}$. The boundary portions Γ_u and Γ_σ are disjoint (i.e., do not overlap), and their sum is the total boundary Γ. Let $\mathbf{u}$ be the displacement vector corresponding to the equilibrium configuration of the body, and let σ and ε be the stress and strain tensors, respectively. The set of admissible configurations are defined by sufficiently differentiable functions that satisfy the geometric boundary conditions $\mathbf{u} = \hat{\mathbf{u}}$ on Γ_u.

If the body is in equilibrium, then of all admissible configurations, the actual one corresponding to the equilibrium configuration makes the total virtual work done zero. In order to determine the equations governing the equilibrium configuration $\mathcal{C}$, we let the body experience a virtual displacement $\delta\mathbf{u}$ from the true configuration $\mathcal{C}$. The virtual displacements are arbitrary, continuous functions except that they satisfy the homogeneous form of geometric boundary conditions, i.e., they must belong to the set of admissible variations.

The principle of virtual displacements can be stated as follows: *if a continuous body is in equilibrium, the virtual work of all actual forces in moving through a virtual displacement is zero:*

$$\delta U + \delta V \equiv \delta W = 0 \qquad (2.2.3)$$

Just as we derived the Euler–Lagrange equations associated with the statement $\delta I = 0$, we can derive them for the statement in Eq. (2.2.3). However, first we must identify δU and δV for a given problem. The principle of virtual work is independent of any constitutive law and applies to both elastic (linear and nonlinear) and inelastic continua.

For a solid body, the external and internal virtual work expressions are given in Eqs. (2.1.2) and (2.1.5), respectively. The principle can be expressed as

$$\int_{\Omega_0} \boldsymbol{\sigma} : \delta\boldsymbol{\varepsilon} \, dv - \int_{\Omega_0} \rho_0 \mathbf{f} \cdot \delta\mathbf{u} \, dv - \int_{\Gamma_\sigma} \mathbf{T} \cdot \delta\mathbf{u} \, ds = 0 \qquad (2.2.4)$$

where Ω_0 is the volume of the undeformed body, and dv and ds denote the volume and surface elements of Ω_0. Writing in terms of the Cartesian rectangular components, Eq. (2.2.4) takes the form

$$\int_{\Omega_0} (\sigma_{ij}\delta\varepsilon_{ij} - \rho_0 f_i \delta u_i) \, dv - \int_{\Gamma_\sigma} T_i \delta u_i \, ds = 0 \qquad (2.2.5)$$

where the summation on repeated subscripts is implied.

We will show that the Euler–Lagrange equations associated with the statement (2.2.4) or (2.2.5) of the principle of virtual displacements are nothing but the equilibrium equations of the 3-D elasticity [see Eq. (1.4.24)]. First we use the rectangular Cartesian component form to establish the result.

The virtual strains $\delta\varepsilon_{ij}$ are related to the virtual displacements δu_i by

$$\delta\varepsilon_{ij} = \frac{1}{2}(\delta u_{i,j} + \delta u_{j,i}), \quad u_{i,j} \equiv \frac{\partial u_i}{\partial x_j} \qquad (2.2.6)$$

Substituting $\delta\varepsilon_{ij}$ from the above equation into Eq. (2.2.5) and using the divergence theorem to transfer differentiation from δu_i to its coefficient, one obtains

$$\begin{aligned} 0 &= \int_{\Omega_0}\left[\frac{1}{2}\sigma_{ij}(\delta u_{i,j}+\delta u_{j,i})-\rho_0 f_i \delta u_i\right]dv - \int_{\Gamma_\sigma}T_i\delta u_i\,ds \\ &= \int_{\Omega_0}(\sigma_{ij}\delta u_{ij}-\rho_0 f_i\delta u_i)\,dv - \int_{\Gamma_\sigma}T_i\delta u_i\,ds \\ &= -\int_{\Omega_0}(\sigma_{ij,j}+\rho_0 f_i)\delta u_i\,dv - \int_{\Gamma_\sigma}T_i\delta u_i\,ds + \oint_\Gamma \sigma_{ij}n_j\delta u_i\,ds \quad (2.2.7) \end{aligned}$$

Since $\delta u_i = 0$ on Γ_u, we have

$$0 = -\int_{\Omega_0}(\sigma_{ij,j}+\rho_0 f_i)\delta u_i\,dv + \int_{\Gamma_\sigma}(\sigma_{ij}n_j - T_i)\delta u_i\,ds \quad (2.2.8)$$

Because the virtual displacements are arbitrary in Ω_0 and on Γ_σ, Eq. (2.2.8) yields the following Euler–Lagrange equations

$$\sigma_{ij,j} + \rho_0 f_i = 0 \text{ in } \Omega_0 \quad (2.2.9)$$

$$\sigma_{ij}n_j - T_i = 0 \text{ on } \Gamma_\sigma \quad (2.2.10)$$

Equations (2.2.9) and (2.2.10) are the Euler–Lagrange equations associated with the principle of virtual displacements for a body undergoing small strains and displacements. The boundary conditions in Eq. (2.2.10) are the natural boundary conditions. The principle of virtual displacements is applicable to any continuous body with arbitrary constitutive behavior (i.e., elastic or inelastic).

Example 2.2.1

Consider the internal and external virtual works in Eqs. (a) and (b) of Example 2.1.1 associated with an Euler–Bernoulli beam on an elastic foundation (modulus k) and subjected to an applied transverse load $q(x)$ at the top. If we assume that the strains are of the form given in Eq. (1.4.14), the principle of virtual work requires

$$\begin{aligned} 0 = \int_0^L \int_A &\left[\sigma_{xx}\left(\frac{d\delta u_0}{dx}+\frac{dw_0}{dx}\frac{d\delta w_0}{dx}-z\frac{d^2\delta w_0}{dx^2}\right)\right. \\ &\left.+\sigma_{zz}\left(\frac{dw_0}{dx}\frac{d\delta w_0}{dx}\right)\right]dAdx - \int_0^L (q-kw_0)\,\delta w_0\,dx \end{aligned}$$

$$= \int_0^L \left[N_{xx} \left(\frac{d\delta u_0}{dx} + \frac{dw_0}{dx} \frac{d\delta w_0}{dx} \right) - M_{xx} \frac{d^2 \delta w_0}{dx^2} + N_{zz} \left(\frac{dw_0}{dx} \frac{d\delta w_0}{dx} \right) \right] dx$$
$$- \int_0^L (q - kw_0) \, \delta w_0 \, dx \quad \text{(a)}$$

$$= \int_0^L \left[-\frac{dN_{xx}}{dx} \delta u_0 - \frac{d}{dx} \left(\frac{dw_0}{dx} N_{xx} \right) \delta w_0 - \frac{d^2 M_{xx}}{dx^2} \delta w_0 \right.$$
$$\left. - \frac{d}{dx} \left(\frac{dw_0}{dx} N_{zz} \right) \delta w_0 \right] dx - \int_0^L (q - kw_0) \, \delta w_0 \, dx$$
$$+ \left[N_{xx} \delta u_0 + \left(\frac{dw_0}{dx} N_{xx} + \frac{dM_{xx}}{dx} \frac{dw_0}{dx} N_{zz} \right) \delta w_0 - M_{xx} \frac{d\delta w_0}{dx} \right]_0^L \quad \text{(b)}$$

where N_{xx}, N_{zz}, and M_{xx} are the stress resultants defined by

$$N_{xx} = \int_A \sigma_{xx} dA, \quad N_{zz} = \int_A \sigma_{zz} dA, \quad M_{xx} = \int_A \sigma_{xx} z \, dA \quad \text{(c)}$$

Hence, the Euler–Lagrange equations are

$$\delta u_0: \quad -\frac{dN_{xx}}{dx} = 0 \quad \text{(d)}$$

$$\delta w_0: \quad -\frac{d^2 M_{xx}}{dx^2} - \frac{d}{dx}\left[\frac{dw_0}{dx} (N_{xx} + N_{zz}) \right] + kw_0 - q = 0 \quad \text{(e)}$$

in $0 < x < L$ and the following boundary conditions at $x = 0, L$ are:

$$N_{xx} = 0, \quad \frac{dM_{xx}}{dx} + \frac{dw_0}{dx}(N_{xx} + N_{zz}) = 0, \quad M_{xx} = 0 \quad \text{(f)}$$

2.2.3 Hamilton's Principle

The principle of virtual displacements is limited to static equilibrium of solid bodies. Hamilton's principle is a generalization of the principle of virtual displacements to dynamics. The dynamics version of the principle of virtual displacements, i.e., Hamilton's principle, is based under the assumption that a dynamical system is characterized by two energy functions: a *kinetic energy* K and a *potential energy* W.

Newton's second law of motion expresses the global statement of the principle of conservation of linear momentum. However, Newton's second law of motion is not sufficient to determine the motion $\mathbf{u} = \mathbf{u}(\mathbf{x}, t)$. The kinematic conditions and constitutive equations derived in Chapter 1 are needed to completely determine the motion. The actual path $\mathbf{u} = \mathbf{u}(\mathbf{x}, t)$ followed by a material particle in position $\mathbf{x}$ in the body is varied, consistent with kinematic (essential) boundary conditions, to $\mathbf{u} + \delta\mathbf{u}$, where $\delta\mathbf{u}$ is the admissible variation (or virtual displacement) of the path. We suppose that the varied path differs from the actual path except at initial and final times, t_1 and t_2, respectively. Thus, an admissible variation $\delta\mathbf{u}$ satisfies the conditions

$$\delta\mathbf{u} = 0 \text{ on } \Gamma_u \text{ for all } t$$

$$\delta\mathbf{u}(\mathbf{x}, t_1) = \delta\mathbf{u}(\mathbf{x}, t_2) = 0 \text{ for all } \mathbf{x} \qquad (2.2.11)$$

where Γ_u denotes the portion of the surface of the body where the displacement vector $\mathbf{u}$ is specified.

Hamilton's principle states that *of all possible paths that a material particle could travel from its position at time t_1 to its position at time t_2, its actual path will be one for which the integral*

$$\int_{t_1}^{t_2} (K - W) \, dt \qquad (2.2.12)$$

is an extremum. The difference between the kinetic and potential energies is called the *Lagrangian* function

$$L = K - W = K - W_I - W_E \qquad (2.2.13)$$

Thus, Hamilton's principle can be expressed as

$$0 = \int_{t_1}^{t_2} (\delta K - \delta W_I - \delta W_E) \, dt \qquad (2.2.14)$$

The kinetic energy of a continuous body is given by

$$K = \frac{1}{2} \int_{\Omega_0} \rho \frac{\partial \mathbf{u}}{\partial t} \cdot \frac{\partial \mathbf{u}}{\partial t} \, dv \qquad (2.2.15)$$

where $\frac{\partial \mathbf{u}}{\partial t}$ denotes the velocity of the material particle.

The requirement that $\delta\mathbf{u}$ be zero at the final time when $\mathbf{u}(\mathbf{x}, t_2)$ is not known or specified is often a point of discussion, and for this reason Hamilton's principle is sometimes referred to as a pseudo variational principle. The principle gives the equations of motion but not the initial conditions (see Oden and Reddy [1] for additional discussion). An application of Hamilton's principle is presented in the next example.

Example 2.2.2: The Timoshenko beam theory

The Euler–Bernoulli beam theory is based on the assumption that a straight line transverse to the axis of the beam before deformation remains (i) straight, (ii) inextensible, and (iii) normal to the midplane after deformation. In the Timoshenko beam theory, the first two assumptions are kept but the normality condition is relaxed by assuming that the rotation is independent of the slope ($w_{0,x}$) of the beam. Using these assumptions, the displacement field of the beam can be expressed as

$$u_1(x, z, t) = u_0(x, t) + z\phi(x, t)$$
$$u_2 = 0$$
$$u_3 = w_0(x, t) \qquad (a)$$

where (u_1, u_2, u_3) are the displacements of a point along the (x, y, z) coordinates, (u_0, w_0) are the displacements of a point on the mid-plane of an undeformed beam, and ϕ is the rotation (about the y−axis) of a transverse normal line. We wish to derive the equations of motion of the Timoshenko beam theory using Hamilton's principle for the case of infinitesimal strains.

The only nonzero strains are

$$\varepsilon_{xx} = \frac{\partial u_1}{\partial x} = \frac{\partial u_0}{\partial x} + z\frac{\partial \phi}{\partial x} \equiv \varepsilon_{xx}^0 + z\varepsilon_{xx}^1$$
$$\gamma_{xz} = \frac{\partial u_1}{\partial z} + \frac{\partial u_3}{\partial x} = \phi + \frac{\partial w_0}{\partial x} \equiv \gamma_{xz}^0 \qquad (b)$$

The virtual strains are

$$\delta\varepsilon_{xx}^0 = \frac{\partial \delta u_0}{\partial x}, \quad \delta\varepsilon_{xx}^1 = \frac{\partial \delta\phi}{\partial x}, \quad \delta\gamma_{xz} = \delta\phi + \frac{\partial \delta w_0}{\partial x}$$

Next, we compute the virtual energies required in Eq. (2.2.14). We have

$$\delta W_I = \int_V \sigma_{ij}\delta\varepsilon_{ij}\, dV$$
$$\delta W_E = \int_0^L q\delta w_0\, dx$$
$$\delta K = \frac{1}{2}\delta \int_V \rho\left[(\dot{u}_1)^2 + (\dot{u}_3)^2\right] dV$$

where ρ is the density and q is the distributed transverse load. We have

$$\delta W_I = \int_0^L \int_A (\sigma_{xx}\delta\varepsilon_{xx} + \sigma_{xz}\delta\gamma_{xz})\,dA\,dx$$

$$= \int_0^L \int_A \left[\sigma_{xx}\left(\delta\varepsilon_{xx}^0 + z\delta\varepsilon_{xx}^1\right) + \sigma_{xz}\delta\gamma_{xz}^0\right]dA\,dx$$

$$= \int_0^L \left(N_{xx}\delta\varepsilon_{xx}^0 + M_{xx}\delta\varepsilon_{xx}^1 + Q_x\delta\gamma_{xz}^0\right)dx \tag{c}$$

$$\delta W_E = -\int_0^L q\delta w_0\,dx \tag{d}$$

$$\delta K = \int_0^L \int_A (\dot{u}_1\delta\dot{u}_1 + \dot{u}_3\delta\dot{u}_3)\,dA\,dx$$

$$= \int_0^L \int_A \left[\left(\dot{u}_0 + z\dot{\phi}\right)\left(\delta\dot{u}_0 + z\delta\dot{\phi}\right) + \dot{w}_0\delta\dot{w}_0\right]dA\,dx$$

$$= \int_0^L \left[I_0\dot{u}_0\delta\dot{u}_0 + I_1\left(\dot{u}_0\delta\dot{\phi} + \dot{\phi}\delta\dot{u}_0\right) + I_2\dot{\phi}\delta\dot{\phi} + I_0\dot{w}_0\delta\dot{w}_0\right]dx \tag{e}$$

where a superposed dot indicates differentiation with respect to time t, and

$$N_{xx} = \int_A \sigma_{xx}\,dA, \quad M_{xx} = \int_A \sigma_{xx}z\,dA, \quad Q_x = K_s\int_A \sigma_{xz}\,dA$$

$$I_i = \int_A \rho\,(z)^i\,dA \tag{f}$$

and K_s is the *shear correction factor* introduced to account for the difference between the actual shear force and that predicted by the Timoshenko beam theory on account of constant state of shear stress through the beam height.

The remaining procedure is the same as that followed for the static problems, i.e., integrate by parts (in space as well as time) to relieve δu_0 and δw_0 of any differentiations, collect like terms together, and use the fundamental lemma to arrive at the Euler–Lagrange equations and the natural boundary conditions. Note also that

$$\delta u_0(x,t_1) = \delta u_0(x,t_2) = \delta w_0(x,t_1) = \delta w_0(x,t_2) = 0$$

Otherwise δu_0 and δw_0 are arbitrary in $0 \leq x \leq L$. We have

$$0 = \int_0^T \int_0^L \left[I_0\dot{u}_0\delta\dot{u}_0 + I_1\left(\dot{u}_0\delta\dot{\phi} + \dot{\phi}\delta\dot{u}_0\right) + I_2\dot{\phi}\delta\dot{\phi} + I_0\dot{w}_0\delta\dot{w}_0\right]dx\,dt$$

$$- \int_{t_1}^{t_2} \int_0^L \left[N_{xx} \frac{\partial \delta u_0}{\partial x} + M_{xx} \frac{\partial \delta \phi}{\partial x} + Q_x \left(\delta \phi + \frac{\partial \delta w_0}{\partial x} \right) - q\, \delta w_0 \right] dx dt$$

$$= \int_0^T \int_0^L \left[\frac{\partial N_{xx}}{\partial x} - \left(I_0 \ddot{u}_0 + I_1 \ddot{\phi} \right) \right] \delta u_0\, dx dt$$

$$+ \int_0^T \int_0^L \left[\frac{\partial M_{xx}}{\partial x} - Q_x - \left(I_1 \ddot{u}_0 + I_2 \ddot{\phi} \right) \right] \delta \phi\, dx dt$$

$$+ \int_0^T \int_0^L \left(\frac{\partial Q_x}{\partial x} + q - I_0 \ddot{w}_0 \right) \delta w_0\, dx dt$$

$$- [N_{xx} \cdot \delta u_0]_0^L - [M_{xx} \cdot \delta \phi]_0^L - [Q_x \cdot \delta w_0]_0^L \tag{g}$$

where it is assumed that ρ is independent of time. We note that

$$I_1 = \int_A \rho z\, dA = b \int_{-\frac{h}{2}}^{\frac{h}{2}} \rho z\, dz = 0$$

whenever ρ is independent of z, where b is the width of the beam (when it is assumed to be of a rectangular cross section).

The Euler–Lagrange equations are given by

$$\delta u_0: \quad \frac{\partial N_{xx}}{\partial x} - I_0 \frac{\partial^2 u_0}{\partial t^2} = 0 \tag{h}$$

$$\delta \phi: \quad \frac{\partial M_{xx}}{\partial x} - Q_x - I_2 \frac{\partial^2 \phi}{\partial t^2} = 0 \tag{i}$$

$$\delta w_0: \quad \frac{\partial Q_x}{\partial x} + q - I_0 \frac{\partial^2 w_0}{\partial t^2} = 0 \tag{j}$$

The force (natural) boundary conditions for the Timoshenko beam theory involve specifying the following secondary variables:

$$N_{xx}, \quad Q_x, \quad \text{and} \quad M_{xx} \quad \text{at } x = 0, L \tag{k}$$

The geometric boundary conditions involve specifying the following primary variables:

$$u_0, \quad w_0, \quad \text{and} \quad \phi \quad \text{at } x = 0, L \tag{l}$$

Thus the pairing of the primary and secondary variables is as follows:

$$(u_0, N_{xx}), \quad (w_0, Q_x), \quad (\phi, M_{xx}) \tag{m}$$

Only one member of each pair may be specified at a point in the beam.

2.2.4 The Principle of Minimum Total Potential Energy

A special case of the principle of virtual displacements that deals with linear as well as nonlinear elastic bodies is known as the *principle of minimum total potential energy*. For elastic bodies (in the absence of temperature variations) there exists a *strain energy density* function U_0 such that

$$\sigma = \frac{\partial U_0}{\partial \varepsilon} \quad \text{or} \quad \sigma_{ij} = \frac{\partial U_0}{\partial \varepsilon_{ij}} \quad (2.2.16)$$

Equation (2.2.16) represents the constitutive equation of an hyperelastic material. The strain energy density U_0 is a single-valued function of strains at a point, and is assumed to be positive definite. The statement of the principle of virtual displacements, Eqs. (2.2.4) and (2.2.5), can be expressed in terms of the strain energy density U_0:

$$\int_{\Omega_0} \frac{\partial U_0}{\partial \varepsilon} : \delta\varepsilon \, dv - \left[\int_{\Omega_0} \rho_0 \mathbf{f} \cdot \delta\mathbf{u} \, dv + \int_{\Gamma_\sigma} \mathbf{T} \cdot \delta\mathbf{u} \, ds \right] = 0 \quad (2.2.17a)$$

or, in component form,

$$\int_{\Omega_0} \frac{\partial U_0}{\partial \varepsilon_{ij}} \delta\varepsilon_{ij} \, dv - \left[\int_{\Omega_0} \rho_0 f_i \delta u_i \, dv + \int_{\Gamma_\sigma} T_i \delta u_i \, ds \right] = 0 \quad (2.2.17b)$$

The first integral is equal to

$$\int_{\Omega_0} \delta U_0 \, dv = \delta U \quad (2.2.18)$$

where U is the internal strain energy functional

$$U = \int_{\Omega_0} U_0 \, dv \quad (2.2.19)$$

Suppose that there exists a potential V whose first variation is

$$\delta V = - \left[\int_{\Omega_0} \rho_0 \mathbf{f} \cdot \delta\mathbf{u} \, dv + \int_{\Gamma_\sigma} \mathbf{T} \cdot \delta\mathbf{u} \, ds \right]$$

$$= - \left[\int_{\Omega_0} \rho_0 f_i \delta u_i \, dv + \int_{\Gamma_\sigma} T_i \delta u_i \, ds \right] \quad (2.2.20)$$

Then the principle of virtual work takes the form

$$\delta U + \delta V = \delta(U + V) \equiv \delta \Pi = 0 \quad (2.2.21)$$

The sum $U+V = \Pi$ is called the *total potential energy* of the elastic body. The statement in Eq. (2.2.21) is known as the *principle of minimum total potential energy*. It states that *of all admissible displacements, those that satisfy the equilibrium equations make the total potential energy a minimum*:

$$\Pi(\mathbf{u}) \leq \Pi(\bar{\mathbf{u}}) \qquad (2.2.22)$$

where $\mathbf{u}$ is the true solution and $\bar{\mathbf{u}}$ is any admissible displacement field. The equality holds only if $\mathbf{u} = \bar{\mathbf{u}}$.

Example 2.2.3

Consider the Timoshenko beam theory of Example 2.2.2. For the static case, the principle of virtual displacements requires that

$$\delta W \equiv \delta W_I + \delta W_E$$

where δW_I and δW_E are given by Eqs. (c) and (d) of Example 2.2.2. Thus we have

$$0 = \int_0^L \left(N_{xx}\delta\varepsilon_{xx}^0 + M_{xx}\delta\varepsilon_{xx}^1 + Q_x\delta\gamma_{xz}^0 - q\delta w_0 \right) dx \qquad (a)$$

Using the uniaxial constitutive relations

$$\sigma_{xx} = E\varepsilon_{xx} = E\left(\frac{\partial u_0}{\partial x} + z\frac{\partial \phi}{\partial x}\right), \quad \sigma_{xz} = G\gamma_{xz} = G\left(\phi + \frac{\partial w_0}{\partial x}\right) \qquad (b)$$

in Eq. (f) of Example 2.2.2 and evaluating the integrals, the force and moment resultants can be evaluated in terms of the displacements as

$$N_{xx} = EA\frac{\partial u_0}{\partial x}, \quad M_{xx} = EI\frac{\partial \phi}{\partial x}, \quad Q_x = GAK_s\left(\phi + \frac{\partial w_0}{\partial x}\right) \qquad (c)$$

where A denotes the area of cross section and I the moment of inertia of the beam.

Rewriting the statement in Eq. (a) in terms of the displacements, we obtain

$$0 = \int_0^L \left[EA\frac{\partial u_0}{\partial x}\frac{\partial \delta u_0}{\partial x} + EI\frac{\partial \phi}{\partial x}\frac{\partial \delta \phi}{\partial x} \right.$$
$$\left. + GAK_s\left(\phi + \frac{\partial w_0}{\partial x}\right)\left(\delta\phi + \frac{\partial \delta w_0}{\partial x}\right) - q\delta w_0 \right] dx \qquad (d)$$

The final step involves pulling the variational symbol out of the entire expression by taking into consideration its operations on the dependent variables u_0, w_0 and ϕ. We obtain

$$0 = \delta \int_0^L \left[\frac{EA}{2} \left(\frac{\partial u_0}{\partial x}\right)^2 + \frac{EI}{2} \left(\frac{\partial \phi}{\partial x}\right)^2 + \frac{GAK_s}{2} \left(\phi + \frac{\partial w_0}{\partial x}\right)^2 - qw_0 \right] dx$$
$$= \delta \Pi(u_0, w_0, \phi). \tag{e}$$

Thus the total potential energy functional for the Timoshenko beam theory is given by

$$\Pi = \int_0^L \left[\frac{EA}{2} \left(\frac{\partial u_0}{\partial x}\right)^2 + \frac{EI}{2} \left(\frac{\partial \phi}{\partial x}\right)^2 + \frac{GAK_s}{2} \left(\phi + \frac{\partial w_0}{\partial x}\right)^2 - qw_0 \right] dx \tag{f}$$

Then the principle of virtual displacements for the linear elastic case becomes the principle of minimum total potential energy, $\delta \Pi = 0$.

2.3 Variational Methods

2.3.1 Introduction

In section 2.2, we have seen how the principles of virtual work and minimum total potential energy can be used to obtain governing differential equations and associated natural boundary conditions. In this section, we study the use of the variational statements, $\delta W = 0$ or $\delta \Pi = 0$, in the solution of the underlying equations of these statements. The methods to be described here are known as the *classical* variational methods. In these methods, we seek an approximate solution to the problem in terms of *adjustable* parameters that are determined by substituting the assumed solution into a variational statement equivalent to the governing equations of the problem. Such solution methods are called *direct methods* because the approximate solutions are obtained directly by applying the same variational principle that was used to derive the governing (i.e., Euler–Lagrange) equations. In the classical variational methods, we seek solution over the total domain, and therefore the methods yield continuous solutions throughout the domain. In the finite element method, the same methods can be used over a sub-domain, called a *finite element*, to set up algebraic relations among the unknown parameters, as will be seen later in this book.

The assumed solutions in the variational methods are in the form of a finite linear combination of *undetermined parameters* and appropriately chosen functions. This amounts to representing a continuous system by a finite set of coordinate functions. Since the solution of a continuum problem, in general, cannot be represented by a finite set of functions, error is introduced into the solution. Therefore, the solution obtained is an approximation to the true solution of the equations describing a physical problem. As the number of linearly independent terms in the assumed solution is increased, the error in the approximation will be reduced and the assumed solution converges to the exact solution, provided certain conditions are met by the assumed solution.

It should be understood that the equations governing a physical problem are themselves approximate. The approximations are introduced through several sources, including representation of the geometry and boundary conditions, loads, and material behavior. Therefore, when one thinks of permissible error in an approximate solution, it is understood to be relative to exact solutions of the governing equations that inherently contain various approximations. The variational methods of approximation to be described here are limited to the Rayleigh–Ritz method and the weighted-residual methods (e.g., the Galerkin method, least-squares method, collocation method, sub-domain method, and so on). Only the Rayleigh–Ritz and Galerkin methods of the weighted-residual family of methods will be discussed in detail here. Readers interested in other methods may consult the references at the end of the chapter for additional details.

2.3.2 The Rayleigh–Ritz Method

There exist a number of approximate methods that can be used to solve differential equations (e.g., finite difference methods, perturbation methods, and so on). The most direct methods are those that bypass the derivation of the Euler–Lagrange equations and go directly from a variational statement of the problem to the solution of the Euler–Lagrange equations. One such direct method was proposed by Lord Rayleigh, and a generalization of the method was proposed independently by Ritz. Thus, the Rayleigh–Ritz method is necessarily based on variational statements, such as those provided by the principles of virtual work or their variants, that contain the natural boundary conditions as a part of the statements.

In the Rayleigh–Ritz method we approximate a dependent unknown (e.g., the displacement) u by a finite linear combination U_N of the form

$$u \approx U_N = \sum_{j=1}^{N} c_j \varphi_j + \varphi_0 \qquad (2.3.1)$$

and then determine the parameters c_j by requiring that the principle of virtual displacements or the minimum total potential energy hold for the approximate problem. In Eq. (2.3.1), c_j denote undetermined parameters, and φ_0 and φ_j are called *approximation functions*, which are appropriately selected functions of position. Equation (2.3.1) can be viewed as a representation of u in a component form; the parameters c_j are the *components* (or coordinates) and φ_j are the *coordinate functions*. Another interpretation of Eq. (2.3.1) is provided by the finite Fourier series, where c_j are known as the *Fourier coefficients*.

Properties of Approximation Functions

In order to ensure that the algebraic equations resulting from the Rayleigh–Ritz approximation have a solution and that the approximate solution converges to the true solution of the problem as the number of parameters N is increased, we must choose φ_j ($j = 1, 2, 3, ..., N$) and φ_0 such that they meet certain requirements. Formal definitions of convergence and completeness require a background in functional analysis (see Reddy [4]), and therefore we give some elementary and intuitive definitions of these concepts.

Convergence. A sequence $\{U_N\}$ of functions is said to *converge* to u_0 if for each $\epsilon > 0$ there is a number $M > 0$, depending on ϵ, such that

$$\|u_N(\mathbf{x}) - u_0(\mathbf{x})\| < \epsilon \quad \text{for all } N > M \qquad (2.3.2)$$

where $\|\cdot\|$ denotes a *norm* or measure of the magnitude of the enclosed quantity and u_0 is called the limit of the sequence. An example of the norm of a function $u(\mathbf{x})$ with $\mathbf{x}$ in the domain Ω is provided by

$$\|u\|^2 = \int_\Omega |u(\mathbf{x})|^2 d\mathbf{x} \qquad (2.3.3)$$

In the above statement U_N can be thought of as the approximate solution and u_0 as the true solution. The statement implies that the N-parameter approximate solution U_N can be made as close to u_0 as we wish, say within ϵ, by choosing N to be greater than $M = M(\epsilon)$, provided that the approximate solution is convergent. While there is no formula to determine M, a series of trials will help determine the value of N for which the approximate solution U_N is within the tolerance.

Completeness. The concept of convergence of a sequence involves a limit of the sequence. If the limit is not a part of the sequence $\{U_N\}_{N=1}^{\infty}$, then there is no hope of attaining convergence. For example, if the true solution to a certain problem is of the form $u_0(x) = ax^2 + bx^3$, where a and b are constants, then the sequence of approximations

$$U_1 = c_1 x^3, \ U_2 = c_1 x^3 + c_2 x^4, \ \cdots, \ U_N = c_1 x^3 + c_2 x^4 + \cdots + c_N x^{N+2}$$

will not converge to the true solution because the sequence does not contain the x^2 term. The sequence is said to be incomplete. As a rule, in selecting an approximate solution one should include all terms up to the highest order term.

Linear Independence. If a function U can be expressed as $U = \sum c_i \varphi_i$, where c_i are real numbers, we say that U is a *linear combination* of φ_i's. An expression of the form $\sum c_i \varphi_i = 0$ is called a *linear relation* among the φ_i's. A linear relation with all $c_i = 0$ is called a *trivial linear relation*. If at least one c_i nonzero, the relation is called a *nontrivial relation*. A set of vectors $\{\varphi_i\}$ is said to be *linearly independent* if there exists no nontrivial relation among them. In other words, no φ_i is expressible in terms of others in the set.

We now list the requirements of a convergent Rayleigh–Ritz approximation (1):

1. $\varphi_j (j = 1, 2, ..., N)$ should satisfy the following three conditions:

    (a) Each φ_j is *continuous*, as is required in the variational statement; i.e., φ_j should be such that U_N has a nonzero contribution to the weak form.

    (b) Each φ_j satisfies the *homogeneous form* of the specified essential boundary conditions.

    (c) The set $\{\varphi_j\}$ is *linearly independent* and *complete*.

2. φ_0 has the primary role of satisfying the specified *essential* (or geometric) boundary conditions associated with the variational formulation; φ_0 plays the role of the *particular solution*. It is necessarily zero when the specified essential boundary conditions are homogeneous.

For polynomial approximations functions, the linear independence and completeness properties require φ_j to be increasingly higher-order polynomials. For example, if φ_1 is a linear polynomial, φ_2 should be a quadratic polynomial, φ_3 should be a cubic polynomial, and so on (but each φ_j need not be a complete polynomial by itself). The completeness property is essential for the convergence of the Rayleigh–Ritz approximation (see Reddy [4], p. 262).

Since the natural boundary conditions of the problem are included in the variational statements, we require the Rayleigh–Ritz approximation U_N to satisfy only the specified essential boundary conditions of the problem. This is done by selecting φ_i to satisfy the homogeneous form and φ_0 to satisfy the actual form of the essential boundary conditions. For instance, if u is specified to be $\hat{u}$ at a point $\mathbf{x} = \mathbf{x}_0$, then using $U_N(x_0) = \hat{u}$ and $\varphi_0(\mathbf{x}_0) = \hat{u}$ in

$$u(\mathbf{x}_0) \approx U_N(\mathbf{x}_0) = \sum_{j=1}^{N} c_j \varphi_j(\mathbf{x}_0) + \varphi_0(\mathbf{x}_0)$$

one can conclude that

$$\sum_{j=1}^{N} c_j \varphi_j(\mathbf{x}_0) = 0$$

Since the above condition must hold for any independent set of parameters c_j, it follows that

$$\varphi_j(\mathbf{x}_0) = 0 \quad \text{for} \quad j = 1, 2, \cdots, N \tag{2.3.4}$$

The requirements on φ_j provide guidelines for selecting the coordinate functions; they do not give any formula for generating the functions. As a general rule, coordinate functions should be selected from a set of linearly independent functions from the lowest order to a desirable order without missing any intermediate admissible functions in the representation of U_N (i.e., satisfy the completeness property). The function φ_0 has no other role to play than to satisfy specified (non-homogeneous) essential boundary conditions; there are no continuity conditions on φ_0. Therefore, one should select the lowest order φ_o that satisfies the essential boundary conditions.

Rayleigh–Ritz Equations for the Parameters

Once the approximation functions φ_0 and φ_i are selected, the parameters c_j in Eq. (2.3.1) are determined by requiring U_N to minimize

ENERGY PRINCIPLES AND VARIATIONAL METHODS   77

the total potential energy functional Π or satisfy the principle of virtual displacements of the problem. First consider the principle of minimum total potential energy. Upon substitution of the approximation $u \approx U_N$, $\Pi(U_N)$ becomes a function of $c_1, c_2, \cdots, c_N$. Hence minimization of the functional $\Pi(c_1, c_2, \cdots, c_N)$ is reduced to the minimization of a function of several variables:

$$0 = \delta\Pi(U_N) = \sum_{i=1}^{N} \frac{\partial \Pi}{\partial c_i} \delta c_i \quad \text{or} \quad \frac{\partial \Pi}{\partial c_i} = 0 \qquad (2.3.5)$$

The same procedure applies to the principle of virtual displacements. We have [see Eq. (2.1.9b)]

$$0 = \delta W(U_N) = \sum_{i=1}^{N} \delta_{c_i} W = \sum_{i=1}^{N} \frac{\partial W}{\partial c_i} \delta c_i \quad \text{or} \quad \frac{\partial W}{\partial c_i} = 0$$

which is the same as Eq. (2.3.5).

Equation (2.3.5) gives N algebraic equations in the N coefficients $(c_1, c_2, ..., c_N)$,

$$0 = \frac{\partial \Pi}{\partial c_i} = \sum_{j=1}^{N} R_{ij} c_j - F_i \quad \text{or} \quad [R]\{c\} = \{F\} \qquad (2.3.6)$$

where R_{ij} and F_i are known coefficients that depend on the problem parameters (e.g., geometry, material coefficients, and loads) and the approximation functions. These coefficients will be defined for each problem discussed in the sequel. Equation (2.3.6) is then solved for $\{c\}$ and substituted back into Eq. (2.3.1) to obtain the N parameter Rayleigh–Ritz solution. Some general features of the Rayleigh–Ritz method are listed below:

1. If the coordinate functions φ_i are selected to satisfy the required conditions, the assumed approximation for the displacements converges to the true solution with an increase in the number of parameters (i.e., as $N \to \infty$). A mathematical proof of such an assertion can be found in [4].

2. For increasing values of N, the previously computed coefficients of the algebraic equations (2.3.6) remain unchanged, provided the previously selected coordinate functions are not changed. One must add only the newly computed coefficients to the system of equations.

3. If the resulting algebraic equations are symmetric, one needs to compute only upper or lower diagonal elements in the coefficient matrix $[R]$. For all plate problems considered here, the coefficient matrix is symmetric whenever the principles of virtual displacements or the minimum total potential energy are used.

4. If the variational (or virtual work) statement is nonlinear in u, then the resulting algebraic equations are also nonlinear in the parameters c_i. To solve such nonlinear equations, a variety of numerical methods are available (e.g., Newton's method, the Newton–Raphson method, the Picard method), which will be discussed later in this book.

5. Since the strains are computed from an approximate displacement field, the strains and stresses are generally less accurate than the displacements.

6. The equilibrium equations of the problem are satisfied only in the energy sense, not in the differential equation sense. Therefore, the displacements obtained from the Rayleigh–Ritz approximation in general do not satisfy the equations of equilibrium point-wise, unless the solution converged to the exact solution.

7. Since a continuous system is approximated by a finite number of coordinates (or degrees of freedom), the approximate system is less flexible than the actual system. Consequently, the displacements obtained using the principle of minimum total potential energy and the Rayleigh–Ritz method converge to the exact displacements from below:

$$U_1 < U_2 < \ldots < U_N < U_M \ldots < u(\text{exact}), \quad \text{for } M > N$$

where U_N denotes the N parameter Rayleigh–Ritz approximation of u obtained from the principle of virtual displacements or the minimum total potential energy principle. It should be noted that the displacements obtained from the Rayleigh–Ritz method based on the total complementary energy principle provide the upper bound.

8. The Rayleigh–Ritz method can be applied, in principle, to any physical problem that can be cast in a *weak form*—a form that contains the natural boundary conditions of the problem. The principles of virtual displacements ($\delta W = 0$) and minimum total potential energy ($\delta \Pi = 0$) provide the weak forms directly. In particular, the Rayleigh–Ritz method can be applied to all plate

Example 2.3.1

Consider axial deformation of an isotropic, elastic, non-uniform cross section bar of length L, fixed at $x = 0$ and subjected to an axial force P_0 at $x = L$ (see Figure 2.3.1). In addition, suppose that the bar experiences a body force $f(x)$. We wish to determine the axial displacement $u(x)$ using the Rayleigh–Ritz method.

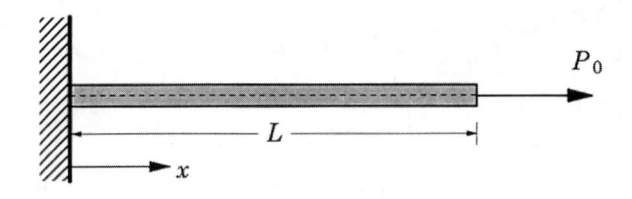

Figure 2.3.1. Axial deformation of a bar with an end load.

First, we write the boundary conditions of the problem:

$$u(0) = 0, \quad \left(EA\frac{du}{dx}\right)_{x=L} = P_0 \tag{a}$$

where E is the modulus and A is the area of cross section of the bar. The first one is a geometric (essential) boundary condition, while the second one is a force (natural) boundary condition.

The principle of virtual displacements for the problem at hand can be written as

$$0 = \int_0^L (EA\frac{du}{dx}\frac{d\delta u}{dx} - f\delta u)dx - P_0\delta u(L) \tag{b}$$

and the total potential energy of the bar is

$$\Pi(u) = \int_0^L \left[\frac{EA}{2}\left(\frac{du}{dx}\right)^2 - fu\right]dx - P_0 u(L) \tag{c}$$

The expression in (b) is equivalent to the statement $\delta\Pi = 0$.

Since the specified geometric boundary condition is homogeneous, we have $\varphi_0 = 0$. The approximation functions $\varphi_i(x)$ must satisfy the condition $\varphi_i(0) = 0$. Clearly, $\varphi_i(x) = x^i$ satisfies the condition and it has a nonzero first derivative, as required in (b) or (c).

Substituting

$$u \approx u_n = \sum_{j=1}^{n} c_j \varphi_j(x) + \varphi_0(x) \tag{d}$$

into the statement (b), we obtain

$$\begin{aligned}0 &= \int_0^L \left[EA \left(\sum_{j=1}^n c_j \frac{d\varphi_j}{dx} + \frac{d\varphi_0}{dx} \right) \left(\sum_{i=1}^n \delta c_i \frac{d\varphi_i}{dx} \right) - f \left(\sum_{i=1}^n \delta c_i \varphi_i \right) \right] dx \\ &\quad - P_0 \left(\sum_{i=1}^n \delta c_i \varphi_i(L) \right) \\ &= \sum_{i=1}^n \delta c_i \left\{ \sum_{j=1}^n \left(\int_0^L EA \frac{d\varphi_i}{dx} \frac{d\varphi_j}{dx} dx \right) c_j + \int_0^L EA \frac{d\varphi_0}{dx} \frac{d\varphi_i}{dx} dx \right. \\ &\quad \left. - \int_0^L f \varphi_i \, dx - P_0 \varphi_i(L) \right\} = \sum_{i=1}^n \delta c_i \left\{ \sum_{j=1}^n R_{ij} c_j - F_i \right\}\end{aligned}$$

where

$$R_{ij} = \int_0^L EA \frac{d\varphi_i}{dx} \frac{d\varphi_j}{dx} dx, \quad F_i = \int_0^L f\varphi_i \, dx + P_0 \varphi_i(L) - \int_0^L EA \frac{d\varphi_0}{dx} \frac{d\varphi_i}{dx} dx$$

Since δc_i are arbitrary and linearly independent, we can write

$$0 = \sum_{j=1}^n R_{ij} c_j - F_i \quad \text{or} \quad [R]\{c\} = \{F\} \tag{e}$$

The result in Eq. (e) can also be obtained from the principle of minimum total potential energy. Substituting Eq. (d) into Eq. (c), we obtain

$$\begin{aligned}\Pi = \frac{1}{2} \int_0^L &\left[EA \left(\sum_{j=1}^n c_j \frac{d\varphi_j}{dx} + \frac{d\varphi_0}{dx} \right) \left(\sum_{i=1}^n c_i \frac{d\varphi_i}{dx} + \frac{d\varphi_0}{dx} \right) \right. \\ &\left. - f \left(\sum_{i=1}^n c_i \varphi_i \right) \right] dx - P_0 \left(\sum_{i=1}^n c_i \varphi_i(L) \right)\end{aligned}$$

ENERGY PRINCIPLES AND VARIATIONAL METHODS   81

Thus Π is a function of $c_1, c_2, \cdots, c_n$, and a necessary condition for its minimum is

$$0 = \sum_{i=1}^{n} \frac{\partial \Pi}{\partial c_i} \delta c_i \quad \text{or} \quad \frac{\partial \Pi}{\partial c_i} = 0 \quad (i = 1, 2, \cdots, n)$$

and

$$\frac{\partial \Pi}{\partial c_i} = \sum_{j=1}^{n} \left(\int_0^L EA \frac{d\varphi_i}{dx} \frac{d\varphi_j}{dx} dx \right) c_j + \int_0^L EA \frac{d\varphi_0}{dx} \frac{d\varphi_i}{dx} dx$$
$$- \int_0^L f\varphi_i \, dx - P_0 \varphi_i(L)$$

which is the same as Eq. (e).

From the earlier discussions, we have (the approximation functions are nondimensionalized)

$$\varphi_0 = 0, \quad \varphi_i = \left(\frac{x}{L}\right)^i, \quad \frac{d\varphi_i}{dx} = \frac{i}{L}\left(\frac{x}{L}\right)^{i-1}$$

In addition, suppose that the modulus E is constant, the area of cross section A varies as $A = A_0 + (x/L)A_1$, and the body force varies as $f(x) = f_0 + (x/L)f_1$. Then

$$R_{ij} = \int_0^L E\left(A_0 + \frac{x}{L}A_1\right) \frac{ij}{L^2}\left(\frac{x}{L}\right)^{i+j-2} dx$$
$$= \frac{E}{L}\left[\left(\frac{ij}{i+j-1}\right)A_0 + \left(\frac{ij}{i+j}\right)A_1\right]$$
$$F_i = \int_0^L \left(f_0 + \frac{x}{L}f_1\right)\left(\frac{x}{L}\right)^i dx + P_0 = L\left(\frac{f_0}{i+1} + \frac{f_1}{i+2}\right) + P_0 \quad \text{(f)}$$

In particular, we choose, for example, $A_0 = 2a_0$, $A_1 = -a_0$, and $f_1 = 0$, we obtain

$$R_{ij} = \frac{Ea_0}{L} \frac{ij(1+i+j)}{(i+j-1)(i+j)}, \quad F_i = \frac{f_0 L}{i+1} + P_0$$

For $n = 2$, we obtain

$$\frac{Ea_0}{6L}\begin{bmatrix} 9 & 8 \\ 8 & 10 \end{bmatrix}\begin{Bmatrix} c_1 \\ c_2 \end{Bmatrix} = \frac{f_0 L}{6}\begin{Bmatrix} 3 \\ 2 \end{Bmatrix} + P_0\begin{Bmatrix} 1 \\ 1 \end{Bmatrix} \quad \text{(g)}$$

82   THEORY AND ANALYSIS OF ELASTIC PLATES

$$c_1 = \frac{7f_0L^2 + 6P_0L}{13Ea_0}, \quad c_2 = \frac{3(P_0L - f_0L^2)}{13Ea_0}$$

$$u_2(x) = \frac{7f_0L^2 + 6P_0L}{13Ea_0}\left(\frac{x}{L}\right) + \frac{3(P_0L - f_0L^2)}{13Ea_0}\left(\frac{x^2}{L^2}\right) \qquad \text{(h)}$$

The exact solution is

$$u(x) = \frac{f_0L}{Ea_0}x + \frac{2f_0L^2}{Ea_0}\log(1 - \frac{x}{2L}) + \frac{(f_0L - P_0)L}{Ea_0}\log(1 - \frac{x}{2L}) \qquad \text{(i)}$$

Table 2.3.1 contains comparison of the Ritz coefficients $\bar{c}_i = c_i P_0 L / Ea_0$ and the end deflection $\bar{u}(L) = (P_0L/Ea_0)u(L) = \bar{c}_1 + \bar{c}_2 + \cdots + \bar{c}_n$ for $n = 1, 2, \ldots, 5$ with the exact coefficients and solution from the expansion

$$u(x) = -\frac{P_0L}{Ea_0}\log(1 - \frac{x}{2L}) \approx \frac{P_0L}{Ea_0}\left(\frac{1}{2}\frac{x}{L} + \frac{1}{8}\frac{x^2}{L^2} + \frac{1}{24}\frac{x^3}{L^3} + \frac{1}{64}\frac{x^4}{L^4} + \cdots\right) \qquad \text{(j)}$$

Clearly the Rayleigh–Ritz solution converges to the exact solution as n increases. This completes the example.

Table 2.3.1. The Ritz coefficients[†] and the end deflection for the axial deformation of a nonuniform, isotropic bar fixed at $x = 0$ and subjected to force P_0 at $x = L$.

n	$\bar{c}_1$	$\bar{c}_2$	$\bar{c}_3$	$\bar{c}_4$	$\bar{c}_5$	$\bar{u}(L)$
1	0.6667					0.6667
2	0.4615	0.2308				0.6923
3	0.5079	0.0794	0.1058			0.6931
4	0.4984	0.1402	0.0000	0.0545		0.6931
5	0.5003	0.1206	0.0610	-0.0187	0.0299	0.6931
Exact	0.5000	0.1250	0.0416	0.0156	0.0062	0.6931

[†] $\bar{c}_i = c_i(P_0L/Ea_0)$, $\bar{u}(L) = u(L)(P_0L/Ea_0)$.

Example 2.3.2

This example is concerned with dynamic study of bars. Consider a uniform cross-section bar of length L with the left end fixed and the right end connected to a rigid support via a linear elastic spring with spring constant k (see Figure 2.3.2). In addition, the member is subjected to a body force $f(x,t)$. The kinetic and potential energies associated with the axial motion of the member are given by

$$K = \int_0^L \frac{\rho A}{2} \left(\frac{\partial u}{\partial t}\right)^2 dx$$

$$U = \int_0^L \frac{EA}{2} \left(\frac{\partial u}{\partial x}\right)^2 dx + \frac{k}{2}[u(L,t)]^2$$

$$V = -\int_0^L fu \, dx \tag{a}$$

Substituting for K, U, and V from Eq. (a) in Hamilton's principle, we obtain

$$0 = \delta \int_0^{t_0} (K - U - V) \, dt$$

$$= \delta \int_0^{t_0} \left\{ \int_0^L \left[\frac{\rho A}{2}\left(\frac{\partial u}{\partial t}\right)^2 - \frac{EA}{2}\left(\frac{\partial u}{\partial x}\right)^2 + fu \right] dx - \frac{k}{2}[u(L,t)]^2 \right\} dt \tag{b}$$

The Euler–Lagrange equations associated with Eq. (b) are

$$\frac{\partial}{\partial x}\left(EA\frac{\partial u}{\partial x}\right) - \frac{\partial}{\partial t}\left(\rho A\frac{\partial u}{\partial t}\right) + f = 0, \quad 0 < x < L; \quad t > 0 \tag{c}$$

$$\left(EA\frac{\partial u}{\partial x} + ku\right)\bigg|_{x=L} = 0 \quad \text{for all} \quad t \geq 0 \tag{d}$$

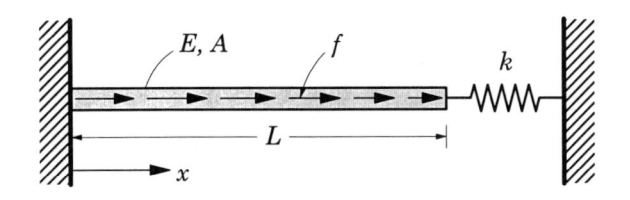

Figure 2.3.2. Axial deformation of a bar with an end spring.

Natural Vibration

First, suppose that we are interested in finding the periodic motion (or natural vibration) of the form (when $f = 0$)

$$u(x,t) = u_0(x)e^{i\omega t}, \quad i = \sqrt{-1} \tag{e}$$

where ω is the *frequency* of natural vibration and $u_0(x)$ is the amplitude. Substituting Eq. (e) into Eqs. (c) and (d), we obtain

$$\frac{d}{dx}\left(EA\frac{du_0}{dx}\right) + \rho A \omega^2 u_0 = 0, \quad 0 < x < L \tag{f}$$

$$\left(EA\frac{du_0}{dx} + ku_0\right)\Big|_{x=L} = 0 \tag{g}$$

The variational statement (or weak form) associated with Eqs. (f) and (g) is given by

$$\begin{aligned} 0 &= \int_0^L \delta u_0 \left[\frac{d}{dx}\left(EA\frac{du_0}{dx}\right) + \rho A \omega^2 u_0\right] \\ &= \int_0^L \left(-EA\frac{d\delta u_0}{dx}\frac{du_0}{dx} + \rho A \omega^2 \delta u_0 u_0\right) dx - k\delta u_0(L) u_0(L) \end{aligned} \tag{h}$$

We use Eq. (h) and the Rayleigh–Ritz method to determine the values of ω. The continuum system described by Eqs. (f) and (g) have an infinite number of frequencies. Any approximate method reduces a continuum problem to a discrete problem, which has a finite number of degrees of freedom and, hence, a finite number of frequencies. The number of natural frequencies obtainable by the Rayleigh–Ritz method is equal to the number of parameters n. It is convenient to nondimensionalize the variables. Let ($\bar{x} = x/L$, $\bar{u} = u_0/L$). Then

$$0 = \int_0^1 \left(\lambda \delta u u - \frac{d\delta u}{dx}\frac{du}{dx}\right) dx - \mu \delta u(1) u(1) \tag{i}$$

$$\mu = \frac{kL}{EA}, \quad \lambda = \frac{\omega^2 \rho L^2}{E} \tag{j}$$

where the bar over the nondimensional variables is omitted in the interest of brevity.

ENERGY PRINCIPLES AND VARIATIONAL METHODS 85

Following the procedure outlined in the previous example, we select (based on the condition $\varphi_i(0) = 0$)

$$\varphi_0 = 0, \quad \varphi_1 = x, \quad \varphi_2 = x^2, \quad \cdots, \quad \varphi_n = x^n \tag{k}$$

Substituting $u_0 \approx \sum_{j=1}^n c_j x^j$ into Eq. (i), we obtain

$$0 = \int_0^1 \left[\lambda \left(\sum_{i=1}^n \delta c_i \varphi_i \right) \left(\sum_{j=1}^n c_j \varphi_j \right) - \left(\sum_{i=1}^n \delta c_i \frac{d\varphi_i}{dx} \right) \left(\sum_{j=1}^n c_j \frac{d\varphi_j}{dx} \right) \right] dx$$

$$- \mu \left(\sum_{i=1}^n \delta c_i \varphi_i(1) \right) \left(\sum_{j=1}^n c_j \varphi_j(1) \right)$$

$$= \sum_{i=1}^n \delta c_i \left\{ \sum_{j=1}^n \left[\int_0^1 \left(\lambda \varphi_i \varphi_j - \frac{d\varphi_i}{dx} \frac{d\varphi_j}{dx} \right) dx - \mu \varphi_i(1) \varphi_j(1) \right] c_j \right\}$$

Since δc_i are arbitrary and linearly independent, we obtain

$$0 = \sum_{j=1}^n (\lambda M_{ij} - R_{ij}) c_j \quad \text{or} \quad (\lambda [M] - [R])\{c\} = \{0\} \tag{l}$$

where

$$M_{ij} = \int_0^1 \varphi_i \varphi_j \, dx = \frac{1}{i+j+1}$$

$$R_{ij} = \int_0^1 \frac{d\varphi_i}{dx} \frac{d\varphi_j}{dx} dx + \mu \varphi_i(1) \varphi_j(1) = \frac{ij}{i+j-1} + \mu \tag{m}$$

For $n = 2$, we have

$$M_{11} = \frac{1}{3}, \quad M_{12} = \frac{1}{4} = M_{21}, \quad M_{22} = \frac{1}{5}$$

$$R_{11} = 1 + \mu, \quad R_{12} = 1 + \mu = R_{21}, \quad R_{22} = \frac{4}{3} + \mu$$

$$\left(\begin{bmatrix} 1+\mu & 1+\mu \\ 1+\mu & \frac{4}{3}+\mu \end{bmatrix} - \lambda \begin{bmatrix} \frac{1}{3} & \frac{1}{4} \\ \frac{1}{4} & \frac{1}{5} \end{bmatrix} \right) \begin{Bmatrix} c_1 \\ c_2 \end{Bmatrix} = \begin{Bmatrix} 0 \\ 0 \end{Bmatrix} \tag{n}$$

A nontrivial solution ($c_1 \neq 0$, $c_2 \neq 0$) to Eq. (n) exists, but only if the determinant of the coefficients matrix is zero:

$$\begin{vmatrix} 1+\mu-\frac{\lambda}{3} & 1+\mu-\frac{\lambda}{4} \\ 1+\mu-\frac{\lambda}{4} & \frac{4}{3}+\mu-\frac{\lambda}{5} \end{vmatrix} = 0$$

86   THEORY AND ANALYSIS OF ELASTIC PLATES

For $\mu = 1$, we obtain

$$\frac{2}{3} - \frac{8}{45}\lambda + \frac{1}{240}\lambda^2 = 0 \quad \text{or} \quad 15\lambda^2 - 640\lambda + 2400 = 0$$

whose roots are

$$\lambda_1 = \frac{64}{3} - \frac{4}{3}\sqrt{166} \approx 4.155, \quad \lambda_2 = \frac{64}{3} + \frac{4}{3}\sqrt{166} \approx 38.512$$

and the first two natural frequencies are $[\omega = (1/L)\sqrt{E\lambda/\rho}]$

$$\omega_1 = \frac{2.038}{L}\sqrt{\frac{E}{\rho}}, \quad \omega_2 = \frac{6.206}{L}\sqrt{\frac{E}{\rho}}$$

The exact values of λ (with $\mu = 1$) are the roots of the transcendental equation

$$\lambda + \tan\lambda = 0 \tag{o}$$

whose first two roots are

$$\omega_1 = \frac{2.029}{L}\sqrt{\frac{E}{\rho}}, \quad \omega_2 = \frac{4.913}{L}\sqrt{\frac{E}{\rho}}$$

Note that the higher approximation frequency is in larger error.

Transient Response

When one is interested in determining the time-dependent solution $u(x,t)$ under applied load $f(x,t)$, the Rayleigh–Ritz solution is sought in the form

$$u(x,t) \approx \sum_{j=1}^{n} c_j(t)\varphi_j(x) + \varphi_0(x) \tag{p}$$

where c_j are now time-dependent parameters to be determined for all times $t > 0$. Substituting (p) into (b), we obtain

$$0 = -\sum_{j=1}^{n}\left[\left(\int_0^1 \varphi_i\varphi_j\,dx\right)\frac{d^2c_j}{dt^2} + \left(\int_0^1 \frac{d\varphi_i}{dx}\frac{d\varphi_j}{dx}dx + \varphi_i(1)\varphi_j(1)\right)c_j\right]$$

$$+ \frac{1}{f_0}\int_0^1 \phi_i f\,dx$$

$$= -\sum_{j=1}^{n}\left(M_{ij}\frac{d^2c_j}{dt^2} + R_{ij}c_j\right) + F_i \tag{q}$$

where the first term of Eq. (b) was integrated by parts in time, x, u, and t are nondimensionalized as

$$\bar{x} = \frac{x}{L}, \quad \bar{u} = \frac{u}{(f_0 L^2/EA)}, \quad \bar{t} = \frac{t}{L\sqrt{\rho/E}}$$

f_0 being a constant, and (omitting the bars over the variables)

$$M_{ij} = \int_0^1 \varphi_i \varphi_j \, dx, \quad R_{ij} = \int_0^1 \frac{d\varphi_i}{dx}\frac{d\varphi_j}{dx} dx + \varphi_i(1)\varphi_j(1)$$
$$F_i = \frac{1}{f_0}\int_0^1 \varphi_i f(x,t) \, dx \tag{r}$$

For $n = 1$ and $f = f_0$ (step loading), we have

$$M_{11}\frac{d^2 c_1}{dt^2} + R_{11} c_1 = F_1 \quad \text{or} \quad \frac{1}{3}\frac{d^2 c_1}{dt^2} + 2c_1 = \frac{1}{2}$$

The solution to the second-order differential equation is

$$c_1(t) = A \sin \sqrt{6}t + B \cos \sqrt{6}t + \frac{1}{4}$$

Hence, the one-parameter Rayleigh–Ritz solution is given by

$$u_1(x,t) = \left(A \sin \sqrt{6}t + B \cos \sqrt{6}t + \frac{1}{4}\right) x$$

where A and B are constants to be determined using the initial conditions. For zero initial conditions $u(x,0) = 0$ and $\dot{u}(x,0) = 0$, we have $B = -1/4$ and $A = 0$, and the solution becomes

$$u_1(x,t) = \frac{1}{4}\left(1 - \cos \sqrt{6}t\right) x$$

For $n \geq 2$, the resulting system of differential equations in time may be solved for $c_i(t)$ using a numerical method. In that case, the solution may be obtained only for discrete values of time. At time $t = t_s = \Delta s$ we have

$$u_n(x, t_s) = \sum_{j=1}^n c_j(t_s)\varphi_j(x)$$

Example 2.3.3

Consider a simply-supported straight beam of length L. We wish to find a Rayleigh–Ritz approximation for the transverse deflection w_0 of the beam under uniformly distributed transverse load of intensity q_0. The principle of virtual displacements for the problem is

$$0 = \int_0^L \left(EI \frac{d^2 \delta w_0}{dx^2} \frac{d^2 w_0}{dx^2} - \delta w_0 q_0 \right) dx \tag{a}$$

The geometric boundary conditions are

$$w_0(0) = w_0(L) = 0$$

The remaining two boundary conditions are both natural and they are $M(0) = M(L) = 0$. In the Rayleigh–Ritz method, these enter the statement of the virtual displacement (a) (work done by the zero bending moments in moving through the nonzero rotations at $x = 0$ and $x = L$ is zero).

We choose a two-parameter approximation

$$w_0(x) \approx w_2(x) = c_1 \varphi_1 + c_2 \varphi_2 + \varphi_0 \tag{b}$$

with

$$\varphi_0 = 0, \quad \varphi_1 = x(L-x), \quad \varphi_2 = x^2(L-x) \tag{c}$$

Note that $\varphi_i(0) = \varphi_i(L) = 0$. Substituting Eq. (b) into Eq. (a), we obtain

$$0 = \int_0^L \left[EI \left(\delta c_1 \varphi_1'' + \delta c_2 \varphi_2'' \right) \left(c_1 \varphi_1'' + c_2 \varphi_2'' \right) - (\delta c_1 \varphi_1 + \delta c_2 \varphi_2) q_0 \right] dx$$

$$= \sum_{i=1}^2 \left(\sum_{j=1}^2 R_{ij} c_j - F_i \right) \delta c_i$$

or

$$0 = \sum_{j=1}^2 R_{ij} c_j - F_i \tag{d}$$

where

$$R_{ij} = \int_0^L EI \frac{d^2 \varphi_i}{dx^2} \frac{d^2 \varphi_j}{dx^2} \, dx, \quad F_i = \int_0^L q_0 \varphi_i \, dx \tag{e}$$

For the particular choice of φ_i in Eq.(c), we have

$$EIL \begin{bmatrix} 4 & 2L \\ 2L & 4L^2 \end{bmatrix} \begin{Bmatrix} c_1 \\ c_2 \end{Bmatrix} = \frac{q_0 L^3}{12} \begin{Bmatrix} 2 \\ L \end{Bmatrix}$$

whose solution is

$$c_1 = \frac{q_0 L^2}{24 EI}, \quad c_2 = 0$$

and the two-parameter Rayleigh–Ritz solution becomes

$$w_2(x) = \frac{q_0 L^3}{24 EI} x - \frac{q_0 L^2}{24 EI} x^2 \tag{f}$$

The exact solution is given by

$$w_0(x) = \frac{q_0 L^4}{24 EI} \left[\left(\frac{x}{L}\right) - 2\left(\frac{x}{L}\right)^3 + \left(\frac{x}{L}\right)^4 \right] \tag{g}$$

The three-parameter approximation with $\varphi_3 = x^3(L-x)$ yields

$$EIL \begin{bmatrix} 4 & 2L & 2L^2 \\ 2L & 4L^2 & 4L^3 \\ 2L^2 & 4L^3 & 4.8L^4 \end{bmatrix} \begin{Bmatrix} c_1 \\ c_2 \\ c_3 \end{Bmatrix} = \frac{q_0 L^3}{12} \begin{Bmatrix} 2 \\ L \\ 0.6L^2 \end{Bmatrix} \tag{h}$$

$$c_1 = c_2 L = -c_3 L^2 = \frac{q_0 L^2}{24 EI} \tag{i}$$

and the three-parameter Rayleigh–Ritz approximation coincides with the exact solution.

Instead of the uniformly distributed load, suppose that the beam is subjected to a transverse point load F_0 at $x = L/2$. The virtual work statement for this case is given by

$$0 = \int_0^L EI \frac{d^2 \delta w_0}{dx^2} \frac{d^2 w_0}{dx^2} dx - F_0 \delta w_0 \left(\frac{L}{2}\right) \tag{j}$$

Using the three-parameter approximation with $\varphi_i(x) = x^i(x-L)$, we obtain the coefficient matrix as in Eq. (h), with the right-hand side given by

$$\frac{F_0 L^2}{4} \begin{Bmatrix} 1 \\ \frac{L}{2} \\ \frac{L^2}{4} \end{Bmatrix} \tag{k}$$

and the solution of these equations is given by

$$c_1 = \frac{F_0 L}{12EI}, \quad c_2 = -\frac{F_0}{12EI}, \quad c_3 = -\frac{5F_0}{64EIL} \quad (1)$$

The exact solution for this case is

$$w_0(x) = \begin{cases} \frac{F_0 L}{12EI}x(L-x) + \frac{F_0}{12EI}x^2(L-x) - \frac{F_0 x L^2}{48EI} \\ \frac{F_0 L}{12EI}x(L-x) + \frac{F_0}{12EI}x^2(L-x) - \frac{F_0 x L^2}{48EI} + \frac{F_0(2x-L)^3}{48EI} \end{cases} \quad (m)$$

The first expression is valid for $0 \leq x \leq L/2$ and the second expression for $L/2 \leq x \leq L$. The maximum deflection predicted by the three-parameter Rayleigh–Ritz solution is

$$w_3\left(\frac{L}{2}\right) = \frac{7}{384}\frac{F_0 L^3}{EI} = \frac{F_0 L^3}{54.86 EI}$$

and the exact value is

$$w_0\left(\frac{L}{2}\right) = \frac{F_0 L^3}{48 EI}$$

The Ritz solution is defined by a single expression, whereas the exact solution is defined by two expressions in two parts of the beam due to the discontinuity of the applied load. Consequently, the exact shear force and bending moment are also discontinuous at $x = L/2$. Irrespective of the number of terms used in the approximation, the Rayleigh–Ritz solution based on the total potential energy functional (j) does not yield the exact solution (m). Obviously, the variational method should also be applied separately in each part, $0 < x < L/2$ and $L/2 < x < L$, and then the continuity of the primary variables at the point $x = L/2$ should be imposed to obtain the exact solution. The *finite element method* is a technique in which the domain is divided into parts and then a variational method is used in each interval to develop relationships among the undetermined parameters (see Reddy [5]). The method will be discussed in the sequel.

One can also use trigonometric polynomials in place of algebraic polynomials for the coordinate functions. However, such functions can make the evaluation of R_{ij} and/or F_i a difficult task. Further, for non-trigonometric loads, the Rayleigh–Ritz solution does not converge to the true solution for a finite number of parameters. The reason is that loads represented by algebraic polynomials (e.g., $q(x) = a_1 + a_2 x + a_3 x^2 + \ldots$)

cannot be represented by a finite trigonometric series. For instance, the simply-supported beam under uniform loading can be modeled by the approximation

$$w_N(x) = c_1 \sin \frac{\pi x}{L} + c_2 \sin \frac{3\pi x}{L} + \ldots + c_N \sin \frac{(2N+1)\pi x}{L} \quad \text{(n)}$$

The functions $\varphi_i = \sin(2i+1)\pi x/L$ are linearly independent and complete if all lower functions up to $\sin(2N+1)\pi x/L$ are included. However, the approximation will not coincide with the exact solution for any finite value of N because the sine-series representation of a uniform load is an infinite series:

$$q_0 = \sum_{i=1,3,\ldots}^{\infty} \frac{16 q_0}{i\pi} \sin \frac{i\pi x}{L}$$

The approximate solution w_N in Eq. (n) converges as the number of parameters N are increased, giving almost an exact solution for only finitely large values of N. For example, a one-term approximation of a simply supported beam with uniformly distributed load of intensity q_0 and point load F_0 at the midspan is

$$w_1 = \frac{2L^3(2q_0 L + \pi F_0)}{EI\pi^5} \sin \frac{\pi x}{L}$$

Evaluating at $x = L/2$, we get

$$w_1\left(\frac{L}{2}\right) = \frac{q_0 L^4}{76.5 EI} + \frac{F_0 L^3}{48.7 EI}$$

whereas the exact solution is

$$w_0\left(\frac{L}{2}\right) = \frac{q_0 L^4}{76.8 EI} + \frac{F_0 L^3}{48 EI}$$

The center deflection is 0.39% in error in the case of the uniform load and 1.46% in error for the point load at the center. The two-term approximation gives the deflection

$$w_2(x) = w_1(x) + \left(\frac{4 f_0 L^4}{243 EI\pi^5} - \frac{2PL^3}{81 EI\pi^5}\right) \sin \frac{3\pi x}{L}$$

Evaluating w_2 at $x = L/2$, we find

$$w_2\left(\frac{L}{2}\right) = \frac{q_0 L^4}{76.8 EI} + \frac{F_0 L^3}{48.1 EI}$$

Thus, the two-term approximation gives a center deflection that is exact for uniform load and 0.21% in error for point load at the center.

2.3.3 The Galerkin Method

Consider an operator equation of the form

$$A(u) = f \quad \text{in } \Omega \tag{2.3.7}$$

subjected to boundary conditions

$$B_1(u) = \hat{u} \quad \text{on } \Gamma_1, \quad B_2(u) = \hat{g} \quad \text{on } \Gamma_2 \tag{2.3.8}$$

where A is a linear or nonlinear differential operator, u is the dependent variable, f is a given force term in the domain Ω, B_1 and B_2 are boundary operators associated with essential and natural boundary conditions of the operator A, and $\hat{u}$ and $\hat{g}$ are specified values on the portions Γ_1 and Γ_2 of the boundary Γ of the domain. An example of Eqs. (2.3.7) and (2.3.8) is given by

$$A(u) = -\frac{d}{dx}\left(a\frac{du}{dx}\right), \quad B_1(u) = \hat{u}, \quad B_2(u) = a\frac{du}{dx}$$

Γ_1 is the point $x = 0$, Γ_2 is the point $x = L$

We seek an approximate solution of u in the form

$$U_N = \sum_{j=1}^{N} c_j \varphi_j + \varphi_0 \tag{2.3.9}$$

where the parameters c_j are determined by requiring that the residual of the approximation

$$\mathcal{R}_N = A\left(\sum_{j=1}^{N} c_j \varphi_j + \varphi_0\right) - f \neq 0 \tag{2.3.10}$$

be orthogonal to a set of N linearly independent *weight functions* ψ_i :

$$\int_\Omega \psi_i \mathcal{R}_N(c_i, \varphi_i, f) d\mathbf{x} = 0, \ i = 1, 2, \cdots, N \qquad (2.3.11)$$

The method based on this procedure is called, for obvious reason, a *weighted-residual method*.

The coordinate function φ_0 and φ_i in a weighted-residual method must satisfy the following conditions:

1. $\varphi_j (j = 1, 2, \cdots, N)$ should satisfy the following three conditions:

   (a) Each φ_j is *continuous* as required in the weighted-residual statement; i.e., φ_j should be such that U_N yields a nonzero value of $A(U_N)$.

   (b) Each φ_j satisfies the *homogeneous form* of all specified (i.e., essential as well as natural) boundary conditions.

   (c) The set $\{\varphi_j\}$ is *linearly independent* and *complete*.

2. φ_0 has the main purpose of satisfying *all* specified boundary conditions associated with the equation. It is necessarily zero when the specified boundary conditions are homogeneous.

3. ψ_i should be linearly independent.

There are two main differences between the approximation functions used in the Rayleigh–Ritz method and those used in weighted-residual methods: (1) *Continuity*. The approximation functions used in the weighted-residual methods are required to have the same differentiability as in the differential equation, whereas those used in the Rayleigh–Ritz method must be differentiable as required by the weak form. (2) *Boundary Conditions*. The approximation functions used in the Rayleigh–Ritz method must satisfy the homogeneous form of both geometric and force boundary conditions. Both of these differences require φ_i to be of a higher order than those used in the Rayleigh–Ritz method.

Equation (2.3.11) provides N linearly independent equations for the determination of the parameters c_i. If A is a nonlinear operator, the resulting algebraic equations will be nonlinear. Whenever A is linear, we can write

$$A \left(\sum_{j=1}^N c_j \varphi_j + \varphi_0 \right) = \sum_{j=1}^N c_j A(\varphi_j) + A(\varphi_0) \qquad (2.3.12)$$

and Eq. (2.3.11) becomes

$$\sum_{j=1}^{N}\left[\int_{\Omega}\psi_i A(\varphi_j)d\mathbf{x}\right]c_j - \int_{\Omega}\psi_i\left[f - A(\varphi_0)\right]d\mathbf{x} = 0$$

$$\sum_{j=1}^{N} G_{ij}c_j - q_i = 0, \quad i = 1, 2, \cdots, N \qquad (2.3.13a)$$

where

$$G_{ij} = \int_{\Omega}\psi_i A(\varphi_j)d\mathbf{x}, \quad q_i = \int_{\Omega}\psi_i\left[f - A(\varphi_0)\right]d\mathbf{x} \qquad (2.3.13b)$$

Note that G_{ij} is not symmetric in general, even when $\psi_i = \varphi_i$ (Galerkin's method). It is symmetric when A is a linear operator and $\psi_i = A(\varphi_i)$ (the least-squares method).

Various special cases of the weighted-residual method differ from each other due to the choice of the weight function, ψ_i. The most commonly used weight functions are:

Galerkin's method: $\psi_i = \varphi_i$.

Least-Squares method: $\psi_i = A(\varphi_i)$.

Collocation method: $\psi_i = \delta(\mathbf{x} - \mathbf{x}_i)$.

Here $\delta(\cdot)$ denotes the Dirac delta function. The weighted-residual method in the general form (2.3.11) (with $\psi_i \neq \varphi_i$) is known as the *Petrov–Galerkin method*.

The Galerkin method is a special case of the Petrov–Galerkin method in which the coordinate functions and the weighted functions are the same ($\varphi_i = \psi_i$). If the Galerkin method is used for second-order or higher-order equations, it would involve the use of higher-order coordinate functions and the solution of non-symmetric equations.

The Rayleigh–Ritz and Galerkin methods yield the same set of algebraic equations for the following two cases:

1. The specified boundary conditions of the problem are all essential type, and therefore the requirements on φ_i in both methods are the same.

2. The problem has both essential and natural boundary conditions, but the coordinate functions used in the Galerkin method are also used in the Rayleigh–Ritz method.

Example 2.3.4

Consider the bar problem of Example 2.3.2. In the weighted-residual method, φ_i must also satisfy the condition $\varphi'_i(1) + \varphi_i(1) = 0$ (for the nondimensionalized problem). The lowest-order function that satisfies the condition is

$$\varphi_1 = 3x - 2x^2 \qquad (a)$$

The one-parameter Galerkin's solution for the natural frequency can be computed using

$$0 = \int_0^1 \varphi_1 \left(\lambda c_1 \varphi_1 + c_1 \frac{d^2\varphi_1}{dx^2} \right) dx \quad \text{or} \quad \left(\frac{4}{5}\lambda - \frac{10}{3} \right) c_1 = 0 \qquad (b)$$

which gives $\lambda_1 = 50/12 = 4.167$. If the same function is used for φ_1 in the one-parameter Rayleigh–Ritz solution, we obtain the same result as in the one-parameter Galerkin's solution.

For the transient response we have

$$0 = \int_0^1 \varphi_1 \left(c_1 \frac{d^2\varphi_1}{dx^2} - \frac{d^2c_1}{dt^2}\varphi_1 + 1 \right) dx \quad \text{or} \quad \frac{4}{5}\frac{d^2c_1}{dt^2} + \frac{10}{3}c_1 = \frac{5}{6} \qquad (c)$$

whose solution is ($\sqrt{50/12} \approx 2.0412$)

$$c_1(t) = A \sin 2.04t + B \cos 2.04t + \frac{1}{4}$$

For zero initial conditions, the total solution becomes

$$u_1(x,t) = \frac{1}{4}(1 - \cos 2.04t)\left(3x - 2x^2\right)$$

Example 2.3.5

Consider the simply supported beam problem of Example 2.3.3. Since all specified boundary conditions are homogeneous, again we have $\varphi_0 = 0$. In the Galerkin method, φ_i must satisfy the homogeneous form of all specified boundary conditions ($w_0 = M_{xx} = 0$ at $x = 0, L$):

$$\varphi_i = 0, \quad \frac{d^2\varphi_i}{dx^2} = 0 \qquad (a)$$

For the choice of algebraic polynomials, we assume a five-parameter polynomial because there are four conditions in Eq. (a):

$$\varphi_1(x) = a + bx + cx^2 + dx^3 + ex^4$$

Using the boundary conditions we find that

$$a = c = 0, \quad bL + dL^3 + eL^4 = 0, \quad 6dL + 12eL^2 = 0$$

Thus we have $b = eL^3$ and $d = -2eL$. The function φ_1 is given by (taking $eL^4 = 1$)

$$\varphi_1 = \frac{x}{L}\left(1 - 2\frac{x^2}{L^2} + \frac{x^4}{L^4}\right) \tag{b}$$

Substituting the one-parameter Galerkin approximation $W_1 = c_1\varphi_1$ into the residual

$$\mathcal{R} = EI\frac{d^4 W_1}{dx^4} - q_0 = \frac{24EI}{L^4}c_1 - q_0 \tag{c}$$

and integrating over 0 to L yields $c_1 = q_0 L^4/(24EI)$. Hence, the solution becomes

$$W_1(x) = \frac{q_0 L^4}{24 EI}\left[\left(\frac{x}{L}\right) - 2\left(\frac{x}{L}\right)^3 + \left(\frac{x}{L}\right)^4\right] \tag{d}$$

which is the same as the exact solution.

It can be shown that a one-parameter Rayleigh–Ritz solution with φ_1 given by Eq. (b) gives the exact solution. From Eq. (e) of Example 2.3.3 we have the result

$$R_{11} = \int_0^L EI\frac{d^2\phi_1}{dx^2}\frac{d^2\phi_1}{dx^2}\,dx = 144\frac{EI}{L^4}\int_0^L \left(-\frac{x}{L} + \frac{x^2}{L^2}\right)^2 dx = \frac{24EIL^5}{5}$$

$$F_1 = \int_0^L q_0\phi_1\,dx = q_0\int_0^L \left[\left(\frac{x}{L}\right) - 2\left(\frac{x}{L}\right)^3 + \left(\frac{x}{L}\right)^4\right]dx = \frac{q_0 L}{5}$$

Hence, we obtain $c_1 = q_0 L^4/(24EI)$.

2.4 Summary

In this chapter, an introduction to the principle of virtual displacements and its special case, the principle of minimum total potential energy, are presented. The variational/energy principles provide a means for the derivation of the governing equations of plates, provided one can write the internal and external virtual work expressions for the problem. They also yield the natural boundary conditions and give the form of the essential and natural boundary conditions.

An introduction to the Rayleigh–Ritz and Galerkin methods is also presented in this chapter. The Rayleigh–Ritz method makes use of the weak form provided by the principle of virtual displacements or the principles of minimum potential energy in determining the approximate solutions. The Galerkin method, on the other hand, does not require any variational statements. It is based on a weighted-integral statement of the governing equation(s). However, the Galerkin solution is required to satisfy all specified boundary conditions, whereas the Rayleigh–Ritz solution need only satisfy the specified geometric boundary conditions.

The single most difficult step in each method presented in this chapter is the selection of the approximation functions. The requirements on the approximation functions merely provide the guidelines for their selection. The selection of coordinate functions becomes more difficult for problems with irregular domains (i.e., noncircular and nonrectangular) or discontinuous data (i.e., loading or geometry). Further, the generation of coefficient matrices for the resulting algebraic equations cannot be automated for a *class* of problems that differ from each other only in the geometry of the domain, boundary conditions, or loading. These limitations of the classical variational methods can be overcome by representing a given domain as a collection of geometrically simple subdomains for which we can systematically generate the coordinate functions. One such technique is the *finite element method*, which is discussed later in this book.

Exercises

Find the Euler–Lagrange equations and the natural boundary conditions associated with each of the functionals given in Exercises 2.1–2.6. The dependent variables are listed as the arguments of the functional, and only the geometric boundary conditions are given. No other variables are functions

2.1 Axial deformation of a spring-supported bar: $u_0(0) = 0$

$$\Pi(u_0) = \int_0^L \left[\frac{EA}{2} \left(\frac{du_0}{dx} \right)^2 - fu_0 \right] dx + \frac{k}{2}[u_0(L)]^2 - Pu_0(L)$$

2.2 Bending of a beam on elastic foundation: $w_0(0) = w_0(L) = 0$

$$\Pi(w_0) = \int_0^L \left[\frac{EI}{2} \left(\frac{d^2 w_0}{dx^2} \right)^2 + \frac{k}{2} w_0^2 - q w_0 \right] dx$$

2.3 Nonlinear bending of a beam

$$\Pi(u_0, w_0) = \int_0^L \left\{ \frac{EA}{2} \left[\frac{du_0}{dx} + \frac{1}{2} \left(\frac{dw_0}{dx} \right)^2 \right]^2 + \frac{EI}{2} \left(\frac{d^2 w_0}{dx^2} \right)^2 \right\} dx$$
$$- F_0 w_0(L) - P u_0(L)$$

$$u_0(0) = 0, \quad w_0(0) = 0, \quad \frac{dw_0}{dx}(0) = 0$$

2.4 Transverse deflection of a membrane

$$I(u) = \int_\Omega \left[\frac{a_{11}}{2} \left(\frac{\partial u}{\partial x} \right)^2 + \frac{a_{22}}{2} \left(\frac{\partial u}{\partial y} \right)^2 - fu \right] dxdy - \int_{\Gamma_\sigma} tu\, ds$$

2.5 Bending of an orthotropic plate

$$\Pi(w_0) = \int_\Omega \left[\frac{D_{11}}{2} \left(\frac{\partial^2 w_0}{\partial x^2} \right)^2 + D_{12} \frac{\partial^2 w_0}{\partial x^2} \frac{\partial^2 w_0}{\partial y^2} + 2 D_{66} \left(\frac{\partial^2 w_0}{\partial x \partial y} \right)^2 \right.$$
$$\left. + \frac{D_{22}}{2} \left(\frac{\partial^2 w_0}{\partial y^2} \right)^2 - q w_0 \right] dxdy$$

$$w_0 = 0, \quad \frac{\partial w_0}{\partial n} = 0 \quad \text{on the boundary } \Gamma$$

2.6 Plane elasticity

$$I(u,v) = \int_\Omega \left\{ \frac{1}{2} \left[c_{11} \left(\frac{\partial u}{\partial x} \right)^2 + c_{22} \left(\frac{\partial v}{\partial y} \right)^2 + 2 c_{12} \frac{\partial u}{\partial x} \frac{\partial v}{\partial y} \right. \right.$$
$$\left. \left. + c_{33} \left(\frac{\partial u}{\partial y} + \frac{\partial v}{\partial x} \right)^2 \right] - f_1 u - f_2 v \right\} dxdy - \int_{\Gamma_2} (t_1 u + t_2 v) ds$$

$$u = \hat{u}, \quad v = \hat{v} \quad \text{on} \quad \Gamma_1, \quad \Gamma_1 + \Gamma_2 = \Gamma$$

2.7 Show that the force and moment resultants N_{xx}, M_{xx}, and Q_x can be expressed in terms of the generalized displacements (u_0, w_0, ϕ) of the Timoshenko beam theory as

$$N_{xx} = EA\left(\frac{\partial u_0}{\partial x} - \alpha T^0\right), \quad M_{xx} = EI\left(\frac{\partial \phi}{\partial x} - \alpha T^1\right)$$

$$Q_x = GAK_s\left(\phi + \frac{\partial w_0}{\partial x}\right)$$

where α is the thermal expansion coefficient, E the modulus, G the shear modulus, A the area of cross section, I the moment of inertia, and $\Delta T = T^0 + zT^1$ is the temperature change. Use the thermoelastic constitutive relations $\sigma_{xx} = E(\varepsilon_{xx} - \alpha \Delta T)$ and $\sigma_{xz} = G\gamma_{xz}$.

2.8 (A Third-Order Beam Theory) Consider the displacement field

$$u(x, z, t) = u_0(x, t) + z\phi(x, t) - c_1 z^3 \left(\phi + \frac{\partial w_0}{\partial x}\right)$$
$$w(x, z, t) = w_0(x, t) \tag{1}$$

where $c_1 = 4/(3h^2)$. The displacement field accommodates quadratic variation of transverse shear strains and stresses and vanishing of transverse shear stress on the top and bottom planes of a beam.

(a) Show that the nonzero linear strains associated with the displacement field (1) are:

$$\varepsilon_{xx} = \varepsilon_{xx}^{(0)} + z\varepsilon_{xx}^{(1)} + z^3 \varepsilon_{xx}^{(3)}$$
$$\gamma_{xz} = \gamma_{xz}^{(0)} + z^2 \gamma_{xz}^{(2)} \tag{2a}$$

where

$$\varepsilon_{xx}^{(0)} = \frac{\partial u_0}{\partial x}, \quad \varepsilon_{xx}^{(1)} = \frac{\partial \phi}{\partial x}, \quad \varepsilon_{xx}^{(3)} = -c_1\left(\frac{\partial \phi}{\partial x} + \frac{\partial^2 w_0}{\partial x^2}\right)$$

$$\gamma_{xz}^{(0)} = \phi + \frac{\partial w_0}{\partial x}, \quad \gamma_{xz}^{(2)} = -c_2\left(\phi + \frac{\partial w_0}{\partial x}\right) \tag{2b}$$

and $c_2 = 4/h^2$.

(b) Using the dynamic version of the principle of virtual displacements, show that the equations of motion of the theory are given by

$$\frac{\partial N_{xx}}{\partial x} = I_0 \frac{\partial^2 u_0}{\partial t^2} \tag{3}$$

$$\frac{\partial \bar{Q}_x}{\partial x} + c_1 \frac{\partial^2 P_{xx}}{\partial x^2} + q$$
$$= I_0 \frac{\partial^2 w_0}{\partial t^2} + c_1 \left(J_4 \frac{\partial^3 \phi}{\partial x \partial t^2} - c_1 I_6 \frac{\partial^4 w_0}{\partial x^2 \partial t^2} \right) \tag{4}$$

$$\frac{\partial \bar{M}_{xx}}{\partial x} - \bar{Q}_x = K_2 \frac{\partial^2 \phi}{\partial t^2} - c_1 J_4 \frac{\partial^3 w_0}{\partial x \partial t^2} \tag{5}$$

where

$$\left\{ \begin{array}{c} N_{xx} \\ M_{xx} \\ P_{xx} \end{array} \right\} = \int_{-h/2}^{h/2} \left\{ \begin{array}{c} 1 \\ z \\ z^3 \end{array} \right\} \sigma_{xx}\, dz, \quad \left\{ \begin{array}{c} Q_x \\ R_x \end{array} \right\} = \int_{-h/2}^{h/2} \left\{ \begin{array}{c} 1 \\ z^2 \end{array} \right\} \sigma_{xz}\, dz \tag{6}$$

$$\bar{M}_{xx} = M_{xx} - c_1 P_{xx}, \quad \bar{Q}_x = Q_x - c_2 R_x \tag{7}$$

$$J_i = I_i - c_1 I_{i+2}, \quad K_2 = I_2 - 2c_1 I_4 + c_1^2 I_6, \quad I_i = \int_{-h/2}^{h/2} \rho(z)^i\, dz \tag{8}$$

and (P_{xx}, R_x) denote the higher-order stress resultants. Note that the primary and secondary variables of the theory are:

$$\text{Primary Variables:} \quad u_0,\ w_0,\ \frac{\partial w_0}{\partial x},\ \phi \tag{9a}$$

$$\text{Secondary Variables:} \quad N_{xx},\ V_x,\ P_{xx},\ \bar{M}_{xx} \tag{9b}$$

where

$$V_x \equiv \bar{Q}_x + N_{xx}\frac{\partial w_0}{\partial x} + c_1 \left[\frac{\partial P_{xx}}{\partial x} - \left(I_3 \ddot{u}_0 + J_4 \ddot{\phi}_x - c_1 I_6 \frac{\partial \ddot{w}_0}{\partial x} \right) \right] \tag{10}$$

2.9 Suppose that the displacement field of the Timoshenko beam theory (see Example 2.2.2 on page 67) can be expressed in the form

$$u(x,z) = z\left(-\frac{dw^b}{dx} + \phi\right), \quad w(x,z) = w^b(x) + w^s(x) \tag{a}$$

where w^b and w^s denote the bending and shear components, respectively, of the total transverse deflection w, and ϕ denotes the shear rotation, in addition to the bending rotation, of a transverse normal about the y axis. Use the principle of virtual displacements to obtain the following Euler–Lagrange equations:

$$\delta\phi: \quad -\frac{dM_x}{dx} + Q_x = 0 \tag{b}$$

$$\delta w^b: \quad -\frac{d^2 M_x}{dx^2} - q = 0 \tag{c}$$

$$\delta w^s: \quad -\frac{dQ_x}{dx} - q = 0 \tag{d}$$

where q is the distributed transverse load (measured per unit length), and M_x and Q_x are the resultants defined in Eq. (f) on page 68.

2.10 (**Unit Dummy Force Method**) The principle of complementary virtual work can be used to find point forces (i.e., reactions) and point displacements of simple structures. Let U_0 be the point displacement at a point 'O' in an elastic structure and e_{ij} be the actual strains in the structure. Let us prescribe a virtual point force δF_0 at point 'O' that induces virtual stresses $\delta\sigma_{ij}^0$ in the structure. Then the principle of complementary virtual work gives

$$0 = \int_V \delta\sigma_{ij}^0 \varepsilon_{ij} \, dV - \delta F_0 U_0 \tag{a}$$

or

$$\delta F_0 U_0 = \int_V \delta\sigma_{ij}^0 \varepsilon_{ij} \, dV \tag{b}$$

Since δF_0 is arbitrary, it can be taken as unity. Then the above equation reduces to

$$U_0 = \int_V \delta\sigma_{ij}^0 \varepsilon_{ij} \, dV \tag{c}$$

where $\delta\sigma_{ij}^0$ are the virtual stresses in the structure due to unit virtual force at point 'O'.

For the bending of (the Euler–Bernoulli) beams, show that Eq. (c) reduces to

$$w_0 = \int_0^L \delta M^0 \frac{M}{EI} \, dx \tag{d}$$

where δM^0 is the bending moment due to unit virtual force at point 'O', M is the actual bending moment, and w_0 is the transverse deflection at point 'O'.

2.11 Use the Unit-Dummy-Force method of Exercise 2.10 to determine the center deflection of a simply-supported beam under uniformly distributed transverse load, f. Note that the point of interest 'O' is at $x = L/2$.

2.12 Use the Unit-Dummy-Force (moment) method to determine the rotation at $x = 0$ in Exercise 2.11.

2.13 (**Castigliano's Second Theorem**) An equivalent procedure to that of the Unit-Dummy-Force method of Exercise 2.10 but based on the total complementary potential energy principle is known as Castigliano's second theorem. It can be derived directly from Eq. (c) of Exercise 2.10:

$$U_0 = \int_V \delta\sigma_{ij}^0 \varepsilon_{ij} \, dV = \int_V \delta U_0^* \, dV = \frac{\partial U^*}{\partial F_0} \qquad (a)$$

where U^* is the complementary strain energy of the structure with virtual load F_0 and actual loads included.

Equation (a) can be derived from the principle of total complementary energy. Suppose that we are interested in determining the displacement U_0 at point 'O' of an elastic structure that is subjected to point loads as well as distributed forces. The total complementary energy is given by

$$\Pi^* = U^* + V^*$$

which is a function of the actual load F_0 (if there is a point load at the point 'O') or the virtual load F_0 (if there is no point load applied at the point 'O'). Then $\delta\Pi^* = 0$ gives, for arbitrary δF_0, the result

$$-\frac{\partial V^*}{\partial F_0} = \frac{\partial U^*}{\partial F_0}$$

Since F_0 is a point load, we have $V^* = -F_0 \cdot U_0 +$ other terms not dependent on F_0. Hence

$$\frac{\partial V^*}{\partial F_0} = -U_0 = \left(\frac{\partial U^*}{\partial F_0}\right)$$

which is same as Eq. (a). When F_0 is a virtual load, $\partial U^*/\partial F_0$ should be evaluated as $F_0 = 0$ before using in Eq. (a). Show that for the bending of beams, Eq. (a) can be expressed as

$$w_0 = \int_0^L \frac{M}{EI} \frac{\partial M}{\partial F_0} \, dx \qquad (b)$$

where w_0 and F_0 are the transverse displacement and force, respectively, at point 'O'. Use Eq. (b) to solve the problem of Exercise 2.11.

2.14 Use Castigliano's second theorem to determine the compressive force and displacements in the linear elastic spring (spring constant, k) supporting the free end of a cantilevered beam under triangular distributed load (see Figure P2.14).

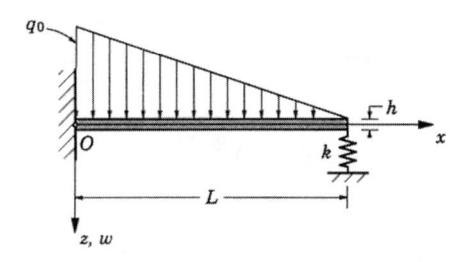

Figure P2.14

2.15 Use Castigliano's second theorem to determine the vertical and horizontal deflections at the free end of the curved beam shown in Figure P2.15.

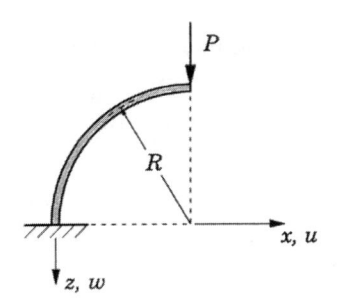

Figure P2.15

2.16 Give the approximation functions φ_1 and φ_0 required in the (i) Rayleigh–Ritz and (ii) Galerkin methods to solve the following problems:

(a) A beam clamped at both ends.

(b) A clamped solid circular plate with uniform transverse load.

2.17 Consider a uniform beam fixed at one end and supported by an elastic spring (spring constant k) in the vertical direction. Assume that the beam is loaded by uniformly distributed load q_0. Determine one-parameter Rayleigh–Ritz solution using algebraic functions.

2.18 Use the total potential energy functional to determine a one-parameter Rayleigh–Ritz solution of a clamped circular plate subjected to a transverse point load F_0 at the center.

2.19 Repeat Exercise 2.18 using the Galerkin method.

2.20 Consider the buckling of a uniform beam according to the Timoshenko beam theory. The total potential energy functional for the problem can be written as

$$\Pi(w_0, \phi_x) = \frac{1}{2} \int_0^L \left[D \left(\frac{d\phi_x}{dx}\right)^2 + S \left(\frac{dw_0}{dx} + \phi_x\right)^2 - N \left(\frac{dw_0}{dx}\right)^2 \right] dx$$

where $w_0(x)$ is the transverse deflection, ϕ_x is the rotation, D is the flexural stiffness, S is the shear stiffness, and N is the axial compressive load. We wish to determine the critical buckling load N_{cr} of a simply supported beam using the Rayleigh–Ritz method. Assume one-parameter approximation of the form

$$w_0(x) = c_1 \sin \frac{\pi x}{L}, \quad \phi_x(x) = c_2 \cos \frac{\pi x}{L}$$

and determine the critical buckling load.

2.21 Determine the n-parameter Rayleigh–Ritz solution for the natural frequencies of a simply supported beam. Use trigonometric functions for $\varphi_i(x)$.

2.22 Determine the n-parameter Rayleigh–Ritz solution for the transient response of a simply supported beam under step loading $q(x,t) = q_0 H(t - t_0)$, where $H(t)$ denotes the Heaviside step function. Use trigonometric functions for $\varphi_i(x)$.

2.23 Show that the two-parameter Rayleigh–Ritz solution for the transient response of the bar considered in Example 2.3.2 yields the equations

$$\begin{bmatrix} \frac{1}{3} & \frac{1}{4} \\ \frac{1}{4} & \frac{1}{5} \end{bmatrix} \begin{Bmatrix} \ddot{c}_1 \\ \ddot{c}_2 \end{Bmatrix} + \begin{bmatrix} 2 & 2 \\ 2 & \frac{7}{3} \end{bmatrix} \begin{Bmatrix} c_1 \\ c_2 \end{Bmatrix} = \begin{Bmatrix} \frac{1}{2} \\ \frac{1}{3} \end{Bmatrix} \quad \text{(a)}$$

Use the Laplace transform method to determine the solution of these equations.

References for Additional Reading

1. Oden, J. T. and Reddy, J. N., *Variational Methods in Theoretical Mechanics*, 2nd Edition, Springer–Verlag, Berlin (1982).
2. Oden, J. T. and Ripperger, E. A., *Mechanics of Elastic Structures*, 2nd ed., Hemisphere, New York (1981).
3. Reddy, J. N., *Energy and Variational Methods in Applied Mechanics*, John Wiley, New York (1984).
4. Reddy, J. N., *Applied Functional Analysis and Variational Methods in Engineering*, McGraw–Hill, New York, 1986; reprinted by Krieger, Melbourne, Florida (1992).
5. Reddy, J. N., *An Introduction to the Finite Element Method*, Second Edition, McGraw–Hill, New York (1993).
6. Reddy, J. N., *Mechanics of Laminated Composite Plates: Theory and Analysis*, CRC Press, Boca Raton, Florida (1997).
7. Reddy, J. N. and Rasmussen, M. L., *Advanced Engineering Analysis*, John Wiley, New York (1982); reprinted by Krieger, Melbourne, Florida (1991).
8. Washizu, K., *Variational Methods in Elasticity and Plasticity*, 3rd Edition, Pergamon Press, New York (1982).
9. Timoshenko, S. P., "On the Transverse Vibrations of Bars of Uniform Cross Section", *Philosophical Magazine*, **43**, 125-131 (1922).
10. Timoshenko, S. P. and Woinowsky-Krieger, S., *Theory of Plates and Shells*, McGraw–Hill, Singapore (1970).
11. Lanczos, C., *The Variational Principles of Mechanics*, The University of Toronto Press, Toronto (1964).
12. Langhaar, H. L., *Energy Methods in Applied Mechanics*, John Wiley, New York (1962).
13. Mikhlin, S. G., *Variational Methods in Mathematical Physics*, (translated from the 1957 Russian edition by T. Boddington) The MacMillan Company, New York (1964).
14. Mikhlin, S. G., *The Problem of the Minimum of a Quadratic Functional* (translated from the 1952 Russian edition by A. Feinstein), Holden-Day, San Francisco (1965).

Chapter Three

The Classical Theory of Plates

3.1 Introduction

The objective of this chapter is to develop the governing equations of the classical theory of plates. The principle of virtual displacements is used to derive the equations of motion and identify the form of the geometric and force boundary conditions.

A *plate* is a structural element with planform dimensions that are large compared to its thickness and is subjected to loads that cause bending deformation in addition to stretching. In most cases, the thickness is no greater than one-tenth of the smallest in-plane dimension. Because of the smallness of thickness dimension, it is often not necessary to model them using 3D elasticity equations. Simple 2D plate theories can be developed to study the deformation and stresses in plate structures.

Plate theories are developed by assuming the form of the displacement or stress field as a linear combination of unknown functions and the thickness coordinate:

$$\varphi_i(x,y,z,t) = \sum_{j=0}^{N}(z)^j \varphi_i^j(x,y,t) \qquad (3.1.1)$$

where φ_i is the ith component of displacement or stress, (x,y) are the in-plane coordinates, z is the thickness coordinate, t denotes the time, and φ_i^j are functions to be determined.

When φ_i are displacements, the equations governing φ_i^j are determined by the principle of virtual displacements (or its dynamic version, when time dependency is to be included):

$$0 = \int_0^T (\delta U + \delta V - \delta K)\,dt \qquad (3.1.2)$$

where $\delta U, \delta V$, and δK denote the virtual strain energy, virtual work done by external applied forces, and virtual kinetic energy, respectively. These quantities are determined in terms of actual stresses and virtual strains, which depend on the assumed displacement functions, φ_i, and their variations. For plate structures, the integration over the domain of the plate is represented as the (tensor) product of integration over the plane of the plate and integration over the thickness of the plate (volume integral=integral over the plane × integral over the thickness). This is possible due to the explicit nature of the assumed displacement field in the thickness coordinate. Thus, we can write

$$\int_{Vol.} (\cdot)\, dV = \int_{-\frac{h}{2}}^{\frac{h}{2}} \int_{\Omega_0} (\cdot)\, d\Omega\, dz \qquad (3.1.3)$$

where h denotes the total thickness of the plate and Ω_0 denotes the undeformed mid-plane of the plate, which is assumed to coincide with the xy–plane. Since all undetermined variables are explicit functions of the thickness coordinate, the integration over plate thickness is carried out explicitly, reducing the problem to a two-dimensional one. Consequently, the Euler–Lagrange equations associated with Eq. (3.1.2) consist of differential equations involving the dependent variables $\varphi_i^j(x,y,t)$ and thickness-averaged stress resultants, $R_{ij}^{(m)}$:

$$R_{ij}^{(m)} = \int_{-\frac{h}{2}}^{\frac{h}{2}} (z)^m \sigma_{ij}\, dz \qquad (3.1.4)$$

The resultants can be written in terms of φ_i with the help of the assumed constitutive equations (stress-strain relations) and strain-displacement relations. More complete development of this procedure is forthcoming in this chapter.

The same approach is used when φ_i denote stress components, except that the basis of the derivation of the governing equations is the principle of virtual forces. In the present book, the stress-based theories will receive very little attention. Readers interested in stress-based theories may consult the book by Panc [5].

3.2 Assumptions of the Theory

The *classical plate theory* (CPT) is one in which the displacement field is selected so as to satisfy the *Kirchhoff hypothesis*. The *Kirchhoff hypothesis* has the following three assumptions (see Figure 3.2.1):

(1) Straight lines perpendicular to the mid-surface (i.e., transverse normals) before deformation, remain straight after deformation.

(2) The transverse normals do not experience elongation (i.e., they are in-extensible).

(3) The transverse normals rotate such that they remain perpendicular to the mid-surface after deformation.

Let us examine the consequences of the Kirchhoff assumptions. Consider a plate of uniform thickness h. We shall use the rectangular Cartesian coordinates (x, y, z) with the xy–plane coinciding with the geometric midplane of the plate and the z–coordinate taken positive upward. Suppose that (u, v, w) denote the total displacements of a point along the (x, y, z) coordinates. A material point occupying the position (x, y, z) in the undeformed plate moves to the position $(x+u, y+v, z+w)$ in the deformed plate.

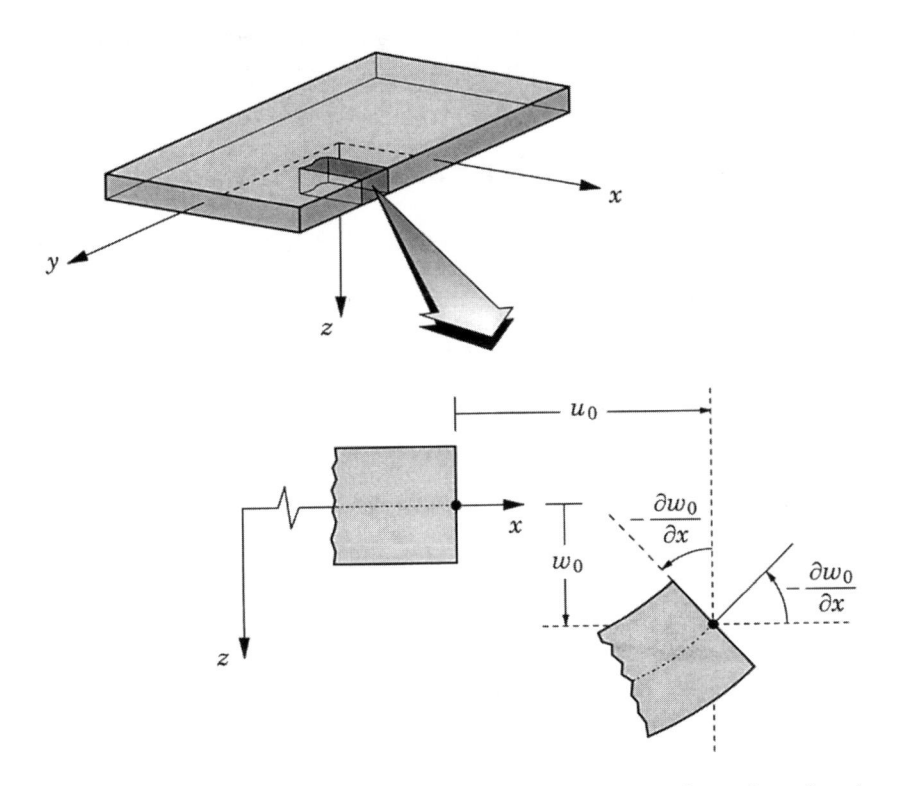

Figure 3.2.1. Undeformed and deformed geometries of an edge of a plate under the Kirchhoff assumptions.

The in-extensibility of a transverse normal (assumptions 1 and 2) implies that the thickness normal strain is zero:

$$\varepsilon_{zz} = \frac{\partial w}{\partial z} = 0 \qquad (3.2.1)$$

which implies that w is independent of z. The third assumption results in zero transverse shear strains

$$\varepsilon_{xz} = \frac{\partial u}{\partial z} + \frac{\partial w}{\partial x} = 0, \quad \varepsilon_{yz} = \frac{\partial v}{\partial z} + \frac{\partial w}{\partial y} = 0 \qquad (3.2.2)$$

These conditions in turn imply, because w is independent of z, that

$$u = -z\frac{\partial w}{\partial x} + u_0(x,y), \quad v = -z\frac{\partial w}{\partial y} + v_0(x,y) \qquad (3.2.3)$$

where u_0 and v_0 represent the values of u and v, respectively, at the point $(x, y, 0)$. In other words, u_0 and v_0 denote the displacements along the x and y coordinates, respectively, of a point on the mid-plane.

3.3 Displacement Field and Strains

Let us denote the undeformed mid-plane of the plate with the symbol Ω_0. The total domain of the plate is the tensor product $\Omega_0 \times (-h/2, h/2)$. The boundary of the total domain consists of the top surface $S_t(z = h/2)$, bottom surfaces $S_b(z = -h/2)$, and the edge $\bar{\Gamma} \equiv \Gamma \times (-h/2, h/2)$. In general, Γ is a curved surface, with outward normal $\hat{\mathbf{n}} = n_x \hat{\mathbf{e}}_x + n_y \hat{\mathbf{e}}_y$, where n_x and n_y are the direction cosines of the unit normal.

The Kirchhoff hypothesis implies, as discussed above, the following form of the displacement field (see Figure 3.2.1)

$$\begin{aligned} u(x,y,z,t) &= u_0(x,y,t) - z\frac{\partial w_0}{\partial x} \\ v(x,y,z,t) &= v_0(x,y,t) - z\frac{\partial w_0}{\partial y} \\ w(x,y,z,t) &= w_0(x,y,t) \end{aligned} \qquad (3.3.1)$$

where (u_0, v_0, w_0) denote the displacements of a material point at $(x, y, 0)$ in (x, y, z) coordinate directions. Note that (u_0, v_0) are associated with extensional deformation of the plate while w_0 denotes the bending deflection.

The strains associated with the displacement field (3.3.1) can be computed using either the nonlinear strain-displacement relations (1.4.4) or the linear strain-displacement relations (1.4.8). The nonlinear strains are given by

$$E_{11} = \frac{\partial u}{\partial x} + \frac{1}{2}\left[\left(\frac{\partial u}{\partial x}\right)^2 + \left(\frac{\partial v}{\partial x}\right)^2 + \left(\frac{\partial w}{\partial x}\right)^2\right]$$

$$E_{12} = \frac{1}{2}\left(\frac{\partial u}{\partial y} + \frac{\partial v}{\partial x} + \frac{\partial u}{\partial x}\frac{\partial u}{\partial y} + \frac{\partial v}{\partial x}\frac{\partial v}{\partial y} + \frac{\partial w}{\partial x}\frac{\partial w}{\partial y}\right)$$

$$E_{13} = \frac{1}{2}\left(\frac{\partial u}{\partial z} + \frac{\partial w}{\partial x} + \frac{\partial u}{\partial x}\frac{\partial u}{\partial z} + \frac{\partial v}{\partial x}\frac{\partial v}{\partial z} + \frac{\partial w}{\partial x}\frac{\partial w}{\partial z}\right)$$

$$E_{22} = \frac{\partial v}{\partial y} + \frac{1}{2}\left[\left(\frac{\partial u}{\partial y}\right)^2 + \left(\frac{\partial v}{\partial y}\right)^2 + \left(\frac{\partial w}{\partial y}\right)^2\right]$$

$$E_{23} = \frac{1}{2}\left(\frac{\partial v}{\partial z} + \frac{\partial w}{\partial y} + \frac{\partial u}{\partial y}\frac{\partial u}{\partial z} + \frac{\partial v}{\partial y}\frac{\partial v}{\partial z} + \frac{\partial w}{\partial y}\frac{\partial w}{\partial z}\right)$$

$$E_{33} = \frac{\partial w}{\partial z} + \frac{1}{2}\left[\left(\frac{\partial u}{\partial z}\right)^2 + \left(\frac{\partial v}{\partial z}\right)^2 + \left(\frac{\partial w}{\partial z}\right)^2\right] \qquad (3.3.2)$$

If the components of the displacement gradients are of the order ϵ, i.e.,

$$\frac{\partial u}{\partial x}, \frac{\partial u}{\partial y}, \frac{\partial v}{\partial x}, \frac{\partial v}{\partial y}, \frac{\partial w}{\partial z} = O(\epsilon) \qquad (3.3.3)$$

then the small strain assumption implies that terms of the order ϵ^2 are omitted in the strains. Terms of order ϵ^2 are:

$$\left(\frac{\partial u}{\partial x}\right)^2, \left(\frac{\partial u}{\partial y}\right)^2, \left(\frac{\partial u}{\partial z}\right)^2, \left(\frac{\partial u}{\partial x}\right)\left(\frac{\partial u}{\partial y}\right), \left(\frac{\partial u}{\partial x}\right)\left(\frac{\partial u}{\partial z}\right), \left(\frac{\partial u}{\partial y}\right)\left(\frac{\partial u}{\partial z}\right)$$

$$\left(\frac{\partial v}{\partial x}\right)^2, \left(\frac{\partial v}{\partial y}\right)^2, \left(\frac{\partial v}{\partial z}\right)^2, \left(\frac{\partial v}{\partial x}\right)\left(\frac{\partial v}{\partial y}\right), \left(\frac{\partial v}{\partial x}\right)\left(\frac{\partial v}{\partial z}\right), \left(\frac{\partial v}{\partial y}\right)\left(\frac{\partial v}{\partial z}\right)$$

$$\left(\frac{\partial w}{\partial x}\right)\left(\frac{\partial w}{\partial z}\right), \left(\frac{\partial w}{\partial y}\right)\left(\frac{\partial w}{\partial z}\right), \left(\frac{\partial w}{\partial z}\right)^2 \qquad (3.3.4)$$

If the rotations of transverse normals are moderate (say $10° - 15°$), then the following terms are small but *not* negligible compared to ϵ

$$\left(\frac{\partial w}{\partial x}\right)^2, \left(\frac{\partial w}{\partial y}\right)^2, \frac{\partial w}{\partial x}\frac{\partial w}{\partial y} \qquad (3.3.5)$$

and they should be included in the strain-displacement relations. Thus for small strains and moderate rotations, the strain-displacement relations (3.3.2) take the form

$$\varepsilon_{xx} = \frac{\partial u}{\partial x} + \frac{1}{2}\left(\frac{\partial w}{\partial x}\right)^2, \quad \varepsilon_{xy} = \frac{1}{2}\left(\frac{\partial u}{\partial y} + \frac{\partial v}{\partial x} + \frac{\partial w}{\partial x}\frac{\partial w}{\partial y}\right)$$

$$\varepsilon_{xz} = \frac{1}{2}\left(\frac{\partial u}{\partial z} + \frac{\partial w}{\partial x}\right), \quad \varepsilon_{yy} = \frac{\partial v}{\partial y} + \frac{1}{2}\left(\frac{\partial w}{\partial y}\right)^2$$

$$\varepsilon_{yz} = \frac{1}{2}\left(\frac{\partial v}{\partial z} + \frac{\partial w}{\partial y}\right), \quad \varepsilon_{zz} = \frac{\partial w}{\partial z} \qquad (3.3.6)$$

where, for this special case of geometric non-linearity (i.e., small strains but moderate rotations), the notation ε_{ij} is used in place of E_{ij}. The corresponding Second–Piola Kirchhoff stresses will be denoted σ_{ij}.

For the displacement field in Eq. (3.3.1), $\partial w/\partial z = 0$. In view of the assumptions in Eqs. (3.3.3)–(3.3.5), the strains (3.3.6) for the displacement field (3.3.1) reduce to

$$\varepsilon_{xx} = \frac{\partial u_0}{\partial x} - z\frac{\partial^2 w_0}{\partial x^2} + \frac{1}{2}\left(\frac{\partial w_0}{\partial x}\right)^2$$

$$\varepsilon_{xy} = \frac{1}{2}\left(\frac{\partial u_0}{\partial y} + \frac{\partial v_0}{\partial x} - 2z\frac{\partial^2 w_0}{\partial x \partial y} + \frac{\partial w_0}{\partial x}\frac{\partial w_0}{\partial y}\right)$$

$$\varepsilon_{yy} = \frac{\partial v_0}{\partial y} - z\frac{\partial^2 w_0}{\partial y^2} + \frac{1}{2}\left(\frac{\partial w_0}{\partial y}\right)^2 \qquad (3.3.7a)$$

$$\varepsilon_{xz} = \frac{1}{2}\left(-\frac{\partial w_0}{\partial x} + \frac{\partial w_0}{\partial x}\right) = 0$$

$$\varepsilon_{yz} = \frac{1}{2}\left(-\frac{\partial w_0}{\partial y} + \frac{\partial w_0}{\partial y}\right) = 0$$

$$\varepsilon_{zz} = 0 \qquad (3.3.7b)$$

The strains in Eqs. (3.3.7a,b) are called the *von Kármán strains*, and the associated plate theory is termed the *von Kármán plate theory*. Note that the transverse strains $(\varepsilon_{xz}, \varepsilon_{yz}, \varepsilon_{zz})$ are identically zero in the classical plate theory.

The strains in Eq. (3.3.7a) have the form

$$\varepsilon_{xx} = \varepsilon_{xx}^0 + z\varepsilon_{xx}^1$$
$$\gamma_{xy} \equiv 2\varepsilon_{xy} = \gamma_{xy}^0 + z\gamma_{xy}^1$$
$$\varepsilon_{yy} = \varepsilon_{yy}^0 + z\varepsilon_{yy}^1 \qquad (3.3.8a)$$

where

$$\varepsilon_{xx}^0 = \frac{\partial u_0}{\partial x} + \frac{1}{2}\left(\frac{\partial w_0}{\partial x}\right)^2 , \quad \varepsilon_{xx}^1 = -\frac{\partial^2 w_0}{\partial x^2}$$

$$\varepsilon_{yy}^0 = \frac{\partial v_0}{\partial y} + \frac{1}{2}\left(\frac{\partial w_0}{\partial y}\right)^2 , \quad \varepsilon_{yy}^1 = -\frac{\partial^2 w_0}{\partial y^2}$$

$$\gamma_{xy}^0 = \frac{\partial u_0}{\partial y} + \frac{\partial v_0}{\partial x} + \frac{\partial w_0}{\partial x}\frac{\partial w_0}{\partial y} , \quad \gamma_{xy}^1 = -2\frac{\partial^2 w_0}{\partial x \partial y} \quad (3.3.8b)$$

where $(\varepsilon_{xx}^0, \varepsilon_{yy}^0, \gamma_{xy}^0)$ are the *membrane strains*, and $(\varepsilon_{xx}^1, \varepsilon_{yy}^1, \gamma_{xy}^1)$ are the flexural (bending) strains, known as the *curvatures*. In matrix notation, Eqs. (3.3.8a,b) can be written as

$$\left\{\begin{array}{c}\varepsilon_{xx}\\ \varepsilon_{yy}\\ \gamma_{xy}\end{array}\right\} = \left\{\begin{array}{c}\varepsilon_{xx}^0\\ \varepsilon_{yy}^0\\ \gamma_{xy}^0\end{array}\right\} + z\left\{\begin{array}{c}\varepsilon_{xx}^1\\ \varepsilon_{yy}^1\\ \gamma_{xy}^1\end{array}\right\} \quad (3.3.9a)$$

$$\left\{\begin{array}{c}\varepsilon_{xx}^0\\ \varepsilon_{yy}^0\\ \gamma_{xy}^0\end{array}\right\} = \left\{\begin{array}{c}\frac{\partial u_0}{\partial x} + \frac{1}{2}\left(\frac{\partial w_0}{\partial x}\right)^2\\ \frac{\partial v_0}{\partial y} + \frac{1}{2}\left(\frac{\partial w_0}{\partial y}\right)^2\\ \frac{\partial u_0}{\partial y} + \frac{\partial v_0}{\partial x} + \frac{\partial w_0}{\partial x}\frac{\partial w_0}{\partial y}\end{array}\right\} , \quad \left\{\begin{array}{c}\varepsilon_{xx}^1\\ \varepsilon_{yy}^1\\ \gamma_{xy}^1\end{array}\right\} = \left\{\begin{array}{c}-\frac{\partial^2 w_0}{\partial x^2}\\ -\frac{\partial^2 w_0}{\partial y^2}\\ -2\frac{\partial^2 w_0}{\partial x \partial y}\end{array}\right\} \quad (3.3.9b)$$

3.4 Equations of Motion

Here, the governing equations are derived using the principle of virtual displacements. In the derivations, we account for thermal (and hence, moisture) effects only with the understanding that the material properties are independent of the temperature and that the temperature T is a known function of position (hence, $\delta T = 0$). Thus, the temperature enters the formulation only through constitutive equations.

As noted earlier, the transverse strains $(\gamma_{xz}, \gamma_{yz}, \varepsilon_{zz})$ are identically zero in the classical plate theory. Consequently, the transverse stresses $(\sigma_{xz}, \sigma_{yz}, \sigma_{zz})$ do not enter the formulation because the virtual strain energy of these stresses is zero [due to the fact that kinematically consistent virtual strains must be zero; see Eq. (3.3.7b)]:

$$\delta\varepsilon_{xz} = 0, \quad \delta\varepsilon_{yz} = 0, \quad \delta\varepsilon_{zz} = 0 \quad (3.4.1)$$

Whether the transverse stresses are accounted for or not in a theory, they are present in reality to keep the plate in equilibrium. In addition,

114 THEORY AND ANALYSIS OF ELASTIC PLATES

these stress components may be specified on the boundary. Thus, the transverse stresses do not enter the virtual strain energy expression but must be accounted for in the boundary conditions and equilibrium of forces.

The dynamic version of the principle of virtual displacements [or the minimum total potential energy, $\delta\Pi = \delta(U + V - K)$] is

$$0 = \int_0^T (\delta U + \delta V - \delta K)\, dt \tag{3.4.2}$$

where U is the strain energy (volume integral of U_0), V is the work done by applied forces, and K is the kinetic energy. Suppose that q_b is the distributed force at the bottom ($z = -h/2$) of the plate, q_t is the distributed force at the top ($z = h/2$) of the plate, and $(\hat{\sigma}_{nn}, \hat{\sigma}_{ns}, \hat{\sigma}_{nz})$ are the specified stress components on the portion Γ_σ of the boundary Γ (see Figure 3.4.1). The virtual displacements must be zero on the portion Γ_u on which the displacements are specified. We have $\Gamma = \Gamma_u + \Gamma_\sigma$.

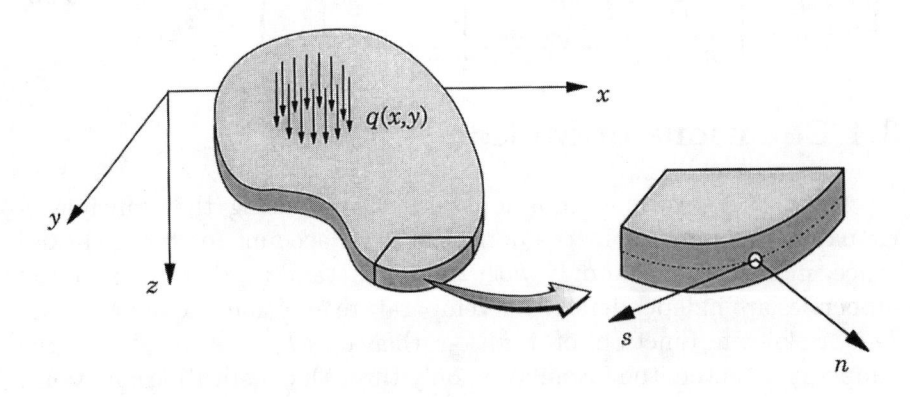

Figure 3.4.1. Geometry of a plate with curved boundary.

The virtual strain energy is given by

$$\delta U = \int_{\Omega_0} \int_{-\frac{h}{2}}^{\frac{h}{2}} (\sigma_{xx}\delta\varepsilon_{xx} + \sigma_{yy}\delta\varepsilon_{yy} + 2\sigma_{xy}\delta\varepsilon_{xy})\, dz\, dx\, dy$$

$$= \int_{\Omega_0} \left\{ \int_{-\frac{h}{2}}^{\frac{h}{2}} \left[\sigma_{xx}\left(\delta\varepsilon_{xx}^0 + z\delta\varepsilon_{xx}^1\right) + \sigma_{yy}\left(\delta\varepsilon_{yy}^0 + z\delta\varepsilon_{yy}^1\right) + \right. \right.$$

THE CLASSICAL THEORY OF PLATES 115

$$+ \sigma_{xy}\left(\delta\gamma_{xy}^0 + z\delta\gamma_{xy}^1\right)\Big] dz \Big\} dxdy$$

$$= \int_{\Omega_0} \left(N_{xx}\delta\varepsilon_{xx}^0 + M_{xx}\delta\varepsilon_{xx}^1 + N_{yy}\delta\varepsilon_{yy}^0 + M_{yy}\delta\varepsilon_{yy}^1\right.$$
$$\left. + N_{xy}\delta\gamma_{xy}^0 + M_{xy}\delta\gamma_{xy}^1\right) dxdy \tag{3.4.3}$$

where (N_{xx}, N_{yy}, N_{xy}) are the *force resultants* and (M_{xx}, M_{yy}, M_{xy}) are the *moment resultants* (see Figure 3.4.2):

$$\left\{\begin{array}{c} N_{xx} \\ N_{yy} \\ N_{xy} \end{array}\right\} = \int_{-\frac{h}{2}}^{\frac{h}{2}} \left\{\begin{array}{c} \sigma_{xx} \\ \sigma_{yy} \\ \sigma_{xy} \end{array}\right\} dz \;,\quad \left\{\begin{array}{c} M_{xx} \\ M_{yy} \\ M_{xy} \end{array}\right\} = \int_{-\frac{h}{2}}^{\frac{h}{2}} \left\{\begin{array}{c} \sigma_{xx} \\ \sigma_{yy} \\ \sigma_{xy} \end{array}\right\} z\, dz \tag{3.4.4}$$

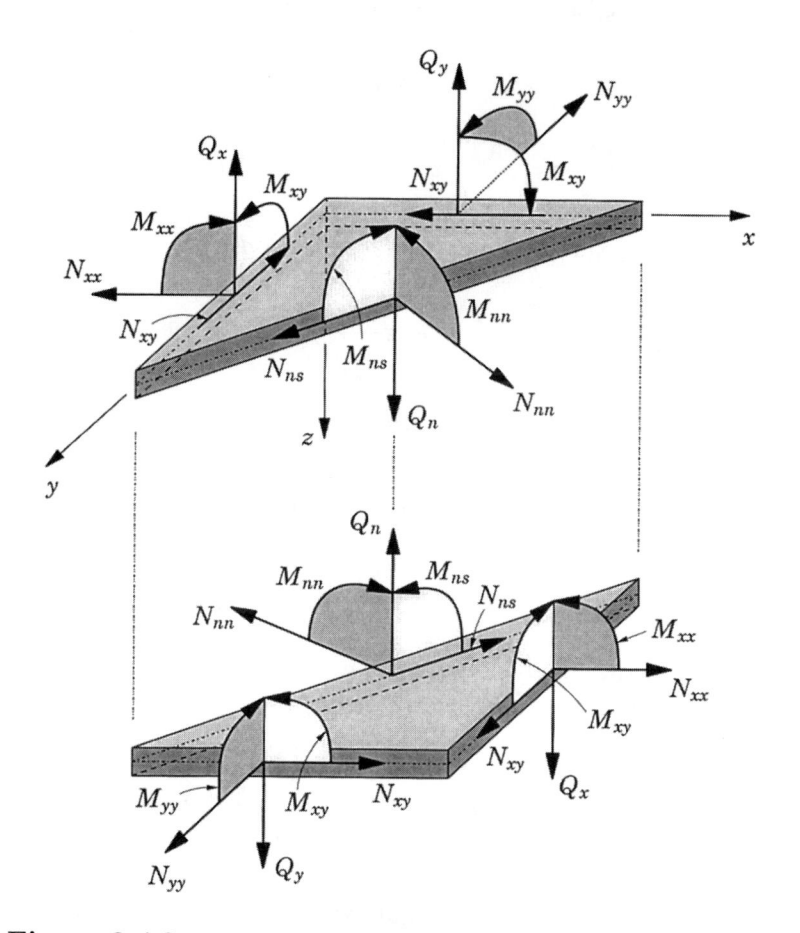

Figure 3.4.2. Force and moment resultants on a plate element.

116 THEORY AND ANALYSIS OF ELASTIC PLATES

The virtual work done by external applied forces can be computed as follows. The external applied forces consist of the distributed transverse load $q_t(x,y)$ applied at the top surface, distributed transverse load $q_b(x,y)$ applied at the bottom surface, and the transverse reaction force of an elastic foundation (if any). In addition, forces and moments due to in-plane normal stress $\hat{\sigma}_{nn}$, in-plane tangential stress $\hat{\sigma}_{ns}$, and the transverse shear stress $\hat{\sigma}_{nz}$, all acting on an edge with normal $\hat{n}$, also do work in moving through corresponding virtual displacements. All stresses are assumed to be positive with respect to the sign convention already established.

The virtual work done by external forces is

$$\delta V = -\int_{\Omega_0} \left[q_t(x,y)\delta w(x,y,\tfrac{h}{2}) + q_b(x,y)\delta w(x,y,-\tfrac{h}{2}) \right] dxdy$$

$$- \int_{\Omega_0} F_s(x,y)\delta w(x,y,\tfrac{h}{2})\, dxdy$$

$$- \int_{\Gamma_\sigma} \int_{-\frac{h}{2}}^{\frac{h}{2}} [\hat{\sigma}_{nn}\delta u_n + \hat{\sigma}_{ns}\delta u_s + \hat{\sigma}_{nz}\delta w]\, dzds \qquad (3.4.5)$$

$$= -\int_{\Omega_0} [q_b(x,y) + q_t(x,y) - kw_0]\,\delta w_0\, dxdy$$

$$- \int_{\Gamma_\sigma}\int_{-\frac{h}{2}}^{\frac{h}{2}} \left[\hat{\sigma}_{nn}\left(\delta u_{0n} - z\frac{\partial \delta w_0}{\partial n}\right) + \hat{\sigma}_{ns}\left(\delta u_{0s} - z\frac{\partial \delta w_0}{\partial s}\right) \right.$$

$$\left. + \hat{\sigma}_{nz}\delta w_0 \right] dzds$$

$$= -\int_{\Omega_0} (q - kw_0)\,\delta w_0\, dxdy - \int_{\Gamma_\sigma} \left(\hat{N}_{nn}\delta u_{0n} - \hat{M}_{nn}\frac{\partial \delta w_0}{\partial n} \right.$$

$$\left. + \hat{N}_{ns}\delta u_{0s} - \hat{M}_{ns}\frac{\partial \delta w_0}{\partial s} + \hat{Q}_n\delta w_0 \right) ds \qquad (3.4.6)$$

where $F_s = -kw_0$, k is the foundation modulus, $q = q_b + q_t$, u_{0n}, u_{0s}, and w_0 are the displacements along the normal, tangential, and transverse directions, respectively, and

$$\left\{ \begin{array}{c} \hat{N}_{nn} \\ \hat{N}_{ns} \end{array} \right\} = \int_{-\frac{h}{2}}^{\frac{h}{2}} \left\{ \begin{array}{c} \hat{\sigma}_{nn} \\ \hat{\sigma}_{ns} \end{array} \right\} dz, \quad \left\{ \begin{array}{c} \hat{M}_{nn} \\ \hat{M}_{ns} \end{array} \right\} = \int_{-\frac{h}{2}}^{\frac{h}{2}} \left\{ \begin{array}{c} \hat{\sigma}_{nn} \\ \hat{\sigma}_{ns} \end{array} \right\} z\, dz \qquad (3.4.7)$$

If the plate is in dynamic equilibrium, the virtual kinetic energy must be calculated. The virtual kinetic energy due to the three velocities is

$$\delta K = \int_{\Omega_0} \int_{-\frac{h}{2}}^{\frac{h}{2}} \rho_0 \left(\dot{u}\delta\dot{u} + \dot{v}\delta\dot{v} + \dot{w}\delta\dot{w} \right) dz \, dxdy$$

$$= \int_{\Omega_0} \int_{-\frac{h}{2}}^{\frac{h}{2}} \rho_0 \left[\left(\dot{u}_0 - z\frac{\partial \dot{w}_0}{\partial x} \right) \left(\delta\dot{u}_0 - z\frac{\partial \delta\dot{w}_0}{\partial x} \right) \right.$$

$$\left. + \left(\dot{v}_0 - z\frac{\partial \dot{w}_0}{\partial y} \right) \left(\delta\dot{v}_0 - z\frac{\partial \delta\dot{w}_0}{\partial y} \right) + \dot{w}_0 \delta\dot{w}_0 \right] dz \, dxdy$$

$$= \int_{\Omega_0} \left[-I_0 \left(\dot{u}_0 \delta u_0 + \dot{v}_0 \delta v_0 + \dot{w}_0 \delta w_0 \right) + I_1 \left(\frac{\partial \delta \dot{w}_0}{\partial x} \dot{u}_0 + \frac{\partial \dot{w}_0}{\partial x} \delta \dot{u}_0 \right. \right.$$

$$\left. \left. + \frac{\partial \delta \dot{w}_0}{\partial y} \dot{v}_0 + \frac{\partial \dot{w}_0}{\partial y} \delta \dot{v}_0 \right) - I_2 \left(\frac{\partial \dot{w}_0}{\partial x} \frac{\partial \delta \dot{w}_0}{\partial x} + \frac{\partial \dot{w}_0}{\partial y} \frac{\partial \delta \dot{w}_0}{\partial y} \right) \right] dxdy$$

(3.4.8)

where the superposed dot on a variable indicates time derivative, $\dot{u}_0 = \partial u_0/\partial t$, and (I_0, I_1, I_2) are the mass moments of inertia

$$\begin{Bmatrix} I_0 \\ I_1 \\ I_2 \end{Bmatrix} = \int_{-\frac{h}{2}}^{\frac{h}{2}} \begin{Bmatrix} 1 \\ z \\ z^2 \end{Bmatrix} \rho_0 \, dz = \rho_0 \begin{Bmatrix} h \\ 0 \\ \frac{h^3}{12} \end{Bmatrix}, \quad \hat{Q}_n = \int_{-\frac{h}{2}}^{\frac{h}{2}} \hat{\sigma}_{nz} \, dz \quad (3.4.9)$$

The virtual displacements are zero on the portion of the boundary where the corresponding actual displacements are specified. For time-dependent problems, the admissible virtual displacements must also vanish at time $t = 0$ and $t = T$. Substituting for $\delta U, \delta V$, and δK from Eqs. (3.4.3), (3.4.6), and (3.4.8) into the virtual work statement in Eq. (3.4.2) and integrating through the thickness of the plate, we obtain

$$0 = \int_0^T \left\{ \int_{\Omega_0} \left[N_{xx}\delta\varepsilon_{xx}^0 + M_{xx}\delta\varepsilon_{xx}^1 + N_{yy}\delta\varepsilon_{yy}^0 + M_{yy}\delta\varepsilon_{yy}^1 + N_{xy}\delta\gamma_{xy}^0 \right. \right.$$

$$+ M_{xy}\delta\gamma_{xy}^1 + kw_0\delta w_0 - I_0 \left(\dot{u}_0 \delta u_0 + \dot{v}_0 \delta v_0 + \dot{w}_0 \delta w_0 \right)$$

$$+ I_1 \left(\frac{\partial \delta \dot{w}_0}{\partial x} \dot{u}_0 + \frac{\partial \dot{w}_0}{\partial x} \delta \dot{u}_0 + \frac{\partial \delta \dot{w}_0}{\partial y} \dot{v}_0 + \frac{\partial \dot{w}_0}{\partial y} \delta \dot{v}_0 \right)$$

$$\left. \left. - I_2 \left(\frac{\partial \dot{w}_0}{\partial x} \frac{\partial \delta \dot{w}_0}{\partial x} + \frac{\partial \dot{w}_0}{\partial y} \frac{\partial \delta \dot{w}_0}{\partial y} \right) - q\delta w_0 \right] dxdy \right\} dt$$

$$- \int_0^T \left\{ \int_{\Gamma_\sigma} \left(\hat{N}_{nn}\delta u_{0n} + \hat{N}_{ns}\delta u_{0s} - \hat{M}_{nn}\frac{\partial \delta w_0}{\partial n} - \hat{M}_{ns}\frac{\partial \delta w_0}{\partial s} \right. \right.$$

$$\left. \left. + \hat{Q}_n\delta w_0 \right) ds \right\} dt \qquad (3.4.10)$$

The virtual strains are known in terms of the virtual displacements:

$$\delta\varepsilon_{xx}^0 = \frac{\partial \delta u_0}{\partial x} + \frac{\partial w_0}{\partial x}\frac{\partial \delta w_0}{\partial x}, \quad \delta\varepsilon_{xx}^1 = -\frac{\partial^2 \delta w_0}{\partial x^2}$$

$$\delta\varepsilon_{yy}^0 = \frac{\partial \delta v_0}{\partial y} + \frac{\partial w_0}{\partial y}\frac{\partial \delta w_0}{\partial y}, \quad \delta\varepsilon_{yy}^1 = -\frac{\partial^2 \delta w_0}{\partial y^2}$$

$$\delta\gamma_{xy}^0 = \frac{\partial \delta u_0}{\partial y} + \frac{\partial \delta v_0}{\partial x} + \frac{\partial \delta w_0}{\partial x}\frac{\partial w_0}{\partial y} + \frac{\partial w_0}{\partial x}\frac{\partial \delta w_0}{\partial y}$$

$$\delta\gamma_{xy}^1 = -2\frac{\partial^2 \delta w_0}{\partial x \partial y} \qquad (3.4.11)$$

Substituting the virtual strains from Eq. (3.4.11) into Eq. (3.4.10) and integrating by parts to relieve the virtual displacements ($\delta u_0, \delta v_0, \delta w_0$) in Ω_0 of any differentiation (so that we can use the fundamental lemma of variational calculus), we obtain

$$0 = \int_0^T \Bigg\{ \int_{\Omega_0} \Bigg[-N_{xx,x}\delta u_0 - \left(N_{xx}\frac{\partial w_0}{\partial x}\right)_{,x} \delta w_0 - M_{xx,xx}\delta w_0 - N_{yy,y}\delta v_0$$

$$- \left(N_{yy}\frac{\partial w_0}{\partial y}\right)_{,y} \delta w_0 - M_{yy,yy}\delta w_0 - N_{xy,y}\delta u_0 - N_{xy,x}\delta v_0$$

$$- \left(N_{xy}\frac{\partial w_0}{\partial y}\right)_{,x} \delta w_0 - \left(N_{xy}\frac{\partial w_0}{\partial x}\right)_{,y} \delta w_0 - 2M_{xy,xy}\delta w_0$$

$$+ (kw_0 - q)\delta w_0 + I_0\left(\ddot{u}_0\delta u_0 + \ddot{v}_0\delta v_0 + \ddot{w}_0\delta w_0\right)$$

$$- I_2\left(\frac{\partial^2 \ddot{w}_0}{\partial x^2} + \frac{\partial^2 \ddot{w}_0}{\partial y^2}\right)\delta w_0 \Bigg] dxdy$$

$$+ \oint_\Gamma \Bigg[N_{xx}n_x\delta u_0 + \left(N_{xx}\frac{\partial w_0}{\partial x}\right) n_x\delta w_0 - M_{xx}n_x\frac{\partial \delta w_0}{\partial x}$$

$$+ M_{xx,x}n_x\delta w_0 + N_{yy}n_y\delta v_0 + \left(N_{yy}\frac{\partial w_0}{\partial y}\right) n_y\delta w_0$$

$$- M_{yy}n_y\frac{\partial \delta w_0}{\partial y} + M_{yy,y}n_y\delta w_0 - M_{xy}n_x\frac{\partial \delta w_0}{\partial y}$$

$$+ M_{xy,x}n_y\delta w_0 - M_{xy}n_y\frac{\partial \delta w_0}{\partial x} + M_{xy,y}n_x\delta w_0 + N_{xy}n_y\delta u_0$$

$$+ N_{xy}n_x\delta v_0 + N_{xy}\frac{\partial w_0}{\partial y}n_x\delta w_0 + N_{xy}\frac{\partial w_0}{\partial x}n_y\delta w_0 \Bigg] ds$$

$$- \int_{\Gamma_\sigma} \left(\hat{N}_{nn}\delta u_{0n} + \hat{N}_{ns}\delta u_{0s} - \hat{M}_{nn}\frac{\partial \delta w_0}{\partial n} - \hat{M}_{ns}\frac{\partial \delta w_0}{\partial s} \right)$$

$$\left. + \hat{Q}_n \delta w_0 \right) ds + \oint_\Gamma I_2 \left(\frac{\partial \ddot{w}_0}{\partial x} n_x + \frac{\partial \ddot{w}_0}{\partial y} n_y \right) \delta w_0 \, ds \bigg\} dt \tag{3.4.12}$$

where a comma followed by subscripts denotes differentiation with respect to the subscripts: $N_{xx,x} = \partial N_{xx}/\partial x$, and so on. Note that both spatial and time integration-by-parts were used in arriving at the last expression. The terms obtained in Ω_0 but evaluated at $t = 0$ and $t = T$ were set to zero because the virtual displacements are zero there.

Collecting the coefficients of each of the virtual displacements $(\delta u_0, \delta v_0, \delta w_0)$ together and noting that the virtual displacements are zero on Γ_u, we obtain

$$0 = \int_0^T \bigg\{ \int_{\Omega_0} \bigg[-(N_{xx,x} + N_{xy,y} - I_0 \ddot{u}_0) \delta u_0 - (N_{xy,x} + N_{yy,y} - I_0 \ddot{v}_0) \delta v_0$$

$$- \bigg(M_{xx,xx} + 2 M_{xy,xy} + M_{yy,yy} + \mathcal{N}(u_0, v_0, w_0)$$

$$- k w_0 + q - I_0 \ddot{w}_0 + I_2 \frac{\partial^2 \ddot{w}_0}{\partial x^2} + I_2 \frac{\partial^2 \ddot{w}_0}{\partial y^2} \bigg) \delta w_0 \bigg] dx dy$$

$$+ \int_{\Gamma_\sigma} \bigg[(N_{xx} n_x + N_{xy} n_y) \delta u_0 + (N_{xy} n_x + N_{yy} n_y) \delta v_0$$

$$+ \bigg(M_{xx,x} n_x + M_{xy,y} n_x + M_{yy,y} n_y + M_{xy,x} n_y$$

$$+ \mathcal{P}(u_0, v_0, w_0) + I_2 \frac{\partial \ddot{w}_0}{\partial x} n_x + I_2 \frac{\partial \ddot{w}_0}{\partial y} n_y \bigg) \delta w_0$$

$$- (M_{xx} n_x + M_{xy} n_y) \frac{\partial \delta w_0}{\partial x} - (M_{xy} n_x + M_{yy} n_y) \frac{\partial \delta w_0}{\partial y} \bigg] ds$$

$$- \int_{\Gamma_\sigma} \bigg(\hat{N}_{nn} \delta u_{0n} + \hat{N}_{ns} \delta u_{0s} - \hat{M}_{nn} \frac{\partial \delta w_0}{\partial n}$$

$$- \hat{M}_{ns} \frac{\partial \delta w_0}{\partial s} + \hat{Q}_n \delta w_0 \bigg) ds \bigg\} dt \tag{3.4.13}$$

where

$$\mathcal{N}(u_0, v_0, w_0) = \frac{\partial}{\partial x} \left(N_{xx} \frac{\partial w_0}{\partial x} + N_{xy} \frac{\partial w_0}{\partial y} \right) + \frac{\partial}{\partial y} \left(N_{xy} \frac{\partial w_0}{\partial x} + N_{yy} \frac{\partial w_0}{\partial y} \right)$$

$$\mathcal{P}(u_0, v_0, w_0) = \left(N_{xx} \frac{\partial w_0}{\partial x} + N_{xy} \frac{\partial w_0}{\partial y} \right) n_x + \left(N_{xy} \frac{\partial w_0}{\partial x} + N_{yy} \frac{\partial w_0}{\partial y} \right) n_y \tag{3.4.14}$$

The Euler–Lagrange equations are obtained by setting the coefficients of δu_0, δv_0, and δw_0 in Ω_0 to zero separately:

$$\delta u_0: \quad \frac{\partial N_{xx}}{\partial x} + \frac{\partial N_{xy}}{\partial y} = I_0 \frac{\partial^2 u_0}{\partial t^2} \quad (3.4.15)$$

$$\delta v_0: \quad \frac{\partial N_{xy}}{\partial x} + \frac{\partial N_{yy}}{\partial y} = I_0 \frac{\partial^2 v_0}{\partial t^2} \quad (3.4.16)$$

$$\delta w_0: \quad \frac{\partial^2 M_{xx}}{\partial x^2} + 2\frac{\partial^2 M_{xy}}{\partial y \partial x} + \frac{\partial^2 M_{yy}}{\partial y^2} + \mathcal{N}(u_0, v_0, w_0) - k w_0 + q$$

$$= I_0 \frac{\partial^2 w_0}{\partial t^2} - I_2 \frac{\partial^2}{\partial t^2}\left(\frac{\partial^2 w_0}{\partial x^2} + \frac{\partial^2 w_0}{\partial y^2}\right) \quad (3.4.17)$$

The terms involving I_2 are called *rotary* (or *rotatory*) inertia terms, and are often neglected in most books. The term can contribute to higher frequencies of vibration.

3.5 Boundary and Initial Conditions

A salient feature of energy principles is that they also yield the natural boundary conditions and identify the variables involved in the specification of the essential boundary conditions of the problem. As a rule, quantities with a variation in the boundary integrals are the *primary variables*, and their specification constitutes the geometric boundary conditions. The expressions that are coefficients of the varied quantities are termed the *secondary variables*, and their specification constitutes the natural boundary conditions. For example, if there are no specified external forces, the last integral expression in Eq. (3.4.13) becomes zero. Then the remaining boundary integral is

$$\int_{\Gamma_\sigma} \Bigg[(N_{xx} n_x + N_{xy} n_y)\, \delta u_0 + (N_{xy} n_x + N_{yy} n_y)\, \delta v_0$$

$$+ \bigg(M_{xx,x} n_x + M_{xy,y} n_x + M_{yy,y} n_y + M_{xy,x} n_y$$

$$+ \mathcal{P}(u_0, v_0, w_0) + I_2 \frac{\partial \ddot{w}_0}{\partial x} n_x + I_2 \frac{\partial \ddot{w}_0}{\partial y} n_y \bigg)\, \delta w_0$$

$$- (M_{xx} n_x + M_{xy} n_y) \frac{\partial \delta w_0}{\partial x} - (M_{xy} n_x + M_{yy} n_y) \frac{\partial \delta w_0}{\partial y} \Bigg]\, ds \quad (3.5.1)$$

By the principle of virtual displacements, since the expression involving integral over the domain (expression in the square brackets) is zero independently, the above expression must be equated to zero. Examination of the above expression indicates that the quantities with variation are $u_0, v_0, w_0, \partial w_0/\partial x$, and $\partial w_0/\partial x$, and therefore they are the primary variables for a plate with edges parallel to the x and y coordinates. The secondary variables associated with the primary variables are

$$\begin{aligned} \delta u_0 : &\quad N_{xx}n_x + N_{xy}n_y = 0 \\ \delta v_0 : &\quad N_{xy}n_x + N_{yy}n_y = 0 \\ \delta w_0 : &\quad (M_{xx,x} + M_{xy,y})\,n_x + (M_{xy,x} + M_{yy,y})\,n_y \\ &\quad + \mathcal{P}(u_0, v_0, w_0) + I_2\frac{\partial \ddot{w}_0}{\partial x}n_x + I_2\frac{\partial \ddot{w}_0}{\partial y}n_y = 0 \\ \delta\left(\frac{\partial w_0}{\partial x}\right) : &\quad -(M_{xx}n_x + M_{xy}n_y) \\ \delta\left(\frac{\partial w_0}{\partial y}\right) : &\quad -(M_{xy}n_x + M_{yy}n_y) \end{aligned} \quad (3.5.2)$$

Next we return to the case in which the external applied edge forces are nonzero and derive the boundary conditions for an edge whose in-plane normal $\hat{\mathbf{n}}$ is oriented at angle θ counterclockwise from the positive x–axis. The displacement vector of a point on this edge can be written in terms of the three displacement components along three coordinate directions (n, s, r) and rotations about n and s coordinates (see Figure 3.4.1). In order to write the components of displacements and force and moment resultants in the (x, y, z) coordinates in terms of the (n, s, r) coordinates, we use the transformation equations (1.3.13b) derived in Chapter 1.

If the unit outward normal vector $\hat{\mathbf{n}}$ is oriented at an angle θ from the positive x–axis, then its direction cosines are: $n_x = \cos\theta$ and $n_y = \sin\theta$. Of course, the transverse normal coordinate r is parallel to the z–axis. Hence, the transformation between the coordinate system (n, s, r) and (x, y, z) is given by

$$\begin{aligned} \hat{\mathbf{e}}_x &= \cos\theta\,\hat{\mathbf{e}}_n - \sin\theta\,\hat{\mathbf{e}}_s = n_x\,\hat{\mathbf{e}}_n - n_y\,\hat{\mathbf{e}}_s \\ \hat{\mathbf{e}}_y &= \sin\theta\,\hat{\mathbf{e}}_n + \cos\theta\,\hat{\mathbf{e}}_s = n_y\,\hat{\mathbf{e}}_n + n_x\,\hat{\mathbf{e}}_s \\ \hat{\mathbf{e}}_z &= \hat{\mathbf{e}}_r \end{aligned} \quad (3.5.3)$$

Hence, the displacements (u_{0n}, u_{0s}) are related to (u_0, v_0) by the same transformation as in Eq. (3.5.3):

$$u_0 = n_x u_{0n} - n_y u_{0s}, \quad v_0 = n_y u_{0n} + n_x u_{0s} \quad (3.5.4a)$$

Similarly, the normal and tangential derivatives $(w_{0,n}, w_{0,s})$ are related to the derivatives $(w_{0,x}, w_{0,y})$ by the relations

$$\frac{\partial w_0}{\partial x} = n_x \frac{\partial w_0}{\partial n} - n_y \frac{\partial w_0}{\partial s}, \quad \frac{\partial w_0}{\partial y} = n_y \frac{\partial w_0}{\partial n} + n_x \frac{\partial w_0}{\partial s} \quad (3.5.4b)$$

Now we can rewrite the boundary expressions in terms of (u_{0n}, u_{0s}) and $(w_{0,n}, w_{0,s})$. We have

$$\begin{aligned}(N_{xx}n_x + N_{xy}n_y)\,\delta u_0 &+ (N_{xy}n_x + N_{yy}n_y)\,\delta v_0 \\ &= (N_{xx}n_x + N_{xy}n_y)(n_x \delta u_{0n} - n_y \delta u_{0s}) \\ &\quad + (N_{xy}n_x + N_{yy}n_y)(n_y \delta u_{0n} + n_x \delta u_{0s}) \\ &= \left(N_{xx}n_x^2 + 2N_{xy}n_x n_y + N_{yy}n_y^2\right)\delta u_{0n} \\ &\quad + \left[-N_{xx}n_x n_y + N_{yy}n_x n_y + N_{xy}\left(n_x^2 - n_y^2\right)\right]\delta u_{0s}\end{aligned} \quad (3.5.5)$$

We recognize that the coefficients of δu_{0n} and δu_{0s} in the right hand side of the above equation are equal to N_{nn} and N_{ns}, respectively. This follows from the fact that the stresses $(\sigma_{nn}, \sigma_{ns})$ are related to $(\sigma_{xx}, \sigma_{yy}, \sigma_{xy})$ by the transformation [see Eq. (1.5.1)]

$$\left\{\begin{array}{c}\sigma_{nn} \\ \sigma_{ns}\end{array}\right\} = \begin{bmatrix} n_x^2 & n_y^2 & 2n_x n_y \\ -n_x n_y & n_x n_y & n_x^2 - n_y^2 \end{bmatrix} \left\{\begin{array}{c}\sigma_{xx} \\ \sigma_{yy} \\ \sigma_{xy}\end{array}\right\} \quad (3.5.6)$$

Hence, we have

$$\left\{\begin{array}{c}N_{nn} \\ N_{ns}\end{array}\right\} = \begin{bmatrix} n_x^2 & n_y^2 & 2n_x n_y \\ -n_x n_y & n_x n_y & n_x^2 - n_y^2 \end{bmatrix} \left\{\begin{array}{c}N_{xx} \\ N_{yy} \\ N_{xy}\end{array}\right\} \quad (3.5.7)$$

and

$$\left\{\begin{array}{c}M_{nn} \\ M_{ns}\end{array}\right\} = \begin{bmatrix} n_x^2 & n_y^2 & 2n_x n_y \\ -n_x n_y & n_x n_y & n_x^2 - n_y^2 \end{bmatrix} \left\{\begin{array}{c}M_{xx} \\ M_{yy} \\ M_{xy}\end{array}\right\} \quad (3.5.8)$$

In view of the above relations, the boundary integrals in Eq. (3.4.13) can be written as

$$0 = \int_0^T \int_{\Gamma_\sigma} \left[\left(N_{nn} - \hat{N}_{nn}\right)\delta u_{0n} + \left(N_{ns} - \hat{N}_{ns}\right)\delta u_{0s} \right.$$

$$+ \left(M_{xx,x}n_x + M_{xy,y}n_x + M_{yy,y}n_y + M_{xy,x}n_y \right.$$
$$\left. + \mathcal{P}(u_0, v_0, w_0) + I_2\frac{\partial \ddot{w}_0}{\partial x}n_x + I_2\frac{\partial \ddot{w}_0}{\partial y}n_y - \hat{Q}_n \right) \delta w_0$$
$$\left. - \left(M_{nn} - \hat{M}_{nn} \right) \frac{\partial \delta w_0}{\partial n} - \left(M_{ns} - \hat{M}_{ns} \right) \frac{\partial \delta w_0}{\partial s} \right] dsdt$$
(3.5.9)

The natural boundary conditions are then given by

$$N_{nn} - \hat{N}_{nn} = 0 \,, \quad N_{ns} - \hat{N}_{ns} = 0$$
$$M_{nn} - \hat{M}_{nn} = 0 \,, \quad M_{ns} - \hat{M}_{ns} = 0 \,, \quad Q_n - \hat{Q}_n = 0 \quad (3.5.10)$$

on Γ_σ, where

$$Q_n \equiv \left(M_{xx,x} + M_{xy,y} + I_2\frac{\partial \ddot{w}_0}{\partial x} \right) n_x$$
$$+ \left(M_{yy,y} + M_{xy,x} + I_2\frac{\partial \ddot{w}_0}{\partial y} \right) n_y + \mathcal{P}(u_0, v_0, w_0) \quad (3.5.11)$$

Thus the primary variables (i.e., generalized displacements) and secondary variables (i.e., generalized forces) of the theory are:

$$\text{primary variables:} \quad u_{0n}, \ u_{0s}, \ w_0, \ \frac{\partial w_0}{\partial n}, \ \frac{\partial w_0}{\partial s}$$
$$\text{secondary variables:} \quad N_{nn}, \ N_{ns}, \ Q_n, \ M_{nn}, \ M_{ns} \quad (3.5.12)$$

In the present case, the generalized displacements are assumed to be specified on Γ_u and the generalized forces are assumed to be specified on Γ_σ.

We note that the equations of motion in Eqs. (3.4.15)–(3.4.17) have the total spatial differential order of eight. In other words, if the equations are expressed in terms of the displacements (u_0, v_0, w_0), they would contain second-order spatial derivatives of u_0 and v_0 and fourth-order spatial derivatives of w_0. Hence, the classical plate theory is said to be an *eighth-order theory*. This implies that there should be only eight (four essential and four natural) boundary conditions, whereas Eq. (3.5.12) shows five essential and five natural boundary conditions, giving a total of ten boundary conditions. To eliminate this discrepancy, one may integrate the tangential derivative term by parts to obtain the boundary term

$$-\oint_\Gamma M_{ns}\frac{\partial \delta w_0}{\partial s}\,ds = \oint_\Gamma \frac{\partial M_{ns}}{\partial s}\delta w_0\,ds - [M_{ns}\delta w_0]_\Gamma \qquad (3.5.13)$$

The term $[M_{ns}\delta w_0]_\Gamma$ is zero when the end points of a closed curve coincide or when $M_{ns} = 0$. If $M_{ns} = 0$ is not specified at corners of the boundary Γ of a polygonal plate, concentrated forces of magnitude

$$F_c = -2M_{ns} \qquad (3.5.14)$$

will be produced at the corners. The factor of 2 appears because M_{ns} from two sides of the corner are added there.

The remaining boundary term in Eq. (3.5.13) is added to the shear force Q_n (because it is a coefficient of δw_0 on Γ) to obtain the effective shear force

$$V_n \equiv Q_n + \frac{\partial M_{ns}}{\partial s} \qquad (3.5.15)$$

which should be balanced by the applied force $\hat{Q}_n$. This boundary condition, $V_n = \hat{Q}_n$, is known as the *Kirchhoff free-edge condition*. The boundary conditions of the classical plate theory are:

Geometric: $u_{0n},\ u_{0s},\ w_0,\ \dfrac{\partial w_0}{\partial n}$;   **Force:** $N_{nn},\ N_{ns},\ V_n,\ M_{nn}$

$$(3.5.16)$$

Thus, at every boundary point one must know u_{0n} or N_{nn}, u_{0s} or N_{ns}, w_0 or V_n, and $\partial w_0/\partial n$ or M_{nn}. On an edge parallel to the x-axis (i.e., $s = x$ and $n = y$), for example, the above boundary conditions become

$$u_{0n} = v_0,\quad u_{0s} = u_0,\quad w_0,\quad \frac{\partial w_0}{\partial n} = \frac{\partial w_0}{\partial y}$$

$$N_{nn} = N_{yy},\quad N_{ns} = N_{yx},\quad V_n,\quad M_{nn} = M_{yy}$$

Next we discuss some common types of boundary conditions for the linear bending of a rectangular plate with edges parallel to the x and y coordinates. Here we use the edge at $y = 0$ ($n_x = 0$ and $n_y = -1$) to discuss the boundary conditions (see Figure 3.4.2). It should be noted that only one element of each of the four pairs may (and should) be specified on an edge of a plate. The force boundary conditions may be expressed in terms of the generalized displacements using the plate constitutive equations discussed in the next section.

Free edge, $y = 0$: A free edge is one which is geometrically not restrained in any way. Hence, we have

$$u_0 \neq 0, \quad v_0 \neq 0, \quad w_0 \neq 0, \quad \frac{\partial w_0}{\partial y} \neq 0 \qquad (3.5.17\text{a})$$

However, the edge may have applied forces and/or moments

$$N_{xy} = \hat{N}_{xy}, \quad N_{yy} = \hat{N}_{yy}, \quad V_n \equiv -Q_y - \frac{\partial M_{xy}}{\partial x} = \hat{V}_n, \quad M_{yy} = \hat{M}_{yy} \qquad (3.5.17\text{b})$$

where quantities with a hat are specified forces/moments. For free rectangular plates, $M_{xy} = 0$, hence no corner forces are developed.

Fixed (or clamped) edge, $y = 0$: A fixed edge is one that is geometrically fully restrained

$$u_0 = 0, \quad v_0 = 0, \quad w_0 = 0, \quad \frac{\partial w_0}{\partial y} = 0 \qquad (3.5.18)$$

Therefore, the forces and moments on a fixed edge are not known a priori (i.e., they are reactions to be determined as a part of the analysis). For clamped rectangular plates, $M_{xy} = 0$, hence no corner forces are developed.

Simply supported edge $y = 0$: The phrase 'simply supported' does not uniquely define the boundary conditions, and one must indicate what it means, especially when both in-plane and bending deflections are involved. Here we *define* two types of simply supported boundary conditions:

$$\textbf{SS-1:} \quad u_0 = 0, \quad w_0 = 0 \qquad (3.5.19\text{a})$$
$$N_{yy} = \hat{N}_{yy}, \quad M_{yy} = \hat{M}_{yy} \qquad (3.5.19\text{b})$$

$$\textbf{SS-2:} \quad v_0 = 0, \quad w_0 = 0 \qquad (3.5.20\text{a})$$
$$N_{xy} = \hat{N}_{xy}, \quad M_{yy} = \hat{M}_{yy} \qquad (3.5.20\text{b})$$

For simply supported rectangular plates, a reacting force of $2M_{xy}$ is developed at each corner of the plate.

When transient response of a plate is of interest, we must know the initial displacement field and velocity field throughout the domain of the plate. The initial conditions of classical plate theory involve specifying the values of the displacements and their first derivatives with respect to time at $t = 0$:

$$u_n = u_n^0, \quad u_s = u_s^0, \quad w_0 = w_0^0$$
$$\dot{u}_n = \dot{u}_n^0, \quad \dot{u}_s = \dot{u}_s^0, \quad \dot{w}_0 = \dot{w}_0^0 \tag{3.5.21}$$

for all points in Ω_0.

3.6 Plate Elastic Stiffnesses

In classical plate theory, all three transverse strain components $(\varepsilon_{zz}, \varepsilon_{xz}, \varepsilon_{yz})$ are zero by definition. Since $\varepsilon_{zz} = 0$, the transverse normal stress σ_{zz}, though not zero identically, does not appear in the virtual work statement and hence in the equations of motion. Consequently, it amounts to neglecting the transverse normal stress. Thus we have, in theory, a case of both plane strain and plane stress. However, from practical considerations, a thin to moderately thick plate is in a state of plane stress because the thickness is small compared to the in-plane dimensions. Hence, the plane stress-reduced constitutive relations may be used.

For an orthotropic material with principal materials axes (x_1, x_2, x_3) coinciding with the plate coordinates (x, y, z), the plane stress-reduced thermoelastic constitutive equations can be expressed as (see Reddy [7])

$$\begin{Bmatrix} \sigma_{xx} \\ \sigma_{yy} \\ \sigma_{xy} \end{Bmatrix} = \begin{bmatrix} Q_{11} & Q_{12} & 0 \\ Q_{12} & Q_{22} & 0 \\ 0 & 0 & Q_{66} \end{bmatrix} \begin{Bmatrix} \varepsilon_{xx} - \alpha_1 \Delta T \\ \varepsilon_{yy} - \alpha_2 \Delta T \\ \gamma_{xy} \end{Bmatrix} \tag{3.6.1}$$

where Q_{ij} are the plane stress-reduced stiffnesses

$$Q_{11} = \frac{E_1}{1 - \nu_{12}\nu_{21}}, \quad Q_{12} = \frac{\nu_{12}E_2}{1 - \nu_{12}\nu_{21}} = \frac{\nu_{21}E_1}{1 - \nu_{12}\nu_{21}} \tag{3.6.2a}$$

$$Q_{22} = \frac{E_2}{1 - \nu_{12}\nu_{21}}, \quad Q_{66} = G_{12} \tag{3.6.2b}$$

and $(\sigma_i, \varepsilon_i)$ are the stress and strain components, respectively, α_1 and α_2 are the coefficients of thermal expansion, and ΔT is the temperature increment from a reference state, $\Delta T = T - T_0$. The moisture strains are similar to thermal strains (i.e., for moisture strains replace ΔT and α_i with the moisture concentration increment and coefficients of hygroscopic expansion, respectively).

If the principal material axes (x_1, x_2, x_3) are such that $x_3 = z$ and the $x_1 x_2$–plane is rotated about the z–axis by an arbitrary angle θ, we must transform the material stiffnesses, as explained in Chapter 1. The stress-strain relations for this case are given by

$$\left\{\begin{array}{c} \sigma_{xx} \\ \sigma_{yy} \\ \sigma_{xy} \end{array}\right\} = \left[\begin{array}{ccc} \bar{Q}_{11} & \bar{Q}_{12} & \bar{Q}_{16} \\ \bar{Q}_{12} & \bar{Q}_{22} & \bar{Q}_{26} \\ \bar{Q}_{16} & \bar{Q}_{26} & \bar{Q}_{66} \end{array}\right] \left(\left\{\begin{array}{c} \varepsilon_{xx} \\ \varepsilon_{yy} \\ \gamma_{xy} \end{array}\right\} - \left\{\begin{array}{c} \alpha_{xx} \\ \alpha_{yy} \\ \alpha_{xy} \end{array}\right\} \Delta T\right) \quad (3.6.3)$$

where

$$\bar{Q}_{11} = Q_{11} \cos^4 \theta + 2(Q_{12} + 2Q_{66}) \sin^2 \theta \cos^2 \theta + Q_{22} \sin^4 \theta$$
$$\bar{Q}_{12} = (Q_{11} + Q_{22} - 4Q_{66}) \sin^2 \theta \cos^2 \theta + Q_{12}(\sin^4 \theta + \cos^4 \theta)$$
$$\bar{Q}_{22} = Q_{11} \sin^4 \theta + 2(Q_{12} + 2Q_{66}) \sin^2 \theta \cos^2 \theta + Q_{22} \cos^4 \theta$$
$$\bar{Q}_{16} = (Q_{11} - Q_{12} - 2Q_{66}) \sin \theta \cos^3 \theta + (Q_{12} - Q_{22} + 2Q_{66}) \sin^3 \theta \cos \theta$$
$$\bar{Q}_{26} = (Q_{11} - Q_{12} - 2Q_{66}) \sin^3 \theta \cos \theta + (Q_{12} - Q_{22} + 2Q_{66}) \sin \theta \cos^3 \theta$$
$$\bar{Q}_{66} = (Q_{11} + Q_{22} - 2Q_{12} - 2Q_{66}) \sin^2 \theta \cos^2 \theta + Q_{66}(\sin^4 \theta + \cos^4 \theta)$$
$$(3.6.4)$$

and α_{xx}, α_{yy}, and α_{xy} are the transformed thermal coefficients of expansion

$$\alpha_{xx} = \alpha_1 \cos^2 \theta + \alpha_2 \sin^2 \theta$$
$$\alpha_{yy} = \alpha_1 \sin^2 \theta + \alpha_2 \cos^2 \theta$$
$$\alpha_{xy} = 2(\alpha_1 - \alpha_2) \sin \theta \cos \theta \quad (3.6.5)$$

Here θ is the angle measured counterclockwise from the x–coordinate to the x_1–coordinate. In most of this book, we consider only orthotropic plates (i.e., $\theta = 0°$ or $90°$).

The plate constitutive equations relate the force and moment resultants in Eq. (3.4.4) to the strains (3.3.9b) of the plate theory. For a plate made of a single anisotropic material (i.e., $\theta \neq 0°$ or $90°$), the plate constitutive relations are obtained using the definitions in Eq. (3.4.4). They are

$$\left\{\begin{array}{c} N_{xx} \\ N_{yy} \\ N_{xy} \end{array}\right\} = \int_{-\frac{h}{2}}^{\frac{h}{2}} \left\{\begin{array}{c} \sigma_{xx} \\ \sigma_{yy} \\ \sigma_{xy} \end{array}\right\} dz =$$

$$= \int_{-\frac{h}{2}}^{\frac{h}{2}} \begin{bmatrix} \bar{Q}_{11} & \bar{Q}_{12} & \bar{Q}_{16} \\ \bar{Q}_{12} & \bar{Q}_{22} & \bar{Q}_{26} \\ \bar{Q}_{16} & \bar{Q}_{26} & \bar{Q}_{66} \end{bmatrix} \begin{Bmatrix} \varepsilon_{xx}^0 + z\varepsilon_{xx}^1 - \alpha_{xx}\Delta T \\ \varepsilon_{yy}^0 + z\varepsilon_{yy}^1 - \alpha_{yy}\Delta T \\ \gamma_{xy}^0 + z\gamma_{xy}^1 - \alpha_{xy}\Delta T \end{Bmatrix} dz$$

$$= \begin{bmatrix} A_{11} & A_{12} & A_{16} \\ A_{12} & A_{22} & A_{26} \\ A_{16} & A_{26} & A_{66} \end{bmatrix} \begin{Bmatrix} \varepsilon_{xx}^0 \\ \varepsilon_{yy}^0 \\ \gamma_{xy}^0 \end{Bmatrix} - \begin{Bmatrix} N_{xx}^T \\ N_{yy}^T \\ N_{xy}^T \end{Bmatrix} \quad (3.6.6)$$

$$\begin{Bmatrix} M_{xx} \\ M_{yy} \\ M_{xy} \end{Bmatrix} = \int_{-\frac{h}{2}}^{\frac{h}{2}} \begin{Bmatrix} \sigma_{xx} \\ \sigma_{yy} \\ \sigma_{xy} \end{Bmatrix} z\, dz$$

$$= \int_{-\frac{h}{2}}^{\frac{h}{2}} \begin{bmatrix} \bar{Q}_{11} & \bar{Q}_{12} & \bar{Q}_{16} \\ \bar{Q}_{12} & \bar{Q}_{22} & \bar{Q}_{26} \\ \bar{Q}_{16} & \bar{Q}_{26} & \bar{Q}_{66} \end{bmatrix} \begin{Bmatrix} \varepsilon_{xx}^0 + z\varepsilon_{xx}^1 - \alpha_{xx}\Delta T \\ \varepsilon_{yy}^0 + z\varepsilon_{yy}^1 - \alpha_{yy}\Delta T \\ \gamma_{xy}^0 + z\gamma_{xy}^1 - \alpha_{xy}\Delta T \end{Bmatrix} z\, dz$$

$$= \begin{bmatrix} D_{11} & D_{12} & D_{16} \\ D_{12} & D_{22} & D_{26} \\ D_{16} & D_{26} & D_{66} \end{bmatrix} \begin{Bmatrix} \varepsilon_{xx}^1 \\ \varepsilon_{yy}^1 \\ \gamma_{xy}^1 \end{Bmatrix} - \begin{Bmatrix} M_{xx}^T \\ M_{yy}^T \\ M_{xy}^T \end{Bmatrix} \quad (3.6.7)$$

where A_{ij} are *extensional stiffnesses* and D_{ij} are *bending stiffnesses*, which are defined in terms of the lamina stiffnesses $\bar{Q}_{ij}$ as

$$(A_{ij}, D_{ij}) = \int_{-\frac{h}{2}}^{\frac{h}{2}} \bar{Q}_{ij}(1, z^2)dz \text{ or } A_{ij} = \bar{Q}_{ij}h \text{ and } D_{ij} = \bar{Q}_{ij}\frac{h^3}{12} \quad (3.6.8)$$

and $\{N^T\}$ and $\{M^T\}$ are thermal stress resultants

$$\begin{Bmatrix} N_{xx}^T \\ N_{yy}^T \\ N_{xy}^T \end{Bmatrix} = \int_{-\frac{h}{2}}^{\frac{h}{2}} \begin{Bmatrix} \bar{Q}_{11}\alpha_{xx} + \bar{Q}_{12}\alpha_{yy} + \bar{Q}_{16}\alpha_{xy} \\ \bar{Q}_{12}\alpha_{xx} + \bar{Q}_{22}\alpha_{yy} + \bar{Q}_{26}\alpha_{xy} \\ \bar{Q}_{16}\alpha_{xx} + \bar{Q}_{26}\alpha_{yy} + \bar{Q}_{66}\alpha_{xy} \end{Bmatrix} \Delta T\, dz \quad (3.6.9a)$$

$$\begin{Bmatrix} M_{xx}^T \\ M_{yy}^T \\ M_{xy}^T \end{Bmatrix} = \int_{-\frac{h}{2}}^{\frac{h}{2}} \begin{Bmatrix} \bar{Q}_{11}\alpha_{xx} + \bar{Q}_{12}\alpha_{yy} + \bar{Q}_{16}\alpha_{xy} \\ \bar{Q}_{12}\alpha_{xx} + \bar{Q}_{22}\alpha_{yy} + \bar{Q}_{26}\alpha_{xy} \\ \bar{Q}_{16}\alpha_{xx} + \bar{Q}_{26}\alpha_{yy} + \bar{Q}_{66}\alpha_{xy} \end{Bmatrix} \Delta T\, z\, dz \quad (3.6.9b)$$

In general, Q's and α's, and therefore A's and D's, can be functions of position (x, y). Relations similar to (3.6.6) and (3.6.7) can be written for hygroscopic effects (i.e., replace ΔT by Δc, where c is the moisture concentration).

For a plate made of a single orthotropic layer, we have

$$\begin{Bmatrix} N_{xx} \\ N_{yy} \\ N_{xy} \end{Bmatrix} = \begin{bmatrix} A_{11} & A_{12} & 0 \\ A_{12} & A_{22} & 0 \\ 0 & 0 & A_{66} \end{bmatrix} \begin{Bmatrix} \varepsilon_{xx}^0 \\ \varepsilon_{yy}^0 \\ \gamma_{xy}^0 \end{Bmatrix} - \begin{Bmatrix} N_{xx}^T \\ N_{yy}^T \\ 0 \end{Bmatrix} \quad (3.6.10a)$$

$$\begin{Bmatrix} M_{xx} \\ M_{yy} \\ M_{xy} \end{Bmatrix} = \begin{bmatrix} D_{11} & D_{12} & 0 \\ D_{12} & D_{22} & 0 \\ 0 & 0 & D_{66} \end{bmatrix} \begin{Bmatrix} \varepsilon_{xx}^1 \\ \varepsilon_{yy}^1 \\ \gamma_{xy}^1 \end{Bmatrix} - \begin{Bmatrix} M_{xx}^T \\ M_{yy}^T \\ 0 \end{Bmatrix} \quad (3.6.10b)$$

where

$$A_{11} = \frac{E_1 h}{1 - \nu_{12}\nu_{21}}, \quad A_{12} = \nu_{21} A_{11}, \quad A_{22} = \frac{E_2}{E_1} A_{11}, \quad A_{66} = \frac{G_{12}}{E_1} A_{11}$$

$$D_{11} = \frac{E_1 h^3}{12(1 - \nu_{12}\nu_{21})}, \quad D_{12} = \nu_{21} D_{11}, \quad D_{22} = \frac{E_2}{E_1} D_{11}, \quad D_{66} = \frac{G_{12} h^3}{12}$$

(3.6.11)

and

$$\left\{ \begin{array}{c} N_{xx}^T \\ N_{yy}^T \end{array} \right\} = \left\{ \begin{array}{c} Q_{11}\alpha_1 + Q_{12}\alpha_2 \\ Q_{12}\alpha_1 + Q_{22}\alpha_2 \end{array} \right\} \int_{-\frac{h}{2}}^{\frac{h}{2}} \Delta T(x,y,z) \, dz \quad (3.6.12a)$$

$$\left\{ \begin{array}{c} M_{xx}^T \\ M_{yy}^T \end{array} \right\} = \left\{ \begin{array}{c} Q_{11}\alpha_1 + Q_{12}\alpha_2 \\ Q_{12}\alpha_1 + Q_{22}\alpha_2 \end{array} \right\} \int_{-\frac{h}{2}}^{\frac{h}{2}} \Delta T(x,y,z) \, z \, dz \quad (3.6.12b)$$

where the temperature change ΔT (above a stress-free temperature) is a known function of position. For isotropic plates, Eqs. (3.6.12a,b) simplify to

$$N_{xx}^T = N_{yy}^T = \frac{N_T}{(1-\nu)}, \quad N_T = E\alpha \int_{-\frac{h}{2}}^{\frac{h}{2}} \Delta T \, dz \quad (3.6.13a)$$

$$M_{xx}^T = M_{yy}^T = \frac{M_T}{(1-\nu)}, \quad M_T = E\alpha \int_{-\frac{h}{2}}^{\frac{h}{2}} \Delta T \, z \, dz \quad (3.6.13b)$$

Plates composed of multiple layers are called *laminated plates* (see Reddy [7] and Figure 3.6.1). A laminated plate composed of multiple orthotropic layers that are *symmetrically disposed*, both from material and geometric properties standpoint, about the midplane of the plate has constitutive equations that are again given by Eqs. (3.6.10a,b), with the laminate stiffnesses A_{ij} and D_{ij} defined by

$$A_{ij} = \sum_{k=1}^{N} \bar{Q}_{ij}^{(k)} (z_{k+1} - z_k), \quad D_{ij} = \frac{1}{3} \sum_{k=1}^{N} \bar{Q}_{ij}^{(k)} (z_{k+1}^3 - z_k^3) \quad (3.6.14)$$

with

$$\bar{Q}_{11}^{(k)} = \frac{E_1^k}{1 - \nu_{12}^k \nu_{21}^k}, \quad \bar{Q}_{12}^{(k)} = \frac{\nu_{21}^k E_1^k}{1 - \nu_{12}^k \nu_{21}^k}, \quad \bar{Q}_{22}^{(k)} = \frac{E_2^k}{1 - \nu_{12}^k \nu_{21}^k}$$

$$\bar{Q}_{16}^{(k)} = 0, \quad \bar{Q}_{26}^{(k)} = 0, \quad \bar{Q}_{66}^{(k)} = G_{12}^k, \quad \bar{Q}_{44}^{(k)} = G_{23}^k, \quad \bar{Q}_{55}^{(k)} = G_{13}^k \quad (3.6.15)$$

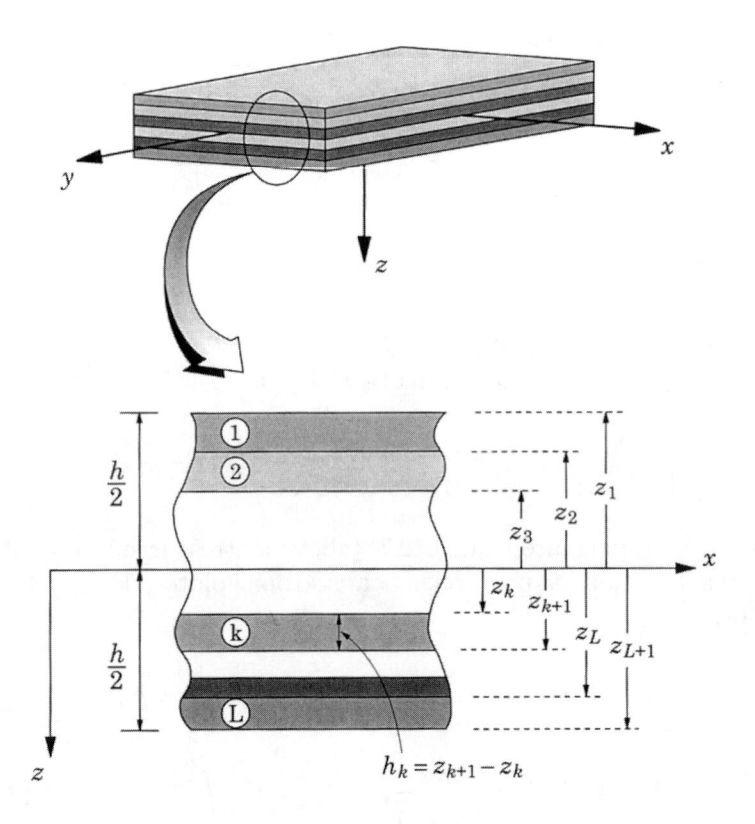

Figure 3.6.1. The layer numbering used for a typical laminated plate.

Here N denotes the number of layers, z_k is the z–coordinate of the bottom of the kth layer, and $(E_i^k, \nu_{ij}^k, G_{ij}^k)$ denote the engineering properties of the kth layer. The layers are numbered from bottom to top, in the positive z direction, as indicated in Figure 3.6.1. For a laminated plate composed of multiple isotropic layers that are *symmetrically disposed*, the laminate stiffnesses A_{ij} and D_{ij} are defined by Eq. (3.6.14) with

$$\bar{Q}_{11}^{(k)} = \bar{Q}_{22}^{(k)} = \frac{E^k}{1-\nu_k^2}, \quad \bar{Q}_{16}^{(k)} = \bar{Q}_{26}^{(k)} = 0$$

$$\bar{Q}_{12}^{(k)} = \frac{\nu_k E^k}{1-\nu_k^2}, \quad \bar{Q}_{44}^{(k)} = \bar{Q}_{55}^{(k)} = \bar{Q}_{66}^{(k)} = \frac{E^k}{2(1+\nu_k)} \qquad (3.6.16)$$

For an analysis of laminated plates, the reader may consult the text book by Reddy [7].

Example 3.6.1

Consider a plate made of a graphite-epoxy material with the following material properties in the principal material coordinates (see Table 1.4.1):

$$E_1 = 20 \text{ msi}, \ E_2 = 1.30 \text{ msi}, \ G_{12} = 1.03 \text{ msi}, \ \nu_{12} = 0.3 \quad (3.6.17)$$

The elastic coefficients Q_{ij} (in msi. $= 10^6$ psi.) in the principal material coordinates can be calculated using Eqs. (3.6.2a,b) as

$$Q_{11} = 20.118, \ Q_{22} = 1.3076, \ Q_{12} = 0.3923, \ Q_{66} = 1.03$$

If the plate is of thickness $h = 0.001$ in., and the plate coordinate axes coincide with the principal material coordinates, the (orthotropic) plate stiffnesses A_{ij} (ksi-in.) and D_{ij} (psi-in^3.) can be calculated using Eq. (3.6.14) as

$$A_{11} = 201.18, \ A_{22} = 13.076, \ A_{12} = 3.9229, \ A_{66} = 10.300$$

$$D_{11} = 1.6765, \ D_{22} = 0.10897, \ D_{12} = 0.03269, \ D_{66} = 0.08583$$

If the principal material axes x_1 and x_2 are oriented at 60° to the plate (x, y) axes and $x_3 = z$, then we have

$$\bar{Q}_{11} = 2.9125, \ \bar{Q}_{22} = 12.318, \ \bar{Q}_{12} = 3.4899, \ \bar{Q}_{66} = 4.1276 \ (\text{msi.})$$

$$A_{11} = 29.125, \ A_{22} = 123.18, \ A_{12} = 34.899, \ A_{66} = 41.276 \ (\text{ksi-in.})$$

$$D_{11} = 0.24271, \ D_{22} = 1.0625, \ D_{12} = 0.29083, \ D_{66} = 0.34397 \ (\text{psi-in}^3.)$$

If the plate is composed of three layers of the same thickness and material but the top and bottom layers oriented at 0° and the middle layer at 90° [denoted (0/90/0) laminate], then the plate stiffnesses are given by

$$A_{11} = 138.48, \ A_{22} = 75.777, \ A_{12} = 3.9229, \ A_{66} = 10.300 \ (\text{ksi-in.})$$

$$D_{11} = 1.6184, \ D_{22} = 0.1670, \ D_{12} = 0.0327, \ D_{66} = 0.0858 \ (\text{psi-in}^3.)$$

Example 3.6.2

Suppose that an orthotropic plate is subjected to loads such that the only nonzero strain at a point (x, y) is $\varepsilon_{xx}^0 = 10^3 \mu = 10^{-3}$ in./in. The material properties are the same as those listed in Eq. (3.6.17). We wish to determine the state of stress $(\sigma_{xx}, \sigma_{yy}, \sigma_{xy})$ in the plate. Assuming that the plate is of thickness 0.01 in. we wish to compute the force and moment resultants.

The stresses (in psi.) are given by

$$\sigma_{xx} = Q_{11}\varepsilon_{xx}^0 = 20,118, \quad \sigma_{yy} = Q_{12}\varepsilon_{xx}^0 = 392, \quad \sigma_{xy} = Q_{16}\varepsilon_{xx}^0 = 0$$

where Q_{ij} are given in Example 3.6.1. The tensile stress σ_{yy} is the reaction of the laminate trying to contract in the y-direction due to the Poisson effect. Since the strain $\varepsilon_{yy} = 0$ by assumption, a tensile stress σ_{yy} is required to maintain the zero strain condition.

The force and moment resultants in the plate are

$$\left\{\begin{array}{c} N_{xx} \\ N_{yy} \\ N_{xy} \end{array}\right\} = \left\{\begin{array}{c} A_{11} \\ A_{12} \\ 0 \end{array}\right\} \varepsilon_{xx}^0 = \left\{\begin{array}{c} 201.18 \\ 3.92 \\ 0 \end{array}\right\} \text{ lbs/in.}, \quad \left\{\begin{array}{c} M_{xx} \\ M_{yy} \\ M_{xy} \end{array}\right\} = \left\{\begin{array}{c} 0 \\ 0 \\ 0 \end{array}\right\} \text{ lbs-in.}$$

If the plate is subjected to a uniform temperature of $T^0 = 250°F$, the thermal forces generated are equal to ($\alpha_1 = 10^{-6}$ in./in/°F and $\alpha_2 = 30 \times 10^{-6}$ in./in/°F)

$$\left\{\begin{array}{c} N_{xx}^T \\ N_{yy}^T \\ N_{xy}^T \end{array}\right\} = \left\{\begin{array}{c} (A_{11}\alpha_1 + A_{12}\alpha_2) \\ (A_{12}\alpha_1 + A_{22}\alpha_2) \\ 0 \end{array}\right\} T^0 = \left\{\begin{array}{c} 79.72 \\ 99.05 \\ 0 \end{array}\right\} \text{ lbs/in.}$$

Example 3.6.3

Consider the case in which the plate of Example 3.6.2 is subjected to loads such that the only nonzero strain at a point (x, y) is the curvature strain $\varepsilon_{xx}^1 = (1/12)$ /in. Hence, $\varepsilon_{xx} = z\varepsilon_{xx}^1 = \frac{z}{12}$ in./in. Then the stresses (in msi.) are given by

$$\sigma_{xx} = zQ_{11}\varepsilon_{xx}^1 = 1.6765z, \quad \sigma_{yy} = zQ_{12}\varepsilon_{xx}^1 = 0.0327z, \quad \sigma_{xy} = 0$$

The stresses σ_{xx} and σ_{yy} are linear through the entire plate thickness, and σ_{xy} is zero everywhere.

The force and moment resultants are ($\varepsilon_{xx}^0 = 0$)

$$\begin{Bmatrix} N_{xx} \\ N_{yy} \\ N_{xy} \end{Bmatrix} = \begin{Bmatrix} 0 \\ 0 \\ 0 \end{Bmatrix} \text{ lbs/in.,} \quad \begin{Bmatrix} M_{xx} \\ M_{yy} \\ M_{xy} \end{Bmatrix} = \begin{Bmatrix} D_{11} \\ D_{12} \\ D_{16} \end{Bmatrix} \varepsilon_{xx}^1 = \begin{Bmatrix} 0.1397 \\ 0.0027 \\ 0 \end{Bmatrix} \text{ lbs-in.}$$

If the plate is subjected to a linear variation of temperature through the thickness of zT^1 with $T^1 = 250°\text{F/in.}$, the thermal moments generated are equal to ($\alpha_1 = 10^{-6}$ in./in./°F and $\alpha_2 = 30 \times 10^{-6}$ in./in./°F)

$$\begin{Bmatrix} M_{xx}^T \\ M_{yy}^T \\ M_{xy}^T \end{Bmatrix} = \begin{Bmatrix} (D_{11}\alpha_1 + D_{12}\alpha_2) \\ (D_{12}\alpha_1 + D_{22}\alpha_2) \\ 0 \end{Bmatrix} T^1 = \begin{Bmatrix} 0.664 \\ 0.826 \\ 0 \end{Bmatrix} 10^{-3} \text{ lbs-in.}$$

3.7 Characterization of Orthotropic Materials

When the engineering constants ($E_1, E_2, \nu_{12}, G_{12}$) of a plate material are available, the plate stiffnesses A_{ij} and D_{ij} may be computed using Eq. (3.6.11) for single-layer plates and Eq. (3.6.14) for symmetrically laminated plates. The engineering parameters E_1, E_2, G_{12}, and ν_{12} of an orthotropic material may be determined experimentally using an appropriate test specimen made of the material. For example, E_1 and ν_{12} of a fiber-reinforced material are measured using a uniaxial test specimen. The specimen consists of several (physical or mathematical) layers of the material with fibers in each layer being aligned with the longitudinal direction. The specimen is then loaded along the longitudinal direction and strains along and perpendicular to the fiber directions are measured using strain gauges. If the load applied is P, then the stress along the fiber direction is $\sigma_{11} = P/A$, where A denotes the cross-sectional area of the specimen. For different values of the load, σ_{11} can be computed and longitudinal and lateral strains ε_{11} and ε_{22} can be recorded. Then E_1 denotes the slope of the experimental relation between σ_{11} and ε_{11}. The slope of the experimental relation between the lateral contraction ε_{22} and longitudinal extension ε_{11} is Poisson's ratio ν_{12}. Similarly, other constants can be determined from appropriate tests. Note that Poisson's ratio ν_{21} is determined using the reciprocal relationship

$$\nu_{ij}E_j = \nu_{ji}E_i \text{ (no sum on } i \text{ and } j\text{);} \quad \nu_{21} = \frac{\nu_{12}E_2}{E_1} \qquad (3.7.1)$$

When it is not possible to determine the engineering constants experimentally, approximate methods based on certain mathematical models are used to determine them. Fiber-reinforced composite materials, for example, are formed by combining two or more materials on a macroscopic scale such that together they have better engineering properties than either of the constituents used alone. Most man-made composite materials are made from two materials: a reinforcement material called *fiber* and a base material called *matrix* material. Reinforced concrete provides an example, in which steel constitutes the fiber and concrete (which itself is a composite) the matrix material. Matrix materials have their usual bulk-form properties whereas fibers have directionally dependent properties.

The theoretical approach, called a *micromechanics approach*, used to determine the engineering constants of a continuous fiber-reinforced composite material is based on the assumptions that (1) perfect bonding exists between fibers and matrix, (2) fibers are parallel and uniformly distributed throughout, (3) the matrix is free of voids or micro-cracks and is initially in a stress-free state, (4) both fibers and matrix are isotropic and obey Hooke's law, and (5) the applied loads are either parallel or perpendicular to the fiber direction. In the micromechanics approach, the moduli and Poisson's ratio of a fiber-reinforced material are expressed in terms of the moduli, Poisson's ratios, and volume fractions of the constituents as (see [10–12])

$$E_1 = E_f v_f + E_m v_m, \quad \nu_{12} = \nu_f v_f + \nu_m v_m$$
$$E_2 = \frac{E_f E_m}{E_f v_m + E_m v_f}, \quad G_{12} = \frac{G_f G_m}{G_f v_m + G_m v_f}$$
$$G_f = \frac{E_f}{2(1+\nu_f)}, \quad G_m = \frac{E_m}{2(1+\nu_m)} \qquad (3.7.2)$$

where E_1 is the longitudinal modulus, E_2 the transverse modulus, ν_{12} the major Poisson's ratio, G_{12} the shear modulus, E_f the modulus of fiber, E_m the modulus of matrix, G_f the shear modulus of fiber, G_m the shear modulus of matrix, ν_f the Poisson's ratio of fiber, ν_m Poisson's ratio of matrix, v_f the fiber volume fraction, and v_m the matrix volume fraction. Here it is assumed that the fiber as well as matrix materials are isotropic. Similarly, the following expressions hold for the thermal coefficients of expansion

$$\alpha_1 = \frac{E_f \alpha_f v_f + E_m \alpha_m v_m}{E_f v_f + E_m v_m}$$
$$\alpha_2 = (1+\nu_m)\alpha_m v_m + (1+\nu_f)\alpha_f v_f - \alpha_1 \nu_{12} \qquad (3.7.3)$$

THE CLASSICAL THEORY OF PLATES   135

Table 3.7.1 summarizes the expressions for bending stiffnesses (or rigidities) for some standard cases (see [1,8–12]). For the steel-reinforced concrete slab, the steel bars are assumed to be placed biaxially along the x and y axes. In the case of stiffened plates, the word 'plate' refers to the portion without stiffeners. If a plate is stiffened by equidistant stiffeners in both x and y directions (Table 3.7.1 shows only along the y–direction), the rigidity D_{11} should be modified as

$$D_{11} = \frac{Eh^3}{12(1-\nu^2)} + \frac{E_s I_s}{s} \qquad (3.7.4)$$

where it is assumed that the stiffeners are identical in geometry and material properties. Otherwise, the additive part, $E_s I_s/s$, should be replaced with $E_x I_x/s_x$ for stiffeners parallel to the x–direction and $E_y I_y/s_y$ for stiffeners in the y–direction, which are added to D_{11} and D_{22}, respectively.

Table 3.7.1. Extensional and bending rigidities for various orthotropic plates.

Plate type	Plate bending stiffnesses D_{ij}
Steel-reinforced concrete slab	$D_{11} = \frac{E_c}{1-\nu_c^2}\left[I_{cx} + \left(\frac{E_s}{E_c} - 1\right)I_{sx}\right]$ $D_{22} = \frac{E_c}{1-\nu_c^2}\left[I_{cy} + \left(\frac{E_s}{E_c} - 1\right)I_{sy}\right]$ $D_{12} = \nu_c\sqrt{D_{11}D_{22}}$ $D_{66} = \frac{1-\nu_c}{2}\sqrt{D_{11}D_{22}}$ ν_c = Poisson's ratio for concrete E_c, E_s = Young's modulus of concrete and steel, respectively.

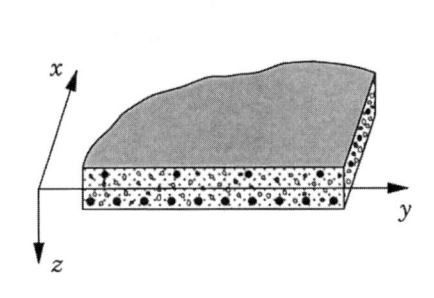

$I_{c\xi}, I_{s\xi}$ = moments of inertia of the concrete and steel bars about the neutral axis in the section ξ = constant ($\xi = x$ or y).

(Table 3.7.1 continued)

Plate reinforced symmetrically by equidistant stiffeners

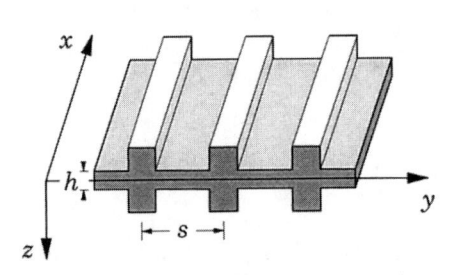

$D_{11} = \frac{Eh^3}{12(1-\nu^2)}, \quad D_{12} = \nu D_{11}$

$D_{22} = \frac{Eh^3}{12(1-\nu^2)} + \frac{E_s I_s}{s}$

$E, E_s =$ Young's modulus of plate and stiffeners, respectively.
$\nu =$ Poisson's ratio of plate.
$s =$ spacing between stiffeners.

$I_s =$ moment of inertia of the stiffener with respect to the mid-plane of the plate.

Plate reinforced by equidistant ribs

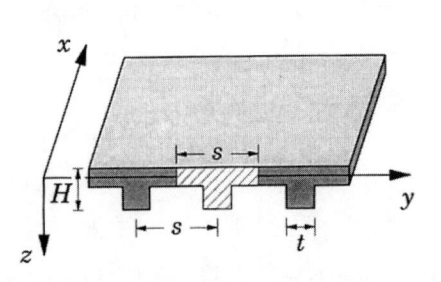

$D_{11} = \frac{Esh^3}{12[s-t+t(h/H)^3]}, \quad D_{22} = \frac{EI}{s}$

$2D_{66} = 2G_{xy} + \frac{C}{s}, \quad D_{12} = 0 \; (\nu = 0)$

$E =$ Young's modulus of the plate
$G_{xy} =$ torsional rigidity of the plate
$C =$ torsional rigidity of one rib
$I_s =$ moment of inertia about neutral axis of a T-section (shaded).

Corrugated plate

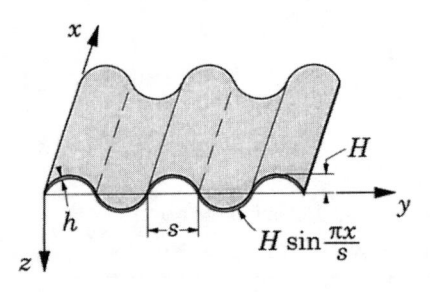

$D_{11} = \frac{s}{\mu} \frac{Eh^3}{12(1-\nu^2)}, \quad D_{22} = EI$

$2D_{66} = \frac{s}{\mu} \frac{Eh^3}{12(1+\nu)}, \quad D_{12} = 0$

$\mu = s\left(1 + \frac{\pi^2 H^2}{4s^2}\right)$

$I = 0.5 H^2 h \left[1 - \frac{0.81}{1+2.5(H/2s)^2}\right]$

$H \sin \frac{\pi x}{s}$

3.8 Equations of Motion in Terms of Displacements

The stress resultants (N's and M's) are related to the displacement gradients and other fields, such as temperature and electric fields. The force and moment resultants in a single-layer orthotropic plate or a laminated plate composed of multiple orthotropic layers that are symmetrically disposed about the mid-plane of the plate can be expressed in terms of the displacements (u_0, v_0, w_0) by the relations

$$\begin{Bmatrix} N_{xx} \\ N_{yy} \\ N_{xy} \end{Bmatrix} = \begin{bmatrix} A_{11} & A_{12} & 0 \\ A_{12} & A_{22} & 0 \\ 0 & 0 & A_{66} \end{bmatrix} \begin{Bmatrix} \frac{\partial u_0}{\partial x} + \frac{1}{2}\left(\frac{\partial w_0}{\partial x}\right)^2 \\ \frac{\partial v_0}{\partial y} + \frac{1}{2}\left(\frac{\partial w_0}{\partial y}\right)^2 \\ \frac{\partial u_0}{\partial y} + \frac{\partial v_0}{\partial x} + \frac{\partial w_0}{\partial x}\frac{\partial w_0}{\partial y} \end{Bmatrix} - \begin{Bmatrix} N_{xx}^T \\ N_{yy}^T \\ 0 \end{Bmatrix} \quad (3.8.1)$$

$$\begin{Bmatrix} M_{xx} \\ M_{yy} \\ M_{xy} \end{Bmatrix} = - \begin{bmatrix} D_{11} & D_{12} & 0 \\ D_{12} & D_{22} & 0 \\ 0 & 0 & D_{66} \end{bmatrix} \begin{Bmatrix} \frac{\partial^2 w_0}{\partial x^2} \\ \frac{\partial^2 w_0}{\partial y^2} \\ 2\frac{\partial^2 w_0}{\partial x \partial y} \end{Bmatrix} - \begin{Bmatrix} M_{xx}^T \\ M_{yy}^T \\ 0 \end{Bmatrix} \quad (3.8.2)$$

where the thermal force and moment resultants are defined by Eqs. (3.6.12a,b). For geometrically linear analysis, the nonlinear parts of strains in Eq. (3.8.1) are omitted.

The equations of motion (3.4.15)–(3.4.17) can be expressed in terms of displacements (u_0, v_0, w_0) by substituting for the force and moment resultants from Eqs. (3.8.1) and (3.8.2). For homogeneous plates (i.e., for plates with constant A's and D's), the equations of motion (3.4.15)–(3.4.17) take the form

$$A_{11}\left(\frac{\partial^2 u_0}{\partial x^2} + \frac{\partial w_0}{\partial x}\frac{\partial^2 w_0}{\partial x^2}\right) + A_{12}\left(\frac{\partial^2 v_0}{\partial x \partial y} + \frac{\partial w_0}{\partial y}\frac{\partial^2 w_0}{\partial x \partial y}\right)$$
$$+ A_{66}\left(\frac{\partial^2 u_0}{\partial y^2} + \frac{\partial^2 v_0}{\partial x \partial y} + \frac{\partial^2 w_0}{\partial x \partial y}\frac{\partial w_0}{\partial y} + \frac{\partial w_0}{\partial x}\frac{\partial^2 w_0}{\partial y^2}\right)$$
$$- \left(\frac{\partial N_{xx}^T}{\partial x} + \frac{\partial N_{xy}^T}{\partial y}\right) = I_0 \frac{\partial^2 u_0}{\partial t^2} \quad (3.8.3)$$

$$A_{66}\left(\frac{\partial^2 u_0}{\partial x \partial y} + \frac{\partial^2 v_0}{\partial x^2} + \frac{\partial^2 w_0}{\partial x^2}\frac{\partial w_0}{\partial y} + \frac{\partial w_0}{\partial x}\frac{\partial^2 w_0}{\partial x \partial y}\right)$$
$$+ A_{12}\left(\frac{\partial^2 u_0}{\partial x \partial y} + \frac{\partial w_0}{\partial x}\frac{\partial^2 w_0}{\partial x \partial y}\right) + A_{22}\left(\frac{\partial^2 v_0}{\partial y^2} + \frac{\partial w_0}{\partial y}\frac{\partial^2 w_0}{\partial y^2}\right)$$

$$-\left(\frac{\partial N_{xy}^T}{\partial x} + \frac{\partial N_{yy}^T}{\partial y}\right) = I_0 \frac{\partial^2 v_0}{\partial t^2} \qquad (3.8.4)$$

$$-D_{11}\frac{\partial^4 w_0}{\partial x^4} - 2(D_{12} + 2D_{66})\frac{\partial^4 w_0}{\partial x^2 \partial y^2} - D_{22}\frac{\partial^4 w_0}{\partial y^4} - kw_0$$

$$-\left(\frac{\partial^2 M_{xx}^T}{\partial x^2} + 2\frac{\partial^2 M_{xy}^T}{\partial y \partial x} + \frac{\partial^2 M_{yy}^T}{\partial y^2}\right) + \mathcal{N} + q$$

$$= I_0 \frac{\partial^2 w_0}{\partial t^2} - I_2 \frac{\partial^2}{\partial t^2}\left(\frac{\partial^2 w_0}{\partial x^2} + \frac{\partial^2 w_0}{\partial y^2}\right) \qquad (3.8.5)$$

where $\mathcal{N}(u_0, v_0, w_0)$ was defined in Eq. (3.4.14), and it contains the thermal force resultants.

The nonlinear partial differential equations (3.8.3)–(3.8.5) can be simplified for the linear, static analysis. The coupling between the in-plane displacements (u_0, v_0) and transverse deflection w_0 disappears in the case of linear analysis, and Eq. (3.8.5) can be solved separately from Eqs. (3.8.3) and (3.8.4).

Example 3.8.1: Cylindrical Bending

If a plate is infinitely long in one direction, the plate becomes a *plate strip*. Consider a plate strip that has a finite dimension along the x−axis and is subjected to a transverse load $q(x)$ that is uniform at any section parallel to the x−axis. In such a case, the deflection w_0 and displacements (u_0, v_0) of the plate are functions of only x. Therefore, all derivatives with respect to y are zero. In such cases, the deflected surface of the plate strip is cylindrical, and it is referred to as the *cylindrical bending*. For this case, the governing equations (3.8.3)–(3.8.5) reduce to

$$A_{11}\left(\frac{\partial^2 u_0}{\partial x^2} + \frac{\partial w_0}{\partial x}\frac{\partial^2 w_0}{\partial x^2}\right) - \frac{\partial N_{xx}^T}{\partial x} = I_0 \frac{\partial^2 u_0}{\partial t^2} \qquad (3.8.6)$$

$$A_{66}\frac{\partial^2 v_0}{\partial x^2} - \frac{\partial N_{xy}^T}{\partial x} = I_0 \frac{\partial^2 v_0}{\partial t^2} \qquad (3.8.7)$$

$$-D_{11}\frac{\partial^4 w_0}{\partial x^4} + \frac{\partial}{\partial x}\left(N_{xx}\frac{\partial w_0}{\partial x}\right) - kw_0 + q - \frac{\partial^2 M_{xx}^T}{\partial x^2}$$

$$= I_0 \frac{\partial^2 w_0}{\partial t^2} - I_2 \frac{\partial^4 w_0}{\partial x^2 \partial t^2} \qquad (3.8.8)$$

Note that for the nonlinear case, the three equations are coupled through the nonlinear term (N_{xx}) in Eq. (3.8.8) and are uncoupled for the linear case. Hence they can be solved independently of each other.

Example 3.8.2: Pure Bending of Plates

For the case of small strains and small displacements, the nonlinear terms in Eqs. (3.8.3)–(3.8.5) can be neglected. In that case, the equation (3.8.5) governing the transverse deflection is uncoupled from Eqs. (3.8.3) and (3.8.4). We have

$$D_{11}\frac{\partial^4 w_0}{\partial x^4} + 2(D_{12} + 2D_{66})\frac{\partial^4 w_0}{\partial x^2 \partial y^2} + D_{22}\frac{\partial^4 w_0}{\partial y^4} + k w_0$$
$$= -\left(\frac{\partial^2 M_{xx}^T}{\partial x^2} + 2\frac{\partial^2 M_{xy}^T}{\partial x \partial y} + \frac{\partial^2 M_{yy}^T}{\partial y^2}\right) + q$$
$$- I_0 \ddot{w}_0 + I_2\left(\frac{\partial^2 \ddot{w}_0}{\partial x^2} + \frac{\partial^2 \ddot{w}_0}{\partial y^2}\right) \tag{3.8.9}$$

$$D_{11} = \frac{E_1 h^3}{12(1 - \nu_{12}\nu_{21})}, \quad D_{22} = \frac{E_2 h^3}{12(1 - \nu_{12}\nu_{21})}$$
$$D_{12} = \frac{\nu_{12} E_2 h^3}{12(1 - \nu_{12}\nu_{21})}, \quad D_{66} = \frac{G_{12} h^3}{12} \tag{3.8.10}$$

For the static case, Eq. (3.8.9) becomes

$$D_{11}\frac{\partial^4 w_0}{\partial x^4} + 2(D_{12} + 2D_{66})\frac{\partial^4 w_0}{\partial x^2 \partial y^2} + D_{22}\frac{\partial^4 w_0}{\partial y^4} + k w_0$$
$$= q - \left(\frac{\partial^2 M_{xx}^T}{\partial x^2} + \frac{\partial^2 M_{yy}^T}{\partial y^2}\right) \tag{3.8.11}$$

For isotropic plates [$D_{11} = D_{22}$, $D_{12} = \nu D$, and $2D_{66} = (1-\nu)D$, $\alpha_1 = \alpha_2 = \alpha$], Eq. (3.8.11) takes the simpler form

$$D\left(\frac{\partial^4 w_0}{\partial x^4} + 2\frac{\partial^4 w_0}{\partial x^2 \partial y^2} + \frac{\partial^4 w_0}{\partial y^4}\right) + k w_0 = q - \frac{1}{1-\nu}\left(\frac{\partial^2 M_T}{\partial x^2} + \frac{\partial^2 M_T}{\partial y^2}\right) \tag{3.8.12}$$

where

$$M_T = E\alpha \int_{-\frac{h}{2}}^{\frac{h}{2}} \Delta T \, z \, dz \tag{3.8.13}$$

and $D = Eh^3/[12(1-\nu^2)]$. Using the Laplace operator ∇^2, we can express the above equation as

$$D\nabla^4 w_0 + k w_0 = q - \frac{1}{1-\nu}\nabla^2 M_T \tag{3.8.14}$$

where $\nabla^4 = \nabla^2\nabla^2$. Equation (3.8.14) is valid in any coordinate system (i.e., invariant form).

Equation (3.8.9) or (3.8.11) must be solved in conjunction with appropriate boundary conditions of the problem for the desired response. The boundary conditions at any point on the boundary are of the form

$$\text{specify:} \quad w_0 \quad \text{or} \quad V_n \equiv Q_n + \frac{\partial M_{ns}}{\partial s} \qquad (3.8.15)$$

$$\text{specify:} \quad \frac{\partial w_0}{\partial n} \quad \text{or} \quad M_{nn} \qquad (3.8.16)$$

where

$$\frac{\partial w_0}{\partial n} = \frac{\partial w_0}{\partial x}n_x + \frac{\partial w_0}{\partial y}n_y, \quad Q_n = Q_x n_x + Q_y n_y \qquad (3.8.17)$$

$$M_{nn} = M_{xx}n_x^2 + M_{yy}n_y^2 + 2M_{xy}n_x n_y$$
$$= -\left(D_{11}\frac{\partial^2 w_0}{\partial x^2} + D_{12}\frac{\partial^2 w_0}{\partial y^2} + M_{xx}^T\right)n_x^2$$
$$- \left(D_{12}\frac{\partial^2 w_0}{\partial x^2} + D_{22}\frac{\partial^2 w_0}{\partial y^2} + M_{yy}^T\right)n_y^2 - 2D_{66}\frac{\partial^2 w_0}{\partial x \partial y}n_x n_y \qquad (3.8.18)$$

$$M_{ns} = (M_{yy} - M_{xx})n_x n_y + \left(n_x^2 - n_y^2\right)M_{xy}$$
$$= \left[(D_{11} - D_{12})\frac{\partial^2 w_0}{\partial x^2} - (D_{22} - D_{12})\frac{\partial^2 w_0}{\partial y^2}\right]n_x n_y$$
$$- 2D_{66}\frac{\partial^2 w_0}{\partial x \partial y}\left(n_x^2 - n_y^2\right) - \left(M_{yy}^T - M_{xx}^T\right)n_x n_y \qquad (3.8.19)$$

Note that the moment resultants M_{nn} and M_{ns} contain thermal parts. The expressions can be simplified for isotropic plates.

The following boundary conditions will be of interest in the present study:

$$\textbf{Simply Supported:} \quad w_0 = 0, \quad M_{nn} = 0 \qquad (3.8.20)$$
$$\textbf{Clamped:} \quad w_0 = 0, \quad \frac{\partial w_0}{\partial n} = 0 \qquad (3.8.21)$$
$$\textbf{Free:} \quad V_n = 0, \quad M_{nn} = 0 \qquad (3.8.22)$$

Exercises

3.1 Starting with a linear distribution of the displacements through the plate thickness in terms of unknown functions $(u_0, v_0, w_0, F_1, F_2, F_3)$

$$u(x,y,z,t) = u_0(x,y,t) + zF_1(x,y,t)$$
$$v(x,y,z,t) = v_0(x,y,t) + zF_2(x,y,t)$$
$$w(x,y,z,t) = w_0(x,y,t) + zF_3(x,y,t)$$

determine the functions (F_1, F_2, F_3) such that the Kirchhoff hypothesis holds:

$$\frac{\partial w}{\partial z} = 0, \quad \frac{\partial u}{\partial z} = -\frac{\partial w}{\partial x}, \quad \frac{\partial v}{\partial z} = -\frac{\partial w}{\partial y}$$

3.2 Construct the total potential energy functional for an orthotropic plate in pure bending. Assume small strains, displacements, and rotations (i.e., neglect non-linearity), and that the only nonzero applied loading is the distributed load, q. Use the classical plate theory.

3.3 Derive the equation of equilibrium for the linear pure bending case using the principle of minimum total potential energy.

3.4 Consider an orthotropic plate and assume that the material coordinates coincide with the plate coordinates. Compute the stresses $(\sigma_{xx}, \sigma_{yy}, \sigma_{xy})$ for the pure bening case using the constitutive equations, and then use the equilibrium equations of the three-dimensional elasticity theory to determine the transverse stresses $(\sigma_{xz}, \sigma_{yz}, \sigma_{zz})$ as a function of the thickness coordinate.

3.5 Consider the equations of motion of 3-D elasticity [see Eq. (1.4.26)] in the absence of body forces:

$$\frac{\partial \sigma_{xx}}{\partial x} + \frac{\partial \sigma_{xy}}{\partial y} + \frac{\partial \sigma_{xz}}{\partial z} = \rho_0 \frac{\partial^2 u}{\partial t^2}$$

$$\frac{\partial \sigma_{xy}}{\partial x} + \frac{\partial \sigma_{yy}}{\partial y} + \frac{\partial \sigma_{yz}}{\partial z} = \rho_0 \frac{\partial^2 v}{\partial t^2}$$

$$\frac{\partial \sigma_{xz}}{\partial x} + \frac{\partial \sigma_{yz}}{\partial y} + \frac{\partial \sigma_{zz}}{\partial z} = \rho_0 \frac{\partial^2 w}{\partial t^2}$$

Integrate the above equations with respect to z over the interval $(-h/2, h/2)$ and express the results in terms of the force resultants (N_{xx}, N_{yy}, N_{xy}). Use the following boundary conditions:

$$\sigma_{xz}(x, y, -\frac{h}{2}) = 0, \quad \sigma_{xz}(x, y, \frac{h}{2}) = 0, \quad \sigma_{yz}(x, y, -\frac{h}{2}) = 0$$

$$\sigma_{yz}(x,y,\frac{h}{2}) = 0, \ \sigma_{zz}(x,y,-\frac{h}{2}) = -q_b, \ \sigma_{zz}(x,y,\frac{h}{2}) = q_t$$

Next, multiply the equations of motion with z, integrate with respect to z over the interval $(-h/2, h/2)$, and express the results in terms of the moment resultants (M_{xx}, M_{yy}, M_{xy}).

References for Additional Reading

1. Timoshenko, S. P. and Woinowsky-Krieger, S., *Theory of Plates and Shells*, McGraw–Hill, Singapore (1970).
2. Ugural, A. C., *Stresses in Plates and Shells*, McGraw–Hill, New York (1981).
3. Szilard, R., *Theory and Analysis of Plates. Classical and Numerical Methods*, Prentice–Hall, Englewood Cliffs, New Jersey (1974).
4. McFarland, D., Smith, B. L., and Bernhart, W. D., *Analysis of Plates*, Spartan Books, Philadelphia, PA (1972).
5. Panc, V., *Theories of Elastic Plates*, Noordhoff, Leyden, The Netherlands (1975).
6. Reddy, J. N., *Energy and Variational Methods in Applied Mechanics*, John Wiley and Sons, New York (1984).
7. Reddy, J. N., *Mechanics of Laminated Composite Plates: Theory and Analysis*, CRC Press, Boca Raton, FL (1997).
8. Troitsky, M. S., *Stiffened Plates-Bending, Stability, and Vibrations*, Elsevier, New York (1976).
9. Huffington, H. J., "Theoretical Determination of Rigidity Properties of Orthotropic Stiffened Plates", *Journal of Applied Mechanics*, **23**, 15–18 (1956).
10. Gibson, R. F., *Principles of Composite Material Mechanics*, McGraw–Hill, New York (1994).

Chapter Four

Analysis of Plate Strips

4.1 Introduction

There are two cases of rectangular plates that can be treated as one-dimensional problems: (1) beams, and (2) cylindrical bending of plate strips. When the width b (i.e., length along the y−axis) of a plate is very small compared to the length along the x−axis, it is treated as a beam. In cylindrical bending, the plate is assumed to be a plate strip that is very long along the y−axis and has a finite dimension a along the x−axis (see Figure 4.1.1). The transverse load $q(x)$ is assumed to be uniform at any section parallel to the x−axis. In such a case, the deflection w_0 and displacements (u_0, v_0) of the plate are functions of only x, and all derivatives with respect to y are zero. The cylindrical bending problem is a *plane strain* problem, whereas the beam problem is a *plane stress* problem. In this chapter we consider cylindrical bending, buckling, and natural vibrations of plate strips.

4.2 Governing Equations

Consider an orthotropic rectangular plate strip, and let the x and y coordinates be parallel to the edges of the strip. Suppose that the plate is long in the y−direction, has a finite dimension along the x−direction, and is subjected to a transverse load $q(x)$ that is uniform at any section parallel to the x−axis. In such a case, the deflection w_0 and displacements (u_0, v_0) of the plate are functions of only x. Therefore, all derivatives with respect to y are zero, and the plate bends into a cylindrical surface. For this *cylindrical bending* problem (see Figure 4.1.1), the governing equations of motion according to the linear classical plate theory (CPT) are given by

144   THEORY AND ANALYSIS OF ELASTIC PLATES

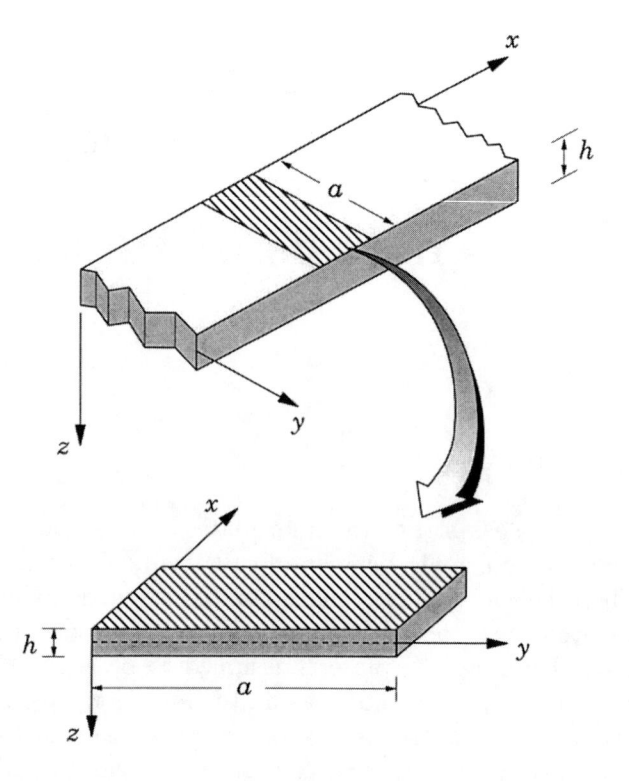

Figure 4.1.1. Geometry of a plate strip in cylindrical bending.

$$A_{11}\frac{\partial^2 u_0}{\partial x^2} - \frac{\partial N_{xx}^T}{\partial x} = I_0\frac{\partial^2 u_0}{\partial t^2} \qquad (4.2.1)$$

$$A_{66}\frac{\partial^2 v_0}{\partial x^2} - \frac{\partial N_{xy}^T}{\partial x} = I_0\frac{\partial^2 v_0}{\partial t^2} \qquad (4.2.2)$$

$$-D_{11}\frac{\partial^4 w_0}{\partial x^4} + \frac{\partial}{\partial x}\left(\hat{N}_{xx}\frac{\partial w_0}{\partial x}\right) - \frac{\partial^2 M_{xx}^T}{\partial x^2} + q = I_0\frac{\partial^2 w_0}{\partial t^2} - I_2\frac{\partial^4 w_0}{\partial x^2 \partial t^2} \qquad (4.2.3)$$

where $\hat{N}_{xx}$ is an applied axial (tensile) load, and

$$D_{11} = \frac{E_1 h^3}{12(1-\nu_{12}\nu 21)}, \quad (I_0, I_2) = \int_{-\frac{h}{2}}^{\frac{h}{2}} (1, z^2)\rho_0\, dz \qquad (4.2.4)$$

The three equations are uncoupled from each other, and therefore each equation can be solved independently of the other two. Here we are

ANALYSIS OF PLATE STRIPS   145

primarily concerned with the bending [i.e., Eq. (4.2.3)] of plate strips. Equation (4.2.3) has the same form as that for a beam. Therefore, the solutions known for various types of beams are also valid for cylindrical bending with the appropriate change of the coefficients ($EI = D_{11} \equiv D$). Also, it should be noted that the modulus E_2 has no direct effect on the behavior of the plate strip. However, its influence is felt through the Poisson ratio ν_{21}, which is computed from the reciprocal relationship $\nu_{21} = \nu_{12} E_2 / E_1$. Thus, the solutions are influenced by three material parameters: E_1, ν_{12}, and E_2/E_1. In the next few pages, we study the analytical solutions for static bending, buckling under in-plane constant compressive force $(-\hat{N}_{xx})$, and natural (or free) flexural vibrations of plate strips in cylindrical bending.

4.3 Bending Analysis
4.3.1 Without Axial Force

For linear static bending analysis in the absence of axial force $\hat{N}_{xx}$ and elastic foundation, Eq. (4.2.3) reduces to ($D = D_{11}$)

$$D \frac{d^4 w_0}{dx^4} = -\frac{d^2 M_{xx}^T}{dx^2} + q \tag{4.3.1}$$

Equation (4.3.1) can be integrated, given thermal and mechanical loads, to obtain $w_0(x)$. We have

$$D \frac{d^2 w_0}{dx^2} = -M_{xx}^T + \int^x \int^\xi q(\eta)\, d\eta\, d\xi + c_1 x + c_2 \tag{4.3.2}$$

$$D \frac{d w_0}{dx} = -\int^x M_{xx}^T(\xi)\, d\xi + \int^x \int^\xi \int^\eta q(\zeta)\, d\zeta\, d\eta\, d\xi$$
$$+ c_1 \frac{x^2}{2} + c_2 x + c_3 \tag{4.3.3}$$

$$D w_0(x) = -\int^x \int^\xi M_{xx}^T(\eta) d\eta\, d\xi + \int^x \int^\xi \int^\eta \int^\zeta q(\mu) d\mu\, d\zeta\, d\eta\, d\xi$$
$$+ c_1 \frac{x^3}{6} + c_2 \frac{x^2}{2} + c_3 x + c_4 \tag{4.3.4}$$

where c_1 through c_4 are constants of integration.

If the temperature distribution in the plate is of the form

$$\Delta T(x, z) = T_0(x) + z T_1(x) \tag{4.3.5}$$

where T_0 and T_1 are functions of x only, then we have

$$M_{xx}^T = \int_{-\frac{h}{2}}^{\frac{h}{2}} (\bar{Q}_{11}\alpha_{xx} + \bar{Q}_{12}\alpha_{yy}) T_1 z^2 \, dz = D^T T_1 \qquad (4.3.6)$$

$$D^T = \int_{-\frac{h}{2}}^{\frac{h}{2}} (\bar{Q}_{11}\alpha_{xx} + \bar{Q}_{12}\alpha_{yy}) z^2 \, dz \qquad (4.3.7)$$

For uniformly distributed load $q = q_0$ and constant temperatures T_0 and T_1, Eq. (4.3.4) yields

$$Dw_0(x) = -D^T T_1 \frac{x^2}{2} + q_0 \frac{x^4}{24} + c_1 \frac{x^3}{6} + c_2 \frac{x^2}{2} + c_3 x + c_4 \qquad (4.3.8)$$

The constants of integration c_i can be determined using the boundary conditions.

Simply supported plate under uniformly distributed load

For a plate strip with simply supported edges at $x = 0$ and $x = a$, the boundary conditions are (see Table 4.3.1)

$$w_0 = 0, \quad M_{xx} \equiv -D \frac{d^2 w_0}{dx^2} - M_{xx}^T = 0 \qquad (4.3.9)$$

Using boundary conditions (4.3.9) in Eqs. (4.3.2) and (4.3.4), we obtain

$$c_2 = 0, \quad c_4 = 0, \quad c_1 = -\frac{q_0 a}{2}, \quad c_3 = \frac{D^T T_1 a}{2} + \frac{q_0 a^3}{24} \qquad (4.3.10)$$

and the solution becomes

$$w_0(x) = \frac{q_0 a^4}{24D} \left[\left(\frac{x}{a}\right)^4 - 2\left(\frac{x}{a}\right)^3 + \left(\frac{x}{a}\right) \right] - \frac{D^T T_1 a^2}{2D} \left[\left(\frac{x}{a}\right)^2 - \left(\frac{x}{a}\right) \right]$$

$$(4.3.11a)$$

$$M_{xx} = -\frac{q_0 a^2}{2} \left[\left(\frac{x}{a}\right)^2 - \left(\frac{x}{a}\right) \right] \qquad (4.3.11b)$$

The maximum transverse deflection and bending moments occur at $x = a/2$, and are given by

$$w_{max} = \frac{5 q_0 a^4}{384 D} + \frac{D^T T_1 a^2}{8D}, \quad M_{max} = \frac{q_0 a^2}{8} \qquad (4.3.12)$$

ANALYSIS OF PLATE STRIPS   147

Table 4.3.1. Conventional boundary conditions in the classical theory of plate strips.

Type of support	Geometric B.C.	Force B.C.
free	None	$N_{xx} = 0$, $N_{xy} = 0$ $M_{xx} = 0$, $\dfrac{dM_{xx}}{dx} = 0$
roller	$w_0 = 0$ $\dfrac{dv_0}{dx} = 0$	$N_{xx} = 0$ $M_{xx} = 0$
simple support	$u_0 = 0$, $w_0 = 0$ $\dfrac{dv_0}{dx} = 0$	$M_{xx} = 0$
clamped	$u_0 = 0$, $v_0 = 0$ $w_0 = 0$, $\dfrac{dw_0}{dx} = 0$	None

Clamped plate under uniformly distributed load

For a plate strip with clamped edges at $x = 0$ and $x = a$ (and free at $y = 0, \infty$), the boundary conditions are

$$w_0 = 0, \quad \frac{dw_0}{dx} = 0 \tag{4.3.13}$$

Using boundary conditions (4.3.13) in Eqs. (4.3.3) and (4.3.4), we obtain

$$c_3 = 0, \quad c_4 = 0, \quad c_1 = -\frac{q_0 a}{2}, \quad c_2 = D^T T_1 + \frac{q_0 a^2}{12} \tag{4.3.14}$$

and the solution becomes

$$w_0(x) = \frac{q_0 a^4}{24 D} \left[\left(\frac{x}{a}\right)^4 - 2\left(\frac{x}{a}\right)^3 + \left(\frac{x}{a}\right)^2 \right] \tag{4.3.15a}$$

$$M_{xx} = -\frac{q_0 a^2}{12}\left[1 - 6\left(\frac{x}{a}\right) + 6\left(\frac{x}{a}\right)^2\right] - D^T T_1 \quad (4.3.15b)$$

The maximum transverse deflection occurs at $x = a/2$ and maximum bending moments occurs at $x = 0$:

$$w_{max} = \frac{q_0 a^4}{384 D}, \quad M_{max} = -\frac{q_0 a^2}{12} - D^T T_1 \quad (4.3.16)$$

4.3.2 With Axial Force

When the beam is subjected to axial (tensile) force $\hat{N}_{xx}$ but without an elastic foundation, Eq. (4.2.3) reduces to

$$D\frac{d^4 w_0}{dx^4} - \hat{N}_{xx}\frac{d^2 w_0}{dx^2} = -\frac{d^2 M_{xx}^T}{dx^2} + q \quad (4.3.17)$$

Integrating twice with respect to x, we arrive at

$$D\frac{d^2 w_0}{dx^2} - \hat{N}_{xx} w_0 = -M_{xx}^T + \int^x \left(\int^\xi q(\eta)\, d\eta\right) d\xi + c_1 x + c_2 \quad (4.3.18)$$

The general solution to the second-order equation consists of the homogeneous part

$$w_0^h(x) = c_3 \sinh\frac{2\mu x}{a} + c_4 \cosh\frac{2\mu x}{a} \quad (4.3.19)$$

and the nonhomogeneous part

$$w_0^p(x) = \frac{a}{4D\mu}\left(e^{2\mu(\frac{x}{a})}\int^x e^{-2\mu(\frac{\xi}{a})} f(\xi) d\xi - e^{-2\mu(\frac{x}{a})}\int^x e^{i\mu(\frac{\xi}{a})} f(\xi) d\xi\right) \quad (4.3.20)$$

where $f(x)$ is defined by

$$f(x) = -M_{xx}^T + \int^x \left(\int^\xi q(\eta)\, d\eta\right) d\xi + c_1 x + c_2 \quad (4.3.21)$$

and μ is defined to be

$$\mu^2 = \frac{\hat{N}_{xx} a^2}{4D} \quad (4.3.22)$$

For uniformly distributed load and constant temperatures T_0 and T_1, we have

$$w_0^p(x) = \frac{D^T T_1 a^2}{4D\mu^2} - \frac{q_0}{16D}\left[2\left(\frac{ax}{\mu}\right)^2 + \left(\frac{a}{\mu}\right)^4\right] - \frac{c_1 x}{4D}\left(\frac{a}{\mu}\right)^2 - \frac{c_2 a^2}{4D\mu^2} \quad (4.3.23)$$

and the complete solution becomes

$$w_0(x) = \frac{D^T T_1 a^2}{4D\mu^2} - \frac{q_0}{16D}\left[2\left(\frac{ax}{\mu}\right)^2 + \left(\frac{a}{\mu}\right)^4\right] - \frac{c_1 x}{4D}\left(\frac{a}{\mu}\right)^2$$
$$+ c_3 \sinh\frac{2\mu x}{a} + c_4 \cosh\frac{2\mu x}{a} - \frac{c_2 a^2}{4D\mu^2} \quad (4.3.24)$$

Simply supported plate under uniformly distributed load

Using the boundary conditions $w_0 = 0$ and $M_{xx} = 0$ at $x = 0, a$ in Eq. (4.3.18), we obtain

$$c_2 = 0, \quad c_1 = -\frac{q_0 a}{2} \quad (4.3.25)$$

and Eq. (4.3.24) for the deflection w_0 becomes

$$w_0(x) = \frac{D^T T_1 a^2}{4D\mu^2} - \frac{q_0}{16D}\left[2\left(\frac{ax}{\mu}\right)^2 + \left(\frac{a}{\mu}\right)^4\right] + \frac{q_0 ax}{8D}\left(\frac{a}{\mu}\right)^2$$
$$+ c_3 \sinh\frac{2\mu x}{a} + c_4 \cosh\frac{2\mu x}{a} \quad (4.3.26)$$

Next, using the boundary condition $w_0 = 0$ at $x = 0, a$, we obtain

$$c_4 = -\frac{D^T T_1 a^2}{4D\mu^2} + \frac{q_0 a^4}{16D\mu^4} \equiv \beta, \quad c_3 = c_4\left(\frac{1 - \cosh 2\mu}{\sinh 2\mu}\right) \quad (4.3.27)$$

and the expression (4.3.26) for the deflection w_0 becomes

$$w_0(x) = \beta\left(\frac{1 - \cosh 2\mu}{\sinh 2\mu}\sinh\frac{2\mu x}{a} + \cosh\frac{2\mu x}{a} - 1\right) + \left(\frac{q_0 a^2}{8D\mu^2}\right)x(a-x)$$
$$= \beta\left[\frac{\cosh\mu(1-\frac{2x}{a})}{\cosh\mu} - 1\right] + \left(\frac{q_0 a^2}{8D\mu^2}\right)x(a-x) \quad (4.3.28)$$

where the last step is arrived at by using the hyperbolic identities

$$\cosh 2\mu = \cosh^2 \mu + \sinh^2 \mu, \quad \sinh 2\mu = 2 \sinh \mu \cosh \mu$$

$$\cosh^2 \mu = 1 + \sinh^2 \mu, \quad \cosh \mu \cosh \frac{2\mu x}{a} - \sinh \mu \sinh \frac{2\mu x}{a} = \cosh \mu \left(1 - \frac{2x}{a}\right)$$

The maximum deflection occurs at $x = a/2$ and is given by

$$w_{max} = \frac{5q_0 a^4}{384 D} f(\mu) + \frac{D^T T_1 a^2}{8D} g(\mu) \tag{4.3.29a}$$

$$f(\mu) = \frac{12}{5\mu^4} \left(2 \operatorname{sech} \mu - 2 + \mu^2\right), \quad g(\mu) = \frac{2}{5\mu^2} (1 - \operatorname{sech} \mu) \tag{4.3.29b}$$

The axial stress for this case consists of two parts: bending stress proportional to bending moment and tensile stress proportional to $\hat{N}_{xx}/h$. The maximum bending moment is given by

$$M_{max} = -D \left(\frac{d^2 w}{dx^2}\right)_{x=\frac{a}{2}} - M_{xx}^T = \frac{q_0 a^2}{8} \psi_0(\mu) + D^T T_1 \phi_0(\mu) \tag{4.3.30a}$$

$$\psi_0(\mu) = \frac{2}{\mu^2}(1 - \operatorname{sech}\mu), \quad \phi_0(\mu) = (1 - \operatorname{sech}\mu) \tag{4.3.30b}$$

Hence, the maximum axial stress occurs at $x = a/2$ and $z = h/2$ and it is given by

$$\sigma_{max} = \frac{\hat{N}_{xx}}{h} + \frac{6 M_{max}}{h^2} = \frac{4\mu^2 D}{a^2 h} + \frac{6}{h^2} \left(\frac{q_0 a^2}{8} \psi_0(\mu) + D^T T_1 \phi_0(\mu)\right) \tag{4.3.31}$$

From Eq. (4.3.31) it is clear that the deflection of a plate strip depends on μ, which in turn depends on the axial force $\hat{N}_{xx}$. The axial force can be determined from the condition that the axial displacement u_0 be zero at $x = 0, a$. By definition, we have [from Eqs. (3.6.10a) and (3.3.8b) when $u_0 = v_0 = 0$]

$$N_{xx} = \frac{1}{2} A_{11} \left(\frac{dw_0}{dx}\right)^2 = \frac{E_1 h}{(1 - \nu_{12}\nu_{21})} \left[\frac{1}{2}\left(\frac{dw_0}{dx}\right)^2\right] \tag{4.3.32}$$

We define the axial force in the plate strip to be the average

$$\hat{N}_{xx} = A_{11} \left[\frac{1}{2a} \int_0^a \left(\frac{dw_0}{dx}\right)^2 dx\right] \tag{4.3.33}$$

which is constant along the length of the plate strip. Substituting for the deflection w_0 from Eq. (4.3.28) into Eq. (4.3.33) and integrating with respect to x, we obtain

$$\frac{a\hat{N}_{xx}}{A_{11}} = \frac{q_0^2 a^7}{D^2} \left[\frac{\mu D^2 \beta^2}{q_0^2 a^8} \left(\tanh \mu + \mu \operatorname{sech}^2 \mu \right) + \frac{D\beta}{4q_0 a^4 \mu^3} \left(\tanh \mu - \mu \right) \right.$$
$$\left. + \frac{1}{384 \mu^4} \right] \tag{4.3.34}$$

where

$$\beta = \left(\frac{q_0 a^4}{16 D \mu^4} - \frac{D^T T_1 a^2}{4 D \mu^2} \right) \tag{4.3.35}$$

Using the definition in Eq. (4.3.22), we can write Eq. (4.3.34) as

$$\frac{E_1^2 s^2}{(1 - \nu_{12}\nu_{21})^2 q_0^2} = \frac{432 D^2 \beta^2}{q_0^2 a^8 \mu} \left(\tanh \mu + \mu \operatorname{sech}^2 \mu \right)$$
$$+ \frac{D\beta}{108 q_0 a^4 \mu^5} \left(\tanh \mu - \mu \right) + \frac{9}{8 \mu^6} \tag{4.3.36}$$

where $s = h/a$. For an isotropic plate strip without thermal loads, Eq. (4.3.36) simplifies to

$$\frac{E^2 s^2}{(1 - \nu^2)^2 q_0^2} = \frac{135}{16} \frac{\tanh \mu}{\mu^7} + \frac{27}{16} \frac{\tanh^2 \mu}{\mu^8} - \frac{135}{16 \mu^8} + \frac{9}{8 \mu^6} \tag{4.3.37}$$

For a given material, a given ratio $s = h/a$, and a given load q_0, the nonlinear (transcendental) equation (4.3.36) for μ can be solved by an iterative method.

The analytical solutions developed above for the deflection, bending moment, and stress involve the determination of μ from the nonlinear equation (4.3.36), which is complicated. If we can determine μ by some approximate means, then Eqs. (4.3.29a,b)–(4.3.31) can be used to determine the maximum deflection, bending moment, and stress. Here we discuss the use of the Rayleigh–Ritz method to determine μ. Suppose that

$$w_0(x) \approx c_1 \sin \frac{\pi x}{a} \tag{4.3.38}$$

which satisfies the geometric boundary conditions $w_0(0) = w_0(a) = 0$. Substituting into the virtual work statement

$$0 = \int_0^a \left[\left(D \frac{d^2 w_0}{dx^2} + M_{xx}^T \right) \frac{d^2 \delta w_0}{dx^2} + \hat{N}_{xx} \frac{dw_0}{dx} \frac{d\delta w_0}{dx} - q \delta w_0 \right] dx \tag{4.3.39}$$

we obtain

$$c_1 = \frac{w^c}{(1+\alpha)}, \quad w^c = \frac{4a^4}{D\pi^5}\left(q_0 + \frac{M_{xx}^T \pi^2}{a^2}\right), \quad \alpha = \hat{N}_{xx}\frac{a^2}{\pi^2 D} \quad (4.3.40)$$

Here, w^c denotes the center deflection of the beam when $\hat{N}_{xx} = 0$, and α denotes the ratio of the axial force $\hat{N}_{xx}$ to the critical buckling load $N_{cr} = \pi^2 D/a^2$ (see section 4.4). Note that the exact value of w^c is given by Eq. (4.3.12)

$$w_0^c = \frac{5a^4}{384D}\left(q_0 + 9.6\frac{M_{xx}^T}{a^2}\right) \quad (4.3.41)$$

Thus using

$$w_0(x) \approx \frac{w^c}{(1+\alpha)}\sin\frac{\pi x}{a} \quad (4.3.42)$$

in Eq. (4.3.33) and carrying out the integration, we obtain

$$\frac{a\hat{N}_{xx}}{A_{11}} = \frac{\pi^2 (w^c)^2}{4a(1+\alpha)^2}$$

The left side of the above equation can be expressed in terms of α using Eq. (4.3.40)

$$\frac{a\hat{N}_{xx}}{A_{11}} = \frac{\pi^2 a^2 \hat{N}_{xx} h^2}{12\pi^2 aD} = \alpha\frac{\pi^2 h^2}{12a}$$

Hence, we have

$$\alpha(1+\alpha)^2 = \frac{3(w^c)^2}{h^2} \quad (4.3.43)$$

which can be used to calculate α. Then μ is given by (4.3.22)

$$\mu^2 = \frac{\hat{N}_{xx} a^2}{4D} = \frac{\pi^2}{4}\alpha \quad (4.3.44)$$

Clamped plate under uniformly distributed load

For the case of a plate with clamped edges, the boundary conditions are $w_0 = 0$ and $dw_0/dx = 0$ at $x = 0, a$. Using these conditions in Eq. (4.3.24), we obtain

$$c_1 = -\frac{q_0 a}{2}, \quad c_2 = \frac{q_0 a^2}{4\mu^2}(\mu\coth\mu - 1) + D^T T_1$$

ANALYSIS OF PLATE STRIPS

$$c_3 = -\frac{q_0 a^4}{16D\mu^3}, \quad c_4 = \frac{q_0 a^4}{16D\mu^3}\coth\mu \quad (4.3.45)$$

and Eq. (4.3.24) for the deflection w_0 becomes

$$w_0(x) = -\frac{q_0 a^4}{16D\mu^3}\sinh\frac{2\mu x}{a} + \frac{q_0 a^4}{16D\mu^3}\coth\mu\cosh\frac{2\mu x}{a}$$

$$-\frac{q_0}{16D}\left[2\left(\frac{ax}{\mu}\right)^2 + \left(\frac{a}{\mu}\right)^4\right] + \frac{q_0 ax}{8D}\left(\frac{a}{\mu}\right)^2 - \frac{q_0 a^4}{16D\mu^4}(\mu\coth\mu - 1)$$

$$= \left(\frac{q_0 a^4}{16D\mu^3 \tanh\mu}\right)\left[\frac{\cosh\mu(1-\frac{2x}{a})}{\cosh\mu} - 1\right] + \frac{q_0 a^2 x(a-x)}{8D\mu^2} \quad (4.3.46)$$

The maximum deflection occurs at $x = a/2$ and is given by

$$w_{max} = \frac{q_0 a^4}{384D}g(\mu), \quad g(\mu) = \frac{12}{\mu^4}\left(\frac{\mu^2}{2} + \frac{\mu}{\sinh\mu} - \frac{\mu}{\tanh\mu}\right) \quad (4.3.47)$$

The axial stress for this case consists of two parts: bending stress proportional to bending moment and tensile stress proportional to $\hat{N}_{xx}/h$. The maximum bending moment is at $x = 0$ and is given by

$$M_{max} = \frac{q_0 a^2}{4\mu^2}(1 - \mu\coth\mu) - D^T T_1 \quad (4.3.48)$$

Hence, the maximum axial stress is given by

$$\sigma_{max} = \frac{4\mu^2 D}{a^2 h} + \frac{6M_{max}}{h^2} \quad (4.3.49)$$

The axial force in the plate strip can be calculated from

$$\hat{N}_{xx} = \frac{E_1 h q_0^2 a^6}{256 D^2 (1-\nu_{12}\nu_{21})}\left(-\frac{3}{\mu^5 \tanh\mu} - \frac{1}{\mu^4 \sinh^2\mu} + \frac{4}{\mu^6} + \frac{2}{3\mu^4}\right) \quad (4.3.50)$$

4.3.3 Plate Strip on an Elastic Foundation

Here we consider the problem of the bending of a uniformly loaded plate strip supported over the entire bottom surface by an elastic foundation. In the absence of the axial force, the governing equation is given by

$$D\frac{d^2 w_0}{dx^2} + kw_0 = q_0 \quad (4.3.51)$$

where k is the foundation modulus (force per unit surface area) and q_0 is the intensity of the transverse load. The general solution of this equation is

$$w_0(x) = \left(c_1 \sinh \frac{2\beta x}{a} + c_2 \cosh \frac{2\beta x}{a}\right) \sin \frac{2\beta x}{a}$$
$$+ \left(c_3 \sinh \frac{2\beta x}{a} + c_4 \cosh \frac{2\beta x}{a}\right) \cos \frac{2\beta x}{a} + \frac{q_0}{k} \quad (4.3.52)$$

where

$$\beta^4 = \frac{ka^4}{64D} \quad (4.3.53)$$

and c_1 through c_4 are constants to be determined using the boundary conditions.

For a plate strip with both edges simply supported, the deflection is symmetrical with respect to the center of the strip. Hence, the slope and shear force at the center must be zero. Suppose that the origin of the coordinate system is taken at the center of the plate. Then, the boundary conditions are

$$\frac{dw_0}{dx} = \frac{d^3 w_0}{dx^3} = 0 \text{ at } x = 0; \quad w_0 = \frac{d^2 w_0}{dx^2} = 0 \text{ at } x = a/2 \quad (4.3.54)$$

The first two boundary conditions give $c_2 = c_3 = 0$. The next two boundary conditions yield

$$c_1 \sin \beta \sinh \beta + c_4 \cos \beta \cosh \beta + \frac{q_0}{k} = 0$$

$$c_1 \cos \beta \cosh \beta + c_4 \sin \beta \sinh \beta = 0$$

from which is obtained

$$c_1 = -\frac{2q_0}{k}\left(\frac{\sin \beta \sinh \beta}{\cos 2\beta + \cosh 2\beta}\right), \quad c_4 = -\frac{2q_0}{k}\left(\frac{\cos \beta \cosh \beta}{\cos 2\beta + \cosh 2\beta}\right) \quad (4.3.55)$$

The deflection in Eq. (4.3.52) takes the form

$$w_0(x) = \frac{q_0 a^4}{64 D \beta^4}\left[1 - \left(\frac{2 \sin \beta \sinh \beta}{\cos 2\beta + \cosh 2\beta}\right) \sin \frac{2\beta x}{a} \sinh \frac{2\beta x}{a}\right.$$
$$\left. - \left(\frac{2 \cos \beta \cosh \beta}{\cos 2\beta + \cosh 2\beta}\right) \cos \frac{2\beta x}{a} \cosh \frac{2\beta x}{a}\right] \quad (4.3.56)$$

The mximum deflection occurs at $x = 0$

$$w_{max} = w_0(\frac{a}{2}) = \frac{5q_0 a^4}{384D}\varphi_1(\beta), \quad \varphi_1(\beta) = \frac{6}{5\beta^4}\left(1 - \frac{2\cos\beta\cosh\beta}{\cos 2\beta + \cosh 2\beta}\right)$$
(4.3.57)

The rotation at $x = -a/2$ is given by

$$\theta_{max} = \frac{dw_0}{dx}\big|_{x=a/2} = \frac{q_0 a^3}{24D}\varphi_2(\beta), \quad \varphi_2(\beta) = \frac{3}{4\beta^3}\left(\frac{2\sinh 2\beta - \sin 2\beta}{\cos 2\beta + \cosh 2\beta}\right)$$
(4.3.58)

The maximum bending moment occurs at $x = 0$, and it is given by

$$M_{max} = \left(-D\frac{d^2 w_0}{dx^2}\right)_{x=0} = \frac{q_0 a^2}{8}\varphi_3(\beta), \quad \varphi_3(\beta) = \frac{2}{\beta^2}\left(\frac{\sinh\beta\sin\beta}{\cos 2\beta + \cosh 2\beta}\right)$$
(4.3.59)

4.4 Buckling Under In-plane Compressive Load

4.4.1 Introduction

A long plate strip subjected to compressive load $\hat{N}_{xx} = -N_{xx}^0$ along its width remains straight but shortens as the load increases from zero to a certain magnitude. If a small additional axial or lateral disturbance applied to the plate keeps it in equilibrium, then the plate is said to be *stable*. If the small additional disturbance results in a large response and the plate does not return to its original equilibrium configuration, the plate is said be *unstable*. The onset of instability is called *buckling* (see Figure 4.4.1). The magnitude of the compressive axial load at which the plate becomes unstable is termed the *critical buckling load*. If the load is increased beyond this critical buckling load, it results in a large deflection and the plate seeks another equilibrium configuration. Thus, the load at which a plate becomes unstable is of practical importance in design. Here we determine critical buckling loads for plate strips in cylindrical bending.

The equilibrium of the plate strip under the applied in-plane compressive load $\hat{N}_{xx} = -N_{xx}^0$ can be obtained from Eqs. (4.2.3) by omitting the inertia terms and thermal resultants

$$D\frac{d^4 w}{dx^4} + N_{xx}^0 \frac{d^2 w}{dx^2} = 0 \qquad (4.4.1)$$

where w denotes the transverse deflection measured from the prebuckling equilibrium state.

156 THEORY AND ANALYSIS OF ELASTIC PLATES

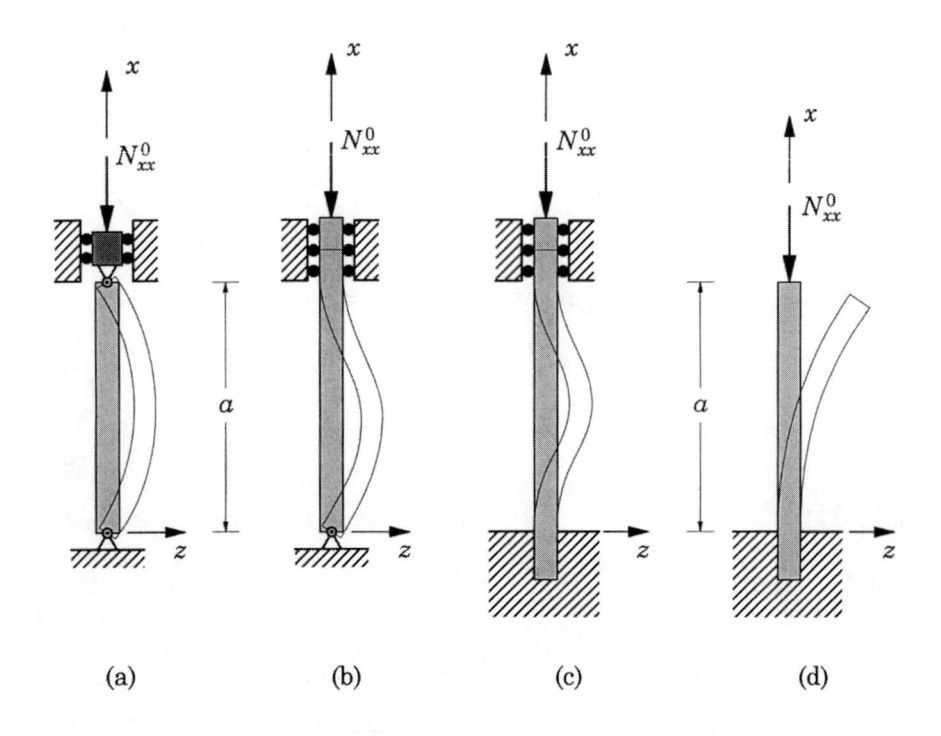

Figure 4.4.1. Buckling of plate strips under various boundary conditions (edge view): (a) hinged-hinged, (b) hinged-clamped, (c) clamped-clamped, and (d) clamped-free.

Integrating Eq. (4.4.1) twice with respect to x, we obtain

$$D\frac{d^2w}{dx^2} + N_{xx}^0 w = K_1 x + K_2 \quad (4.4.2)$$

where K_1 and K_2 are constants. Next we consider the solution of Eq. (4.4.2) in two parts: homogeneous solution and particular solution. The homogeneous solution of Eq. (4.4.2) is obtained from

$$D\frac{d^2w}{dx^2} + N_{xx}^0 w = 0 \quad \text{or} \quad \frac{d^2w}{dx^2} + \lambda^2 w = 0 \quad (4.4.3)$$

where

$$\lambda^2 = \frac{N_{xx}^0}{D} \quad \text{or} \quad N_{xx}^0 = D\lambda^2 \quad (4.4.4)$$

The homogeneous solution of the second-order equation (4.4.3) is given by
$$w_h(x) = c_1 \sin \lambda x + c_2 \cos \lambda x \qquad (4.4.5a)$$
The particular solution is given by
$$w_p(x) = \frac{1}{\lambda^2}(K_1 \sin \lambda x + c_2 \cos \lambda x) \qquad (4.4.5b)$$
Hence, the complete solution is
$$w(x) = c_1 \sin \lambda x + c_2 \cos \lambda x + c_3 x + c_4 \qquad (4.4.6)$$
where $c_3 = K_1/\lambda^2$ and $c_4 = K_2/\lambda^2$. Three of the four constants c_1, c_2, c_3, c_4, and λ (or N_{xx}^0) are determined using (four) boundary conditions of the problem. Once λ is known, the buckling load can be determined using Eq. (4.4.4).

4.4.2 Simply Supported Plate Strips

For a simply supported plate strip (at $x = 0, a$), we have
$$w = 0, \quad \frac{d^2 w}{dx^2} = 0 \qquad (4.4.7)$$
Use of the boundary conditions on w gives

$$w(0) = 0 : \quad c_2 + c_4 = 0 \text{ or } c_4 = -c_2$$
$$\frac{d^2 w}{dx^2}\bigg|_{x=0} = 0 : \quad -\lambda^2 c_2 = 0 \text{ or } c_2 = 0$$
$$w(a) = 0 : \quad c_1 \sin \lambda a + c_3 a = 0$$
$$\frac{d^2 w}{dx^2}\bigg|_{x=a} = 0 : \quad -\lambda^2 c_1 \sin \lambda a = 0 \qquad (4.4.8a)$$

Thus we have
$$c_1 \sin \lambda a = 0, \quad c_2 = 0, \quad c_3 = 0, \quad c_4 = 0 \qquad (4.4.8b)$$
The first equation implies that either $c_1 = 0$ or (and) $\sin \lambda a = 0$. If $c_1 = 0$ then the buckling deflection of Eq. (4.4.3) is zero, implying that the plate did not begin to buckle. For nonzero deflection w, we must have
$$\sin \lambda a \equiv \sin\left(\sqrt{\frac{N_{xx}^0}{D}}\right) = 0 \quad \text{or} \quad N_{xx}^0 = D\left(\frac{n\pi}{a}\right)^2 \qquad (4.4.9)$$

158   THEORY AND ANALYSIS OF ELASTIC PLATES

The critical buckling load (i.e., the smallest value of N_{xx}^0 at which the plate buckles) is given by $(n = 1)$

$$N_{cr} = \frac{D\pi^2}{a^2} \qquad (4.4.10)$$

For an orthotropic plate with $E_1/E_2 = 25$ and $\nu_{12} = 0.25$, we have

$$N_{cr} = \frac{100\pi^2}{1197} \frac{E_1 h^3}{a^2} = 0.8245 \frac{E_1 h^3}{a^2} = 20.6125 \frac{E_2 h^3}{a^2} \qquad (4.4.11)$$

For an isotropic plate $(D_{22} = D_{11} = D)$ with $\nu = 0.25$ the buckling load becomes

$$N_{cr} = \frac{100\pi^2}{1125} \frac{E h^3}{a^2} = 0.8773 \frac{E h^3}{a^2} \qquad (4.4.12)$$

4.4.3 Clamped Plate Strips

For a clamped plate strip (at $x = 0, a$), we have

$$w = 0, \quad \frac{dw}{dx} = 0 \qquad (4.4.13)$$

Use of the boundary conditions on w gives

$$w(0) = 0: \quad c_2 + c_4 = 0 \text{ or } c_4 = -c_2$$
$$\frac{dw}{dx}\bigg|_{x=0} = 0: \quad \lambda c_1 + c_3 = 0 \text{ or } c_3 = -\lambda c_1$$
$$w(a) = 0: \quad c_1 \sin \lambda a + c_2 \cos \lambda a + c_3 a + c_4 = 0$$
$$\frac{dw}{dx}\bigg|_{x=a} = 0: \quad \lambda (c_1 \cos \lambda a - c_2 \sin \lambda a) + c_3 = 0 \qquad (4.4.14)$$

Using the first two equations, c_3 and c_4 can be eliminated from the last two equations. We have

$$c_1 (\sin \lambda a - \lambda a) + c_2 (\cos \lambda a - 1) = 0, \quad c_1 (\cos \lambda a - 1) - c_2 \sin \lambda a = 0$$

For nonzero tranverse deflection (i.e., for nonzero values of c_2 and c_3), we require that the determininant of the above pair of equations be zero:

$$\begin{vmatrix} \sin \lambda a - \lambda a & \cos \lambda a - 1 \\ \cos \lambda a - 1 & -\sin \lambda a \end{vmatrix} = 0$$

$$\lambda a \sin \lambda a + 2 \cos \lambda a - 2 = 0 \tag{4.4.15}$$

This nonlinear (transcendental) equation can be solved by an iterative (Newton's) method for various roots of the equation. The smallest root of this equation is $\lambda = 2\pi$, and the critical buckling load becomes

$$N_{cr} = \frac{4D\pi^2}{a^2} \tag{4.4.16}$$

Thus, the buckling load of a clamped plate strip is four times that of a simply supported plate strip.

For an orthotropic plate with $E_1/E_2 = 25$ and $\nu_{12} = 0.25$, we have

$$N_{cr} = \frac{100\pi^2}{1197} \frac{E_1 h^3}{a^2} = 3.298 \frac{E_1 h^3}{a^2} = 82.45 \frac{E_2 h^3}{a^2}$$

For an isotropic plate ($D_{22} = D_{11} = D$) with $\nu = 0.25$, the buckling load becomes

$$N_{cr} = \frac{100\pi^2}{1125} \frac{E h^3}{a^2} = 3.5092 \frac{E h^3}{a^2}$$

4.4.4 Other Boundary Conditions

The procedure illustrated in the two previous sections to determine the buckling load can be used for any set of boundary conditions. The analytical expressions for various boundary conditions are listed in Eqs. (4.4.17)–(4.4.21) below. The relations among the constants c_1, c_2, c_3, and c_4 of Eq. (4.4.6) and the equation governing λ are listed for various combinations of classical boundary conditions in Table 4.4.1. These can be used to readily determine the critical buckling loads of plate strips.

Free:

$$M_{xx} = 0 \quad \text{and} \quad V_x = Q_x - N_{xx}^0 \frac{d^2 w}{dx^2} = 0$$

or

$$\frac{d^2 w}{dx^2} = 0 \quad \text{and} \quad D \frac{d^3 w}{dx^3} - N_{xx}^0 \frac{d^2 w}{dx^2} = 0 \tag{4.4.17}$$

Hinged (or) Simply Supported:

$$w = 0, \quad M_{xx} = 0 \quad \text{or} \quad \frac{d^2 w}{dx^2} = 0 \tag{4.4.18}$$

Table 4.4.1. Values of the constants and eigenvalues for buckling of plate strips with various boundary conditions ($\lambda^2 \equiv N_{xx}^0/D = (e_n/a)^2$). The classical plate theory is used.

End conditions at $x = 0$ and $x = a$	Constants[†]	Characteristic equation and values[*] of $e_n \equiv \lambda_n a$
• Hinged-Hinged	$c_1 \neq 0$ $c_2 = c_3 = c_4 = 0$	$\sin e_n = 0,\ e_n = n\pi$
• Fixed-Fixed	$c_1 = 1/(\sin e_n - e_n)$ $c_3 = -1/\lambda_n,\ c_2 = -c_4$ $c_2 = 1/(\cos e_n - 1)$	$e_n \sin e_n = 2(1 - \cos e_n)$ $e_n = 2\pi, 8.987, 4\pi, \cdots$
• Fixed-Free	$c_1 = c_3 = 0$ $c_2 = -c_4 \neq 0$	$\cos e_n = 0$ $e_n = (2n-1)\pi/2$
• Free-Free	$c_1 = c_3 = 0$ $c_2 \neq 0,\ c_4 \neq 0$	$\sin e_n = 0$ $e_n = n\pi$
• Hinged-Fixed	$c_1 = 1/(e_n \cos e_n)$ $c_3 = -1,\ c_2 = c_4 = 0$	$\tan e_n = e_n$ $e_n = 4.493, 7.725, \cdots$

[†] See Eq. (4.4.6): $w(x) = c_1 \sin \lambda x + c_2 \cos \lambda x + c_3 x + c_4$.
[*] For critical buckling load, only the first (minimum) value of $e = \lambda a$ is needed.

Fixed (or) Clamped:

$$w = 0,\quad \frac{dw}{dx} = 0 \qquad (4.4.19)$$

Elastically Simply Supported:

$$w = \alpha F_0,\quad M_{xx} = 0 \text{ or } \frac{d^2w}{dx^2} = 0 \qquad (4.4.20)$$

where F_0 is the vertical reacting force of the elastic support and α is the inverse of the elastic (spring) constant. When $\alpha = 0$ (i.e., the support is rigid), we recover the conventional simply supported boundary condition. When α is very large, the boundary condition approaches that of a free edge.

Elastically Clamped:

$$w = 0, \quad \frac{dw}{dx} = -\beta M_0 \qquad (4.4.21)$$

where M_0 is the reacting moment (clockwise) of the elastic support and β is the torsional spring constant. When $\beta = 0$ (i.e., the restraint is rigid), we recover the conventional clamped boundary condition. On the other hand, if β is very large (i.e., the restraint is very flexible), the condition approaches that of a simply supported case.

4.5 Free Vibration
4.5.1 General Formulation

In the absence of any applied transverse mechanical and thermal loads, Eq. (4.2.3) reduces to

$$D\frac{\partial^4 w_0}{\partial x^4} = I_2 \frac{\partial^4 w_0}{\partial x^2 \partial t^2} - I_0 \frac{\partial^2 w_0}{\partial t^2} + \hat{N}_{xx} \frac{\partial^2 w_0}{\partial x^2} \qquad (4.5.1)$$

For natural flexural vibration (i.e., periodic motion), the solution of Eq. (4.5.1) is assumed to be periodic

$$w_0(x, t) = w(x) \cos \omega t \qquad (4.5.2)$$

where ω is the natural frequency of vibration and $w(x)$ is the mode shape. Substituting for w_0 from Eq. (4.5.2) into Eq. (4.5.1) and cancelling the factor $\cos \omega t$ (because the result should hold for any time t), we obtain

$$D\frac{d^4 w}{dx^4} - \hat{N}_{xx}\frac{d^2 w}{dx^2} = I_0 \omega^2 w - I_2 \omega^2 \frac{d^2 w}{dx^2} \qquad (4.5.3)$$

Equation (4.5.3) can be expressed in the general form

$$a\frac{d^4 w}{dx^4} + b\frac{d^2 w}{dx^2} - cw = 0 \qquad (4.5.4a)$$

where

$$a = D(= D_{11}), \quad b = \omega^2 I_2 - \hat{N}_{xx}, \quad c = \omega^2 I_0 \qquad (4.5.4b)$$

Assume solution of the homogeneous equation (4.5.4a) in the form

$$w(x) = Ae^{rx} \quad (4.5.5)$$

and substitute it into Eq. (4.5.4a) to obtain

$$ar^4 + br^2 - c = 0 \quad \text{or} \quad as^2 + bs - c = 0 \ (s = r^2) \quad (4.5.6)$$

The roots of the above equation are

$$s_1 = \frac{1}{2a}\left(-b - \sqrt{b^2 + 4ac}\right) \equiv -\lambda^2, \quad s_2 = \frac{1}{2a}\left(-b + \sqrt{b^2 + 4ac}\right) \equiv \mu^2 \quad (4.5.7)$$

Hence, the solution of Eq. (4.5.4a) can be expressed as

$$w(x) = c_1 \sin \lambda x + c_2 \cos \lambda x + c_3 \sinh \mu x + c_4 \cosh \mu x \quad (4.5.8)$$

where

$$\lambda = \sqrt{\frac{1}{2a}\left(b + \sqrt{b^2 + 4ac}\right)}, \quad \mu = \sqrt{\frac{1}{2a}\left(-b + \sqrt{b^2 + 4ac}\right)} \quad (4.5.9)$$

and c_1, c_2, c_3, and c_4 are integration constants, which are to be determined using the boundary conditions. In reality, the four boundary conditions (two at each edge of the plate strip) are used to determine λ and only three of the four constants.

The frequency of vibration is determined from λ (or μ) as follows. From Eq. (4.5.7) we have

$$(2a\lambda - b)^2 = b^2 + 4ac \quad \text{or} \quad a\lambda^4 - b\lambda^2 - c = 0$$

Using the definitions of a, b, and c, we obtain

$$D\lambda^4 - \left(\omega^2 I_2 - \hat{N}_{xx}\right)\lambda^2 - \omega^2 I_0 = 0$$

from which we have

$$\left(I_0 + \lambda^2 I_2\right)\omega^2 = D\lambda^4 + \hat{N}_{xx}\lambda^2$$

or

$$\omega^2 = \frac{D\lambda^4 + \hat{N}_{xx}\lambda^2}{I_0 + \lambda^2 I_2} \quad (4.5.10)$$

If the applied in-plane force is zero, the natural frequency of vibration, with rotary inertia included, is given by

$$\omega^2 = \frac{D\lambda^4}{I_0 + \lambda^2 I_2} = \frac{D}{I_0}\left(1 - \frac{I_2\lambda^2}{I_0 + \lambda^2 I_2}\right)\lambda^2 \qquad (4.5.11)$$

In addition, if rotary inertia is neglected, we have

$$\omega = \lambda^2 \sqrt{\frac{D}{I_0}} \qquad (4.5.12)$$

The result in Eq. (4.5.12), when the rotary inertia is neglected and there is no applied in-plane force, can be obtained directly from Eq. (4.5.3), which reduces to

$$D\frac{d^4 w}{dx^4} - I_0 \omega^2 w = 0$$

The solution to this equation is also given by Eq. (4.5.8) with $\lambda = \mu$ and

$$\lambda^2 = \omega \sqrt{\frac{I_0}{D}} \qquad (4.5.13)$$

In the following sections, plates with both ends simply supported or both ends clamped are considered to illustrate the procedure to evaluate the constants c_1 through c_4, and, more importantly, to determine λ and find ω. The smallest frequency is known as the *fundamental frequency*. For other boundary conditions, the reader is referred to Table 4.5.1.

4.5.2 Simply Supported Plate Strips

For a simply supported plate strip, we use the boundary conditions

$$w = 0 \quad M_{xx} = D\frac{d^2 w}{dx^2} = 0 \quad \text{at} \quad x = 0, a$$

and obtain

$$c_1 \sin \lambda a = 0, \quad c_2 = c_3 = c_4 = 0$$

Hence $\lambda = \frac{n\pi}{a}$, and from Eq. (4.5.10) it follows that

$$\omega_n = \left(\frac{n\pi}{a}\right)^2 \left[\left(\frac{D}{I_0}\right)\left(1 + \frac{\hat{N}_{xx}}{(\frac{n\pi}{a})^2 D}\right)\left(\frac{1}{1 + (\frac{n\pi}{a})^2 \frac{I_2}{I_0}}\right)\right]^{\frac{1}{2}} \qquad (4.5.14)$$

Table 4.5.1. Values of the constants and eigenvalues for natural vibration of plate strips with various boundary conditions ($\lambda_n^4 \equiv \omega_n^2 I_0/D = (e_n/a)^4$). The classical plate theory *without* rotary inertia is used.

End conditions at $x = 0$ and $x = a$	Constants†	Characteristic equation and values of $e_n \equiv \lambda_n a$
• Hinged-Hinged	$c_1 \neq 0$ $c_2 = c_3 = c_4 = 0$	$\sin e_n = 0, \ e_n = n\pi$
• Fixed-Fixed	$c_1 = -c_3 = 1/p_n$ $-c_2 = c_4 = 1/q_n$ $p_n = (\sin e_n - \sinh e_n)$	$\cos e_n \cosh e_n - 1 = 0$ $e_n = 4.730, 7.853, \cdots$ $q_n = (\cos e_n - \cosh e_n)$
• Fixed-Free	$c_1 = -c_3 = 1/p_n$ $-c_2 = c_4 = 1/q_n$ $p_n = (\sin e_n + \sinh e_n)$	$\cos e_n \cosh e_n + 1 = 0$ $e_n = 1.875, 4.694, \cdots$ $q_n = (\cos e_n + \cosh e_n)$
• Free-Free	$c_1 = c_3 = 1/p_n$ $c_2 = c_4 = -1/q_n$ $p_n = (\sin e_n - \sinh e_n)$	$\cos e_n \cosh e_n - 1 = 0$ $e_n = 4.730, 7.853, \cdots$ $q_n = (\cos e_n - \cosh e_n)$
• Hinged-Fixed	$c_1 = 1/\sin e_n$ $c_2 = c_4 = 0$ $c_3 = 1/\sinh e_n$	$\tan e_n = \tanh e_n$ $e_n = 3.927, 7.069, \cdots$
• Hinged-Free	$c_1 = 1/\sin e_n$ $c_2 = c_4 = 0$ $c_3 = -1/\sinh e_n$	$\tan e_n = \tanh e_n$ $e_n = 3.927, 7.069, \cdots$

† See Eq. (4.5.8): $W(x) = c_1 \sin \lambda x + c_2 \cos \lambda x + c_3 \sinh \mu x + c_4 \cosh \mu x$.

If the rotary inertia is neglected, we obtain

$$\omega_n = \left(\frac{n\pi}{a}\right)^2 \left[\left(\frac{D}{I_0}\right)\left(1 + \frac{\hat{N}_{xx}}{\left(\frac{n\pi}{a}\right)^2 D}\right)\right]^{\frac{1}{2}} \qquad (4.5.15)$$

Thus, the effect of the in-plane tensile force $\hat{N}_{xx}$ is to increase the natural frequencies. If we have a very flexible plate, say a membrane under large tension, the second term under the radical in Eq. (4.5.15) becomes very large in comparison with unity; if n is not large, we have

$$\omega_n \approx \frac{n\pi}{a}\sqrt{\frac{\hat{N}_{xx}}{I_0}} \qquad (4.5.16)$$

which are natural frequencies of a stretched membrane. We also note from Eq. (4.5.15) that frequencies of natural vibration decrease when a compressive force instead of a tensile force is acting on the plate strip.

When $\hat{N}_{xx} = 0$, we obtain from Eq. (4.5.14)

$$\omega_n = \left(\frac{n\pi}{a}\right)^2 \left[\frac{D}{I_0}\left(\frac{I_0}{I_0 + (\frac{n\pi}{a})^2 I_2}\right)\right]^{\frac{1}{2}}$$

$$= \left(\frac{n\pi}{a}\right)^2 \left[\frac{D}{I_0}\left(1 - \frac{I_2(\frac{n\pi}{a})^2}{(\frac{n\pi}{a})^2 I_2 + I_0}\right)\right]^{\frac{1}{2}} \qquad (4.5.17)$$

Thus, rotatory inertia decreases frequencies of natural vibration. If the rotatory inertia is neglected, we obtain

$$\omega_n = \left(\frac{n\pi}{a}\right)^2 \sqrt{\frac{D}{I_0}} \qquad (4.5.18)$$

For a simply supported orthotropic plate with $E_1/E_2 = 25$, $\nu_{12} = 0.25$, $a/h = 10$ and rotatory inertia included, we obtain

$$\omega_1 = \frac{14.205}{a^2}\sqrt{\frac{E_2 h^3}{I_0}} = 14.205 \frac{h}{a^2}\sqrt{\frac{E_2}{\rho}}$$

4.5.3 Clamped Plate Strips

For a plate strip clamped at both ends, the boundary conditions

$$w = 0, \quad \frac{dw}{dx} = 0 \text{ at } x = 0, a$$

lead to the relations

$$c_2 + c_4 = 0, \quad \lambda c_1 + \mu c_3 = 0 \qquad (4.5.19a)$$

$$\begin{bmatrix} \sin \lambda a - \left(\frac{\lambda}{\mu}\right) \sinh \mu a & \cos \lambda a - \cosh \mu a \\ \cos \lambda a - \cosh \mu a & -\sin \lambda a - \left(\frac{\mu}{\lambda}\right) \sinh \mu a \end{bmatrix} \begin{Bmatrix} c_1 \\ c_2 \end{Bmatrix} = \begin{Bmatrix} 0 \\ 0 \end{Bmatrix} \tag{4.5.19b}$$

where relations (4.5.19a) were used to eliminate c_3 and c_4 in writing the relations in (4.5.19b). For nonzero c_1 and c_2, we require the determinant of the coefficient matrix of the above equations to vanish. This leads to the characteristic polynomial

$$-2 + 2\cos \lambda a \cosh \mu a + \left(\frac{\lambda}{\mu} - \frac{\mu}{\lambda}\right) \sin \lambda a \sinh \mu a = 0 \tag{4.5.20}$$

The transcendental equation (4.5.20) needs to be solved iteratively for λ and μ, in conjunction with

$$\lambda^2 = \mu^2 + \frac{b}{a} = \mu^2 + \frac{I_2 \omega^2 - \hat{N}_{xx}}{D} \tag{4.5.21}$$

For natural vibration without rotatory inertia, Eq. (4.5.20) takes the simpler form

$$\cos \lambda a \cosh \lambda a - 1 = 0 \tag{4.5.22}$$

The roots of Eq. (4.5.22) are

$$\lambda_1 a = 4.730, \quad \lambda_2 a = 7.853, \quad \cdots, \quad \lambda_n a \approx (n + \tfrac{1}{2})\pi \tag{4.5.23}$$

In general, the roots of Eq. (4.5.22) are not the same as those of Eq. (4.5.20). If one approximates Eq. (4.5.22) as (4.5.20) (i.e., $\lambda \approx \mu$), the roots in Eq. (4.5.23) can be used to determine the natural frequencies of vibration *with rotary inertia* from Eq. (4.5.17). When rotary inertia is neglected, the frequencies are given by Eq. (4.5.18) with λ given by Eq. (4.5.23). The frequencies obtained from Eq. (4.5.17) with the values of λ from Eq. (4.5.22) are only an approximation of the frequencies with rotary inertia.

For a clamped orthotropic plate with $E_1/E_2 = 25$, $\nu_{12} = 0.25$, $a/h = 10$ and rotatory inertia included, we obtain

$$\omega_1 = \frac{32.169}{a^2} \sqrt{\frac{E_2 h^3}{I_0}} = 32.169 \frac{h}{a^2} \sqrt{\frac{E_2}{\rho}} \tag{4.5.24}$$

4.6 Transient Analysis
4.6.1 Preliminary Comments

In this section, we study the transient response of plate strips. The Rayleigh–Ritz method or Navier's method may be used to reduce the partial differential equation (4.2.3) to an ordinary differential equation in time. In the Navier method, the deflection as well as the load are expanded using the Fourier series. The expansions are selected such that they satisfy all of the boundary conditions of the problem for any choice of the Fourier coefficients. The expansions are then substituted into the governing differential equation to obtain a differential equation in time. The Navier method may be used only for simply supported boundary conditions.

Suppose that the solution $w_0(x,t)$ of Eq. (4.2.3), either exact or approximate, is of the form

$$w_0(x,t) \approx \sum_{n=1}^{N} A_n(t)\varphi_n(x) \qquad (4.6.1)$$

where $A_n(t)$ are parameters to be determined and $\varphi_n(x)$ are known functions of x. In the N-parameter Rayleigh–Ritz solution, $\varphi_n(x)$ represent the approximation functions that are selected to satisfy at least the geometric boundary conditions (see section 2.3.2) of the problem for any time. In the Navier method, φ_n satisfy all of the boundary conditions of the problem. Further, N may be infinity (i.e., Eq. (4.6.1) represents an infinite series). In both methods, an ordinary differential equation of the form

$$M_n \frac{d^2 A_n}{dt^2} + K_n A_n = F_n \qquad (4.6.2)$$

is obtained, which is then solved using initial conditions of the problem. The exact form of the coefficients M_n, K_n, and F_n depend on the method used.

Suppose that we wish to solve the equation

$$D\frac{\partial^4 w_0}{\partial x^4} = q - I_0 \frac{\partial^2 w_0}{\partial t^2} + I_2 \frac{\partial^4 w_0}{\partial x^2 \partial t^2} \qquad (4.6.3)$$

subjected to the simply-supported boundary conditions

$$w_0 = 0, \quad M \equiv -D\frac{\partial^2 w_0}{\partial x^2} = 0 \qquad (4.6.4)$$

168   THEORY AND ANALYSIS OF ELASTIC PLATES

at $x = 0, a$. Here we discuss both the Navier method and the Rayleigh–Ritz method of reducing Eq. (4.6.3) to Eq. (4.6.2).

4.6.2 The Navier Solution

The first step in the Navier solution procedure is to expand the dependent unknown $w_0(x,t)$ in the Fourier series

$$w_0(x,t) = \sum_{n=1}^{\infty} A_n(t) \sin \alpha_n x + \sum_{n=0}^{\infty} B_n(t) \cos \alpha_n x \qquad (4.6.5)$$

where (A_n, B_n) are Fourier coefficients to be determined for any time t, and $\alpha_n = n\pi/a$. The expansion in Eq. (4.6.3) must be such that w_0 satisfies the boundary conditions in Eq. (4.6.4). Clearly, the expansion in (4.6.5) satisfies the boundary conditions (4.6.4) for any A_n but $B_n = 0$.

The next step is to substitute the expansion (4.6.5) into Eq. (4.6.3)

$$\sum_{n=1}^{\infty} \left\{ \left[D\alpha_n^4 \right] A_n(t) + \left[I_0 + I_2 \alpha_n^2 \right] \ddot{A}_n(t) \right\} \sin \alpha_n x = q \qquad (4.6.6)$$

where the superposed dots on A_n denote derivatives with respect to time. Since the left side of Eq. (4.6.4) represents a Fourier sine series, the load q is also expanded in a sine series

$$q(x,t) = \sum_{n=1}^{\infty} Q_n(t) \sin \alpha_n x \qquad (4.6.7)$$

Since q is a known function, it is possible to determine the coefficients Q_n from the equation

$$Q_n(t) = \frac{2}{a} \int_0^a q(x,t) \sin \alpha_n x \, dx \qquad (4.6.8)$$

For example, if the beam is suddenly subjected to uniformly distributed load of intensity q_0 at time $t = 0$, we have

$$Q_n = \frac{2}{a} \int_0^a q_0 \sin \alpha_n x \, dx = \frac{4q_0}{a\alpha_n} \quad \text{when } n = 1, 3, 5, \cdots \qquad (4.6.9)$$

and $Q_n = 0$ when n is even.

Substituting the load expansion (4.6.7) into Eq. (4.6.6), we obtain

$$\sum_{n=1}^{\infty} \left[K_n A_n + M_n \ddot{A}_n - F_n \right] \sin \alpha_n x = 0 \qquad (4.6.10a)$$

where K_n, M_n, and F_n are the parameters (independent of x)

$$K_n = D\alpha_n^4, \quad M_n = I_0 + I_2 \alpha_n^2, \quad F_n = Q_n \qquad (4.6.10b)$$

Since Eq. (4.6.10a) must hold for every x and $\sin \alpha_n x$ is not zero for all x, it follows that the coefficient in the square brackets of Eq. (4.6.10a) must be zero:

$$M_n \frac{d^2 A_n}{dt^2} + K_n A_n - F_n = 0 \qquad (4.6.11)$$

This completes the application of the Navier solution approach. Note that if there is no time dependence, we set the term containing $\ddot{A}_n$ to zero and solve for $A_n = F_n/K_n$. Then the static solution is given by (which is exact)

$$w_0(x) = \sum_{n=1}^{\infty} \frac{F_n}{D\alpha_n^4} \sin \alpha_n x \qquad (4.6.12)$$

4.6.3 The Rayleigh–Ritz Solution

In the Rayleigh–Ritz method the approximation functions $\varphi_n(x)$ are chosen to satisfy the geometric boundary conditions $w_0(0,t) = w_0(a,t) = 0$. If we were to choose trigonometric functions, we set

$$w_0(x,t) \approx W_N(x,t) = \sum_{n=1}^{N} A_n(t) \varphi_n(x) \qquad (4.6.13a)$$

$$\varphi_n(x) = \sin \alpha_n x, \quad \alpha_n = \frac{n\pi}{a} \qquad (4.6.13b)$$

The function $\varphi_0(x)$ is zero in the present case because the geometric boundary conditions are homogeneous.

Next, we substitute the approximation W_N in Eq. (4.6.13a) into the statement of the principle of virtual displacements or the principle of the minimum total potential energy and determine the equation governing $A_n(t)$. The expression for the total virtual work is given by

$$\delta W = \int_0^T \int_0^a \left(D \frac{\partial^2 W_N}{\partial x^2} \frac{\partial^2 \delta W_N}{\partial x^2} - q W_N - I_0 \dot{W}_N \delta \dot{W}_N \right.$$
$$\left. - I_2 \frac{\partial^2 W_N}{\partial x \partial t} \frac{\partial^2 \delta W_N}{\partial x \partial t} \right) dx dt \qquad (4.6.14)$$

Substituting Eq. (4.6.13a) into Eq. (4.6.14) and invoking the principle of virtual displacements $\delta W = 0$, we obtain

$$0 = \int_0^T \left[\sum_{n=1}^N \left(M_{mn} \frac{d^2 A_n}{dt^2} + K_{mn} A_n \right) - F_n \right] dt \qquad (4.6.15a)$$

for $m = 1, 2, \cdots, N$, where

$$K_{mn} = \int_0^a D \frac{\partial^2 \varphi_m}{\partial x^2} \frac{\partial^2 \varphi_n}{\partial x^2} dx$$
$$= \begin{cases} \frac{a D \alpha_n^4}{2} & \text{for } m = n \\ 0 & \text{for } m \neq n \end{cases}$$

$$M_{mn} = \int_0^a \left(I_0 \varphi_m \varphi_n + I_2 \frac{\partial \varphi_m}{\partial x} \frac{\partial \varphi_n}{\partial x} \right) dx$$
$$= \begin{cases} \frac{a}{2} (I_0 + I_2 \alpha_n^2) & \text{for } m = n \\ 0 & \text{for } m \neq n \end{cases}$$

$$F_m = \int_0^a q \varphi_m dx \qquad (4.6.15b)$$

Since Eq. (4.6.15a) must hold for all times $t > 0$, we have

$$\sum_{n=1}^N \left(M_{mn} \frac{d^2 A_n}{dt^2} + K_{mn} A_n \right) = F_m \quad \text{for } m = 1, 2, \cdots, N \qquad (4.6.16)$$

Equation (4.6.16) is similar to Eq. (4.6.11), except that Eq. (4.6.16) is a matrix equation, which, in view of the fact that the off-diagonal values are zero [see Eq. (4.6.15b)], reduces to a set of N uncoupled equations similar to Eq. (4.6.11):

$$M_n \frac{d^2 A_n}{dt^2} + K_n A_n = F_n \quad \text{no sum on } n \qquad (4.6.17)$$

ANALYSIS OF PLATE STRIPS   171

for $n = 1, 2, \cdots, N$, where we set $M_{nn} = M_n$ and $K_{nn} = K_n$. When $N = \infty$, Eqs. (4.6.11) and (4.6.17) are the same.

4.6.4 Transient Response

Equation (4.6.11) or (4.6.17) represents a second-order ordinary differential equation in time for each n. The general solution of (4.6.11) is given by

$$A_n(t) = A_n^0 \sin \lambda_n t + B_n^0 \cos \lambda_n t + \frac{F_n}{K_n}, \quad \lambda_n^2 = \frac{K_n}{M_n} \qquad (4.6.18)$$

where A_n^0 and B_n^0 are constants to be determined using the initial conditions of the problem. The complete solution is then given by

$$w_0(x,t) = \sum_{n=1}^{\infty} \left[A_n^0 \sin \lambda_n t + B_n^0 \cos \lambda_n t + \frac{Q_n}{K_n} \right] \sin \alpha_n x \qquad (4.6.19)$$

In the case of the Rayleigh–Ritz solution, we write

$$W_N(x,t) = \sum_{n=1}^{N} \left[A_n^0 \sin \lambda_n t + B_n^0 \cos \lambda_n t + \frac{F_n}{K_n} \right] \sin \alpha_n x \qquad (4.6.20)$$

Let the initial conditions on the displacement and velocity be

$$w_0(x,0) = d_0(x) = \sum_{n=1}^{\infty} D_n^0 \sin \alpha_n x, \quad \frac{\partial w_0}{\partial t}(x,0) = v_0 = \sum_{n=1}^{\infty} V_n^0 \sin \alpha_n x$$
$$(4.6.21)$$

where D_n^0 and V_n^0 have the same form as Q_n in Eq. (4.6.8). These conditions give

$$B_n^0 = D_n^0 - \frac{F_n}{K_n}, \quad A_n^0 \lambda_n = V_n^0 - \frac{1}{K_n} \frac{dF_n}{dt} \qquad (4.6.22)$$

Then the solution becomes

$$w_0(x,t) = \sum_{n=1}^{\infty} \left[D_n^0 \sin \lambda_n t + \frac{V_n^0}{\lambda_n} \cos \lambda_n t \right.$$
$$\left. + \frac{F_n}{K_n}(1 - \cos \lambda_n t) - \frac{1}{K_n} \frac{dF_n}{dt} \sin \lambda_n t \right] \sin \alpha_n x \quad (4.6.23)$$

172 THEORY AND ANALYSIS OF ELASTIC PLATES

When the initial deflection and velocity are zero, we obtain

$$w_0(x,t) = \sum_{n=1}^{\infty} \left[\frac{Q_n}{K_n}(1 - \cos \lambda_n t) - \frac{1}{K_n}\frac{dQ_n}{dt}\sin \lambda_n t \right] \sin \alpha_n x \quad (4.6.24)$$

For a beam simply supported at both ends and subjected to a suddenly applied uniformly distributed load $q = q_0 H(t)$, the solution becomes

$$w_0(x,t) = \frac{4q_0}{\pi}\sum_{n=1}^{\infty}\frac{1}{nK_n}(1 - \cos \lambda_n t)\sin \alpha_n x \quad (4.6.25)$$

for n odd. The static deflection can be obtained by setting the expression $(1 - \cos \lambda_n t)$ to unity.

Equation (4.6.11) can also be solved using the Laplace transform method. The Laplace transform of a function $f(t)$ is defined by

$$\overline{f}(s) \equiv \mathcal{L}[f(t)] = \int_0^\infty e^{-st} f(t)\,dt \quad (4.6.26)$$

The Laplace transform (or inverse transform) of some typical functions are given in Table 4.6.1. To solve Eq. (4.6.11) by the Laplace transform method, let $\mathcal{L}[A_n(t)] = \overline{A}_n(s)$. Then Eq. (4.6.11) transforms to

$$M_n\left(s^2\overline{A}_n(s) - sA_n(0) - \dot{A}_n(0)\right) + K_n\overline{A}_n(s) - \overline{F}_n(s) = 0 \quad (4.6.27a)$$

or

$$(s^2 + \lambda_n^2)\overline{A}_n(s) = \left(sA_n(0) + \dot{A}_n(0) + \frac{\overline{F}_n}{M_n}\right) \quad (4.6.27b)$$

Now suppose that the applied load is uniformly distributed step loading. Then

$$\overline{F}_n(s) = \left(\frac{4q_0}{a\alpha_n}\right)\frac{1}{s} \quad (4.6.28)$$

Substituting the above expression into Eq. (4.6.27b) and using the inverse Laplace transform (see Table 4.6.1), we obtain

$$A_n(t) = \left[A_n(0)\cos \lambda_n t + \frac{\dot{A}_n(0)}{\lambda_n}\sin \lambda_n t\right] + \left(\frac{4q_0}{a\alpha_n K_n}\right)(1 - \cos \lambda_n t) \quad (4.6.29)$$

Table 4.6.1. The Laplace transform pairs of typical functions[†].

$F(t)$	$\overline{F}(s)$	$F(t)$	$\overline{F}(s)$
1	$\frac{1}{s}$	$\frac{df}{dt}$	$s\overline{f}(s) - f(0)$
e^{at}	$\frac{1}{s-a}$	$\frac{d^2 f}{dt^2}$	$s^2\overline{f}(s) - sf(0) - f'(0)$
$\sin at$	$\frac{a}{s^2+a^2}$	$\cos at$	$\frac{s}{s^2+a^2}$
$\frac{1}{a^2}(1-\cos at)$	$\frac{1}{s(s^2+a^2)}$	$\frac{1}{a^3}(at-\sin at)$	$\frac{1}{s^2(s^2+a^2)}$
$\frac{1}{2a}\sin at$	$\frac{s}{(s^2+a^2)^2}$	$\frac{1}{2a^3}(\sin at - at\cos at)$	$\frac{1}{(s^2+a^2)^2}$
$t\cos at$	$\frac{s^2-a^2}{(s^2+a^2)^2}$	$\frac{1}{2a}(\sin at + at\cos at)$	$\frac{s^2}{(s^2+a^2)^2}$
$\frac{1}{b}e^{at}\sin bt$	$\frac{1}{(s-a)^2+b^2}$	$\frac{\cos at - \cos bt}{b^2-a^2}$	$\frac{s}{(s^2+a^2)(s^2+b^2)}$
$\sinh at$	$\frac{a}{s^2-a^2}$	$\cosh at$	$\frac{s}{s^2-a^2}$
$t^n f(t)$	$(-)^n \frac{d\overline{f}}{ds}$	$tf(t)$	$-\frac{d}{ds}\overline{f}(s)$
$\frac{1}{t}f(t)$	$\int_s^\infty \overline{f}(r)dr$	$\int_0^t f(\tau)d\tau$	$\frac{1}{s}\overline{f}(s)$
$e^{at}f(t)$	$\overline{f}(s-a)$	$f*g$	$\overline{f}(s)\overline{g}(s)$

[†] a and b are constants; $f*g = \int_0^t f(\tau)g(t-\tau)d\tau$.

where we used the identity

$$\frac{1}{s(s^2+\lambda_n^2)} = \frac{1}{\lambda_n^2}\left(\frac{1}{s} - \frac{s}{s^2+\lambda_n^2}\right)$$

For $A_n(0) = D_n^0$ and $\dot{A}_n(0) = V_n^0$, Eq. (4.6.29) becomes

$$A_n(t) = \left[D_n^0 \cos\lambda_n t + \frac{V_n^0}{\lambda_n}\sin\lambda_n t\right] + \left(\frac{4q_0}{a\alpha_n K_n}\right)(1-\cos\lambda_n t) \quad (4.6.30)$$

If the initial conditions are zero, then $D_n^0 = V_n^0 = 0$ and $A_n(t)$ become

$$A_n(t) = \left(\frac{4q_0}{a\alpha_n K_n}\right)(1-\cos\lambda_n t) \quad (4.6.31)$$

Equation (4.6.31) is the same as the solution obtained earlier. The Laplace transform method can also be applied to a set of (matrix) differential equations.

Exercises

4.1 Consider a simply supported plate strip under point loads F_0 at $x = a/4$ and $x = 3a/4$ (the so-called *four-point bending*). Use the symmetry about $x = a/2$ to determine the deflection $w_0(x)$ using the classical plate theory. (*Ans:* The maximum deflection is $w_{max} = -11F_0 a^3/384D$.)

4.2 Determine the static deflection of a clamped plate strip under uniformly distributed load q_0 and a center point load F_0 using the classical plate theory.

4.3 Determine the static deflection of a plate strip with elastically built-in edges and under uniformly distributed load q_0 using the classical plate theory. Assume that there is no axial force. Note that the boundary conditions in Eq. (4.3.13) are now replaced with

$$w_0 = 0, \quad \frac{dw_0}{dx} = -\beta M_0$$

where β is a constant that defines the degree of the rigidity of the restraint against rotation. The larger the value of β, the more flexible is the restraint, and the boundary condition approaches that of a simply supported edge.

4.4 Repeat Exercise 4.3 for the case in which there is an axial force.

4.5 Show that the critical buckling load of a clamped-free plate strip using the classical plate theory is given by $N_{cr} = D\pi^2/(4a^2)$.

4.6 Show that the characteristic equation governing the buckling of a clamped-hinged plate strip using the classical plate theory is given by

$$\sin \lambda a - \lambda a \cos \lambda a = 0$$

4.7 Show that the characteristic equation governing the natural vibration of a clamped-free plate strip using the classical plate theory is given by

$$\cos \lambda a \cosh \lambda a + 1 = 0$$

4.8 Show that the characteristic equation governing the natural vibration of a clamped-hinged plate strip using the classical plate theory when rotary inertia is neglected is given by

$$\sin \lambda a \cosh \lambda a - \cos \lambda a \sinh \lambda a = 0$$

4.9 Derive the characteristic equation governing the natural vibration of a clamped-hinged plate strip using the classical plate theory when rotary inertia is *not* neglected.

4.10 Use the total potential energy functional

$$\Pi(w_0) = \frac{1}{2}\int_0^a \left[D\left(\frac{d^2w_0}{dx^2}\right)^2 - N_{xx}^0\left(\frac{dw_0}{dx}\right)^2 - I_0\omega^2 w_0^2 \right] dx$$

to construct a one-parameter Rayleigh–Ritz solution to determine the natural frequency of vibration, ω, of a *simply supported* plate strip with compressive load N_{xx}^0. Use algebraic polynomials for the approximate functions. (*Ans*: $\omega = (1/a)\sqrt{(10/I_0)[(12D/a^2) - N_{xx}^0]}$.)

4.11 Repeat Exercise 4.10 for fixed-free boundary conditions (fixed at $x = 0$ and free at $x = a$). (*Ans*: $\omega = (1/a)\sqrt{(5/3I_0)[(12D/a^2) - 4N_{xx}^0]}$.)

4.12 Use the total potential energy functional

$$\Pi(w_0) = \int_0^a \left[\frac{D_{11}}{2}\left(\frac{d^2w_0}{dx^2}\right)^2 - qw_0 \right] dx$$

to construct a one-parameter Ritz solution of w_0 for a simply supported plate strip. Use algebraic polynomials for the approximate functions. (*Ans*: $c_1 = q_0 a^2/24 D_{11}$.)

4.13 Repeat Exercise 4.12 for a plate strip with clamped boundary conditions at $x = 0$ and free boundary conditions at $x = a$. (*Ans*: $c_1 = q_0 a^2/12 D_{11}$.)

4.14 Use the total potential energy functional

$$\Pi(w) = \frac{1}{2}\int_0^a \left[D_{11}\left(\frac{d^2w}{dx^2}\right)^2 - N_{xx}^0\left(\frac{dw}{dx}\right)^2 \right] dx$$

to construct a one-parameter (for each variable) Ritz solution to determine the critical buckling load N_{cr} of a plate strip with clamped boundary conditions at $x = 0$ and free boundary conditions at $x = a$. Use algebraic polynomials for the approximate functions. (*Ans*: $N_{cr} = 3D_{11}/a^2$.)

4.15 Use the total potential energy functional

$$\Pi(w_0) = \frac{1}{2}\int_0^a \left[D_{11}\left(\frac{d^2w}{dx^2}\right)^2 - N_{xx}^0\left(\frac{dw}{dx}\right)^2 - I_0\omega^2 w^2 \right] dx$$

to construct a one-parameter (for each variable) Ritz solution to determine the natural frequency of vibration, ω, of a simply supported

plate strip with edge compressive load N_{xx}^0. Use algebraic polynomials for the approximate functions.
(Ans: $\omega = (1/a)\sqrt{(10/I_0)[(12D_{11}/a^2) - N_{xx}^0]}$.)

4.16 Repeat Exercise 4.15 for a plate strip with clamped boundary condition at $x = 0$ and free at $x = a$.
(Ans: $\omega = (1/a)\sqrt{(20/3I_0)[(3D_{11}/a^2) - N_{xx}^0]}$.)

4.17 The deflection, bending moment, and shear force of the first order plate theory for cylindrical bending can be expressed in terms of the corresponding quantities of the classical plate theory. In order to establish these relationships, we use the following equations of the two theories:

$$-D_{11}\frac{d^4w_0^c}{dx^4} = -q(x), \quad D_{11}\frac{d^3\phi^s}{dx^3} = -q(x)$$

$$GhK_s\left(\frac{d\phi^s}{dx} + \frac{d^2w_0^s}{dx^2}\right) = -q(x)$$

where K_s is the shear correction coefficient, and superscripts c and s on variables refer to the classical and shear deformation plate theories. Show that

$$D_{11}w_0^s(x) = D_{11}w_0^c(x) + \frac{D_{11}}{GhK_s}M_{xx}^c(x) + C_1\left(\frac{D_{11}}{GhK_s}x - \frac{x^3}{6}\right)$$
$$- C_2\frac{x^2}{2} - C_3x - C_4 \quad (1)$$

$$D_{11}\phi^s(x) = -D_{11}\frac{dw_0^c}{dx} + C_1\frac{x^2}{2} + C_2x + C_3 \quad (2)$$

$$M_{xx}^s(x) = M_{xx}^c(x) + C_1x + C_2 \quad (3)$$

$$Q_x^s(x) = Q_x^c(x) + C_1 \quad (4)$$

where h is the plate thickness and $C_1, C_2, C_3,$ and C_4 are constants of integration, which are to be determined using the boundary conditions of the particular beam.

4.18 Show that for simply supported beams all C_i of Exercise 4.17 are zero.

4.19 Show that for cantilevered beams all C_i except C_4 of Exercise 4.17 are zero [$C_4 = M_{xx}^c(0)D_{11}/(GhK_s)$].

4.20 Consider the bending of a plate strip of length a, clamped (or fixed) at the left end and simply supported at the right, that is subjected to a uniformly distributed transverse load q_0. The boundary conditions of the classical (CPT) and shear deformation (SDT) theories for the problem are as follows:

$$\text{CPT}: \quad w_0^c(0) = w_0^c(a) = \frac{dw_0^c}{dx}(0) = M_{xx}^c(a) = 0 \quad (1)$$

$$\text{SDT}: \quad w_0^s(0) = w_0^s(a) = \phi^s(0) = M_{xx}^s(a) = 0 \quad (2)$$

Show that the constants of integration in Exercise 4.14 are given by

$$C_1 = \frac{3\Omega}{(1+3\Omega)L}[M_{xx}^c(0) - M_{xx}^c(a)]$$

$$C_2 = -C_1 a, \quad C_3 = 0, \quad C_4 = \Omega M_{xx}^c(0)a^2 \quad (3)$$

where $\Omega = D_{11}/(GhK_s a^2)$.

References for Additional Reading

1. Clough, R. W. and Penzien, J., *Dynamics of Structures*, McGraw–Hill, New York (1975).
2. Pipes, L. A. and Harvill, L. R., *Applied Mathematics for Engineers and Physicists*, Third Edition, McGraw–Hill, New York (1970).
3. Reddy, J. N., *Energy and Variational Methods in Applied Mechanics*, John Wiley, New York (1984).
4. Reddy, J. N., *Mechanics of Laminated Composite Plates: Theory and Analysis*, CRC Press, Boca Raton, FL (1997).
5. Timoshenko, S. P. and Gere, J. P., *Theory of Elastic Stability*, Second Edition, McGraw–Hill, New York (1959).
6. Timoshenko, S. P., "On the Transverse Vibrations of Bars of Uniform Cross Section", *Philosophical Magazine*, **43**, 125–131 (1922).
7. Timoshenko, S. P. and Woinowsky-Krieger, S., *Theory of Plates and Shells*, McGraw–Hill, Singapore (1970).
8. Weaver, W., Jr., Timoshenko, S. P., and Young, D. H., *Vibration Problems in Engineering*, Fifth Edition, John Wiley, New York (1990).
9. Wang, C. M., "Timoshenko Beam-Bending Solutions in Terms of Euler-Bernoulli Solutions", *Journal of Engineering Mechanics*, ASCE, **121**(6), 763–765 (1995).

10. Reddy, J. N., Wang, C. M., and Lee, K. H., "Relationships Between Bending Solutions of Classical and Shear Deformation Beam Theories", *International Journal of Solids and Structures,* **34**(26), 3373–3384 (1997).
11. Wang, C. M. and Reddy, J. N., "Buckling Load Relationship Between Reddy and Kirchhoff Plates of Polygonal Shape with Simply Supported Edges", *Mechanics Research Communications,* **24**(1), 103–108 (1997).
12. Reddy, J. N., Wang, C. M., and Lam, K. Y., "Unified Beam Finite Elements Based on the Classical and Shear Deformation Theories of Beams and Axisymmetric Circular Plates", *Communications in Numerical Methods in Engineering,* **13**, 495–510 (1997).
13. Reddy, J. N., "On Locking–Free Shear Deformable Beam Finite Elements", *Computer Methods in Applied Mechanics and Engineering,* **149**, 113–132 (1997).
14. Reddy, J. N., "On the Dynamic Behavior of the Timoshenko Beam Finite Elements", *Sadhana* (Journal of the Indian Academy of Sciences), to appear.

Chapter Five

Analysis of Circular Plates

5.1 Introduction

Circular plates are found in end plates, manhole covers, closures of pressure vessels, pump diaphragms, clutches, and turbine disks, to name a few. Therefore, it is of interest to study stretching and bending of circular plates. The state of rotational symmetry (or axisymmetry with respect to the z-axis, which is taken normal to the plane of the plate with the origin at the center of the plate) exists (i.e., deformation is independent of the angular coordinate) only when the boundary conditions, material properties, and loading are also rotationally symmetric. Here we consider bending of circular plates under both axisymmetric and asymmetric transverse loads.

5.2 Governing Equations

5.2.1 Transformation of Equations from Rectangular Coordinates to Polar Coordinates

The equations governing circular plates may be obtained either (a) using the transformation between the polar coordinates (r, θ) and the rectangular Cartesian coordinates (x, y), or (b) by direct derivation in the polar coordinates. The former is mathematical while the latter provides more physical insight. To illustrate the first (transformation) method, we consider the Poisson equation

$$\nabla^2 u = f \qquad (5.2.1)$$

governing a function u in a two-dimensional domain with a source f. Eq. (5.2.1) is valid in any coordinate system, and we only have to define ∇^2 in that coordinate system. In the rectangular coordinates, we have

180 THEORY AND ANALYSIS OF ELASTIC PLATES

[see Eq. (1.2.22)]

$$\begin{aligned}\nabla^2 &= \nabla \cdot \nabla \\ &= \left(\hat{\mathbf{e}}_x \frac{\partial}{\partial x} + \hat{\mathbf{e}}_y \frac{\partial}{\partial y}\right) \cdot \left(\hat{\mathbf{e}}_x \frac{\partial}{\partial x} + \hat{\mathbf{e}}_y \frac{\partial}{\partial y}\right) \\ &= \frac{\partial^2}{\partial x^2} + \frac{\partial^2}{\partial y^2}\end{aligned} \quad (5.2.2)$$

and in the polar coordinates, we have [see Eqs. (1.2.30)–(1.2.33)]

$$\begin{aligned}\nabla^2 &= \left(\hat{\mathbf{e}}_r \frac{\partial}{\partial r} + \frac{\hat{\mathbf{e}}_\theta}{r} \frac{\partial}{\partial \theta}\right) \cdot \left(\hat{\mathbf{e}}_r \frac{\partial}{\partial r} + \frac{\hat{\mathbf{e}}_\theta}{r} \frac{\partial}{\partial \theta}\right) \\ &= \frac{1}{r} \frac{\partial}{\partial r}\left(r \frac{\partial}{\partial r}\right) + \frac{1}{r^2} \frac{\partial^2}{\partial \theta^2}\end{aligned} \quad (5.2.3)$$

Therefore, the Poisson equation (5.2.1) has the form

$$\frac{\partial^2 u}{\partial x^2} + \frac{\partial^2 u}{\partial y^2} = f(x,y) \quad (5.2.4a)$$

in the rectangular coordinate system, and

$$\frac{1}{r} \frac{\partial}{\partial r}\left(r \frac{\partial u}{\partial r}\right) + \frac{1}{r^2} \frac{\partial^2 u}{\partial \theta^2} = f(r,\theta) \quad (5.2.4b)$$

in the polar coordinate system (see Figure 5.2.1).

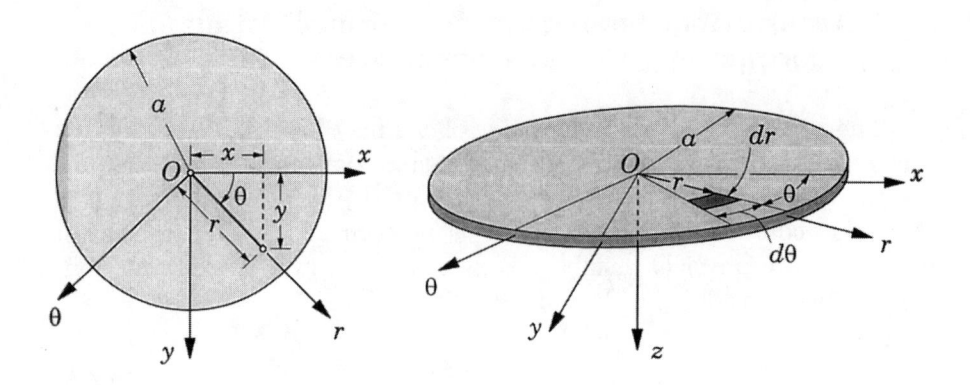

Figure 5.2.1. Transformation between rectangular and polar coordinate systems.

ANALYSIS OF CIRCULAR PLATES   181

The result in Eq. (5.2.3) can also be obtained from Eq. (5.2.2) using the transformation relations [see Eq. (1.2.23) and Figure 5.2.1]

$$x = r\cos\theta, \quad y = r\sin\theta \tag{5.2.5a}$$

$$\frac{\partial}{\partial x} = \frac{\partial r}{\partial x}\frac{\partial}{\partial r} + \frac{\partial\theta}{\partial x}\frac{\partial}{\partial\theta}, \quad \frac{\partial}{\partial y} = \frac{\partial r}{\partial y}\frac{\partial}{\partial r} + \frac{\partial\theta}{\partial y}\frac{\partial}{\partial\theta} \tag{5.2.5b}$$

where [from $r^2 = x^2 + y^2$ and Eqs. (5.2.5a,b)]

$$\frac{\partial r}{\partial x} = \frac{x}{r} = \cos\theta, \quad \frac{\partial r}{\partial y} = \frac{y}{r} = \sin\theta \tag{5.2.6a}$$

$$\frac{\partial\theta}{\partial x} = -\frac{\sin\theta}{r}, \quad \frac{\partial\theta}{\partial y} = \frac{\cos\theta}{r} \tag{5.2.6b}$$

Hence

$$\frac{\partial}{\partial x} = \cos\theta\frac{\partial}{\partial r} - \frac{\sin\theta}{r}\frac{\partial}{\partial\theta}, \quad \frac{\partial}{\partial y} = \sin\theta\frac{\partial}{\partial r} + \frac{\cos\theta}{r}\frac{\partial}{\partial\theta} \tag{5.2.7}$$

The second derivatives with respect to x and y can now be computed using the above expression

$$\frac{\partial^2}{\partial x^2} = \cos\theta\frac{\partial}{\partial r}\left(\frac{\partial}{\partial x}\right) - \frac{\sin\theta}{r}\frac{\partial}{\partial\theta}\left(\frac{\partial}{\partial x}\right)$$

$$= \cos\theta\frac{\partial}{\partial r}\left(\cos\theta\frac{\partial}{\partial r} - \frac{\sin\theta}{r}\frac{\partial}{\partial\theta}\right) - \frac{\sin\theta}{r}\frac{\partial}{\partial\theta}\left(\cos\theta\frac{\partial}{\partial r} - \frac{\sin\theta}{r}\frac{\partial}{\partial\theta}\right)$$

$$= \cos^2\theta\frac{\partial^2}{\partial r^2} + \frac{\sin^2\theta}{r}\frac{\partial}{\partial r} - \frac{\sin 2\theta}{r}\frac{\partial^2}{\partial r\partial\theta} + \frac{\sin 2\theta}{r^2}\frac{\partial}{\partial\theta} + \frac{\sin^2\theta}{r^2}\frac{\partial^2}{\partial\theta^2} \tag{5.2.8}$$

$$\frac{\partial^2}{\partial y^2} = \sin^2\theta\frac{\partial^2}{\partial r^2} + \frac{\cos^2\theta}{r}\frac{\partial}{\partial r} + \frac{\sin 2\theta}{r}\frac{\partial^2}{\partial r\partial\theta} - \frac{\sin 2\theta}{r^2}\frac{\partial}{\partial\theta} + \frac{\cos^2\theta}{r^2}\frac{\partial^2}{\partial\theta^2} \tag{5.2.9}$$

$$\frac{\partial^2}{\partial x\partial y} = \frac{\sin 2\theta}{2}\frac{\partial^2}{\partial r^2} - \frac{\sin 2\theta}{2r}\frac{\partial}{\partial r} + \frac{\cos 2\theta}{r}\frac{\partial^2}{\partial r\partial\theta} - \frac{\cos 2\theta}{r^2}\frac{\partial}{\partial\theta} - \frac{\sin 2\theta}{2r^2}\frac{\partial^2}{\partial\theta^2} \tag{5.2.10}$$

Adding Eqs. (5.2.8) and (5.2.9), we arrive at the result in Eq. (5.2.3).

With the above relations in hand, one can write the equation of equilibrium governing the linear bending of an isotropic plate on elastic foundation of modulus k [see Eq. (3.8.14)]

$$D\nabla^4 w_0 + k w_0 = q \tag{5.2.11}$$

where w_0 is the transverse deflection, q the distributed transverse load, and D the flexural rigidity, as

$$D\left(\frac{\partial^4 w_0}{\partial x^4} + 2\frac{\partial^4 w_0}{\partial x^2 \partial y^2} + \frac{\partial^4 w_0}{\partial y^4}\right) + k w_0 = q(x,y) \tag{5.2.12}$$

in the Cartesian rectangular coordinate system and as

$$D\left[\frac{1}{r}\frac{\partial}{\partial r}\left(r\frac{\partial}{\partial r}\right) + \frac{1}{r^2}\frac{\partial^2}{\partial \theta^2}\right]\left[\frac{1}{r}\frac{\partial}{\partial r}\left(r\frac{\partial w_0}{\partial r}\right) + \frac{1}{r^2}\frac{\partial^2 w_0}{\partial \theta^2}\right] + k w_0 = q(r,\theta) \tag{5.2.13}$$

in the Cartesian polar coordinate system. The former equation is more suitable for polygonal plates, while the latter equation is more suitable for circular plates.

For axisymmetric deformation, Eq. (5.2.13) reduces to

$$D\left[\frac{1}{r}\frac{\partial}{\partial r}\left(r\frac{\partial}{\partial r}\right)\right]\left[\frac{1}{r}\frac{\partial}{\partial r}\left(r\frac{\partial w_0}{\partial r}\right)\right] + k w_0(r) = q(r), \quad b < r < a \tag{5.2.14}$$

where r is taken radially outward from the center of the circular plate, and a and b denote the outer and inner radii of an annular plate. For solid circular plates, we set $b = 0$.

In order to derive the strain-displacement equations, equations of motion, and plate constitutive equations in polar coordinates from corresponding equations in the rectangular coordinates, one must first use the transformation relations between the strain components and stress components (hence, force and moment resultants) of the two coordinate systems. Then the transformation equations (5.2.7) through (5.2.10) may be used to replace the derivatives with respect to x and y in terms of the derivatives with respect to r and θ. Thus, the procedure involves quite a bit of mathematical manipulation. For example, the moment-displacement and shear force-displacement relations for an

ANALYSIS OF CIRCULAR PLATES   183

isotropic plate are given by (see Figure 5.2.2)

$$M_{rr} = -D\left[\frac{\partial^2 w_0}{\partial r^2} + \nu\left(\frac{1}{r}\frac{\partial w_0}{\partial r} + \frac{1}{r^2}\frac{\partial^2 w_0}{\partial \theta^2}\right)\right] \quad (5.2.15a)$$

$$M_{\theta\theta} = -D\left[\nu\frac{\partial^2 w_0}{\partial r^2} + \frac{1}{r}\frac{\partial w_0}{\partial r} + \frac{1}{r^2}\frac{\partial^2 w_0}{\partial \theta^2}\right] \quad (5.2.15b)$$

$$M_{r\theta} = -(1-\nu)D\left(\frac{1}{r}\frac{\partial^2 w_0}{\partial r \partial \theta} - \frac{1}{r^2}\frac{\partial w_0}{\partial \theta}\right) \quad (5.2.15c)$$

$$Q_r = -D\frac{\partial}{\partial r}\left(\nabla^2 w_0\right), \quad Q_\theta = -D\frac{1}{r}\frac{\partial}{\partial \theta}\left(\nabla^2 w_0\right) \quad (5.2.16)$$

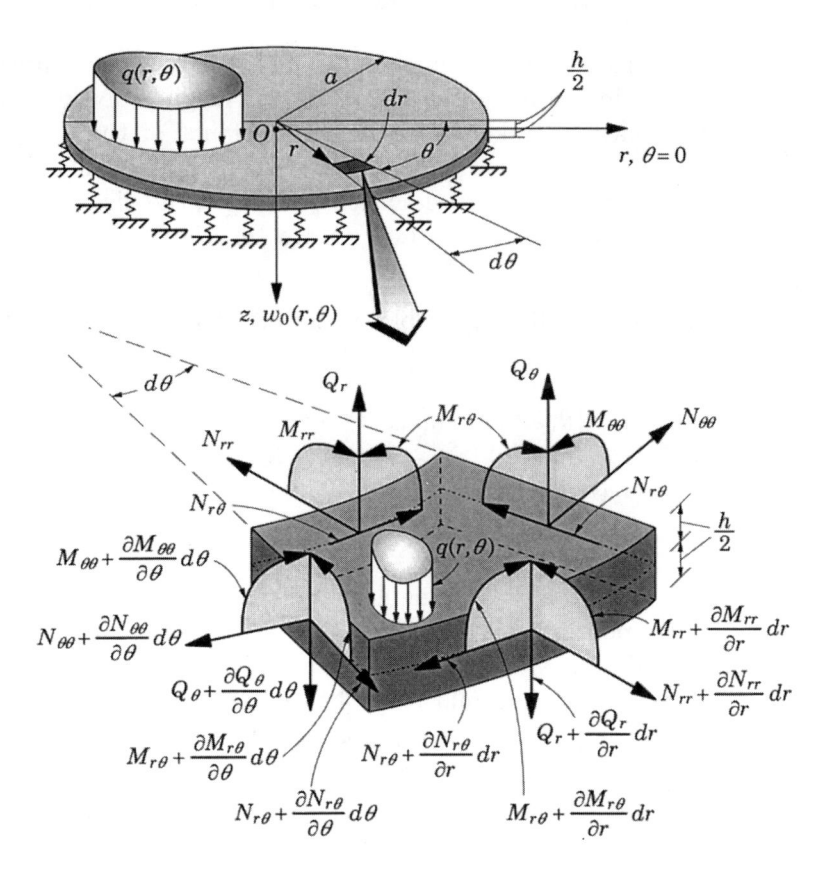

Figure 5.2.2. Moment and shear force resultants on an element of a circular plate.

184   THEORY AND ANALYSIS OF ELASTIC PLATES

The effective transverse shear forces for fixed r and θ are given by V_r and V_θ, respectively:

$$V_r \equiv Q_r + \frac{1}{r}\frac{\partial M_{r\theta}}{\partial \theta}$$

$$= -D\left[\frac{\partial}{\partial r}\left(\nabla^2 w_0\right) + (1-\nu)\frac{1}{r}\frac{\partial}{\partial \theta}\left(\frac{1}{r}\frac{\partial^2 w_0}{\partial r \partial \theta} - \frac{1}{r^2}\frac{\partial w_0}{\partial \theta}\right)\right] \quad (5.2.17a)$$

$$V_\theta \equiv Q_\theta + \frac{\partial M_{r\theta}}{\partial r}$$

$$= -D\left[\frac{1}{r}\frac{\partial}{\partial \theta}\left(\nabla^2 w_0\right) + (1-\nu)\frac{\partial}{\partial r}\left(\frac{1}{r}\frac{\partial^2 w_0}{\partial r \partial \theta} - \frac{1}{r^2}\frac{\partial w_0}{\partial \theta}\right)\right] \quad (5.2.17b)$$

5.2.2 Direct Derivation of Equations in Polar Coordinates

Let the r-coordinate be taken radially outward from the center of the plate, z-coordinate along the thickness (or height) of the plate, and the θ-coordinate be taken along a circumference of the plate, as shown in Figure 5.2.2. In a general case where applied loads and geometric boundary conditions are not axisymmetric, the displacements (u_r, u_θ, u_z) along the coordinates (r, θ, z) are functions of $r, \theta,$ and z coordinates.

We begin with the following displacement field for the classical plate theory (CPT):

$$u_r(r,\theta,z,t) = u_0(r,\theta,t) - z\frac{\partial w_0}{\partial r}$$

$$u_\theta(r,\theta,z,t) = v_0(r,\theta,t) - z\left(\frac{1}{r}\frac{\partial w_0}{\partial \theta}\right)$$

$$u_z(r,\theta,z,t) = w_0(r,\theta,t) \quad (5.2.18)$$

where (u_0, v_0, w_0) are the radial, angular, and transverse displacements, respectively, of a point on the midplane (i.e., $z = 0$) of the plate. The displacement field (5.2.18) is based on the Kirchhoff hypothesis that straight lines normal to the mid-plane before deformation remain (i) in-extensible, (ii) straight, and (iii) normal to the mid-surface after deformation.

The nonlinear strains are given by Eq. (1.4.6). Accounting only for the von Kármán non-linearity, we obtain

$$\varepsilon_{rr} = \frac{\partial u_r}{\partial r} + \frac{1}{2}\left(\frac{\partial u_z}{\partial r}\right)^2$$

ANALYSIS OF CIRCULAR PLATES

$$\varepsilon_{\theta\theta} = \frac{u_r}{r} + \frac{1}{r}\frac{\partial u_\theta}{\partial \theta} + \frac{1}{2}\left(\frac{1}{r}\frac{\partial u_z}{\partial \theta}\right)^2$$

$$\varepsilon_{zz} = \frac{\partial u_z}{\partial z}$$

$$\varepsilon_{r\theta} = \frac{1}{2}\left(\frac{1}{r}\frac{\partial u_r}{\partial \theta} + \frac{\partial u_\theta}{\partial r} - \frac{u_\theta}{r}\right)$$

$$\varepsilon_{z\theta} = \frac{1}{2}\left(\frac{\partial u_\theta}{\partial z} + \frac{1}{r}\frac{\partial u_z}{\partial \theta}\right)$$

$$\varepsilon_{rz} = \frac{1}{2}\left(\frac{\partial u_r}{\partial z} + \frac{\partial u_z}{\partial r}\right) + \frac{1}{2}\frac{\partial u_z}{\partial r}\frac{\partial u_z}{\partial z} \quad (5.2.19)$$

For the choice of the displacement field in Eq. (5.2.18), the only nonzero strains are

$$\varepsilon_{rr} = \varepsilon_{rr}^{(0)} + z\varepsilon_{rr}^{(1)}, \quad \varepsilon_{\theta\theta} = \varepsilon_{\theta\theta}^{(0)} + z\varepsilon_{\theta\theta}^{(1)}, \quad 2\varepsilon_{r\theta} = \gamma_{r\theta}^{(0)} + z\gamma_{r\theta}^{(1)} \quad (5.2.20)$$

where

$$\varepsilon_{rr}^{(0)} = \frac{\partial u_0}{\partial r} + \frac{1}{2}\left(\frac{\partial w_0}{\partial r}\right)^2, \quad \varepsilon_{rr}^{(1)} = -\frac{\partial^2 w_0}{\partial r^2}$$

$$\varepsilon_{\theta\theta}^{(0)} = \frac{u_0}{r} + \frac{1}{r}\frac{\partial v_0}{\partial \theta} + \frac{1}{2r^2}\left(\frac{\partial w_0}{\partial \theta}\right)^2, \quad \varepsilon_{\theta\theta}^{(1)} = -\frac{1}{r}\left(\frac{\partial w_0}{\partial r} + \frac{1}{r}\frac{\partial^2 w_0}{\partial \theta^2}\right)$$

$$\gamma_{r\theta}^{(0)} = \frac{1}{r}\frac{\partial u_0}{\partial \theta} + \frac{\partial v_0}{\partial r} - \frac{v_0}{r}, \quad \gamma_{r\theta}^{(1)} = -\frac{2}{r}\left(\frac{\partial^2 w_0}{\partial r \partial \theta} - \frac{1}{r}\frac{\partial w_0}{\partial \theta}\right) \quad (5.2.21)$$

The principle of virtual displacements for the bending of circular plates is given by $\delta W = 0$, where δW denotes the total virtual work done

$$\delta W = \int_0^T \int_{\Omega_0} \int_{-\frac{h}{2}}^{\frac{h}{2}} (\sigma_{rr}\delta\varepsilon_{rr} + \sigma_{\theta\theta}\delta\varepsilon_{\theta\theta} + \sigma_{r\theta}\delta\gamma_{r\theta})\, dz\, dr\, d\theta\, dt$$

$$- \int_0^T \int_{\Omega_0} \int_{-\frac{h}{2}}^{\frac{h}{2}} \rho\,(\dot{u}_r\delta\dot{u}_r + \dot{u}_\theta\delta\dot{u}_\theta + \dot{u}_z\delta\dot{u}_z)\, dz\, dr\, d\theta\, dt$$

$$- \int_0^T \int_{\Omega_0} (q + F_s)\delta w_0 \, r\, dr\, d\theta\, dt \quad (5.2.22)$$

where h is the plate thickness, $q = q(r, \theta, t)$ is the distributed transverse load, F_s is the reaction force of an elastic foundation, Ω_0 denotes the midplane of the plate, and the superposed dot indicates the time derivative.

186 THEORY AND ANALYSIS OF ELASTIC PLATES

The virtual work done by any applied nonzero edge loads and moments should be added to the expression in Eq. (5.2.22).

Substituting for the virtual strains $\delta\varepsilon_{rr}$ and $\delta\varepsilon_{\theta\theta}$ from Eq. (5.2.21) into Eq. (5.2.22) and integrating over the plate thickness, we obtain

$$0 = \int_0^T \int_{\Omega_0} \left[N_{rr} \left(\frac{\partial \delta u_0}{\partial r} + \frac{\partial w_0}{\partial r}\frac{\partial \delta w_0}{\partial r} \right) + N_{r\theta} \left(\frac{1}{r}\frac{\partial \delta u_0}{\partial \theta} + \frac{\partial \delta v_0}{\partial r} - \frac{\delta v_0}{r} \right) \right.$$

$$+ N_{\theta\theta} \left(\frac{\delta u_0}{r} + \frac{1}{r}\frac{\partial \delta v_0}{\partial \theta} + \frac{1}{r^2}\frac{\partial w_0}{\partial \theta}\frac{\partial \delta w_0}{\partial \theta} \right) - M_{rr}\frac{\partial^2 \delta w_0}{\partial r^2}$$

$$- M_{\theta\theta}\frac{1}{r}\left(\frac{\partial \delta w_0}{\partial r} + \frac{1}{r}\frac{\partial^2 \delta w_0}{\partial \theta^2} \right) - 2M_{r\theta}\frac{1}{r}\left(\frac{\partial^2 \delta w_0}{\partial r \partial \theta} - \frac{1}{r}\frac{\partial \delta w_0}{\partial \theta} \right)$$

$$- I_0(\dot{u}_0 \delta \dot{u}_0 + \dot{v}_0 \delta \dot{v}_0 + \dot{w}_0 \delta \dot{w}_0) - I_2 \left(\frac{\partial \dot{w}_0}{\partial r}\frac{\partial \delta \dot{w}_0}{\partial r} + \frac{1}{r^2}\frac{\partial \dot{w}_0}{\partial \theta}\frac{\partial \delta \dot{w}_0}{\partial \theta} \right)$$

$$\left. - q\delta w_0 + kw_0 \delta w_0 \right] r \, dr \, d\theta \, dt \qquad (5.2.23)$$

where $F_s = -kw_0$ is used, and the force and moment resultants are defined by

$$N_{rr} = \int_{-\frac{h}{2}}^{\frac{h}{2}} \sigma_{rr}\, dz, \quad N_{\theta\theta} = \int_{-\frac{h}{2}}^{\frac{h}{2}} \sigma_{\theta\theta}\, dz, \quad N_{r\theta} = \int_{-\frac{h}{2}}^{\frac{h}{2}} \sigma_{r\theta}\, dz \qquad (5.2.24)$$

$$M_{rr} = \int_{-\frac{h}{2}}^{\frac{h}{2}} \sigma_{rr} z\, dz, \quad M_{\theta\theta} = \int_{-\frac{h}{2}}^{\frac{h}{2}} \sigma_{rr} z\, dz, \quad M_{r\theta} = \int_{-\frac{h}{2}}^{\frac{h}{2}} \sigma_{r\theta} z\, dz \quad (5.2.25)$$

The Euler–Lagrange equations are given by

$$\delta u_0 : \quad \frac{1}{r}\left[\frac{\partial}{\partial r}(rN_{rr}) + \frac{\partial N_{r\theta}}{\partial \theta} - N_{\theta\theta} \right] = I_0 \frac{\partial^2 u_0}{\partial t^2} \qquad (5.2.26)$$

$$\delta v_0 : \quad \frac{1}{r}\left[\frac{\partial}{\partial r}(rN_{r\theta}) + \frac{\partial N_{\theta\theta}}{\partial \theta} + N_{r\theta} \right] = I_0 \frac{\partial^2 v_0}{\partial t^2} \qquad (5.2.27)$$

$$\delta w_0 : \quad \frac{1}{r}\left[\frac{\partial^2}{\partial r^2}(rM_{rr}) - \frac{\partial M_{\theta\theta}}{\partial r} + \frac{1}{r}\frac{\partial^2 M_{\theta\theta}}{\partial \theta^2} + 2\frac{\partial^2 M_{r\theta}}{\partial r \partial \theta} + \frac{2}{r}\frac{\partial M_{r\theta}}{\partial \theta} \right.$$

$$\left. + \frac{\partial}{\partial r}\left(rN_{rr}\frac{\partial w_0}{\partial r} \right) + \frac{1}{r}\frac{\partial}{\partial \theta}\left(N_{\theta\theta}\frac{\partial w_0}{\partial \theta} \right) \right] + q - kw_0$$

$$= I_0 \frac{\partial^2 w_0}{\partial t^2} - I_2 \frac{\partial^2}{\partial t^2}\left[\frac{1}{r}\frac{\partial}{\partial r}\left(r\frac{\partial w_0}{\partial r} \right) + \frac{1}{r^2}\frac{\partial^2 w_0}{\partial \theta^2} \right] \qquad (5.2.28)$$

Introducing the transverse shear forces acting on the rz–plane and θz–plane as

$$Q_r = \frac{1}{r}\left[\frac{\partial}{\partial r}(rM_{rr}) + \frac{\partial M_{r\theta}}{\partial \theta} - M_{\theta\theta}\right] \quad (5.2.29a)$$

$$Q_\theta = \frac{1}{r}\left[\frac{\partial}{\partial r}(rM_{r\theta}) + \frac{\partial M_{\theta\theta}}{\partial \theta} + M_{r\theta}\right] \quad (5.2.29b)$$

the third equation of motion, Eq. (5.2.28), can be expressed as

$$\frac{1}{r}\left[\frac{\partial}{\partial r}(rQ_r) + \frac{\partial Q_\theta}{\partial \theta} + \frac{\partial}{\partial r}\left(rN_{rr}\frac{\partial w_0}{\partial r}\right) + \frac{1}{r}\frac{\partial}{\partial \theta}\left(N_{\theta\theta}\frac{\partial w_0}{\partial \theta}\right)\right] + q - kw_0$$

$$= I_0\frac{\partial^2 w_0}{\partial t^2} - I_2\frac{\partial^2}{\partial t^2}\left[\frac{1}{r}\frac{\partial}{\partial r}\left(r\frac{\partial w_0}{\partial r}\right) + \frac{1}{r^2}\frac{\partial^2 w_0}{\partial \theta^2}\right] \quad (5.2.30)$$

The natural boundary conditions involve specifying the following expressions on an edge with unit normal $\hat{n} = n_r\hat{e}_r + n_\theta\hat{e}_\theta$, where (n_r, n_θ) are such that $n_r = 1$ when $n_\theta = 0$ and vice versa:

$$\delta u_0: \quad rN_{rr}n_r + N_{r\theta}n_\theta \quad (5.2.31a)$$

$$\delta v_0: \quad rN_{r\theta}n_r + N_{\theta\theta}n_\theta \quad (5.2.31b)$$

$$\delta w_0: \quad Q_n + \frac{\partial M_{ns}}{\partial s} + rI_2\frac{\partial \ddot{w}_0}{\partial n} + \mathcal{P}(w_0) \quad (5.2.31c)$$

$$\frac{\partial \delta w_0}{\partial n}: \quad M_{nn} \quad (5.2.31d)$$

where

$$Q_n = rQ_rn_r + Q_\theta n_\theta \quad (5.2.32)$$

$$\mathcal{P}(w_0) = rN_{rr}\frac{\partial w_0}{\partial r}n_r + \frac{1}{r}N_{\theta\theta}\frac{\partial w_0}{\partial \theta}n_\theta \quad (5.2.33)$$

$$M_{ns} = M_{r\theta}(rn_r^2 - n_\theta^2) \quad (5.2.34a)$$

$$M_{nn} = rM_{rr}n_r^2 + M_{\theta\theta}n_\theta^2 \quad (5.2.34b)$$

$$\frac{\partial}{\partial n} = n_r\frac{\partial}{\partial r} + n_\theta\frac{1}{r}\frac{\partial}{\partial \theta}, \quad \frac{\partial}{\partial s} = -n_\theta\frac{\partial}{\partial r} + n_r\frac{1}{r}\frac{\partial}{\partial \theta} \quad (5.2.35)$$

and the Kirchhoff free edge condition is incorporated in Eq. (5.2.31c). The boundary condition in Eq. (5.2.31c) can be expressed in terms of the equivalent shear force V_n as

$$\delta w_0: \quad \bar{V}_n \equiv V_n + rI_2\frac{\partial \ddot{w}_0}{\partial n} + \left(rN_{rr}\frac{\partial w_0}{\partial r}n_r + \frac{1}{r}N_{\theta\theta}\frac{\partial w_0}{\partial \theta}n_\theta\right) \quad (5.2.36)$$

where

$$V_n = rV_r n_r + V_\theta n_\theta, \quad V_r = Q_r + \frac{1}{r}\frac{\partial M_{r\theta}}{\partial \theta}, \quad V_\theta = Q_\theta + \frac{\partial M_{r\theta}}{\partial r} \quad (5.2.37)$$

The boundary conditions for *circular plates* involve specifying one quantity in each of the following pairs on positive r- and θ-planes:

At $r = \hat{r}$, constant:

$$u_0 = \hat{u}_0 \quad \text{or} \quad \hat{r}N_{rr} = \hat{r}\hat{N}_{rr} \quad (5.3.38a)$$

$$v_0 = \hat{v}_0 \quad \text{or} \quad \hat{r}N_{r\theta} = \hat{r}\hat{N}_{r\theta} \quad (5.2.38b)$$

$$w_0 = \hat{w}_0 \quad \text{or} \quad \hat{r}\bar{V}_r = \hat{r}\hat{\bar{V}}_r \quad (5.3.38c)$$

$$\frac{\partial w_0}{\partial r} = \frac{\partial \hat{w}_0}{\partial r} \quad \text{or} \quad \hat{r}M_{rr} = \hat{r}\hat{M}_{rr} \quad (5.2.38d)$$

At $\theta = \hat{\theta}$, constant:

$$u_0 = \hat{u}_0 \quad \text{or} \quad N_{r\theta} = \hat{N}_{r\theta} \quad (5.2.39a)$$

$$v_0 = \hat{v}_0 \quad \text{or} \quad N_{\theta\theta} = \hat{N}_{\theta\theta} \quad (5.2.39b)$$

$$w_0 = \hat{w}_0 \quad \text{or} \quad \bar{V}_\theta = \hat{\bar{V}}_\theta \quad (5.2.39c)$$

$$\frac{1}{r}\frac{\partial w_0}{\partial \theta} = \frac{1}{r}\frac{\partial \hat{w}_0}{\partial \theta} \quad \text{or} \quad M_{\theta\theta} = \hat{M}_{\theta\theta} \quad (5.2.39d)$$

where

$$\bar{V}_r = Q_r + \frac{1}{r}\frac{\partial M_{r\theta}}{\partial \theta} + I_2 \frac{\partial \ddot{w}_0}{\partial r} + N_{rr}\frac{\partial w_0}{\partial r} \quad (5.2.40a)$$

$$\bar{V}_\theta = Q_\theta + \frac{\partial M_{r\theta}}{\partial r} + I_2 \frac{\partial \ddot{w}_0}{\partial \theta} + \frac{1}{r^2}N_{\theta\theta}\frac{\partial w_0}{\partial \theta} \quad (5.2.40b)$$

For the linear case, Eq. (5.2.28) is uncoupled from Eqs. (5.2.26) and (5.2.27). For this case, Eq. (5.2.28) takes the form

$$\frac{1}{r}\left[\frac{\partial^2}{\partial r^2}(rM_{rr}) - \frac{\partial M_{\theta\theta}}{\partial r} + \frac{1}{r}\frac{\partial^2 M_{\theta\theta}}{\partial \theta^2} + 2\frac{\partial^2 M_{r\theta}}{\partial r \partial \theta} + \frac{2}{r}\frac{\partial M_{r\theta}}{\partial \theta}\right] + q - kw_0$$

$$= I_0 \frac{\partial^2 w_0}{\partial t^2} - I_2 \frac{\partial^2}{\partial t^2}\left[\frac{1}{r}\frac{\partial}{\partial r}\left(r\frac{\partial w_0}{\partial r}\right) + \frac{1}{r^2}\frac{\partial^2 w_0}{\partial \theta^2}\right] \quad (5.2.41)$$

and for the static case it reduces to
$$\frac{1}{r}\left[\frac{\partial^2}{\partial r^2}(rM_{rr}) - \frac{\partial M_{\theta\theta}}{\partial r} + \frac{1}{r}\frac{\partial^2 M_{\theta\theta}}{\partial \theta^2} + 2\frac{\partial^2 M_{r\theta}}{\partial r\partial \theta} + \frac{2}{r}\frac{\partial M_{r\theta}}{\partial \theta}\right] + q - kw_0 = 0 \tag{5.2.42}$$

In the preceding discussion, the force and moment resultants are assumed to contain contributions of mechanical, thermal, and electrical effects.

The force and moment resultants can be expressed in terms of the strains of Eq. (5.2.20) once the constitutive relations are assumed. For a linear thermoelastic case (assuming that the elastic stiffnesses are independent of the temperature), we have [see Eq. (3.6.1)]

$$\begin{Bmatrix} \sigma_{rr} \\ \sigma_{\theta\theta} \\ \sigma_{r\theta} \end{Bmatrix} = \begin{bmatrix} Q_{11} & Q_{12} & 0 \\ Q_{12} & Q_{22} & 0 \\ 0 & 0 & Q_{66} \end{bmatrix} \begin{Bmatrix} \varepsilon_{rr} - \alpha_1 \Delta T \\ \varepsilon_{\theta\theta} - \alpha_2 \Delta T \\ 2\varepsilon_{r\theta} \end{Bmatrix} \tag{5.2.43}$$

where $\Delta T(r, \theta, z)$ is the temperature increment, α_1 and α_2 are the coefficients of thermal expansion, and Q_{ij} are the elastic stiffness coefficients. The principal material axes are assumed to coincide with the polar coordinates (r, θ). Then we find that

$$N_{rr} = \int_{-\frac{h}{2}}^{\frac{h}{2}} \sigma_{rr}\, dz = A_{11}\varepsilon_{rr}^{(0)} + A_{12}\varepsilon_{\theta\theta}^{(0)} - N_{rr}^T$$

$$N_{\theta\theta} = \int_{-\frac{h}{2}}^{\frac{h}{2}} \sigma_{\theta\theta}\, dz = A_{12}\varepsilon_{rr}^{(0)} + A_{22}\varepsilon_{\theta\theta}^{(0)} - N_{\theta\theta}^T$$

$$N_{r\theta} = \int_{-\frac{h}{2}}^{\frac{h}{2}} \sigma_{r\theta}\, dz = A_{66}\gamma_{r\theta}^{(0)} \tag{5.2.44a}$$

$$M_{rr} = \int_{-\frac{h}{2}}^{\frac{h}{2}} \sigma_{rr}z\, dz = D_{11}\varepsilon_{rr}^{(1)} + D_{12}\varepsilon_{\theta\theta}^{(1)} - M_{rr}^T$$

$$M_{\theta\theta} = \int_{-\frac{h}{2}}^{\frac{h}{2}} \sigma_{rr}z\, dz = D_{12}\varepsilon_{rr}^{(1)} + D_{22}\varepsilon_{\theta\theta}^{(1)} - M_{\theta\theta}^T$$

$$M_{r\theta} = \int_{-\frac{h}{2}}^{\frac{h}{2}} \sigma_{r\theta}z\, dz = D_{66}\gamma_{r\theta}^{(1)} \tag{5.2.44b}$$

where A_{ij} and D_{ij} denote the extensional and bending stiffnesses of polar orthotropic plates

$$(A_{ij}, D_{ij}) = \int_{-\frac{h}{2}}^{\frac{h}{2}} Q_{ij}(1, z)dz = Q_{ij}(h, \frac{h^3}{12}) \tag{5.2.45}$$

190   THEORY AND ANALYSIS OF ELASTIC PLATES

and N_{rr}^T, $N_{\theta\theta}^T$, M_{rr}^T, and $M_{\theta\theta}^T$ are the thermal forces and moments [see Eqs. (3.6.12a,b)]

$$N_{rr}^T = \int_{-\frac{h}{2}}^{\frac{h}{2}} (Q_{11}\alpha_1 + Q_{12}\alpha_2)\,\Delta T\,dz$$

$$N_{\theta\theta}^T = \int_{-\frac{h}{2}}^{\frac{h}{2}} (Q_{12}\alpha_1 + Q_{22}\alpha_2)\,\Delta T\,dz$$

$$M_{rr}^T = \int_{-\frac{h}{2}}^{\frac{h}{2}} (Q_{11}\alpha_1 + Q_{12}\alpha_2)\,\Delta T\,z\,dz$$

$$M_{\theta\theta}^T = \int_{-\frac{h}{2}}^{\frac{h}{2}} (Q_{12}\alpha_1 + Q_{22}\alpha_2)\,\Delta T\,z\,dz \qquad (5.2.46)$$

The moment resultants can be expressed in terms of the deflection w_0 using the strain-displacement relations. For a geometrically linear case, we have

$$M_{rr} = -\left[D_{11}\frac{\partial^2 w_0}{\partial r^2} + D_{12}\frac{1}{r}\left(\frac{\partial w_0}{\partial r} + \frac{1}{r}\frac{\partial^2 w_0}{\partial \theta^2}\right)\right] - M_{rr}^T \qquad (5.2.47a)$$

$$M_{\theta\theta} = -\left[D_{12}\frac{\partial^2 w_0}{\partial r^2} + D_{22}\frac{1}{r}\left(\frac{\partial w_0}{\partial r} + \frac{1}{r}\frac{\partial^2 w_0}{\partial \theta^2}\right)\right] - M_{\theta\theta}^T \qquad (5.2.47b)$$

$$M_{r\theta} = -2D_{66}\frac{1}{r}\left[\frac{\partial^2 w_0}{\partial r \partial \theta} - \frac{1}{r}\frac{\partial w_0}{\partial \theta}\right] \qquad (5.2.47c)$$

Equation of motion (5.2.41) can be written in terms of the displacement with the aid of Eqs. (5.2.47a–c).

For isotropic plates, we set $D_{11} = D_{22} = D$, $D_{12} = \nu D$ and $2D_{66} = (1 - \nu)D$, and the bending moment-deflection relationships (5.2.47a–c) reduce to [c.f., Eqs. (5.2.15a–c)]

$$M_{rr} = -D\left[\frac{\partial^2 w_0}{\partial r^2} + \frac{\nu}{r}\left(\frac{\partial w_0}{\partial r} + \frac{1}{r}\frac{\partial^2 w_0}{\partial \theta^2}\right)\right] - \frac{M_T}{1-\nu} \qquad (5.2.48a)$$

$$M_{\theta\theta} = -D\left[\nu\frac{\partial^2 w_0}{\partial r^2} + \frac{1}{r}\left(\frac{\partial w_0}{\partial r} + \frac{1}{r}\frac{\partial^2 w_0}{\partial \theta^2}\right)\right] - \frac{M_T}{1-\nu} \qquad (5.2.48b)$$

$$M_{r\theta} = -(1-\nu)D\frac{1}{r}\left(\frac{\partial^2 w_0}{\partial r \partial \theta} - \frac{1}{r}\frac{\partial w_0}{\partial \theta}\right) \qquad (5.2.48c)$$

where
$$M_T = E\alpha \int_{-\frac{h}{2}}^{\frac{h}{2}} \Delta T(x,z)\, z\, dz \qquad (5.2.49)$$

The equation of motion (5.2.41) can be expressed in terms of the deflection. To this end, let us introduce the moment sum, called the *Marcus moment*

$$\mathcal{M} \equiv \frac{M_{rr} + M_{\theta\theta}}{1+\nu}$$
$$= -D\left[\frac{1}{r}\frac{\partial}{\partial r}\left(r\frac{\partial w_0}{\partial r}\right) + \frac{1}{r^2}\frac{\partial^2 w_0}{\partial \theta^2}\right] = -D\nabla^2 w_0 \qquad (5.2.50)$$

Then we have

$$\left[\frac{1}{r}\frac{\partial}{\partial r}\left(r\frac{\partial \mathcal{M}}{\partial r}\right) + \frac{1}{r^2}\frac{\partial^2 \mathcal{M}}{\partial \theta^2}\right] - \frac{1}{1-\nu}\left[\frac{1}{r}\frac{\partial}{\partial r}\left(r\frac{\partial M_T}{\partial r}\right) + \frac{1}{r^2}\frac{\partial^2 M_T}{\partial \theta^2}\right]$$
$$+ q - kw_0 = I_0\frac{\partial^2 w_0}{\partial t^2} - I_2\frac{\partial^2}{\partial t^2}\left[\frac{1}{r}\frac{\partial}{\partial r}\left(r\frac{\partial w_0}{\partial r}\right) + \frac{1}{r^2}\frac{\partial^2 w_0}{\partial \theta^2}\right] \qquad (5.2.51)$$

and in terms of the deflection, we have

$$D\left[\frac{1}{r}\frac{\partial}{\partial r}\left(r\frac{\partial}{\partial r}\right) + \frac{1}{r^2}\frac{\partial^2}{\partial \theta^2}\right]\left[\frac{1}{r}\frac{\partial}{\partial r}\left(r\frac{\partial w_0}{\partial r}\right) + \frac{1}{r^2}\frac{\partial^2 w_0}{\partial \theta^2}\right] + kw_0$$
$$= q - \frac{1}{1-\nu}\left[\frac{1}{r}\frac{\partial}{\partial r}\left(r\frac{\partial M_T}{\partial r}\right) + \frac{1}{r^2}\frac{\partial^2 M_T}{\partial \theta^2}\right]$$
$$- I_0\frac{\partial^2 w_0}{\partial t^2} + I_2\frac{\partial^2}{\partial t^2}\left[\frac{1}{r}\frac{\partial}{\partial r}\left(r\frac{\partial w_0}{\partial r}\right) + \frac{1}{r^2}\frac{\partial^2 w_0}{\partial \theta^2}\right] \qquad (5.2.52)$$

Using the Laplace operator of Eq. (5.2.3), Eq. (5.2.52) can be written simply as

$$D\nabla^2\nabla^2 w_0 + kw_0 = q - \frac{1}{1-\nu}\nabla^2 M_T - I_0\frac{\partial^2 w_0}{\partial t^2} + I_2\frac{\partial^2}{\partial t^2}\left(\nabla^2 w_0\right) \qquad (5.2.53)$$

For the static case, we have [c.f., Eqs. (5.2.11) and (5.2.13)]

$$D\nabla^2\nabla^2 w_0 + kw_0 = q - \frac{1}{1-\nu}\nabla^2 M_T \qquad (5.2.54)$$

5.3 Axisymmetric Bending
5.3.1 Governing Equations

When isotropic or polar orthotropic circular (solid or annular) plates are subjected to rotationally symmetric (or axisymmetric) loads and the edge conditions are also axisymmetric, then all variables are only functions of the radial coordinate r (and possibly time t) and u_θ is identically zero. In this section, we consider linear bending of circular plates with axisymmetric loads and edge conditions. Since u_0 is uncoupled from w_0 for the linear case, we focus on pure bending deformation only.

The nonzero linear strains, stresses, bending moments, and shear force for the axisymmetric bending are given by

$$\varepsilon_{rr} = -z\frac{d^2w_0}{dr^2} - \alpha_1 \Delta T, \quad \varepsilon_{\theta\theta} = -z\frac{1}{r}\frac{dw_0}{dr} - \alpha_2 \Delta T \qquad (5.3.1)$$

$$\sigma_{rr} = -z\left(Q_{11}\frac{d^2w_0}{dr^2} + Q_{12}\frac{1}{r}\frac{dw_0}{dr}\right) - (Q_{11}\alpha_1 + Q_{12}\alpha_2)\Delta T \qquad (5.3.2a)$$

$$\sigma_{\theta\theta} = -z\left(Q_{12}\frac{d^2w_0}{dr^2} + Q_{22}\frac{1}{r}\frac{dw_0}{dr}\right) - (Q_{12}\alpha_1 + Q_{22}\alpha_2)\Delta T \qquad (5.3.2b)$$

$$M_{rr} = -\left(D_{11}\frac{d^2w_0}{dr^2} + D_{12}\frac{1}{r}\frac{dw_0}{dr}\right) - M_{rr}^T \qquad (5.3.3a)$$

$$M_{\theta\theta} = -\left(D_{12}\frac{d^2w_0}{dr^2} + D_{22}\frac{1}{r}\frac{dw_0}{dr}\right) - M_{\theta\theta}^T \qquad (5.3.3b)$$

$$Q_r \equiv \frac{1}{r}\left[\frac{d}{dr}(rM_{rr}) - M_{\theta\theta}\right]$$
$$= -D_{11}\frac{1}{r}\frac{d}{dr}\left(r\frac{d^2w_0}{dr^2}\right) + D_{22}\frac{1}{r^2}\frac{dw_0}{dr} - \frac{1}{r}\left[\frac{d}{dr}(rM_{rr}^T) - M_{\theta\theta}^T\right]$$
$$(5.3.4)$$

The equation of equilibrium for the axisymmetric case can be deduced from Eq. (5.2.30) by setting the nonlinear and time-derivative terms and terms involving differentiation with respect to θ to zero, i.e.,

$$-\frac{1}{r}\frac{d}{dr}(rQ_r) + kw_0 = q \qquad (5.3.5)$$

Using Eq. (5.3.4) for rQ_r in Eq. (5.3.5), we find that

$$\frac{1}{r}\frac{d}{dr}\left[D_{11}\frac{d}{dr}\left(r\frac{d^2w_0}{dr^2}\right) - D_{22}\frac{1}{r}\frac{dw_0}{dr}\right] + kw_0 = q - \frac{1}{r}\frac{d}{dr}\left[\frac{d}{dr}\left(rM_{rr}^T\right) - M_{\theta\theta}^T\right] \quad (5.3.6)$$

In view of the identity

$$\frac{d}{dr}\left(r\frac{d^2w_0}{dr^2}\right) = r\frac{d}{dr}\left[\frac{1}{r}\frac{d}{dr}\left(r\frac{dw_0}{dr}\right)\right] + \frac{1}{r}\frac{dw_0}{dr} \quad (5.3.7)$$

we can rewrite Eq. (5.3.6) as

$$D_{11}\frac{1}{r}\frac{d}{dr}\left\{r\frac{d}{dr}\left[\frac{1}{r}\frac{d}{dr}\left(r\frac{dw_0}{dr}\right)\right] + \left(\frac{D_{11} - D_{22}}{D_{11}}\right)\frac{1}{r}\frac{dw_0}{dr}\right\} + kw_0$$

$$= q - \frac{1}{r}\frac{d}{dr}\left[\frac{d}{dr}\left(rM_{rr}^T\right) - M_{\theta\theta}^T\right] \quad (5.3.8)$$

Equation (5.3.8) must be solved with appropriate boundary conditions [see Eqs. (5.2.38c,d)] on w_0 or Q_r and $\frac{dw_0}{dr}$ or M_{rr}.

For isotropic plates $[D_{11} = D_{22} = D;\ M_{rr}^T = M_{\theta\theta}^T = M_T/(1-\nu)]$, Eqs. (5.3.2)–(5.3.4) become

$$\sigma_{rr} = -\frac{Ez}{(1-\nu^2)}\left(\frac{d^2w_0}{dr^2} + \frac{\nu}{r}\frac{dw_0}{dr}\right) - \frac{E\alpha}{1-\nu}\Delta T \quad (5.3.9a)$$

$$\sigma_{\theta\theta} = -\frac{Ez}{(1-\nu^2)}\left(\nu\frac{d^2w_0}{dr^2} + \frac{1}{r}\frac{dw_0}{dr}\right) - \frac{E\alpha}{1-\nu}\Delta T \quad (5.3.9b)$$

$$M_{rr} = -D\left(\frac{d^2w_0}{dr^2} + \frac{\nu}{r}\frac{dw_0}{dr}\right) - \frac{M_T}{1-\nu} \quad (5.3.10a)$$

$$M_{\theta\theta} = -D\left(\nu\frac{d^2w_0}{dr^2} + \frac{1}{r}\frac{dw_0}{dr}\right) - \frac{M_T}{1-\nu} \quad (5.3.10b)$$

$$Q_r = -D\frac{d}{dr}\left[\frac{1}{r}\frac{d}{dr}\left(r\frac{dw_0}{dr}\right)\right] - \frac{1}{1-\nu}\frac{dM_T}{dr} \quad (5.3.11a)$$

$$M_T = E\alpha\int_{-\frac{h}{2}}^{\frac{h}{2}}\Delta T(r,z)z\,dz \quad (5.3.11b)$$

The governing equation (5.3.8) becomes

$$\frac{D}{r}\frac{d}{dr}\left\{r\frac{d}{dr}\left[\frac{1}{r}\frac{d}{dr}\left(r\frac{dw_0}{dr}\right)\right]\right\} + kw_0 = q - \frac{1}{1-\nu}\frac{1}{r}\frac{d}{dr}\left(r\frac{dM_T}{dr}\right) \quad (5.3.12)$$

5.3.2 Analytical Solutions

In this section we develop the general analytical solution of isotropic plates and the general Rayleigh–Ritz formulation of polar orthotropic plates (i.e., for plates whose principal material coordinates coincide with the polar coordinates). In the subsequent sections these general solutions are specialized to plates with various boundary conditions and loads.

We assume that the plate is isotropic and *not* resting on an elastic foundation (i.e., $k = 0$). Circular plates on elastic foundation and polar orthotropic plates will be considered towards the end of the chapter.

Multiplying Eq. (5.3.5) throughout with r and integrating with respect to r, we obtain

$$-rQ(r) = \int rq(r)dr + c_1 \tag{5.3.13}$$

or, in view of Eq. (5.3.11a),

$$Dr\frac{d}{dr}\left[\frac{1}{r}\frac{d}{dr}\left(r\frac{dw_0}{dr}\right)\right] + \frac{r}{1-\nu}\frac{dM_T}{dr} = \int rq(r)dr + c_1 \tag{5.3.14}$$

Successive integrations lead to the relations

$$D\frac{d}{dr}\left(r\frac{dw_0}{dr}\right) = r\int\frac{1}{r}\int rq(r)drdr - \frac{rM_T}{1-\nu} + c_1 r \log r + c_2 r \tag{5.3.15}$$

$$D\frac{dw_0}{dr} = F'(r) - G'(r) + \frac{r}{4}(2\log r - 1)c_1 + \frac{r}{2}c_2 + \frac{1}{r}c_3 \tag{5.3.16}$$

$$Dw_0(r) = F(r) - G(r) + \frac{r^2}{4}(\log r - 1)c_1 + \frac{r^2}{4}c_2 + c_3 \log r + c_4 \tag{5.3.17}$$

$$F(r) = \int\frac{1}{r}\int r\int\frac{1}{r}\int rq(r)drdrdrdr, \quad G(r) = \frac{1}{(1-\nu)}\int\frac{1}{r}\int rM_T\,drdr \tag{5.3.18}$$

and c_i ($i = 1, 2, 3, 4$) are constants of integration that will be evaluated using boundary conditions. Typical boundary conditions for solid circular plates and annular plates are listed in Table 5.3.1 (also see Figure 5.3.1). Since the second derivative of w_0 is required in the force boundary conditions [see Eqs. (5.3.10a,b) and (5.3.11a)], it can be obtained by differentiating Eq. (5.3.16) with respect to r

$$D\frac{d^2w_0}{dr^2} = F'' - G'' + \frac{1}{4}(2\log r + 1)c_1 + \frac{1}{2}c_2 - \frac{1}{r^2}c_3 \tag{5.3.19}$$

Table 5.3.1. Typical boundary conditions for solid circular and annular plates†.

Plate Type/Edge	Free	Hinged	Clamped
Circular Plate $r = 0$ (for all cases) $r = a$	$\frac{dw_0}{dr} = 0,$ $Q_r = Q_a$ $M_{rr} = M_a$	$2\pi r Q_r = -Q_0$ $w_0 = 0$ $M_{rr} = M_a$	$w_0 = 0$ $\frac{dw_0}{dr} = 0$
Annular Plate $r = b$ $r = a$	$Q_r = Q_b$ $M_{rr} = M_b$ $Q_r = Q_a$ $M_{rr} = M_a$	$w_0 = 0$ $M_{rr} = M_b$ $w_0 = 0$ $M_{rr} = M_a$	$w_0 = 0$ $\frac{dw_0}{dr} = 0$ $w_0 = 0$ $\frac{dw_0}{dr} = 0$

†Q_a, Q_b, M_a, and M_b are distributed edge forces and moments and Q_0 is a concentrated force.

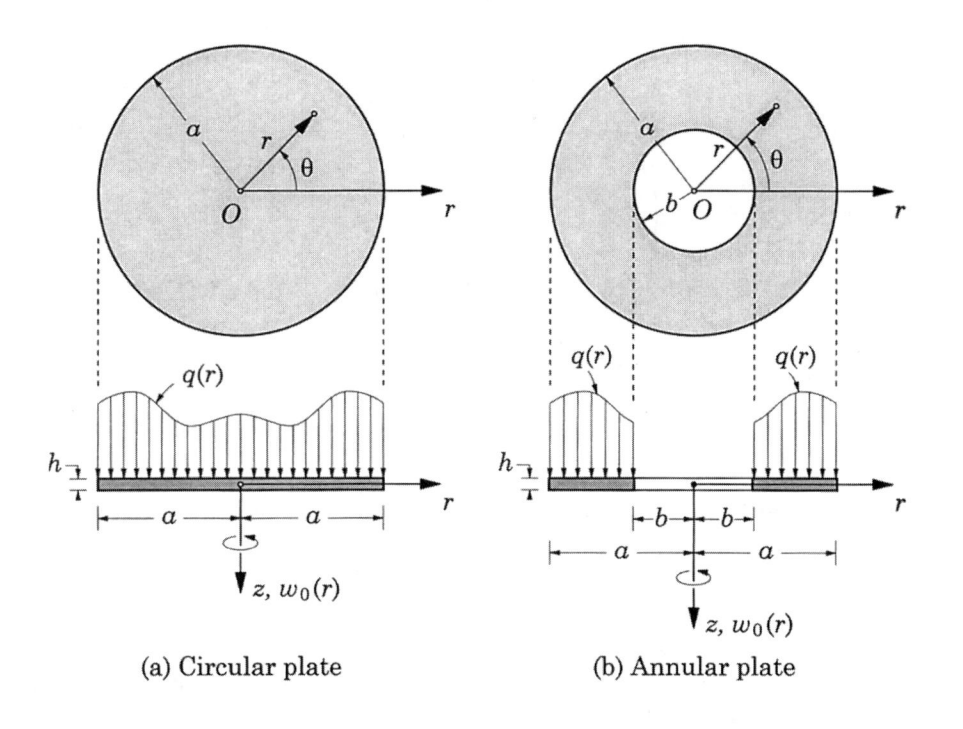

(a) Circular plate

(b) Annular plate

Figure 5.3.1. Axisymmetric circular and annular plates.

Using Eqs. (5.3.16) and (5.3.19) in Eqs. (5.3.9a,b) and (5.3.10a,b), we arrive at the results

$$\sigma_{rr}(r,z) = -\frac{Ez}{1-\nu^2}\left(\frac{d^2w_0}{dr^2} + \nu\frac{1}{r}\frac{dw_0}{dr}\right)$$
$$= -\frac{12z}{h^3}\left\{\left(H'' + \frac{\nu}{r}H'\right) + \left[\frac{1+\nu}{2}\log r + \frac{1}{4}(1-\nu)\right]c_1 \right.$$
$$\left. +\frac{1+\nu}{2}c_2 - \frac{1-\nu}{r^2}c_3\right\} \qquad (5.3.20)$$

$$\sigma_{\theta\theta}(r,z) = -\frac{Ez}{1-\nu^2}\left(\nu\frac{d^2w_0}{dr^2} + \frac{1}{r}\frac{dw_0}{dr}\right)$$
$$= -\frac{12z}{h^3}\left\{\left(\nu H'' + \frac{1}{r}H'\right) + \left[\frac{1+\nu}{2}\log r - \frac{1}{4}(1-\nu)\right]c_1 \right.$$
$$\left. +\frac{1+\nu}{2}c_2 + \frac{1-\nu}{r^2}c_3\right\} \qquad (5.3.21)$$

$$M_{rr}(r) = -D\left(\frac{d^2w_0}{dr^2} + \frac{\nu}{r}\frac{dw_0}{dr}\right)$$
$$= -\left(H'' + \frac{\nu}{r}H'\right) - \left[\frac{1+\nu}{2}\log r + \frac{1}{4}(1-\nu)\right]c_1$$
$$-\frac{1+\nu}{2}c_2 + \frac{1-\nu}{r^2}c_3 \qquad (5.3.22)$$

$$M_{\theta\theta}(r) = -D\left(\nu\frac{d^2w_0}{dr^2} + \frac{1}{r}\frac{dw_0}{dr}\right)$$
$$= -\left(\nu H'' + \frac{1}{r}H'\right) - \left[\frac{1+\nu}{2}\log r - \frac{1}{4}(1-\nu)\right]c_1$$
$$-\frac{1+\nu}{2}c_2 - \frac{1-\nu}{r^2}c_3 \qquad (5.3.23)$$

where $H = F - G$ and primes on F, G, and H denote derivatives with respect to r.

In general, *for a (solid) circular plate*, the boundary condition

$$\frac{dw_0}{dr} = 0 \qquad (5.3.24)$$

requires that c_3 be zero (because $r \log r = 0$ as $r \to 0$) for boundedness of the slope. In addition, the shear force at $r = 0$ should equal the externally applied force Q_0

$$2\pi(rQ_r) = -Q_0 \quad \text{at} \quad r = 0 \qquad (5.3.25)$$

which gives, in view of Eq. (5.3.13), the result
$$c_1 = Q_0/(2\pi) \tag{5.3.26}$$
In the *absence* of a point load at the center (i.e., $Q_0 = 0$), we have the boundary condition
$$c_1 = 0. \tag{5.3.27}$$

The fact that c_1 is not equal to zero in the case of a nonzero central point load presents a problem because of the presence of the logarithmic term in the solution. In particular, the deflection, bending moments, and stresses at $r = 0$ are unbounded. This is due to the fact that in the classical theory of plates, we assumed that the transverse shear strain is zero, an assumption that does not hold in the vicinity of a point load. To alleviate the problem, the deflection, bending moments, and stresses at the center of a circular plate loaded at the center are computed by omitting the logarithmic term.

For the case of a circular plate under arbitrarily distributed transverse load $q(r)$ and temperature distribution $\Delta T(r, z)$, we have $c_1 = c_3 = 0$. The deflection, bending moments, and stresses become

$$Dw_0(r) = H(r) + c_2 \frac{r^2}{4} + c_4 \tag{5.3.28}$$

$$D\frac{dw_0}{dr} = H' + c_2 \frac{r}{2} \tag{5.3.29}$$

$$M_{rr} = -\left(H'' + \frac{\nu}{r}H'\right) - \frac{1+\nu}{2}c_2 \tag{5.3.30}$$

$$M_{\theta\theta} = -\left(\nu H'' + \frac{1}{r}H'\right) - \frac{1+\nu}{2}c_2 \tag{5.3.31}$$

$$\sigma_{rr} = -\frac{12z}{h^3}\left[\left(H'' + \frac{\nu}{r}H'\right) + \frac{1+\nu}{2}c_2\right] \tag{5.3.32}$$

$$\sigma_{\theta\theta} = -\frac{12z}{h^3}\left[\left(\nu H'' + \frac{1}{r}H'\right) + \frac{1+\nu}{2}c_2\right] \tag{5.3.33}$$

The maximum values of these quantities occur at $r = 0$ and are given by

$$w_{max} = w_0(0) = \frac{c_4}{D} \tag{5.3.34}$$

$$M_{max} = M_{rr}(0) = M_{\theta\theta} = -\frac{1+\nu}{2}c_2 \tag{5.3.35}$$

$$\sigma_{max} = \sigma_{rr}(0, \frac{h}{2}) = \sigma_{\theta\theta}(0, \frac{h}{2}) = -\frac{3(1+\nu)}{h^2}c_2 \tag{5.3.36}$$

198   THEORY AND ANALYSIS OF ELASTIC PLATES

The constants c_1 through c_4 for annular plates with various types of mechanical and thermal loads and circular plates with central point load, and constants c_2 and c_4 for circular plates with distributed transverse loads and temperature will be determined in the subsequent sections. It is reminded that the analytical solutions presented are limited to isotropic plates. The Rayleigh–Ritz formulation presented next is for orthotropic plates.

5.3.3 Rayleigh–Ritz Formulation

Here we consider Rayleigh–Ritz solutions of polar orthotropic circular plates, possibly on elastic foundation. Towards this end, we first write the variational statement of the problem for the general case. From the virtual work statement in Eq. (5.2.23), adding terms due to virtual work done by applied generalized forces and restricting to the axisymmetric case, we have

$$
\begin{aligned}
0 = & -2\pi \int_b^a \left(M_{rr} \frac{d^2 \delta w_0}{dr^2} + M_{\theta\theta} \frac{1}{r} \frac{d\delta w_0}{dr} - kw_0 \delta w_0 + q\delta w_0 \right) r\, dr \\
& + 2\pi \left[aQ_a \delta w_0(a) - bQ_b \delta w_0(b) - aM_a \left(\frac{d\delta w_0}{dr}\right)_a + bM_b \left(\frac{d\delta w_0}{dr}\right)_b \right] \\
= & \; 2\pi \int_b^a \left[\left(D_{11} \frac{d^2 w_0}{dr^2} + D_{12} \frac{1}{r} \frac{dw_0}{dr} \right) \frac{d^2 \delta w_0}{dr^2} \right. \\
& + \left(D_{12} \frac{d^2 w_0}{dr^2} + D_{22} \frac{1}{r} \frac{dw_0}{dr} \right) \frac{1}{r} \frac{d\delta w_0}{dr} + kw_0 \delta w_0 - q\delta w_0 \\
& \left. + M_{rr}^T \frac{d^2 \delta w_0}{dr^2} + M_{\theta\theta}^T \frac{1}{r} \frac{d\delta w_0}{dr} \right] r\, dr + 2\pi \left[aQ_a \delta w_0(a) - bQ_b \delta w_0(b) \right] \\
& + 2\pi \left[-aM_a \left(\frac{d\delta w_0}{dr}\right)_a + bM_b \left(\frac{d\delta w_0}{dr}\right)_b \right] \quad (5.3.37)
\end{aligned}
$$

where a and b denote the outer and inner radii of the plate, k is the spring constant of the linear elastic foundation, D_{11}, D_{12}, and D_{22} the bending stiffnesses of a polar orthotropic material, q the distributed load, Q_a and Q_b the intensities of line loads at the outer and inner edges, respectively, and M_a and M_b the distributed edge moments at the outer and inner edges, respectively. When $b = 0$ (for solid circular plate), we have $M_b = 0$ and $2\pi b Q_b = Q_0$, Q_0 being the point load at the center of the plate.

Assume a solution of the form (assuming that all specified geometric boundary conditions are homogeneous so that we can take $\varphi_0(r) = 0$)

$$w_0(r) \approx W_N(r) = \sum_{j=1}^{N} c_j \varphi_j(r) \tag{5.3.38}$$

Substituting into Eq. (5.3.37), we obtain

$$0 = 2\pi \sum_{j=1}^{N} c_j \int_b^a \left[\left(D_{11} \frac{d^2\varphi_j}{dr^2} + D_{12} \frac{1}{r} \frac{d\varphi_j}{dr} \right) \frac{d^2\phi_i}{dr^2} \right.$$
$$\left. + \left(D_{12} \frac{d^2\varphi_j}{dr^2} + D_{22} \frac{1}{r} \frac{d\varphi_j}{dr} \right) \frac{1}{r} \frac{d\varphi_i}{dr} + k\varphi_i\varphi_j \right] r \, dr$$
$$- 2\pi \int_b^a \left(q - M_{rr}^T \frac{d^2\varphi_i}{dr^2} - M_{\theta\theta}^T \frac{1}{r} \frac{d\varphi_i}{dr} \right) r \, dr$$
$$+ 2\pi \left[aQ_a \varphi_i(a) - bQ_b \varphi_i(b) - aM_a \left(\frac{d\varphi_i}{dr} \right)_a + bM_b \left(\frac{d\varphi_i}{dr} \right)_b \right]$$
$$= \sum_{j=1}^{N} R_{ij} c_j - F_i \tag{5.3.39a}$$

where

$$R_{ij} = 2\pi \int_b^a \left[D_{11} \frac{d^2\varphi_i}{dr^2} \frac{d^2\varphi_j}{dr^2} + D_{12} \frac{1}{r} \left(\frac{d\varphi_i}{dr} \frac{d^2\varphi_j}{dr^2} + \frac{d^2\varphi_j}{dr^2} \frac{d\varphi_i}{dr} \right) \right.$$
$$\left. + D_{22} \frac{1}{r^2} \frac{d\varphi_i}{dr} \frac{d\varphi_j}{dr} + k\varphi_i\varphi_j \right] r \, dr \tag{5.3.39b}$$

$$F_i = 2\pi \int_b^a \left[q\varphi_i - \left(M_{rr}^T \frac{d^2\varphi_i}{dr^2} + M_{\theta\theta}^T \frac{1}{r} \frac{d\varphi_i}{dr} \right) \right] r \, dr$$
$$- 2\pi \left[aQ_a \varphi_i(a) - bQ_b \varphi_i(b) - aM_a \left(\frac{d\varphi_i}{dr} \right)_a + bM_b \left(\frac{d\varphi_i}{dr} \right)_b \right] \tag{5.3.39c}$$

In matrix form, Eq. (5.3.39a) can be written as

$$[R]\{c\} = \{F\} \tag{5.3.40}$$

The approximation functions φ_i are determined using the geometric boundary conditions of the problem. Once φ_i are known, one can determine R_{ij} and F_i, and Eq. (5.3.40) can be solved for c_i ($i = 1, 2, \cdots, N$) and the solution is given by Eq. (5.3.38).

5.3.4 Simply Supported Circular Plate Under Distributed Load

The boundary conditions associated with the simply supported outer edges are $w_0 = 0$ and $M_{rr} = 0$ at $r = a$. Using these conditions in Eqs. (5.3.28) and (5.3.30), we obtain

$$c_2 = -\frac{2}{(1+\nu)}\left(F_0'' + \frac{\nu}{a}F_0'\right), \quad c_4 = -\left(F_0 + \frac{a^2}{4}c_2\right) \quad (5.3.41)$$

where F_0, for example, denotes $F_0 = F(a)$.

For uniformly distributed load of intensity q_0 (see Figure 5.3.2), we have

$$F(r) = \frac{q_0 r^4}{64}, \quad F'(r) = \frac{q_0 r^3}{16}, \quad F''(r) = \frac{3q_0 r^2}{16} \quad (5.3.42a)$$

$$c_2 = -\frac{(3+\nu)}{8(1+\nu)}q_0 a^2, \quad c_4 = \frac{(5+\nu)}{64(1+\nu)}q_0 a^4 \quad (5.3.42b)$$

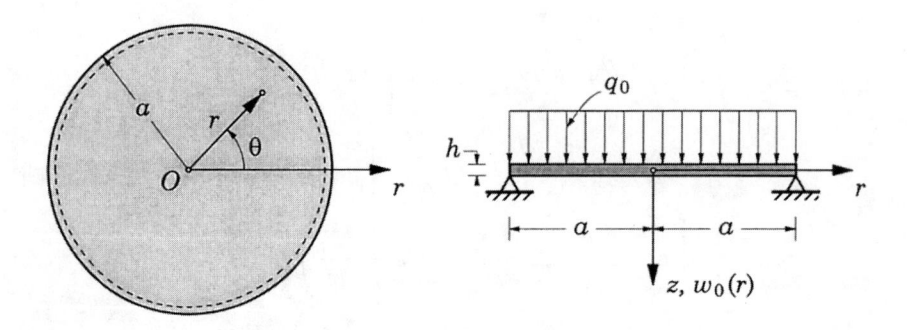

Figure 5.3.2. A simply supported circular plate under uniformly distributed load.

The deflection, bending moment, and stresses in a simply supported circular plate under uniformly distributed load are

$$w_0(r) = \frac{q_0 a^4}{64D}\left[\left(\frac{r}{a}\right)^4 - 2\left(\frac{3+\nu}{1+\nu}\right)\left(\frac{r}{a}\right)^2 + \left(\frac{5+\nu}{1+\nu}\right)\right] \quad (5.3.43)$$

$$M_{rr}(r) = (3+\nu)\frac{q_0 a^2}{16}\left[1 - \left(\frac{r}{a}\right)^2\right] \tag{5.3.44a}$$

$$M_{\theta\theta}(r) = \frac{q_0 a^2}{16}\left[(3+\nu) - (1+3\nu)\left(\frac{r}{a}\right)^2\right] \tag{5.3.44b}$$

$$\sigma_{rr}(r,z) = 3(3+\nu)\frac{q_0 a^2}{4h^2}\frac{z}{h}\left[1 - \left(\frac{r}{a}\right)^2\right] \tag{5.3.45a}$$

$$\sigma_{\theta\theta}(r,z) = 3\frac{q_0 a^2}{4h^2}\frac{z}{h}\left[(3+\nu) - (1+3\nu)\left(\frac{r}{a}\right)^2\right] \tag{5.3.45b}$$

The solution in Eq. (5.3.43) for the deflection can also be obtained with a three-parameter Rayleigh–Ritz approximation, as will be shown in the sequel.

The maximum deflection, bending moments, and stresses occur at $r = 0$, and they are

$$w_{max} = \left(\frac{5+\nu}{1+\nu}\right)\frac{q_0 a^4}{64D}, \quad M_{max} = (3+\nu)\frac{q_0 a^2}{16}, \quad \sigma_{max} = 3(3+\nu)\frac{q_0 a^2}{8h^2} \tag{5.3.46}$$

For a linearly distributed load of the type (see Figure 5.3.3)

$$q(r) = q_0 + \frac{q_1 - q_0}{a}r \tag{5.3.47}$$

we obtain

$$F(r) = \frac{q_0 r^4}{64} + \left(\frac{q_1 - q_0}{a}\right)\frac{r^5}{225}, \quad F'(r) = \frac{q_0 r^3}{16} + \left(\frac{q_1 - q_0}{a}\right)\frac{r^4}{45}$$

$$F''(r) = \frac{3q_0 r^2}{16} + \left(\frac{q_1 - q_0}{a}\right)\frac{4r^3}{45} \tag{5.3.48}$$

$$c_2 = -\frac{2a^2}{(1+\nu)}\left[\frac{3+\nu}{16}q_0 + \frac{4+\nu}{45}(q_1 - q_0)\right]$$

$$c_4 = \frac{a^4}{(1+\nu)}\left[\frac{5+\nu}{64}q_0 + \frac{6+\nu}{150}(q_1 - q_0)\right] \tag{5.3.49}$$

The deflection, bending moments, and stresses are given by Eqs. (5.3.28)–(5.3.33), and their maximum values are given by Eqs. (5.3.34)–(5.3.36).

202   THEORY AND ANALYSIS OF ELASTIC PLATES

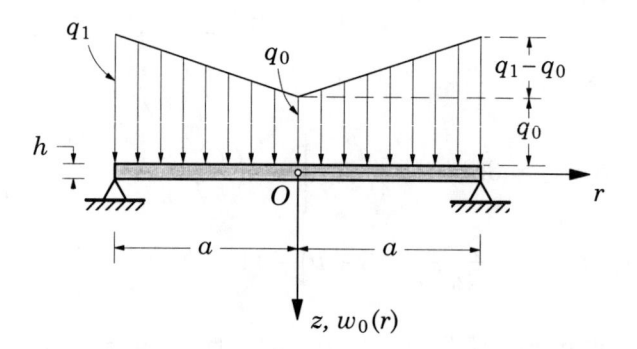

Figure 5.3.3. A simply supported circular plate under linearly varying distributed load.

The Rayleigh–Ritz solution of a simply supported polar orthotropic circular plate under uniformly distributed load can be obtained using the formulation of section 5.3.3. The primary task is to select admissible approximation functions $\varphi_i(r)$. Recall that for this problem the geometric (or essential) boundary conditions are

$$\frac{dw_0}{dr}(0) = 0, \quad w_0(a) = 0$$

and the force (or natural) boundary conditions are given by

$$M_{rr} = 0 \text{ at } r = a, \quad Q_r = \frac{1}{r}\left[\frac{d}{dr}(rM_{rr}) - M_{\theta\theta}\right] = 0 \text{ at } r = 0$$

The natural boundary conditions will have no bearing on the selection of φ_i in the Rayleigh–Ritz method. Each φ_i must satisfy the conditions

$$\frac{d\varphi_i}{dr}(0) = 0, \quad \varphi_i(a) = 0$$

For the choice of algebraic polynomials, we assume

$$\varphi_1(r) = \alpha_1 + \alpha_2 r + \alpha_3 r^2$$

and determine, using the above conditions, that $\alpha_2 = 0$ and $\alpha_1 + \alpha_3 a^2 = 0$. Thus we have (for the choice of $\alpha_1 = 1$)

$$\varphi_1 = 1 - \frac{r^2}{a^2}$$

ANALYSIS OF CIRCULAR PLATES

The procedure can be used to obtain a linearly independent and complete set of functions

$$\varphi_1 = 1 - \frac{r^2}{a^2}, \quad \varphi_2 = 1 - \frac{r^3}{a^3}, \quad \ldots, \quad \varphi_j = 1 - \left(\frac{r}{a}\right)^{j+1} \tag{5.3.50}$$

Substituting for φ_j from the above approximation into Eqs. (5.3.39b,c), and noting that $k = 0$, we obtain

$$R_{ij} = \frac{2\pi}{a^2} [ijD_{11} + D_{12}(i+j) + D_{22}] \frac{(i+1)(j+1)}{(i+j)} \left(1 - \beta^{i+j}\right)$$

$$F_i = 2\pi \left\{ q_0 a^2 \left[(i+1) - (i+3)\beta^2 + 2\beta^{i+3}\right] \frac{1}{2(i+3)} + bQ_b \left(1 - \beta^{i+1}\right) \right.$$
$$\left. + \left(1 - \beta^{i+1}\right) \left(iM_{rr}^T + M_{\theta\theta}^T\right) - (i+1)M_a + (i+1)\beta^{i+1}M_b \right\} \tag{5.3.51}$$

where $\beta = b/a$.

Consider the case of a circular plate ($b = 0$ or $\beta = 0$) made of a material for which we have $D_{11}/D_{22} = 10$ and $D_{12}/D_{22} = 1/3$, and subjected to only uniformly distributed load of intensity q_0. Then

$$R_{ij} = \frac{2\pi D_{22}}{3a^2} (30ij + i + j + 3) \frac{(i+1)(j+1)}{(i+j)}, \quad F_i = 2\pi q_0 a^2 \frac{(i+1)}{2(i+2)}$$

For $N = 1$, we have

$$R_{11} = \frac{140 D_{22} \pi}{3a^2}, \quad F_1 = \frac{2\pi q_0 a^2}{4}, \quad c_1 = \frac{3 q_0 a^4}{280 D_{22}}$$

and the one-parameter Rayleigh–Ritz solution becomes

$$W_1(r) = \frac{3 q_0 a^4}{280 D_{22}} \left(1 - \frac{r^2}{a^2}\right)$$

For $N = 2$, we have

$$\frac{D_{22}}{a^2} \begin{bmatrix} \frac{70}{3} & 44 \\ 44 & \frac{381}{4} \end{bmatrix} \begin{Bmatrix} c_1 \\ c_2 \end{Bmatrix} = \frac{q_0 a^2}{20} \begin{Bmatrix} 5 \\ 6 \end{Bmatrix}$$

and the solution is given by

$$W_2(r) = \frac{q_0 a^4}{D_{22}} \left[0.03704 \left(1 - \frac{r^2}{a^2}\right) - 0.01396 \left(1 - \frac{r^3}{a^3}\right) \right]$$

For $N = 3$, we have

$$\frac{D_{22}}{a^2} \begin{bmatrix} \frac{70}{3} & 44 & \frac{194}{3} \\ 44 & 381 & \frac{752}{5} \\ \frac{194}{3} & \frac{752}{5} & 248 \end{bmatrix} \begin{Bmatrix} c_1 \\ c_2 \\ c_3 \end{Bmatrix} = q_0 a^2 \begin{Bmatrix} \frac{1}{4} \\ \frac{3}{10} \\ \frac{1}{3} \end{Bmatrix} \qquad (5.3.52)$$

and the solution is given by

$$W_3(r) = \frac{q_0 a^4}{D_{22}} \left[0.05841 \left(1 - \frac{r^2}{a^2}\right) - 0.04493 \left(1 - \frac{r^3}{a^3}\right) \right.$$
$$\left. + 0.01336 \left(1 - \frac{r^4}{a^4}\right) \right]$$

For an isotropic circular plate, the coefficients R_{ij} become

$$R_{ij} = \frac{2\pi D}{a^2} \left(\frac{ij+1}{i+j} + \nu \right) (i+1)(j+1)$$

For $N = 3$ we have the algebraic equations

$$\frac{D}{a^2} \begin{bmatrix} 4(1+\nu) & 6(1+\nu) & 8(1+\nu) \\ 6(1+\nu) & 9(1.25+\nu) & 12(1.4+\nu) \\ 8(1+\nu) & 12(1.4+\nu) & 16(\frac{5}{3}+\nu) \end{bmatrix} \begin{Bmatrix} c_1 \\ c_2 \\ c_3 \end{Bmatrix} = \frac{q_0 a^2}{60} \begin{Bmatrix} 15 \\ 18 \\ 20 \end{Bmatrix}$$
$$(5.3.53a)$$

The one-, two- and three-parameter Rayleigh–Ritz solutions are

$$W_1(r) = \frac{q_0 a^4}{16D(1+\nu)} \left(1 - \frac{r^2}{a^2}\right)$$

$$W_2(r) = \frac{q_0 a^4}{80D} \left(\frac{9+4\nu}{1+\nu}\right) \left(1 - \frac{r^2}{a^2}\right) - \frac{q_0 a^4}{30D} \left(1 - \frac{r^3}{R_0^3}\right)$$

$$W_3(r) = \frac{q_0 a^4}{64D} \left(\frac{6+2\nu}{1+\nu}\right) \left(1 - \frac{r^2}{a^2}\right) - \frac{q_0 a^4}{64D} \left(1 - \frac{r^4}{a^4}\right)$$

$$= \frac{q_0 a^4}{64D} \left[\frac{5+\nu}{1+\nu} - 2\left(\frac{3+\nu}{1+\nu}\right) \left(\frac{r}{a}\right)^2 + \left(\frac{r}{a}\right)^4 \right] \qquad (5.3.53b)$$

The three-parameter solution coincides with the exact solution (5.3.43).

5.3.5 Simply Supported Circular Plate Under Central Point Load

For this case, we have $c_1 = Q_0/2\pi$ and $c_3 = 0$ (see Figure 5.3.4). Using the boundary condition $w_0(a) = 0$, we obtain from Eq. (5.3.17) the result

$$\frac{Q_0 a^2}{8\pi}(\log a - 1) + \frac{a^2}{4}c_2 + c_4 = 0$$

If there is no applied bending moment at $r = a$, Eq. (5.3.22) yields the result

$$-\frac{Q_0}{2\pi}\left[\frac{1+\nu}{2}\log a + \frac{1-\nu}{4}\right] - \frac{1+\nu}{2}c_2 = 0$$

Solving for the constants, we obtain

$$c_2 = -\frac{Q_0}{2\pi}\left[\log a + \frac{1}{2}\left(\frac{1-\nu}{1+\nu}\right)\right], \quad c_4 = \left(\frac{3+\nu}{1+\nu}\right)\frac{Q_0 a^2}{16\pi} \qquad (5.3.54)$$

The solution for any $r \neq 0$ is given by

$$w_0(r) = \frac{Q_0 a^2}{16\pi D}\left[\left(\frac{3+\nu}{1+\nu}\right)\left(1 - \frac{r^2}{a^2}\right) + 2\left(\frac{r}{a}\right)^2 \log\left(\frac{r}{a}\right)\right] \qquad (5.3.55)$$

$$\sigma_{rr}(r) = -\frac{3zQ_0(1+\nu)}{h^3\pi}\log\left(\frac{r}{a}\right) \qquad (5.3.56a)$$

$$\sigma_{\theta\theta}(r) = -\frac{3zQ_0}{h^3\pi}\left[(1+\nu)\log\left(\frac{r}{a}\right) - (1-\nu)\right] \qquad (5.3.56b)$$

$$M_{rr}(r) = -\frac{Q_0(1+\nu)}{4\pi}\log\left(\frac{r}{a}\right) \qquad (5.3.57a)$$

$$M_{\theta\theta}(r) = -\frac{Q_0}{4\pi}\left[(1+\nu)\log\left(\frac{r}{a}\right) - (1-\nu)\right] \qquad (5.3.57b)$$

The maximum deflection is given by

$$w_{max} = w_0(0) = \frac{Q_0 a^2}{16\pi D}\left(\frac{3+\nu}{1+\nu}\right) \qquad (5.3.58)$$

The maximum stresses and bending moments cannot be calculated using Eqs.(5.3.56a,b) and (5.3.57a,b) due to the logarithmic singularity. The maximum finite stresses produced by load Q_0 on a very small circular area of radius r_c can be calculated using the so-called equivalent radius r_e in Eqs. (5.3.56a,b) and (5.3.57a,b) (see Roark and Young [3])

$$r_e = \sqrt{1.6 r_c^2 + h^2} - 0.675h \quad \text{when } r_c < 1.7h \qquad (5.3.59a)$$

$$r_e = r_c \quad \text{when } r_c \geq 1.7h \qquad (5.3.59b)$$

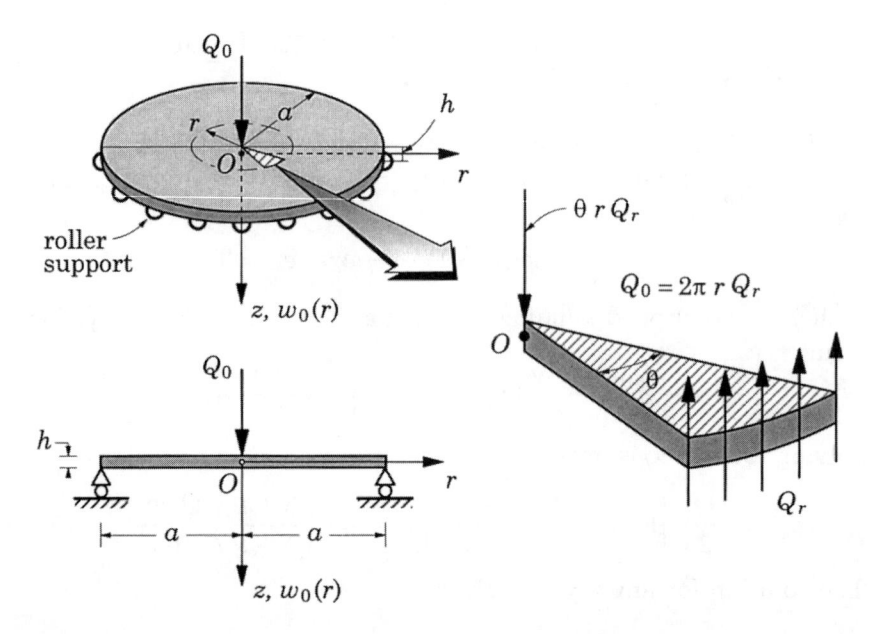

Figure 5.3.4. A simply supported circular plate under central point load.

This problem can also be solved using the Rayleigh–Ritz method. Although the exact solution is logarithmic, we can seek an algebraic approximation of the solution. The approximation functions φ_i given in Eq. (5.3.50) are admissible for this problem. Using the three-parameter approximation, we obtain the system of algebraic equations given in Eq. (5.3.53a). Suppose that the circular plate is subjected to moment M_a at edge $r = a$, point load $Q_0 = 2\pi b Q_b$ at the center, and uniformly distributed load q_0. Then from Eq. (5.3.51) we have $R_{11} = 8\pi D(1+\nu)/a^2$ and $F_1 = \pi q_0 a^2/2 + Q_0 + 4\pi M_a$. Hence, the one-parameter Rayleigh–Ritz solution is given by

$$W_1(r) = \left[\frac{q_0 a^4}{16D(1+\nu)} + \frac{Q_0 a^2}{8\pi D(1+\nu)} + \frac{M_a a^2}{2D(1+\nu)} \right] \left(1 - \frac{r^2}{a^2}\right) \quad (5.3.60)$$

Note that the deflection due to the edge moment M_a is exact, the maximum deflection due to the distributed load q_0 is in 24.5% error, and the maximum deflection due to the point load Q_0 is in 40% error. It was already shown in Eq. (5.3.53b) that the three-parameter Raleigh–Ritz solution is exact for uniformly distributed load. Hence, we consider only the point load in evaluating the two- and three-parameter solutions.

For $N = 2$, we obtain

$$c_1 = \frac{5.25 + 3\nu}{18(1+\nu)}\left(\frac{Q_0 a^2}{\pi D}\right), \quad c_2 = -\frac{Q_0 a^2}{9\pi D} \quad (5.3.61)$$

and the solution is given by

$$W_2(r) = \frac{Q_0 a^2}{18\pi D}\left[\left(\frac{5.25 + 3\nu}{1+\nu}\right)\left(1 - \frac{r^2}{a^2}\right) - 2\left(1 - \frac{r^3}{a^3}\right)\right] \quad (5.3.62)$$

The maximum deflection predicted by the two-parameter approximation is 4.38% in error. Lastly, for $N = 3$, we obtain

$$c_1 = \frac{2.96 + 2\nu}{1+\nu}\Omega, \quad c_2 = -2.133\Omega, \quad c_3 = 0.6\Omega, \quad \Omega = \frac{Q_0 a^2}{7.68\pi D}$$

and the solution becomes

$$W_3(r) = \Omega\left[\left(\frac{2.96 + 2\nu}{1+\nu}\right)\left(1 - \frac{r^2}{a^2}\right) - 2.133\left(1 - \frac{r^3}{a^3}\right) + 0.6\left(1 - \frac{r^4}{a^4}\right)\right]$$

$$(5.3.63)$$

The maximum deflection predicted by the three-parameter Rayleigh–Ritz solution is only one percent in error compared to the exact deflection in Eq. (5.3.58).

It should be noted that the method of superposition may be used to obtain the solution to a problem on a variety of loads. For example, the center deflection of a simply supported plate subjected to uniformly distributed load as well as a point load at the center may be obtained by adding the expressions in Eqs. (5.3.43) and (5.3.55).

5.3.6 Annular Plate with Simply Supported Outer Edge

Circular plates with a concentric circular hole of radius b are called annular plates (see Figure 5.3.1b). For this case, the solutions given in Eqs. (5.3.17) and (5.3.20)–(5.3.23) are valid for $b \leq r \leq a$. For annular plates, the edge $r = b$ can also be subjected to various geometric and/or force boundary conditions. We consider several cases here.

Suppose that the inner edge is subjected to a bending moment M_b while the outer edge is subjected to a bending moment M_a (see Figure 5.3.5a). The boundary conditions are then

$$\text{At } r = b: \quad M_{rr} = M_b, \quad rQ_r = 0 \quad (5.3.64a)$$
$$\text{At } r = a: \quad w_0 = 0, \quad M_{rr} = M_a \quad (5.3.64b)$$

208   THEORY AND ANALYSIS OF ELASTIC PLATES

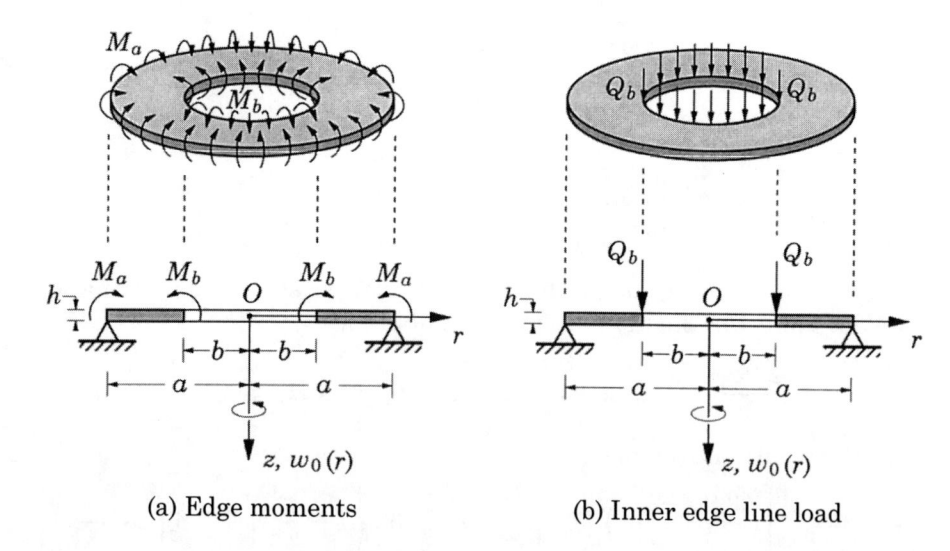

Figure 5.3.5. A simply supported annular plate with (a) edge moments or (b) inner edge line load.

Hence, we obtain from Eq. (5.3.13) $c_1 = 0$ and

$$-\frac{1+\nu}{2}c_2 + \frac{1-\nu}{b^2}c_3 = M_b, \quad -\frac{1+\nu}{2}c_2 + \frac{1-\nu}{a^2}c_3 = M_a$$

$$\frac{a^2}{4}c_2 + c_3 \log a + c_4 = 0 \tag{5.3.65}$$

which give

$$c_3 = \frac{M_b - M_a}{1-\nu}\left(\frac{b^2 a^2}{a^2 - b^2}\right), \quad c_2 = \frac{2}{1+\nu}\left(\frac{M_b b^2 - M_a a^2}{a^2 - b^2}\right)$$

$$c_4 = -\frac{a^2}{2(1+\nu)}\left(\frac{M_b b^2 - M_a a^2}{a^2 - b^2}\right) - \frac{(M_b - M_a)\log a}{1-\nu}\left(\frac{b^2 a^2}{a^2 - b^2}\right)$$

$$\tag{5.3.66}$$

The deflection of Eq. (5.3.17) becomes

$$w_0 = -\left(\frac{a^2 - r^2}{a^2 - b^2}\right)\left(\frac{M_b b^2 - M_a a^2}{2(1+\nu)D}\right) + \left(\frac{b^2 a^2}{a^2 - b^2}\right)\left(\frac{M_a - M_b}{(1-\nu)D}\right)\log \bar{r}$$

$$= -\frac{a^2}{2D(1-\beta^2)} \left[\beta^2 \left(\frac{1-\bar{r}^2}{1+\nu} - \frac{2\log \bar{r}}{1-\nu} \right) M_b \right.$$
$$\left. - \left(\frac{1-\bar{r}^2}{1+\nu} - \frac{2\beta^2 \log \bar{r}}{1-\nu} \right) M_a \right] \tag{5.3.67}$$

$$M_{rr} = \frac{1}{1-\beta^2} \left[\beta^2 \left(\frac{a^2}{r^2} - 1 \right) M_b + \left(1 - \beta^2 \frac{a^2}{r^2} \right) M_a \right] \tag{5.3.68}$$

$$M_{\theta\theta} = \frac{1}{1-\beta^2} \left[-\beta^2 \left(\frac{a^2}{r^2} + 1 \right) M_b + \left(1 + \beta^2 \frac{a^2}{r^2} \right) M_a \right] \tag{5.3.69}$$

where $\bar{r} = r/a$ and $\beta = b/a$. The above results can be specialized to the case in which $M_b = 0$ or $M_a = 0$ as well as to the case of solid circular plate with applied moment at edge $r = a$ by setting $\beta = 0$.

Next consider an annular plate with the inner edge subjected to transverse line load Q_b (see Figure 5.3.5b). The boundary conditions are then

$$\text{At } r = b: \quad M_{rr} = 0, \quad (rQ_r) = -bQ_b \tag{5.3.70a}$$
$$\text{At } r = a: \quad w_0 = 0, \quad M_{rr} = 0 \tag{5.3.70b}$$

Hence, we have from Eq. (5.3.13) $c_1 = bQ_b$ and

$$-\left[\frac{1+\nu}{2} \log b + \frac{1}{4}(1-\nu) \right] bQ_b - \frac{1+\nu}{2} c_2 + \frac{1-\nu}{b^2} c_3 = 0$$

$$-\left[\frac{1+\nu}{2} \log a + \frac{1}{4}(1-\nu) \right] bQ_b - \frac{1+\nu}{2} c_2 + \frac{1-\nu}{a^2} c_3 = 0$$

$$\frac{a^2}{4} (\log a - 1) bQ_b + \frac{a^2}{4} c_2 + c_3 \log a + c_4 = 0 \tag{5.3.71}$$

from which we obtain

$$c_2 = bQ_b \left[\frac{b^2}{a^2-b^2} \log \left(\frac{b}{a} \right) - \log a - \frac{1}{2}\left(\frac{1-\nu}{1+\nu} \right) \right]$$

$$c_3 = bQ_b \frac{1}{2} \left(\frac{1+\nu}{1-\nu} \right) \frac{b^2 a^2}{a^2-b^2} \log \left(\frac{b}{a} \right)$$

$$c_4 = -bQ_b \left[\frac{a^2}{4} \left(\frac{b^2}{a^2-b^2} \log \left(\frac{b}{a} \right) - 1 \right) - \frac{a^2}{8} \left(\frac{1-\nu}{1+\nu} \right) \right.$$
$$\left. + \frac{1}{2} \left(\frac{1+\nu}{1-\nu} \right) \frac{b^2 a^2}{a^2-b^2} \log \left(\frac{b}{a} \right) \log a \right] \tag{5.3.72}$$

210 THEORY AND ANALYSIS OF ELASTIC PLATES

Hence, the solution becomes

$$w_0(r) = \frac{Q_b b a^2}{8D}\left[\left(1 - \frac{r^2}{a^2}\right)\left(\frac{3+\nu}{1+\nu} - 2\kappa\right) + 2\frac{r^2}{a^2}\log\left(\frac{r}{a}\right)\right.$$
$$\left. + 4\kappa\left(\frac{1+\nu}{1-\nu}\right)\log\left(\frac{r}{a}\right)\right] \tag{5.3.73}$$

$$M_{rr}(r) = -\frac{bQ_b(1+\nu)}{2}\left[\log\left(\frac{r}{a}\right) + \kappa\left(1 - \frac{r^2}{a^2}\right)\right] \tag{5.3.74}$$

$$M_{\theta\theta}(r) = -\frac{bQ_b(1+\nu)}{2}\left[\log\left(\frac{r}{a}\right) - \frac{1-\nu}{1+\nu} + \kappa\left(1 + \frac{r^2}{a^2}\right)\right] \tag{5.3.75}$$

where

$$\kappa = \frac{\beta^2}{1-\beta^2}\log\beta, \quad \beta = \frac{b}{a} \tag{5.3.76}$$

Note that if the radius b of the hole becomes infinitesimally small, $b^2 \log(b/a)$ vanishes, and Eqs. (5.3.73)–(5.3.75), by letting $2\pi b Q_b = Q_0$ [or $Q_b = Q_0/(2\pi b)$], reduce to Eqs. (5.3.55), (5.3.57a), and (5.3.57b), respectively.

Finally, we consider a simply supported annular plate under uniformly distributed load of intensity q_0. The boundary conditions are

$$\text{At } r = b: \quad M_{rr} = 0, \quad (rQ_r) = 0 \tag{5.3.77a}$$
$$\text{At } r = a: \quad w_0 = 0, \quad M_{rr} = 0 \tag{5.3.77b}$$

Hence, we have $c_1 = -q_0 b^2/2$ and

$$-\left(\frac{3+\nu}{16}\right)q_0 b^2 + \left[\frac{1+\nu}{4}\log b + \frac{1-\nu}{8}\right]b^2 q_0 - \frac{1+\nu}{2}c_2 + \frac{1-\nu}{b^2}c_3 = 0$$

$$-\left(\frac{3+\nu}{16}\right)q_0 a^2 + \left[\frac{1+\nu}{4}\log a + \frac{1}{8}(1-\nu)\right]b^2 q_0 - \frac{1+\nu}{2}c_2 + \frac{1-\nu}{a^2}c_3 = 0$$

$$\frac{q_0 a^4}{64} - \frac{a^2}{8}(\log a - 1)q_0 b^2 + \frac{a^2}{4}c_2 + c_3\log a + c_4 = 0 \tag{5.3.78}$$

We obtain

$$c_2 = -\left(\frac{1+3\nu}{1+\nu}\right)\frac{q_0 b^2}{8} - \left(\frac{3+\nu}{1+\nu}\right)\frac{q_0 a^2}{8} + \frac{q_0 b^2}{2}\log a - \frac{q_0 b^4}{2(a^2-b^2)}\log\beta$$

$$c_3 = -\left(\frac{3+\nu}{1-\nu}\right)\frac{q_0 b^2 a^2}{16} - \left(\frac{1+\nu}{1-\nu}\right)\frac{q_0 a^2 b^4}{4(a^2-b^2)}\log\beta$$

$$c_4 = -\frac{q_0 a^4}{64} + \left(\frac{3+\nu}{1+\nu}\right)\frac{q_0 a^2(a^2-b^2)}{32} + \left(\frac{3+\nu}{1-\nu}\right)\frac{q_0 b^2 a^2}{16}\log a$$

$$+ \frac{q_0 b^4 a^2}{8(a^2-b^2)}\log\beta + \left(\frac{1+\nu}{1-\nu}\right)\frac{q_0 b^4 a^2}{4(a^2-b^2)}\log a \log\beta \qquad (5.3.79)$$

The deflection and bending moments become

$$w_0 = \frac{q_0 a^4}{64D}\left\{-\left[1-\left(\frac{r}{a}\right)^4\right] + \frac{2\alpha_1}{1+\nu}\left[1-\left(\frac{r}{a}\right)^2\right] - \frac{4\alpha_2\beta^2}{1-\nu}\log\left(\frac{r}{a}\right)\right\}$$

$$M_{rr} = \frac{q_0 a^2}{16}\left\{(3+\nu)\left[1-\left(\frac{r}{a}\right)^2\right] - \beta^2(3+\nu)\left[1+\left(\frac{r}{a}\right)^2\right]\right.$$

$$\left. + 4(1+\nu)\beta^2\kappa\left[1-\left(\frac{r}{a}\right)^2\right] + 4(1+\nu)\beta^2\log\left(\frac{r}{a}\right)\right\}$$

$$M_{\theta\theta} = \frac{q_0 a^2}{16}\left\{(3+\nu)\left[1-\left(\frac{r}{a}\right)^2\right] + \beta^2\left[(5\nu-1)+(3+\nu)\left(\frac{r}{a}\right)^2\right]\right.$$

$$\left. ++4(1+\nu)\beta^2\kappa\left[1+\left(\frac{r}{a}\right)^2\right]\right\} \qquad (5.3.80a)$$

where

$$\alpha_1 = (3+\nu)(1-\beta^2) - 4(1+\nu)\beta^2\kappa, \quad \alpha_2 = (3+\nu)+4(1+\nu)\kappa \quad (5.3.80b)$$

and β and κ are defined Eq. (5.3.76).

The Rayleigh–Ritz formulation presented in Eqs. (5.3.37)–(5.3.40) is also applicable to annular plates. The approximation functions of Eq. (5.3.50) were derived using the conditions $w_0(a) = 0$ and $dw_0/dr = 0$ at $r = 0$. While $w_0(a) = 0$ is also valid for simply supported annular plates, we no longer require that $dw_0/dr = 0$ at $r = 0$, which is not even a boundary point. The lowest order function $\varphi_1(r)$ that meets the condition $\varphi_1(a) = 0$ is $(1 - r/a)$, which does not meet the continuity requirement. Hence, the next choice is $\varphi_1 = (1 - r^2/a^2)$. Thus, the functions in Eq. (5.3.50) can be used for simply supported annular plates. For an isotropic annular plate with edge moments and edge forces, and subjected to distributed load, Eq. (5.3.51) yields

$$R_{ij} = \frac{2\pi D}{a^2}[ij+(1+\nu)(i+j)+1]\frac{(i+1)(j+1)}{(i+j)}\left(1-\beta^{i+j}\right)$$

$$F_i = 2\pi \left\{ q_0 a^2 \left[(i+1) - (i+3)\beta^2 + 2\beta^{i+3}\right] \frac{1}{2(i+3)} \right.$$
$$\left. + bQ_b \left(1 - \beta^{i+1}\right) - (i+1)M_a + (i+1)\beta^{i+1} M_b \right\} \quad (5.3.81)$$

where $\beta = b/a$. For instance, the one-parameter solution is

$$c_1 = \frac{1}{D(1+\nu)} \left[\frac{q_0 a^4 (1 - \beta^2)}{16} + \frac{Q_b a^2 b}{4} - \frac{a^2}{2(1-\beta^2)} \left(\beta^2 M_b - M_a\right)\right]$$

Of course, although the Rayleigh–Ritz approximation for this choice of φ_i will not give the exact solution (because of the lack of logarithmic terms), it can give a very accurate solution as the number of terms is increased. Alternatively, one may choose an approximation of the form

$$w_0(r) \approx c_1 \log r + c_2 r^2 \log r + \sum_{j=3}^{N} c_j \left(1 - \frac{r^{j+1}}{a^{j+1}}\right)$$

and obtain the exact solution. The first term is needed to capture the response due to the edge moment and distributed load, and the second term is needed to capture the response due to the point load.

5.3.7 Clamped Circular Plate Under Distributed Load

The boundary conditions associated with the clamped outer edges are $w_0 = 0$ and $\frac{dw_0}{dr} = 0$ at $r = a$ (see Figure 5.3.6). Using these conditions in Eqs. (5.3.28) and (5.3.29), we obtain

$$c_2 = -\frac{2F_0'}{a}, \quad c_4 = -F_0 + \frac{F_0' a}{2} \quad (5.3.82)$$

For uniformly distributed load q_0, we have $c_2 = -q_0 a^2/8$ and $c_4 = q_0 a^4/64$. Hence, the deflection (5.3.28) becomes

$$w_0(r) = \frac{q_0 a^4}{64D} \left(1 - \frac{r^2}{a^2}\right)^2 \quad (5.3.83)$$

and the maximum deflection occurs at the center of the plate ($r = 0$):

$$w_{max} = \frac{q_0 a^4}{64D} \quad (5.3.84)$$

Expressions for the bending moments and stresses can be obtained from Eqs. (5.3.30)–(5.3.33):

$$M_{rr}(r) = \frac{q_0 a^2}{16}\left[(1+\nu) - (3+\nu)\frac{r^2}{a^2}\right] \qquad (5.3.85)$$

$$M_{\theta\theta}(r) = \frac{q_0 a^2}{16}\left[(1+\nu) - (1+3\nu)\frac{r^2}{a^2}\right] \qquad (5.3.86)$$

$$\sigma_{rr}(r,z) = \frac{3q_0 a^2 z}{4h^3}\left[(1+\nu) - (3+\nu)\frac{r^2}{a^2}\right] \qquad (5.3.87)$$

$$\sigma_{\theta\theta}(r,z) = \frac{3q_0 a^2 z}{4h^3}\left[(1+\nu) - (1+3\nu)\frac{r^2}{a^2}\right] \qquad (5.3.88)$$

The maximum (in magnitude) values of the bending moments are found to be at the fixed edge:

$$M_{rr}(a) = -\frac{q_0 a^2}{8}, \quad M_{\theta\theta}(a) = -\frac{\nu q_0 a^2}{8} \qquad (5.3.89)$$

Hence, the maximum stress is given by

$$\sigma_{rr}(a, -\frac{h}{2}) = -\frac{6M_{rr}(a)}{h^2} = \frac{3q_0}{4}\left(\frac{a}{h}\right)^2 \qquad (5.3.90)$$

Note that the one-parameter Rayleigh–Ritz or Galerkin solution with $\varphi_1 = (1 - r^2/a^2)^2$ will yield the same solution as in Eq. (5.3.83); see Exercises 12–14 at the end of the chapter.

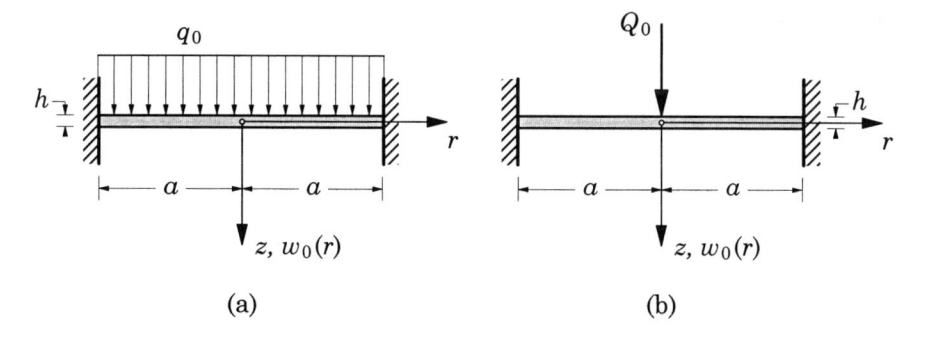

Figure 5.3.6. A clamped circular plate under a uniformly distributed load.

5.3.8 Clamped Circular Plate Under Central Point Load

For this case, we have $c_1 = Q_0/2\pi$ and $c_3 = 0$, where Q_0 is the magnitude of the central point load. The boundary conditions of the clamped edge give

$$\frac{Q_0 a}{8\pi}(2\log a - 1) + \frac{a}{2}c_2 = 0, \quad \frac{Q_0 a^2}{8\pi}(\log a - 1) + \frac{a^2}{4}c_2 + c_4 = 0$$

from which we obtain

$$c_2 = -\frac{Q_0}{4\pi}(2\log a - 1), \quad c_4 = \frac{Q_0 a^2}{16\pi} \qquad (5.3.91)$$

Hence, the solution becomes

$$w_0(r) = \frac{Q_0 a^2}{16\pi D}\left[1 - \frac{r^2}{a^2} + 2\frac{r^2}{a^2}\log\left(\frac{r}{a}\right)\right] \qquad (5.3.92)$$

$$M_{rr}(r) = -\frac{Q_0}{4\pi}\left[1 + (1+\nu)\log\left(\frac{r}{a}\right)\right] \qquad (5.3.93)$$

$$M_{\theta\theta}(r) = -\frac{Q_0}{4\pi}\left[\nu + (1+\nu)\log\left(\frac{r}{a}\right)\right] \qquad (5.3.94)$$

The maximum deflection occurs at $r = 0$ and is given by

$$w_{max} = \frac{Q_0 a^2}{16\pi D} \qquad (5.3.95)$$

Let us consider a Rayleigh–Ritz solution to the problem. The geometric boundary conditions are

$$\frac{dw}{dr} = 0 \text{ at } r = 0 \quad \text{and} \quad w = \frac{dw}{dr} = 0 \text{ at } r = a \qquad (5.3.96)$$

The known natural boundary condition is that the shear force $2\pi(rQ_r)$ at $r = 0$ is Q_0, which enters the variational statement [see Eq. (5.3.37)].

For $N = 2$, and with the approximation functions [the reader is asked to verify that the following functions satisfy the conditions in Eq. (5.3.96)]

$$\varphi_1(r) = 1 - 3\left(\frac{r}{a}\right)^2 + 2\left(\frac{r}{a}\right)^3, \quad \varphi_2(r) = \left[1 - \left(\frac{r}{a}\right)^2\right]^2 \qquad (5.3.97)$$

we obtain from Eq. (5.3.39b,c), with $q = 0$, $k = 0$, $b = 0$, $Q_a = 0$, $2\pi b Q_b = Q_0$, and $D_{11} = D_{22} = D$ the coefficients and the solution

$$R_{11} = \frac{18\pi D}{a^2}, \quad R_{22} = \frac{64\pi}{3a^2}, \quad R_{12} = R_{21} = \frac{96\pi}{5a^2} \tag{5.3.98}$$

$$F_1 = F_2 = Q_0, \quad c_1 = \frac{5Q_0 a^2}{36\pi D}, \quad c_2 = -\frac{5Q_0 a^2}{64\pi D} \tag{5.3.99}$$

$$W_2(r) = \frac{5Q_0 a^2}{576\pi D}\left[7 - 30\left(\frac{r}{a}\right)^2 + 32\left(\frac{r}{a}\right)^3 - 9\left(\frac{r}{a}\right)^4\right] \tag{5.3.100}$$

The one-parameter solution is in 11.2% error while the two-parameter solution is in 3.8% error compared to the exact solution (5.3.95).

5.3.9 Annular Plates with Clamped Outer Edges

Suppose that the inner edge is subjected to a bending moment M_b while the outer edge is clamped (see Figure 5.3.7). The boundary conditions are then

$$\text{At } r = b: \quad M_{rr} = M_b, \quad (rQ_r) = 0 \tag{5.3.101a}$$

$$\text{At } r = a: \quad w_0 = 0, \quad \frac{dw_0}{dr} = 0 \tag{5.3.101b}$$

(a) Inner edge moments

(b) Inner edge line load

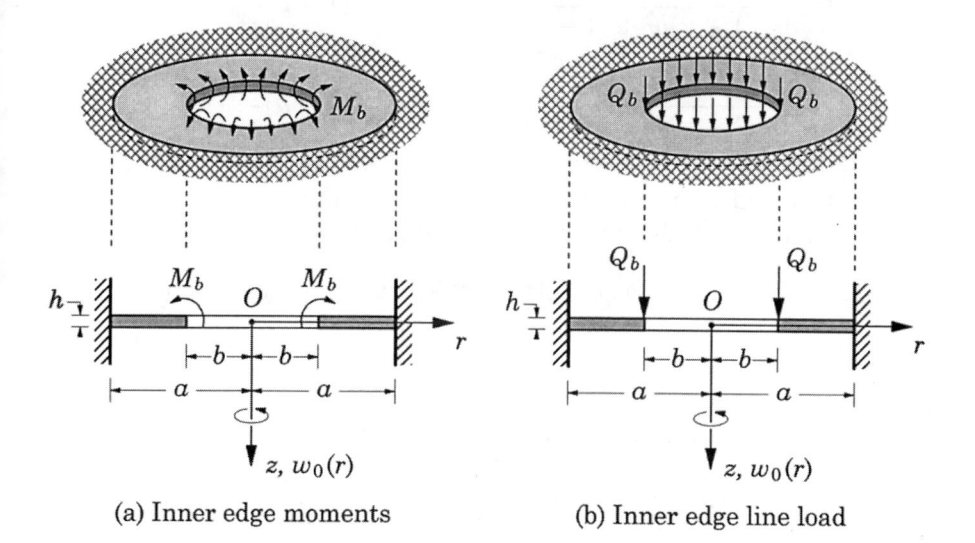

Figure 5.3.7. A clamped annular plate with the inner edge subjected to a (a) bending moment or (b) line load.

216   THEORY AND ANALYSIS OF ELASTIC PLATES

Hence, we obtain from Eq. (5.3.13) $c_1 = 0$ and from Eqs. (5.3.16) and (5.3.17) the relations

$$-\frac{1+\nu}{2}c_2 + \frac{1-\nu}{b^2}c_3 = M_b, \quad \frac{a}{2}c_2 + \frac{1}{a}c_3 = 0$$

$$\frac{a^2}{4}c_2 + c_3 \log a + c_4 = 0 \tag{5.3.102}$$

which give

$$c_2 = -\frac{2M_b b^2}{(1+\nu)b^2 + (1-\nu)a^2}, \quad c_3 = \frac{M_b b^2 a^2}{(1+\nu)b^2 + (1-\nu)a^2}$$

$$c_4 = \frac{1}{2}\left(\frac{M_b b^2 a^2}{(1+\nu)b^2 + (1-\nu)a^2}\right)(1 - 2\log a) \tag{5.3.103}$$

Hence, the deflection and bending moments of Eqs. (5.3.17), (5.3.22) and (5.3.23) for the present problem become

$$w_0(r) = \frac{M_b b^2 a^2}{2[(1+\nu)b^2 + (1-\nu)a^2]}\left[1 - \frac{r^2}{a^2} + 2\log\left(\frac{r}{a}\right)\right] \tag{5.3.104}$$

$$M_{rr}(r) = \frac{M_b b^2}{(1+\nu)b^2 + (1-\nu)a^2}\left[(1+\nu) + (1-\nu)\frac{a^2}{r^2}\right] \tag{5.3.105}$$

$$M_{\theta\theta}(r) = \frac{M_b b^2}{(1+\nu)b^2 + (1-\nu)a^2}\left[(1+\nu) - (1-\nu)\frac{a^2}{r^2}\right] \tag{5.3.106}$$

Next consider an annular plate with the inner edge subjected to transverse line load Q_b. The boundary conditions are then

$$\text{At } r = b: \quad M_{rr} = 0, \quad rQ_r = -bQ_b \tag{5.3.107a}$$

$$\text{At } r = a: \quad w_0 = 0, \quad \frac{dw_0}{dr} = 0 \tag{5.3.107b}$$

Hence, we have from Eq. (5.3.13) $c_1 = bQ_b$ and

$$-\left[\frac{1+\nu}{2}\log b + \frac{1}{4}(1-\nu)\right]bQ_b - \frac{1+\nu}{2}c_2 + \frac{1-\nu}{b^2}c_3 = 0 \tag{5.3.108a}$$

$$\frac{a^2}{4}(\log a - 1)bQ_b + \frac{a^2}{4}c_2 + c_3 \log a + c_4 = 0 \tag{5.3.108b}$$

$$\frac{a}{4}(2\log a - 1)bQ_b + \frac{a}{2}c_2 + \frac{c_3}{a} = 0 \tag{5.3.108c}$$

which gives

$$c_2 = -\frac{bQ_b}{\Delta}\left[(1-\nu)\log a + (1+\nu)\beta^2 \log b - \frac{1}{2}(1-\nu)(1-\beta^2)\right]$$

$$c_3 = -\frac{a^2}{4}[(2\log a - 1)bQ_b + 2c_2]$$

$$c_4 = -\frac{a^2}{4}[(\log a - 1)bQ_b + 2c_2] - c_3 \log a$$

$$\Delta = (1-\nu) + (1+\nu)\beta^2, \quad \beta = \frac{b}{a} \tag{5.3.109}$$

The deflections and bending moments can be obtained from Eqs. (5.3.17), (5.3.22), and (5.3.23).

Now consider an annular plate under uniformly distributed load of intensity q_0 and with a free inner edge. The boundary conditions are

$$\text{At } r = b: \quad M_{rr} = 0, \quad rQ_r = 0 \tag{5.3.110a}$$

$$\text{At } r = a: \quad w_0 = 0, \quad \frac{dw_0}{dr} = 0 \tag{5.3.110b}$$

Hence, we have $c_1 = -q_0 b^2/2$ and

$$-\left(\frac{3+\nu}{16}\right)q_0 b^2 + \left[\frac{1+\nu}{4}\log b + \frac{1-\nu}{8}\right]b^2 q_0 - \frac{1+\nu}{2}c_2 + \frac{1-\nu}{b^2}c_3 = 0$$

$$\frac{q_0 a^4}{64} - \frac{q_0 a^2 b^2}{8}(\log a - 1) + \frac{a^2}{4}c_2 + c_3 \log a + c_4 = 0$$

$$\frac{q_0 a^3}{16} - \frac{q_0 a b^2}{8}(2\log a - 1) + \frac{c_2 a}{2} + \frac{c_3}{a} = 0$$

We obtain

$$c_2 = -\frac{q_0 a^2}{8\Delta}\Big[(1-\nu)(1+2\beta^2) + (1+3\nu)\beta^4 - 4(1-\nu)\beta^2 \log a$$
$$- 4(1+\nu)\beta^4 \log b\Big]$$

$$c_3 = -\frac{q_0 a^4}{16}\left[1 - 2\beta^2(2\log a - 1)\right] - \frac{a^2}{2}c_2$$

$$c_4 = -\frac{q_0 a^4}{64}\left[1 - 8\beta^2(\log a - 1)\right] - \frac{a^2}{4}c_2 - c_3 \log a \tag{5.3.111}$$

where Δ and β are as defined in Eq. (5.3.109). Once again, the deflection and bending moments can be obtained using Eqs. (5.3.17), (5.3.22), and (5.3.23).

218   THEORY AND ANALYSIS OF ELASTIC PLATES

Next, we consider a circular plate clamped to a rigid circular inclusion of radius b at the center, and subjected to a point load Q_0 at the center (see Figure 5.3.8). The boundary conditions are

$$\text{At } r = b: \quad \frac{dw_0}{dr} = 0, \quad 2\pi(rQ_r) = -Q_0 \quad (5.3.112a)$$

$$\text{At } r = a: \quad w_0 = 0, \quad \frac{dw_0}{dr} = 0 \quad (5.3.112b)$$

The boundary conditions yield $c_1 = Q_0/2\pi$ and

$$\frac{b}{8\pi}(2\log b - 1)Q_0 + \frac{b}{2}c_2 + \frac{1}{b}c_3 = 0$$

$$\frac{a^2}{8\pi}(\log a - 1)Q_0 + \frac{a^2}{4}c_2 + \log a \, c_3 + c_4 = 0$$

$$\frac{a}{8\pi}(2\log a - 1)Q_0 + \frac{a}{2}c_2 + \frac{1}{a}c_3 = 0$$

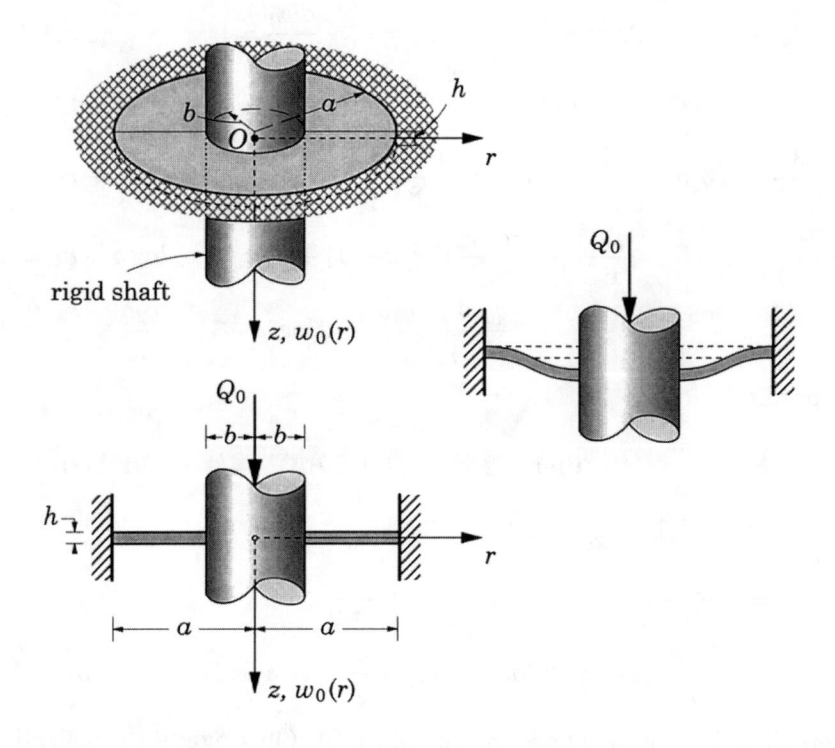

Figure 5.3.8. An annular plate with a rigid shaft at the center and subjected to a central point load.

ANALYSIS OF CIRCULAR PLATES   219

which give ($\beta = b/a$)

$$c_2 = \frac{Q_0}{4\pi}\left(1 + \frac{2\beta^2}{1-\beta^2}\log b - \frac{2}{1-\beta^2}\log a\right)$$

$$c_3 = -\frac{Q_0 b^2}{4\pi}\frac{\log \beta}{1-\beta^2}$$

$$c_4 = \frac{Q_0 a^2}{16\pi}\left[1 - \frac{2\beta^2}{1-\beta^2}\log \beta\,(1 + 2\log a)\right] \qquad (5.3.113)$$

The deflection and bending moments are given by Eqs. (5.3.17), (5.3.22), and (5.3.23).

Finally, we consider a circular plate clamped to a rigid circular inclusion of radius b at the center (see Figure 5.3.9), and subjected to uniformly distributed load q_0. In addition, suppose that it is also clamped at $r = a$. The boundary conditions are

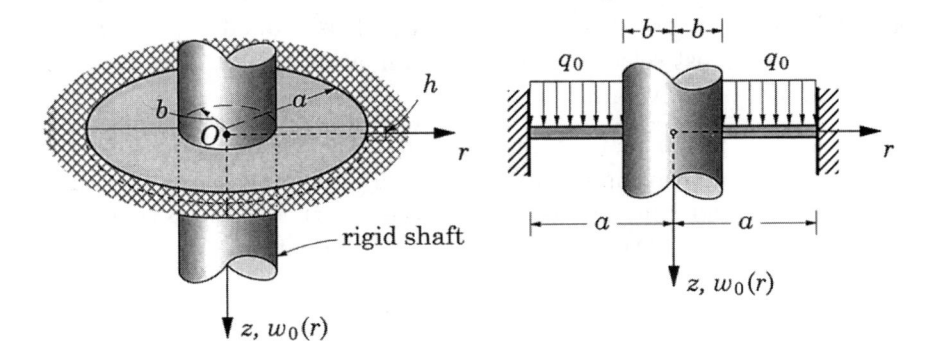

$$\text{At } r = b: \quad \frac{dw_0}{dr} = 0, \quad rQ_r = 0 \qquad (5.3.114a)$$

$$\text{At } r = a: \quad w_0 = 0, \quad \frac{dw_0}{dr} = 0 \qquad (5.3.114b)$$

Figure 5.3.9. An annular plate with a rigid shaft at the center and subjected to uniformly distributed load.

The boundary conditions yield $c_1 = -q_0 b^2/2$ and

$$\frac{q_0 b^3}{16} - \frac{q_0 b^3}{8}(2\log b - 1) + \frac{c_2 b}{2} + \frac{c_3}{b} = 0$$

$$\frac{q_0 a^4}{64} - \frac{q_0 a^4}{8}(\log a - 1) + \frac{a^2}{4}c_2 + c_3 \log a + c_4 = 0$$

$$\frac{q_0 a^3}{16} - \frac{q_0 b^2 a}{8}(2\log a - 1) + \frac{c_2 a}{2} + \frac{c_3}{a} = 0$$

or

$$c_2 = -\frac{q_0 a^2}{8(1-\beta^2)}\left[1 + 2\beta^2(1 - 2\log a) + \beta^4(-3 + 4\log b)\right]$$

$$c_3 = \frac{q_0 a^2 b^2}{16}\left(1 - \frac{4\beta^2}{1-\beta^2}\log \beta\right)$$

$$c_4 = -\frac{q_0 a^4}{64}[1 - 8(\log a - 1)] - \frac{a^2}{4}c_2 + c_3 \log a \quad (5.3.115)$$

5.3.10 Circular Plates on Elastic Foundation

The exact solutions of circular plates resting on an elastic foundation can be developed in terms of the Bessel functions (see Timoshenko and Woinowsky-Krieger [1], Chapter 8). Here we consider Rayleigh–Ritz solutions of circular plates on elastic foundations. The weak form and the Rayleigh–Ritz formulation presented in Eqs. (5.3.37)–(5.3.40) are valid here. We must select suitable approximation functions φ_i for each set of boundary conditions.

For a *free* circular plate on an elastic foundation, the approximation functions for the Rayleigh–Ritz method are required to satisfy only the continuity and completeness conditions, as there are no boundary conditions to satisfy. We begin with the approximation ($N \geq 2$)

$$W_N(r) = c_0 + c_1 r + c_2 r^2 + \cdots + c_N r^N \quad (5.3.116)$$

and note that the c_1–term presents difficulty (logarithmic singularity) when $b = 0$. Hence, we omit it for circular plates but include for annular plates. For solid circular plates, we have from Eq. (5.3.39b,c) the result $R_{ij} = R_{ij}^1 + R_{ij}^2$, where

$$R_{ij}^1 = 2\pi k \int_0^a \varphi_i \varphi_j \, r dr = \frac{2\pi k a^{i+j+2}}{i+i+2} \quad (5.3.117a)$$

when $i = 0, j = 0, 2, \cdots, N$ and when $j = 0, i = 0, 2, \cdots, N$, and

$$R_{ij}^2 = 2\pi \int_b^a \left[D_{11} \frac{d^2\varphi_i}{dr^2} \frac{d^2\varphi_j}{dr^2} + D_{12} \frac{1}{r} \left(\frac{d\varphi_i}{dr} \frac{d^2\varphi_j}{dr^2} + \frac{d^2\varphi_j}{dr^2} \frac{d\varphi_i}{dr} \right) \right.$$
$$\left. + D_{22} \frac{1}{r^2} \frac{d\varphi_i}{dr} \frac{d\varphi_j}{dr} \right] r dr$$
$$= 2\pi \left[ij(i-1)(j-1) D_{11} + ij(i+j-2) D_{12} + ij D_{22} \right] \frac{a^{i+j-2}}{i+j-2}$$
(5.3.117b)

when $i, j = 2, 3, \cdots, N$. Also, we have

$$F_i = \frac{2\pi q_0 a^{i+1}}{i+2} + \hat{Q} \delta_{i0} \qquad (5.3.118)$$

Consider as an example a circular plate resting freely on an elastic foundation and subjected to a center point load Q_0 (see Figure 5.3.10). For $N = 2$ (i.e., $w_0(r) \approx c_0 + c_2 r^2$) and $\hat{Q} = Q_0$, we have

$$2\pi \begin{bmatrix} \frac{ka^2}{2} & \frac{ka^4}{4} \\ \frac{ka^4}{4} & \frac{ka^6}{6} + \alpha \end{bmatrix} \begin{Bmatrix} c_0 \\ c_2 \end{Bmatrix} = Q_0 \begin{Bmatrix} 1 \\ 0 \end{Bmatrix}$$

where

$$\alpha = (2D_{11} + 4D_{12} + 2D_{22}) a^2$$

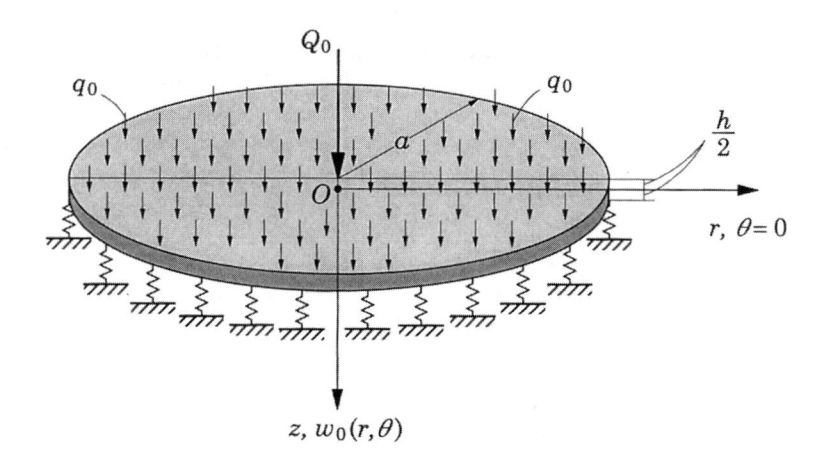

Figure 5.3.10. A circular plate on elastic foundation and subjected to a central point load.

The solution is given by

$$c_0 = \frac{4Q_0}{k\pi a^2}\left(\frac{ka^6 + 6\alpha}{ka^6 + 24\alpha}\right), \quad c_2 = -\frac{6Q_0 a^2}{\pi(ka^6 + 24\alpha)} \quad (5.3.119a)$$

For an isotropic plate, we have [$\alpha = 4(1+\nu)Da^2$ and $\hat{D} = D/(ka^4)$]

$$c_0 = \frac{4Q_0}{\pi ka^2}\left[\frac{1 + 24(1+\nu)\hat{D}}{1 + 96(1+\nu)\hat{D}}\right], \quad c_2 = -\frac{6Q_0}{\pi ka^4[1 + 96(1+\nu)\hat{D}]} \quad (5.3.119b)$$

The maximum deflection occurs at $r = 0$ and it is given by

$$w_{max} = c_0 = \frac{4Q_0}{k\pi a^2}\left[\frac{1 + 24(1+\nu)\hat{D}}{1 + 96(1+\nu)\hat{D}}\right] \quad (5.3.120)$$

Suppose that

$$a = 5 \text{ in.}, \quad \hat{D} \equiv \frac{D}{ka^4} = 1.0, \quad \frac{Q_0}{\pi ka^2} = 408 \times 10^{-4}, \quad \nu = 0.3 \quad (5.3.121)$$

Then we find that $w_{max} = c_0 = 0.04177$ in. This solution is about 3% less than that obtained using the analytical solution based on the Bessel functions (see [1], pp. 263–265).

Although the deflection obtained using the approximation $w_0 \approx c_0 + c_2 r^2$ is very accurate, the bending moments and stresses computed will not be accurate for a circular plate under a central point load. This is due to the fact that the assumed approximation does not contain a term that significantly contributes to the stress calculation. Note from Eqs. (5.3.17) and (5.3.26) that the solution of a circular plate under a central point load consists of a term of the form $r^2 \log r$. Hence, to obtain accurate stresses around the point load, one must include this term in the approximate solution sought

$$w_0(r) \approx c_0 + c_2 r^2 + c_3 r^2 \log r \quad (5.3.122)$$

For circular plates on elastic foundation and with various boundary conditions, one may select the approximation developed earlier. For a simply supported plate on an elastic foundation and subjected to a central point load, for example, approximation functions of the form used in Eq. (5.3.50) may be employed. The coefficient matrix R_{ij}^2 of

ANALYSIS OF CIRCULAR PLATES   223

Eq. (5.3.117b) is valid for the problem, and the coefficients R_{ij}^1 can be computed with

$$\varphi_i(r) = 1 - \left(\frac{r}{a}\right)^{i+1} \tag{5.3.123}$$

We find

$$R_{ij}^1 = 2\pi k a^2 \left[\frac{1}{2} + \frac{1}{i+j+4} - \frac{i+j+6}{(i+3)(j+3)}\right] \tag{5.3.124}$$

For $N = 1$, we have

$$\left[\frac{8\pi(1+\nu)D}{a^2} + \frac{\pi k a^2}{3}\right] c_1 = Q_0$$

or

$$c_1 = \left(\frac{3Q_0}{\pi k a^2}\right) \frac{1}{1 + 24(1+\nu)\hat{D}} \tag{5.3.125}$$

The maximum deflection with the parameters $a = 5$ in., $\hat{D} = 1.0$, $Q_0/\pi k a^2 = 408 \times 10^{-4}$, and $\nu = 0.3$ is

$$w_{max} = c_1 = 0.0038 \text{ in.}$$

5.3.11 Bending of Circular Plates Under Thermal Loads

As discussed in section 5.3.2, Eq. (5.3.17) gives the deflection of a circular plate (solid or annular) under arbitrary thermal loading. We have

$$Dw_0(r) = -G(r) + \frac{r^2}{4}(\log r - 1)c_1 + \frac{r^2}{4}c_2 + c_3 \log r + c_4 \tag{5.3.126a}$$

where

$$G(r) = \frac{1}{(1-\nu)} \int \frac{1}{r} \int r M_T \, dr dr \tag{5.3.126b}$$

The constants c_1 through c_4 must be determined using the boundary conditions. For solid circular plates we necessarily have $c_1 = c_3 = 0$ and the solution becomes

$$Dw_0(r) = -G(r) + c_2 \frac{r^2}{4} + c_4 \tag{5.3.127}$$

Once the deflection is known, the stresses and bending moments may be computed using Eqs. (5.3.9a,b) and (5.3.10a,b).

The boundary conditions for the three standard cases are listed below:

Hinged: $w_0 = 0$, $\quad D\left(\dfrac{d^2 w_0}{dr^2} + \dfrac{\nu}{r}\dfrac{dw_0}{dr}\right) + \dfrac{M_T}{1-\nu} = 0$ $\quad$ (5.3.128)

Fixed: $w_0 = 0$, $\quad \dfrac{dw_0}{dr} = 0$ $\qquad\qquad\qquad\qquad$ (5.3.129)

Free: $D\left(\dfrac{d^2 w_0}{dr^2} + \dfrac{\nu}{r}\dfrac{dw_0}{dr}\right) + \dfrac{M_T}{1-\nu} = 0$, and

$$D\dfrac{d}{dr}\left[\dfrac{1}{r}\dfrac{d}{dr}\left(r\dfrac{dw_0}{dr}\right)\right] + \dfrac{1}{1-\nu}\dfrac{dM_T}{dr} = 0 \quad (5.3.130)$$

The force boundary conditions for free thermal problems are not homogeneous because of the thermal moment M_T.

The constants for typical boundary conditions are given below. In these equations, the notation $M_T(a)$, for example, means that $M_T(r)$ is evaluated at $r = a$.

Circular plate, simply supported at $r = a$:

$$c_2 = \dfrac{2}{1+\nu}\left[G''(a) + \dfrac{\nu}{a}G'(a) - \dfrac{M_T(a)}{1-\nu}\right]$$

$$c_4 = G(a) - \dfrac{a^2}{2(1+\nu)}\left[G''(a) + \dfrac{\nu}{a}G'(a) - \dfrac{M_T(a)}{1-\nu}\right] \quad (5.3.131)$$

Circular plate, fixed at $r = a$:

$$c_2 = \dfrac{2}{a}G'(a), \quad c_4 = G(a) - \dfrac{a}{2}G'(a) \quad (5.3.132)$$

Annular plate, fixed at $r = b$ and $r = a$:

$$c_1 = \dfrac{a_{22}b_1 - a_{12}b_2}{a_{11}a_{22} - a_{12}a_{21}}, \quad c_2 = \dfrac{a_{11}b_2 - a_{21}b_1}{a_{11}a_{22} - a_{12}a_{21}}$$

$$a_{11} = \dfrac{1}{4}\left[b^2(2\log b - 1) - a^2(2\log a - 1)\right]$$

$$a_{12} = \dfrac{1}{4}\left(b^2 - a^2\right), \quad a_{22} = \dfrac{1}{4}\left[b^2 - a^2(1 + \log \beta)\right]$$

$$a_{21} = \frac{1}{4}\left[b^2\left(\log b - 1\right) - a^2\left(\log a - 1\right)\left(1 + \log \beta\right)\right]$$
$$b_1 = bG'(b) - aG'(a), \quad b_2 = G'(b) - G'(a) - aG'(a)\log \beta$$
$$c_3 = aG'(a) - \frac{a^2}{4}\left[(2\log a - 1)c_1 + c_2\right], \quad \beta = \frac{b}{a}$$
$$c_4 = G(a) - aG'(a)\log a + \frac{a^2}{4}\{[2\log a(\log a - 1) + 1]c_1$$
$$+ (\log a - 1)c_2\} \tag{5.3.133}$$

For uniform temperature $\Delta T = T^0$, we find $M_T = 0$; however, N_T will not be zero. For temperature linearly varying through the thickness, $\Delta T = zT^1$, we have $M_T = E\alpha T^1 h^3/12$ and $N_T = 0$.

The Rayleigh–Ritz formulation presented in section 5.3.3 can also be used to determine solutions to thermal problems. Since the force boundary conditions do not enter the construction of the approximation functions, the functions derived earlier for mechanical loading may be used. In fact, solutions developed for circular and annular plates under mechanical loads can be extended/modified for thermal loads.

5.4 Asymmetrical Bending

5.4.1 General Solution

In this section, we consider the bending of circular plates under Non-axisymmetric loads. In other words, the geometry and boundary conditions are axisymmetric but the loading is not axisymmetric. To determine solutions under such loads, we return to the governing equation (5.2.13) developed in section 5.2. For static bending of an isotropic plate, not resting on an elastic foundation and without thermal loads, Eq. (5.2.13) becomes

$$D\left[\frac{1}{r}\frac{\partial}{\partial r}\left(r\frac{\partial}{\partial r}\right) + \frac{1}{r^2}\frac{\partial^2}{\partial \theta^2}\right]\left[\frac{1}{r}\frac{\partial}{\partial r}\left(r\frac{\partial w_0}{\partial r}\right) + \frac{1}{r^2}\frac{\partial^2 w_0}{\partial \theta^2}\right] = q(r,\theta) \tag{5.4.1}$$

The solution w_0 to this equation consists of two parts: a homogeneous solution w_h such that

$$D\left[\frac{1}{r}\frac{\partial}{\partial r}\left(r\frac{\partial}{\partial r}\right) + \frac{1}{r^2}\frac{\partial^2}{\partial \theta^2}\right]\left[\frac{1}{r}\frac{\partial}{\partial r}\left(r\frac{\partial w_h}{\partial r}\right) + \frac{1}{r^2}\frac{\partial^2 w_h}{\partial \theta^2}\right] = 0 \tag{5.4.2a}$$

226 THEORY AND ANALYSIS OF ELASTIC PLATES

and a particular solution w_p so that

$$w_0 = w_h + w_p \tag{5.4.2b}$$

The homogeneous solution may be expressed in the following form

$$w_h(r,\theta) = \sum_{n=0}^{\infty} a_n(r)\cos n\theta + \sum_{n=1}^{\infty} b_n(r)\sin n\theta \tag{5.4.3}$$

where a_n and b_n are functions of r only. Substitution of Eq. (5.4.3) into Eq. (5.4.2a) leads to the following set of equations for a_n and b_n:

$$a_0 = A_0 + B_0 r^2 + C_0 \log r + D_0 r^2 \log r$$
$$a_1 = A_1 r + B_1 r^3 + C_1 r^{-1} + D_1 r \log r$$
$$b_1 = E_1 r + F_1 r^3 + G_1 r^{-1} + H_1 r \log r$$
$$a_n = A_n r^n + B_n r^{-n} + C_n r^{n+2} + D_n r^{-n+2}$$
$$b_n = E_n r^n + F_n r^{-n} + G_n r^{n+2} + H_n r^{-n+2} \tag{5.4.4}$$

for $n = 2, 3, \cdots$ where $A_n, \cdots, H_n$ are constants that are determined using boundary conditions. The solution $w_h = a_0$, which is independent of the angle θ, represents symmetrical bending of circular plates.

5.4.2 General Solution of Circular Plates Under Linearly Varying Asymmetric Loading

Consider a circular plate of radius a and subjected to an asymmetric loading of the type (see Figure 5.4.1)

$$q(r,\theta) = q_0 + q_1 \frac{r}{a} \cos\theta \tag{5.4.5}$$

where q_0 represents the uniform part of the load for which the solution was already determined previously for various boundary conditions. The particular solution w_{p1} due to the uniform load q_0 for this case is given by [see Eq. (5.3.18)]

$$D w_{p1} = \int \frac{1}{r} \int r \int \frac{1}{r} \int r q_0 \, dr dr dr dr = \frac{q_0 r^4}{64} \tag{5.4.6}$$

Similarly, the particular solution due to the second portion of the load is given by

$$D w_{p2} = A \frac{q_1}{a} \cos\theta \left(\int \frac{1}{r} \int r \int \frac{1}{r} \int r^2 \, dr dr dr dr \right)$$
$$= A \frac{q_1}{a} \cos\theta \left(\frac{r^5}{225} \right) \tag{5.4.7}$$

ANALYSIS OF CIRCULAR PLATES   227

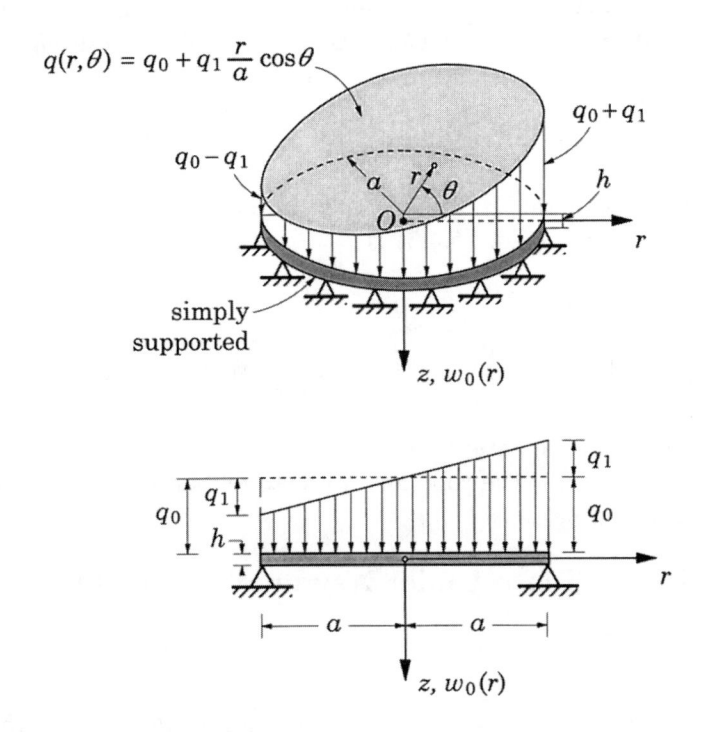

Figure 5.4.1. A circular plate subjected to an asymmetric loading.

where A is a constant to be determined by substituting into

$$D\left[\frac{1}{r}\frac{\partial}{\partial r}\left(r\frac{\partial}{\partial r}\right)+\frac{1}{r^2}\frac{\partial^2}{\partial \theta^2}\right]\left[\frac{1}{r}\frac{\partial}{\partial r}\left(r\frac{\partial w_{p2}}{\partial r}\right)+\frac{1}{r^2}\frac{\partial^2 w_{p2}}{\partial \theta^2}\right]=q_1\frac{r}{a}\cos\theta \quad (5.4.8)$$

We obtain $A = 225/192$. Thus, the particular solution is

$$w_p = w_{p1} + w_{p2} = \frac{q_0 r^4}{64D} + \frac{q_1 r^5}{192 D a}\cos\theta \quad (5.4.9)$$

The fourth-order derivative with respect to θ appearing in Eq. (5.4.2a) and the nature of the load distribution (5.4.5) implies that the homogeneous solution w_h is of the form (see [1,5,6])

$$w_h(r) = \sum_{n=0}^{\infty} a_n(r)\cos n\theta \quad (5.4.10)$$

We take the first two terms of the series. Since the deflection must be finite at $r = 0$, it follows that $C_0 = D_0 = C_1 = D_1 = 0$ in a_0 and a_1. Hence, the homogeneous solution is of the form

$$w_h = A_0 + B_0 r^2 + \left(A_1 r + B_1 r^3 \right) \cos \theta \tag{5.4.11}$$

The complete general solution of a circular plate subjected to asymmetric loading in Eq. (5.4.5) is

$$w_0(r, \theta) = w_p + w_h$$
$$= A_0 + B_0 r^2 + \frac{q_0 r^4}{64D} + \left(\frac{q_1 r^5}{192 D a} + A_1 r + B_1 r^3 \right) \cos \theta \tag{5.4.12}$$

The four constants are determined using the boundary conditions. The procedure to determine the constants for clamped and simply supported plates is discussed next.

5.4.3 Clamped Plate Under Asymmetric Loading

The boundary conditions of a clamped circular plate are

$$w_0 = 0, \quad \frac{\partial w_0}{\partial r} = 0 \quad \text{at} \quad r = a \text{ and for all } \theta \tag{5.4.13}$$

Substituting the deflection expression from Eq. (5.4.12) into the boundary conditions in Eq. (5.4.13), we arrive at

$$\frac{q_0 a^4}{64D} + A_0 + B_0 a^2 + \left(\frac{q_1 a^4}{192D} + A_1 a + B_1 a^3 \right) \cos \theta = 0$$

$$\frac{4 q_0 a^3}{64D} + 2 B_0 a + \left(\frac{5 q_1 a^3}{192D} + A_1 + 3 B_1 a^2 \right) \cos \theta = 0$$

Since the above conditions must hold for all values of θ, it follows that

$$\frac{q_0 a^4}{64D} + A_0 + B_0 a^2 = 0, \quad \frac{q_1 a^4}{192D} + A_1 a + B_1 a^3 = 0$$
$$\frac{4 q_0 a^3}{64D} + 2 B_0 a = 0, \quad \frac{5 q_1 a^3}{192D} + A_1 + 3 B_1 a^2 = 0$$

We obtain

$$A_0 = \frac{q_0 a^4}{64D}, \quad B_0 = -\frac{q_0 a^2}{32D}, \quad A_1 = \frac{q_1 a^3}{192D}, \quad B_1 = -\frac{q_1 a}{96D} \tag{5.4.14}$$

ANALYSIS OF CIRCULAR PLATES 229

The deflection of a clamped plate under asymmetric loading becomes

$$w_0(r,\theta) = \left[\frac{q_0 a^4}{64D} + \frac{q_1 a^4}{192D}\left(\frac{r}{a}\right)\cos\theta\right]\left[1-\left(\frac{r}{a}\right)^2\right]^2 \qquad (5.4.15)$$

The bending moments can be calculated using Eq. (5.2.15a–c) and they are given by

$$M_{rr} = \frac{q_0 a^2}{16}\left[(1+\nu) - (3+\nu)\left(\frac{r}{a}\right)^2\right]$$

$$-\frac{q_1 a^2}{48}\left[(5+\nu)\left(\frac{r}{a}\right)^3 - (3+\nu)\left(\frac{r}{a}\right)\right]\cos\theta \qquad (5.4.16)$$

$$M_{\theta\theta} = \frac{q_0 a^2}{16}\left[(1+\nu) - (1+3\nu)\left(\frac{r}{a}\right)^2\right]$$

$$-\frac{q_1 a^2}{48}\left[(1+5\nu)\left(\frac{r}{a}\right)^3 - (1+3\nu)\left(\frac{r}{a}\right)\right]\cos\theta \qquad (5.4.17)$$

$$M_{r\theta} = -\frac{(1-\nu)q_1 a^2}{48}\left[1-\left(\frac{r}{a}\right)^2\right]\sin\theta \qquad (5.4.18)$$

Note that when $q_1 = 0$, all of the above results reduce to those of a clamped circular plate under a uniformly distributed load [see Eqs. (5.3.83), (5.3.85), and (5.3.86)].

5.4.4 Simply Supported Plate Under Asymmetric Loading

The boundary conditions of a simply supported circular plate are

$$w_0 = 0, \quad M_{rr} = 0 \quad \text{at } r = a \text{ and for all } \theta \qquad (5.4.19)$$

where the expression for M_{rr} in terms of the deflection is given in Eq. (5.2.15a). Substituting the deflection expression from Eq. (5.4.12) into the boundary conditions in Eq. (5.4.19), we arrive at

$$\frac{q_0 a^4}{64D} + A_0 + B_0 a^2 = 0, \quad \frac{q_1 a^3}{192D} + A_1 + B_1 a^2 = 0$$

$$\frac{4(3+\nu)q_0 a^3}{64D} + 2(1+\nu)B_0 a = 0, \quad \frac{4(5+\nu)q_1 a^2}{192D} + 2(3+\nu)B_1 a = 0$$

$$(5.4.20)$$

and

$$A_0 = \left(\frac{5+\nu}{1+\nu}\right)\frac{q_0 a^4}{64D}, \quad B_0 = -\left(\frac{3+\nu}{1+\nu}\right)\frac{q_0 a^2}{32D}$$
$$A_1 = \left(\frac{7+\nu}{3+\nu}\right)\frac{q_1 a^3}{192D}, \quad B_1 = -\left(\frac{5+\nu}{3+\nu}\right)\frac{q_1 a}{96D} \quad (5.4.21)$$

Then the deflection of a simply supported plate under the asymmetric loading becomes

$$w_0(r,\theta) = \frac{q_0 a^4}{64D}\left(1-\frac{r^2}{a^2}\right)\left(\frac{5+\nu}{1+\nu}-\frac{r^2}{a^2}\right) + \frac{q_1 a^4}{(3+\nu)192D}\left(\frac{r}{a}\right)\left(1-\frac{r^2}{a^2}\right)$$
$$\times \left[7+\nu-(3+\nu)\frac{r^2}{a^2}\right]\cos\theta \quad (5.4.22)$$

The bending moments can be calculated using Eq. (5.2.15a–c) as before

$$M_{rr} = \frac{(3+\nu)q_0 a^2}{16}\left[1-\left(\frac{r}{a}\right)^2\right] + \frac{(5+\nu)q_1 a^2}{48}\left[\frac{r}{a}-\left(\frac{r}{a}\right)^3\right]\cos\theta$$
$$(5.4.23)$$

$$M_{\theta\theta} = \frac{(3+\nu)q_0 a^2}{16}\left[1-\left(\frac{1+3\nu}{3+\nu}\right)\left(\frac{r}{a}\right)^2\right]$$
$$+ \frac{(1+3\nu)q_1 a^2}{48}\left[\left(\frac{5+\nu}{3+\nu}\right)\frac{r}{a}-\left(\frac{1+5\nu}{1+3\nu}\right)\left(\frac{r}{a}\right)^2\right]\cos\theta \quad (5.4.24)$$

$$M_{r\theta} = -\frac{(1-\nu)q_1 a^2}{48}\left[\frac{5+\nu}{3+\nu}-\left(\frac{r}{a}\right)^2\right]\sin\theta \quad (5.4.25)$$

5.4.5 Circular Plates Under Non-central Point Load

Consider an isotropic circular plate with a *point load* Q_0 applied at point A at $r = b$. The solution can be sought by dividing the plate into two regions: the inside circle of radius b (shown with dashed line in Figure 5.4.2), and the annulus of inner radius b and outer radius a. Then Eq. (5.4.3) can be applied to each region separately. If the angle θ is measured from the radial line passing through point A, then the solution (5.4.3) containing $\cos n\theta$ should be retained because the solution is symmetric about $\theta = 0$ radial line.

ANALYSIS OF CIRCULAR PLATES 231

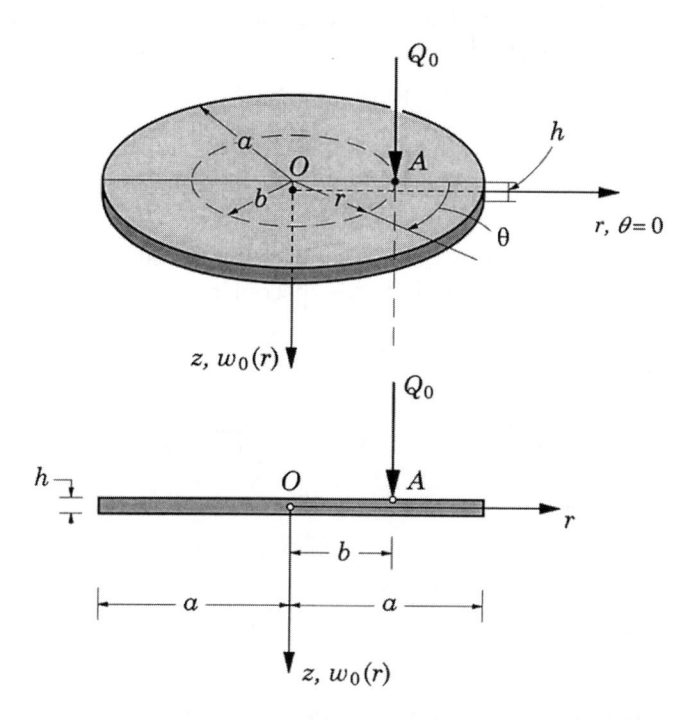

Figure 5.4.2. A circular plate subjected to an asymmetric point load.

For the outer (annulus) region, we have

$$w_0(r,\theta) = \sum_{n=0}^{\infty} a_n \cos n\theta \qquad (5.4.26)$$

where a_0 and a_n are as defined in Eq. (5.4.4). Similarly, the solution for the inner region is

$$w_i(r,\theta) = \sum_{n=0}^{\infty} \hat{a}_n \cos n\theta \qquad (5.4.27)$$

where $\hat{a}_n$ have expressions similar to a_n. The requirement that the deflection, slope, and moments be finite at the center of the plate forces the following constants to be zero:

$$\hat{C}_0 = \hat{D}_0 = \hat{C}_1 = \hat{D}_1 = \hat{B}_n = \hat{D}_n = 0$$

for $n > 1$. Therefore, four constants for each term of the series (5.4.26) and two constants for each term of the series (5.4.27) must be

232   THEORY AND ANALYSIS OF ELASTIC PLATES

determined. The six constants require six conditions for each n. These conditions are developed next.

The two solutions w_0 and w_i must be continuous at $r = b$

$$w_0 = w_i, \quad \frac{\partial w_0}{\partial r} = \frac{\partial w_i}{\partial r}, \quad \frac{\partial^2 w_0}{\partial r^2} = \frac{\partial^2 w_i}{\partial r^2} \qquad (5.4.28)$$

In addition, the shear force is also continuous at all points of the circle except at point A. To represent this discontinuity, first the applied load Q_0 is represented in the form

$$Q_0 = \sum_{n=0}^{\infty} Q_n^0 \cos n\theta = \frac{Q_0}{2\pi} \left(\frac{1}{2} + \sum_{n=1}^{\infty} \cos n\theta \right) \qquad (5.4.29)$$

Next the condition for the discontinuity of shear force is written as [see Eq. (5.2.17a)]

$$D\left(\nabla^2 w_0\right)_{r=b} - D\left(\nabla^2 w_i\right)_{r=b} = \frac{Q_0}{2\pi}\left(\frac{1}{2} + \sum_{n=1}^{\infty} \cos n\theta\right) \qquad (5.4.30)$$

The outer region also has two specified boundary conditions, the nature of which depends on the type of support. For instance, if the plate is clamped, we have

$$w_0(a) = 0, \quad \left(\frac{\partial w_0}{\partial r}\right)_{r=a} = 0 \qquad (5.4.31)$$

Equations (5.4.28), (5.4.30), and (5.4.31) provide the necessary six conditions to determine the six constants.

For a clamped circular plate, the functions $a_0(r), \cdots, \hat{a}_n(r)$ are obtained as (see [1])

$$a_0 = \frac{Q_0 a^2}{16\pi D}\left[2\left(\beta^2 + \bar{r}^2\right)\log \bar{r} + \left(1 + \beta^2\right)\left(1 - \bar{r}^2\right)\right]$$

$$\hat{a}_0 = \frac{Q_0 a^2}{16\pi D}\left[2\left(\beta^2 + \bar{r}^2\right)\log \beta + \left(1 - \beta^2\right)\left(1 + \bar{r}^2\right)\right]$$

$$a_1 = -\frac{Q_0 ab}{16\pi D}\left[\frac{\beta^2}{\bar{r}} + 2\left(1 - \beta^2\right)\bar{r} - \left(2 - \beta^2\right)\bar{r}^3 - 4\bar{r}\log\frac{1}{\bar{r}}\right]$$

$$\hat{a}_1 = -\frac{Q_0 ab}{16\pi D}\left[2\left(1-\beta^2\right)\bar{r} + \frac{(1-\beta^2)^2}{\beta^2}\bar{r}^3 - 4\bar{r}\log\frac{1}{\beta}\right]$$

$$a_n = \frac{Q_0 a^2 \beta^n}{8n(n-1)\pi D}\left\{\bar{r}^n\left[(n-1)\beta^2 - n + (n-1)\bar{r}^2 - \frac{n(n-1)}{n+1}\beta^2\bar{r}^2\right]\right.$$
$$\left. + \frac{1}{\bar{r}^n}\left(\bar{r}^2 - \frac{n-1}{n+1}\beta^2\right)\right\}$$

$$\hat{a}_n = \frac{Q_0 a^2 \beta^n}{8n(n-1)\pi D}\left\{\bar{r}^n\left[(n-1)\beta^2 - n + \frac{1}{\beta^{n-2}}\right]\right.$$
$$\left. + \frac{n-1}{n+1}\bar{r}^{n+2}\left[1 + n\left(1-\beta^2\right) - \frac{1}{\beta^{2n}}\right]\right\} \quad (5.4.32)$$

where $\bar{r} = r/a$ and $\beta = b/a$. The deflection under the load is given by

$$w_0(b,0) = \frac{Q_0 a^2}{16\pi D}(1-\beta^2)^2 \quad (5.4.33)$$

Solutions to circular plates under other loads and boundary conditions can be developed using the procedure discussed above. See Chapter 9 of Timoshenko and Woinowski-Krieger [1] and sections 1.13 and 1.14 of Szilard [4] for additional details.

5.4.6 Rayleigh–Ritz Formulation

Here, we present the Rayleigh–Ritz formulation for the bending of of circular plates under general loading and boundary conditions. The weak formulation or the statement of virtual work for this case was already presented in Eq. (5.2.23). For the linear case, the in-plane deformation is uncoupled from the bending deformation. For static bending case, Eq. (5.2.23) simplifies to

$$0 = \int_\Omega\left[-M_{rr}\frac{\partial^2 \delta w_0}{\partial r^2} - M_{\theta\theta}\left(\frac{1}{r}\frac{\partial \delta w_0}{\partial r} + \frac{1}{r^2}\frac{\partial^2 \delta w_0}{\partial \theta^2}\right)\right.$$
$$\left. - 2M_{r\theta}\left(\frac{1}{r}\frac{\partial^2 \delta w_0}{\partial r \partial \theta} - \frac{1}{r^2}\frac{\partial \delta w_0}{\partial \theta}\right) - (q - kw_0)\delta w_0\right]r\,dr\,d\theta \quad (5.4.34)$$

where Ω denotes the r-θ plane ($z = 0$). Rewriting Eq. (5.4.34) in terms of the displacements by employing the plate constitutive equations (5.2.47a–c), we obtain

$$0 = \int_\Omega\left\{\left[D_{11}\frac{\partial^2 w_0}{\partial r^2} + D_{12}\left(\frac{1}{r}\frac{\partial w_0}{\partial r} + \frac{1}{r^2}\frac{\partial^2 w_0}{\partial \theta^2}\right) + M_{rr}^T\right]\frac{\partial^2 \delta w_0}{\partial r^2}\right.$$

234 THEORY AND ANALYSIS OF ELASTIC PLATES

$$+ \left[D_{12} \frac{\partial^2 w_0}{\partial r^2} + D_{22} \left(\frac{1}{r} \frac{\partial w_0}{\partial r} + \frac{1}{r^2} \frac{\partial^2 w_0}{\partial \theta^2} \right) + M_{\theta\theta}^T \right]$$

$$\times \left(\frac{1}{r} \frac{\partial \delta w_0}{\partial r} + \frac{1}{r^2} \frac{\partial^2 \delta w_0}{\partial \theta^2} \right)$$

$$+ 4D_{66} \left(\frac{1}{r} \frac{\partial^2 w_0}{\partial r \partial \theta} - \frac{1}{r^2} \frac{\partial w_0}{\partial \theta} \right) \left(\frac{1}{r} \frac{\partial^2 \delta w_0}{\partial r \partial \theta} - \frac{1}{r^2} \frac{\partial \delta w_0}{\partial \theta} \right) \Big\} r \, dr \, d\theta$$

$$- \int_\Omega (q - k w_0) \, \delta w_0 \, r \, dr \, d\theta \tag{5.4.35a}$$

Rearranging the terms, we arrive at

$$0 = \int_\Omega \Big[D_{11} \frac{\partial^2 w_0}{\partial r^2} \frac{\partial^2 \delta w_0}{\partial r^2} + D_{12} \frac{1}{r} \left(\frac{\partial w_0}{\partial r} \frac{\partial^2 \delta w_0}{\partial r^2} + \frac{\partial \delta w_0}{\partial r} \frac{\partial^2 w_0}{\partial r^2} \right.$$

$$\left. + \frac{1}{r} \frac{\partial^2 w_0}{\partial \theta^2} \frac{\partial^2 \delta w_0}{\partial r^2} + \frac{1}{r} \frac{\partial^2 \delta w_0}{\partial \theta^2} \frac{\partial^2 w_0}{\partial r^2} \right)$$

$$+ D_{22} \left(\frac{1}{r} \frac{\partial w_0}{\partial r} + \frac{1}{r^2} \frac{\partial^2 w_0}{\partial \theta^2} \right) \left(\frac{1}{r} \frac{\partial \delta w_0}{\partial r} + \frac{1}{r^2} \frac{\partial^2 \delta w_0}{\partial \theta^2} \right)$$

$$+ 4D_{66} \left(\frac{1}{r} \frac{\partial^2 w_0}{\partial r \partial \theta} - \frac{1}{r^2} \frac{\partial w_0}{\partial \theta} \right) \left(\frac{1}{r} \frac{\partial^2 \delta w_0}{\partial r \partial \theta} - \frac{1}{r^2} \frac{\partial \delta w_0}{\partial \theta} \right)$$

$$+ M_{rr}^T \frac{\partial^2 \delta w_0}{\partial r^2} + M_{\theta\theta}^T \left(\frac{1}{r} \frac{\partial \delta w_0}{\partial r} + \frac{1}{r^2} \frac{\partial^2 \delta w_0}{\partial \theta^2} \right)$$

$$- (q - k w_0) \delta w_0 \Big] r \, dr \, d\theta \tag{5.4.35b}$$

Now it is easy to identify the total potential energy functional $\Pi(w_0)$ from Eq. (5.4.35b). It is given by

$$\Pi = \int_\Omega \Big[\frac{D_{11}}{2} \left(\frac{\partial^2 w_0}{\partial r^2} \right)^2 + D_{12} \left(\frac{1}{r} \frac{\partial w_0}{\partial r} \frac{\partial^2 w_0}{\partial r^2} + \frac{1}{r^2} \frac{\partial^2 w_0}{\partial \theta^2} \frac{\partial^2 w_0}{\partial r^2} \right)$$

$$+ \frac{D_{22}}{2} \left(\frac{1}{r} \frac{\partial w_0}{\partial r} + \frac{1}{r^2} \frac{\partial^2 w_0}{\partial \theta^2} \right)^2 + 2 D_{66} \left(\frac{1}{r} \frac{\partial^2 w_0}{\partial r \partial \theta} - \frac{1}{r^2} \frac{\partial w_0}{\partial \theta} \right)^2$$

$$+ M_{rr}^T \frac{\partial^2 w_0}{\partial r^2} + M_{\theta\theta}^T \left(\frac{1}{r} \frac{\partial w_0}{\partial r} + \frac{1}{r^2} \frac{\partial^2 w_0}{\partial \theta^2} \right) - q w_0 + \frac{k}{2} w_0^2 \Big] r \, dr \, d\theta$$

$$\tag{5.4.36}$$

For an isotropic plate, the total potential energy functional is given by

$$\Pi = \int_\Omega \left\{ \frac{D}{2} \left[\left(\frac{\partial^2 w_0}{\partial r^2}\right)^2 + 2\frac{\nu}{r}\left(\frac{\partial w_0}{\partial r} + \frac{1}{r}\frac{\partial^2 w_0}{\partial \theta^2}\right)\frac{\partial^2 w_0}{\partial r^2} \right.\right.$$
$$\left. + \left(\frac{1}{r}\frac{\partial w_0}{\partial r} + \frac{1}{r^2}\frac{\partial^2 w_0}{\partial \theta^2}\right)^2 + 2(1-\nu)\left(\frac{1}{r}\frac{\partial^2 w_0}{\partial r \partial \theta} - \frac{1}{r^2}\frac{\partial w_0}{\partial \theta}\right)^2 \right]$$
$$\left. + \frac{M_T}{1-\nu}\left(\frac{\partial^2 w_0}{\partial r^2} + \frac{1}{r}\frac{\partial w_0}{\partial r} + \frac{1}{r^2}\frac{\partial^2 w_0}{\partial \theta^2}\right) - qw_0 + \frac{k}{2}w_0^2 \right\} r\,dr\,d\theta$$
(5.4.37)

where

$$M_T(r,\theta) = E\alpha \int_{-\frac{h}{2}}^{\frac{h}{2}} \Delta T(r,\theta,z)\, z\, dz \qquad (5.4.38)$$

Assume an N-parameter Rayleigh–Ritz approximation of the form

$$w_0(r,\theta) \approx \sum_{j=1}^{N} c_j \varphi_j(r,\theta) + \varphi_0(r,\theta) \qquad (5.4.39)$$

Substituting Eq. (5.4.39) into Eq. (5.4.35b), we find

$$0 = \sum_{j=1}^{N} \left(R_{ij}^1 + R_{ij}^2\right) c_j - \left(F_j^1 + F_i^2\right) \qquad (5.4.40a)$$

where

$$R_{ij}^1 = \int_\Omega \left[D_{11}\frac{\partial^2 \varphi_i}{\partial r^2}\frac{\partial^2 \varphi_j}{\partial r^2} + D_{22}\left(\frac{1}{r}\frac{\partial \varphi_i}{\partial r} + \frac{1}{r^2}\frac{\partial^2 \varphi_i}{\partial \theta^2}\right)\left(\frac{1}{r}\frac{\partial \varphi_j}{\partial r} + \frac{1}{r^2}\frac{\partial^2 \varphi_j}{\partial \theta^2}\right) \right.$$
$$+ \frac{D_{12}}{r}\left(\frac{\partial \varphi_i}{\partial r}\frac{\partial^2 \varphi_j}{\partial r^2} + \frac{\partial \varphi_j}{\partial r}\frac{\partial^2 \varphi_i}{\partial r^2} + \frac{1}{r}\frac{\partial^2 \varphi_i}{\partial \theta^2}\frac{\partial^2 \varphi_j}{\partial r^2} + \frac{1}{r}\frac{\partial^2 \varphi_j}{\partial \theta^2}\frac{\partial^2 \varphi_i}{\partial r^2}\right)$$
$$\left. + 4D_{66}\left(\frac{1}{r}\frac{\partial^2 \varphi_i}{\partial r \partial \theta} - \frac{1}{r^2}\frac{\partial \varphi_i}{\partial \theta}\right)\left(\frac{1}{r}\frac{\partial^2 \varphi_j}{\partial r \partial \theta} - \frac{1}{r^2}\frac{\partial \varphi_j}{\partial \theta}\right) \right] r\,dr\,d\theta$$
(5.4.40b)

$$R_{ij}^2 = k\int_\Omega \varphi_i \varphi_j\, r\,dr\,d\theta \qquad (5.4.40c)$$

$$F_i^1 = \int_\Omega \left[-D_{11}\frac{\partial^2 \varphi_i}{\partial r^2}\frac{\partial^2 \varphi_0}{\partial r^2} - D_{22}\left(\frac{1}{r}\frac{\partial \varphi_i}{\partial r} + \frac{1}{r^2}\frac{\partial^2 \varphi_i}{\partial \theta^2}\right)\left(\frac{1}{r}\frac{\partial \varphi_0}{\partial r} + \frac{1}{r^2}\frac{\partial^2 \varphi_0}{\partial \theta^2}\right) \right.$$

$$-\frac{D_{12}}{r}\left(\frac{\partial\varphi_i}{\partial r}\frac{\partial^2\varphi_0}{\partial r^2} + \frac{\partial\varphi_0}{\partial r}\frac{\partial^2\varphi_i}{\partial r^2} + \frac{1}{r}\frac{\partial^2\varphi_i}{\partial\theta^2}\frac{\partial^2\varphi_0}{\partial r^2} + \frac{1}{r}\frac{\partial^2\varphi_0}{\partial\theta^2}\frac{\partial^2\varphi_i}{\partial r^2}\right)$$
$$-4D_{66}\left(\frac{1}{r}\frac{\partial^2\varphi_i}{\partial r\partial\theta} - \frac{1}{r^2}\frac{\partial\varphi_i}{\partial\theta}\right)\left(\frac{1}{r}\frac{\partial^2\varphi_0}{\partial r\partial\theta} - \frac{1}{r^2}\frac{\partial\varphi_0}{\partial\theta}\right) + q\varphi_i\bigg]rdrd\theta$$
(5.4.40d)

$$F_i^2 = -\frac{1}{1-\nu}\int_\Omega M_T(r,\theta)\left[\frac{\partial^2\varphi_i}{\partial r^2} + \left(\frac{1}{r}\frac{\partial\varphi_i}{\partial r} + \frac{1}{r^2}\frac{\partial^2\varphi_i}{\partial\theta^2}\right)\right]rdrd\theta \quad (5.4.40e)$$

This completes the formulation of the Rayleigh–Ritz equations for an arbitrarily loaded, orthotropic circular plate.

As an example, consider a clamped circular plate on elastic foundation and subjected to asymmetric loading of the form (see section 5.4.2)

$$q(r,\theta) = q_1\frac{r}{a}\cos\theta \quad (5.4.41a)$$

and with a temperature distribution of the form

$$\Delta T(r,\theta,z) = T(z)\cos\theta \quad (5.4.41b)$$

so that $M_T = M_T^0\cos\theta$, where

$$M_T^0 = E\alpha\int_{-\frac{h}{2}}^{\frac{h}{2}} T(z)\,z\,dz \quad (5.4.41c)$$

For this case, the geometric boundary conditions are

$$w_0(r,\theta) = \frac{\partial w_0}{\partial r} = 0 \text{ at } r = a \text{ and } \frac{\partial w_0}{\partial r} = 0 \text{ at } r = 0 \quad (5.4.42)$$

for any θ. For one-parameter Rayleigh–Ritz solution we can assume

$$w_0(r,\theta) \approx c_1\varphi_1(r,\theta) \equiv c_1 f_1(r)f_2(\theta) \quad (5.4.43)$$

where f_1 is only a function of r and f_2 is only a function of θ. Therefore, the conditions on the approximation functions

$$\varphi_i(r,\theta) = \frac{\partial\varphi_i}{\partial r} = 0 \text{ at } r = a \text{ and } \frac{\partial\varphi_i}{\partial r} = 0 \text{ at } r = 0 \quad (5.4.44)$$

translate into the conditions $f_1(a) = f_1'(a) = f_1'(0) = 0$. Clearly, the choice

$$f_1(r) = 1 - 3\left(\frac{r}{a}\right)^2 + 2\left(\frac{r}{a}\right)^3 \quad (5.4.45)$$

satisfies the conditions. The form of the load suggests that $f_2 = \cos\theta$. Since
$$f_1' = -\frac{6r}{a^2} + \frac{6r^2}{a^3}, \quad f_1'' = -\frac{6}{a^2} + \frac{12r}{a^3}$$
it is clear that the integrals
$$\int_0^a \frac{1}{r} f_1 f_1'' \, dr, \quad \int_0^a \frac{1}{r^3} f_1 f_1 \, dr \tag{5.4.46}$$
required in R_{11}^1 of Eq. (5.4.40a) do not exist because of the logarithmic singularity. Thus $f_1(r)$ defined in Eq. (5.4.45) is not admissible. The next function that satisfies the boundary conditions $f_1(a) = f_1'(a) = f_1'(0) = 0$ is
$$f_1(r) = \left[1 - \left(\frac{r}{a}\right)^2\right]^2 = 1 - 2\left(\frac{r}{a}\right)^2 + \left(\frac{r}{a}\right)^4 \tag{5.4.47}$$
However, this too does not permit the evaluation of the integrals in Eq. (5.4.46). The next admissible function is
$$f_1(r) = \frac{r}{a}\left[1 - \left(\frac{r}{a}\right)^2\right]^2 = \frac{r}{a} - 2\left(\frac{r}{a}\right)^3 + \left(\frac{r}{a}\right)^5 \tag{5.4.48}$$
which does not present any problem with the evaluation of the necessary integrals. Substituting $\varphi_1 = f_1(r)\cos\theta$ into R_{11}^1 of Eq. (5.4.40b) with $D_{11} = D_{22} = D$, $D_{12} = \nu D$, and $2D_{66} = (1-\nu)D$ and into F_1^1 and F_1^2 of Eqs. (5.4.40d,e), we obtain
$$R_{11}^1 = D \int_0^{2\pi} \int_0^a \left\{\left[f_1'' f_1'' + \left(\frac{1}{r}f_1' - \frac{1}{r^2}f_1\right)^2 + \frac{2\nu}{r}\left(f_1'f_1'' - \frac{1}{r}f_1 f_1''\right)\right.\right.$$
$$\left.\left. + \frac{M_T^0}{1-\nu}\right]\cos^2\theta + 2(1-\nu)\left(\frac{1}{r}f' - \frac{1}{r^2}f_1\right)^2 \sin^2\theta\right\} r\, dr\, d\theta$$
$$= \frac{D\pi}{a^2}\left[6 + \frac{2}{3} - \nu + \frac{7\nu}{3} + \frac{4(1-\nu)}{3}\right] = \frac{8D\pi}{a^2} \tag{5.4.49a}$$
$$R_{11}^2 = k \int_0^{2\pi}\int_0^a f_1 f_1 \cos^2\theta\, r\, dr\, d\theta = \frac{ka^2\pi}{60} \tag{5.4.49b}$$
$$F_1^1 = \int_0^{2\pi}\int_0^a \left(q_1\frac{r}{a}\cos\theta\right)f_1\cos\theta\, r\, dr\, d\theta = \frac{q_1 a^2\pi}{24} \tag{5.4.49c}$$
$$F_1^2 = -\frac{M_T^0}{1-\nu}\int_0^{2\pi}\int_0^a \left(f_1'' + \frac{1}{r}f_1' - \frac{1}{r^2}f_1\right)\cos^2\theta\, r\, dr\, d\theta$$
$$= \frac{8M_T^0\pi}{15(1-\nu)} \tag{5.4.49d}$$

where we used the identities

$$\cos^2\theta = \frac{1}{2}(1+\cos 2\theta), \quad \sin^2\theta = \frac{1}{2}(1-\cos 2\theta)$$

to evaluate the integrals. Thus, we have

$$\left(R_{11}^1 + R_{11}^2\right)c_1 = F_1^1 + F_1^2$$

which gives

$$c_1 = \frac{1}{\left(1 + \frac{ka^4}{480D}\right)}\left[\frac{q_1 a^4}{192D} + \frac{M_T^0 a^2}{15(1-\nu)D}\right]$$

and the Rayleigh–Ritz solution becomes

$$w_0(r,\theta) = \frac{1}{\left(1+\frac{ka^4}{480D}\right)}\left[\frac{q_1 a^4}{192D} + \frac{M_T^0 a^2}{15(1-\nu)D}\right]\left\{\frac{r}{a}\left[1-\left(\frac{r}{a}\right)^2\right]^2\right\}\cos\theta \quad (5.4.50)$$

which coincides with the exact solution in Eq. (5.4.15) for the case $q_0 = 0$, $k = 0$, and $M_T^0 = 0$.

5.5 Free Vibration
5.5.1 Introduction

In the preceding sections of this chapter, we considered bending of circular and annular plates. Here we consider natural vibration. The equation of motion of an isotropic plate is given by Eq. (5.2.53)

$$D\nabla^2\nabla^2 w_0 + kw_0 + I_0\frac{\partial^2 w_0}{\partial t^2} - I_2\frac{\partial^2}{\partial t^2}\left(\nabla^2 w_0\right) = 0 \quad (5.5.1)$$

where $I_0 = \rho h$ and $I_2 = \rho h^3/12$ are the principal and rotatory inertias, respectively. When free vibration is assumed, the deflection is periodic and can be expressed as

$$w_0(r,\theta,t) = W(r,\theta)\cos\omega t \quad (5.5.2)$$

where ω is the circular frequency of vibration (radians per unit time), and W is a function of only r and θ. Substituting Eq. (5.5.2) into Eq. (5.5.1), we obtain

$$D\nabla^2\nabla^2 W + kW - I_0\omega^2 W + I_2\omega^2\nabla^2 W = 0 \quad (5.5.3)$$

The presence of the rotatory inertia I_2 presents difficulties while it contributes little to the frequencies, especially to the fundamental frequency. Hence, we neglect the rotatory inertia term. Equation (5.5.3) becomes

$$\left(\nabla^4 - \beta^4\right) W = 0 \tag{5.5.4}$$

where

$$\beta^4 = \frac{I_0 \omega^2 - k}{D} \tag{5.5.5}$$

Equation (5.5.4) can be factored into

$$\left(\nabla^2 + \beta^2\right)\left(\nabla^2 - \beta^2\right) W = 0 \tag{5.5.6}$$

so that the complete solution to Eq. (5.5.6) can be obtained by superimposing the solutions of the equations

$$\nabla^2 W_1 + \beta^2 W_1 = 0, \quad \nabla^2 W_2 - \beta^2 W_2 = 0 \tag{5.5.7}$$

5.5.2 General Analytical Solution

We assume solution to Eq. (5.5.4) in the form of the general Fourier series

$$W(r, \theta) = \sum_{n=0}^{\infty} W_n(r) \cos n\theta + \sum_{n=1}^{\infty} W_n^*(r) \sin n\theta \tag{5.5.8}$$

Substitution of Eq. (5.5.8) into Eq. (5.5.7) yields

$$\frac{d^2 W_{n1}}{dr^2} + \frac{1}{r}\frac{dW_{n1}}{dr} - \left(\frac{n^2}{r^2} - \beta^2\right) W_{n1} = 0 \tag{5.5.9a}$$

$$\frac{d^2 W_{n2}}{dr^2} + \frac{1}{r}\frac{dW_{n2}}{dr} - \left(\frac{n^2}{r^2} - \beta^2\right) W_{n2} = 0 \tag{5.5.9b}$$

and two identical equations for W_n^* (W_{n1}^* and W_{n2}^*). Equations (5.5.9a,b) are a form of Bessel's equations, which have the solutions

$$W_{n1} = A_n J_n(\beta r) + B_n Y_n(\beta r), \quad W_{n2} = C_n I_n(\beta r) + D_n K_n(\beta r) \tag{5.5.10}$$

respectively, where J_n and Y_n are the Bessel functions of first and second kind, respectively, and I_n and K_n are the modified Bessel functions

240 THEORY AND ANALYSIS OF ELASTIC PLATES

of the first and second kind, respectively (see [9]). The coefficients A_n, B_n, C_n, and D_n, which determine the mode shapes, are solved using the boundary conditions. Thus, the general solution of Eq. (5.5.4) is

$$W(r,\theta) = \sum_{n=0}^{\infty} [A_n J_n(\beta r) + B_n Y_n(\beta r) + C_n I_n(\beta r) + D_n K_n(\beta r)] \cos n\theta$$
$$+ \sum_{n=1}^{\infty} [A_n^* J_n(\beta r) + B_n^* Y_n(\beta r) + C_n^* I_n(\beta r) + D_n^* K_n(\beta r)] \sin n\theta$$

(5.5.11)

For solid circular plates, the terms involving Y_n and K_n in the solution (5.5.11) must be discarded in order to avoid singularity of deflections and stresses (i.e., avoid infinite values) at the origin, $r = 0$. In addition, if the boundary conditions are symmetrically applied about a diameter of the plate, then the second expression containing $\sin n\theta$ is not needed to represent the solution. Then, the nth term of Eq. (5.5.11) becomes

$$W_n(r,\theta) = [A_n J_n(\beta r) + C_n I_n(\beta r)] \cos n\theta \qquad (5.5.12)$$

for $n = 0, 1, \cdots, \infty$. A *nodal line* is one which has zero deflection (i.e., $W_n = 0$). For circular plates nodal lines are either concentric circles or diameters. The nodal diameters are determined by $n\theta = \pi/2, 3\pi/2, \cdots$.

Next we apply boundary conditions (which are homogeneous) to obtain a pair of algebraic equations among A_n and C_n for any n. For a nontrivial solution of the pair, the determinant of the matrix representing the two equations is set to zero. The resulting equation is known as the characteristic polynomial. In the following paragraphs, we consider several typical boundary conditions.

5.5.3 Clamped Circular Plates

The boundary conditions are

$$W_n = 0 \text{ and } \frac{\partial W_n}{\partial r} = 0 \text{ at } r = a \text{ for any } \theta \qquad (5.5.13)$$

Using Eq. (5.5.12) in Eq. (5.5.13), we obtain

$$\begin{bmatrix} J_n(\lambda) & I_n(\lambda) \\ J_n'(\lambda) & I_n'(\lambda) \end{bmatrix} \begin{Bmatrix} A_n \\ C_n \end{Bmatrix} = \begin{Bmatrix} 0 \\ 0 \end{Bmatrix} \qquad (5.5.14)$$

where $\lambda = \beta a$ and the prime denotes differentiation with respect to the argument, βr. For nontrivial solution we set the determinant of the coefficient matrix in Eq. (5.5.14) to zero

$$\begin{vmatrix} J_n(\lambda) & I_n(\lambda) \\ J'_n(\lambda) & I'_n(\lambda) \end{vmatrix} = 0 \qquad (5.5.15)$$

Expanding the determinant and using the recursion relations

$$\lambda J'_n(\lambda) = nJ_n(\lambda) - \lambda J_{n+1}(\lambda), \quad \lambda I'_n(\lambda) = nI_n(\lambda) + \lambda I_{n+1}(\lambda) \qquad (5.5.16)$$

we find

$$J_n(\lambda)I_{n+1}(\lambda) + I_n(\lambda)J_{n+1}(\lambda) = 0 \qquad (5.5.17a)$$

or

$$\frac{J_{n+1}(\lambda)}{J_n(\lambda)} + \frac{I_{n+1}(\lambda)}{I_n(\lambda)} = 0 \qquad (5.5.17b)$$

which is called the *frequency equation*. The roots λ of Eq. (5.5.17), called the eigenvalues, are used to determine the frequencies ω [see Eq. (5.5.5)]

$$\omega^2 = \frac{D\beta^4 + k}{I_0} = \frac{D\lambda^4 + ka^4}{a^4 I_0} \qquad (5.5.18)$$

When $k = 0$, Eq. (5.5.18) reduces to

$$\omega^2 = \frac{D\lambda^4}{a^4 I_0} \quad \text{or} \quad \lambda^2 = \omega a^2 \sqrt{I_0/D} \qquad (5.5.19)$$

Note that the frequencies of a clamped circular plate do not depend on Poisson's ratio.

There are an infinite number of roots λ of Eq. (5.5.17) for each value of n, which represents the number of nodal diameters. For instance, when $n = 0$ (i.e., when the only nodal diameter is the boundary circle) the roots in order of magnitude correspond with $1, 2, \cdots, m$ nodal circles. The mode shape associated with λ is determined using Eq. (5.5.15)

$$\frac{A_n}{C_n} = -\frac{I_n(\lambda)}{J_n(\lambda)} \qquad (5.5.20)$$

where λ is the solution (i.e., root) of Eq. (5.5.17). The radii of nodal circles $\xi = r/a$ are determined from Eqs. (5.5.20) and (5.5.12)

$$\frac{J_n(\lambda\xi)}{J_n(\lambda)} = \frac{I_n(\lambda\xi)}{I_n(\lambda)} \qquad (5.5.21)$$

Values of λ^2 taken from [8–10] are presented in Table 5.5.1, where n denotes the number of nodal diameters and m is the number of nodal circles, not including the circle $r = a$. Figure 5.5.1 shows typical nodal patterns for a clamped circular plate.

Table 5.5.1. Values of $\lambda^2 = \omega a^2 \sqrt{I_0/D}$ for a clamped circular plate.

$n \rightarrow$ $m \downarrow$	0	1	2	3	4	5
0	10.2158	21.26	34.88	51.04	69.6659	90.7390
1	39.771	60.82	84.58	111.01	140.1079	171.8029
2	89.104	120.08	153.81	190.30	229.5185	271.4283
3	158.183	199.06	242.71	289.17	338.4113	390.3896

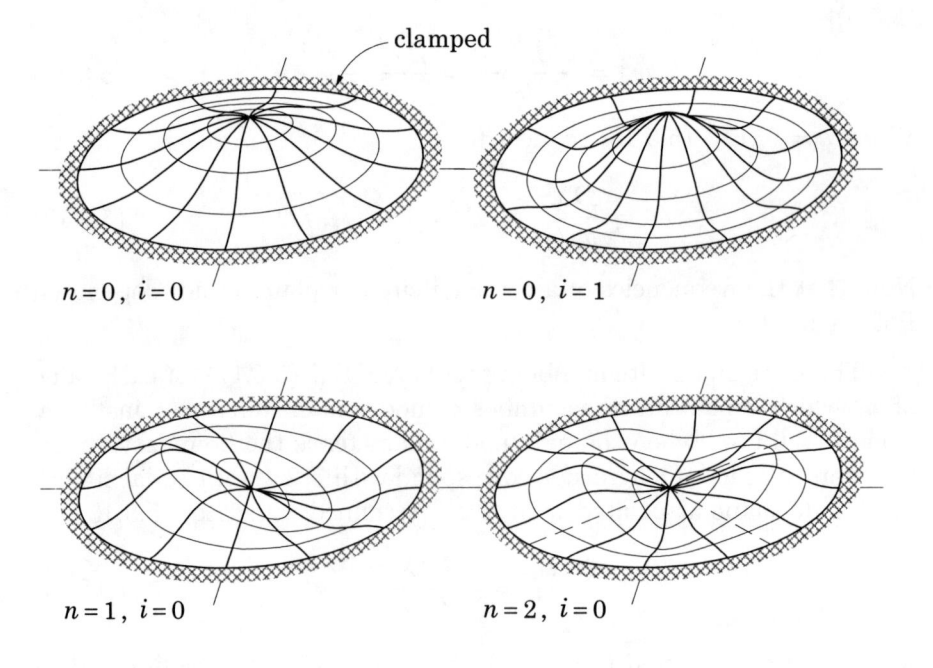

Figure 5.5.1. Typical modes of free vibration of a clamped circular plate, showing the nodal diameters and nodal circles.

5.5.4 Simply Supported Circular Plates

The boundary conditions are

$$W_n = 0 \quad \text{and} \quad M_{rr} = 0 \quad \text{at } r = a \text{ for any } \theta \tag{5.5.22}$$

In addition, we note that $\partial^2 w_0/\partial\theta^2 = 0$ on the boundary. Use of Eq. (5.5.12) in Eq. (5.5.21) results in the following two equations:

$$A_n J_n(\lambda) + C_n I_n(\lambda) = 0 \tag{5.5.23a}$$

$$A_n \left[J_n''(\lambda) + \frac{\nu}{\lambda} J_n'(\lambda) \right] + C_n \left[I_n''(\lambda) + \frac{\nu}{\lambda} I_n'(\lambda) \right] = 0 \tag{5.5.23b}$$

These equations lead to the frequency equation

$$\frac{J_{n+1}(\lambda)}{J_n(\lambda)} + \frac{I_{n+1}(\lambda)}{I_n(\lambda)} = \frac{2\lambda}{1-\nu} \tag{5.5.24}$$

The mode shape is now determined using Eq. (5.5.23a,b)

$$\frac{A_n}{C_n} = -\frac{I_n(\lambda)}{J_n(\lambda)} \tag{5.5.25}$$

where λ is a solution of Eq. (5.5.24).

Table 5.5.2 contains values of λ^2 for various values of n and m and Poisson's ratio $\nu = 0.3$.

Table 5.5.2. Values of $\lambda^2 = \omega a^2 \sqrt{I_0/D}$ for a simply supported circular plate ($\nu = 0.3$).

$n \rightarrow$ $m \downarrow$	0	1	2
0	4.977	13.94	25.65
1	29.76	48.51	70.14
2	74.20	102.80	134.33
3	138.34	176.84	218.24

5.5.5 Rayleigh–Ritz Solutions

Natural frequencies of solid circular plates with other boundary conditions and annular plates with various boundary conditions can be found in the papers listed at the end of the chapter. In the general case, analytical solutions of free vibration lead to mathematically complex equations, often in terms of Bessel functions, and the solutions are difficult to obtain, requiring the use of approximate methods of solution. Here, the Rayleigh–Ritz formulation for free vibration of circular plates is presented.

The equation of motion in Eq. (5.2.41), with M_{rr}, $M_{r\theta}$, and $M_{\theta\theta}$ given by Eqs. (5.2.47a–c) and no applied loads, is valid for free vibration analysis. The solution to this equation is again sought in the form of Eq. (5.5.2). The variational (or weak) form of the resulting equation is the same as that in Eq. (5.4.35b) without the thermal and mechanical loads but with the addition of appropriate terms corresponding to the inertia terms. We find that

$$0 = \int_\Omega \left\{ D_{11} \frac{\partial^2 W}{\partial r^2} \frac{\partial^2 \delta W}{\partial r^2} + D_{12} \frac{1}{r} \left(\frac{\partial W}{\partial r} \frac{\partial^2 \delta W}{\partial r^2} + \frac{\partial \delta W}{\partial r} \frac{\partial^2 W}{\partial r^2} \right. \right.$$
$$\left. + \frac{1}{r} \frac{\partial^2 W}{\partial \theta^2} \frac{\partial^2 \delta W}{\partial r^2} + \frac{1}{r} \frac{\partial^2 \delta W}{\partial \theta^2} \frac{\partial^2 W}{\partial r^2} \right) + kW\delta W$$
$$+ D_{22} \left(\frac{1}{r} \frac{\partial W}{\partial r} + \frac{1}{r^2} \frac{\partial^2 W}{\partial \theta^2} \right) \left(\frac{1}{r} \frac{\partial \delta W}{\partial r} + \frac{1}{r^2} \frac{\partial^2 \delta W}{\partial \theta^2} \right)$$
$$+ 4D_{66} \left(\frac{1}{r} \frac{\partial^2 W}{\partial r \partial \theta} - \frac{1}{r^2} \frac{\partial W}{\partial \theta} \right) \left(\frac{1}{r} \frac{\partial^2 \delta W}{\partial r \partial \theta} - \frac{1}{r^2} \frac{\partial \delta W}{\partial \theta} \right)$$
$$\left. - \omega^2 \left[I_0 W \delta W + I_2 \left(\frac{\partial W}{\partial r} \frac{\partial \delta W}{\partial r} + \frac{1}{r^2} \frac{\partial W}{\partial \theta} \frac{\partial \delta W}{\partial \theta} \right) \right] \right\} r \, dr \, d\theta$$
$$(5.5.26)$$

where ω is the frequency of free vibration.

Assume an N-parameter Rayleigh–Ritz approximation of the form

$$W(r,\theta) \approx \sum_{j=1}^{N} c_j \varphi_j(r,\theta) \tag{5.5.27}$$

Substituting Eq. (5.5.27) into Eq. (5.5.26), we obtain

ANALYSIS OF CIRCULAR PLATES 245

$$0 = \sum_{j=1}^{N} \left(R_{ij}^1 + R_{ij}^2 - \omega^2 M_{ij} \right) c_j \tag{5.5.28}$$

where R_{ij}^1 and R_{ij}^2 are defined in Eqs. (5.4.40b,c), and M_{ij} is defined as

$$M_{ij} = \int_\Omega \left[I_0 \varphi_i \varphi_j + I_2 \left(\frac{\partial \varphi_i}{\partial r} \frac{\partial \varphi_j}{\partial r} + \frac{1}{r^2} \frac{\partial \varphi_i}{\partial \theta} \frac{\partial \varphi_j}{\partial \theta} \right) \right] r dr d\theta \tag{5.5.29}$$

In matrix notation, Eq. (5.5.28) has the form of an eigenvalue problem

$$\left([R^1 + R^2] - \omega^2 [M] \right) \{c\} = \{0\} \tag{5.5.30}$$

For a non-trivial solution, the coefficient matrix in Eq. (5.5.30) should be singular, i.e.,

$$|([R^1] + [R^2] - \omega^2 [M]| = 0 \tag{5.5.31}$$

The above formulation is valid also for annular plates.

As an example, consider free vibration of a clamped solid circular plate. For a one-parameter ($N = 1$) Rayleigh–Ritz solution, let [see Eqs. (5.4.42) through (5.4.48)]

$$\varphi_1(r, \theta) = f_1(r) \cos n\theta \tag{5.5.32}$$

and compute R_{11}^1, R_{11}^2, and M_{11}

$$R_{11}^1 = \int_0^{2\pi} \int_0^a \left\{ \left[D_{11} f_1'' f_1'' + D_{22} \left(\frac{1}{r} f_1' - \frac{n^2}{r^2} f_1 \right)^2 \right. \right.$$
$$\left. + 2 D_{12} \frac{1}{r} \left(f_1' f_1'' - \frac{n^2}{r} f_1 f_1'' \right) \right] \cos^2 n\theta$$
$$\left. + 4 D_{66} \left(\frac{n}{r} f' - \frac{n}{r^2} f_1 \right)^2 \sin^2 n\theta \right\} r dr d\theta \tag{5.5.33a}$$

$$R_{11}^2 = k \int_0^{2\pi} \int_0^a f_1 f_1 \cos^2 n\theta \, r dr d\theta \tag{5.5.33b}$$

$$M_{ij} = \int_0^{2\pi} \int_0^a \left[(I_0 f_1 f_1 + f_1' f_1') \cos^2 n\theta \right.$$
$$\left. + I_2 \frac{n^2}{r^2} f_1 f_1 \sin^2 n\theta \right] r dr d\theta \tag{5.5.33c}$$

246   THEORY AND ANALYSIS OF ELASTIC PLATES

In particular, for $n = 0$, we select the function in Eq. (5.4.47) and obtain

$$R_{11}^1 = 2\pi \int_0^a \left(D_{11} f_1'' f_1'' + D_{22} \frac{1}{r^2} f_1' f_1' + 2D_{12} \frac{1}{r} f_1' f_1'' \right) r dr$$
$$= \frac{2\pi}{a^2} \left(8D_{11} + \frac{8}{3} D_{22} \right) \qquad (5.5.34a)$$

$$R_{11}^2 = 2\pi k \int_0^a f_1 f_1 \, r dr = 2\pi k \frac{a^2}{10} \qquad (5.5.34b)$$

$$M_{11} = 2\pi \int_0^a \left(I_0 f_1 f_1 + I_2 f_1' f_1' \right) r dr = 2\pi \left(\frac{a^2}{10} I_0 + \frac{3}{a^2} I_2 \right) \quad (5.5.34c)$$

We find that

$$\lambda^2 = \omega a^2 \sqrt{\frac{I_0}{D_{22}}} = \left[\frac{\frac{80}{3} \left(3 \frac{D_{11}}{D_{22}} + 1 \right) + \frac{a^4 k}{80 D_{22}}}{1 + \frac{30 I_2}{a^2 I_0}} \right]^{\frac{1}{2}} \qquad (5.5.35)$$

Clearly, rotatory inertia has the effect of reducing the frequency of vibration while the elastic foundation modulus increases it. For a clamped isotropic plate without elastic foundation, the frequency becomes ($I_2 = I_0 h^2 / 12$)

$$\lambda^2 = \omega a^2 \sqrt{\frac{I_0}{D}} = 10.328 \left[\frac{1}{1 + 2.5 \frac{h^2}{a^2}} \right]^{\frac{1}{2}} \qquad (5.5.36)$$

For very thin plates, say $h/a = 0.01$, the effect of rotatory inertia is negligible. Even for $h/a = 0.1$, the effect is less than one percent. Note that the one-parameter Rayleigh–Ritz solution differs from that listed in Table 5.5.1 (for $m = n = 0$) by less than one percent!

For $n = 1$, the function in Eq. (5.4.47) is not admissible and we select the function in Eq. (5.4.48) and obtain

$$R_{11}^1 = \frac{\pi}{a^2} \left[6 D_{11} + \frac{2}{3} D_{22} - 2 \left(\frac{1}{2} - \frac{7}{6} \right) D_{12} + 4 \frac{2}{3} D_{66} \right] \quad (5.5.37a)$$

$$R_{11}^2 = \frac{\pi a^2}{60} k, \quad M_{11} = \frac{a^2}{60} I_0 + \frac{1}{6} I_2 \qquad (5.5.37b)$$

The frequency is given by

$$\lambda^2 = \omega a^2 \sqrt{\frac{I_0}{D_{22}}} = \left[\frac{6 \frac{D_{11}}{D_{22}} + \frac{2}{3} + \frac{4}{3} \frac{D_{12} + 2 D_{66}}{D_{22}} + \frac{a^4 k}{60 D_{22}}}{1 + \frac{10 I_2}{a^2 I_0}} \right]^{\frac{1}{2}} \quad (5.5.38)$$

ANALYSIS OF CIRCULAR PLATES   247

For a clamped isotropic plate without elastic foundation, the frequency for the case $m = 0$, $n = 1$ becomes ($I_2 = I_0 h^2/12$)

$$\lambda^2 = \omega a^2 \sqrt{\frac{I_0}{D}} = 21.91 \left[\frac{1}{1 + \frac{5}{6}\frac{h^2}{a^2}}\right]^{\frac{1}{2}} \tag{5.5.39}$$

When rotatory inertia is neglected, the frequency predicted by Eq. (5.5.39) differs from the value given in Table 5.5.1 only by three percent.

For other values of n, one must select functions that are admissible (i.e., allow evaluation of the integrals). Generally, higher values of n require higher-order functions $f_i(r)$.

Exercises

5.1 Verify the strain-displacement relations in Eqs. (5.2.21).

5.2 Use the 3-D equations of equilibrium in terms of stresses (with the body forces equal to zero) in the cylindrical coordinate system to derive the equilibrium equations in terms of the force and moment resultants of Eqs. (5.2.24) and (5.2.25) of a circular plate (see Exercise 3.5).

5.3 Show that

$$\frac{\partial M_{r\theta}}{\partial r} = \frac{1}{r}\frac{\partial F}{\partial \theta} \tag{a}$$

$$\frac{\partial \mathcal{M}}{\partial r} = \frac{1}{r}\left[\frac{\partial}{\partial r}(rM_{rr}) - M_{\theta\theta} + \frac{\partial M_{r\theta}}{\partial \theta}\right] \tag{b}$$

where

$$\mathcal{M} = \frac{M_{rr} + M_{\theta\theta}}{1+\nu}, \quad F = -(1-\nu)Dr\frac{\partial}{\partial r}\left(\frac{1}{r}\frac{\partial w_0}{\partial r} - \frac{w_0}{r^2}\right) \tag{c}$$

5.4 Show that the expression for the deflection of a simply supported circular plate subjected to uniformly distributed transverse load and applied bending moment $M_{rr} = M_a$ at $r = a$ is

$$w_0(r) = \bar{w}_0(r) + \frac{M_a a^2}{2D(1+\nu)}\left(1 - \frac{r^2}{a^2}\right)$$

where $\bar{w}_0$ is the deflection given in Eq. (5.3.43).

5.5 Determine the deflection and bending moments of a circular plate under uniformly distributed transverse load q_0 when the edge $r = a$ is elastically built-in. The boundary conditions are

$$w_0(a) = 0, \quad M_{rr} = \beta \frac{dw_0}{dr} \quad \text{at} \quad r = a$$

where β denotes rotational stiffness constant. *Answer:*

$$w_0 = \frac{q_0 a^4}{64D} \left\{ \frac{(5+\nu)D + \beta a}{(1+\nu)D + \beta a} - 2 \left[\frac{(3+\nu)D + \beta a}{(1+\nu)D + \beta a} \right] \left(\frac{r}{a}\right)^2 + \left(\frac{r}{a}\right)^4 \right\}$$

$$M_{rr} = \frac{q_0 a^2}{16} \left[(1+\nu) \frac{(3+\nu)D + \beta a}{(1+\nu)D + \beta a} - (3+\nu) \left(\frac{r}{a}\right)^2 \right]$$

$$M_{\theta\theta} = \frac{q_0 a^2}{16} \left[(1+\nu) \frac{(3+\nu)D + \beta a}{(1+\nu)D + \beta a} - (1+3\nu) \left(\frac{r}{a}\right)^2 \right]$$

5.6 Show that the maximum deflection and bending moment of a simply supported circular plate under linearly varying load $q = q_0(1 - r/a)$ are

$$w_{max} = \frac{q_0 a^4}{4800 D} \left(\frac{183 + 43\nu}{1+\nu} \right), \quad M_{max} = q_0 a^2 \left(\frac{71 + 29\nu}{720} \right)$$

5.7 Show that the maximum deflection and bending moment of a simply supported circular plate under linearly varying load $q = q_1(r/a)$ are

$$w_{max} = \frac{q_1 a^4}{150 D} \left(\frac{6+\nu}{1+\nu} \right), \quad M_{max} = q_1 a^2 \left(\frac{4+\nu}{45} \right)$$

5.8 Show that the deflection of a clamped circular plate under the load $q = q_0(r^2/a^2)$ is given by

$$w_0(r) = \frac{q_0 a^4}{576 D} \left[2 - 3 \left(\frac{r}{a}\right)^2 + \left(\frac{r}{a}\right)^6 \right]$$

5.9 Show that the expressions for the deflection and bending moments of an annular plate of inner radius a and outer radius b, simply supported at the inner edge $r = a$ and subjected to bending moment $M_{rr} = M_b$

ANALYSIS OF CIRCULAR PLATES   249

at $r = b$, are the same as those given in Eqs. (5.3.67)–(5.3.69) with $M_a = 0$.

5.10 Show that the expressions for the deflection and bending moments of an annular plate of inner radius a and outer radius b, simply supported at the inner edge $r = a$ and subjected to bending moment $M_{rr} = M_a$ at $r = a$, are the same as those given in Eqs. (5.3.67)–(5.3.69) with $M_b = 0$.

5.11 Determine the deflection and bending moments of an annular plate of inner radius b and outer radius a, clamped at the inner edge $r = b$ and subjected to uniformly distributed load q_0.

5.12 Show that the expressions for the deflection and bending moments of an annular plate of inner radius a and outer radius b, simply supported at the inner edge $r = a$ and subjected to shear force $rQ_r = bQ_b$ at $r = b$, are the same as those given in Eqs. (5.3.73)–(5.3.75).

5.13 Show that the expression for the deflection of a clamped (at the outer edge) circular plate under linearly varying load, $q = q_0(1 - r/a)$ is

$$w_0(r) = \frac{q_0 a^4}{14400 D}\left(129 - 290\frac{r^2}{a^2} + 225\frac{r^4}{a^4} - 64\frac{r^5}{a^5}\right)$$

5.14 Show that the expressions for the deflection and bending moments of a clamped (at the outer edge) circular plate under linearly varying load $q = q_1(r/a)$ are

$$w_0(r) = \frac{q_1 a^4}{450 D}\left(3 - 5\frac{r^2}{a^2} + 2\frac{r^5}{a^5}\right)$$

$$M_{rr}(r) = \frac{q_1 a^2}{45}\left[(1+\nu) - (4+\nu)\frac{r^3}{a^3}\right]$$

$$M_{\theta\theta}(r) = \frac{q_1 a^2}{45}\left[(1+\nu) - (1+4\nu)\frac{r^3}{a^3}\right]$$

5.15 Show that the one-parameter Rayleigh–Ritz solution for the deflection of an isotropic clamped (at the outer edge) circular plate under uniformly distributed load q_0 is given by

$$w_1(r) = c_1\phi_1(r) = \frac{q_0 a^4}{60 D}\left[1 - 3\left(\frac{r}{a}\right)^2 + 2\left(\frac{r}{a}\right)^3\right]$$

5.16 Show that the two-parameter Rayleigh–Ritz solution for the deflection of an isotropic, clamped (at the outer edge) circular plate under uniformly distributed load q_0 coincides with the exact solution.

5.17 Show that the one-parameter Galerkin's solution for the deflection of an isotropic clamped (at the outer edge) circular plate under uniformly distributed load q_0 coincides with the exact solution.

5.18 Use the two-parameter Rayleigh–Ritz solution of the form

$$W_2(r,\theta) = c_1\varphi_1(r) + c_2\varphi_2(r)\cos\theta$$

and determine the coefficients c_1 and c_2 for a clamped circular plate subjected to asymmetric loading given in Eq. (5.4.5). You must determine φ_1 and φ_2 first (see Exercise 5.16 above).

5.19 Determine the fundamental frequency of a simply supported circular plate using a one-parameter Rayleigh–Ritz approximation.

5.20 (**First-order shear deformation theory of circular plates**) The first order shear deformation plate theory (FST) is the simplest theory that accounts for nonzero transverse shear strain. It is based on the displacement field (for axisymmetric bending)

$$u_r(r,z) = z\phi(r), \quad u_z(r,z) = w_0(r) \qquad (a)$$

where ϕ denotes rotation of a transverse normal in the plane $\theta =$ constant. Use the principle of virtual work to show that the governing equations are given by

$$-\frac{1}{r}\left[\frac{d}{dr}(rM_{rr}) + M_{\theta\theta}\right] + Q_r = 0 \qquad (b)$$

$$-\frac{1}{r}\frac{d}{dr}(rQ_r) = q \qquad (c)$$

5.21 (**Third-order shear deformation theory of circular plates**) The third-order plate theory of Reddy [18] is based on the displacement field (for axisymmetric bending)

$$u_r(r,z) = z\phi(r) - c_1 z^3\left(\phi + \frac{dw_0}{dr}\right), \quad u_z(r,z) = w_0(r) \qquad (a)$$

where $c_1 = 4/(3h^2)$ and h denotes the total thickness of the plate. Show that the governing equations are given by

$$-\frac{1}{r}\left[\frac{d}{dr}(rM_{rr}) - M_{\theta\theta}\right] + Q_r + c_1\frac{1}{r}\left[\frac{d}{dr}(rP_{rr}) - P_{\theta\theta}\right] - c_2 R_r = 0$$

$$-\frac{1}{r}\frac{d}{dr}(rQ_r) + c_2\frac{1}{r}\frac{d}{dr}(rR_r) - c_1\frac{1}{r}\left[\frac{d^2}{dr^2}(rP_{rr}) - \frac{dP_{\theta\theta}}{dr}\right] = q$$

where $(P_{rr}, P_{\theta\theta}, R_r)$ are the higher-order stress resultants and $c_2 = 4/h^2$.

References for Additional Reading

1. Timoshenko, S. P. and Woinowsky-Krieger, S., *Theory of Plates and Shells*, McGraw–Hill, Singapore (1970).
2. Ugural, A. C., *Stresses in Plates and Shells*, McGraw–Hill, New York (1981).
3. Roark, J. R. and Young, W. C., *Formulas for Stress and Strain*, McGraw–Hill, New York (1975).
4. Szilard, R., *Theory and Analysis of Plates. Classical and Numerical Methods*, Prentice–Hall, Englewood Cliffs, New Jersey (1974).
5. McFarland, D., Smith, B. L., and Bernhart, W. D., *Analysis of Plates*, Spartan Books, Philadelphia, PA (1972).
6. Sokolnikoff, I. S. and Redheffer, R. M., *Mathematics of Physics and Modern Engineering*, McGraw–Hill, New York (1966).
7. Panc, V., *Theories of Elastic Plates*, Noordhoff, Leyden, The Netherlands (1975).
8. Leissa, A. W., *Vibration of Plates*, NASA, Washington, D. C. (1969).
9. McLachlan, N., *Bessel Functions for Engineers*, Oxford University Press, London (1948).
10. Kantham, C. L., "Bending and Vibration of Elastically Restrained Circular Plates", *Journal Franklin Institute*, **265**(6), 483-491 (1958).
11. Bartlett, C.C., "The Vibration and Buckling of a Circular Plate Clamped on Part of its Boundary and Simply Supported on the Remainder", *Quarterly Journal of Mechanics and Applied Mathematics*, **16**, 431–440 (1963).
12. Southwell, R. V., "On the Free Transverse Vibrations of a Uniform Circular Disc Clamped at its Center and on the Effect of Rotation", *Proceedingfs of the Royal Society (London)*, Series A, **101**, 133–153 (1922).

13. Roberson, R. E., "Vibrations of a Clamped Circular Plate Carrying Concentrated Mass", *Journal of Applied Mechanics,* **18**(4), 349–352 (1951).
14. Raju, P. N., "Vibrations of Annular Plates", *Journal of Aeronautical Society of India,* **14**(2), 37–52 (1962).
15. Joga Rao, C. V. and Pickett, G., "Vibrations of Plates of Irregular Shapes and Plates with Holes", *Journal of Aeronautical Society of India,* **13**(3), 83–88 (1961).
16. Reddy, J. N. and Wang, C. M., "Relationships Between Classical and Shear Deformation Theories of Axisymmetric Bending of Circular Plates", *AIAA Journal,* **35**(12), 1862–1868 (1997).
17. Reddy, J. N., Wang, C. M., and Lam, K. Y., "Unified Beam Finite Elements Based on the Classical and Shear Deformation Theories of Beams and Axisymmetric Circular Plates", *Communications in Numerical Methods in Engineering,* **13**, 495–510 (1997).
18. Reddy, J. N., *Mechanics of Laminated Composite Plates: Theory and Analysis,* CRC, Boca Raton, Florida (1997).

Chapter Six

Bending of Simply Supported Rectangular Plates

6.1 Introduction

6.1.1 Governing Equations

In this chapter, analytical solutions for deflections and stresses of simply supported rectangular plates are developed using the Navier method, the Lévy method with the state-space approach, and the Rayleigh–Ritz method. The Navier solutions can be developed for rectangular plates when all four edges are simply supported. The Lévy solutions can be developed for rectangular plates with two opposite edges simply supported and the remaining two edges having any possible combination of *free, simply supported,* and *clamped/fixed* boundary conditions. The Rayleigh–Ritz method can be used to determine approximate solutions for more general boundary conditions, provided that suitable approximation functions that satisfy the geometric boundary conditions can be constructed.

The governing equations of rectangular plates are best described using Cartesian rectangular coordinates, as discussed in Chapter 3. For the linear analysis of plates studied here, we consider plates subjected to transverse distributed load q and thermal loads due to temperature change ΔT. When in-plane forces $(\hat{N}_{xx}, \hat{N}_{xx}, \hat{N}_{xy})$ and the nonlinear terms are $(\mathcal{N})$ omitted, the equation governing the bending, Eq. (3.8.5), becomes uncoupled from Eqs. (3.8.3) and (3.8.4). For an orthotropic plate we have

$$D_{11}\frac{\partial^4 w_0}{\partial x^4} + 2\hat{D}_{12}\frac{\partial^4 w_0}{\partial x^2 \partial y^2} + D_{22}\frac{\partial^4 w_0}{\partial y^4} + kw_0$$
$$= q - \left(\frac{\partial^2 M_{xx}^T}{\partial x^2} + 2\frac{\partial^2 M_{xy}^T}{\partial x \partial y} + \frac{\partial^2 M_{yy}^T}{\partial y^2}\right) \qquad (6.1.1)$$

where
$$\hat{D}_{12} = (D_{12} + 2D_{66}) \qquad (6.1.2)$$

and the thermal resultants M_{xx}^T, M_{xy}^T, and M_{yy}^T are defined by [see Eq. (3.6.12b)]

$$\left\{\begin{array}{c} M_{xx}^T \\ M_{yy}^T \\ M_{xy}^T \end{array}\right\} = \left\{\begin{array}{c} Q_{11}\alpha_1 + Q_{12}\alpha_2 \\ Q_{12}\alpha_1 + Q_{22}\alpha_2 \\ 0 \end{array}\right\} \int_{-\frac{h}{2}}^{\frac{h}{2}} \Delta T(x,y,z)\, z\, dz \qquad (6.1.3)$$

where Q_{ij} and α_i are the elastic stiffnesses and thermal coefficients of expansion, respectively, of an orthotropic plate, and $\Delta T(x,y,z)$ is the temperature increment above a reference temperature (at which the plate is stress free). Note that M_{xy}^T is identically zero for isotropic or orthotropic plates.

For isotropic plates ($D_{11} = D_{22} = \hat{D}_{12} = D$, $\alpha_1 = \alpha_2 = \alpha$), Eq. (6.1.1) simplifies to

$$D\left(\frac{\partial^4 w_0}{\partial x^4} + 2\frac{\partial^4 w_0}{\partial x^2 \partial y^2} + \frac{\partial^4 w_0}{\partial y^4}\right) + kw_0$$
$$= q - \frac{1}{1-\nu}\left(\frac{\partial^2 M_T}{\partial x^2} + \frac{\partial^2 M_T}{\partial y^2}\right) \qquad (6.1.4)$$

where
$$M_T = E\alpha \int_{-\frac{h}{2}}^{\frac{h}{2}} \Delta T(x,y,z)\, z\, dz \qquad (6.1.5)$$

Equation (6.1.4) can be expressed in terms of the Laplace operator ∇^2 as

$$D\nabla^2\nabla^2 w_0 + kw_0 = q - \frac{1}{(1-\nu)}\nabla^2 M_T \qquad (6.1.6)$$

6.1.2 Boundary Conditions

Simply supported boundary conditions on all four edges of a rectangular plate can be expressed as

$$w_0(0,y) = 0, \quad w_0(a,y) = 0, \quad w_0(x,0) = 0, \quad w_0(x,b) = 0 \qquad (6.1.7)$$

$$M_{xx}(0,y) = 0, \quad M_{xx}(a,y) = 0, \quad M_{yy}(x,0) = 0, \quad M_{yy}(x,b) = 0 \qquad (6.1.8)$$

where the bending moments are related to the transverse deflection by

$$M_{xx} = -\left(D_{11}\frac{\partial^2 w_0}{\partial x^2} + D_{12}\frac{\partial^2 w_0}{\partial y^2}\right) - M_{xx}^T \tag{6.1.9a}$$

$$M_{yy} = -\left(D_{12}\frac{\partial^2 w_0}{\partial x^2} + D_{22}\frac{\partial^2 w_0}{\partial y^2}\right) - M_{yy}^T \tag{6.1.9b}$$

$$M_{xy} = -2D_{66}\frac{\partial^2 w_0}{\partial x \partial y} \tag{6.1.9c}$$

$$D_{11} = \frac{E_1 h^3}{12(1-\nu_{12}\nu_{21})}, \quad D_{22} = \frac{E_2 h^3}{12(1-\nu_{12}\nu_{21})}$$

$$D_{12} = \frac{\nu_{12} E_2 h^3}{12(1-\nu_{12}\nu_{21})}, \quad D_{66} = \frac{G_{12} h^3}{12} \tag{6.1.10}$$

and a and b denote the in-plane dimensions along the x and y coordinate directions of a rectangular plate. The origin of the coordinate system is taken at the upper left corner of the mid-plane (see Figure 6.1.1).

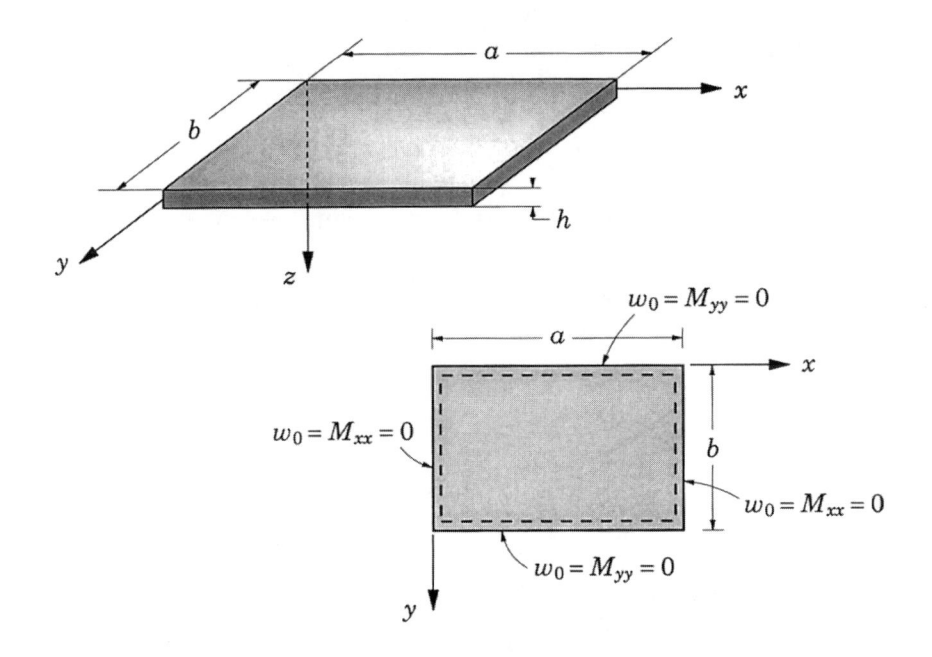

Figure 6.1.1. Geometry and coordinate system for a rectangular plate.

6.2 The Navier Solutions

6.2.1 Solution Procedure

The solution of Eq. (6.1.1) for rectangular plates with simply supported boundary conditions can be obtained using Navier's method. In Navier's method the displacement w_0, mechanical load q, and thermal loads M_{xx}^T and M_{yy}^T are expanded in trigonometric series. The choice of the trigonometric functions used for the deflection and thermal moments is restricted to those that satisfy the boundary conditions of the problem. Substitution of the expansion for the deflection into the governing equation (6.1.1) will dictate the choice of the expansions used for the mechanical and thermal loads.

The simply supported boundary conditions in Eq. (6.1.7) are met by the following form of the transverse deflection

$$w_0(x,y) = \sum_{n=1}^{\infty}\sum_{m=1}^{\infty} W_{mn} \sin\frac{m\pi x}{a} \sin\frac{n\pi y}{b} \quad (6.2.1)$$

where W_{mn} are coefficients to be determined such that Eq. (6.1.1) is satisfied everywhere in the domain of the plate. Substituting Eq. (6.2.1) into Eq. (6.1.1), we obtain

$$\sum_{n=1}^{\infty}\sum_{m=1}^{\infty}\left[D_{11}\left(\frac{m\pi}{a}\right)^4 + 2\hat{D}_{12}\left(\frac{m\pi}{a}\right)^2\left(\frac{n\pi}{b}\right)^2 \right.$$

$$\left. + D_{22}\left(\frac{n\pi}{b}\right)^4 + k\right] W_{mn} \sin\frac{m\pi x}{a} \sin\frac{n\pi y}{b}$$

$$= q - \left(\frac{\partial^2 M_{xx}^T}{\partial x^2} + 2\frac{\partial^2 M_{xy}^T}{\partial x \partial y} + \frac{\partial^2 M_{yy}^T}{\partial y^2}\right) \quad (6.2.2)$$

Equation (6.2.2) suggests that the right-hand side also be expanded in double sine series

$$q(x,y) = \sum_{n=1}^{\infty}\sum_{m=1}^{\infty} q_{mn} \sin\frac{m\pi x}{a} \sin\frac{n\pi y}{b} \quad (6.2.3)$$

$$M_{xx}^T(x,y) = \sum_{n=1}^{\infty}\sum_{m=1}^{\infty} M_{mn}^1 \sin\frac{m\pi x}{a} \sin\frac{n\pi y}{b} \quad (6.2.4a)$$

$$M_{yy}^T(x,y) = \sum_{n=1}^{\infty}\sum_{m=1}^{\infty} M_{mn}^2 \sin\frac{m\pi x}{a} \sin\frac{n\pi y}{b} \quad (6.2.4b)$$

and that $M_{xy}^T = 0$. The coefficients q_{mn}, for example, are calculated from

$$q_{mn} = \frac{4}{ab} \int_0^b \int_0^a q(x,y) \sin\frac{m\pi x}{a} \sin\frac{n\pi y}{b} \, dx\, dy \qquad (6.2.5)$$

Similar expressions hold for M_{mn}^1 and M_{mn}^2. For example, when $q = q_0$, we find from Eq. (6.2.5) that

$$\begin{aligned} q_{mn} &= \frac{4}{ab} \int_0^b \int_0^a q_0 \sin\frac{m\pi x}{a} \sin\frac{n\pi y}{b} \, dx\, dy \\ &= \frac{4}{ab} \left[-\frac{\cos\frac{m\pi x}{a}}{\frac{m\pi}{a}}\right]_0^a \left[-\frac{\cos\frac{n\pi y}{b}}{\frac{n\pi}{b}}\right]_0^b \\ &= \frac{4}{mn\pi^2}(\cos m\pi - 1)(\cos n\pi - 1)\end{aligned}$$

for $m, n = 1, 2, 3, \cdots$. Note that the above expression is identically zero for the even values of m and n, and it is equal to

$$q_{mn} = \frac{16}{mn\pi^2}, \quad \text{for } m, n = 1, 3, 5, \cdots \qquad (6.2.6)$$

Similarly, for a point load Q_0 applied at $x = x_0$ and $y = y_0$, we set $q = Q_0 \delta(x - x_0)\delta(y - y_0)$ and find

$$\begin{aligned}q_{mn} &= \frac{4}{ab} \int_0^b \int_0^a Q_0 \delta(x - x_0)\delta(y - y_0) \sin\frac{m\pi x}{a} \sin\frac{n\pi y}{b} \, dx\, dy \\ &= \frac{4Q_0}{ab} \sin\frac{m\pi x_0}{a} \sin\frac{n\pi y_0}{b} \end{aligned} \qquad (6.2.7)$$

for $m, n = 1, 2, 3, \cdots$, where $\delta(\cdot)$ is the Dirac delta function. For $x_0 = a/2$ and $y_0 = b/2$, we have

$$q_{mn} = \frac{4Q_0}{ab}, \quad m, n = 1, 3, 5, \cdots. \qquad (6.2.8)$$

The coefficients q_{mn} for other types of loads are listed in Table 6.2.1.

We note that the boundary conditions (6.1.8) on bending moments are satisfied by the double sine series expansions selected for the deflection and thermal moments [see Eqs. (6.1.9a,b)]. Therefore, the exact analytical solution of Eq. (6.1.1) is given by Eq. (6.2.1), where

Table 6.2.1. Coefficients in the double trigonometric series expansion of loads in the Navier method.

Loading	Coefficients q_{mn}
	Uniform load, $q(x,y) = q_0$ $$q_{mn} = \frac{16q_0}{\pi^2 mn}$$ $(m, n = 1, 3, 5, \cdots)$
	Hydrostatic load, $q(x,y) = q_0 \frac{y}{b}$ $$q_{mn} = \frac{8q_0}{\pi^2 mn}(-1)^{n+1}$$ $(m = 1, 3, 5, \cdots)$ $(n = 1, 2, 3, \cdots)$
	Point load, [i.e., Q_0 at (x_0, y_0)] $$q_{mn} = \frac{4Q_0}{ab} \sin \frac{m\pi x_0}{a} \sin \frac{n\pi y_0}{b}$$ $(m, n = 1, 2, 3, \cdots)$
	Line load, $q(x,y) = q_0\, \delta(y - y_0)$ $$q_{mn} = \frac{8q_0}{\pi b m} \sin \frac{n\pi y_0}{b}$$ $(m = 1, 3, 5, \cdots)$ $(n = 1, 2, 3, \cdots)$

the coefficients W_{mn} are determined from Eq. (6.2.2) by substituting the expansions for the load and thermal moments. We find

$$\sum_{n=1}^{\infty}\sum_{m=1}^{\infty}\left\{W_{mn}\left[D_{11}\left(\frac{m\pi}{a}\right)^4+2\hat{D}_{12}\left(\frac{m\pi}{a}\right)^2\left(\frac{n\pi}{b}\right)^2+D_{22}\left(\frac{n\pi}{b}\right)^4+k\right]\right.$$
$$\left.-q_{mn}-\left[\left(\frac{m\pi}{a}\right)^2 M_{mn}^1+\left(\frac{n\pi}{b}\right)^2 M_{mn}^2\right]\right\}\sin\frac{m\pi x}{a}\sin\frac{n\pi y}{b}=0$$

Since the equation must hold for all points (x, y) of the domain $0 < x < a$ and $0 < y < b$, the coefficient of $\sin\frac{m\pi x}{a}\sin\frac{n\pi y}{b}$ must be zero for any m and n. This yields

$$W_{mn}=\frac{1}{d_{mn}}\left[q_{mn}+\left(\frac{m\pi}{a}\right)^2 M_{mn}^1+\left(\frac{n\pi}{b}\right)^2 M_{mn}^2\right] \quad (6.2.9)$$

$$d_{mn}=\frac{\pi^4}{b^4}\left[D_{11}m^4 s^4+2\hat{D}_{12}m^2 n^2 s^2+D_{22}n^4\right]+k \quad (6.2.10)$$

where s denotes the plate aspect ratio, $s = b/a$. The deflection in Eq. (6.1.1) becomes

$$w_0(x,y)=\sum_{n=1}^{\infty}\sum_{m=1}^{\infty}\frac{1}{d_{mn}}\left[q_{mn}+\left(\frac{m\pi}{a}\right)^2 M_{mn}^1+\left(\frac{n\pi}{b}\right)^2 M_{mn}^2\right]$$
$$\times \sin\frac{m\pi x}{a}\sin\frac{n\pi y}{b} \quad (6.2.11)$$

For an isotropic plate, we have

$$w_0(x,y)=\sum_{n=1}^{\infty}\sum_{m=1}^{\infty}\left\{\frac{q_{mn}+\pi^2\left(\frac{m^2}{a^2}+\frac{n^2}{b^2}\right)M_{mn}^T}{\pi^4 D\left(\frac{m^2}{a^2}+\frac{n^2}{b^2}\right)^2+k}\right\}$$
$$\times \sin\frac{m\pi x}{a}\sin\frac{n\pi y}{b} \quad (6.2.12)$$

where

$$M_{mn}^T=\frac{4}{ab(1-\nu)}\int_0^b\int_0^a M_T(x,y)\sin\frac{m\pi x}{a}\sin\frac{n\pi y}{b} \quad (6.2.13)$$

and M_T is defined in Eq. (6.1.5).

6.2.2 Calculation of Bending Moments, Shear Forces, and Stresses

The bending moments can be calculated using Eqs. (6.1.9a–c) and (6.2.1) as

$$M_{xx} = \sum_{n=1}^{\infty}\sum_{m=1}^{\infty} \left(B_{xx}W_{mn} - M_{mn}^1\right) \sin\frac{m\pi x}{a} \sin\frac{n\pi y}{b}$$

$$M_{yy} = \sum_{n=1}^{\infty}\sum_{m=1}^{\infty} \left(B_{yy}W_{mn} - M_{mn}^2\right) \sin\frac{m\pi x}{a} \sin\frac{n\pi y}{b}$$

$$M_{xy} = -\sum_{n=1}^{\infty}\sum_{m=1}^{\infty} B_{xy}W_{mn} \cos\frac{m\pi x}{a} \cos\frac{n\pi y}{b} \qquad (6.2.14a)$$

where

$$B_{xx} = D_{11}\left(\frac{m\pi}{a}\right)^2 + D_{12}\left(\frac{n\pi}{b}\right)^2$$

$$B_{yy} = D_{12}\left(\frac{m\pi}{a}\right)^2 + D_{22}\left(\frac{n\pi}{b}\right)^2$$

$$B_{xy} = 2D_{66}\left(\frac{m\pi}{a}\right)\left(\frac{n\pi}{b}\right) \qquad (6.2.14b)$$

The transverse shear forces Q_x and Q_y are defined by

$$Q_x = \frac{\partial M_{xx}}{\partial x} + \frac{\partial M_{xy}}{\partial y}$$

$$= \sum_{n=1}^{\infty}\sum_{m=1}^{\infty} \left[S_{xx}W_{mn} - \left(\frac{m\pi}{a}\right)M_{mn}^1\right] \cos\frac{m\pi x}{a} \sin\frac{n\pi y}{b} \qquad (6.2.15a)$$

$$Q_y = \frac{\partial M_{xy}}{\partial x} + \frac{\partial M_{yy}}{\partial y}$$

$$= \sum_{n=1}^{\infty}\sum_{m=1}^{\infty} \left[S_{yy}W_{mn} - \left(\frac{n\pi}{b}\right)M_{mn}^2\right] \sin\frac{m\pi x}{a} \cos\frac{n\pi y}{b} \qquad (6.2.15b)$$

where

$$S_{xx} = D_{11}\left(\frac{m\pi}{a}\right)^3 + \hat{D}_{12}\left(\frac{m\pi}{a}\right)\left(\frac{n\pi}{b}\right)^2$$

$$S_{yy} = D_{22}\left(\frac{n\pi}{b}\right)^3 + \hat{D}_{12}\left(\frac{m\pi}{a}\right)^2\left(\frac{n\pi}{b}\right) \qquad (6.2.15c)$$

Reaction forces V_x and V_y along the simply supported edges $x = a$ and $y = b$, respectively, can be calculated using the definition in Eq. (3.5.15) as follows:

$$V_x(a, y) = Q_x + \frac{\partial M_{xy}}{\partial y} = \frac{\partial M_{xx}}{\partial x} + 2\frac{\partial M_{xy}}{\partial y}$$

$$= \sum_{n=1}^{\infty}\sum_{m=1}^{\infty} (-1)^m \left[\hat{S}_{xx} W_{mn} - \left(\frac{m\pi}{a}\right) M_{mn}^1\right] \sin\frac{n\pi y}{b} \quad (6.2.16)$$

$$V_y(x, b) = Q_y + \frac{\partial M_{xy}}{\partial x} = 2\frac{\partial M_{xy}}{\partial x} + \frac{\partial M_{yy}}{\partial y}$$

$$= \sum_{n=1}^{\infty}\sum_{m=1}^{\infty} (-1)^n \left[\hat{S}_{yy} W_{mn} - \left(\frac{n\pi}{b}\right) M_{mn}^2\right] \sin\frac{m\pi x}{a} \quad (6.2.17)$$

where

$$\hat{S}_{xx} = D_{11}\left(\frac{m\pi}{a}\right)^3 + (D_{12} + 4D_{66})\left(\frac{m\pi}{a}\right)\left(\frac{n\pi}{b}\right)^2$$

$$\hat{S}_{yy} = D_{22}\left(\frac{n\pi}{b}\right)^3 + (D_{12} + 4D_{66})\left(\frac{m\pi}{a}\right)^2\left(\frac{n\pi}{b}\right) \quad (6.2.18)$$

Thus, the distribution of the reaction forces along the edges follows a sinusoidal form. Similar expressions hold for $V_x(0, y)$ and $V_y(x, 0)$.

In addition to the reactions in Eqs. (6.2.16) and (6.2.17) along the edges, the plate experiences concentrated forces [see Eq. (3.5.14)] at the corners of a rectangular plate due to the twisting moment M_{xy} per unit length (which has the dimensions of a force). The concentrated force at the corner $x = a$ and $y = b$ is given by

$$F_c = -2M_{xy} = 4D_{66}\frac{\partial^2 w_0}{\partial x \partial y}$$

$$= 4D_{66} \sum_{n=1}^{\infty}\sum_{m=1}^{\infty} \left(\frac{m\pi}{a}\right)\left(\frac{n\pi}{b}\right)(-1)^{m+n} W_{mn} \quad (6.2.19)$$

We assume that the temperature variation through thickness is of the form

$$\Delta T(x, y, z) = T^0(x, y) + zT^1(x, y)$$

$$\equiv \sum_{n=1}^{\infty}\sum_{m=1}^{\infty} \left(T_{mn}^0 + zT_{mn}^1\right) \sin\frac{m\pi x}{a} \sin\frac{n\pi y}{b} \quad (6.2.20)$$

where the coefficients T_{mn}^0 and T_{mn}^1 are computed using equations of the form (6.2.5). Using Eq. (6.2.20) in the definitions of the thermal moments, Eq. (6.1.3), we find that [c.f., Eqs. (6.2.4a,b)]

$$M_{xx}^T = \int_{-\frac{h}{2}}^{\frac{h}{2}} (Q_{11}\alpha_1 + Q_{12}\alpha_2) \Delta T \, z \, dz$$

$$= (D_{11}\alpha_1 + D_{12}\alpha_2) \sum_{n=1}^{\infty} \sum_{m=1}^{\infty} T_{mn}^1 \sin \frac{m\pi x}{a} \sin \frac{n\pi y}{b}$$

$$\equiv \sum_{n=1}^{\infty} \sum_{m=1}^{\infty} M_{mn}^1 \sin \frac{m\pi x}{a} \sin \frac{n\pi y}{b} \qquad (6.2.21a)$$

$$M_{yy}^T = \int_{-\frac{h}{2}}^{\frac{h}{2}} (Q_{12}\alpha_1 + Q_{22}\alpha_2) \Delta T \, z \, dz$$

$$= (D_{12}\alpha_1 + D_{22}\alpha_2) \sum_{n=1}^{\infty} \sum_{m=1}^{\infty} T_{mn}^1 \sin \frac{m\pi x}{a} \sin \frac{n\pi y}{b}$$

$$\equiv \sum_{n=1}^{\infty} \sum_{m=1}^{\infty} M_{mn}^2 \sin \frac{m\pi x}{a} \sin \frac{n\pi y}{b} \qquad (6.2.21b)$$

$$M_{mn}^1 = (D_{11}\alpha_1 + D_{12}\alpha_2) T_{mn}^1, \quad M_{mn}^2 = (D_{12}\alpha_1 + D_{22}\alpha_2) T_{mn}^1 \quad (6.2.22)$$

For the pure bending case considered here, i.e., when the membrane strains are zero, the stresses in a simply supported plate are given by

$$\begin{Bmatrix} \sigma_{xx} \\ \sigma_{yy} \\ \sigma_{xy} \end{Bmatrix} = -z \begin{bmatrix} Q_{11} & Q_{12} & 0 \\ Q_{12} & Q_{22} & 0 \\ 0 & 0 & Q_{66} \end{bmatrix} \begin{Bmatrix} \frac{\partial^2 w_0}{\partial x^2} + \alpha_1 T^1 \\ \frac{\partial^2 w_0}{\partial y^2} + \alpha_2 T^1 \\ 2\frac{\partial^2 w_0}{\partial x \partial y} \end{Bmatrix}$$

$$= z \sum_{n=1}^{\infty} \sum_{m=1}^{\infty} \begin{Bmatrix} (R_{xx} W_{mn} - \mathcal{M}_{mn}^1) \sin \frac{m\pi x}{a} \sin \frac{n\pi y}{b} \\ (R_{yy} W_{mn} - \mathcal{M}_{mn}^2) \sin \frac{m\pi x}{a} \sin \frac{n\pi y}{b} \\ -R_{xy} W_{mn} \cos \frac{m\pi x}{a} \cos \frac{n\pi y}{b} \end{Bmatrix} \qquad (6.2.23)$$

where

$$R_{xx} = \frac{\pi^2}{b^2} \left(Q_{11} m^2 s^2 + Q_{12} n^2 \right), \quad R_{yy} = \frac{\pi^2}{b^2} \left(Q_{12} m^2 s^2 + Q_{22} n^2 \right)$$

$$R_{xy} = 2mns \frac{\pi^2}{b^2} Q_{66} \qquad (6.2.24)$$

$$\mathcal{M}_{mn}^1 = (Q_{11}\alpha_1 + Q_{12}\alpha_2) T_{mn}^1, \quad \mathcal{M}_{mn}^2 = (Q_{12}\alpha_1 + Q_{22}\alpha_2) T_{mn}^1$$

BENDING OF SIMPLY SUPPORTED RECTANGULAR PLATES 263

The maximum stresses occur at $(x, y, z) = (a/2, b/2, \pm h/2)$, depending on the nature of the thermomechanical loads.

In the classical plate theory, the transverse stresses $(\sigma_{xz}, \sigma_{yz}, \sigma_{zz})$ are identically zero when computed from the constitutive equations because the transverse shear strains are zero. However, they can be computed using the 3-D stress equilibrium equations for any $-h/2 \leq z \leq h/2$:

$$\sigma_{xz} = -\int_{-\frac{h}{2}}^{z} \left(\frac{\partial \sigma_{xx}}{\partial x} + \frac{\partial \sigma_{xy}}{\partial y} \right) dz + C_1(x,y)$$

$$\sigma_{yz} = -\int_{-\frac{h}{2}}^{z} \left(\frac{\partial \sigma_{xy}}{\partial x} + \frac{\partial \sigma_{yy}}{\partial y} \right) dz + C_2(x,y)$$

$$\sigma_{zz} = -\int_{-\frac{h}{2}}^{z} \left(\frac{\partial \sigma_{xz}}{\partial x} + \frac{\partial \sigma_{yz}}{\partial y} \right) dz + C_3(x,y) \quad (6.2.25)$$

where the stresses σ_{xx}, σ_{xy}, and σ_{yy} are known from Eq. (6.2.23) and C_i are functions to be determined using the conditions $\sigma_{xz}(x, y, -h/2) = \sigma_{yz}(x, y, -h/2) = \sigma_{zz}(x, y, -h/2) = 0$. We obtain $C_i = 0$ and

$$\sigma_{xz} = \frac{h^2}{8}\left[1 - \left(\frac{2z}{h}\right)^2\right] \sum_{n=1}^{\infty}\sum_{m=1}^{\infty} S_{mn}^{xz} \cos\frac{m\pi x}{a} \sin\frac{n\pi y}{b} \quad (6.2.26a)$$

$$\sigma_{yz} = \frac{h^2}{8}\left[1 - \left(\frac{2z}{h}\right)^2\right] \sum_{n=1}^{\infty}\sum_{m=1}^{\infty} S_{mn}^{yz} \sin\frac{m\pi x}{a} \cos\frac{n\pi y}{b} \quad (6.2.26b)$$

$$\sigma_{zz} = -\frac{h^3}{48}\left\{\left[1 + \left(\frac{2z}{h}\right)^3\right] - 3\left[1 + \left(\frac{2z}{h}\right)\right]\right\}$$
$$\times \sum_{n=1}^{\infty}\sum_{m=1}^{\infty} S_{mn}^{zz} \sin\frac{m\pi x}{a} \sin\frac{n\pi y}{b} \quad (6.2.26c)$$

where

$$S_{mn}^{xz} = \left(\frac{m\pi}{a}\right)\left(R_{xx}W_{mn} - \mathcal{M}_{mn}^1\right) + \left(\frac{n\pi}{b}\right)R_{xy}W_{mn}$$
$$= S_{13}W_{mn} - T_{13}T_{mn}^1$$
$$S_{mn}^{yz} = \left(\frac{n\pi}{b}\right)\left(R_{yy}W_{mn} - \mathcal{M}_{mn}^2\right) + \left(\frac{m\pi}{a}\right)R_{xy}W_{mn}$$
$$= S_{23}W_{mn} - T_{23}T_{mn}^1$$
$$S_{mn}^{zz} = \left(\frac{m\pi}{a}\right)S_{mn}^{xz} + \left(\frac{n\pi}{b}\right)S_{mn}^{yz} = S_{33}W_{mn} - T_{33}T_{mn}^1 \quad (6.2.27)$$

with S_{ij} defined by

$$S_{13} = \left(\frac{m\pi}{a}\right)^3 Q_{11} + \left(\frac{m\pi}{a}\right)\left(\frac{n\pi}{b}\right)^2 (Q_{12} + 2Q_{66})$$

$$S_{23} = \left(\frac{n\pi}{b}\right)^3 Q_{22} + \left(\frac{m\pi}{a}\right)^2 \left(\frac{n\pi}{b}\right)(Q_{12} + 2Q_{66})$$

$$S_{33} = \left(\frac{m\pi}{a}\right)^4 Q_{11} + 2\left(\frac{m\pi}{a}\right)^2\left(\frac{n\pi}{b}\right)^2 (Q_{12} + 2Q_{66}) + \left(\frac{n\pi}{b}\right)^4 Q_{22}$$

$$T_{13} = \left(\frac{m\pi}{a}\right)(Q_{11}\alpha_1 + Q_{12}\alpha_2), \quad T_{23} = \left(\frac{n\pi}{b}\right)(Q_{12}\alpha_1 + Q_{22}\alpha_2)$$

$$T_{33} = \left(\frac{m\pi}{a}\right)^2 Q_{11}\alpha_1 + \left[\left(\frac{m\pi}{a}\right)^2 \alpha_2 + \left(\frac{n\pi}{b}\right)^2 \alpha_1\right] Q_{12} + \left(\frac{n\pi}{b}\right)^2 Q_{22}\alpha_2$$

(6.2.28)

Note that σ_{xz} and σ_{yz} are zero and $\sigma_{zz} = q$ at the top surface of the plate ($z = h/2$). The transverse shear stress σ_{xz} is the maximum at $(x,y,z) = (0,b/2,0)$, σ_{yz} is the maximum at $(x,y,z) = (a/2,0,0)$, and the transverse normal stress σ_{zz} is the maximum at $(x,y,z) = (a/2,b/2,h/2)$.

All developments presented above can be specialized to isotropic plates by setting $\alpha_1 = \alpha_2 = \alpha$ and

$$D_{11} = D_{22} = D, \quad D_{12} = \nu D, \quad 2D_{66} = (1-\nu)D, \quad D = \frac{Eh^3}{12(1-\nu^2)}$$

(6.2.29)

The thermal resultants take the form

$$M_{mn}^1 = M_{mn}^2 = D(1+\nu)\alpha T_{mn}^1 = \frac{Eh^3\alpha}{12(1-\nu)}T_{mn}^1 \qquad (6.2.30)$$

The transverse deflection becomes

$$w_0(x,y) = \sum_{n=1}^{\infty}\sum_{m=1}^{\infty} W_{mn} \sin\frac{m\pi x}{a} \sin\frac{n\pi y}{b} \qquad (6.2.31)$$

where $s = b/a$ is the plate aspect ratio and

$$W_{mn} = \frac{b^4}{D\pi^4}\left[\frac{q_{mn} + \delta_T(m^2 s^2 + n^2)}{(m^2 s^2 + n^2)^2 + \hat{k}}\right] \qquad (6.2.32a)$$

$$\hat{k} = \frac{kb^4}{D\pi^4}, \quad \delta_T = \frac{T_{mn}^1 \alpha D(1+\nu)\pi^2}{b^2} \qquad (6.2.32b)$$

BENDING OF SIMPLY SUPPORTED RECTANGULAR PLATES

The bending moments become

$$M_{xx} = D \sum_{n=1}^{\infty} \sum_{m=1}^{\infty} \left[\frac{\pi^2}{b^2} \left(m^2 s^2 + \nu n^2 \right) W_{mn} - (1+\nu)\alpha T_{mn}^1 \right]$$
$$\times \sin \frac{m\pi x}{a} \sin \frac{n\pi y}{b}$$

$$M_{yy} = D \sum_{n=1}^{\infty} \sum_{m=1}^{\infty} \left[\frac{\pi^2}{b^2} \left(\nu m^2 s^2 + n^2 \right) W_{mn} - (1+\nu)\alpha T_{mn}^1 \right]$$
$$\times \sin \frac{m\pi x}{a} \sin \frac{n\pi y}{b}$$

$$M_{xy} = -(1-\nu) \frac{\pi^2}{b^2} Ds \sum_{n=1}^{\infty} \sum_{m=1}^{\infty} mn\, W_{mn} \cos \frac{m\pi x}{a} \cos \frac{n\pi y}{b}$$

(6.2.33)

The transverse shear forces Q_x and Q_y can be calculated using Eqs. (6.2.15a,b) with

$$S_{xx} = D \left[\left(\frac{m\pi}{a}\right)^3 + \left(\frac{m\pi}{a}\right) \left(\frac{n\pi}{b}\right)^2 \right]$$

$$S_{yy} = D \left[\left(\frac{n\pi}{b}\right)^3 + \left(\frac{m\pi}{a}\right)^2 \left(\frac{n\pi}{b}\right) \right] \quad (6.2.34)$$

6.2.3 Sinusoidally Loaded Plates

The Navier solution for a sinusoidally distributed transverse load

$$q(x,y) = q_0 \sin \frac{\pi x}{a} \sin \frac{\pi y}{b} \quad (6.2.35)$$

is a one-term solution ($m = n = 1$ and $q_{11} = q_0$), and therefore it is a *closed-form solution*. For this case, the deflection in Eq. (6.2.11) becomes

$$w_0(x,y) = \frac{q_0 b^4}{\pi^4 (D_{11} s^4 + 2\hat{D}_{12} s^2 + D_{22}) + k b^4} \sin \frac{\pi x}{a} \sin \frac{\pi y}{b} \quad (6.2.36)$$

where $s = b/a$ is the plate aspect ratio. The maximum deflection occurs at $x = a/2$ and $y = b/2$:

$$w_{max} = w_0(\frac{a}{2}, \frac{b}{2}) = \frac{q_0 b^4}{\pi^4 (D_{11} s^4 + 2\hat{D}_{12} s^2 + D_{22}) + k b^4} \quad (6.2.37)$$

If $D_{11} = 25D_{22}$ and $2\hat{D}_{12} = 2.5D_{22}$ (typical of a graphite-epoxy fiber-reinforced material), $k = 0$, and $a = b$, we have

$$w_{max} = \frac{q_0 b^4}{28.5 D_{22} \pi^4} = 0.0003602 \frac{q_0 b^4}{D_{22}} = 0.0043115 \frac{q_0 b^4}{E_2 h^3} \quad (6.2.38)$$

The bending moments for the case when $k = 0$ are given by

$$M_{xx} = \frac{q_0 b^2 (s^2 D_{11} + D_{12})}{\pi^2 (s^4 D_{11} + 2s^2 \hat{D}_{12} + D_{22})} \sin\frac{\pi x}{a} \sin\frac{\pi y}{b}$$

$$M_{yy} = \frac{q_0 b^2 (s^2 D_{12} + D_{22})}{\pi^2 (s^4 D_{11} + 2s^2 \hat{D}_{12} + D_{22})} \sin\frac{\pi x}{a} \sin\frac{\pi y}{b}$$

$$M_{xy} = -\frac{2s q_0 b^2 D_{66}}{\pi^2 (s^4 D_{11} + 2s^2 \hat{D}_{12} + D_{22})} \cos\frac{\pi x}{a} \cos\frac{\pi y}{b} \quad (6.2.39)$$

The maximum values of the bending moments M_{xx} and M_{yy} occur at $x = a/2$ and $y = b/2$, and the maximum of M_{xy} occurs at $x = y = 0$.

The in-plane stresses are

$$\begin{Bmatrix} \sigma_{xx} \\ \sigma_{yy} \\ \sigma_{xy} \end{Bmatrix} = z \frac{w_s \pi^2}{b^2} \begin{Bmatrix} (s^2 Q_{11} + Q_{12}) \sin\frac{\pi x}{a} \sin\frac{\pi y}{b} \\ (s^2 Q_{12} + Q_{22}) \sin\frac{\pi x}{a} \sin\frac{\pi y}{b} \\ -2s Q_{66} \cos\frac{\pi x}{a} \cos\frac{\pi y}{b} \end{Bmatrix} \quad (6.2.40)$$

where

$$w_s = \frac{q_0 b^4}{\pi^4 (s^4 D_{11} + 2s^2 \hat{D}_{12} + D_{22}) + k b^4} \quad (6.2.41)$$

The maximum normal stresses occur at $(x, y, z) = (a/2, b/2, h/2)$, and the shear stress is maximum at $(x, y, z) = (a, b, -h/2)$.

The transverse shear and normal stresses $(\sigma_{xz}, \sigma_{yz}, \sigma_{zz})$ computed from the equilibrium equations are

$$\sigma_{xz} = \frac{h^2}{8}\left[1 - \left(\frac{2z}{h}\right)^2\right] S_{13} \, w_s \cos\frac{\pi x}{a} \sin\frac{\pi y}{b}$$

$$\sigma_{yz} = \frac{h^2}{8}\left[1 - \left(\frac{2z}{h}\right)^2\right] S_{23} \, w_s \sin\frac{\pi x}{a} \cos\frac{\pi y}{b}$$

$$\sigma_{zz} = -\frac{h^3}{48}\left\{\left[1 + \left(\frac{2z}{h}\right)^3\right] - 3\left[1 + \left(\frac{2z}{h}\right)\right]\right\} S_{33} \, w_s \sin\frac{\pi x}{a} \sin\frac{\pi y}{b}$$

$$(6.2.42)$$

where

$$S_{13} = \frac{\pi^3}{b^3}\left[s^3 Q_{11} + s^2 (Q_{12} + 2Q_{66})\right]$$

$$S_{23} = \frac{\pi^3}{b^3}\left[s^3 Q_{22} + s^2 (Q_{12} + 2Q_{66})\right]$$

$$S_{33} = \frac{\pi^4}{b^4}\left[s^4 Q_{11} + 2s^2 (Q_{12} + 2Q_{66}) + Q_{22}\right] \qquad (6.2.43)$$

Clearly, σ_{xz} is maximum at $(x, y, z) = (0, b/2, 0)$, σ_{yz} is maximum at $(x, y, z) = (a/2, 0, 0)$, and σ_{zz} is maximum at $(x, y, z) = (a/2, b/2, h/2)$.

For isotropic plates with $k = 0$, the deflection becomes

$$w_0(x, y) = \frac{q_0 b^4}{\pi^4 D(s^2 + 1)^2} \sin\frac{\pi x}{a} \sin\frac{\pi y}{b} \qquad (6.2.44)$$

The maximum deflection occurs at $x = a/2$ and $y = b/2$

$$w_{max} = \frac{q_0 b^4}{\pi^4 D(s^2 + 1)^2} \qquad (6.2.45)$$

and for a square ($s = b/a = 1$) isotropic plate, we have

$$w_{max} = \frac{q_0 a^4}{4D\pi^4} \approx 0.002566 \frac{q_0 a^4}{D} \qquad (6.2.46)$$

For rectangular isotropic plates, the maximum bending moments become

$$(M_{xx})_{max} = \frac{q_0 b^2 (s^2 + \nu)}{\pi^2 (s^2 + 1)^2}, \quad (M_{yy})_{max} = \frac{q_0 b^2 (\nu s^2 + 1)}{\pi^2 (s^2 + 1)^2} \qquad (6.2.47)$$

$$(M_{xy})_{max} = -\frac{q_0 s b^2 (1 - \nu)}{\pi^2 (s^2 + 1)^2} \qquad (6.2.48)$$

and for a square isotropic plate with $\nu = 0.3$ we have

$$(M_{xx})_{max} = (M_{yy})_{max} = \frac{(1 + \nu) q_0 b^2}{4\pi^2} = 0.03293 q_0 a^2 \qquad (6.2.49)$$

$$(M_{xy})_{max} = -\frac{(1 - \nu) q_0 b^2}{4\pi^2} = -0.01773 q_0 a^2 \qquad (6.2.50)$$

268 THEORY AND ANALYSIS OF ELASTIC PLATES

The reaction force at the corner $x = a$ and $y = b$ of an isotropic rectangular plate is given by

$$R_c = \frac{2q_0 s b^2 (1-\nu)}{\pi^2 (s^2 + 1)^2} \qquad (6.2.51)$$

which acts in the direction of the applied load. In other words, if the load is applied downward (i.e., q_0 is positive), the corners have the tendency to lift, but this is prevented by the concentrated downward reactions at the corners. For a square isotropic plate ($\nu = 0.3$) the value is $R_c = 0.03546 q_0 a^2$.

The in-plane normal stresses in a rectangular isotropic plate are maximum at the center of the plate. Substituting $x = a/2$, $y = b/2$ and $z = h/2$ in Eq. (40), we find

$$\left\{ \begin{matrix} \sigma_{xx} \\ \sigma_{yy} \end{matrix} \right\}_{max} = \frac{6 q_0 b^2}{\pi^2 h^2 (s^2 + 1)^2} \left\{ \begin{matrix} (s^2 + \nu) \\ (\nu s^2 + 1) \end{matrix} \right\} \qquad (6.2.52)$$

For $a > b$, $(\sigma_{xx})_{max} > (\sigma_{yy})_{max}$. The shear stress σ_{xy} is maximum at $(x, y, z) = (a, b, -h/2)$ and $(x, y, z) = (0, 0, -h/2)$. We have

$$(\sigma_{xy})_{max} = \frac{6 q_0 b^2 s (1-\nu)}{\pi^2 h^2 (s^2 + 1)^2} \qquad (6.2.53)$$

The maximum transverse shear stress σ_{yz}, for example, occurs at $(x, y, z) = (a/2, 0, 0)$ and its value is given by

$$(\sigma_{yz})_{max} = \frac{3 q_0 b s^2}{2 \pi h (s^2 + 1)} \qquad (6.2.54)$$

If we use the formula $\tilde{\sigma}_{yz} = 3 V_y / 2h$, we obtain

$$(\tilde{\sigma}_{yz})_{max} = \frac{3 q_0 b s^2 [1 + (2-\nu) s^2]}{2 \pi h (s^2 + 1)^2} \qquad (6.2.55)$$

which is larger in value. For a square isotropic ($\nu = 0.3$) plate the values are

$$(\sigma_{xx})_{max} = (\sigma_{yy})_{max} = 0.1976 \frac{q_0 a^2}{h^2}, \quad (\sigma_{xy})_{max} = 0.1064 \frac{q_0 a^2}{h^2}$$

$$(\sigma_{yz})_{max} = 0.2387 \frac{q_0 a}{h} \quad \text{and} \quad (\tilde{\sigma}_{yz})_{max} = 0.3223 \frac{q_0 a}{h}$$

6.2.4 Plates with Distributed and Point Loads

For loads other than sinusoidal load, the Navier solution is a series solution that can be evaluated for sufficient number of terms in the series. In particular, for *uniformly distributed load* $q(x,y) = q_0$, a constant, we have

$$q_{mn} = \frac{16q_0}{\pi^2 mn} \quad \text{for } m, n, \text{ odd} \tag{6.2.56}$$

and the solution becomes

$$w_0(x,y) = \frac{16q_0}{\pi^2} \sum_{n=1,3,\cdots}^{\infty} \sum_{m=1,3,\cdots}^{\infty} \frac{1}{mn d_{mn}} \sin \frac{m\pi x}{a} \sin \frac{n\pi y}{b} \tag{6.2.57}$$

where d_{mn} is defined in Eq. (6.2.10).

For the *hydrostatic load* $q(x,y) = q_0(y/b)$, we have

$$q_{mn} = \frac{8q_0}{\pi^2 mn}(-1)^{n+1} \quad \text{for } m = 1, 3, \cdots; n = 1, 2, 3, \cdots \tag{6.2.58}$$

and the solution becomes

$$w_0(x,y) = \frac{8q_0}{\pi^2} \sum_{n=1,2,\cdots}^{\infty} \sum_{m=1,3,\cdots}^{\infty} \frac{(-1)^{n+1}}{mn d_{mn}} \sin \frac{m\pi x}{a} \sin \frac{n\pi y}{b} \tag{6.2.59}$$

For a *point load* Q_0 located at (x_0, y_0), the load coefficients are given by Eq. (6.2.7)

$$q_{mn} = \frac{4Q_0}{ab} \sin \frac{m\pi x_0}{a} \sin \frac{n\pi y_0}{b} \tag{6.2.60}$$

and the solution becomes

$$w_0(x,y) = \frac{4Q_0}{ab} \sum_{n=1}^{\infty} \sum_{m=1}^{\infty} \frac{1}{d_{mn}} \sin \frac{m\pi x_0}{a} \sin \frac{n\pi y_0}{b} \sin \frac{m\pi x}{a} \sin \frac{n\pi y}{b} \tag{6.2.61}$$

The series converges rapidly in the case of a uniform load, but it is slower for a point load.

The bending moments can be calculated from

$$M_{xx} = \frac{\pi^2}{b^2} \sum_{n=1}^{\infty} \sum_{m=1}^{\infty} \left(m^2 s^2 D_{11} + n^2 D_{12}\right) W_{mn} \sin \frac{m\pi x}{a} \sin \frac{n\pi y}{b}$$

$$M_{yy} = \frac{\pi^2}{b^2} \sum_{n=1}^{\infty} \sum_{m=1}^{\infty} \left(m^2 s^2 D_{12} + n^2 D_{22}\right) W_{mn} \sin \frac{m\pi x}{a} \sin \frac{n\pi y}{b}$$

$$M_{xy} = -\frac{2s\pi^2 D_{66}}{b^2} \sum_{n=1}^{\infty} \sum_{m=1}^{\infty} mn W_{mn} \cos \frac{m\pi x}{a} \cos \frac{n\pi y}{b} \tag{6.2.62}$$

and the in-plane stresses are given by

$$\left\{\begin{array}{c}\sigma_{xx}\\ \sigma_{yy}\\ \sigma_{xy}\end{array}\right\} = z\frac{\pi^2}{b^2}\sum_{n=1}^{\infty}\sum_{m=1}^{\infty} W_{mn} \left\{\begin{array}{c}(s^2m^2Q_{11}+n^2Q_{12})\sin\frac{m\pi x}{a}\sin\frac{n\pi y}{b}\\ (s^2m^2Q_{12}+n^2Q_{22})\sin\frac{m\pi x}{a}\sin\frac{n\pi y}{b}\\ -2smnQ_{66}\cos\frac{m\pi x}{a}\cos\frac{n\pi y}{b}\end{array}\right\} \quad (6.2.63)$$

where $W_{mn} = q_{mn}/d_{mn}$. The maximum normal stresses occur at $(x,y,z) = (a/2, b/2, h/2)$, and the shear stress is maximum at $(x,y,z) = (a, b, -h/2)$.

The transverse stresses $(\sigma_{xz}, \sigma_{yz}, \sigma_{zz})$ computed using the 3-D stress equilibrium equations for any $-h/2 \le z \le h/2$ are

$$\sigma_{xz} = \frac{h^2}{8}\left[1-\left(\frac{2z}{h}\right)^2\right]\sum_{n=1}^{\infty}\sum_{m=1}^{\infty} S_{13}W_{mn}\cos\frac{m\pi x}{a}\sin\frac{n\pi y}{b}$$

$$\sigma_{yz} = \frac{h^2}{8}\left[1-\left(\frac{2z}{h}\right)^2\right]\sum_{n=1}^{\infty}\sum_{m=1}^{\infty} S_{23}W_{mn}\sin\frac{m\pi x}{a}\cos\frac{n\pi y}{b}$$

$$\sigma_{zz} = -\frac{h^3}{48}\left\{\left[1+\left(\frac{2z}{h}\right)^3\right] - 3\left[1+\left(\frac{2z}{h}\right)\right]\right\}$$

$$\times \sum_{n=1}^{\infty}\sum_{m=1}^{\infty} S_{33}W_{mn}\sin\frac{m\pi x}{a}\sin\frac{n\pi y}{b} \quad (6.2.64a)$$

where

$$S_{13} = \left(\frac{m\pi}{a}\right)^3 Q_{11} + \left(\frac{m\pi}{a}\right)\left(\frac{n\pi}{b}\right)^2 (Q_{12}+2Q_{66})$$

$$S_{23} = \left(\frac{n\pi}{b}\right)^3 Q_{22} + \left(\frac{m\pi}{a}\right)^2\left(\frac{n\pi}{b}\right) (Q_{12}+2Q_{66})$$

$$S_{33} = \left(\frac{m\pi}{a}\right)^4 Q_{11} + 2\left(\frac{m\pi}{a}\right)^2\left(\frac{n\pi}{b}\right)^2 (Q_{12}+2Q_{66}) + \left(\frac{n\pi}{b}\right)^4 Q_{22}$$

(6.2.64b)

In integrating the stress-equilibrium equations, it is assumed that the stresses $(\sigma_{xz}, \sigma_{yz}, \sigma_{zz})$ are zero at the bottom surface of the plate $(z = -h/2)$. Because of the assumptions of the plate theory, they will also be zero at $z = h/2$. The transverse shear stress σ_{xz} is the maximum at $(x,y,z) = (0, b/2, 0)$, σ_{yz} is the maximum at $(x,y,z) = (a/2, 0, 0)$, and the transverse normal stress σ_{zz} is the maximum at $(x,y,z) = (a/2, b/2, h/2)$.

For an isotropic rectangular plate with $k = 0$ and subjected to a uniformly distributed load, the deflection is given by

$$w_0(x,y) = \frac{16q_0 b^4}{D\pi^6} \sum_{n=1,3,\cdots}^{\infty} \sum_{m=1,3,\cdots}^{\infty} \frac{1}{mn(m^2 s^2 + n^2)^2} \sin\frac{m\pi x}{a} \sin\frac{n\pi y}{b} \qquad (6.2.65)$$

The center deflection becomes

$$w_0\left(\frac{a}{2},\frac{b}{2}\right) = \frac{16q_0 b^4}{D\pi^6} \sum_{n=1,3,\cdots}^{\infty} \sum_{m=1,3,\cdots}^{\infty} \frac{(-1)^{\frac{m+n}{2}-1}}{mn(m^2 s^2 + n^2)^2} \qquad (6.2.66)$$

For a square plate ($b = a$) it reduces to

$$w_{max} = \frac{16q_0 a^4}{D\pi^6} \sum_{n=1,3,\cdots}^{\infty} \sum_{m=1,3,\cdots}^{\infty} \frac{(-1)^{\frac{m+n}{2}-1}}{mn(m^2 + n^2)^2} \qquad (6.2.67)$$

A one-term solution is given by

$$w_{max} = \frac{4q_0 a^4}{D\pi^6} = 0.00416\frac{q_0 a^4}{D} \qquad (6.2.68)$$

which is about 2.4% in error compared to the solution obtained with $m, n = 1, 3, \cdots, 9$. Thus, the series in Eq. (6.2.66) converges rapidly. The expressions for bending moments and stresses, which involve second derivatives of w_0, do not converge as rapidly. The first-term values are ($\nu = 0.3$)

$$(M_{xx})_{max} = 0.05338 q_0 a^2, \quad (\sigma_{xx})_{max} = 0.3203\frac{q_0 a^2}{h^2}$$

The first four terms of the series (i.e., $m, n = 1, 3$) yields

$$w_{max} = 0.004056\frac{q_0 a^4}{D}$$

$$(M_{xx})_{max} = 0.0469 q_0 a^2, \quad (\sigma_{xx})_{max} = 0.2816\frac{q_0 a^2}{h^2}$$

For an isotropic rectangular plate ($k = 0$) under a point load Q_0 at (x_0, y_0), the deflection is given by

$$w_0(x,y) = \frac{4Q_0 b^2 s}{D\pi^4} \sum_{n=1,3,\cdots}^{\infty} \sum_{m=1,3,\cdots}^{\infty} \frac{\sin\frac{m\pi x_0}{a} \sin\frac{n\pi y_0}{b}}{(m^2 s^2 + n^2)^2} \sin\frac{m\pi x}{a} \sin\frac{n\pi y}{b} \qquad (6.2.69)$$

272 THEORY AND ANALYSIS OF ELASTIC PLATES

The center deflection when the load is applied at the center is given by

$$w_0\left(\frac{a}{2},\frac{b}{2}\right) = \frac{4Q_0 b^2 s}{D\pi^4} \sum_{n=1,3,\cdots}^{\infty} \sum_{m=1,3,\cdots}^{\infty} \frac{1}{(m^2 s^2 + n^2)^2} \qquad (6.2.70)$$

In the case of a square plate, the center deflection becomes

$$w_{max} = \frac{4Q_0 a^2}{D\pi^4} \sum_{n=1,3,\cdots}^{\infty} \sum_{m=1,3,\cdots}^{\infty} \frac{1}{(m^2 + n^2)^2} \qquad (6.2.71)$$

The first term of the series yields ($\nu = 0.3$)

$$w_{max} = 0.01027 \frac{Q_0 a^2}{D} = 0.1121 \frac{Q_0 a^2}{Eh^3}$$

$$(M_{xx})_{max} = 0.1317 Q_0 a, \quad (\sigma_{xx})_{max} = 0.7903 \frac{Q_0 a}{h^2}$$

Taking the first four terms (i.e., $m, n = 1, 3$) of the series, we obtain

$$w_{max} = 0.01121 \frac{Q_0 a^2}{D} = 0.1225 \frac{Q_0 a^2}{Eh^3}$$

$$(M_{xx})_{max} = 0.199 Q_0 a, \quad (\sigma_{xx})_{max} = 1.194 \frac{Q_0 a}{h^2}$$

The deflection is about 3.4% in error compared to the solution 0.0116 $(Q_0 a^2/D)$ obtained using $m, n = 1, 3, \cdots, 19$.

Table 6.2.2 contains the nondimensionalized maximum transverse deflections and stresses of square and rectangular plates under various types of loads when $k = 0$. The transverse deflection and stresses are nondimensionalized as follows:

$$\bar{w} = w_0(0,0)\left(\frac{E_2 h^3}{a^4 q_0}\right); \quad \bar{\sigma}_{xx} = \sigma_{xx}(a/2, b/2, h/2)\left(\frac{h^2}{a^2 q_0}\right)$$

$$\bar{\sigma}_{yy} = \sigma_{yy}(a/2, b/2, h/2)\left(\frac{h^2}{a^2 q_0}\right); \quad \bar{\sigma}_{xy} = \sigma_{xy}(a, b, -h/2)\left(\frac{h^2}{a^2 q_0}\right)$$

$$\bar{\sigma}_{xz} = \sigma_{xz}(0, b/2, 0)\left(\frac{h}{a q_0}\right); \quad \bar{\sigma}_{yz} = \sigma_{yz}(a/2, 0, 0)\left(\frac{h}{a q_0}\right)$$

Table 6.2.2. Transverse deflections and stresses in isotropic ($\nu = 0.3$) and orthotropic ($E_1/E_2 = 25$, $G_{12} = G_{13} = 0.5E_2$, $\nu_{12} = 0.25$) square and rectangular plates subjected to distributed loads.

Load	$\bar{w}$	$\bar{\sigma}_{xx}$	$\bar{\sigma}_{yy}$	$\bar{\sigma}_{xy}$	$\bar{\sigma}_{xz}$	$\bar{\sigma}_{yz}$
Square Plates ($b/a = 1$)						
Isotropic plates ($\nu = 0.3$)						
SL	0.0280	0.1976	0.1976	0.1064	0.2387	0.2387
UL (19)[†]	0.0444	0.2873	0.2873	0.1946	0.4909	0.4909
HL (19)	0.0222	0.1436	0.1436	0.0775	0.2455	0.1353
PL (29)	0.1266	2.4350	2.4350	0.3658	1.0010	1.0010
Orthotropic plates						
SL	0.0043	0.5387	0.0267	0.0213	0.4398	0.0376
UL (19)	0.0065	0.7866	0.0244	0.0463	0.7758	0.1811
HL (19)	0.0032	0.3933	0.0122	0.0105	0.3879	0.0092
PL (29)	0.0232	6.0075	0.8670	0.0409	1.8150	0.1338
Rectangular Plates ($b/a = 3$)						
Isotropic plates ($\nu = 0.3$)						
SL	0.0908	0.5088	0.2024	0.1149	0.4297	0.1432
UL (19)	0.1336	0.7130	0.2433	0.2830	0.7221	0.5110
HL (19)	0.0668	0.3565	0.1217	0.0579	0.3610	0.0636
PL (29)	0.1845	2.3523	1.8828	0.0566	0.9032	0.2257
Orthotropic plates						
SL	0.0048	0.6016	0.0087	0.0080	0.4746	0.0085
UL (19)	0.0062	0.7494	0.0071	0.0447	0.7310	0.1509
HL (19)	0.0031	0.3747	0.0035	0.0034	0.3655	0.0044
PL (29)	0.0227	4.8157	0.5648	0.0000	1.4820	0.0864

[†] The number in parenthesis denotes the maximum values of m and n used to evaluate the series.

The load consists of only sinusoidal (SL), uniform (UL), hydrostatic (HL), or central point load (PL). Since convergence for derivatives of a function is slower than the function itself, the convergence is slower for stresses, which are calculated using the derivatives of the deflection. In the case of the point load, convergence for stresses will not be reached due to the stress singularity at the center of the plate.

274   THEORY AND ANALYSIS OF ELASTIC PLATES

Deflections for other load cases can also be determined by knowing q_{mn}. For example, a uniform load of intensity q_0 is applied on a rectangular patch of size $2c \times 2d$ (see Figure 6.2.2) whose center is located at (x_0, y_0). Then, from Eq. (6.2.5) we have

$$q_{mn} = \frac{4q_0}{ab} \int_{y_0-d}^{y_0+d} \int_{x_0-c}^{x_0+c} \sin\frac{m\pi x}{a} \sin\frac{n\pi y}{b}\, dx dy$$

$$= \frac{16q_0}{\pi^2 mn} \sin\frac{m\pi x_0}{a} \sin\frac{n\pi y_0}{b} \sin\frac{m\pi c}{a} \sin\frac{n\pi d}{b} \quad (6.2.72)$$

Similarly, if a point load Q_0 is applied at point (x_0, y_0), then q_{mn} can be computed from Eq. (6.2.72) by setting $4q_0 = Q_0/(cd)$ and letting c and d approach zero. We obtain

$$q_{mn} = \frac{4Q_0}{ab} \sin\frac{m\pi x_0}{a} \sin\frac{n\pi y_0}{b} \quad (6.2.73)$$

and the deflection is given by Eq. (6.2.61).

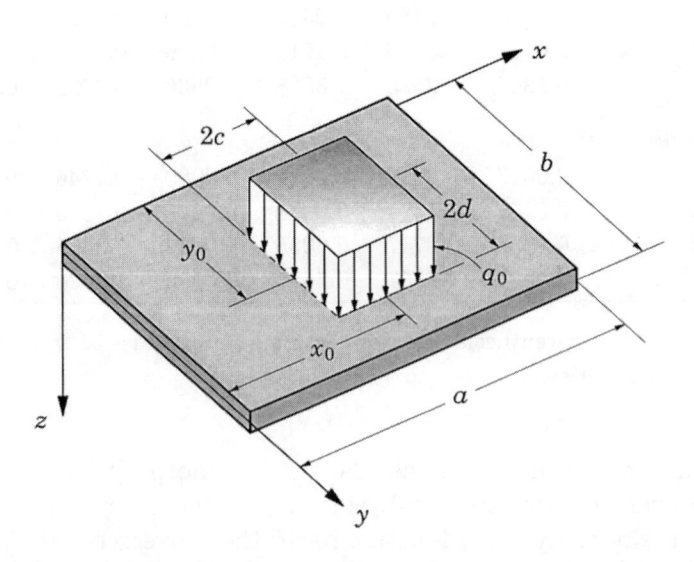

Figure 6.2.2. Rectangular plate with a patch loading.

6.2.5 Plates with Thermal Loads

If an orthotropic plate is subjected to temperature field that varies linearly through the thickness $\Delta T = zT_1$ and $q = k = 0$, then the solution (6.2.11) becomes

$$w_0(x,y) = \sum_{n=1}^{\infty}\sum_{m=1}^{\infty} W_{mn} \sin\frac{m\pi x}{a} \sin\frac{n\pi y}{b} \qquad (6.2.74a)$$

where [see Eqs. (6.2.9) and (6.2.10)]

$$W_{mn} = \frac{T_{mn}^1 b^2}{\pi^2} \frac{\left[m^2 s^2 (D_{11}\alpha_1 + D_{12}\alpha_2) + n^2 (D_{12}\alpha_1 + D_{22}\alpha_2)\right]}{(m^4 s^4 D_{11} + 2m^2 n^2 s^2 \hat{D}_{12} + n^4 D_{22})} \qquad (6.2.74b)$$

and T_{mn}^1 is defined by equation of the form (6.2.5). For a rectangular isotropic plate, the deflection is given by Eq. (6.2.74a) with

$$W_{mn} = \frac{(1+\nu)\alpha b^2}{\pi^2} \frac{T_{mn}^1}{(m^2 s^2 + n^2)} \qquad (6.2.74c)$$

The bending moments can be calculated using Eqs. (6.2.14a,b) and stresses by Eqs. (6.2.23) and (6.2.26a–c), with W_{mn} defined above and M_{mn}^1 and M_{mn}^2 in Eq. (6.2.22).

Sinusoidally distributed temperature field

For sinusoidally distributed temperature field (but linear through the plate thickness)

$$T_1(x,y) = T_0^1 \sin\frac{m\pi x}{a} \sin\frac{n\pi y}{b} \qquad (6.2.75)$$

the coefficients T_{mn}^1 are equal to T_0^1 and the deflection becomes

$$w_0(x,y) = \hat{w} \sin\frac{\pi x}{a} \sin\frac{\pi y}{b} \qquad (6.2.76a)$$

where

$$\hat{w} = \frac{b^2 \left[s^2(D_{11}\alpha_1 + D_{12}\alpha_2) + (D_{12}\alpha_1 + D_{22}\alpha_2)\right] T_0^1}{\pi^2 (s^4 D_{11} + 2s^2 \hat{D}_{12} + D_{22})} \qquad (6.2.76b)$$

For a square plate, we have

$$w_0(x,y) = \frac{b^2\left[(D_{11}+D_{12})\alpha_1 + (D_{12}+D_{22})\alpha_2\right]T_0^1}{\pi^2(D_{11}+2\hat{D}_{12}+D_{22})} \sin\frac{\pi x}{a} \sin\frac{\pi y}{b} \tag{6.2.77}$$

If $D_{11} = 25D_{22}$, $2\hat{D}_{12} = 2.5D_{22}$, $\alpha_1 = 3\alpha_2$, and $a = b$, the maximum deflection becomes

$$w_{max} = \frac{77}{28.5\pi^2} T_0^1 \alpha_2 b^2 \approx 0.2738\, T_0^1 \alpha_2 a^2 \tag{6.2.78}$$

The bending moments are

$$M_{xx} = \frac{\pi^2}{b^2}\left[\left(D_{11}s^2 + D_{12}\right)\hat{w} - \left(Q_{11}s^2 + Q_{12}\right)T_0^1\right]\sin\frac{\pi x}{a}\sin\frac{\pi y}{b}$$

$$M_{yy} = \frac{\pi^2}{b^2}\left[\left(D_{12}s^2 + D_{22}\right)\hat{w} - \left(Q_{12}s^2 + Q_{22}\right)T_0^1\right]\sin\frac{\pi x}{a}\sin\frac{\pi y}{b}$$

$$M_{xy} = -2\frac{\pi^2}{ab}D_{66}\hat{w}\cos\frac{\pi x}{a}\cos\frac{\pi y}{b} \tag{6.2.79}$$

For an isotropic square plate, the deflection becomes

$$w_0(x,y) = \frac{(1+\nu)\alpha a^2 T_0^1}{2\pi^2}\sin\frac{\pi x}{a}\sin\frac{\pi y}{b} \tag{6.2.80}$$

The maximum deflection is $w_{max} = 0.0633(\alpha a^2 T_0^1)$ when $\nu = 0.25$ and $w_{max} = 0.06586(\alpha a^2 T_0^1)$ when $\nu = 0.3$. The bending moments become

$$M_{xx} = \frac{\pi^2 D}{a^2}\left[\frac{(1+\nu)^2 \alpha a^2 T_0^1}{2\pi^2} - \frac{E\alpha T_0^1}{1-\nu}\right]\sin\frac{\pi x}{a}\sin\frac{\pi y}{a}$$

$$M_{yy} = \frac{\pi^2 D}{a^2}\left[\frac{(1+\nu)^2 \alpha b^2 T_0^1}{2\pi^2} - \frac{E\alpha T_0^1}{1-\nu}\right]\sin\frac{\pi x}{a}\sin\frac{\pi y}{a}$$

$$M_{xy} = -\frac{E\alpha T_0^1 h^3}{24}\cos\frac{\pi x}{a}\cos\frac{\pi y}{a} \tag{6.2.81}$$

The maximum bending moments are

$$(M_{xx})_{max} = (M_{yy})_{max} = (M_{xy})_{max} = -(h^3 E\alpha T_0^1/24) \tag{6.2.82a}$$

and the maximum stresses (at $z = h/2$) are

$$(\sigma_{xx})_{max} = (\sigma_{yy})_{max} = (\sigma_{xy})_{max} = -(hE\alpha T_0^1/4) \tag{6.2.82b}$$

Uniformly distributed temperature field

For uniformly distributed temperature field, $T_1(x,y) = T_0^1$, the coefficients T_{mn}^1 are equal to $16T_0^1/\pi^2 mn$. Hence, the deflection becomes

$$w_0(x,y) = \sum_{n=1}^{\infty}\sum_{m=1}^{\infty} W_{mn} \sin\frac{m\pi x}{a} \sin\frac{n\pi y}{b} \qquad (6.2.83)$$

where

$$W_{mn} = 16T_0^1 \frac{[m^2 s^2 (D_{11}\alpha_1 + D_{12}\alpha_2) + n^2 (D_{12}\alpha_1 + D_{22}\alpha_2)]}{mn\pi^4(m^4 s^4 D_{11} + 2m^2 n^2 s^2 \hat{D}_{12} + n^4 D_{22})} \qquad (6.2.84)$$

For an isotropic rectangular plate, the deflection becomes

$$w_0(x,y) = \frac{16T_0^1 \alpha(1+\nu)b^2}{\pi^4} \sum_{n=1}^{\infty}\sum_{m=1}^{\infty} \frac{1}{(m^2 s^2 + n^2)mn} \sin\frac{m\pi x}{a} \sin\frac{n\pi y}{b} \qquad (6.2.85)$$

In general, the series for thermal problems converge more slowly than the corresponding ones for mechanical loads. This can be seen by examining the denominators of the series, for example, in Eqs. (6.2.65) and (6.2.85).

Table 6.2.3 contains the nondimensionalized maximum transverse deflections and stresses

$$\hat{w} = w_0(0,0)/(\alpha_2 T_0^1 a^2), \quad \hat{\sigma} = \sigma \times \beta, \quad \beta = (1/hE_2\alpha_2 T_0^1)$$

of square plates under two types of temperature variations, of which both are assumed to vary linearly through the thickness. The first one is a sinusoidal distribution of the form

$$\Delta T = zT_0^1 \sin\frac{\pi x}{a} \sin\frac{\pi y}{b} \qquad (6.2.86)$$

and the second one is the uniform temperature of the form

$$\Delta T = zT_0^1 \qquad (6.2.87)$$

In the case of uniform temperature field (with respect to x and y), the double trigonometric series are evaluated using $m, n = 1, 3, \cdots, 99$.

Table 6.2.3. Transverse deflections and stresses in isotropic ($\nu = 0.25$) and orthotropic ($E_1/E_2 = 25$, $G_{12} = G_{13} = 0.5E_2$, $\nu_{12} = 0.25$, $\alpha_1 = 3\alpha_2$) square plates subjected to distributed temperature fields of the form $\Delta T = zT^1(x,y)$.

Load	$\hat{w}$	$\hat{\sigma}_{xx}$	$\hat{\sigma}_{yy}$	$\hat{\sigma}_{xy}$
Isotropic plates				
SL	0.0633	-0.0025	-0.0025	0.0025
UL(1)[†]	0.1027	0.0000	0.0000	0.0000
UL(3)	0.0902	0.0030	0.0030	0.0061
UL(9)	0.0922	-0.0019	-0.0019	0.0090
UL(49)	0.0921	-0.0024	-0.0024	0.0141
UL(99)	0.0921	0.0041	0.0041	0.0163
UL(199)	0.0921	0.0041	0.0041	0.0185
Orthotropic plates				
SL	0.2738	-0.0352	0.0082	0.0135
UL(1)	0.4438	0.1772	0.0187	0.0219
UL(3)	0.3924	0.3615	0.0098	0.0328
UL(9)	0.3977	0.0538	0.0533	0.0464
UL(49)	0.3974	0.0334	0.0046	0.0707
UL(99)	0.3974	0.4033	0.0132	0.0812
UL(199)	0.3974	0.4045	0.0132	0.0917

[†] The number in parenthesis denotes the maximum value of m and n used to evaluate the series.

6.3 Lévy's Solutions

6.3.1 Solution Procedure

Here we consider an alternate procedure to determine the solution to simply supported rectangular plates. The solution to the problem of a rectangular plate with two opposite edges simply supported and the other two edges having arbitrary boundary conditions can be represented in terms of single Fourier series as

$$w_0(x,y) = \sum_{n=1}^{\infty} W_n(x) \sin \frac{n\pi y}{b} \qquad (6.3.1)$$

which satisfies the following simply supported boundary conditions on edges $y = 0$ and $y = b$ (see Figure 6.3.1):

BENDING OF SIMPLY SUPPORTED RECTANGULAR PLATES 279

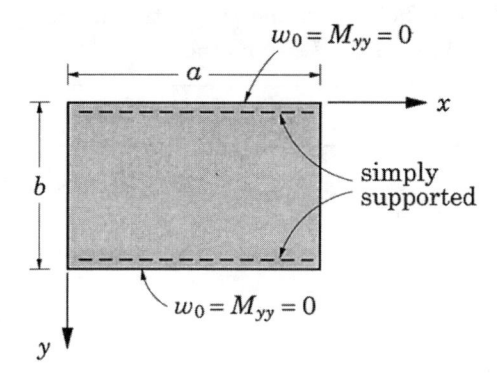

Figure 6.3.1. A rectangular plate with simply supported edges at $y = 0, b$.

$$w_0(x,0) = w_0(x,b) = 0, \quad M_{yy}(x,0) = M_{yy}(x,b) = 0 \qquad (6.3.2)$$

Similarly, the load q and thermal moments M_{xx}^T and M_{yy}^T are represented as

$$q(x,y) = \sum_{n=1}^{\infty} q_n(x) \sin \frac{n\pi y}{b} \qquad (6.3.3)$$

$$M_{xx}^T(x,y) = \sum_{n=1}^{\infty} M_n^1(x) \sin \frac{n\pi y}{b} \qquad (6.3.4a)$$

$$M_{yy}^T(x,y) = \sum_{n=1}^{\infty} M_n^2(x) \sin \frac{n\pi y}{b} \qquad (6.3.4b)$$

where $q_n(x)$, for example, is defined by

$$q_n(x) = \frac{2}{b} \int_0^b q(x,y) \sin \frac{n\pi y}{b} \, dy \qquad (6.3.5)$$

The coefficients for various distributions of $q(x,y)$ are given in Table 6.3.1.

Substitution of Eqs. (6.3.1), (6.3.3), and (6.3.4a,b) into Eq. (6.1.1) results in

$$\sum_{n=1}^{\infty} \left\{ D_{11} \frac{d^4 W_n}{dx^4} - 2 \left(\frac{n\pi}{b} \right)^2 \hat{D}_{12} \frac{d^2 W_n}{dx^2} + \left[\left(\frac{n\pi}{b} \right)^4 D_{22} + k \right] W_n \right.$$
$$\left. + \frac{d^2 M_n^1}{dx^2} - \left(\frac{n\pi}{b} \right)^2 M_n^2 - q_n \right\} \sin \frac{n\pi y}{b} = 0 \qquad (6.3.6)$$

Table 6.3.1. Coefficients in the single trigonometric series expansion of loads in the Lévy method.

Loading	Coefficients $q_n(x)$
	Uniform load, $q = q_0$ $q_n = \frac{4q_0}{n\pi}$ $(n = 1, 3, 5, \cdots)$
	Hydrostatic load $q(x,y) = q_0 \frac{y}{b}$ $q_n = \frac{2q_0}{n\pi}(-1)^{n+1}$ $(n = 1, 2, 3 \cdots)$
	Point load, Q_0 at (x_0, y_0) $q_n = \frac{2Q_0}{b}\delta(x - x_0)\sin\frac{n\pi y_0}{b}$ $(n = 1, 2, 3, \cdots)$
	Line load, $q(x,y) = q_0\delta(y - y_0)$ $q_n = \frac{2q_0}{b}\sin\frac{n\pi y_0}{b}$ $(n = 1, 2, 3, \cdots)$

Since the result must hold for any y, it follows that the expression inside the braces must be zero:

$$D_{11}\frac{d^4W_n}{dx^4} - 2\left(\frac{n\pi}{b}\right)^2 \hat{D}_{12}\frac{d^2W_n}{dx^2} + \left[\left(\frac{n\pi}{b}\right)^4 D_{22} + k\right]W_n$$
$$= q_n - \frac{d^2 M_n^1}{dx^2} + \left(\frac{n\pi}{b}\right)^2 M_n^2 \equiv \bar{q}_n \qquad (6.3.7)$$

The fourth-order differential equation in Eq. (6.3.7) can be solved either analytically or by an approximate method. Analytically, Eq. (6.3.7) can be solved directly or by the *state-space approach* used in control theory. As for approximate methods, the Rayleigh–Ritz, finite difference, and finite element methods are good candidates. Here the analytical and Rayleigh–Ritz solutions are discussed.

6.3.2 Analytical Solution

The general form of the analytical (exact) solution to the fourth-order differential equation (6.3.7) consists of two parts: homogeneous and nonhomogeneous (or particular) solutions. The homogeneous solution is of the form

$$W_n^h(x) = C \exp(\lambda x) \qquad (6.3.8)$$

where λ denotes a root of the algebraic equation

$$D_{11}\lambda^4 - 2\left(\frac{n\pi}{b}\right)^2 (D_{12} + 2D_{66})\lambda^2 + \left(\frac{n\pi}{b}\right)^4 D_{22} + k = 0 \qquad (6.3.9)$$

The roots of this equation are

$$\lambda_1^2 = \beta_n^2 \left(\frac{D_{12} + 2D_{66}}{D_{11}}\right) - \beta_n^2 \left[\left(\frac{D_{12} + 2D_{66}}{D_{11}}\right)^2 - \left(\frac{D_{22}}{D_{11}} + \frac{k}{D_{11}\beta_n^4}\right)\right]^{\frac{1}{2}}$$

$$\lambda_2^2 = \beta_n^2 \left(\frac{D_{12} + 2D_{66}}{D_{11}}\right) + \beta_n^2 \left[\left(\frac{D_{12} + 2D_{66}}{D_{11}}\right)^2 - \left(\frac{D_{22}}{D_{11}} + \frac{k}{D_{11}\beta_n^4}\right)\right]^{\frac{1}{2}}$$
$$(6.3.10)$$

where

$$\beta_n = \frac{n\pi}{b} \qquad (6.3.11)$$

Since there are four roots $(\lambda_1, -\lambda_1, \lambda_2, -\lambda_2)$, the solution (6.3.7) can be written as a linear combination of functions of these four roots. The true form of the solution depends on the nature of the roots, i.e., real or complex and equal or distinct. Three different cases are discussed below.

Case 1: Roots are real and distinct

When $(D_{12} + 2D_{66})^2 > D_{11}(D_{22} + k/\beta_n^4)$, the roots are real and unequal: $\lambda_1, -\lambda_1, \lambda_2,$ and $-\lambda_2$, where λ_1 and λ_2 are given by Eq. (6.3.10). Then the homogeneous part of the solution for this case is

$$W_n^h(x) = A_n \cosh \lambda_1 x + B_n \sinh \lambda_1 x + C_n \cosh \lambda_2 x + D_n \sinh \lambda_2 x \quad (6.3.12)$$

Case 2: Roots are real and equal

When $(D_{12} + 2D_{66})^2 = D_{11}(D_{22} + k/\beta_n^4)$, the roots are real but equal:

$$\lambda_1 = \lambda_2 = -\lambda_3 = -\lambda_4 \equiv \lambda, \quad \lambda^2 = \beta_n^2 \left(\frac{D_{12} + 2D_{66}}{D_{11}} \right) \quad (6.3.13)$$

and the homogeneous part of the solution is

$$W_n^h(x) = (A_n + B_n x) \cosh \lambda x + (C_n + D_n x) \sinh \lambda x \quad (6.3.14)$$

Case 3: Roots are complex

When $(D_{12} + 2D_{66})^2 < D_{11}(D_{22} + k/\beta_n^4)$, the roots are complex and they appear in complex conjugate pairs $\lambda_1 \pm i\lambda_2$ and $-\lambda_1 \pm i\lambda_2$ ($i = \sqrt{-1}, \lambda_1 > 0, \lambda_2 > 0$):

$$(\lambda_1)^2 = \frac{\beta_n^2}{2D_{11}} \left[\sqrt{D_{11}\left(D_{22} + \frac{k}{\beta_n^4}\right)} + (D_{12} + 2D_{66}) \right]$$

$$(\lambda_2)^2 = \frac{\beta_n^2}{2D_{11}} \left[\sqrt{D_{11}\left(D_{22} + \frac{k}{\beta_n^4}\right)} - (D_{12} + 2D_{66}) \right] \quad (6.3.15)$$

The homogeneous part of the solution is

$$W_n^h(x) = (A_n \cos \lambda_2 x + B_n \sin \lambda_2 x) \cosh \lambda_1 x$$
$$+ (C_n \cos \lambda_2 x + D_n \sin \lambda_2 x) \sinh \lambda_1 x \quad (6.3.16)$$

BENDING OF SIMPLY SUPPORTED RECTANGULAR PLATES 283

For isotropic plates, we have $D_{11} = D_{22} = D$ and $D_{12} + 2D_{66} = D$. Hence, Case 2 applies when $k = 0$ and Case 3 applies when $k \neq 0$. For isotropic plates *without* elastic foundation, the homogeneous solution is given by

$$W_n^h(x) = (A_n + B_n x) \cosh \beta_n x + (C_n + D_n x) \sinh \beta_n x \qquad (6.3.17)$$

For isotropic plates on elastic foundation, Case 3 applies and we have

$$W_n^h(x) = (A_n \cos \lambda_2 x + B_n \sin \lambda_2 x) \cosh \lambda_1 x \\ + (C_n \cos \lambda_2 x + D_n \sin \lambda_2 x) \sinh \lambda_1 x \qquad (6.3.18)$$

where

$$(\lambda_1)^2 = \frac{\beta_n^2}{2}\left[\sqrt{\left(1 + \frac{k}{D\beta_n^4}\right)} + 1\right], \quad (\lambda_2)^2 = \frac{\beta_n^2}{2}\left[\sqrt{\left(1 + \frac{k}{D\beta_n^4}\right)} - 1\right] \qquad (6.3.19)$$

The particular solution W_n^p of the fourth-order differential equation (6.3.7) in the general case in which q_n, M_n^1, and M_n^2 are functions of x can be determined by expanding the solution and the loads in Fourier series:

$$W_n^p(x) = \sum_{m=1}^{\infty} W_{mn} \sin \frac{m\pi x}{a}$$

$$q_n(x) = \sum_{m=1}^{\infty} q_{mn} \sin \frac{m\pi x}{a}$$

$$M_n^1(x) = \sum_{m=1}^{\infty} M_{mn}^1 \sin \frac{m\pi x}{a}$$

$$M_n^2(x) = \sum_{m=1}^{\infty} M_{mn}^2 \sin \frac{m\pi x}{a} \qquad (6.3.20)$$

where the coefficients q_{mn}, M_{mn}^1, and M_{mn}^2 have the same meaning as in Eqs. (6.2.3), (6.2.4a), and (6.2.4b), respectively. Substituting these expressions into

$$D_{11}\frac{d^4 W_n^p}{dx^4} - 2\beta_n^2 \hat{D}_{12}\frac{d^2 W_n^p}{dx^2} + \left(\beta_n^4 D_{22} + k\right) W_n^p \\ = q_n - \frac{d^2 M_n^1}{dx^2} + \beta_n^2 M_n^2 \qquad (6.3.21)$$

we obtain [c.f., Eq. (6.2.9)]

$$W_n^p(x) = \sum_{m=1}^{\infty} \frac{1}{d_{mn}} \left[q_{mn} + \left(\frac{m\pi}{a}\right)^2 M_{mn}^1 + \left(\frac{n\pi}{b}\right)^2 M_{mn}^2 \right] \sin \frac{m\pi x}{a} \quad (6.3.22)$$

where d_{mn} is defined by Eq. (6.2.10). The complete solution is given by

$$w_0(x,y) = \sum_{n=1}^{\infty} \left(W_n^h(x) + W_n^p(x) \right) \sin \beta_n y \quad (6.3.23)$$

where W_n^h for the three cases are given in Eqs. (6.3.12), (6.3.14), and (6.3.16), and W_n^p for all cases is given by Eq. (6.3.22). The four constants $A_n, B_n, C_n,$ and D_n present in the expressions for W_n^h are determined by using the four boundary conditions associated with the boundary points $x = 0, a$ (two at each point). In particular, when the edges at $x = 0, a$ are also simply supported, the constants $A_n, B_n, C_n,$ and D_n will be identically zero, and we obtain the solution in Eq. (6.2.11).

Alternatively, if q_n, $d^2 M_n^1/dx^2$, and M_n^2 are at most linear functions of x, the particular solution is also a constant or a linear function of x, which is determined by substituting it into Eq. (6.3.21). The particular solution is given by

$$W_n^p(x) = \frac{q_n - \frac{d^2 M_n^1}{dx^2} + \beta_n^2 M_n^2}{(D_{22}\beta_n^4 + k)} \equiv \hat{q}_n \quad (6.3.24)$$

and the complete solution becomes

$$w_0(x,y) = \sum_{n=1}^{\infty} \left(W_n^h(x) + \hat{q}_n \right) \sin \beta_n y \quad (6.3.25)$$

In this case, even for the case of simply supported boundary conditions on edges $x = 0, a$, the constants $A_n, B_n, C_n,$ and D_n will not be zero.

The particular solution in Eq. (6.3.24) is valid for any $\bar{q} = q_n - (d^2 M_n^1/dx^2) + \beta_n^2 M_n^2$ that is a constant or a linear function of x. Thus, $q(x,y)$ can be an arbitrary function of y but must be a linear function of x in order for the particular solution in Eq. (6.3.24) to be valid. Note that a point load can be expressed as

$$q(x,y) = Q_0\, \delta(x - x_0)\, \delta(y - y_0) \quad (6.3.26a)$$

BENDING OF SIMPLY SUPPORTED RECTANGULAR PLATES 285

and therefore

$$q_n = \frac{2}{b}\int_0^b q(x,y)\sin\frac{n\pi y}{b}\,dy = \frac{2Q_0}{b}\delta(x-x_0)\sin\frac{n\pi y_0}{b}$$

Similarly, hydrostatic load of the type

$$q(x,y) = q_0\left(\frac{x}{a}\right) \qquad (6.3.26\text{b})$$

yields

$$q_n = \frac{2}{b}\int_0^b q(x,y)\sin\frac{n\pi y}{b}\,dy = \frac{2q_0}{n\pi}\left(\frac{x}{a}\right)$$

For these cases, the particular solutions are given by

$$W_n^p(x) = \frac{\frac{2Q_0}{b}\delta(x-x_0)\sin\frac{n\pi y_0}{b}}{(D_{22}\beta_n^4 + k)}, \qquad W_n^p(x) = \frac{\frac{2q_0}{n\pi}\left(\frac{x}{a}\right)}{(D_{22}\beta_n^4 + k)}$$

When $\hat{q}_n$ is a quadratic or higher-order function of x, one must use Eq. (6.3.22). In the following discussion, we assume that the loads are distributed such that $\bar{q}_n$ is at most a linear function of x so that we can use the particular solution in Eq. (6.3.24).

6.3.3 Plates Under Distributed Transverse Loads

Consider a simply supported plate subjected to a distributed transverse load $q(x,y)$ such that $\bar{q}_n$ is independent of x. Since the simply supported boundary conditions on edges $y = 0$ and $y = b$ are identically satisfied by the assumed solution (6.3.1), the remaining simply supported boundary conditions on edges $x = 0$ and $x = a$ are

$$w_0 = 0, \quad M_{xx} \equiv -\left(D_{11}\frac{\partial^2 w_0}{\partial x^2} + D_{12}\frac{\partial^2 w_0}{\partial y^2}\right) = 0 \qquad (6.3.27\text{a})$$

Because of the form of the solution in Eq. (6.3.1), Eq. (6.3.27a) implies that

$$W_n(0) = 0, \quad -D_{11}\left(\frac{d^2 W_n}{dx^2}\right)_{x=0} + D_{12}\beta_n^2 W_n(0) = 0 \qquad (6.3.27\text{b})$$

for any n. Thus, if $w_0 = 0$ at a point, then $M_{xx} = 0$ necessarily implies that

$$\frac{\partial^2 w_0}{\partial x^2} = 0 \text{ at } x = 0 \qquad (6.3.27\text{c})$$

Case 1: $(D_{12} + 2D_{66})^2 > D_{11}(D_{22} + k/\beta_n^4)$. Only certain orthotropic plates fall into this category. The solution for this case is given by

$$w_0(x,y) = \sum_{n=1}^{\infty} (A_n \cosh \lambda_1 x + B_n \sinh \lambda_1 x + C_n \cosh \lambda_2 x$$
$$+ D_n \sinh \lambda_2 x + \hat{q}_n) \sin \beta_n y \qquad (6.3.28a)$$

where λ_1 and λ_2 are defined by Eq. (6.3.10) and $\hat{q}_n$ is defined by Eq. (6.3.24). The second derivative of w_0 with respect to x is

$$\frac{\partial^2 w_0}{\partial x^2} = \sum_{n=1}^{\infty} \left[\lambda_1^2 (A_n \cosh \lambda_1 x + B_n \sinh \lambda_1 x) + \lambda_2^2 (C_n \cosh \lambda_2 x \right.$$
$$\left. + D_n \sinh \lambda_2 x) \right] \sin \beta_n y \qquad (6.3.28b)$$

Substituting Eqs. (6.3.28a,b) into Eq. (6.3.27a), the following algebraic equations are obtained:

$$\begin{bmatrix} 1 & 0 & 1 & 0 \\ \cosh \bar{\lambda}_1 & \sinh \bar{\lambda}_1 & \cosh \bar{\lambda}_2 & \sinh \bar{\lambda}_2 \\ \lambda_1^2 & 0 & \lambda_2^2 & 0 \\ \lambda_1^2 \cosh \bar{\lambda}_1 & \lambda_1^2 \sinh \bar{\lambda}_1 & \lambda_2^2 \cosh \bar{\lambda}_2 & \lambda_2^2 \sinh \bar{\lambda}_2 \end{bmatrix} \begin{Bmatrix} A_n \\ B_n \\ C_n \\ D_n \end{Bmatrix} = - \begin{Bmatrix} \hat{q}_n \\ \hat{q}_n \\ 0 \\ 0 \end{Bmatrix}$$
$$(6.3.29)$$

where $\bar{\lambda}_i = \lambda_i a$. These equations can be solved by Cramer's rule (see section 1.2.5). The determinant of the 4 × 4 coefficient matrix in Eq. (6.3.29) is

$$\Delta = (\lambda_2^2 - \lambda_1^2)^2 \sinh \bar{\lambda}_1 \sinh \bar{\lambda}_2$$

and the solution of the matrix equation (6.3.29) is given by

$$A_n = -\frac{\lambda_2^2}{(\lambda_2^2 - \lambda_1^2)} \hat{q}_n, \quad B_n = -\frac{\lambda_2^2 (1 - \cosh \bar{\lambda}_1)}{(\lambda_2^2 - \lambda_1^2) \sinh \bar{\lambda}_1} \hat{q}_n$$
$$C_n = \frac{\lambda_1^2}{(\lambda_2^2 - \lambda_1^2)} \hat{q}_n, \quad D_n = \frac{\lambda_1^2 (1 - \cosh \bar{\lambda}_2)}{(\lambda_2^2 - \lambda_1^2) \sinh \bar{\lambda}_2} \hat{q}_n \qquad (6.3.30)$$

Case 2: $(D_{12} + 2D_{66})^2 = D_{11}(D_{22} + k/\beta_n^4)$. Isotropic plates without elastic foundation ($k = 0$) and certain orthotropic plates (with or without k) fall into this category. The solution for this case is given by

$$w_0(x,y) = \sum_{n=1}^{\infty} [(A_n + B_n x) \cosh \lambda x + (C_n + D_n x) \sinh \lambda x$$
$$+ \hat{q}_n] \sin \beta_n y \qquad (6.3.31a)$$

where λ is defined by Eq. (6.3.13). The second derivative of w_0 with respect to x is

$$\frac{\partial^2 w_0}{\partial x^2} = \sum_{n=1}^{\infty} \left[\lambda^2 (A_n + B_n a) \cosh \lambda x + \lambda^2 (C_n + D_n x) \sinh \lambda x \right.$$
$$\left. + 2\lambda \left(B_n \sinh \lambda x + D_n \cosh \lambda x \right) \right] \sin \beta_n y \quad (6.3.31b)$$

Substituting Eqs. (6.3.31a,b) into the boundary conditions (6.3.27a,b), we obtain

$$A_n + \hat{q}_n = 0, \quad A_n \lambda^2 + 2D_n \lambda = 0$$
$$(A_n + B_n a) \cosh \lambda a + (C_n + D_n a) \sinh \lambda a + \hat{q}_n = 0$$
$$\lambda^2 (A_n + B_n a) \cosh \lambda a + \lambda^2 (C_n + D_n a) \sinh \lambda a$$
$$+ 2\lambda (B_n \sinh \lambda a + D_n \cosh \lambda a) = 0$$

The solution of the above equations is

$$A_n = -\hat{q}_n, \quad D_n = \frac{1}{2} \lambda \hat{q}_n, \quad B_n = \frac{1}{2} \left(\frac{1 - \cosh \lambda a}{\sinh \lambda a} \right) \lambda \hat{q}_n$$
$$C_n = \frac{1}{2} \left(\frac{1 - \cosh \lambda a}{\sinh \lambda a} \right) \left(\frac{\lambda a - 2 \sinh \lambda a}{\sinh \lambda a} \right) \hat{q}_n \quad (6.3.32)$$

Case 3: $(D_{12} + 2D_{66})^2 < D_{11}(D_{22} + k/\beta_n^4)$. Isotropic plates on elastic foundation ($k \neq 0$) and orthotropic plates (with or without k) fall into this category. The solution for this case is given by

$$w_0(x,y) = \sum_{n=1}^{\infty} \left[(A_n \cos \lambda_2 x + B_n \sin \lambda_2 x) \cosh \lambda_1 x + (C_n \cos \lambda_2 x \right.$$
$$\left. + D_n \sin \lambda_2 x) \sinh \lambda_1 x + \hat{q}_n \right] \sin \beta_n y \quad (6.3.33a)$$

where λ_1 and λ_2 are defined by Eq. (6.3.15). Note that

$$\frac{\partial^2 w_0}{\partial x^2} = \sum_{n=1}^{\infty} \left[-\lambda_2^2 (A_n \cos \lambda_2 x + B_n \sin \lambda_2 x) \cosh \lambda_1 x \right.$$
$$- \lambda_2^2 (C_n \cos \lambda_2 x + D_n \sin \lambda_2 x) \sinh \lambda_1 x$$
$$+ \lambda_1^2 (A_n \cos \lambda_2 x + B_n \sin \lambda_2 x) \cosh \lambda_1 x$$
$$+ \lambda_1^2 (C_n \cos \lambda_2 x + D_n \sin \lambda_2 x) \sinh \lambda_1 x$$
$$+ 2\lambda_1 \lambda_2 (-A_n \sin \lambda_2 x + B_n \cos \lambda_2 x) \sinh \lambda_1 x$$
$$\left. + 2\lambda_1 \lambda_2 (-C_n \sin \lambda_2 x + D_n \cos \lambda_2 x) \cosh \lambda_1 x \right] \sin \beta_n y$$
$$(6.3.33b)$$

Substituting Eqs. (6.3.33a,b) into the boundary conditions (6.3.27), we obtain

$$A_n + \hat{q}_n = 0, \quad \left(\lambda_1^2 - \lambda_2^2\right) A_n + 2\lambda_1\lambda_2 D_n = 0$$
$$(A_n \cos \lambda_2 a + B_n \sin \lambda_2 a) \cosh \lambda_1 a$$
$$+ (C_n \cos \lambda_2 a + D_n \sin \lambda_2 a) \sinh \lambda_1 a + \hat{q}_n = 0$$
$$\left(\lambda_1^2 - \lambda_2^2\right) [(A_n \cos \lambda_2 a + B_n \sin \lambda_2 a) \cosh \lambda_1 a$$
$$+ (C_n \cos \lambda_2 a + D_n \sin \lambda_2 a) \sinh \lambda_1 a]$$
$$+ 2\lambda_1\lambda_2 [(-A_n \sin \lambda_2 a + B_n \cos \lambda_2 a) \sinh \lambda_1 a$$
$$+ (-C_n \sin \lambda_2 a + D_n \cos \lambda_2 a) \cosh \lambda_1 a] = 0$$

The solution of these equations is

$$A_n = -\hat{q}_n, \quad D_n = \left(\frac{\lambda_1^2 - \lambda_2^2}{2\lambda_1\lambda_2}\right) \hat{q}_n$$

$$B_n = \frac{1}{\Delta} \left[\sin \lambda_2 a \left(\cosh \lambda_1 a - \cos \lambda_2 a\right) \hat{q}_n \right.$$
$$\left. + \sinh \lambda_1 a \left(\cos \lambda_2 a + \cosh \lambda_1 a\right) D_n\right]$$

$$C_n = \frac{1}{\Delta} \left[\sinh \lambda_1 a \left(\cos \lambda_2 a - \cosh \lambda_1 a\right) \hat{q}_n \right.$$
$$\left. - \sin \lambda_2 a \left(\cosh \lambda_1 a + \cos \lambda_2 a\right) D_n\right]$$

$$\Delta = -\left(\sin^2 \lambda_2 a \cosh^2 \lambda_1 a + \cos^2 \lambda_2 a \sinh^2 \lambda_1 a\right) \quad (6.3.34)$$

Consider a simply supported plate under a distributed transverse load. For *uniformly distributed load*, we have

$$q(x,y) = q_0, \quad \hat{q}_n = \frac{4q_0 b^4}{n^5 \pi^5 D}$$

and for the case of *hydrostatic loading*, we find

$$q(x,y) = q_0 \frac{y}{b}, \quad \hat{q}_n = \frac{2q_0 b^4}{n^5 \pi^5 D}(-1)^{n+1}$$

An *isotropic plate without elastic foundation* ($k = 0$) falls into Case 2 (real but equal roots). The solution is given by ($\lambda = \beta_n$)

$$w_0(x,y) = \sum_{n=1}^{\infty} [(A_n + B_n x) \cosh \beta_n x + (C_n + D_n x) \sinh \beta_n x$$
$$+ \hat{q}_n] \sin \beta_n y \quad (6.3.35)$$

where [c.f., Eq.(6.3.32)]

$$A_n = -\hat{q}_n, \quad D_n = \frac{1}{2}\beta_n\hat{q}_n, \quad B_n = \frac{1}{2}\left(\frac{1-\cosh\beta_n a}{\sinh\beta_n a}\right)\beta_n\hat{q}_n$$
$$C_n = \frac{1}{2}\left(\frac{1-\cosh\beta_n a}{\sinh\beta_n a}\right)\left(\frac{\beta_n a - 2\sinh\beta_n a}{\sinh\beta_n a}\right)\hat{q}_n \qquad (6.3.36)$$

Bending moments and stresses can be computed using the expression given for w_0 in Eq. (6.3.35).

An *isotropic plate on elastic foundation* ($k \neq 0$) falls into Case 3. The solution is given by Eq. (6.3.33a) with the constants defined in Eq. (6.3.34). The roots λ_1 and λ_2 are given by Eq. (6.3.19). We note that

$$\lambda_1^2 - \lambda_2^2 = \beta_n^2, \quad 2\lambda_1\lambda_2 = \sqrt{\frac{k}{D}} \qquad (6.3.37)$$

Orthotropic plates for which $E_1 > E_2$, with or without elastic foundation, also fall into Case 3. The solution is again given by Eqs. (6.3.33a) and (6.3.34), with λ_1 and λ_2 defined by Eq. (6.3.15). In this case, we have

$$\lambda_1^2 - \lambda_2^2 = \beta_n^2\left(\frac{D_{12} + 2D_{66}}{D_{11}}\right), \quad 2\lambda_1\lambda_2 = \sqrt{\frac{D_{22}\beta_n^4 + k}{D_{11}} - \frac{\hat{D}_{12}^2}{D_{11}^2}\beta_n^4} \qquad (6.3.38)$$

Table 6.3.2 contains numerical results for the nondimensionalized center deflection $\bar{w} = w_0(a/2, b/2)(D/q_0 a^4) \times 10^2$ and bending moments $\bar{M} = M(a/2, b/2) \times 10/(q_0 a^2)$ obtained in the Navier and Lévy methods for various simply supported plates under uniformly distributed load (UDL) or hydrostatic load (HSL). The series was evaluated using a final value of 9 for m and n. Note that the number of actual terms used in evaluating the series for deflection and bending moment for a given final value of n is different in the two methods. For example, $m, n = 1, 2, 3$ in the Navier method involves four terms ($m = n = 1; m = 1, n = 3; m = 3, n = 1; m = n = 3$) of the series, while the Lévy method involves only two terms ($n = 1$ and $n = 3$). Thus, the number of terms used from each series is vastly different for increasing n.

The center deflection due to the hydrostatic load is one-half the center deflection due to uniform load, as it should be. The maximum deflection of a plate under hydrostatic loading (HSL) occurs away from

Table 6.3.2. Nondimensional center deflections ($\bar{w}$) and bending moments $\bar{M}_{xx}$ and $\bar{M}_{xx}$ of various simply supported plates subjected to distributed transverse loads.

Method	Variable	$\frac{b}{a}=1$ HSL	UDL	$\frac{b}{a}=1.5$ HSL	UDL	$\frac{b}{a}=2$ HSL	UDL
Isotropic plates ($k=0$)							
Navier	$\bar{w}$	0.2031	0.4062	0.3862	0.7724	0.5065	1.0130
	$\bar{M}_{xx}$	0.2398	0.4797	0.4063	0.8126	0.5091	1.0182
	$\bar{M}_{yy}$	0.2398	0.4797	0.2500	0.5000	0.2331	0.4661
Lévy	$\bar{w}$	0.2031	0.4062	0.3862	0.7724	0.5065	1.0130
	$\bar{M}_{xx}$	0.2395	0.4791	0.4060	0.8120	0.5088	1.0176
	$\bar{M}_{yy}$	0.2397	0.4795	0.2499	0.4998	0.2330	0.4659
Isotropic plates on elastic foundation ($k=100$)							
Navier	$\bar{w}$	0.0503	0.1007	0.0536	0.1073	0.0524	0.1048
	$\bar{M}_{xx}$	0.0465	0.0930	0.0401	0.0802	0.0369	0.0738
	$\bar{M}_{yy}$	0.0465	0.0930	0.0176	0.0351	0.0110	0.0221
Lévy	$\bar{w}$	0.0503	0.1007	0.0536	0.1073	0.0524	0.1048
	$\bar{M}_{xx}$	0.0462	0.0923	0.0398	0.0796	0.0366	0.0732
	$\bar{M}_{yy}$	0.0464	0.0928	0.0175	0.0349	0.0109	0.0219
Orthotropic plates ($k=0$)							
Navier	$\bar{w}$	0.0271	0.0543	0.0269	0.0538	0.0262	0.0524
	$\bar{M}_{xx}$	0.6560	1.3119	0.6458	1.2917	0.6291	1.2583
	$\bar{M}_{yy}$	0.0207	0.0414	0.0068	0.0136	0.0066	0.0132
Lévy	$\bar{w}$	0.0271	0.0543	0.0269	0.0538	0.0262	0.0524
	$\bar{M}_{xx}$	0.6556	1.3113	0.6455	1.2910	0.6288	1.2576
	$\bar{M}_{yy}$	0.0207	0.0414	0.0068	0.0136	0.0066	0.0132
Orthotropic plates ($k=100$)							
Navier	$\bar{w}$	0.0185	0.0371	0.0178	0.0356	0.0175	0.0350
	$\bar{M}_{xx}$	0.4418	0.8835	0.4219	0.8438	0.4137	0.8273
	$\bar{M}_{yy}$	0.0115	0.0229	0.0037	0.0075	0.0049	0.0098
Lévy	$\bar{w}$	0.0185	0.0371	0.0178	0.0356	0.0175	0.0350
	$\bar{M}_{xx}$	0.4415	0.8829	0.4216	0.8432	0.4133	0.8267
	$\bar{M}_{yy}$	0.0115	0.0229	0.0037	0.0075	0.0049	0.0098

the center. For isotropic plates, it occurs at $y = 0.557b$. However, the value is not significantly different from the center deflection. The point of maximum deflection approaches the center of the plate as the aspect ratio a/b increases. The maximum bending moments also occur away from the center of the plate. Of course, the effect of elastic foundation is to decrease the deflections as well as bending moments.

6.3.4 Plates with Distributed Edge Moments

Consider a simply supported (on all four edges) rectangular plate subjected to distributed bending moments along edges $x = 0, a$ (see Figure 6.3.2). We wish to determine the deflections and bending moments throughout the plate using the Lévy solution procedure. The general solutions developed earlier in Eqs. (6.3.28a), (6.3.31a), and (6.3.33a) for rectangular plates with simply supported edges $y = 0, b$ are valid here except that the distributed load is zero: $q = 0$ (hence, $\hat{q}_n = 0$). Instead, we have nonzero bending moments applied on edges $x = 0, a$. The constants A_n, B_n, C_n, and D_n now must be determined subjected to the boundary conditions

$$w_0(0, y) = 0, \quad M_{xx}(0, y) = f_1(y), \quad w_0(a, y) = 0, \quad M_{xx}(a, y) = f_2(y) \tag{6.3.39}$$

where f_1 and f_2 are the applied bending moment distributions along the edges $x = 0$ and $x = a$, respectively. Here we consider isotropic plates (Case 2) and orthotropic plates (Case 3) without elastic foundation ($k = 0$).

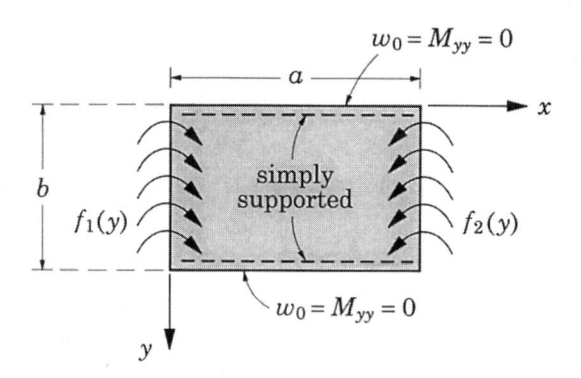

Figure 6.3.2. A rectangular plate with simply supported edges at $y = 0, b$ and subjected to distributed moments along edges $x = 0, a$.

Isotropic plates. The solution for this case is given by Eq. (6.3.31a) with $\hat{q}_n = 0$ and $\lambda = \beta_n$. The boundary condition $w_0(0, y) = 0$ yields $A_n = 0$. The condition $M_{xx}(0, y) = f_1(y)$ together with $w_0(0, y) = 0$ yields

$$-2D \sum_{n=1}^{\infty} \beta_n D_n \sin \beta_n y = f_1(y) \qquad (6.3.40)$$

The form of Eq. (6.3.40) suggests that we must express $f_1(y)$ in the series form

$$M_{xx}(0, y) = f_1(y) = \sum_{n=1}^{\infty} F_n^1 \sin \beta_n y \qquad (6.3.41)$$

where F_n^1 is calculated in the same way as before [see Eq. (6.3.5)]

$$F_n^1 = \frac{2}{b} \int_0^b f_1(y) \sin \beta_n y \, dy \qquad (6.3.42)$$

Then Eq. (6.3.40) yields

$$D_n = -\frac{F_n^1}{2D\beta_n} \qquad (6.3.43)$$

Using the remaining two boundary conditions, we obtain

$$aB_n \cosh \beta_n a + (C_n + D_n a) \sinh \beta_n a = 0$$
$$-2D\beta_n (B_n \sinh \beta_n a + D_n \cosh \beta_n a) = F_n^2$$

where $f_2(y)$ is also expressed in the same form as $f_1(y)$ so that

$$F_n^2 = \frac{2}{b} \int_0^b f_2(y) \sin \beta_n y \, dy \qquad (6.3.44)$$

Solving the above two equations, we find

$$B_n = \frac{1}{2D\beta_n \sinh \beta_n a} \left(F_n^1 \cosh \beta_n a - F_n^2 \right)$$
$$C_n = \frac{a}{2D\beta_n \sinh^2 \beta_n a} \left(F_n^2 \cosh \beta_n a - F_n^1 \right) \qquad (6.3.45)$$

Orthotropic plates. The solution for this case is given by Eq. (6.3.33a) with $\hat{q}_n = 0$. The boundary conditions give

$$A_n = 0, \quad D_n = -\frac{F_n^1}{2D_{11}\lambda_1\lambda_2} \qquad (6.3.46a)$$

$$B_n = \frac{\sinh \lambda_1 a}{2D_{11}\lambda_1\lambda_2\Delta} \left(F_n^2 \cos \lambda_2 a - F_n^1 \cosh \lambda_1 a\right)$$

$$C_n = \frac{\sin \lambda_2 a}{2D_{11}\lambda_1\lambda_2\Delta} \left(F_n^1 \cos \lambda_2 - F_n^2 \cosh \lambda_1 a\right)$$

$$\Delta = -\left(\sin^2 \lambda_2 a \cosh^2 \lambda_1 a + \cos^2 \lambda_2 a \sinh^2 \lambda_1 a\right) \quad (6.3.46b)$$

Nondimensional deflections and bending moments of simply supported rectangular plates with uniformly distributed moments of intensity $f_1 = f_2 = M_0$ are presented in Table 6.3.3 for various aspect ratios. The results were computed using $n = 1, 3, \cdots, 9$. The nondimensionalizations used are as follows:

For $\dfrac{a}{b} \leq 1$: $\bar{w} = w_0(a/2, b/2)\dfrac{D_{22}}{M_0 a^2}$, $\bar{M} = M(a/2, b/2)\dfrac{1}{M_0}$

For $\dfrac{a}{b} \geq 1$: $\bar{w} = w_0(a/2, b/2)\dfrac{D_{22}}{M_0 b^2}$, $\bar{M} = M(a/2, b/2)\dfrac{1}{M_0}$

$$(6.3.47)$$

Table 6.3.3. Nondimensional center deflections and bending moments of simply supported plates subjected to *uniformly distributed bending moments* along edges $x = 0, a$.

Variable $a/b \to$	0.5	0.75	1.0	1.5	2.0	3.0
Isotropic plates ($\nu = 0.3$)						
$\bar{w}$	0.0965	0.0620	0.0368	0.0280	0.0174	0.0055
$\bar{M}_{xx}$	0.7701	0.4764	0.2562	0.0465	-0.0103	-0.0148
$\bar{M}_{yy}$	0.3873	0.4240	0.3938	0.2635	0.1530	0.0446
Orthotropic plates ($E_1/E_2 = 25$, $G_{12} = 0.5E_2$, $\nu = 0.25$)						
$\bar{w}$	0.0050	0.0052	0.0052	0.0097	0.0121	0.0099
$\bar{M}_{xx}$	0.9999	1.0553	1.0606	0.8560	0.5282	0.0523
$\bar{M}_{yy}$	0.0094	0.0139	0.0378	0.0927	0.1209	0.0990

6.3.5 An Alternate Form of the Lévy Solution

The choice of coordinate system can simplify the development of Lévy solutions. Here we illustrate this by considering a rectangular plate with the (x, y) coordinate system such that $-b/2 \leq y \leq b/2$ and $0 \leq x \leq a$ (see Figure 6.3.3). Suppose that the edges $x = 0, a$ are simply supported. The Lévy solution for the plate can be written as

$$w_0(x, y) = \sum_{m=1}^{\infty} W_m(y) \sin \frac{m\pi x}{a} \qquad (6.3.48)$$

which satisfies the simply supported boundary conditions at $x = 0$ and $x = a$.

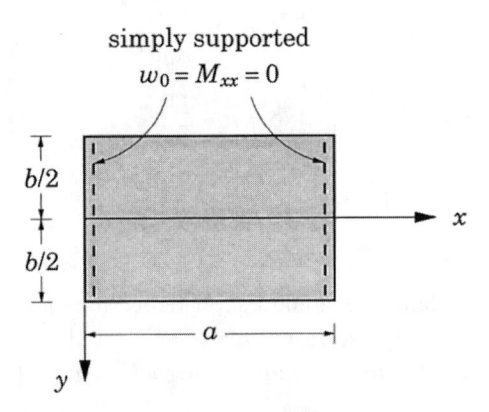

Figure 6.3.3. Geometry and coordinate system for a rectangular plate with sides $x = 0, a$ simply supported.

Following the procedure outlined in section 6.3.1, we find that the complete solution of the governing equation

$$D_{11} \frac{\partial^4 w_0}{\partial x^4} + 2\hat{D}_{12} \frac{\partial^4 w_0}{\partial x^2 \partial y^2} + D_{22} \frac{\partial^4 w_0}{\partial y^4} + k w_0 = q \qquad (6.3.49)$$

where $\hat{D}_{12} = (D_{12} + 2D_{66})$, is given by

$$w_0(x, y) = \sum_{m=1}^{\infty} \left(W_m^h(y) + W_m^p(y) \right) \sin \alpha_m x \qquad (6.3.50)$$

BENDING OF SIMPLY SUPPORTED RECTANGULAR PLATES   295

Here W_m^h denotes the homogeneous solution and W_m^p the particular solution of

$$\left(D_{11}\alpha_m^4 + k\right) W_m - 2\alpha_m^2 \hat{D}_{12}\frac{d^2 W_m}{dy^2} + D_{22}\frac{d^4 W_m}{dy^4} = q_m \qquad (6.3.51)$$

and q_m is defined by

$$q_m(y) = \frac{2}{a}\int_0^a q(x,y) \sin \alpha_m x \, dx, \quad \alpha_m = \frac{m\pi}{a} \qquad (6.3.52)$$

The homogeneous solution W_m^h is taken in the form [see Eq. (6.3.16)]

$$\begin{aligned}W_m^h(y) &= (A_m \cos \lambda_2 y + B_m \sin \lambda_2 y) \cosh \lambda_1 y \\ &+ (C_m \cos \lambda_2 y + D_m \sin \lambda_2 y) \sinh \lambda_1 y\end{aligned} \qquad (6.3.53)$$

where λ_1 and λ_2 are the roots of the equation [assuming that $(D_{12} + 2D_{66})^2 < D_{22}(D_{11} + k/\beta_n^4)$]

$$D_{22}\lambda^4 - 2\alpha_m^2 \hat{D}_{12}\lambda^2 + D_{11}\alpha_m^4 + k = 0 \qquad (6.3.54)$$

and they are given by

$$\begin{aligned}(\lambda_1)^2 &= \frac{\alpha_m^2}{2D_{22}}\left[\sqrt{D_{22}\left(D_{11} + \frac{k}{\alpha_m^4}\right)} + \hat{D}_{12}\right] \\ (\lambda_2)^2 &= \frac{\alpha_m^2}{2D_{22}}\left[\sqrt{D_{22}\left(D_{11} + \frac{k}{\alpha_m^4}\right)} - \hat{D}_{12}\right]\end{aligned} \qquad (6.3.55)$$

The particular solution in a general case is given by the solution of

$$\left(D_{11}\alpha_m^4 + k\right) W_m^p - 2\alpha_m^2 \hat{D}_{12}\frac{d^2 W_m^p}{dy^2} + D_{22}\frac{d^4 W_m^p}{dy^4} = q_m \qquad (6.3.56)$$

Assuming solution of the form

$$W_m^p(y) = \sum_{n=1}^{\infty} W_{mn} \sin \frac{n\pi y}{b} \qquad (6.3.57)$$

and expanding the load q_m also in the same form

$$q_m(y) = \sum_{n=1}^{\infty} q_{mn} \sin \frac{n\pi y}{b} \qquad (6.3.58)$$

296 THEORY AND ANALYSIS OF ELASTIC PLATES

and substituting into Eq. (6.3.56), we obtain

$$W_{mn} = \frac{q_{mn}}{d_{mn}}, \quad d_{mn} = \left(D_{11}\alpha_m^4 + k\right) + 2\alpha_m^2\beta_n^2 \hat{D}_{12} + D_{22}\beta_n^4 \quad (6.3.59)$$

The particular solution becomes

$$W_m^p(y) = \sum_{n=1}^{\infty} \frac{q_{mn}}{d_{mn}} \sin \frac{n\pi y}{b} \quad (6.3.60)$$

When q_m is a constant or a linear function of y, the particular solution is given by

$$W_m^p(y) = \frac{q_m}{D_{11}\alpha_m^4 + k} \equiv \hat{q}_m \quad (6.3.61)$$

The complete solution becomes

$$w_0(x,y) = \sum_{m=1}^{\infty} [(A_m \cos \lambda_2 y + B_m \sin \lambda_2 y)\cosh \lambda_1 y + (C_m \cos \lambda_2 y + D_m \sin \lambda_2 y)\sinh \lambda_1 y + W_m^p\,]\sin \alpha_m x \quad (6.3.62)$$

where W_m^p is given by Eq. (6.3.60), in the general case of loading, or by Eq. (6.3.61), when the loading is such that q_m is a constant or a linear function of y. The constants A_m, B_m, C_m, and D_m must be determined for the particular set of boundary conditions on edges $y = \pm b/2$.

When the applied load and boundary conditions are symmetric about the x axis, then the deflection w_0 is also symmetric about the x axis. Note that the choice of the coordinate system [with $-b/2 \leq y \leq b/2$] made it possible to arrive at this conclusion. Thus, only even functions of y should be kept in the expression (6.3.62), requiring the constants B_m and C_m to be zero. Of course, this can be shown by actual calculation of the constants. We have

$$w_0(x,y) = \sum_{m=1}^{\infty} (A_m \cos \lambda_2 y \cosh \lambda_1 y + D_m \sin \lambda_2 y \sinh \lambda_1 y + \hat{q}_m)\sin \alpha_m x$$
$$(6.3.63)$$

Once w_0 is known, the bending moments and shear forces can be computed using the relations

$$M_{xx} = -D_{11}\frac{\partial^2 w_0}{\partial x^2} - D_{12}\frac{\partial^2 w_0}{\partial y^2}$$

BENDING OF SIMPLY SUPPORTED RECTANGULAR PLATES

$$M_{yy} = -D_{12}\frac{\partial^2 w_0}{\partial x^2} - D_{22}\frac{\partial^2 w_0}{\partial y^2}$$

$$M_{xy} = -2D_{66}\frac{\partial^2 w_0}{\partial x \partial y} \tag{6.3.64}$$

$$Q_x = -D_{11}\frac{\partial^3 w_0}{\partial x^3} - \hat{D}_{12}\frac{\partial^2 w_0}{\partial x \partial y^2}$$

$$Q_y = -\hat{D}_{12}\frac{\partial^2 w_0}{\partial x^2 \partial y} - D_{22}\frac{\partial^3 w_0}{\partial y^3} \tag{6.3.65}$$

$$V_x = -D_{11}\frac{\partial^3 w_0}{\partial x^3} - \bar{D}_{12}\frac{\partial^2 w_0}{\partial x \partial y^2}$$

$$V_y = -\bar{D}_{12}\frac{\partial^2 w_0}{\partial x^2 \partial y} - D_{22}\frac{\partial^3 w_0}{\partial y^3} \tag{6.3.66}$$

where $\hat{D}_{12} = D_{12} + 2D_{66}$, $\bar{D}_{12} = D_{12} + 4D_{66}$, and

$$\frac{\partial^2 w_0}{\partial x^2} = \sum_{m=1}^{\infty} (-\alpha_m^2)\left[A_m f_1(y) + D_m f_2(y) + \hat{q}_m\right] \sin \alpha_m x \tag{6.3.67a}$$

$$\frac{\partial^2 w_0}{\partial y^2} = \sum_{m=1}^{\infty} \left\{ (\lambda_1^2 - \lambda_2^2)\left[A_m f_1(y) + D_m f_2(y)\right] \right.$$
$$\left. + 2\lambda_1 \lambda_2 \left[-A_m f_2(y) + D_m f_1(y)\right] \right\} \sin \alpha_m x \tag{6.3.67b}$$

$$\frac{\partial^2 w_0}{\partial x \partial y} = \sum_{m=1}^{\infty} \alpha_m \left[A_m f_1'(y) + D_m f_2'(y)\right] \cos \alpha_m x \tag{6.3.67c}$$

$$\frac{\partial^3 w_0}{\partial x^2 \partial y} = \sum_{m=1}^{\infty} (-\alpha_m^2)\left[A_m f_1'(y) + D_m f_2'(y)\right] \sin \alpha_m x \tag{6.3.67d}$$

$$\frac{\partial^2 w_0}{\partial x \partial y^2} = \sum_{m=1}^{\infty} \alpha_m \left\{ (\lambda_1^2 - \lambda_2^2)\left[A_m f_1(y) + D_m f_2(y)\right] \right.$$
$$\left. + 2\lambda_1 \lambda_2 \left[-A_m f_2(y) + D_m f_1(y)\right] \right\} \cos \alpha_m x \tag{6.3.67e}$$

$$\frac{\partial^3 w_0}{\partial x^3} = \sum_{m=1}^{\infty} (-\alpha_m^3)\left[A_m f_1(y) + D_m f_2(y) + \hat{q}_m\right] \cos \alpha_m x \tag{6.3.67f}$$

$$\frac{\partial^3 w_0}{\partial y^3} = \sum_{m=1}^{\infty} \left\{ (\lambda_1^2 - \lambda_2^2)\left[A_m f_1'(y) + D_m f_2'(y)\right] \right.$$
$$\left. + 2\lambda_1 \lambda_2 \left[-A_m f_2'(y) + D_m f_1'(y)\right] \right\} \sin \alpha_m x \tag{6.3.67g}$$

$$f_1 = \cos \lambda_2 y \cosh \lambda_1 y, \quad f_2 = \sin \lambda_2 y \sinh \lambda_1 y$$
$$f_1' = \lambda_1 q_1 - \lambda_2 g_2, \quad f_2' = \lambda_2 g_1 + \lambda_1 g_2$$
$$g_1 = \cos \lambda_2 y \sinh \lambda_1 y, \quad g_2 = \sin \lambda_2 y \cosh \lambda_1 y \tag{6.3.67h}$$

As a specific example, consider a rectangular plate with all its edges simply supported and subjected to distributed loads that are independent of y. For this case, the deflection of the plate is symmetric about the x axis (or symmetric in y), and Eq. (6.3.63) is valid here. The constants A_m and D_m are determined using the boundary conditions

$$w_0(x, \pm\frac{b}{2}) = 0, \quad M_{yy}(x, \pm\frac{b}{2}) = 0 \tag{6.3.68}$$

From Eqs. (6.3.63) and (6.3.64), it follows that the conditions in Eq. (6.3.68) are equivalent to

$$w_0 = 0, \quad \frac{\partial^2 w_0}{\partial y^2} = 0 \tag{6.3.69}$$

at $y = \pm b/2$. Using Eqs. (6.3.63) and (6.3.67b) in Eq. (6.3.69), we obtain

$$A_m \cos \bar{\lambda}_2 \cosh \bar{\lambda}_1 + D_m \sin \bar{\lambda}_2 \sinh \bar{\lambda}_1 = -\hat{q}_m \tag{6.3.70a}$$
$$-A_m \sin \bar{\lambda}_2 \sinh \bar{\lambda}_1 + D_m \cos \bar{\lambda}_2 \cosh \bar{\lambda}_1 = \gamma \hat{q}_m \tag{6.3.70b}$$

where

$$\bar{\lambda}_i = \frac{\lambda_i b}{2}, \quad \gamma = \frac{\lambda_1^2 - \lambda_2^2}{2\lambda_1 \lambda_2} \tag{6.3.71}$$

Equations (6.3.70a,b) give

$$A_m = -\frac{\hat{q}_m}{\Delta} \left(\gamma \sinh \bar{\lambda}_1 \sin \bar{\lambda}_2 + \cosh \bar{\lambda}_1 \cos \bar{\lambda}_2 \right)$$
$$D_m = \frac{\hat{q}_m}{\Delta} \left(\gamma \cosh \bar{\lambda}_1 \cos \bar{\lambda}_2 - \sinh \bar{\lambda}_1 \sin \bar{\lambda}_2 \right)$$
$$\Delta = \cosh^2 \bar{\lambda}_1 \cos^2 \bar{\lambda}_2 + \sinh^2 \bar{\lambda}_1 \sin^2 \bar{\lambda}_2 \tag{6.3.72}$$

The same procedure applies to simply supported plates with applied edge moments (see Exercise 6.11).

Table 6.3.4 contains nondimensionalized deflections, bending moments, and shear forces

$$\bar{w} = w_0(\frac{a}{2}, 0) \times 10^2 \frac{D}{q_0 a^4}, \quad \bar{M}_{\xi\xi} = M_{\xi\xi}(\frac{a}{2}, 0) \frac{10}{q_0 a^2} \tag{6.3.73a}$$

BENDING OF SIMPLY SUPPORTED RECTANGULAR PLATES   299

Table 6.3.4. Maximum nondimensional deflections, bending moments, and shear forces in simply supported isotropic ($\nu = 0.3$) rectangular plates subjected to *uniformly distributed load*.

$\frac{b}{a}$	$\bar{w}$	$\bar{M}_{xx}$	$\bar{M}_{yy}$	$\bar{M}_{xy}$	$\bar{Q}_x$	$\bar{Q}_y$	$\bar{V}_x$	$\bar{V}_y$
1.0(399)[†]	0.4062	0.4789	0.4789	0.3248	0.3372	0.3377	0.4200	0.4205
1.5(249)	0.7724	0.8116	0.4984	0.4293	0.4230	0.3640	0.4848	0.4796
2.0(199)	1.0129	1.0168	0.4635	0.4627	0.4640	0.3697	0.5023	0.4958
2.5(149)	1.1496	1.1294	0.4295	0.4718	0.4827	0.3709	0.5046	0.4999
3.0(149)	1.2233	1.1886	0.4063	0.4741	0.4914	0.3712	0.5034	0.5009
3.5 (99)	1.2619	1.2191	0.3922	0.4747	0.4947	0.3712	0.5010	0.5011
4.0 (99)	1.2819	1.2346	0.3842	0.4748	0.4965	0.3712	0.4998	0.5011
4.5 (99)	1.2920	1.2424	0.3798	0.4749	0.4973	0.3712	0.4990	0.5011
5.0 (89)	1.2971	1.2462	0.3775	0.4749	0.4974	0.3712	0.4983	0.5012
10.0(39)	1.3021	1.2500	0.3750	0.4748	0.4949	0.3732	0.4949	0.5010

[†] The number in parenthesis denotes the final value of m used to evaluate the quantities.

$$\bar{M}_{xy} = M_{xy}(0, \frac{b}{2})\frac{10}{q_0 a^2}, \quad \bar{Q}_x = Q_x(0,0)\frac{1}{q_0 a}, \quad \bar{Q}_y = -Q_y(\frac{a}{2}, \frac{b}{2})\frac{1}{q_0 a}$$
(6.3.73b)

$$\bar{V}_x = \left(Q_x + \frac{\partial M_{xy}}{\partial y}\right)\frac{1}{q_0 a}, \quad \bar{V}_y = -\left(Q_y + \frac{\partial M_{xy}}{\partial x}\right)\frac{1}{q_0 a} \quad (6.3.73c)$$

for simply supported rectangular plates under uniform loading. The effective shear forces V_x and V_y were also evaluated at the same points as Q_x and Q_y, respectively. The computation requires evaluation of hyperbolic functions [see Eq. (6.3.72)] that can cause overflow problems for large aspect ratios (b/a) and/or large values of m. Hence the largest value of m is chosen for each aspect ratio that would not cause overflow problems in a Pentium PC. The numerical values for the shear forces may differ slightly depending on the final value of m used to evaluate them. Since the Lévy solution is not symmetric in x and y (whereas the Navier solution is), the equalities $\bar{Q}_x = \bar{Q}_y$ and $\bar{V}_x = \bar{V}_y$ for square plates and plate strips (i.e., $b/a \gg 1$) are not predicted, as evidenced by the numerical results of Table 6.3.4.

6.3.6 The Rayleigh–Ritz Solutions

Equation (6.3.7) can also be solved using the Rayleigh–Ritz method. In the Rayleigh–Ritz method, the solution of Eq. (6.3.7) is sought in the form

$$W_n(x) \approx \sum_{j=1}^{N} c_j^{(n)} \varphi_j(x) \tag{6.3.74}$$

where $\varphi_j(x)$ are approximation functions that must meet the continuity and completeness conditions and satisfy the homogeneous form of the geometric boundary conditions (see Chapter 2). The parameters $c_j^{(n)}$ are then determined by requiring that the variational or weak form of Eq. (6.3.7) be satisfied. The weak form of Eq. (6.3.7) is given by

$$0 = \int_0^a \left[D_{11} \frac{d^2 W_n}{dx^2} \frac{d^2 \delta W_n}{dx^2} + 2\hat{D}_{12}\beta_n^2 \frac{dW_n}{dx} \frac{d\delta W_n}{dx} \right.$$
$$\left. + \left(D_{22}\beta_n^4 + k \right) W_n \delta W_n - \bar{q}_n \delta W_n \right] dx \tag{6.3.75}$$

where δW_n denotes the virtual variation in W_n and $\beta_n = n\pi/b$. In arriving at the weak form (6.3.75), it is assumed that the edges $x = 0, a$ are simply supported and that there are no applied bending moments. See Exercise 6.13 for the weak form when moments are applied on edges $x = 0, a$.

Substituting Eq. (6.3.74) into (6.3.75), we obtain

$$0 = \sum_{j=1}^{N} c_j^{(n)} \int_0^a \left[D_{11} \frac{d^2 \varphi_j}{dx^2} \frac{d^2 \varphi_i}{dx^2} + 2\hat{D}_{12}\beta_n^2 \frac{d\varphi_j}{dx} \frac{d\varphi_i}{dx} \right.$$
$$\left. + \left(D_{22}\beta_n^4 + k \right) \varphi_j \varphi_i \right] dx - \int_0^a \bar{q}_n \varphi_i dx \tag{6.3.76}$$

and in matrix form we can write

$$0 = \sum_{j=1}^{N} R_{ij}^{(n)} c_j^{(n)} - F_i^{(n)} \quad \text{or} \quad [R]^{(n)}\{c\}^{(n)} = \{F\}^{(n)} \tag{6.3.77a}$$

where ($R_{ij}^{(n)} = R_{ji}^{(n)}$, symmetric)

$$R_{ij}^{(n)} = \int_0^a \left[D_{11} \frac{d^2 \varphi_i}{dx^2} \frac{d^2 \varphi_j}{dx^2} + 2\hat{D}_{12}\beta_n^2 \frac{d\varphi_i}{dx} \frac{d\varphi_j}{dx} + \left(D_{22}\beta_n^4 + k \right) \varphi_i \varphi_j \right] dx$$

$$F_i^{(n)} = \int_0^a \bar{q}_n \varphi_i \, dx \tag{6.3.77b}$$

Equation (6.3.77a) represents a set of N algebraic equations among the $c_i^{(n)}$ for each n. The complete solution is given by [see Eq. (6.3.1)]

$$w_0(x,y) = \sum_{n=1}^{\infty}\sum_{j=1}^{N} c_j^{(n)} \varphi_j(x) \sin \beta_n y \qquad (6.3.78)$$

The bending moments and stresses can be computed using the deflection of Eq. (6.3.78). For example, the bending moments are given by

$$M_{xx} = \sum_{n=1}^{\infty}\sum_{j=1}^{N} c_j^{(n)} \left[-D_{11}\frac{d^2\varphi_j}{dx^2} + D_{12}\beta_n^2 \varphi_j \right] \sin \beta_n y$$

$$M_{yy} = \sum_{n=1}^{\infty}\sum_{j=1}^{N} c_j^{(n)} \left[-D_{12}\frac{d^2\varphi_j}{dx^2} + D_{22}\beta_n^2 \varphi_j \right] \sin \beta_n y$$

$$M_{xy} = -2D_{66} \sum_{n=1}^{\infty}\sum_{j=1}^{N} \beta_n c_j^{(n)} \frac{d\varphi_j}{dx} \cos \beta_n y \qquad (6.3.79)$$

For a rectangular plate with edges $x = 0, a$ simply supported, the geometric boundary conditions are given by

$$W_n(0) = 0, \quad W_n(a) = 0 \qquad (6.3.80)$$

Hence, the approximation functions φ_i must be selected such that $\varphi_i = 0$ at $x = 0, a$. If an algebraic polynomial is to be selected, one may use the polynomial

$$\varphi_i(x) = \left(\frac{x}{a}\right)^i \left(1 - \frac{x}{a}\right) = \left(\frac{x}{a}\right)^i - \left(\frac{x}{a}\right)^{i+1} \qquad (6.3.81)$$

For the choice of $\varphi_i(x)$ in Eq. (6.3.81), we have

$$R_{ij} = \frac{D_{11}}{a^3}\left[\frac{ij(i-1)(j-1)}{i+j-3} - \frac{2ij(ij-1)}{i+j-2} + \frac{ij(i+1)(j+1)}{i+j-1}\right]$$
$$+ \frac{2\hat{D}_{12}}{a}\left[\frac{ij}{i+j-1} - \frac{2ij+i+j}{i+j} + \frac{(i+1)(j+1)}{i+j+1}\right]\beta_n^2$$
$$+ a\left(\beta_n^4 D_{22} + k\right)\left[\frac{1}{i+j+1} - \frac{2}{i+j+2} + \frac{1}{i+j+3}\right] \qquad (6.3.82)$$

The load vector $\{F^{(n)}\}$ can be evaluated for different loads.

302   THEORY AND ANALYSIS OF ELASTIC PLATES

Uniform and hydrostatic loads ($\bar{q}_n$ =constant):

$$F_i^{(n)} = \frac{\bar{q}_n a}{(i+1)(i+2)} \tag{6.3.83}$$

Point load $[\bar{q}_n = \frac{2Q_0}{b}\delta(x-x_0)\sin\frac{n\pi y_0}{b}]$:

$$F_i^{(n)} = \frac{2Q_0}{b}\sin\frac{n\pi y_0}{b}\varphi_i(x_0) = \frac{2Q_0}{b}\sin\frac{n\pi y_0}{b}\left(\frac{x_0}{a}\right)^i\left(1-\frac{x_0}{a}\right) \tag{6.3.84}$$

As a specific example, consider $N = 3$. The constants $c_i^{(n)}$ are determined by solving the equations

$$\left(\frac{D_{11}}{a^3}\begin{bmatrix} 4 & 2 & 2 \\ 2 & 4 & 4 \\ 2 & 4 & \frac{4}{5} \end{bmatrix} + \frac{2\beta_n^2 \hat{D}_{12}}{a}\begin{bmatrix} \frac{1}{3} & \frac{1}{6} & \frac{1}{5} \\ \frac{1}{6} & \frac{2}{15} & \frac{1}{5} \\ \frac{1}{5} & \frac{1}{5} & \frac{3}{35} \end{bmatrix}\right.$$

$$\left. + a(\beta_n^4 \hat{D}_{22} + k)\begin{bmatrix} \frac{1}{30} & \frac{1}{60} & \frac{1}{105} \\ \frac{1}{60} & \frac{1}{105} & \frac{1}{168} \\ \frac{1}{105} & \frac{1}{168} & \frac{1}{252} \end{bmatrix}\right)\begin{Bmatrix} c_1^{(n)} \\ c_2^{(n)} \\ c_3^{(n)} \end{Bmatrix} = \bar{q}_n a \begin{Bmatrix} \frac{1}{6} \\ \frac{1}{12} \\ \frac{1}{20} \end{Bmatrix}$$

For a square orthotropic plate with the material properties

$$E_1 = 25E_2, \quad G_{12} = 0.5E_2, \quad \nu_{12} = 0.25, \quad E_2 = 10^6 \text{ psi} \tag{6.3.85}$$

and $k = 0$, the above equation reduces, for instance when $n = 1$, to

$$\frac{1}{a^3}\begin{bmatrix} 9.311 & 4.656 & 4.460 \\ 4.656 & 8.706 & 8.608 \\ 4.460 & 8.608 & 10.23 \end{bmatrix}\begin{Bmatrix} c_1^{(1)} \\ c_2^{(1)} \\ c_3^{(1)} \end{Bmatrix} = q_1 a \begin{Bmatrix} 0.1667 \\ 0.0833 \\ 0.0500 \end{Bmatrix}$$

For uniformly distributed load ($\bar{q}_1 = q_1 = 4q_0/\pi$), the maximum deflection and stresses are (at $x = y = a/2$)

$$w_{max} = 0.6963 \times 10^{-2}\frac{q_0 a^4}{E_2 h^3}$$

$$(\sigma_{xx})_{max} = 0.8452\frac{q_0 a^2}{h^2}, \quad (\sigma_{yy})_{max} = 0.0428\frac{q_0 a^2}{h^2} \tag{6.3.86}$$

BENDING OF SIMPLY SUPPORTED RECTANGULAR PLATES   303

Table 6.3.5 contains a comparison of the deflections obtained through the Rayleigh–Ritz method with those obtained through Navier's method for isotropic and orthotropic, simply supported square plates under uniformly distributed load (UDL), hydrostatic load (HSL), and central point load (CPL). The same nondimensionalization as in Eq. (6.2.72) is used. Clearly, the Rayleigh–Ritz solution converges with increasing values of N. It is found that even values of N do not contribute to the solution over that obtained using the preceding odd value of N. Table 6.3.6 contains comparison of the deflections and stresses. The subscript 'N' denotes Navier's solution and 'R' denotes the Rayleigh–Ritz solution. The Rayleigh–Ritz solution is not symmetric in x and y (it is algebraic in x and trigonometric in y), and hence may not yield $\sigma_{xx} = \sigma_{yy}$ for an isotropic plate when n is small. However, with increasing $n = N$, we do obtain $\sigma_{xx} = \sigma_{yy}$. As noted earlier, the convergence is slower for the point load case.

Table 6.3.5. Transverse deflections $\bar{w} = w_0(a/2, b/2)(E_2 h^3/q_0 a^4)$ in simply supported isotropic and orthotropic square plates subjected to various types of loads.

Load	Navier solution $N=1$	Rayleigh–Ritz solution $N=3$	$N=5$	$N=9$	$N=9^{\dagger}$
Isotropic plates ($\nu = 0.3$)						
UDL (1)[‡]	0.0454	0.0419	0.0451	0.0449	0.0449	0.0115
UDL (9)	0.0444	0.0416	0.0445	0.0444	0.0444	0.0110
CPL (1)	0.1121	0.0987	0.1131	0.1161	0.1172	0.0342
CPL (9)	0.1260	0.1037	0.1194	0.1233	0.1253	0.0417
Orthotropic plates ($E_1/E_2 = 25$, $G_{12} = 0.5E_2$, $\nu_{12} = 0.25$)						
UDL (1)	0.0070	0.0057	0.0069	0.0069	0.0069	0.0048
UDL (9)	0.0065	0.0053	0.0065	0.0065	0.0065	0.0044
CPL (1)	0.0173	0.0134	0.0172	0.0175	0.0175	0.0123
CPL (9)	0.0229	0.0177	0.0222	0.0227	0.0229	0.0174

[†] This column corresponds to a plate on elastic foundation, $k = 100$.

[‡] The number in parenthesis denotes the maximum value of $m = n$ and N used to evaluate the series in the Navier and Rayleigh–Ritz methods.

Table 6.3.6. Transverse deflections $\bar{w} = w_0(a/2, b/2)(E_2 h^3/q_0 a^4)$ and stresses $\bar{\sigma} = \sigma(h^2/q_0 a^2)$ in simply supported isotropic and orthotropic square plates subjected to various types of loads.

Load	$\bar{w}_N$	$\bar{w}_R$	$(\bar{\sigma}_{xx})_N$	$(\bar{\sigma}_{xx})_R$	$(\bar{\sigma}_{yy})_N$	$(\bar{\sigma}_{yy})_R$
Isotropic plates ($\nu = 0.3$)						
UDL (9)[†]	0.0444	0.0444	0.2878	0.2874	0.2878	0.2877
UDL(29)	0.0444	0.0444	0.2873	0.2873	0.2873	0.2873
HSL (9)	0.0222	0.0222	0.1439	0.1437	0.1439	0.1438
HSL (29)	0.0222	0.0222	0.1437	0.1437	0.1437	0.1437
CPL (9)	0.1260	0.1253	1.7546	1.5342	1.7546	1.5802
CPL(29)	0.1266	0.1262	1.4350	1.8993	1.4350	1.8996
Orthotropic plates ($E_1/E_2 = 25$, $G_{12} = G_{13} = 0.5 E_2$, $\nu_{12} = 0.25$)						
UDL (9)	0.0065	0.0065	0.7872	0.7868	0.0248	0.0248
UDL(29)	0.0065	0.0065	0.7867	0.7867	0.0245	0.0245
HSL (9)	0.0325	0.0325	0.3936	0.3934	0.0124	0.0124
HSL (29)	0.0325	0.0325	0.3933	0.3934	0.0122	0.0122
CPL (9)	0.0229	0.0229	4.4530	4.0844	0.5559	0.5445
CPL(29)	0.0232	0.0232	6.0075	5.2142	0.8670	0.8430

[†] The number in parenthesis denotes the maximum value of m and n used to evaluate the series in Navier's solution; it also denotes the values used for $n = N$ in the Rayleigh–Ritz solution.

If we were to select trigonometric functions for φ_j of Eq. (6.3.74), we would take

$$\varphi_j(x) = \sin \frac{j\pi x}{a}, \quad j = 1, 2, \cdots N$$

because it satisfies the simply supported boundary conditions in Eq. (6.3.80). Substituting the following Rayleigh–Ritz approximation

$$W_n(x) \approx W_n^{(N)}(x) = \sum_{j=1}^{N} c_j^{(n)} \sin \frac{j\pi x}{a} \tag{6.3.87}$$

into Eq. (6.3.77b) yields

$$R_{ij}^{(n)} = \begin{cases} \frac{a}{2} \left(D_{11} \alpha_i^4 + 2\hat{D}_{12} \alpha_i^2 \beta_n^2 + D_{22} \beta_n^4 + k \right) & \text{for } i = j \\ 0 & \text{for } i \neq j \end{cases} \tag{6.3.88a}$$

where $\alpha_i = (i\pi/b)$. For the case in which $\bar{q}_n$ is a constant, we can evaluate $F_i^{(n)}$ using definitions in Eq. (6.3.77b).

$$F_i^{(n)} = -\frac{\bar{q}_n}{\alpha_i}\left[(-1)^i - 1\right] = \frac{2a\bar{q}_n}{i\pi} \quad \text{for } i = 1, 3, \cdots, N \quad (6.3.88\text{b})$$

Thus, Eq. (6.3.88a) in the present case is a diagonal system of equations (because $[R^{(n)}]$ is a diagonal matrix) and $c_i^{(n)}$ are nontrivial only when i is an odd integer. We have

$$c_i^{(n)} = \frac{2a\bar{q}_n}{i\pi R_{ii}^{(n)}} \quad \text{(no sum on } i\text{)} \quad (6.3.89)$$

The complete solution becomes

$$w_0(x, y) = \sum_{n=1}^{\infty}\sum_{i=1}^{N} c_i^{(n)} \sin\alpha_i x \sin\beta_n y = \sum_{n=1}^{\infty}\sum_{m=1}^{N} c_m^{(n)} \sin\alpha_m x \sin\beta_n y \quad (6.3.90)$$

where

$$c_m^{(n)} = \frac{q_{mn}}{d_{mn}}, \quad q_{mn} = \frac{4\bar{q}_n}{m\pi}$$

$$d_{mn} = D_{11}\left(\frac{m\pi}{a}\right)^4 + 2\hat{D}_{12}\left(\frac{m\pi}{a}\right)^2\left(\frac{n\pi}{b}\right)^2 + D_{22}\left(\frac{n\pi}{b}\right)^4 + k \quad (6.3.91)$$

Comparing Eqs. (6.3.90) and (6.3.91) with Eqs. (6.2.1), (6.2.9), and (6.2.10), we note that the Rayleigh–Ritz solution coincides with the Navier solution for any finite values of m and n.

It should be noted that the Lévy solution procedure and the Rayleigh–Ritz formulation described in this chapter are also valid for rectangular plates with two parallel edges simply supported and other edges with *any* boundary conditions, as discussed in the next chapter. The Rayleigh–Ritz formulation for plates with arbitrary boundary conditions will be discussed in the sequel.

Exercises

6.1 Verify the expressions in Eqs. (6.2.26a–c) for σ_{xz}, σ_{yz}, and σ_{zz}.

6.2 Determine the maximum transverse deflection, bending moments, and stresses of an isotropic ($\nu = 0.25$), simply supported, square plate with an elastic foundation ($ka^4/D = 1125$) and subjected to a sinusoidally distributed transverse load of intensity q_0. Ans.:

$$w_{max} = 0.66 \times 10^{-3}(q_0 a^4/D)$$

$$(M_{xx})_{max} = 0.008145 q_0 a^2, \quad (\sigma_{xx})_{max} = 0.04887(q_0 a^2/h^2)$$

6.3 Determine the maximum transverse deflection of an isotropic ($\nu = 0.25$), simply supported square plate with an elastic foundation ($ka^4/D = 1125$) and subjected to uniformly distributed transverse load of intensity q_0. Use the first term of the series to evaluate the deflection. Ans.: $w_{max}^{(1)} = 0.01204(q_0 a^4/Eh^3)$, $w_{max}^{(3)} = 0.01098(q_0 a^4/Eh^3)$.

6.4 Repeat Exercise 6.3 when the plate is subjected to a central point load of intensity Q_0. Use the first term of the series to evaluate the deflection. Ans.: $w_{max}^{(1)} = 0.02971(Q_0 a^3/Eh^3)$, $w_{max}^{(3)} = 0.03937(Q_0 a^3/Eh^3)$.

6.5 Determine the deflection of a simply supported plate ($k = 0$) when the plate is subjected to a load of the type

$$q(x, y) = q_0 \sin \frac{\pi x}{a}$$

where q_0 is a constant.

6.6 Determine the deflection of a simply supported plate ($k = 0$) when the plate is subjected to a line load of the type

$$q(x, y) = q^1 \sin \frac{\pi x}{a} \delta(y - y_0)$$

where q^1 is a constant and $\delta(y)$ is the Dirac delta function.

6.7 Determine the deflection of a simply supported isotropic plate when the plate is subjected to a concentrated force Q_0 at $(x, y) = (x_0, y_0)$

$$q(x, y) = Q_0 \, \delta(x - x_0) \, \delta(y - y_0)$$

and uniform compressive in-plane force N_{xx}^0. Ans: The deflection is given by Eq. (6.2.1) with

$$W_{mn} = \frac{4Q_0 \sin \frac{m\pi x_0}{a} \sin \frac{n\pi y_0}{b}}{abD\pi^4 \left[\left(\frac{m^2}{a^2} + \frac{n^2}{b^2}\right)^2 + k - \frac{m^2 N_{xx}^0}{a^2 \pi^2 D}\right]}$$

6.8 Verify Eqs. (6.3.29) and (6.3.30).

6.9 Verify the expressions for $A_n, B_n, C_n,$ and D_n given in Eq. (6.3.32).

6.10 Verify the expressions for $A_n, B_n, C_n,$ and D_n given in Eq. (6.3.34).

6.11 Consider a simply supported plate subjected to uniformly distributed bending moments on edges $y = \pm b/2$ (see Figure 6.3.3). Determine the constants A_m and D_m of Eq. (6.3.63) (with $\hat{q}_m = 0$) so that the boundary conditions

$$w_0(x, \pm\frac{b}{2}) = 0, \quad M_{yy}(x, \pm\frac{b}{2}) = M_0 \tag{a}$$

are satisfied, where M_0 is the value of the applied bending moment. In particular, show that

$$A_m = \frac{M_m \sinh \bar{\lambda}_1 \sin \bar{\lambda}_2}{2D_{22}\bar{\lambda}_1\bar{\lambda}_2\Delta}, \quad D_m = -\frac{M_m \cosh \bar{\lambda}_1 \cos \bar{\lambda}_2}{2D_{22}\bar{\lambda}_1\bar{\lambda}_2\Delta}$$

$$\Delta = \cosh^2 \bar{\lambda}_1 \cos^2 \bar{\lambda}_2 + \sinh^2 \bar{\lambda}_1 \sin^2 \bar{\lambda}_2 \tag{b}$$

where $M_m = (4M_0/m\pi)$ and $\bar{\lambda}_i = \lambda_i b/2$.

6.12 Evaluate the Rayleigh–Ritz equations for a square isotropic ($\nu = 0.25$ and $k = 0$) plate under uniform loading when $n = 1$ and for $N = 1, 2,$ and 3, and determine the maximum deflection [see Eqs. (6.3.77a,b), (6.3.82), and (6.3.83)]. *Ans.:* For $N = 3$ we have

$$\frac{1}{a^3}\begin{bmatrix} 1.2290 & 0.6145 & 0.4357 \\ 0.6145 & 0.6720 & 0.5826 \\ 0.4357 & 0.5826 & 0.6114 \end{bmatrix} \begin{Bmatrix} c_1^{(1)} \\ c_2^{(1)} \\ c_3^{(1)} \end{Bmatrix} = q_0 a \begin{Bmatrix} 0.2122 \\ 0.1061 \\ 0.0637 \end{Bmatrix}$$

$$w_{max} = 0.04644 \frac{q_0 a^4}{Eh^3}$$

6.13 Show that the weak form for a simply supported rectangular plate with edge moments is given by [note that $W_n(0) = W_n(a) = 0$]

$$0 = \int_0^a \left[D_{11} \frac{d^2W_n}{dx^2} \frac{d^2\delta W_n}{dx^2} + 2\hat{D}_{12}\beta_n^2 \frac{dW_n}{dx} \frac{d\delta W_n}{dx} + D_{22}\beta_n^4 W_n \delta W_n \right] dx$$

$$- F_n^1 \left(\frac{d\delta W_n}{dx}\right)_{x=0} + F_n^2 \left(\frac{d\delta W_n}{dx}\right)_{x=a} \tag{a}$$

where F_n^1 and F_n^2 are constants defined by Eqs. (6.3.42) and (6.3.44), respectively.

6.14 Formulate the Rayleigh–Ritz equations for a simply supported, isotropic ($\nu = 0.3$), square plate with distributed edge moments of magnitude M_0 at $x = 0, a$, and determine the solution for $N = 3$ and $n = 1$. *Ans:*

$$\frac{1}{a^3}\begin{bmatrix} 1.266 & 0.6331 & 0.4489 \\ 0.6331 & 0.6923 & 0.6002 \\ 0.4489 & 0.6002 & 0.6299 \end{bmatrix} \begin{Bmatrix} c_1^{(1)} \\ c_2^{(1)} \\ c_3^{(1)} \end{Bmatrix} = \frac{4M_0}{a\pi} \begin{Bmatrix} 2 \\ 1 \\ 1 \end{Bmatrix}$$

$$w_0(a/2, b/2) = 0.0367 \frac{M_0 a^2}{D}, \quad M_{xx}(a/2, b/2) = 0.242 M_0$$

$$M_{yy}(a/2, b/2) = 0.397 M_0$$

Compare these results with those in Table 6.3.3.

6.15 Determine the Rayleigh–Ritz solution of a simply supported, isotropic ($\nu = 0.3$), square plate under hydrostatic load $q = q_0(y/b)$ for $N = 3$ and $n = 1$. *Ans:*

$$w_0(a/2, b/2) = 0.002064 \frac{q_0 a^4}{D}, \quad M_{xx}(a/2, b/2) = 0.0254 q_0 a^2$$

$$M_{yy}(a/2, b/2) = 0.0262 q_0 a^2$$

References for Additional Reading

1. Timoshenko, S. P. and Woinowsky-Krieger, S., *Theory of Plates and Shells*, McGraw–Hill, Singapore (1970).
2. McFarland, D., Smith, B. L., and Bernhart, W. D., *Analysis of Plates*, Spartan Books, Philadelphia, PA (1972).
3. Szilard, R., *Theory and Analysis of Plates. Classical and Numerical Methods*, Prentice–Hall, Englewood Cliffs, NJ (1974).
4. Ugural, A. C., *Stresses in Plates and Shells*, McGraw–Hill, New York (1981).
5. Reddy, J. N., *Energy and Variational Methods in Applied Mechanics*, John Wiley and Sons, New York (1984).
6. Reddy, J. N., *Mechanics of Laminated Composite Plates: Theory and Analysis*, CRC Press, Boca Raton, FL (1997).

Chapter Seven

Bending of Rectangular Plates with Various Boundary Conditions

7.1 Introduction

In the previous chapter, analytical solutions using Navier's, Lévy's, and the Rayleigh–Ritz methods were presented for bending of rectangular plates with all four edges simply supported. In this chapter, solutions will be developed for rectangular plates with various boundary conditions. First, we use the Lévy method to determine solutions of rectangular plates with two opposite edges simply supported while each of the remaining two edges are free, simply supported, or clamped. The Lévy method was discussed in section 6.3, and only pertinent equations are repeated here for the sake of completeness. Next, the Rayleigh–Ritz solutions are developed for rectangular plates with general boundary conditions.

7.2 Lévy Solutions

7.2.1 Basic Equations

Consider a rectangular plate with simply supported edges along $y = 0, b$ (see Figure 7.2.1) and subjected to a distributed transverse load q. The other two edges at $x = 0, a$, can each be free, simply supported, or clamped. The solution to the problem is represented as

$$w_0(x,y) = \sum_{n=1}^{\infty} W_n(x) \sin \frac{n\pi y}{b} \qquad (7.2.1)$$

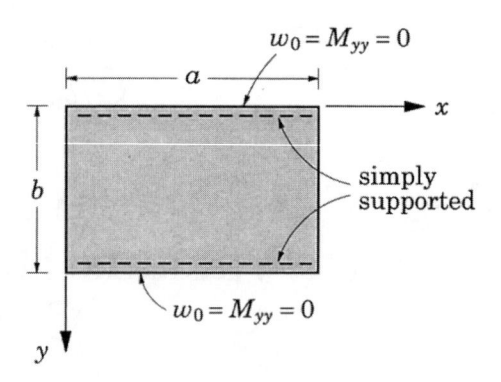

Figure 7.2.1. Geometry and coordinate system for a rectangular plate with sides $y = 0, b$ simply supported.

which satisfies the simply supported boundary conditions at $y = 0$ and $y = b$. Similarly, the load q and thermal moments M_{xx}^T and M_{yy}^T are represented as

$$q(x,y) = \sum_{n=1}^{\infty} q_n(x) \sin \frac{n\pi y}{b} \tag{7.2.2}$$

$$M_{xx}^T(x,y) = \sum_{n=1}^{\infty} M_n^1(x) \sin \frac{n\pi y}{b} \tag{7.2.3a}$$

$$M_{yy}^T(x,y) = \sum_{n=1}^{\infty} M_n^2(x) \sin \frac{n\pi y}{b} \tag{7.2.3b}$$

The coefficents q_n, for example, are given in Table 6.3.1 for various distributions of $q(x,y)$.

Substitution of Eqs. (7.2.1)–(7.2.3) into Eq. (6.1.1) results in the following ordinary differential equation in $W_n(x)$:

$$D_{11} \frac{d^4 W_n}{dx^4} - 2 \left(\frac{n\pi}{b}\right)^2 \hat{D}_{12} \frac{d^2 W_n}{dx^2} + \left[\left(\frac{n\pi}{b}\right)^4 D_{22} + k\right] W_n$$

$$= q_n - \frac{d^2 M_n^1}{dx^2} + \left(\frac{n\pi}{b}\right)^2 M_n^2 \equiv \bar{q}_n \tag{7.2.4}$$

The solution of Eq. (7.2.4) is of the form $W_n(x) = W_n^h(x) + W_n^p(x)$ so that the complete solution is given by

$$w_0(x,y) = \sum_{n=1}^{\infty} \left(W_n^h(x) + W_n^p(x)\right) \sin \beta_n y \qquad (7.2.5)$$

where W_n^h is the homogeneous solution and W_n^p is the particular solution of Eq. (7.2.4). The form of the homogeneous solution depends on the nature of the roots λ of the equation

$$D_{11}\lambda^4 - 2\left(\frac{n\pi}{b}\right)^2 (D_{12} + 2D_{66})\lambda^2 + \left(\frac{n\pi}{b}\right)^4 D_{22} + k = 0 \qquad (7.2.6)$$

The following three cases were discussed in section 6.3 and are summarized below:

Case 1: $(D_{12} + 2D_{66})^2 > D_{11}(D_{22} + k/\beta_n^4)$

$$W_n^h = A_n \cosh \lambda_1 x + B_n \sinh \lambda_1 x + C_n \cosh \lambda_3 x + D_n \sinh \lambda_3 x \qquad (7.2.7a)$$

$$\lambda_1^2 = \beta_n^2 \left(\frac{D_{12} + 2D_{66}}{D_{11}}\right) - \beta_n^2 \left[\left(\frac{D_{12} + 2D_{66}}{D_{11}}\right)^2 - \left(\frac{D_{22}}{D_{11}} + \frac{k}{D_{11}\beta_n^4}\right)\right]^{\frac{1}{2}}$$

$$\lambda_2^2 = \beta_n^2 \left(\frac{D_{12} + 2D_{66}}{D_{11}}\right) + \beta_n^2 \left[\left(\frac{D_{12} + 2D_{66}}{D_{11}}\right)^2 - \left(\frac{D_{22}}{D_{11}} + \frac{k}{D_{11}\beta_n^4}\right)\right]^{\frac{1}{2}}$$

$$(7.2.7b)$$

Case 2: $(D_{12} + 2D_{66})^2 = D_{11}(D_{22} + k/\beta_n^4)$

$$W_n^h(x) = (A_n + B_n x) \cosh \lambda x + (C_n + D_n x) \sinh \lambda x \qquad (7.2.8a)$$

$$\lambda^2 = \beta_n^2 \left(\frac{D_{12} + 2D_{66}}{D_{11}}\right) \qquad (7.2.8b)$$

Case 3: $(D_{12} + 2D_{66})^2 < D_{11}(D_{22} + k/\beta_n^4)$

$$W_n^h(x) = (A_n \cos \lambda_2 x + B_n \sin \lambda_2 x) \cosh \lambda_1 x$$
$$+ (C_n \cos \lambda_2 x + D_n \sin \lambda_2 x) \sinh \lambda_1 x \qquad (7.2.9a)$$

$$(\lambda_1)^2 = \frac{\beta_n^2}{2D_{11}} \left[\sqrt{D_{11}\left(D_{22} + \frac{k}{\beta_n^4}\right)} + (D_{12} + 2D_{66})\right]$$

$$(\lambda_2)^2 = \frac{\beta_n^2}{2D_{11}} \left[\sqrt{D_{11}\left(D_{22} + \frac{k}{\beta_n^4}\right)} - (D_{12} + 2D_{66})\right] \qquad (7.2.9b)$$

The four constants, A_n, B_n, C_n, and D_n, are determined in each case by using the boundary conditions on edges $x = 0, a$.

The particular solution W_n^p may be determined in two alternative ways. In the general case, in which $\bar{q}_n$ is an arbitrary function of x, the particular solution is determined by expanding the solution and the loads in Fourier series

$$W_n^p(x) = \sum_{m=1}^{\infty} W_{mn} \sin \frac{m\pi x}{a}$$

$$q_n(x) = \sum_{m=1}^{\infty} q_{mn} \sin \frac{m\pi x}{a}$$

$$M_n^1(x) = \sum_{m=1}^{\infty} M_{mn}^1 \sin \frac{m\pi x}{a}$$

$$M_n^2(x) = \sum_{m=1}^{\infty} M_{mn}^2 \sin \frac{m\pi x}{a} \qquad (7.2.10)$$

where the coefficients q_{mn}, M_{mn}^1, and M_{mn}^2 have the same meaning as in Eqs. (6.2.3), (6.2.4a), and (6.2.4b), respectively. We have

$$W_n^p(x) = \sum_{m=1}^{\infty} W_{mn} \sin \frac{m\pi x}{a} \qquad (7.2.11)$$

$$W_{mn} = \frac{1}{d_{mn}} \left[q_{mn} + \left(\frac{m\pi}{a}\right)^2 M_{mn}^1 + \left(\frac{n\pi}{b}\right)^2 M_{mn}^2 \right] \qquad (7.2.12a)$$

$$d_{mn} = D_{11} \left(\frac{m\pi}{a}\right)^4 + 2\hat{D}_{12} \left(\frac{m\pi}{a}\right)^2 \left(\frac{n\pi}{b}\right)^2 + D_{22} \left(\frac{n\pi}{b}\right)^4 + k \qquad (7.2.12b)$$

When the applied loads are such that $\bar{q}_n$ is a constant or a linear function of x, the particular solution may be determined directly, as discussed in Eqs. (6.3.24)–(6.3.26). The particular solution is given by

$$W_n^p = \frac{\bar{\bar{q}}_n}{(D_{22}\beta_n^4 + k)} \equiv \hat{q}_n, \quad \bar{\bar{q}}_n = q_n - \frac{d^2 M_n^1}{dx^2} + \beta_n^2 M_n^2 \qquad (7.2.13)$$

where $\beta_n = (n\pi/b)$. The complete solution becomes

$$w_0(x, y) = \sum_{n=1}^{\infty} \left(W_n^h(x) + \hat{q}_n \right) \sin \beta_n y \qquad (7.2.14)$$

7.2.2 Plates with Edges $x = 0, a$ Clamped (CCSS)

Here we consider rectangular plates with the first ($x = 0$) and second ($x = a$) edges clamped (C), and the third ($y = 0$) and fourth ($y = b$) edges simply supported (S) (see Figure 7.2.2). We shall label it as the CCSS plate. The notation SCFS, for example, means that edge $x = 0$ is simply supported (S), edge $x = a$ clamped (C), edge $y = 0$ free (F), and edge $y = b$ simply supported (S). Thus, the first two letters indicate boundary conditions on edges $x = 0, a$ and the next two letters indicate the boundary conditions on edges $y = 0, b$.

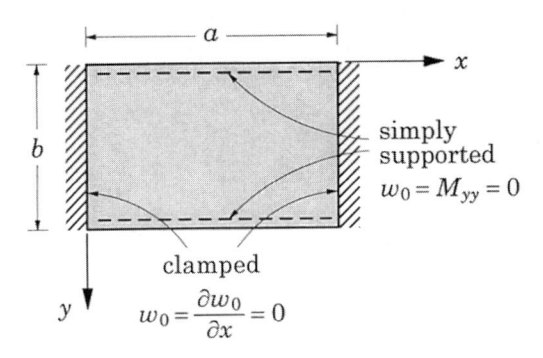

Figure 7.2.2. Geometry and coordinate system for a rectangular plate with sides $x = 0, a$ clamped and $y = 0, b$ simply supported (CCSS).

The clamped boundary conditions on edges $x = 0, a$ are

$$w_0 = 0, \quad \frac{\partial w_0}{\partial x} = 0 \tag{7.2.15}$$

We consider (i) isotropic plates without elastic foundation, and (ii) isotropic plates with elastic foundation or orthotropic plates with $(D_{12} + 2D_{66})^2 < D_{11}(D_{22} + k/\beta_n^4)$ and determine the constants A_n, B_n, C_n, and D_n. The solution to the first problem falls into Case 2 and the second one into Case 3, as discussed earlier. The first problem also includes orthotropic plates on elastic foundation for which we have $(D_{12} + 2D_{66})^2 = D_{11}(D_{22} + k/\beta_n^4)$.

Isotropic plates ($k = 0$)

First, we consider the particular solution in the form of Eq. (7.2.11). The complete solution for this case is given by

$$w_0(x,y) = \sum_{n=1}^{\infty}\left[(A_n + B_n x)\cosh\beta_n x + (C_n + D_n x)\sinh\beta_n x\right.$$
$$\left. + \sum_{m=1}^{\infty} W_{mn} \sin\alpha_m x\right]\sin\beta_n y \qquad (7.2.16)$$

where W_{mn} is given by

$$W_{mn} = \frac{q_{mn} + \alpha_m^2 M_{mn}^1 + \beta_n^2 M_{mn}^2}{D\left(\alpha_m^2 + \beta_n^2\right)^2} \qquad (7.2.17a)$$

and

$$\alpha_m = \frac{m\pi}{a}, \quad \beta_n = \frac{n\pi}{b} \qquad (7.2.17b)$$

Differentiating Eq. (7.2.16), we obtain

$$\frac{\partial w_0}{\partial x} = \sum_{n=1}^{\infty}\left[\beta_n(A_n + B_n x)\sinh\beta_n x + B_n \cosh\beta_n x\right.$$
$$+ \beta_n(C_n + D_n x)\cosh\beta_n x + D_n \sinh\beta_n x$$
$$\left. + \sum_{m=1}^{\infty} \alpha_m W_{mn} \cos\alpha_m x\right]\sin\beta_n y \qquad (7.2.18)$$

Substitution of Eqs. (7.2.16) and (7.2.18) into Eq. (7.2.15) yields

$$A_n = 0, \quad B_n + \beta_n C_n + \sum_{m=1}^{\infty}\alpha_m W_{mn} = 0$$

$$(A_n + B_n a)\cosh\bar{\beta}_n + (C_n + D_n a)\sinh\bar{\beta}_n = 0$$

$$\beta_n(A_n + B_n a)\sinh\bar{\beta}_n + \beta_n(C_n + D_n a)\cosh\bar{\beta}_n$$
$$+ B_n \cosh\bar{\beta}_n + D_n \sinh\bar{\beta}_n + \sum_{m=1}^{\infty}(-1)^m \alpha_m W_{mn} = 0 \qquad (7.2.19)$$

where $\bar{\beta}_n = a\beta_n = n\pi a/b$. The solution of these equations is ($A_n = 0$)

$$B_n = \frac{1}{\Delta_n}\left(\sinh\bar{\beta}_n \hat{q}_n^0 + \bar{\beta}_n \hat{q}_n^1\right)\sinh\bar{\beta}_n, \quad \Delta_n = \bar{\beta}_n^2 - \sinh^2\bar{\beta}_n$$

$$C_n = -\frac{1}{\Delta_n}\left(\bar{\beta}_n \hat{q}_n^0 + \sinh\bar{\beta}_n \hat{q}_n^1\right)$$

$$D_n = -\frac{1}{\Delta_n}\left[\left(-\bar{\beta}_n + \cosh\bar{\beta}_n \sinh\bar{\beta}_n\right)\hat{q}_n^0 + \left(\bar{\beta}_n \cosh\bar{\beta}_n - \sinh\bar{\beta}_n\right)\hat{q}_n^1\right]$$

$$\hat{q}_n^0 = \sum_{m=1}^{\infty} \alpha_m W_{mn}, \quad \hat{q}_n^1 = \sum_{m=1}^{\infty}(-1)^m \alpha_m W_{mn} \qquad (7.2.20)$$

Next we consider the case in which the applied loads are such that $\bar{q}_n$ is a constant. Hence, the complete solution is given by

$$w_0(x,y) = \sum_{n=1}^{\infty}\left[(A_n + B_n x)\cosh\beta_n x + (C_n + D_n x)\sinh\beta_n x + \frac{\bar{q}_n}{D\beta_n^4}\right]\sin\beta_n y \qquad (7.2.21)$$

where $\bar{q}_n$ is defined in Eq. (7.2.4) and q_n, M_n^1, and M_n^2 can be calculated using equations of the form

$$q_n(x) = \frac{2}{b}\int_0^b q(x,y)\sin\beta_n y\, dy \qquad (7.2.22)$$

Using the boundary conditions in Eq. (7.2.15), we obtain

$$A_n + \frac{\bar{q}_n b^4}{Dn^4\pi^4} = 0, \quad B_n + \beta_n C_n = 0$$

$$(A_n + B_n a)\cosh\bar{\beta}_n + (C_n + D_n a)\sinh\bar{\beta}_n + \frac{\bar{q}_n b^4}{Dn^4\pi^4} = 0$$

$$\beta_n(A_n + B_n a)\sinh\bar{\beta}_n + \beta_n(C_n + D_n a)\cosh\bar{\beta}_n$$
$$+ B_n\cosh\bar{\beta}_n + D_n\sinh\bar{\beta}_n = 0 \qquad (7.2.23)$$

Solving these equations, we find that

$$A_n = -\frac{\bar{q}_n b^4}{n^4\pi^4 D}, \qquad B_n = -\beta_n A_n\left(\frac{1-\cosh\bar{\beta}_n}{\bar{\beta}_n + \sinh\bar{\beta}_n}\right)$$

$$C_n = A_n\left(\frac{1-\cosh\bar{\beta}_n}{\bar{\beta}_n + \sinh\bar{\beta}_n}\right), \quad D_n = -\beta_n A_n\left(\frac{\sinh\bar{\beta}_n}{\bar{\beta}_n + \sinh\bar{\beta}_n}\right) \qquad (7.2.24)$$

The bending moments and stresses can be computed using the deflection in Eq. (7.2.21). First note that

$$\frac{\partial^2 w_0}{\partial x^2} = \sum_{n=1}^{\infty} \beta_n \bigg\{ [\beta_n (A_n + B_n x) + 2 D_n] \cosh \beta_n x$$

$$+ [\beta_n (C_n + D_n x) + 2 B_n] \sinh \beta_n x \bigg\} \sin \beta_n y$$

$$\frac{\partial^2 w_0}{\partial y^2} = -\sum_{n=1}^{\infty} \beta_n^2 \bigg[(A_n + B_n x) \cosh \beta_n x + (C_n + D_n x) \sinh \beta_n x$$

$$- A_n \bigg] \sin \beta_n y \qquad (7.2.25)$$

Next we consider several specific load cases.

Uniformly distributed load, q_0. For this case, we have $\bar{q}_n = q_n$ and

$$q_n = \frac{4 q_0}{n \pi}, \quad A_n = -\frac{4 q_0 b^4}{n^5 \pi^5 D} \qquad (7.2.26)$$

The deflection is given by Eq. (7.2.21) with A_n given by Eq. (7.2.26) and B_n, C_n, and D_n defined by Eq. (7.2.24). For $n = 1$ and $b/a = 1$, the center deflection and bending moments are given by

$$w_0(a/2, b/2) = 0.001962 \left(\frac{q_0 a^4}{D} \right), \quad M_{xx}(a/2, b/2) = 0.03461 q_0 a^2$$

$$M_{yy}(a/2, b/2) = 0.028 q_0 a^2$$

The bending moments at the center of the clamped edge are:

$$M_{xx}(0, b/2) = -0.07383 q_0 a^2, \quad M_{yy}(0, b/2) = -0.02215 q_0 a^2$$

Table 7.2.1 contains nondimensionalized deflections and bending moments

$$\hat{w} = w_0(a/2, b/2) \left(\frac{D}{q_0 b^4} \right) \times 10^2, \quad \hat{M} = M \left(\frac{1}{q_0 b^2} \right) \times 10 \qquad (7.2.27)$$

$$\bar{w} = w_0(a/2, b/2) \left(\frac{D}{q_0 a^4} \right) \times 10^2, \quad \bar{M} = M \left(\frac{1}{q_0 a^2} \right) \times 10 \qquad (7.2.28)$$

BENDING OF PLATES WITH VARIOUS BOUNDARY CONDITIONS   317

Table 7.2.1. Maximum nondimensional deflections and bending moments in isotropic ($\nu = 0.3$) rectangular plates with edges $y = 0, b$ simply supported and edges $x = 0, a$ clamped (CCSS), and subjected to *uniformly distributed load*.

Variable	$\frac{b}{a}=\frac{1}{3}$	$\frac{b}{a}=\frac{1}{2}$	$\frac{b}{a}=1$	$\frac{b}{a}=\frac{3}{2}$	$\frac{b}{a}=2$	$\frac{b}{a}=3$
$\hat{w}(a/2, b/2)$	1.1681	0.8445	0.1917	0.2476	0.2612	0.2619
$\hat{M}_{xx}(a/2, b/2)$	0.4214	0.4738	0.3326	0.4067	0.4214	0.4201
$\hat{M}_{yy}(a/2, b/2)$	1.1442	0.8693	0.2445	0.1794	0.1441	0.1302
$-\hat{M}_{xx}(0, b/2)$	1.2467	1.1915	0.6990	0.8233	0.8451	0.8417
$-\hat{M}_{yy}(0, b/2)$	0.3740	0.3574	0.2097	0.2470	0.2535	0.2525

for various aspect ratios of isotropic plates. The first set of nondimensionalization is used for $b/a \leq 1$ (the first three columns of values) and the second set is used for $b/a > 1$ (the last three columns of values). The values are obtained using the first five terms of the series ($n = 1, 3, \cdots, 9$).

Hydrostatic load. The hydrostatic load is assumed to be of the form

$$q(x, y) = q_0 \frac{y}{b} \qquad (7.2.29)$$

Then $\bar{q}_n = q_n$ is given by (see Table 6.3.1)

$$q_n = \frac{2q_0}{n\pi}(-1)^{n+1}, \quad A_n = -\frac{2q_0 b^4}{n^5 \pi^5 D}(-1)^{n+1} \qquad (7.2.30)$$

for $n = 1, 2, 3, \cdots, \infty$. For $n = 1$ and $b/a = 1$, the center deflection and bending moments are given by

$$w_0(a/2, b/2) = 0.000981 \left(\frac{q_0 a^4}{D}\right), \quad M_{xx}(a/2, b/2) = 0.01731 q_0 a^2$$

$$M_{yy}(a/2, b/2) = 0.014 q_0 a^2$$

The bending moments at the center of the clamped edge are:

$$M_{xx}(0, b/2) = -0.03691 q_0 a^2, \quad M_{yy}(0, b/2) = -0.01107 q_0 a^2$$

and the bending moments at $y = 3b/4$ of the clamped edge are:

$$M_{xx}(0, 3b/4) = -0.0261 q_0 a^2, \quad M_{yy}(0, 3b/4) = -0.00783 q_0 a^2$$

Table 7.2.2 contains nondimensionalized deflections and bending moments [see Eqs. (7.2.27) and (7.2.28)] for various aspect ratios of isotropic plates. The values are obtained using the first nine terms of the series ($n = 1, 2, 3, \cdots, 9$).

Table 7.2.2. Maximum nondimensional deflections and bending moments in isotropic ($\nu = 0.3$) rectangular plates with edges $y = 0, b$ simply supported and edges $x = 0, a$ clamped (CCSS), and subjected to *hydrostatic* load.

Variable	$\frac{b}{a} = \frac{1}{3}$	$\frac{b}{a} = \frac{1}{2}$	$\frac{b}{a} = 1$	$\frac{b}{a} = \frac{3}{2}$	$\frac{b}{a} = 2$	$\frac{b}{a} = 3$
$\hat{w}(a/2, b/2)$	0.5841	0.4222	0.0959	0.1238	0.1306	0.1309
$\hat{M}_{xx}(a/2, b/2)$	0.2107	0.2369	0.1663	0.2033	0.2107	0.2100
$\hat{M}_{yy}(a/2, b/2)$	0.5721	0.4346	0.1222	0.0897	0.0721	0.0651
$-\hat{M}_{xx}(0, b/2)$	0.6233	0.5957	0.3495	0.4117	0.4225	0.4209
$-\hat{M}_{yy}(0, b/2)$	0.1870	0.1787	0.1048	0.1235	0.1267	0.1263
$\hat{w}(a/2, 3b/4)$	0.4368	0.3218	0.0841	0.1313	0.1614	0.1884
$\hat{M}_{xx}(a/2, 3b/4)$	0.1803	0.1997	0.1562	0.2276	0.2705	0.3055
$\hat{M}_{yy}(a/2, 3b/4)$	0.5085	0.4097	0.1640	0.1665	0.1509	0.1153
$-\hat{M}_{xx}(0, 3b/4)$	0.5449	0.5254	0.3475	0.4810	0.5566	0.6140
$-\hat{M}_{yy}(0, 3b/4)$	0.1635	0.1576	0.1042	0.1443	0.1670	0.1842

Orthotropic plates

For isotropic plates on elastic foundation or orthotropic plates with $(D_{12} + 2D_{66})^2 < D_{11}(D_{22} + k/\beta_n^4)$, the homogeneous solution is given by Eq. (7.2.9a). When the applied loads are such that $\bar{q}_n$ is a constant or linear function of x, the complete solution is given by

$$w_0(x, y) = \sum_{n=1}^{\infty} \left[\frac{q_n}{D_{22}\beta_n^4 + k} + (A_n \cos \lambda_2 x + B_n \sin \lambda_2 x) \cosh \lambda_1 x \right.$$
$$\left. + (C_n \cos \lambda_2 x + D_n \sin \lambda_2 x) \sinh \lambda_1 x \right] \sin \beta_n y \quad (7.2.31a)$$

BENDING OF PLATES WITH VARIOUS BOUNDARY CONDITIONS

where λ_1 and λ_2 are given by Eq. (7.2.9b). Differentiating Eq. (7.2.31a), we obtain

$$\frac{\partial w_0}{\partial x} = \sum_{n=1}^{\infty} \Big[\lambda_2 \left(-A_n \sin \lambda_2 x + B_n \cos \lambda_2 x \right) \cosh \lambda_1 x$$
$$+ \lambda_1 \left(A_n \cos \lambda_2 x + B_n \sin \lambda_2 x \right) \sinh \lambda_1 x$$
$$+ \lambda_2 \left(-C_n \sin \lambda_2 x + D_n \cos \lambda_2 x \right) \sinh \lambda_1 x$$
$$+ \lambda_1 \left(C_n \cos \lambda_2 x + D_n \sin \lambda_2 x \right) \cosh \lambda_1 x \Big] \sin \beta_n y \quad (7.2.31b)$$

Using Eqs. (7.2.30a,b) in Eq. (7.2.15), we find that

$$A_n + \frac{\bar{q}_n}{D_{22}\beta_n^4 + k} = 0, \quad \lambda_2 B_n + \lambda_1 C_n = 0$$

$$\frac{\bar{q}_n}{D_{22}\beta_n^4 + k} + \left(A_n \cos \lambda_2 a + B_n \sin \lambda_2 a \right) \cosh \lambda_1 a$$
$$+ \left(C_n \cos \lambda_2 a + D_n \sin \lambda_2 a \right) \sinh \lambda_1 a = 0$$

$$\lambda_2 \left(- A_n \sin \lambda_2 a + B_n \cos \lambda_2 a \right) \cosh \lambda_1 a$$
$$+ \lambda_1 \left(A_n \cos \lambda_2 a + B_n \sin \lambda_2 a \right) \sinh \lambda_1 a$$
$$+ \lambda_2 \left(-C_n \sin \lambda_2 a + D_n \cos \lambda_2 a \right) \sinh \lambda_1 a$$
$$+ \lambda_1 \left(C_n \cos \lambda_2 a + D_n \sin \lambda_2 a \right) \cosh \lambda_1 a = 0 \quad (7.2.32)$$

which can be solved for A_n, B_n, C_n, and D_n. We have

$$A_n = -\frac{\bar{q}_n}{D_{22}\beta_n^4 + k}, \quad C_n = \frac{b_1 a_{22} - b_2 a_{12}}{a_{11} a_{22} - a_{12} a_{21}}, \quad D_n = \frac{b_2 a_{11} - b_1 a_{21}}{a_{11} a_{22} - a_{12} a_{21}}$$

$$B_n = -\frac{\lambda_2}{\lambda_1} C_n, \quad b_1 = (1 - \cos \lambda_2 a \, \cosh \lambda_1 a) A_n$$

$$b_2 = (\lambda_2 \sin \lambda_2 a \, \cosh \lambda_1 a - \lambda_1 \cos \lambda_2 a \, \sinh \lambda_1 a) A_n$$

$$a_{11} = \cos \lambda_2 a \, \sinh \lambda_1 a - \frac{\lambda_1}{\lambda_2} \sin \lambda_2 a \, \cosh \lambda_1 a$$

$$a_{12} = \sin \lambda_2 a \, \sinh \lambda_1 a, \quad a_{21} = -\left(\frac{\lambda_2^2 + \lambda_1^2}{\lambda_2} \right) \sin \lambda_2 a \, \sinh \lambda_1 a$$

$$a_{22} = \lambda_1 \sin \lambda_2 a \, \cosh \lambda_1 a + \lambda_2 \cos \lambda_2 a \, \sinh \lambda_1 a \quad (7.2.33)$$

The bending moments can be calculated using Eqs. (6.1.9a–c). Note that the second derivatives of w_0 in Eq. (7.2.31a) with respect to x and y are given by

320   THEORY AND ANALYSIS OF ELASTIC PLATES

$$\frac{\partial^2 w_0}{\partial x^2} = \sum_{n=1}^{\infty} \Big[\left(\lambda_1^2 - \lambda_2^2\right) (A_n \cos \lambda_2 x + B_n \sin \lambda_2 x) \cosh \lambda_1 x$$
$$+ 2\lambda_1 \lambda_2 (-A_n \sin \lambda_2 x + B_n \cos \lambda_2 x) \sinh \lambda_1 x$$
$$+ \left(\lambda_1^2 - \lambda_2^2\right) (C_n \cos \lambda_2 x + D_n \sin \lambda_2 x) \sinh \lambda_1 x$$
$$+ 2\lambda_1 \lambda_2 (-C_n \sin \lambda_2 x + D_n \cos \lambda_2 x) \cosh \lambda_1 x \Big] \sin \beta_n y \quad (7.2.34)$$

$$\frac{\partial^2 w_0}{\partial y^2} = -\sum_{n=1}^{\infty} \beta_n^2 \Big[\hat{q}_n + (A_n \cos \lambda_2 x + B_n \sin \lambda_2 x) \cosh \lambda_1 x$$
$$+ (C_n \cos \lambda_2 x + D_n \sin \lambda_2 x) \sinh \lambda_1 x \Big] \sin \beta_n y \quad (7.2.35)$$

where $\hat{q}_n = \bar{q}_n/(D_{22}\beta_n^4 + k)$.

Table 7.2.3 contains nondimensionalized deflections and bending moments [see Eqs. (7.2.27) and (7.2.28) for the nondimensionalizations] for various aspect ratios of orthotropic and isotropic plates on elastic foundation under uniformly distributed load (UDL) or hydrostatic load (HSL). The values are obtained using the first nine terms of the series ($n = 1, 2, 3, \cdots, 9$). The values in column $k = 0$ were obtained with the procedure of this section but by setting $k = 10^{-6}$.

7.2.3 Plates with Edge $x = 0$ Clamped and Edge $x = a$ Simply Supported (CSSS)

The boundary conditions for this case are (see Figure 7.2.3)

$$w_0 = \frac{\partial w_0}{\partial x} = 0 \text{ at } x = 0 \quad \text{and} \quad w_0 = M_{xx} = 0 \text{ at } x = a \quad (7.2.36)$$

Here we consider Case 3 for orthotropic plates; isotropc plate solutions can be obtained as a special case. The solution is given by Eq. (7.2.31a). Using Eqs. (7.2.31b), (7.2.34), and (7.2.35) in Eq. (7.2.36), we obtain

$$A_n + \hat{q}_n = 0, \quad \lambda_2 B_n + \lambda_1 C_n = 0 \quad (7.2.37a)$$
$$\hat{q}_n + (A_n \cos \lambda_2 a + B_n \sin \lambda_2 a) \cosh \lambda_1 a$$
$$+ (C_n \cos \lambda_2 a + D_n \sin \lambda_2 a) \sinh \lambda_1 a = 0 \quad (7.2.37b)$$
$$[(-A_n \sin \lambda_2 a + B_n \cos \lambda_2 a) \sinh \lambda_1 a + (-C_n \sin \lambda_2 a$$
$$+ D_n \cos \lambda_2 a) \cosh \lambda_1 a] - \hat{q}_n \left(\frac{\lambda_1^2 - \lambda_2^2}{2\lambda_1 \lambda_2}\right) = 0 \quad (7.2.37c)$$

Table 7.2.3. Maximum nondimensional deflections and bending moments in isotropic ($\nu = 0.3$) and orthotropic ($E_1/E_2 = 25$, $G_{12} = G_{13} = 0.5E_2$, $\nu_{12} = 0.25$) square plates with edges $y = 0, b$ simply supported and edges $x = 0, a$ clamped (CCSS), and subjected to distributed loads.

Variable	$k=0$	$k=1$	$k=10$	$k=10^2$	$k=10^3$	$k=10^4$
Isotropic plates on elastic foundation, UDL						
$\hat{w}(a/2, b/2)$	0.1917	0.1892	0.1688	0.0795	0.0103	0.0009
$\hat{M}_{xx}(a/2, b/2)$	0.3326	0.3277	0.2888	0.1199	0.0025	0.0000
$-\hat{M}_{xx}(0, b/2)$	0.6990	0.6908	0.6258	0.3392	0.0978	0.0308
Orthotropic plates on elastic foundation, UDL						
$\hat{w}(a/2, b/2)$	0.0108	0.0108	0.0107	0.0098	0.0053	0.0009
$\hat{M}_{xx}(a/2, b/2)$	0.4331	0.4327	0.4285	0.3909	0.1998	0.0180
$-\hat{M}_{xx}(0, b/2)$	0.8630	0.8622	0.8554	0.7930	0.4748	0.1479
Isotropic plates on elastic foundation, HSL						
$\hat{w}(a/2, b/2)$	0.0958	0.0946	0.0844	0.0397	0.0051	0.0004
$\hat{M}_{xx}(a/2, b/2)$	0.1663	0.1639	0.1444	0.0599	0.0012	0.0000
$\hat{w}(a/2, 3b/4)$	0.0841	0.0831	0.0754	0.0404	0.0075	0.0007
$\hat{M}_{xx}(a/2, 3b/4)$	0.4607	0.4553	0.4119	0.2091	0.0157	0.0005
$-\hat{M}_{xx}(0, 3b/4)$	1.3370	1.3265	1.2416	0.8271	0.2989	0.0608
$-\hat{M}_{yy}(0, 3b/4)$	0.4011	0.3979	0.3725	0.2481	0.0897	0.0183
Orthotropic plates on elastic foundation, HSL						
$\hat{w}(a/2, 3b/4)$	0.0054	0.0054	0.0053	0.0049	0.0027	0.0004
$\hat{M}_{xx}(a/2, 3b/4)$	0.2166	0.2163	0.2143	0.1954	0.0999	0.0090
$\hat{w}(a/2, 3b/4)$	0.0078	0.0078	0.0078	0.0072	0.0042	0.0007
$\hat{M}_{xx}(a/2, 3b/4)$	1.4317	1.4304	1.4188	1.3117	0.7186	0.0602
$-\hat{M}_{xx}(0, 3b/4)$	2.6650	2.6628	2.6432	2.4616	1.4520	0.2867
$-\hat{M}_{yy}(0, 3b/4)$	0.0266	0.0266	0.0264	0.0246	0.0145	0.0028

which can be solved for the constants A_n, B_n, C_n, and D_n, and the deflection is given by Eq. (7.2.31a). It should be noted that the solution of Eq. (7.2.37a–c) includes plates on elastic foundation.

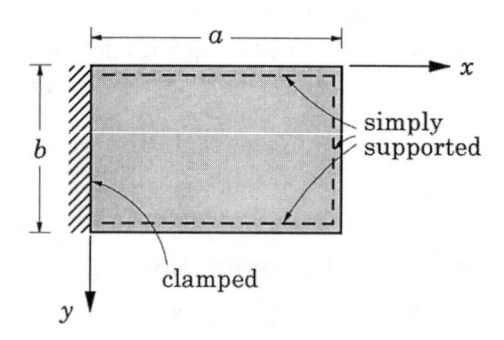

Figure 7.2.3. Geometry and coordinate system for a rectangular plate with side $x = 0$ clamped, side $x = a$ simply supported, and sides $y = 0, b$ simply supported (CSSS).

Table 7.2.4. Nondimensional deflections and bending moments in isotropic and orthotropic rectangular plates with side $x = 0$ clamped, side $x = a$ simply supported, and sides $y = 0, b$ simply supported (CSSS), and subjected to *uniformly distributed load*.

Variable	$\frac{b}{a} = \frac{1}{3}$	$\frac{b}{a} = \frac{1}{2}$	$\frac{b}{a} = 1$	$\frac{b}{a} = \frac{3}{2}$	$\frac{b}{a} = 2$	$\frac{b}{a} = 3$
Isotropic plates ($\nu = 0.3$)						
$\hat{w}(a/2, b/2)$	1.0526	0.9270	0.2786	0.4250	0.4880	0.5193
$\hat{M}_{xx}(a/2, b/2)$	8.4196	0.4689	0.3920	0.5442	0.6021	0.6267
$\hat{M}_{yy}(a/2, b/2)$	-6.8394	0.9419	0.3395	0.2862	0.2373	0.2002
$-\hat{M}_{xx}(0, b/2)$	1.2484	1.2157	0.8394	1.1135	1.2143	1.2562
$-\hat{M}_{yy}(0, b/2)$	0.3745	0.3647	0.2518	0.3341	0.3643	0.3769
Orthotropic plates ($E_1/E_2 = 25$, $G_{12} = 0.5E_2$, $\nu_{12} = 0.25$)						
$\hat{w}(a/2, b/2)$	0.7735	0.2873	0.0221	0.0210	0.0209	0.0212
$\hat{M}_{xx}(a/2, b/2)$	2.6029	2.1967	0.6651	0.6304	0.6262	0.6354
$\hat{M}_{yy}(a/2, b/2)$	0.7524	0.2691	0.0109	0.0059	0.0085	0.0104
$-\hat{M}_{xx}(0, b/2)$	6.2887	4.5522	1.3224	1.2648	1.2605	1.2785
$-\hat{M}_{yy}(0, b/2)$	0.0629	0.0455	0.0132	0.0126	0.0126	0.0128

Nondimensionalized deflections and bending moments [see Eqs. (7.2.27) and (7.2.28)] for various aspect ratios of isotropic and orthotropic plates under uniformly distributed transverse load q_0 are presented in Table 7.2.4. The values are obtained using the first five terms of the series ($n = 1, 3, \cdots, 9$). Table 7.2.5 contains the same quantities for hydrostatic loading. For square isotropic plates under uniform loading, the magnitude of the bending moment at the middle of the clamped edge is $M_{xx}(0, b/2) = -0.0839 q_0 b^2$, which is numerically larger than the corresponding moment $M_{xx}(0, b/2) = -0.0699 q_0 b^2$ for square isotropic plates with two edges clamped (see Table 7.2.3). The deflections and moments of hydrostatically loaded plates are equal to one-half of the corresponding deflections and moments of uniformly loaded plates, as they should be.

Table 7.2.5. Maximum nondimensional deflections and bending moments in isotropic and orthotropic rectangular plates with edges $y = 0, b$ and $x = a$ simply supported and edge $x = 0$ clamped (CSSS), and subjected to hydrostatic load.

Variable	$\frac{b}{a}=\frac{1}{3}$	$\frac{b}{a}=0.5$	$\frac{b}{a}=1$	$\frac{b}{a}=1.5$	$\frac{b}{a}=2$	$\frac{b}{a}=3$
Isotropic plates ($\nu = 0.3$)						
$\hat{w}(a/2, b/2)$	0.5263	0.4635	0.1393	0.2125	0.2440	0.2596
$\hat{M}_{xx}(a/2, b/2)$	4.2098	0.2344	0.1960	0.2721	0.3011	0.3134
$\hat{M}_{yy}(a/2, b/2)$	-3.4197	0.4710	0.1697	0.1431	0.1187	0.1001
$-\hat{M}_{xx}(0, b/2)$	0.6242	0.6079	0.4197	0.5568	0.6072	0.6281
$-\hat{M}_{yy}(0, b/2)$	0.1873	0.1824	0.1259	0.1670	0.1821	0.1884
Orthotropic plates ($E_1/E_2 = 25$, $G_{12} = 0.5 E_2$, $\nu_{12} = 0.25$)						
$\hat{w}(a/2, b/2)$	0.3867	0.1436	0.0110	0.0105	0.0104	0.0106
$\hat{M}_{xx}(a/2, b/2)$	1.3014	1.0984	0.3326	0.3152	0.3131	0.3177
$\hat{M}_{yy}(a/2, b/2)$	0.3762	0.1345	0.0054	0.0030	0.0042	0.0052
$-\hat{M}_{xx}(0, b/2)$	3.1444	2.2761	0.6612	0.6324	0.6302	0.6393
$-\hat{M}_{yy}(0, b/2)$	0.0314	0.0228	0.0066	0.0063	0.0063	0.0064

So far we have considered simply supported and clamped edges only. In the next two sections we shall consider plates with a free edge.

7.2.4 Plates with Edge $x = 0$ Clamped and Edge $x = a$ Free (CFSS)

Here we consider rectangular plates with edge $x = 0$ clamped, edge $x = a$ free, and edges $y = 0, b$ simply supported (see Figure 7.2.4). The boundary conditions at the free edge $x = a$ are

$$M_{xx} = 0, \qquad V_x \equiv \frac{\partial M_{xx}}{\partial x} + 2\frac{\partial M_{xy}}{\partial y} = 0 \qquad (7.2.38a)$$

or, in terms of the deflection,

$$-\left(D_{11}\frac{\partial^2 w_0}{\partial x^2} + D_{12}\frac{\partial^2 w_0}{\partial y^2}\right) = 0, \quad -\left(D_{11}\frac{\partial^3 w_0}{\partial x^3} + \bar{D}_{12}\frac{\partial^3 w_0}{\partial x \partial y^2}\right) = 0$$
$$(7.2.38b)$$

where $\bar{D}_{12} = (D_{12} + 4D_{66})$. Thus, in addition to the second-order derivatives of w_0 as given in Eqs. (7.2.34) and (7.2.35), we need the third-order derivatives of w_0. From Eqs. (7.2.34), (7.2.35), and (7.2.31b), we find

$$\begin{aligned}\frac{\partial^3 w_0}{\partial x^3} = \sum_{n=1}^{\infty}\Big\{ &\left(\lambda_1^2 - \lambda_2^2\right)[\lambda_1\left(A_n \cos \lambda_2 x + B_n \sin \lambda_2 x\right) \sinh \lambda_1 x \\ &+ \lambda_2 \left(-A_n \sin \lambda_2 x + B_n \cos \lambda_2 x\right) \cosh \lambda_1 x] \\ &+ 2\lambda_1^2 \lambda_2 \left(-A_n \sin \lambda_2 x + B_n \cos \lambda_2 x\right) \cosh \lambda_1 x \\ &- 2\lambda_1 \lambda_2^2 \left(A_n \cos \lambda_2 x + B_n \sin \lambda_2 x\right) \sinh \lambda_1 x \\ &+ \left(\lambda_1^2 - \lambda_2^2\right)[\lambda_1 \left(C_n \cos \lambda_2 x + D_n \sin \lambda_2 x\right) \cosh \lambda_1 x \\ &+ \lambda_2 \left(-C_n \sin \lambda_2 x + D_n \cos \lambda_2 x\right) \sinh \lambda_1 x] \\ &+ 2\lambda_1^2 \lambda_2 \left(-C_n \sin \lambda_2 x + D_n \cos \lambda_2 x\right) \sinh \lambda_1 x \\ &- 2\lambda_1 \lambda_2^2 \left(C_n \cos \lambda_2 x + D_n \sin \lambda_2 x\right) \cosh \lambda_1 x \Big\} \sin \beta_n y\end{aligned}$$
$$(7.2.39)$$

$$\begin{aligned}\frac{\partial^3 w_0}{\partial x \partial y^2} = -\sum_{n=1}^{\infty} \beta_n^2 \Big[&\lambda_2 \left(-A_n \sin \lambda_2 x + B_n \cos \lambda_2 x\right) \cosh \lambda_1 x \\ &+ \lambda_1 \left(A_n \cos \lambda_2 x + B_n \sin \lambda_2 x\right) \sinh \lambda_1 x \\ &+ \lambda_2 \left(-C_n \sin \lambda_2 x + D_n \cos \lambda_2 x\right) \sinh \lambda_1 x \\ &+ \lambda_1 \left(C_n \cos \lambda_2 x + D_n \sin \lambda_2 x\right) \cosh \lambda_1 x \Big] \sin \beta_n y \quad (7.2.40)\end{aligned}$$

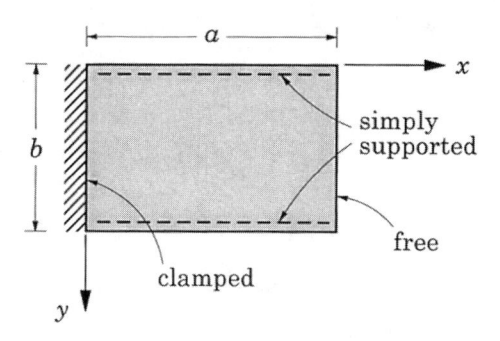

BENDING OF PLATES WITH VARIOUS BOUNDARY CONDITIONS 325

Figure 7.2.4. Geometry and coordinate system for a rectangular plate with sides $y = 0, b$ simply supported, edge $x = 0$ clamped, and edge $x = a$ free (CFSS).

The boundary conditions in Eq. (7.2.38b), coupled with Eq. (7.2.37a) associated with the clamped edge at $x = 0$, yield the following relations among the four constants $[\hat{q}_n = \bar{q}_n/(D_{22}\beta_n^4 + k)]$:

$$A_n + \hat{q}_n = 0, \quad \lambda_2 B_n + \lambda_1 C_n = 0 \quad (7.2.41a)$$

$$D_{11}\bigg[\left(\lambda_1^2 - \lambda_2^2\right)(A_n \cos \lambda_2 a + B_n \sin \lambda_2 a) \cosh \lambda_1 a$$

$$+ 2\lambda_1\lambda_2 (-A_n \sin \lambda_2 a + B_n \cos \lambda_2 a) \sinh \lambda_1 a$$

$$+ \left(\lambda_1^2 - \lambda_2^2\right)(C_n \cos \lambda_2 a + D_n \sin \lambda_2 a) \sinh \lambda_1 a$$

$$+ 2\lambda_1\lambda_2 (-C_n \sin \lambda_2 a + D_n \cos \lambda_2 a) \cosh \lambda_1 a\bigg]$$

$$- D_{12}\beta_n^2 \bigg[\hat{q}_n + (A_n \cos \lambda_2 a + B_n \sin \lambda_2 a) \cosh \lambda_1 a$$

$$+ (C_n \cos \lambda_2 a + D_n \sin \lambda_2 a) \sinh \lambda_1 x\bigg] = 0 \quad (7.2.41b)$$

$$D_{11}\bigg\{\left(\lambda_1^2 - \lambda_2^2\right)[\lambda_1 (A_n \cos \lambda_2 a + B_n \sin \lambda_2 a) \sinh \lambda_1 a$$

$$+ \lambda_2 (-A_n \sin \lambda_2 a + B_n \cos \lambda_2 a) \cosh \lambda_1 a]$$

$$+ 2\lambda_1^2\lambda_2 (-A_n \sin \lambda_2 a + B_n \cos \lambda_2 a) \cosh \lambda_1 a$$

$$- 2\lambda_1\lambda_2^2 (A_n \cos \lambda_2 a + B_n \sin \lambda_2 a) \sinh \lambda_1 a$$

$$+ \left(\lambda_1^2 - \lambda_2^2\right) [\lambda_1 \left(C_n \cos \lambda_2 a + D_n \sin \lambda_2 a\right) \cosh \lambda_1 a$$
$$+ \lambda_2 \left(-C_n \sin \lambda_2 a + D_n \cos \lambda_2 a\right) \sinh \lambda_1 a]$$
$$+ 2\lambda_1^2 \lambda_2 \left(-C_n \sin \lambda_2 a + D_n \cos \lambda_2 a\right) \sinh \lambda_1 a$$
$$- 2\lambda_1 \lambda_2^2 \left(C_n \cos \lambda_2 a + D_n \sin \lambda_2 a\right) \cosh \lambda_1 a \Big\}$$
$$- \bar{D}_{12} \beta_n^2 \Big[\lambda_2 \left(-A_n \sin \lambda_2 a + B_n \cos \lambda_2 a\right) \cosh \lambda_1 a$$
$$+ \lambda_1 \left(A_n \cos \lambda_2 a + B_n \sin \lambda_2 a\right) \sinh \lambda_1 a$$
$$+ \lambda_2 \left(-C_n \sin \lambda_2 a + D_n \cos \lambda_2 a\right) \sinh \lambda_1 a$$
$$+ \lambda_1 \left(C_n \cos \lambda_2 a + D_n \sin \lambda_2 a\right) \cosh \lambda_1 a \Big] = 0 \quad (7.2.41c)$$

Table 7.2.6 contains nondimensionalized deflections and bending moments of isotropic and orthotropic plates under uniformly distributed load and for various aspect ratios. The values were obtained using the first five terms of the series ($n = 1, 3, \cdots, 9$). The deflection is the maximum at the center of the free edge. The bending moments at the center of the clamped edge as well as at the center of the free edge are tabulated. The maximum bending moment also occurs at the center of the free edge.

7.2.5 Plates with Edge $x = 0$ Simply Supported and Edge $x = a$ Free (SFSS)

Here we consider rectangular plates with edges $y = 0, b$ and $x = 0$ simply supported and the edge ($x = a$) free (F) (see Figure 7.2.5). When $\hat{q}_n$ is a constant, the boundary conditions on the edge $x = 0$ yield the following relations for the constants A_n and D_n:

$$A_n = -\hat{q}_n, \quad D_n = -\left(\frac{\lambda_1^2 - \lambda_2^2}{2\lambda_1 \lambda_2}\right) A_n \qquad (7.2.42)$$

The relations among B_n and C_n are provided by Eqs. (7.2.41a–c).

The nondimensional deflections and bending moments at the middle of the free edge of uniformly loaded plates (SFSS) are presented in Table 7.2.7. The last two columns contain bending moments at the center of the plate. The first five terms of the series are used to evaluate the quantities.

BENDING OF PLATES WITH VARIOUS BOUNDARY CONDITIONS   327

Table 7.2.6. Maximum nondimensional deflections and bending moments in rectangular plates (CFSS) and subjected to *uniformly* distributed load.

$\frac{b}{a}$	w_{max}	$-M_{xx}(0,b/2)$	$-M_{yy}(0,b/2)$	$M_{yy}(a,b/2)$
Isotropic plates ($\nu = 0.3$)				
1	$0.0112(\frac{q_0 a^4}{D})$	$0.1185(q_0 a^2)$	$0.0355(q_0 a^2)$	$0.0972(q_0 a^2)$
1.5	$0.0335(\frac{q_0 a^4}{D})$	$0.2271(q_0 a^2)$	$0.0681(q_0 a^2)$	$0.1258(q_0 a^2)$
		$0.1009(q_0 b^2)$	$0.0303(q_0 b^2)$	$0.0559(q_0 b^2)$
2.0	$0.0582(\frac{q_0 a^4}{D})$	$0.3192(q_0 a^2)$	$0.0958(q_0 a^2)$	$0.1173(q_0 a^2)$
		$0.0798(q_0 b^2)$	$0.0239(q_0 b^2)$	$0.0293(q_0 b^2)$
2.5	$0.0788(\frac{q_0 a^4}{D})$	$0.3850(q_0 a^2)$	$0.1155(q_0 a^2)$	$0.0946(q_0 a^2)$
		$0.0616(q_0 b^2)$	$0.0185(q_0 b^2)$	$0.0151(q_0 b^2)$
3.0	$0.0940(\frac{q_0 a^4}{D})$	$0.4285(q_0 a^2)$	$0.1285(q_0 a^2)$	$0.0710(q_0 a^2)$
		$0.0476(q_0 b^2)$	$0.0143(q_0 b^2)$	$0.0079(q_0 b^2)$
Orthotropic plates ($E_1/E_2 = 25$, $G_{12} = G_{13} = 0.5 E_2$, $\nu_{12} = 0.25$)				
1	$0.0039(\frac{q_0 a^4}{D})$	$0.4289(q_0 a^2)$	$0.0043(q_0 a^2)$	$0.0344(q_0 a^2)$
1.5	$0.0051(\frac{q_0 a^4}{D})$	$0.5189(q_0 a^2)$	$0.0052(q_0 a^2)$	$0.0146(q_0 a^2)$
2.0	$0.0053(\frac{q_0 a^4}{D})$	$0.5262(q_0 a^2)$	$0.0053(q_0 a^2)$	$0.0039(q_0 a^2)$
2.5	$0.0052(\frac{q_0 a^4}{D})$	$0.5160(q_0 a^2)$	$0.0052(q_0 a^2)$	$0.0002(q_0 a^2)$
3.0	$0.0051(\frac{q_0 a^4}{D})$	$0.5078(q_0 a^2)$	$0.0051(q_0 a^2)$	$-0.0002(q_0 a^2)$

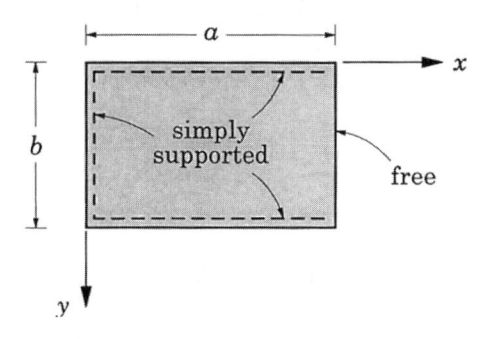

Figure 7.2.5. Geometry and coordinate system for a rectangular plate with sides $y = 0, b$ and $x = 0$ simply supported and edge $x = a$ free (SFSS).

Table 7.2.7. Maximum nondimensional deflections and bending moments in rectangular plates (SFSS) subjected to *uniformly distributed load.*

$\frac{b}{a}$	w_{max}	$M_{yy}(a,b/2)$	$M_{yy}(a/2,b/2)$	$M_{xx}(a/2,b/2)$
3.0	$0.00399(\frac{q_0 b^4}{D})$	$0.0330(q_0 b^2)$	$0.0207(q_0 b^2)$	$0.0125(q_0 b^2)$
2.5	$0.00526(\frac{q_0 b^4}{D})$	$0.0442(q_0 b^2)$	$0.0279(q_0 b^2)$	$0.0166(q_0 b^2)$
2.0	$0.00709(\frac{q_0 b^4}{D})$	$0.0602(q_0 b^2)$	$0.0386\ (q_0 b^2)$	$0.0223(q_0 b^2)$
1.5	$0.00968(\frac{q_0 b^4}{D})$	$0.0833(q_0 b^2)$	$0.0552(q_0 b^2)$	$0.0303(q_0 b^2)$
1.0	$0.01285(\frac{q_0 b^4}{D})$	$0.1185(q_0 b^2)$	$0.0390\ (q_0 b^2)$	$0.0799(q_0 b^2)$
2/3	$0.01461(\frac{q_0 b^4}{D})$	$0.1274(q_0 b^2)$	$0.1011(q_0 b^2)$	$0.0421(q_0 b^2)$
0.5	$0.01507(\frac{q_0 b^4}{D})$	$0.1320(q_0 b^2)$	$0.1127(q_0 b^2)$	$0.0415(q_0 b^2)$
1/3	$0.01521(\frac{q_0 b^4}{D})$	$0.1321(q_0 b^2)$	$0.1209(q_0 b^2)$	$0.0388(q_0 b^2)$

When the plate is loaded by hydrostatic load of the form

$$q(x,y) = q_0 \left(1 - \frac{x}{a}\right) \tag{7.2.43}$$

the particular solution is given by

$$W_n^p(x) = \frac{q_n}{D_{22}\beta_n^4}\left(1 - \frac{x}{a}\right) \tag{7.2.44}$$

The values of A_n and D_n given in Eq. (7.2.42) remain unchanged with $\bar{q}_n = q_n$. However, equations relating B_n and C_n must be modified as

$$\begin{aligned}
D_{11}&\bigg[\left(\lambda_1^2 - \lambda_2^2\right)(A_n\cos\lambda_2 a + B_n\sin\lambda_2 a)\cosh\lambda_1 a \\
&+ 2\lambda_1\lambda_2(-A_n\sin\lambda_2 a + B_n\cos\lambda_2 a)\sinh\lambda_1 a \\
&+ \left(\lambda_1^2 - \lambda_2^2\right)(C_n\cos\lambda_2 a + D_n\sin\lambda_2 a)\sinh\lambda_1 a \\
&+ 2\lambda_1\lambda_2(-C_n\sin\lambda_2 a + D_n\cos\lambda_2 a)\cosh\lambda_1 a\bigg] \\
-D_{12}\beta_n^2&\bigg[(A_n\cos\lambda_2 a + B_n\sin\lambda_2 a)\cosh\lambda_1 a \\
&+ (C_n\cos\lambda_2 a + D_n\sin\lambda_2 a)\sinh\lambda_1 x\bigg] = 0 \tag{7.2.45}
\end{aligned}$$

BENDING OF PLATES WITH VARIOUS BOUNDARY CONDITIONS   329

$$\bar{D}_{11}\Big\{\left(\lambda_1^2-\lambda_2^2\right)[\lambda_1\left(A_n\cos\lambda_2 a+B_n\sin\lambda_2 a\right)\sinh\lambda_1 a$$
$$+\lambda_2\left(-A_n\sin\lambda_2 a+B_n\cos\lambda_2 a\right)\cosh\lambda_1 a]$$
$$+2\lambda_1^2\lambda_2\left(-A_n\sin\lambda_2 a+B_n\cos\lambda_2 a\right)\cosh\lambda_1 a$$
$$-2\lambda_1\lambda_2^2\left(A_n\cos\lambda_2 a+B_n\sin\lambda_2 a\right)\sinh\lambda_1 a$$
$$+\left(\lambda_1^2-\lambda_2^2\right)[\lambda_1\left(C_n\cos\lambda_2 a+D_n\sin\lambda_2 a\right)\cosh\lambda_1 a$$
$$+\lambda_2\left(-C_n\sin\lambda_2 a+D_n\cos\lambda_2 a\right)\sinh\lambda_1 a\,]$$
$$+2\lambda_1^2\lambda_2\left(-C_n\sin\lambda_2 a+D_n\cos\lambda_2 a\right)\sinh\lambda_1 a$$
$$-2\lambda_1\lambda_2^2\left(C_n\cos\lambda_2 a+D_n\sin\lambda_2 a\right)\cosh\lambda_1 a\Big\}$$
$$-\bar{D}_{12}\beta_n^2\bigg[-\frac{q_n}{aD_{22}\beta_n^4}+\lambda_2\left(-A_n\sin\lambda_2 a+B_n\cos\lambda_2 a\right)\cosh\lambda_1 a$$
$$+\lambda_1\left(A_n\cos\lambda_2 a+B_n\sin\lambda_2 a\right)\sinh\lambda_1 a$$
$$+\lambda_2\left(-C_n\sin\lambda_2 a+D_n\cos\lambda_2 a\right)\sinh\lambda_1 a$$
$$+\lambda_1\left(C_n\cos\lambda_2 a+D_n\sin\lambda_2 a\right)\cosh\lambda_1 a\bigg]=0 \qquad (7.2.46)$$

and the solution is given by

$$w_0(x,y)=\sum_{n=1}^{\infty}\bigg[\left(1-\frac{x}{a}\right)\frac{q_n}{D_{22}\beta_n^4}+\left(A_n\cos\lambda_2 x+B_n\sin\lambda_2 x\right)\cosh\lambda_1 x$$
$$+\left(C_n\cos\lambda_2 x+D_n\sin\lambda_2 x\right)\sinh\lambda_1 x\bigg]\sin\beta_n y \qquad (7.2.47)$$

The nondimensional deflections and bending moments

$$\bar{w}=w_0(a,b/2)\frac{D}{q_0 b^4},\quad \hat{w}=w_0(a/2,b/2)\frac{D}{q_0 b^4}$$

$$\bar{M}=M(a,b/2)\frac{1}{q_0 b^2},\quad \hat{M}=M(a/2,b/2)\frac{1}{q_0 b^4}$$

at the middle of the free edge and at the center of hydrostatically loaded plates (SFSS) are presented in Table 7.2.8. Of course, the results for the hydrostatic loading

$$q(x,y)=q_0\left(\frac{y}{b}\right)$$

can be obtained as discussed before. The magnitudes of maximum deflections and bending moments would be one-half of those presented in Table 7.2.8.

Table 7.2.8. Maximum nondimensional deflections and bending moments in rectangular plates with edges $y = 0, b$ and $x = 0$ simply supported and edge $x = a$ free (SFSS) and subjected to hydrostatic load, $q = q_0(1 - x/a)$.

$\dfrac{b}{a}$	$\bar{w}$	$\bar{M}_{yy}$	$\hat{w}$	$\hat{M}_{yy}$	$\hat{M}_{xx}$
2.5	0.00172	0.0145	0.00095	0.0103	0.0085
2.0	0.00229	0.0197	0.00134	0.0146	0.0117
1.5	0.00304	0.0265	0.00199	0.0215	0.0162
1.0	0.00368	0.0325	0.00314	0.0331	0.0214
2/3	0.00347	0.0308	0.00445	0.0455	0.0231
0.5	0.00291	0.0258	0.00533	0.0531	0.0223

We close this section on the Lévy solutions with a note that the solutions for other boundary conditions and loads, within the limitations of the Lévy solutions discussed, may be obtained in the manner illustrated in this section. Solutions of plates with more general boundary conditions (e.g., plates with two adjacent edges clamped while the other edges simply supported) may be obtained using the method of superposition, where the solution of a given problem is represented as the sum of the solutions of plates for which solutions can be either developed or readily available (see Timoshenko and Woinowsky-Krieger [1]), as shown next. In section 7.3, we use the Rayleigh–Ritz method to determine solutions of plates with various boundary conditions.

7.2.6 Solution by the Method of Superposition

The linear solutions of rectangular plates with any edge conditions and loads can be determined using the method of superposition. There are two types of problems that can be used by the method of superposition. (1) Plates with multiple loads, and (2) plates with different boundary conditions. Deflections, bending moments, and stresses in a plate with specific boundary conditions and subjected to several different loads (the first type) can be obtained by simply adding the solutions of plates with the same boundary conditions but subjected to one load at a time. For instance, the deflection of a rectangular plate with all edges simply supported and subjected to hydrostatic load $q = q_0(x/a)$ and distributed bending moment M_0 along the edges $x = 0, a$ can be obtained by simply adding the deflection due to the

BENDING OF PLATES WITH VARIOUS BOUNDARY CONDITIONS 331

hydrostatic load and that due to distributed edge moment. An example of the second type of problems is provided by the plate shown in Figure 7.2.6. The deflection of the problem can be obtained for any applied loading by the superposition of the deflections of two different plate problems: (1) a simply supported plate under the applied loading, and (2) a simply supported plate with applied edge moments M_{yy} necessary to make the rotations $\partial w_0/\partial y$ vanish at the clamped edges. Note that solutions of the two problems whose solutions are superposed are known from the Navier and/or Lévy solution procedures. Of course, the solution to the problem shown in Figure 7.2.6 can be derived directly using the Lévy method (see section 7.2.2). We illustrate the procedure for the SSCC plate.

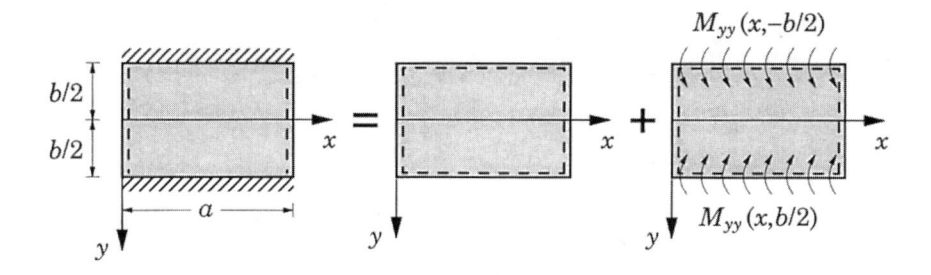

Figure 7.2.6. Application of the method of superposition to a rectangular plate with edges $x = 0, a$ simply supported and edges $y = \pm b/2$ clamped (SSCC).

Consider the SSCC plate shown in Figure 7.2.6 under uniformly distributed transverse load q_0. The deflection of the problem is represented as the sum of the deflection w_1 of a simply supported plate under distributed transverse load and deflection w_2 of a simply supported plate with distributed edge moments M_{yy} at $y = \pm b/2$. The edge moments M_{yy} are determined such that the rotations are zero along the edges $y = \pm b/2$. The general solution of plates with simply supported edges at $x = 0, a$ is given by [see section 6.3.5 and Eq. (6.3.63)]

$$w_0(x,y) = \sum_{m=1}^{\infty} (A_m \cos \lambda_2 y \cosh \lambda_1 y + D_m \sin \lambda_2 y \sinh \lambda_1 y + \hat{q}_m) \sin \alpha_m x$$

(7.2.48)

where
$$\hat{q}_m = \frac{4q_0 a^4}{m^5 \pi^5 D_{11}} \tag{7.2.49}$$

The deflection of a simply supported plate under distributed transverse load is given by

$$w_1(x,y) = \sum_{m=1}^{\infty} \hat{q}_m \left(A_m^{(1)} \cos \lambda_2 y \cosh \lambda_1 y + D_m^{(1)} \sin \lambda_2 y \sinh \lambda_1 y + 1 \right)$$
$$\times \sin \alpha_m x \tag{7.2.50}$$

where $A_m^{(1)}$ and $D_m^{(1)}$ are defined by [see Eqs. (6.3.68)–(6.3.72)]

$$A_m^{(1)} = -\frac{1}{\Delta} \left(\gamma \sinh \bar{\lambda}_1 \sin \bar{\lambda}_2 + \cosh \bar{\lambda}_1 \cos \bar{\lambda}_2 \right)$$

$$D_m^{(1)} = \frac{1}{\Delta} \left(\gamma \cosh \bar{\lambda}_1 \cos \bar{\lambda}_2 - \sinh \bar{\lambda}_1 \sin \bar{\lambda}_2 \right)$$

$$\Delta^{(1)} = \cosh^2 \bar{\lambda}_1 \cos^2 \bar{\lambda}_2 + \sinh^2 \bar{\lambda}_1 \sin^2 \bar{\lambda}_2$$

$$\bar{\lambda}_i = \frac{\lambda_i b}{2}, \quad \gamma = \frac{\lambda_1^2 - \lambda_2^2}{2\lambda_1 \lambda_2} \tag{7.2.51}$$

and the deflection of a simply supported plate with edge moment

$$M_{yy}(x, \pm \frac{b}{2}) = f(x) = \sum_{m=1}^{\infty} F_m \sin \alpha_m x \tag{7.2.52}$$

is given by

$$w_2(x,y) = \sum_{m=1}^{\infty} \frac{F_m}{2 D_{22} \lambda_1 \lambda_2} \left(A_m^{(2)} \cos \lambda_2 y \cosh \lambda_1 y \right.$$
$$\left. + D_m^{(2)} \sin \lambda_2 y \sinh \lambda_1 y \right) \sin \alpha_m x \tag{7.2.53}$$

where $A_m^{(2)}$ and $D_m^{(2)}$ are defined by

$$A_m^{(2)} = \frac{\sinh \bar{\lambda}_1 \sin \bar{\lambda}_2}{\Delta}, \quad D_m^{(2)} = -\frac{\cosh \bar{\lambda}_1 \cos \bar{\lambda}_2}{\Delta}$$

$$\Delta^{(2)} = \cosh^2 \bar{\lambda}_1 \cos^2 \bar{\lambda}_2 + \sinh^2 \bar{\lambda}_1 \sin^2 \bar{\lambda}_2 \tag{7.2.54}$$

The slope of the deflection surface w_1 along the edge $y = b/2$ is

$$\left(\frac{\partial w_1}{\partial x}\right)_{y=\frac{b}{2}} = \sum_{m=1}^{\infty} \hat{q}_m \Big[A_m^{(1)}\left(\lambda_1 \cos \bar{\lambda}_2 \sinh \bar{\lambda}_1 - \lambda_2 \cosh \bar{\lambda}_1 \sin \bar{\lambda}_2\right)$$
$$+ D_m^{(1)}\left(\lambda_1 \sin \bar{\lambda}_2 \cosh \bar{\lambda}_1 + \lambda_2 \sinh \bar{\lambda}_1 \cos \bar{\lambda}_2\right)\Big] \sin \alpha_m x$$
(7.2.55)

The slope of the deflection surface w_2 along the edge $y = b/2$ is

$$\left(\frac{\partial w_2}{\partial x}\right)_{y=\frac{b}{2}} = \sum_{m=1}^{\infty} \frac{F_m}{2 D_{22} \lambda_1 \lambda_2}\Big[A_m^{(2)}\left(\lambda_1 \cos \bar{\lambda}_2 \sinh \bar{\lambda}_1 - \lambda_2 \cosh \bar{\lambda}_1 \sin \bar{\lambda}_2\right)$$
$$+ D_m^{(2)}\left(\lambda_1 \sin \bar{\lambda}_2 \cosh \bar{\lambda}_1 + \lambda_2 \sinh \bar{\lambda}_1 \cos \bar{\lambda}_2\right)\Big] \sin \alpha_m x$$
(7.2.56)

To satisfy the actual boundary conditions along the clamped edges, we equate the negative of the slope produced by the moment distribution F_m to that given in Eq. (7.2.55) and solve for F_m. We obtain

$$F_m = 2\lambda_1 \lambda_2 D_{22} \hat{q}_m \left(\frac{W_m^1}{W_m^2}\right) \tag{7.2.57}$$

$$W_m^1 = A_m^{(1)}\left(\lambda_1 \cos \bar{\lambda}_2 \sinh \bar{\lambda}_1 - \lambda_2 \cosh \bar{\lambda}_1 \sin \bar{\lambda}_2\right)$$
$$+ D_m^{(1)}\left(\lambda_1 \sin \bar{\lambda}_2 \cosh \bar{\lambda}_1 + \lambda_2 \sinh \bar{\lambda}_1 \cos \bar{\lambda}_2\right)$$
$$W_m^2 = A_m^{(2)}\left(\lambda_1 \cos \bar{\lambda}_2 \sinh \bar{\lambda}_1 - \lambda_2 \cosh \bar{\lambda}_1 \sin \bar{\lambda}_2\right)$$
$$+ D_m^{(2)}\left(\lambda_1 \sin \bar{\lambda}_2 \cosh \bar{\lambda}_1 + \lambda_2 \sinh \bar{\lambda}_1 \cos \bar{\lambda}_2\right) \tag{7.2.58}$$

Hence the bending moment along the clamped edge is

$$M_{yy}\left(x, \pm \frac{b}{2}\right) = 2 D_{22} \sum_{m=1}^{\infty} \lambda_1 \lambda_2 \hat{q}_m \left(\frac{W_m^1}{W_m^2}\right) \sin \alpha_m x \tag{7.2.59}$$

The deflection of the SSCC plate is now given by

$$w_0(x, y) = w_1(x, y) - w_2(x, y) \tag{7.2.60}$$

where F_m of Eq. (7.2.53) is defined by Eq. (7.2.57).

7.3 Approximate Solutions by the Rayleigh–Ritz Method

7.3.1 Analysis of the Lévy Plates

We recall from section 6.3.6 that the ordinary differential equation resulting from the application of the Lévy solution procedures, namely Eq. (7.2.4), can be solved by the Rayleigh–Ritz method. The Rayleigh–Ritz solutions of the Lévy equation (7.2.4) were presented for simply supported plates in section 6.3.6. Here we use the method for other boundary conditions. Towards this end, first we construct the weak form of Eq. (7.2.4) for general boundary conditions at $x = 0, a$. We have

$$0 = \int_0^a \left[D_{11} \frac{d^2 W_n}{dx^2} \frac{d^2 \delta W_n}{dx^2} + 2\hat{D}_{12} \beta_n^2 \frac{dW_n}{dx} \frac{d\delta W_n}{dx} \right.$$
$$\left. + \left(D_{22} \beta_n^4 + k \right) W_n \delta W_n - q_n \delta W_n \right] dx$$
$$+ \left[\left(D_{11} \frac{d^3 W_n}{dx^3} - 2\hat{D}_{12} \beta_n^2 \frac{dW_n}{dx} \right) \delta W_n - D_{11} \frac{d^2 W_n}{dx^2} \frac{d\delta W_n}{dx} \right]_0^a$$
(7.3.1)

$\hat{D}_{12} = D_{12} + 2D_{66}$. For a simply supported edge without an applied edge moment $[W_n = 0$ and $(d^2 W_n/dx^2) = 0]$ or a clamped edge $[W_n = 0$ and $(dW_n/dx) = 0]$, the boundary terms in Eq. (7.3.1) vanish identically. However, for a free edge, the vanishing of bending moment and effective shear force require

$$D_{11} \frac{d^2 W_n}{dx^2} - D_{12} \beta_n^2 \, W_n = 0, \quad D_{11} \frac{d^3 W_n}{dx^3} - \bar{D}_{12} \beta_n^2 \frac{dW_n}{dx} = 0 \quad (7.3.2)$$

where $\bar{D}_{12} = D_{12} + 4D_{66}$. Hence, on a free edge, the boundary expression of Eq. (7.3.1) can be simplified to

$$\left[\left(D_{11} \frac{d^3 W_n}{dx^3} - 2\hat{D}_{12} \beta_n^2 \frac{dW_n}{dx} \right) \delta W_n - D_{11} \frac{d^2 W_n}{dx^2} \frac{d\delta W_n}{dx} \right]_0^a$$
$$= -\beta_n^2 D_{12} \left[W_n \frac{d\delta W_n}{dx} + \frac{dW_n}{dx} \delta W_n \right]_0^a \quad (7.3.3)$$

The weak forms for various plates are summarized here.

SSSS, CCSS, and CSSS plates:

$$0 = \int_0^a \left[D_{11} \frac{d^2 W_n}{dx^2} \frac{d^2 \delta W_n}{dx^2} + 2\hat{D}_{12} \beta_n^2 \frac{dW_n}{dx} \frac{d\delta W_n}{dx} \right.$$

$$\left. + \left(D_{22} \beta_n^4 + k \right) W_n \, \delta W_n - q_n \, \delta W_n \right] dx \quad (7.3.4)$$

CFSS and SFSS plates:

$$0 = \int_0^a \left[D_{11} \frac{d^2 W_n}{dx^2} \frac{d^2 \delta W_n}{dx^2} + 2\hat{D}_{12} \beta_n^2 \frac{dW_n}{dx} \frac{d\delta W_n}{dx} \right.$$

$$\left. + \left(D_{22} \beta_n^4 + k \right) W_n \, \delta W_n - q_n \, \delta W_n \right] dx$$

$$- \beta_n^2 D_{12} \left[W_n \frac{d\delta W_n}{dx} + \frac{dW_n}{dx} \delta W_n \right]_0^a \quad (7.3.5)$$

FFSS plate:

$$0 = \int_0^a \left[D_{11} \frac{d^2 W_n}{dx^2} \frac{d^2 \delta W_n}{dx^2} + 2\hat{D}_{12} \beta_n^2 \frac{dW_n}{dx} \frac{d\delta W_n}{dx} \right.$$

$$\left. + \left(D_{22} \beta_n^4 + k \right) W_n \, \delta W_n - q_n \, \delta W_n \right] dx$$

$$- \beta_n^2 D_{12} \left[W_n \frac{d\delta W_n}{dx} + \frac{dW_n}{dx} \delta W_n \right]_0^a \quad (7.3.6)$$

The Rayleigh–Ritz equations are obtained by substituting the approximation

$$W_n(x) \approx \sum_{j=1}^N c_j \varphi_j(x) \quad (7.3.7)$$

into the weak forms (7.3.4)–(7.3.6). We obtain

$$[R + B]^{(n)} \{c\}^{(n)} = \{F\}^{(n)} \quad (7.3.8)$$

where

$$R_{ij}^{(n)} = \int_0^a \left[D_{11} \frac{d^2 \varphi_i}{dx^2} \frac{d^2 \varphi_j}{dx^2} + 2\beta_n^2 \hat{D}_{12} \frac{d\varphi_i}{dx} \frac{d\varphi_j}{dx} + \left(D_{22} \beta_n^4 + k \right) \varphi_i \varphi_j \right] dx$$

$$(7.3.9a)$$

$$B_{ij}^{(n)} = -\beta_n^2 D_{12} \left[\varphi_j \frac{d\varphi_i}{dx} + \frac{d\varphi_j}{dx} \varphi_i \right]_0^a, \quad F_i^{(n)} = \int_0^a q_n \varphi_i \, dx \quad (7.3.9b)$$

336   THEORY AND ANALYSIS OF ELASTIC PLATES

The coefficients $B_{ij}^{(n)}$ are nonzero only for the CFSS, SFSS, and FFSS plates.

Limiting our choice of the approximation functions to algebraic polynomials and using the *geometric* boundary conditions, we can derive φ_i for various plates. The geometric boundary conditions and the approximation functions are presented below.

SSSS Plates:

$$W_n(0) = 0, \quad W_n(a) = 0 \tag{7.3.10}$$

$$\varphi_i = \left(\frac{x}{a}\right)^i - \left(\frac{x}{a}\right)^{i+1}$$

$$\frac{d\varphi_i}{dx} = \frac{1}{a}\left[i\left(\frac{x}{a}\right)^{i-1} - (i+1)\left(\frac{x}{a}\right)^i\right]$$

$$\frac{d^2\varphi_i}{dx^2} = \frac{1}{a^2}\left[i(i-1)\left(\frac{x}{a}\right)^{i-2} - i(i+1)\left(\frac{x}{a}\right)^{i-1}\right] \tag{7.3.11}$$

CCSS Plates:

$$W_n = 0, \quad \frac{dW_n}{dx} = 0 \text{ for } x = 0, a \tag{7.3.12}$$

$$\varphi_i = \left(\frac{x}{a}\right)^{i+1} - 2\left(\frac{x}{a}\right)^{i+2} + \left(\frac{x}{a}\right)^{i+3}$$

$$\frac{d\varphi_i}{dx} = \frac{1}{a}\left[(i+1)\left(\frac{x}{a}\right)^i - 2(i+2)\left(\frac{x}{a}\right)^{i+1} + (i+3)\left(\frac{x}{a}\right)^{i+2}\right]$$

$$\frac{d^2\varphi_i}{dx^2} = \frac{1}{a^2}\left[i(i+1)\left(\frac{x}{a}\right)^{i-1} - 2(i+1)(i+2)\left(\frac{x}{a}\right)^i \right.$$

$$\left. + (i+2)(i+3)\left(\frac{x}{a}\right)^{i+1}\right] \tag{7.3.13}$$

CSSS Plates:

$$W_n(0) = 0, \quad \frac{dW_n}{dx}(0) = 0, \quad W_n(a) = 0 \tag{7.3.14}$$

$$\varphi_i = \left(\frac{x}{a}\right)^{i+1} - \left(\frac{x}{a}\right)^{i+2}$$

BENDING OF PLATES WITH VARIOUS BOUNDARY CONDITIONS 337

$$\frac{d\varphi_i}{dx} = \frac{1}{a}\left[(i+1)\left(\frac{x}{a}\right)^i - (i+2)\left(\frac{x}{a}\right)^{i+1}\right]$$

$$\frac{d^2\varphi_i}{dx^2} = \frac{1}{a^2}\left[i(i+1)\left(\frac{x}{a}\right)^{i-1} - (i+1)(i+2)\left(\frac{x}{a}\right)^i\right] \quad (7.3.15)$$

CFSS Plates:

$$W_n(0) = 0, \quad \frac{dW_n}{dx}(0) = 0 \quad (7.3.16)$$

$$\varphi_i = \left(\frac{x}{a}\right)^{i+1}$$

$$\frac{d\varphi_i}{dx} = \frac{(i+1)}{a}\left(\frac{x}{a}\right)^i$$

$$\frac{d^2\varphi_i}{dx^2} = \frac{1}{a^2}\left[i(i+1)\left(\frac{x}{a}\right)^{i-1}\right] \quad (7.3.17)$$

SFSS Plates:

$$W_n(0) = 0 \quad (7.3.18)$$

$$\varphi_i = \left(\frac{x}{a}\right)^i$$

$$\frac{d\varphi_i}{dx} = \frac{i}{a}\left(\frac{x}{a}\right)^{i-1}$$

$$\frac{d^2\varphi_i}{dx^2} = \frac{1}{a^2}\left[i(i-1)\left(\frac{x}{a}\right)^{i-2}\right] \quad (7.3.19)$$

FFSS Plates: $(N > 1)$

$$\varphi_i = \left(\frac{x}{a}\right)^{i-1}$$

$$\frac{d\varphi_i}{dx} = \frac{(i-1)}{a}\left(\frac{x}{a}\right)^{i-2}$$

$$\frac{d^2\varphi_i}{dx^2} = \frac{1}{a^2}\left[(i-1)(i-2)\left(\frac{x}{a}\right)^{i-3}\right] \quad (7.3.20)$$

Table 7.3.1 contains numerical results obtained using the approximation functions in Eqs. (7.3.11), (7.3.13), (7.3.15), (7.3.17),

(7.3.19), and (7.3.20) for various plates under center point load Q_0 and uniformly distributed transverse load q_0. The results were obtained using $n = 1, 3, \cdots, 9$ and $N = 9$. The deflections and moments are nondimensionalized as follows:

$$\bar{w} = w_0(a/2, b/2)\left(\frac{D}{q_0 a^4}\right), \quad \bar{M} = M(a/2, b/2)\left(\frac{1}{q_0 a^2}\right) \quad (7.3.21a)$$

$$\hat{w} = w_0(a, b/2)\left(\frac{D}{q_0 a^4}\right), \quad \hat{M} = M(a, b/2)\left(\frac{1}{q_0 a^2}\right) \quad (7.3.21b)$$

$$\tilde{M} = M(0, b/2)\left(\frac{1}{q_0 a^2}\right) \quad (7.3.21c)$$

In the case of point load Q_0, the above nondimensionalizations hold with $q_0 a^2 = Q_0$.

Table 7.3.1. Transverse deflections and bending moments in isotropic square plates with various boundary conditions and subjected to central point load (CPL) or uniformly distributed load (UDL).

Plate	$\bar{w}$	$\hat{w}$	$\bar{M}_{xx}$	$\bar{M}_{yy}$	$\hat{M}_{yy}$	$\tilde{M}_{xx}$	$\tilde{M}_{yy}$
Uniformly Distributed Load (UDL) $(q_n = \frac{4q_0}{n\pi})$							
SSSS	0.00406	0.00000	0.0479	0.0479	0.0000	0.0000	0.0000
CCSS	0.00192	0.00000	0.0333	0.0244	-0.0210	-0.0699	-0.0210
CSSS	0.00279	0.00000	0.0392	0.0339	0.0210	-0.0839	-0.0252
CFSS	0.00567	0.01124	0.0280	0.0564	0.0973	-0.1185	-0.0355
SFSS	0.00793	0.01285	0.0390	0.0799	0.1118	0.0000	0.0000
FFSS	0.01309	0.01501	0.0271	0.1226	0.1312	0.0000	0.1312
Central Point Load (CPL) $[q_n = \frac{2Q_0}{b}\delta(x-\frac{a}{2})\sin\frac{n\pi}{2}]$							
SSSS	0.01145	0.00000	0.2557	0.2729	0.0000	0.0000	0.0000
CCSS	0.00695	0.00000	0.2398	0.2309	-0.0601	-0.2002	-0.0601
CSSS	0.00876	0.00000	0.2373	0.2429	0.0000	-0.1566	-0.0470
CFSS	0.01223	0.01363	0.2237	0.2700	0.1385	-0.1980	-0.0594
SFSS	0.01640	0.01664	0.2250	0.3026	0.1381	0.0000	0.0000
FFSS	0.02304	0.01942	0.2095	0.3576	0.1630	0.1630	-0.0494

7.3.2 Formulation for General Plates

In this section, we discuss applications of the Rayleigh–Ritz method to the bending of rectangular plates with various boundary conditions. The virtual work statement (or weak form) and the total potential energy expressions for an orthotropic rectangular plate are [see Eq. (3.4.10)]

$$0 = \int_0^b \int_0^a \left[D_{11} \frac{\partial^2 w_0}{\partial x^2} \frac{\partial^2 \delta w_0}{\partial x^2} + D_{12} \left(\frac{\partial^2 w_0}{\partial y^2} \frac{\partial^2 \delta w_0}{\partial x^2} + \frac{\partial^2 w_0}{\partial x^2} \frac{\partial^2 \delta w_0}{\partial y^2} \right) \right.$$
$$\left. + 4 D_{66} \frac{\partial^2 w_0}{\partial x \partial y} \frac{\partial^2 \delta w_0}{\partial x \partial y} + D_{22} \frac{\partial^2 w_0}{\partial y^2} \frac{\partial^2 \delta w_0}{\partial y^2} - q \delta w_0 \right] dx dy$$
$$- \int_\Gamma \left(-\hat{M}_{nn} \frac{\partial \delta w_0}{\partial n} + \hat{V}_n \delta w_0 \right) ds \qquad (7.3.22)$$

where $\hat{M}_{nn}$ and $\hat{V}_n$ are applied edge moment and effective shear force, respectively, on the boundary Γ. The total potential energy functional is given by

$$\Pi(w_0) = \frac{1}{2} \int_0^b \int_0^a \left[D_{11} \left(\frac{\partial^2 w_0}{\partial x^2} \right)^2 + 2 D_{12} \frac{\partial^2 w_0}{\partial x^2} \frac{\partial^2 w_0}{\partial y^2} + 4 D_{66} \left(\frac{\partial^2 w_0}{\partial x \partial y} \right)^2 \right.$$
$$\left. + D_{22} \left(\frac{\partial^2 w_0}{\partial y^2} \right)^2 - 2 q w_0 \right] dx dy - \int_\Gamma \left(-\hat{M}_{nn} \frac{\partial w_0}{\partial n} + \hat{V}_n w_0 \right) ds$$
$$(7.3.23)$$

We seek an N-parameter Rayleigh–Ritz solution in the form

$$w_0(x, y) \approx \sum_{j=1}^N c_j \varphi_j(x, y) \qquad (7.3.24)$$

and substitute it into Eq. (7.3.22) to obtain

$$[R]\{c\} = \{F\} \qquad (7.3.25)$$

where

$$R_{ij} = \int_0^b \int_0^a \left[D_{11} \frac{\partial^2 \varphi_i}{\partial x^2} \frac{\partial^2 \varphi_j}{\partial x^2} + D_{12} \left(\frac{\partial^2 \varphi_i}{\partial y^2} \frac{\partial^2 \varphi_j}{\partial x^2} + \frac{\partial^2 \varphi_i}{\partial x^2} \frac{\partial^2 \varphi_j}{\partial y^2} \right) \right.$$
$$\left. + 4 D_{66} \frac{\partial^2 \varphi_i}{\partial x \partial y} \frac{\partial^2 \varphi_j}{\partial x \partial y} + D_{22} \frac{\partial^2 \varphi_i}{\partial y^2} \frac{\partial^2 \varphi_j}{\partial y^2} \right] dx dy \qquad (7.3.26a)$$

$$F_i = \int_0^b \int_0^a q \varphi_i \, dx dy + \int_\Gamma \left(-\hat{M}_{nn} \frac{\partial \varphi_i}{\partial n} + \hat{V}_n \varphi_i \right) ds \qquad (7.3.26b)$$

In view of the rectangular geometry, it is convenient to express the Rayleigh–Ritz approximation in the form

$$w_0(x,y) \approx W_{mn}(x,y) = \sum_{i=1}^{N} \sum_{j=1}^{N} c_{ij}\, \varphi_{ij}(x,y) \tag{7.3.27}$$

and write $\varphi_{ij}(x,y)$ as a tensor product of the one-dimensional functions X_i and Y_j

$$\varphi_{ij}(x,y) = X_i(x) Y_j(y) \tag{7.3.28}$$

$i,j = 1, 2, \cdots, N$. The functions given in Eqs. (7.3.11), (7.3.13), (7.3.15), (7.3.17), (7.3.19), and (7.3.20) are candidates for X_i and Y_i. An alternative choice is provided by the characteristic equations of beams (see Table 4.5.1; set $\mu = \lambda$). These sets of functions are summarized below for typical boundary conditions (see Figure 7.3.1). In the case of characteristic polynomials, the roots λ_i are to be determined by solving a nonlinear equation. The first few values of $\lambda_i a$ are given in Table 4.5.1.

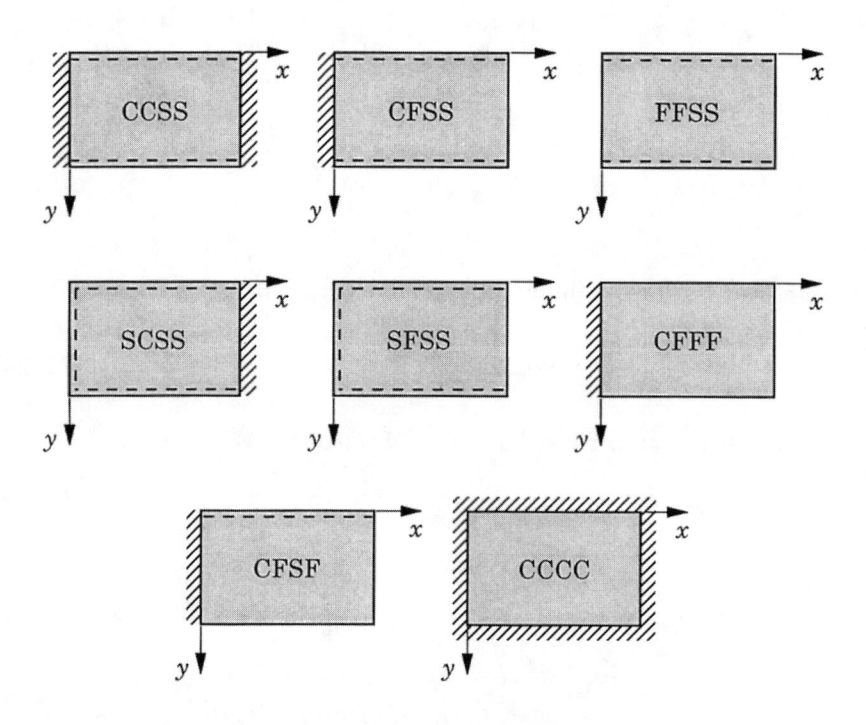

Figure 7.3.1. Rectangular plates with various boundary conditions along the four sides.

BENDING OF PLATES WITH VARIOUS BOUNDARY CONDITIONS   341

CCSS Plates:

Algebraic polynomials

$$X_i(x) = \left(\frac{x}{a}\right)^{i+1} - 2\left(\frac{x}{a}\right)^{i+2} + \left(\frac{x}{a}\right)^{i+3}$$
$$Y_j(x) = \left(\frac{y}{b}\right)^{j} - \left(\frac{y}{b}\right)^{j+1} \tag{7.3.29}$$

Characteristic polynomials

$$X_i(x) = \sin \lambda_i x - \sinh \lambda_i x + \alpha_i (\cosh \lambda_i x - \cos \lambda_i x)$$
$$Y_j(y) = \sin \frac{n\pi y}{b} \tag{7.3.30a}$$

where λ_i are the roots of the characteristic equation

$$\cos \lambda_i a \, \cosh \lambda_i a - 1 = 0 \tag{7.3.30b}$$

and α_i are defined by

$$\alpha_i = \frac{\sinh \lambda_i a - \sin \lambda_i a}{\cosh \lambda_i a - \cos \lambda_i a} \tag{7.3.30c}$$

CFSS Plates:

Algebraic polynomials

$$X_i(x) = \left(\frac{x}{a}\right)^{i+1}, \quad Y_j(x) = \left(\frac{y}{b}\right)^{j} - \left(\frac{y}{b}\right)^{j+1} \tag{7.3.31}$$

Characteristic polynomials

$$X_i(x) = \sin \lambda_i x - \sinh \lambda_i x + \alpha_i (\cosh \lambda_i x - \cos \lambda_i x)$$
$$Y_j(y) = \sin \frac{n\pi y}{b} \tag{7.3.32a}$$

$$\cos \lambda_i a \, \cosh \lambda_i a + 1 = 0, \quad \alpha_i = \frac{\sinh \lambda_i a + \sin \lambda_i a}{\cosh \lambda_i a + \cos \lambda_i a} \tag{7.3.32b}$$

FFSS Plates:

Algebraic polynomials $(M > 1)$

$$X_i(x) = \left(\frac{x}{a}\right)^{i-1}, \quad Y_j(x) = \left(\frac{y}{b}\right)^j - \left(\frac{y}{b}\right)^{j+1} \qquad (7.3.33)$$

Characteristic polynomials

$$\begin{aligned} X_i(x) &= \sin \lambda_i x + \sinh \lambda_i x - \alpha_i \left(\cosh \lambda_i x + \cos \lambda_i x\right) \\ Y_j(y) &= \sin \frac{n\pi y}{b} \end{aligned} \qquad (7.3.34a)$$

$$\cos \lambda_i a \, \cosh \lambda_i a - 1 = 0, \quad \alpha_i = \frac{\sinh \lambda_i a - \sin \lambda_i a}{\cosh \lambda_i a - \cos \lambda_i a} \qquad (7.3.34b)$$

SCSS Plates:

Algebraic polynomials

$$X_i(x) = \left(\frac{x}{a}\right)\left[1 - \left(\frac{x}{a}\right)\right]^{i+1}, \quad Y_j(x) = \left(\frac{y}{b}\right)^j - \left(\frac{y}{b}\right)^{j+1} \qquad (7.3.35)$$

Characteristic polynomials

$$X_i(x) = \sinh \lambda_i a \sin \lambda_i x + \sin \lambda_i a \sinh \lambda_i x, \quad Y_j(y) = \sin \frac{n\pi y}{b} \qquad (7.3.36a)$$

$$\tan \lambda_i a - \tanh \lambda_i a = 0 \qquad (7.3.36b)$$

SFSS Plates:

Algebraic polynomials

$$X_i(x) = \left(\frac{x}{a}\right)^i, \quad Y_j(x) = \left(\frac{y}{b}\right)^j - \left(\frac{y}{b}\right)^{j+1} \qquad (7.3.37)$$

Characteristic polynomials

$$X_i(x) = \sinh \lambda_i a \, \sin \lambda_i x - \sin \lambda_i a \, \sinh \lambda_i x, \quad Y_j(y) = \sin \frac{n\pi y}{b} \qquad (7.3.38a)$$

$$\tan \lambda_i a - \tanh \lambda_i a = 0 \qquad (7.3.38b)$$

CFFF Plates: ($N > 1$)

Algebraic polynomials

$$X_i(x) = \left(\frac{x}{a}\right)^{i+1}, \quad Y_j(y) = \left(\frac{y}{b}\right)^{j-1} \qquad (7.3.39)$$

Characteristic polynomials

$$X_i(x) = \sin \lambda_i x - \sinh \lambda_i x + \alpha_i (\cosh \lambda_i x - \cos \lambda_i x)$$
$$Y_j(y) = \sin \mu_j y + \sinh \mu_j y - \beta_i (\cosh \mu_j y + \cos \mu_j y) \quad (7.3.40a)$$

$$\cos \lambda_i a \cosh \lambda_i a + 1 = 0, \quad \cos \mu_j b \cosh \mu_j b - 1 = 0 \qquad (7.3.40b)$$

$$\alpha_i = \frac{\sinh \lambda_i a + \sin \lambda_i a}{\cosh \lambda_i a + \cos \lambda_i a}, \quad \beta_j = \frac{\sinh \mu_j b - \sin \mu_i b}{\cosh \mu_j b - \cos \mu_j b} \qquad (7.3.40c)$$

CFSF Plates:

Algebraic polynomials

$$X_i(x) = \left(\frac{x}{a}\right)^{i+1}, \quad Y_j(y) = \left(\frac{y}{b}\right)^{j} \qquad (7.3.41)$$

Characteristic polynomials

$$X_i(x) = \sin \lambda_i x - \sinh \lambda_i x + \alpha_i (\cosh \lambda_i x - \cos \lambda_i x)$$
$$Y_j(y) = \sinh \mu_j b \, \sin \mu_j y - \sin \mu_j b \, \sinh \mu_j y \qquad (7.3.42a)$$

$$\cos \lambda_i a \cosh \lambda_i a + 1 = 0, \quad \tan \mu_j b - \tanh \mu_j b = 0 \qquad (7.3.42b)$$

$$\alpha_i = \frac{\sinh \lambda_i a + \sin \lambda_i a}{\cosh \lambda_i a + \cos \lambda_i a} \qquad (7.3.42c)$$

CCCC Plates:

Algebraic polynomials

$$X_i(x) = \left(\frac{x}{a}\right)^{i+1} - 2\left(\frac{x}{a}\right)^{i+2} + \left(\frac{x}{a}\right)^{i+3} \qquad (7.3.43a)$$

$$Y_j(y) = \left(\frac{y}{b}\right)^{j+1} - 2\left(\frac{y}{b}\right)^{j+2} + \left(\frac{y}{b}\right)^{j+3} \qquad (7.3.43b)$$

Characteristic polynomials

$$X_i(x) = \sin \lambda_i x - \sinh \lambda_i x + \alpha_i (\cosh \lambda_i x - \cos \lambda_i x)$$
$$Y_j(y) = \sin \lambda_j y - \sinh \lambda_j y + \alpha_j (\cosh \lambda_j y - \cos \lambda_j y) \quad (7.3.44a)$$

$$\cos \lambda_i a \cosh \lambda_i a - 1 = 0 \quad (7.3.44b)$$

$$\alpha_i = \frac{\sinh \lambda_i a - \sin \lambda_i a}{\cosh \lambda_i a - \cos \lambda_i a} = \frac{\cosh \lambda_i a - \cos \lambda_i a}{\sinh \lambda_i a + \sin \lambda_i a} \quad (7.3.44c)$$

The examples given above indicate how one can construct the approximation functions for any combination of fixed, hinged, and free boundary conditions on the four edges of a rectangular plate. The difficult task is to evaluate the integrals of these functions as required in Eq. (7.3.26a,b). One may use a symbolic manipulator, such as *Mathematica* or *Maple*, to evaluate the integrals. In general, the Rayleigh–Ritz method for general rectangular plates with arbitrary boundary conditions is algebraically more complicated than a numerical method, such as the finite element method.

7.3.3 Clamped Plates (CCCC)

For clamped plates, the Navier or Lévy methods cannot be used. As an example of the application of the Rayleigh–Ritz method to plates with arbitrary boundary conditions, we consider a CCCC plate subjected to distributed transverse load $q(x,y)$. The approximation functions φ_{ij} of Eqs. (7.3.43a,b) or (7.3.44a–c) may be used. Substituting Eq. (7.3.27) into Eq. (7.3.22) (with $\delta w_{0,n} = \delta w_0 = 0$ on Γ), we obtain

$$0 = \sum_{i=1}^{m} \sum_{j=1}^{n} \left\{ \int_0^b \int_0^a \left[D_{11} \frac{d^2 X_i}{dx^2} Y_j \frac{d^2 X_p}{dx^2} Y_q + 4 D_{66} \frac{dX_i}{dx} \frac{dY_j}{dy} \frac{dX_p}{dx} \frac{dY_q}{dy} \right. \right.$$
$$+ D_{12} \left(X_i \frac{d^2 Y_j}{dy^2} \frac{d^2 X_p}{dx^2} Y_q + \frac{d^2 X_i}{dx^2} Y_j X_p \frac{d^2 Y_q}{dy^2} \right)$$
$$\left. \left. + D_{22} X_i \frac{d^2 Y_j}{dy^2} X_p \frac{d^2 Y_q}{dy^2} \right] dx dy \right\} c_{ij}$$
$$- \int_0^b \int_0^a q X_p Y_q \, dx dy \quad (7.3.45)$$

Equation (7.3.45) represents $m \times n$ algebraic equations among the coefficients c_{ij}. Note that all integrals in (7.3.45) are line integrals, and they involve evaluating the following five different integrals:

$$\int_0^a X_i\,dx, \quad \int_0^a X_i X_p\,dx, \quad \int_0^a \frac{dX_i}{dx}\frac{dX_p}{dx}\,dx$$

$$\int_0^a X_i \frac{d^2 X_p}{dx^2}\,dx, \quad \int_0^a \frac{d^2 X_i}{dx^2}\frac{d^2 X_p}{dx^2}\,dx \qquad (7.3.46)$$

First we consider the algebraic functions in Eqs. (7.3.43a,b) with $m = n = 1$ and $q = q_0$ (uniformly distributed load of intensity q_0). The integrals in Eq. (7.3.46) for this case are given by

$$\int_0^a X_1\,dx = \frac{a}{30}, \quad \int_0^a X_1 X_1\,dx = \frac{a}{630}, \quad \int_0^a \frac{dX_1}{dx}\frac{dX_1}{dx}\,dx = \frac{2}{105a}$$

$$\int_0^a X_1 \frac{d^2 X_1}{dx^2}\,dx = -\frac{2}{105a}, \quad \int_0^a \frac{d^2 X_1}{dx^2}\frac{d^2 X_1}{dx^2}\,dx = \frac{4}{5a^3} \qquad (7.3.47)$$

and Eq. (7.3.45) takes the form

$$0 = \left[\left(\frac{4}{5a^3}\right)\left(\frac{b}{630}\right)D_{11} + 4D_{66}\left(\frac{2}{105a}\right)\left(\frac{2}{105b}\right)\right.$$
$$\left. + 2D_{12}\left(-\frac{2}{105a}\right)\left(-\frac{2}{105b}\right) + \left(\frac{a}{630}\right)\left(\frac{4}{5b^3}\right)D_{22}\right]c_{11} - \left(\frac{ab}{900}\right)q_0$$

or

$$\left[\frac{7}{a^4}D_{11} + \frac{4}{a^2 b^2}(D_{12} + 2D_{66}) + \frac{7}{b^4}D_{22}\right]c_{11} = \frac{49}{8}q_0 \qquad (7.3.48)$$

and the one-parameter Rayleigh–Ritz solution becomes

$$W_{11}(x,y) = \left(\frac{49}{8}\right)\frac{q_0 a^4 \left[\frac{x}{a} - \left(\frac{x}{a}\right)^2\right]^2 \left[\frac{y}{b} - \left(\frac{y}{b}\right)^2\right]^2}{7D_{11} + 4(D_{12} + 2D_{66})s^2 + 7D_{22}s^4} \qquad (7.3.49)$$

where $s = a/b$ denotes the plate aspect ratio. The maximum deflection occurs at $x = a/2$ and $y = b/2$:

$$W_{11}\left(\frac{a}{2}, \frac{b}{2}\right) = 0.00342 \frac{q_0 a^4}{D_{11} + 0.5714(D_{12} + 2D_{66})s^2 + D_{22}s^4} \qquad (7.3.50)$$

Next we use the characteristic functions in Eqs. (7.3.44a,b) for $m = n = 1$

$$X_1(x) = \sin\frac{4.73x}{a} - \sinh\frac{4.73x}{a} + 1.0178\left(\cosh\frac{4.73x}{a} - \cos\frac{4.73x}{a}\right)$$

$$Y_1(y) = \sin\frac{4.73y}{b} - \sinh\frac{4.73y}{b} + 1.0178\left(\cosh\frac{4.73y}{b} - \cos\frac{4.73y}{b}\right)$$

$$(7.3.51)$$

Evaluating the integrals in Eq. (7.3.46), we obtain

$$\int_0^a X_1 \, dx = 0.84555a, \quad \int_0^a X_1 X_1 \, dx = 1.035966a$$

$$\int_0^a \frac{dX_1}{dx}\frac{dX_1}{dx} \, dx = \frac{12.7442}{a}, \quad \int_0^a X_1 \frac{d^2 X_1}{dx^2} \, dx = -\frac{12.7442}{a}$$

$$\int_0^a \frac{d^2 X_1}{dx^2}\frac{d^2 X_1}{dx^2} \, dx = \frac{518.531}{a^3}$$

and Eq. (7.3.45) becomes

$$\left[\frac{537.181 b}{a^3} D_{11} + \frac{324.829}{ab}(D_{12} + 2D_{66}) + \frac{537.181 a}{b^3} D_{22}\right] c_{11} = 0.715 q_0 ab$$

The maximum deflection is given by ($X_1(a/2) = Y_1(b/2) = 1.6164$)

$$W_{11}(\frac{a}{2},\frac{b}{2}) = 0.00348 \frac{q_0 a^4}{D_{11} b^4 + 0.6047(D_{12} + 2D_{66})s^2 + D_{22}s^4} \quad (7.3.52)$$

For an isotropic square plate, the maximum deflection in Eq. (7.3.52) becomes $W_{11}(a/2, b/2) = 0.00134(q_0 a^4/D)$, whereas Eq. (7.3.50) gives $W_{11}(a/2, b/2) = 0.00133(q_0 a^4/D)$. The "exact" solution (see [1], p. 202) is $W_{11}(a/2, b/2) = 0.00126(q_0 a^4/D)$.

Exercises

7.1 Write the necessary algebraic equations to determine the constants A_n, B_n, C_n, and D_n in the solution (7.2.31a) of a rectangular plate with edges $x = 0, a$ free and edges $y = 0, b$ simply supported (FFSS), and subjected to loads in which $\hat{q}_n$ is a constant. Note that conditions in Eq. (7.2.38b) must be satisfied at $x = 0, a$. *Ans:* From Eqs. (7.2.34), (7.2.35), (7.2.39), and (7.3.40), we have

$$D_{11}\left[\left(\lambda_1^2 - \lambda_2^2\right) A_n + 2\lambda_1 \lambda_2 D_n\right] - D_{12}\beta_n^2\left[\frac{q_n}{D_{22}\beta_n^4 + k} + A_n\right] = 0$$

$$D_{11}\left[\left(\lambda_1^2 - \lambda_2^2\right)(\lambda_2 B_n + \lambda_1 C_n) + 2\lambda_1 \lambda_2 (\lambda_1 B_n - \lambda_2 C_n)\right]$$

$$-\bar{D}_{12}\beta_n^2\left(\lambda_2 B_n + \lambda_1 C_n\right) = 0$$

BENDING OF PLATES WITH VARIOUS BOUNDARY CONDITIONS 347

The remaining two relations are provided by Eqs. (7.2.41b,c).

7.2 Repeat Exercise 7.1 for the case of plates with hydrostatic loading of the type $q = q_0(x/a)$.

7.3 Determine a one-parameter Rayleigh–Ritz solution of an isotropic rectangular plate with all edges clamped and subjected to uniformly distributed transverse load q_0. Use the approximation

$$W(x,y) = c_{11}\left(1 - \cos\frac{2\pi x}{a}\right)\left(1 - \cos\frac{2\pi y}{b}\right) \quad \text{(a)}$$

Determine the maximum deflection of a square plate. *Ans:*

$$c_{11} = \frac{q_0 a^4}{4\pi^4 D}\left(\frac{1}{3 + 2s^2 + 3s^4}\right), \quad s = \frac{a}{b} \quad \text{(b)}$$

The maximum deflection for $a = b$ is $w_{max} = 0.00128(q_0 a^4/D)$.

7.4 Determine a one-parameter Rayleigh–Ritz solution of an isotropic rectangular plate with all edges clamped and subjected to a central point load Q_0 using approximation in Eq. (a) of Exercise 7.3 and find the maximum deflection of a square plate. *Ans:*

$$c_{11} = \frac{Q_0 a^4}{ab\pi^4 D}\left(\frac{1}{3 + 2s^2 + 3s^4}\right), \quad s = \frac{a}{b} \quad \text{(a)}$$

The maximum deflection for $a = b$ is $w_{max} = 0.00513(Q_0 a^2/D)$.

7.5 Determine a one-parameter Rayleigh–Ritz solution of an isotropic rectangular plate with edges $x = 0, a$ clamped and edges $y = 0, b$ simply supported, and subjected to uniformly distributed transverse load q_0. Use the approximation

$$W(x,y) = c_{11}\left(1 - \cos\frac{2\pi x}{a}\right)\sin\frac{\pi y}{b} \quad \text{(a)}$$

to determine the maximum deflection of a square plate. *Ans:*

$$c_{11} = \frac{8q_0 a^4}{\pi^5 D}\left(\frac{1}{16 + 8s^2 + 3s^4}\right), \quad s = \frac{a}{b} \quad \text{(b)}$$

The maximum deflection for $a = b$ is $w_{max} = 0.00194(q_0 a^4/D)$.

7.6 Determine a one-parameter Rayleigh–Ritz solution of an isotropic rectangular plate with edges $x = 0, a$ clamped and edges $y = 0, b$

simply supported, and subjected to a center point load Q_0 transverse load. Use the approximation in Eq. (a) of Exercise 7.5. Determine the maximum deflection of a square plate. *Ans:*

$$c_{11} = \frac{8Q_0 a^3}{b\pi^4 D}\left(\frac{1}{16 + 8s^2 + 3s^4}\right), \quad s = \frac{a}{b} \qquad (a)$$

The maximum deflection is $w_{max} = 0.00610(Q_0 a^2/D)$.

7.7 Give the functions X_i and Y_j of Eq. (7.3.28) for a rectangular plate with edges $x = 0, a$ and $y = 0$ clamped and edge $y = b$ simply supported.

7.8 Give the functions X_i and Y_j of Eq. (7.3.28) for a rectangular plate with edges $x = 0$ clamped and the remaining edges free.

7.9 Give the functions X_i and Y_j of Eq. (7.3.28) for a rectangular plate with edges $x = 0, a$ and $y = 0$ clamped and edge $y = b$ free.

7.10 Give the functions X_i and Y_j of Eq. (7.3.28) for a rectangular plate with edges $x = 0$ and $y = 0$ clamped and edges $x = a$ and $y = b$ free.

Exercises on Non-Rectangular Plates

7.11 Consider an equilateral triangular plate (see Figure P7.11). Suppose that all edges are simply supported. The equations of the edges, with respect to the coordinate system shown, are

$$\frac{x}{a} + \frac{1}{3} = 0 \quad \text{on BC}, \qquad \frac{1}{\sqrt{3}}\frac{x}{a} + \frac{y}{a} - \frac{2}{3\sqrt{3}} = 0 \quad \text{on AC}$$

$$\frac{1}{\sqrt{3}}\frac{x}{a} - \frac{y}{a} - \frac{2}{3\sqrt{3}} = 0 \quad \text{on AB}$$

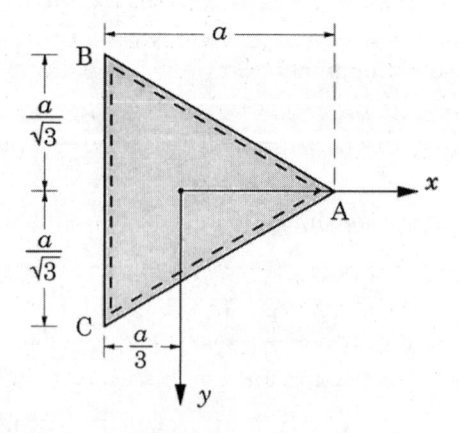

Figure P7.11

Use a one-parameter Rayleigh–Ritz approximation of the form

$$w_0(x,y) = c_1 \left(\frac{x}{a} + \frac{1}{3}\right)\left(\frac{1}{\sqrt{3}}\frac{x}{a} + \frac{y}{a} - \frac{2}{3\sqrt{3}}\right)\left(\frac{1}{\sqrt{3}}\frac{x}{a} - \frac{y}{a} - \frac{2}{3\sqrt{3}}\right)$$

$$= d_1\left[\frac{4}{27} - \left(\frac{x}{a}\right)^2 - \left(\frac{y}{a}\right)^2 - 3\left(\frac{x}{a}\right)\left(\frac{y}{a}\right)^2 + \left(\frac{x}{a}\right)^3\right]$$

$$\equiv d_1 \varphi_1(x,y) \qquad \text{(a)}$$

and determine $d_1 = c_1/3$ when the plate is subjected to uniformly distributed transverse load q_0. The exact deflection of an isotropic plate is given by (can be verified by substitution into the governing equation)

$$w_0(x,y) = \frac{q_0 a^4}{64D}\left[\frac{4}{27} - \left(\frac{x}{a}\right)^2 - \left(\frac{y}{a}\right)^2 - 3\left(\frac{x}{a}\right)\left(\frac{y}{a}\right)^2 + \left(\frac{x}{a}\right)^3\right]$$

$$\times \left[\frac{4}{9} - \left(\frac{x}{a}\right)^2 - \left(\frac{y}{a}\right)^2\right] \qquad \text{(c)}$$

The exact maximum deflection for an isotropic plate is $w_{max} = w_0(0,0) = (q_0 a^4 / 972 D)$.

7.12 Consider the equilateral triangular plate of Figure P7.11. Suppose that all edges are simply supported and loaded by a uniform moment $\hat{M}_{nn} = M_0$ along its edges. Use a one-parameter Rayleigh–Ritz solution of Exercise 7.11 and determine F_1 (and R_{11} is known from Exercise 7.11). Show that

$$w_0(x,y) = \frac{M_0 a^2}{4D}\left[\frac{4}{27} - \left(\frac{x}{a}\right)^2 - \left(\frac{y}{a}\right)^2 - 3\left(\frac{x}{a}\right)\left(\frac{y}{a}\right)^2 + \left(\frac{x}{a}\right)^3\right] \qquad \text{(a)}$$

which is the exact deflection. *Hint:* Evaluate the line integral

$$F_1 = \oint_\Gamma M_0 \left(n_x \frac{\partial \varphi_1}{\partial x} + n_y \frac{\partial \varphi_1}{\partial y}\right) ds$$

where Γ denotes the boundary of the triangle and (n_x, n_y) are the direction cosines of the line segment.

7.13 Determine the deflection surface $w_0(x,y)$ of the simply supported triangular plate of Figure P7.11 using a one-parameter Rayleigh–Ritz approximation in Eq. (a) of Exercise 7.11 when the plate is loaded by a point load Q_0 at the centroid $(x,y) = (0,0)$. Assume that the plate is made of an isotropic material. *Ans:* $d_1 = (Q_0 a^2 / 36\sqrt{3} D)$.

7.14 Consider an isotropic elliptic plate with major and minor axes $2a$ and $2b$, respectively. Obtain a one-parameter Galerkin solution for the case in which the plate is clamped and subjected to uniformly distributed load q_0.

Hint: Take

$$\varphi_1(x,y) = \left(1 - \frac{x^2}{a^2} - \frac{y^2}{b^2}\right)^2 \tag{a}$$

which satisfies the geometric boundary conditions

$$w_0 = \frac{\partial w_0}{\partial x} = \frac{\partial w_0}{\partial y} = 0 \tag{b}$$

Ans: The Galerkin solution, which coincides with the exact solution, is given by

$$w_0(x,y) = \frac{q_0 a^4}{8D(3s^4 + 2s^2 + 3)} \left(1 - \frac{x^2}{a^2} - \frac{y^2}{b^2}\right)^2 \tag{c}$$

where $s = a/b$. When $a = b$, the solution reduces to that of a circular plate.

7.15 Repeat Exercise 7.14 for the case in which the plate is subjected to distributed load $q = q_0(x/a)$. Ans: The solution is

$$w_0(x,y) = \frac{q_0 a^4 x}{24D(s^4 + 2s^2 + 5)} \left(1 - \frac{x^2}{a^2} - \frac{y^2}{b^2}\right)^2 \tag{a}$$

7.16 Determine a one-parameter Rayleigh–Ritz solution of a simply supported, isotropic elliptic plate under uniformly distributed load.

Hint: Use

$$\varphi_1(x,y) = \left(1 - \frac{x^2}{a^2} - \frac{y^2}{b^2}\right) \tag{a}$$

which satisfies the geometric boundary condition $w_0 = 0$. Ans: The Rayleigh–Ritz solution, which *does not* coincide with the exact solution, is given by ($s = a/b$)

$$w_0(x,y) = \frac{q_0 a^4}{8D(1 + 2\nu s^2 + s^4)} \left(1 - \frac{x^2}{a^2} - \frac{y^2}{b^2}\right) \tag{b}$$

7.17 *Solution by State-Space Approach.* An alternative method of solving Eq. (7.2.4) is provided by the state-space approach [6,7]. The approach involves writing a pth-order ordinary differential equation as a set of p first-order equations in matrix form, and then its solution is obtained in terms of the eigenvalues of the matrix operator. In the present case, Eq. (7.2.4) is a fourth-order equation with constant coefficients that can be expressed in terms of four first-order equations. Let

$$Z_1 = W_n, \quad Z_2 = Z_1' = W_n', \quad Z_3 = Z_2' = W_n'', \quad Z_4 = Z_3' = W_n''' \quad \text{(a)}$$

Then we have

$$Z_1' = Z_2, \quad Z_2' = Z_3, \quad Z_3' = Z_4, \quad Z_4' = C_1 Z_1 + C_2 Z_3 + \hat{q}_n \quad \text{(b)}$$

or in matrix form

$$\{Z'\} = [T]\{Z\} + \{F\} \quad \text{(c)}$$

where

$$\{Z\} = \begin{Bmatrix} W_n \\ W_n' \\ W_n'' \\ W_n''' \end{Bmatrix}, \quad [T] = \begin{bmatrix} 0 & 1 & 0 & 0 \\ 0 & 0 & 1 & 0 \\ 0 & 0 & 0 & 1 \\ C_1 & 0 & C_2 & 0 \end{bmatrix}, \quad \{F\} = \begin{Bmatrix} 0 \\ 0 \\ 0 \\ \hat{q}_n \end{Bmatrix} \quad \text{(d)}$$

$$C_1 = -\frac{1}{D_{11}}\left(D_{22}\beta_n^4 + k\right), \quad C_2 = 2\frac{\hat{D}_{12}}{D_{11}}\beta_n^2, \quad \tilde{q}_n = \frac{\bar{q}_n}{D_{11}} \quad \text{(e)}$$

The general solution of Eq. (c) is given by

$$\mathbf{Z}(x) = e^{\mathbf{T}x}\left(\mathbf{K} + \int^x e^{-\mathbf{T}\xi}\,\mathbf{F}(\xi)\,d\xi\right)$$
$$\equiv \mathbf{G}(x)\mathbf{K} + \mathbf{H}(x) \quad \text{(f)}$$

Here $e^{\mathbf{T}x}$ denotes the matrix product

$$e^{\mathbf{T}x} = [E] \begin{bmatrix} e^{\lambda_1 x} & & 0 \\ & \ddots & \\ 0 & & e^{\lambda_4 x} \end{bmatrix} [E]^{-1} \quad \text{(g)}$$

Here $[E]$ is the matrix of distinct eigenvectors of matrix $[T]$, $[E]^{-1}$ denotes its inverse, λ_j ($j = 1, 2, 3, 4$) are the eigenvalues associated with matrix $[T]$, and $\{K\}$ is a vector of constants to be determined using the boundary conditions of the problem.

Consider an orthotropic, simply supported, rectangular plate ($k = 0$) subjected to uniformly distributed load of intensity q_0. The simply supported boundary conditions at $x = 0, a$ imply

$$W_n = 0, \quad -D_{11}W_n'' + D_{12}\beta_n^2 W_n = 0 \tag{h}$$

Show that these four conditions yield the following four algebraic equations among K_i ($i = 1, 2, 3, 4$):

$$\sum_{j=1}^{4} G_{1j}(0)K_j + H_1(0) = 0, \quad \sum_{j=1}^{4} G_{1j}(a)K_j + H_1(a) = 0$$

$$\sum_{j=1}^{4} \left(D_{11}G_{3j}(0) - \beta^2 D_{12}G_{1j}(0)\right) K_j + D_{11}H_3(0) - D_{12}\beta^2 H_1(0) = 0$$

$$\sum_{j=1}^{4} \left(D_{11}G_{3j}(a) - \beta^2 D_{12}G_{1j}(a)\right) K_j + D_{11}H_3(a) - D_{12}\beta^2 H_1(a) = 0$$

References for Additional Reading

1. Timoshenko, S. P. and Woinowsky-Krieger, S., *Theory of Plates and Shells*, McGraw–Hill, Singapore (1970).

2. Szilard, R., *Theory and Analysis of Plates. Classical and Numerical Methods*, Prentice–Hall, Englewood Cliffs, NJ (1974).

3. Ugural, A. C., *Stresses in Plates and Shells*, McGraw–Hill, New York (1981).

4. Reddy, J. N., *Energy and Variational Methods in Applied Mechanics*, John Wiley & Sons, New York (1984).

5. Reddy, J. N., *Mechanics of Laminated Composite Plates: Theory and Analysis*, CRC Press, Boca Raton, FL (1997).

6. Franklin, J. N., *Matrix Theory*, Prentice–Hall, Englewood Cliffs, NJ (1968).

7. Nosier, A. and Reddy, J. N., "Vibration and Stability Analyses of Cross-Ply Laminated Circular Cylindrical Shells", *Journal of Sound and Vibration*, **157**(1), 139–159 (1992).

Chapter Eight

General Buckling of Rectangular Plates

8.1 Buckling of Simply Supported Plates Under Compressive Loads

8.1.1 Governing Equations

When a plate is subjected to uniform compressive forces applied in the middle plane of the plate, and if the forces are sufficiently small, the force-displacement response is linear. The linear relationship holds until a certain load is reached. At that load, called the *buckling load*, the stable state of the plate is disturbed and the plate seeks an alternative equilibrium configuration accompanied by a change in the load-deflection behavior. The phenomenon of changing the equilibrium configuration at the same load and without drastic changes in deformation is termed *bifurcation*. The load-deflection curve for buckled plates is often bilinear. The magnitude of the buckling load depends, as will be shown shortly, on geometry, material properties, and the buckling mode shape, i.e., geometric configuration of the plate at buckling.

In the present study, we assume that the only applied loads are the uniform in-plane forces and that all other mechanical and thermal loads are zero. Since the prebuckling deformation w_0 is that of an equilibrium configuration, it satisfies the equilibrium equations, and the equation governing buckling deflection w is given by

$$D_{11}\frac{\partial^4 w}{\partial x^4} + 2\hat{D}_{12}\frac{\partial^4 w}{\partial x^2 \partial y^2} + D_{22}\frac{\partial^4 w}{\partial y^4} + kw = \hat{N}_{xx}\frac{\partial^2 w}{\partial x^2} + \hat{N}_{yy}\frac{\partial^2 w}{\partial y^2} \quad (8.1.1)$$

where $\hat{D}_{12} = D_{12} + 2D_{66}$, and $\hat{N}_{xx} < 0$ and $\hat{N}_{yy} < 0$ for compressive forces. Here we wish to determine a nonzero deflection w that satisfies

Eq. (8.1.1) when the in-plane forces are (see Figure 8.1.1)

$$\hat{N}_{xx} = -N_0, \quad \hat{N}_{yy} = -\gamma N_0, \quad \gamma = \frac{\hat{N}_{yy}}{\hat{N}_{xx}} \qquad (8.1.2)$$

and the edges are simply supported.

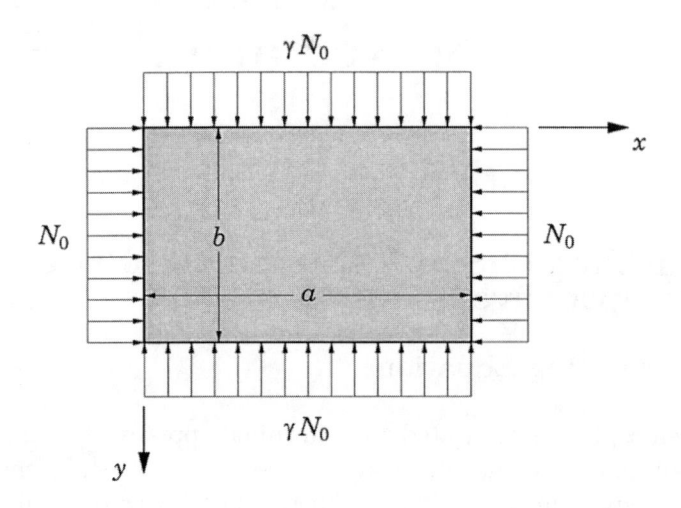

Figure 8.1.1. Biaxial compression of a rectangular plate ($\hat{N}_{xx} = -N_0$ and $N_{yy}^0 = -\gamma N_0$).

8.1.2 The Navier Solution

As in the case of bending, we select an expansion for w that satisfies the boundary conditions

$$w(0,y) = 0, \quad w(a,y) = 0, \quad w(x,0) = 0, \quad w(x,b) = 0 \qquad (8.1.3)$$
$$M_{xx}(0,y) = 0, \quad M_{xx}(a,y) = 0, \quad M_{yy}(x,0) = 0, \quad M_{yy}(x,b) = 0 \qquad (8.1.4)$$

We take

$$w(x,y) = W_{mn} \sin \alpha_m x \, \sin \beta_n y, \quad \alpha_m = \frac{m\pi}{a}, \quad \beta_n = \frac{n\pi}{b} \qquad (8.1.5)$$

Substituting Eq. (8.1.5) into Eq. (8.1.1), we obtain for any m and n the relation

$$0 = \left\{ \left[D_{11}\alpha_m^4 + 2\hat{D}_{12}\alpha_m^2\beta^2 + D_{22}\beta_n^4 + k \right] - (\alpha_m^2 + \gamma\beta_n^2)N_0 \right\}$$
$$\times W_{mn} \, \sin \alpha_m x \, \sin \beta_n y \qquad (8.1.6)$$

Since Eq. (8.1.6) must hold for every point (x,y) of the domain for nontrivial buckling mode $w(x,y)$ (i.e., $W_{mn} \neq 0$), the expression inside the braces should be zero for every m and n. This yields

$$N_0(m,n) = \frac{\pi^2}{b^2}\left(\frac{D_{11}s^4m^4 + 2\hat{D}_{12}s^2m^2n^2 + D_{22}n^4 + \bar{k}}{s^2m^2 + \gamma n^2}\right) \quad (8.1.7)$$

where s is the plate aspect ratio and k is the elastic foundation modulus

$$s = \frac{b}{a}, \quad \bar{k} = \frac{kb^4}{\pi^4} \quad (8.1.8)$$

Thus, for each choice of m and n there corresponds a unique value of N_0. The *critical buckling load* is the smallest of $N_0(m,n)$. For a given plate this value is dictated by a particular combination of the values of m and n, value of γ, plate geometry, and material properties. It is difficult to determine the critical buckling load in the most general case. Next we investigate critical buckling loads of various plates.

8.1.3 Biaxial Compression of a Plate

For a square orthotropic plate subjected to the same magnitude of uniform compressive forces N_{xx} and N_{yy} on both edges (i.e., biaxial compression with $\gamma = 1$; see Figure 8.1.2), Eq. (8.1.7) yields

$$N_0(m,n) = \frac{\pi^2}{a^2}\left(\frac{m^4 D_{11} + 2m^2n^2\hat{D}_{12} + n^4 D_{22} + \bar{k}}{m^2 + n^2}\right) \quad (8.1.9)$$

where $\bar{k} = ka^4/\pi^4$. Now suppose that $D_{11} > D_{22}$. Then $D_{11}m^2$ increases more rapidly than the decrease in D_{22}/m^2 with an increase in m. Thus, the minimum of N_0 occurs when $m = 1$:

$$N_0(1,n) = \left(\frac{\pi^2}{a^2}\right)\left(\frac{D_{11} + 2\hat{D}_{12}n^2 + D_{22}n^4 + \bar{k}}{1 + n^2}\right) \quad (8.1.10)$$

The buckling load is a minimum when n is the nearest positive integer to the real number R

$$R^2 = -1 + \left(1 + \frac{D_{11} - 2\hat{D}_{12} + \bar{k}}{D_{22}}\right)^{\frac{1}{2}} \quad (8.1.11)$$

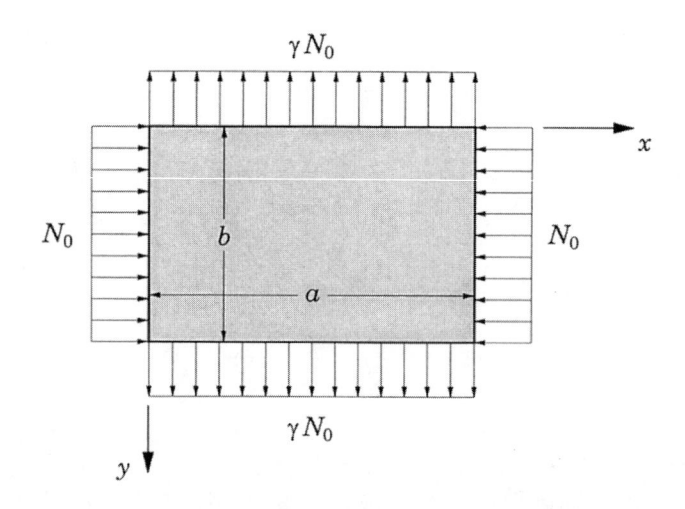

Figure 8.1.2. Plate subjected to uniform compression along x ($\hat{N}_{xx} = -N_0$) and uniform tension along y ($\hat{N}_{yy} = \gamma N_0$).

For example, for modulus ratios of $D_{11}/D_{22} = 10$ and $\hat{D}_{12}/D_{22} = 1$ and $k = 0$, the minimum buckling load occurs at $n = 1$ (because $R = \sqrt{2}$) and it is given by

$$N_{cr} \equiv N_0(1,1) = 6.5 \left(\frac{\pi^2 D_{22}}{a^2} \right) \qquad (8.1.12)$$

For modulus ratios of $D_{11}/D_{22} = 12$ and $\hat{D}_{12}/D_{22} = 1$ and $k = 0$, the value of R is 1.52. Hence, the minimum buckling load occurs for $n = 2$

$$N_{cr} = 7.2 \left(\frac{\pi^2 D_{22}}{a^2} \right) \qquad (8.1.13a)$$

and the mode shape is given by

$$W_{12} = \sin \frac{\pi x}{a} \sin \frac{2\pi y}{a} \qquad (8.1.13b)$$

For a rectangular isotropic [$D_{11} = D_{22} = D$, $D_{12} = \nu D$, $2D_{66} = (1-\nu)D$ or $\hat{D}_{12} = D$] plate under biaxial compression, the buckling load can be calculated using Eq. (8.1.7):

$$N_0(m,n) = \frac{\pi^2 D}{b^2} \left(m^2 s^2 + n^2 + \frac{\hat{k}}{m^2 s^2 + n^2} \right), \quad \hat{k} = \frac{kb^4}{D\pi^4} \qquad (8.1.14)$$

where s is the plate aspect ratio $s = b/a$. Clearly, the critical buckling load for the case in which $k = 0$ occurs at $m = n = 1$ and it is equal to

$$N_{cr} = (1 + s^2)\frac{\pi^2 D}{b^2}, \quad s = \frac{b}{a} \tag{8.1.15a}$$

and for a square plate it reduces to

$$N_{cr} = \frac{2\pi^2 D}{b^2} \tag{8.1.15b}$$

When $k \neq 0$, N_0 is not a minimum for any finite values of m and n.

8.1.4 Biaxial Loading of a Plate

When the edges $x = 0, a$ of a square plate are subjected to compressive load $\hat{N}_{xx} = -N_0$ and the edges $y = 0, b$ are subjected to tensile load $\hat{N}_{yy} = \gamma N_0$ (see Figure 8.1.2), Eq. (8.1.7) becomes

$$N_0(m, n) = \frac{\pi^2}{a^2} \left(\frac{m^4 D_{11} + 2m^2 n^2 \hat{D}_{12} + n^4 D_{22} + \frac{ka^4}{\pi^4}}{m^2 - \gamma n^2} \right) \tag{8.1.16}$$

when $\gamma n^2 < m^2$. For example, when $\gamma = 0.5$ and $k = 0$, the minimum buckling load occurs at $m = 1$ and $n = 1$:

$$N_0(1, 1) = \frac{2\pi^2}{a^2} \left(D_{11} + 2\hat{D}_{12} + D_{22} \right) \tag{8.1.17}$$

If $D_{11}/D_{22} = 10$, $\hat{D}_{12}/D_{22} = 1$, then the critical buckling load becomes

$$N_{cr} = \frac{26\pi^2 D_{22}}{a^2} \tag{8.1.18}$$

For a rectangular isotropic plate (when $k = 0$), the buckling load under biaxial loading ($N_{xx} = -N_0$ and $N_{yy} = \gamma N_0$) becomes

$$N_0(m, n) = \left(\frac{\pi^2 D}{b^2}\right) \frac{(m^2 s^2 + n^2)^2}{m^2 s^2 - \gamma n^2} \tag{8.1.19}$$

and the minimum buckling load occurs for $n = 1$

$$N_0(m, 1) = \left(\frac{\pi^2 D}{b^2}\right) \frac{(m^2 s^2 + 1)^2}{m^2 s^2 - \gamma} \tag{8.1.20}$$

In theory, the minimum of $N_0(m,1)$ occurs when $m^2 s^2 = 1 + 2\gamma$. For a square plate with $\gamma = 0.5$, we find

$$N_0(1,1) = \frac{8\pi^2 D}{a^2}, \quad N_0(2,1) = 7.1429 \frac{\pi^2 D}{a^2} = N_{cr}$$

8.1.5 Uniaxial Compression of a Rectangular Plate

When a rectangular plate with $k = 0$ is subjected to uniform compressive load N_0 on edges $x = 0$ and $x = a$, i.e., when $\gamma = 0$, the buckling load can be calculated using Eq. (8.1.7)

$$N_0(m,n) = \frac{\pi^2}{m^2 s^2 b^2} \left(m^4 s^4 D_{11} + 2s^2 m^2 n^2 \hat{D}_{12} + n^4 D_{22} \right) \quad (8.1.21)$$

An examination of the expression in Eq. (8.1.21) shows that the smallest value of N_0 for any m occurs for $n = 1$:

$$N_0(m,1) = \frac{\pi^2 D_{22}}{b^2} \left(m^2 s^2 \frac{D_{11}}{D_{22}} + 2 \frac{\hat{D}_{12}}{D_{22}} + \frac{1}{m^2 s^2} \right) \quad (8.1.22)$$

Thus the plate buckles in such a way that there can be several ($m \geq 1$) half-waves in the direction of compression but only one ($n = 1$) half-wave in the perpendicular direction. The critical buckling load is then determined by finding the minimum of $N_0 = N_0(m)$ in Eq. (8.1.22) with respect to m. We have

$$\frac{dN_0}{dm} = 0 \quad \text{gives} \quad m_c^4 = \frac{1}{s^4} \frac{D_{22}}{D_{11}}, \quad s = \frac{b}{a} \quad (8.1.23)$$

The second derivative of N_0 with respect to m can be shown to be positive. Since the value of m from Eq. (8.1.23) is not always an integer, the minimum buckling load cannot be predicted by substituting the value of m_c from Eq. (8.1.23) for m into Eq. (8.1.22). The minimum value of N_0 is given by Eq. (8.1.22) when m_c is the nearest integer value given by Eq. (8.1.23). Since the value of m_c depends on the ratio of the principal bending stiffnesses D_{11} and D_{22} as well as plate aspect ratio $s = b/a$, we must investigate the variation of N_0 with aspect ratio s for different values of m_c for a given rectangular plate.

For instance, consider a plate with $D_{11}/D_{22} = 10$ and $a/b = 1.778$. Then we have

$$N_0(m,1) = \frac{\pi^2 D_{22}}{b^2}\left(10m^2 s^2 + 2 + \frac{1}{s^2 m^2}\right) \tag{8.1.24a}$$

$$m_c^4 = \frac{D_{22}}{D_{11}}\left(\frac{a}{b}\right)^4 = 0.1 \times (1.778)^4 = 0.9994 \approx 1 \tag{8.1.24b}$$

In fact, for $(a/b) < 2.66$, we have

$$m_c^4 = \frac{D_{22}}{D_{11}}\left(\frac{a}{b}\right)^4 = 0.1 \times (2.66)^4 \quad \text{or} \quad m_c = 1.496 \tag{8.1.25}$$

Thus the closest integer is $m = 1$. The critical buckling load of a laminate with

$$\frac{D_{11}}{D_{22}} = 10, \quad (D_{12} + 2D_{66}) = D_{22}, \quad \frac{a}{b} < 2.66 \tag{8.1.26}$$

is given by

$$N_{cr} \equiv N_0(1,1) = \frac{\pi^2 D_{22}}{s^2 b^2}\left(10 s^4 + 2 s^2 + 1\right) \tag{8.1.27}$$

For various aspect ratios, we have

$$\frac{a}{b} = 1: \quad N_{cr} = 13\,\frac{\pi^2 D_{22}}{b^2}\,; \quad \frac{a}{b} = 1.5: \quad N_{cr} = 8.69\,\frac{\pi^2 D_{22}}{b^2}$$

$$\frac{a}{b} = 2: \quad N_{cr} = 8.5\,\frac{\pi^2 D_{22}}{b^2}\,; \quad \frac{a}{b} = 2.5: \quad N_{cr} = 9.85\,\frac{\pi^2 D_{22}}{b^2}$$

and the buckling mode is

$$W_{11} = \sin\frac{\pi x}{a}\,\sin\frac{\pi y}{a}$$

It can be shown that if the plate aspect ratio a/b is greater than 2.66 but less than 4.44, the buckling load is the minimum for $n = 1$ and $m = 2$. For example, for $a/b = 3$, we have from Eq. (8.1.24a)

$$N_0(1,1) = \frac{109}{9}\,\frac{\pi^2 D_{22}}{b^2} \approx 12.11\,\frac{\pi^2 D_{22}}{b^2}$$

$$N_0(2,1) = \frac{313}{36}\,\frac{\pi^2 D_{22}}{b^2} \approx 8.69\,\frac{\pi^2 D_{22}}{b^2} = N_{cr}$$

$$N_0(3,1) = 13\,\frac{\pi^2 D_{22}}{b^2}$$

Thus, for aspect ratios between 2.66 and 4.44, the plate buckles into two half-waves in the x-direction and one half-wave in the y-direction. The larger aspect ratios lead to higher modes of buckling. Figure 8.1.3 contains a plot of the nondimensionalized buckling load $\bar{N} = N_0 b^2/(\pi^2 D_{22})$ versus plate aspect ratio a/b for laminates whose material properties are $D_{11}/D_{22} = 10$ and $\hat{D}_{12}/D_{22} = 1$. For aspect ratios less than 2.5, the plate buckles into a single half-wave in the x-direction. As the aspect ratio increases, the plate buckles into more and more half-waves in the x-direction. Note that intersections of two consecutive modes correspond to certain aspect ratios (see Figure 8.1.4). Thus, for each of these aspect ratios, there are two possible buckled mode shapes. The $\bar{N}$ versus a/b curve gets flatter with the increasing aspect ratio, and it approaches the value

$$N_{cr} = \frac{2\pi^2 D_{22}}{b^2} \left(\sqrt{\frac{D_{11}}{D_{22}}} + \frac{\hat{D}_{12}}{D_{22}} \right) \tag{8.1.28}$$

which is obtained from Eq. (8.1.22) after substituting for $m^2 = m_c^2$ from Eq. (8.1.23). For the data in Eq. (8.1.26), this limiting value of the critical buckling load is

$$N_{cr} = 8.325 \left(\frac{\pi^2 D_{22}}{b^2} \right) \tag{8.1.29}$$

For an isotropic plate, Eqs. (8.1.21) and (8.1.22) reduce to

$$N_0(m,n) = \frac{\pi^2 a^2 D}{m^2} \left(\frac{m^2}{a^2} + \frac{n^2}{b^2} \right)^2 \tag{8.1.30}$$

$$N_0(m,1) = \frac{\pi^2 D}{a^2} \left(m + \frac{1}{m} \frac{a^2}{b^2} \right)^2 \tag{8.1.31}$$

For a given aspect ratio, two different modes, m_1 and m_2, will have the same buckling load when $\sqrt{m_1 m_2} = a/b$. In particular, the point of intersection of curves m and $m+1$ occurs for aspect ratios

$$\frac{a}{b} = \sqrt{2}, \sqrt{6}, \sqrt{12}, \sqrt{20}, \cdots, \sqrt{m^2 + m}$$

Thus, there is a mode change at these aspect ratios from m half-waves to $m+1$ half-waves. Putting $m = 1$ in Eq. (8.1.31), we find

$$N_{cr} = \frac{\pi^2 D}{b^2} \left(\frac{a}{b} + \frac{b}{a} \right)^2 \tag{8.1.32}$$

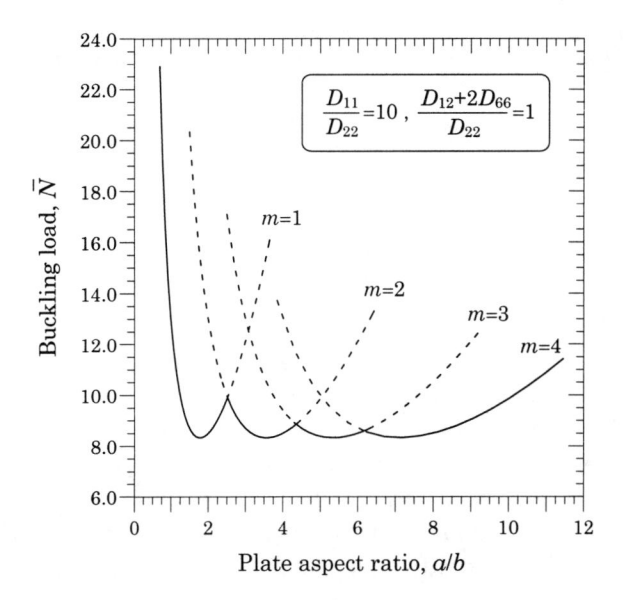

Figure 8.1.3. Nondimensionalized buckling load, $\bar{N} = N_0 b^2/(\pi^2 D_{22})$, versus plate aspect ratio a/b.

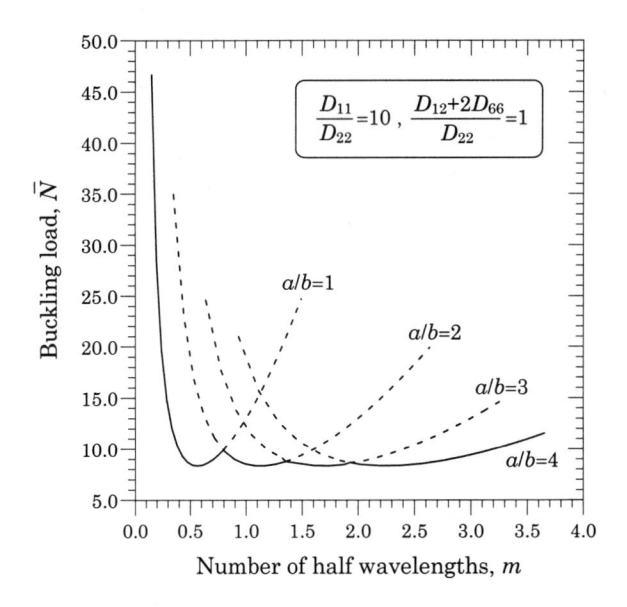

Figure 8.1.4. Nondimensionalized buckling load, $\bar{N}$, versus number of half-wavelengths m in the x–direction.

For a square plate Eq. (8.1.32) yields

$$N_{cr} = \frac{4\pi^2 D}{b^2} \qquad (8.1.33)$$

The critical value of the compressive stress is given by $\sigma_{cr} = N_{cr}/h$, where h is the thickness of the plate.

Table 8.1.1 shows the effect of plate aspect ratio and modulus ratio (orthotropy) on the critical buckling loads $\bar{N} = N_{cr}b^2/(\pi^2 D_{22})$ of rectangular isotropic ($\nu = 0.3$) and orthotropic ($G_{12} = 0.5E_2$, $\nu_{12} = 0.25$) plates under uniform axial compression ($\gamma = 0$) and biaxial compression ($\gamma = 1$). In all cases, the critical buckling mode is $(m, n) = (1, 1)$, except as indicated. The nondimensionalized buckling load increases as the modulus ratio increases. Figures 8.1.5 and 8.1.6 show the effect of aspect ratio and mode on critical buckling loads.

Table 8.1.1. Effect of plate aspect ratio and modulus ratio on the nondimensionalized buckling loads $\bar{N}$ of simply supported (SSSS) rectangular plates under uniform axial compression ($\gamma = 0$) and biaxial compression ($\gamma = 1$).

γ	$\frac{a}{b}$	$\frac{E_1}{E_2}=1$	$\frac{E_1}{E_2}=3$	$\frac{E_1}{E_2}=10$	$\frac{E_1}{E_2}=25$
0	0.5	6.250	14.708	42.737	102.750
	1.0	4.000	6.458	13.488	28.495
	1.5	4.340[(2,1)]†	6.042	9.182	15.856
	2.0	4.000[(2,1)]	6.458[(2,1)]	8.987	12.745
	2.5	4.134[(3,1)]	5.941[(2,1)]	10.338	12.745
	3.0	4.000[(3,1)]	6.042[(2,1)]	9.182[(2,1)]	14.273
1	0.5	5.000	11.767	25.427[(1,3)]	40.784[(1,4)]
	1.0	2.000	3.229	6.744	10.196[(1,2)]
	1.5	1.444	1.859	2.825	4.879
	2.0	1.250	1.442	1.798	2.549
	2.5	1.160	1.267	1.426	1.758
	3.0	1.111	1.179	1.260	1.427

† Denotes mode numbers (m, n) at which the critical buckling load occurred; $(m, n) = (1, 1)$ for all other cases.

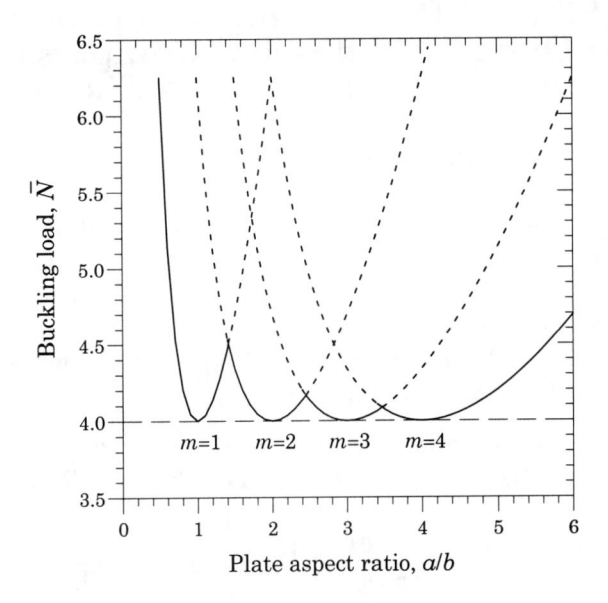

Figure 8.1.5. Nondimensionalized buckling load, $\bar{N} = N_0 b^2/(\pi^2 D_{22})$, versus plate aspect ratio a/b for isotropic SSSS plates.

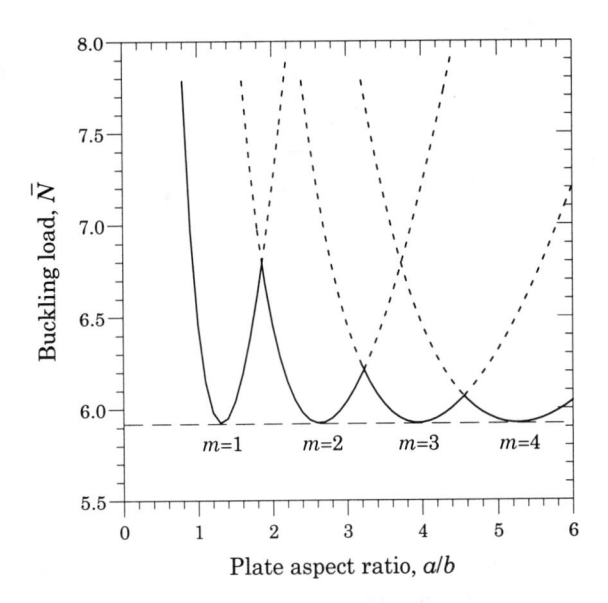

Figure 8.1.6 Nondimensionalized buckling load, $\bar{N} = N_0 b^2/(\pi^2 D_{22})$, versus plate aspect ratio a/b for orthotropic SSSS plates.

8.2 Buckling of Plates Simply Supported Along Two Opposite Sides and Compressed in the Direction Perpendicular to Those Sides

8.2.1 The Lévy Solution

Here we consider buckling of uniformly compressed rectangular plates simply supported along two opposite edges perpendicular to the direction of compression (see Figure 8.2.1) and having various edge conditions along the other two sides. We use the Lévy method of solution to reduce the governing partial differential equation (8.1.1) to an ordinary differential equation in y.

For the case of uniform compression along the x axis, we have $\hat{N}_{xx} = -N_0$ and $\hat{N}_{yy} = 0$, and Eq. (8.1.1) reduces to ($k = 0$)

$$D_{11}\frac{\partial^4 w}{\partial x^4} + 2\hat{D}_{12}\frac{\partial^4 w}{\partial x^2 \partial y^2} + D_{22}\frac{\partial^4 w}{\partial y^4} = -N_0 \frac{\partial^2 w}{\partial x^2} \qquad (8.2.1)$$

where $\hat{D}_{12} = D_{12} + 2D_{66}$. This equation must be solved for the buckling load N_0 and mode shape w for any given boundary conditions.

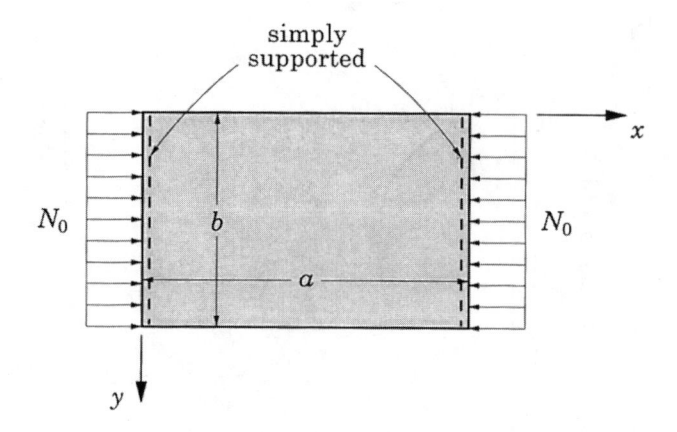

Figure 8.2.1. Uniformly compressed rectangular plates simply supported along two opposite sides ($x = 0, a$) perpendicular to the direction of compression and having various boundary conditions along the other two sides ($y = 0, b$).

We assume solution of Eq. (8.2.1) in the form

$$w(x,y) = W(y) \sin \frac{m\pi x}{a} \qquad (8.2.2)$$

i.e., under the action of compressive forces the plate buckles into m sinusoidal half-waves. The function $W(y)$, which is to be determined later, represents the buckling shape along the y axis. The assumed solution satisfies the boundary conditions along the simply supported edges $x = 0, a$ of the plate:

$$w = 0, \quad M_{xx} \equiv -\left(D_{11}\frac{\partial^2 w}{\partial x^2} + D_{12}\frac{\partial^2 w}{\partial y^2}\right) = 0 \quad \text{at } x = 0, a \qquad (8.2.3)$$

Substituting Eq. (8.2.2) into Eq. (8.2.1), we obtain

$$\left(D_{11}\alpha_m^4 - N_0\alpha_m^2\right) W - 2\alpha_m^2 \hat{D}_{12}\frac{d^2 W}{dy^2} + D_{22}\frac{d^4 W}{dy^4} = 0 \qquad (8.2.4)$$

The form of the solution to Eq. (8.2.4) depends on the nature of the roots λ of the equation

$$D_{22}\lambda^4 - 2\alpha_m^2 \hat{D}_{12}\lambda^2 + \left(\alpha_m^4 D_{11} - \alpha_m^2 N_0\right) = 0 \qquad (8.2.5)$$

Due to the geometric constraints on the edges $y = 0, b$, the buckling load N_0 [see Eq. (8.1.22)] is such that

$$N_0 > \alpha_m^2 D_{11} \qquad (8.2.6)$$

Hence, the solution to Eq. (8.2.4) is of the form

$$W(y) = A \cosh \lambda_1 y + B \sinh \lambda_1 y + C \cos \lambda_2 y + D \sin \lambda_2 y \qquad (8.2.7)$$

where

$$(\lambda_1)^2 = \sqrt{\alpha_m^2 \frac{N_0}{D_{22}} - \alpha_m^4 \frac{D_{11}}{D_{22}} + \alpha_m^4 \left(\frac{\hat{D}_{12}}{D_{22}}\right)^2} + \frac{\hat{D}_{12}}{D_{22}}\alpha_m^2 \qquad (8.2.8a)$$

$$(\lambda_2)^2 = \sqrt{\alpha_m^2 \frac{N_0}{D_{22}} - \alpha_m^4 \frac{D_{11}}{D_{22}} + \alpha_m^4 \left(\frac{\hat{D}_{12}}{D_{22}}\right)^2} - \frac{\hat{D}_{12}}{D_{22}}\alpha_m^2 \qquad (8.2.8b)$$

For isotropic plates, Eqs. (8.2.8a,b) reduce to

$$(\lambda_1)^2 = \sqrt{\alpha_m^2 \frac{N_0}{D} + \alpha_m^2}, \quad (\lambda_2)^2 = \sqrt{\alpha_m^2 \frac{N_0}{D} - \alpha_m^2} \qquad (8.2.8c)$$

8.2.2 Buckling of SSSF Plates

Consider buckling of uniformly compressed rectangular plates with side $y = 0$ simply supported and side $y = b$ free (see Figure 8.2.2). The boundary conditions on the simply supported and free edges are

$$w = 0, \quad M_{yy} = -\left(D_{12}\frac{\partial^2 w}{\partial x^2} + D_{22}\frac{\partial^2 w}{\partial y^2}\right) = 0 \quad \text{at} \quad y = 0 \quad (8.2.9)$$

$$M_{yy} = 0, \quad V_y = -\left(D_{22}\frac{\partial^3 w}{\partial y^3} + \bar{D}_{12}\frac{\partial^3 w}{\partial x^2 \partial y}\right) = 0 \quad \text{at} \quad y = b \quad (8.2.10)$$

where $\bar{D}_{12} = D_{12} + 4D_{66}$. Using the solution (8.2.7) in Eq. (8.2.9) yields $A = C = 0$, and using it in Eq. (8.2.10) yields the following two linear relations among B and D:

$$\left(D_{12}\alpha_m^2 - D_{22}\lambda_1^2\right) B \sinh \lambda_1 b + \left(D_{12}\alpha_m^2 + D_{22}\lambda_2^2\right) D \sin \lambda_2 b = 0$$

$$\lambda_1 \left(\bar{D}_{12}\alpha_m^2 - D_{22}\lambda_1^2\right) B \cosh \lambda_1 b + \lambda_2 \left(\bar{D}_{12}\alpha_m^2 + D_{22}\lambda_2^2\right) D \cos \lambda_2 b = 0$$

$$(8.2.11)$$

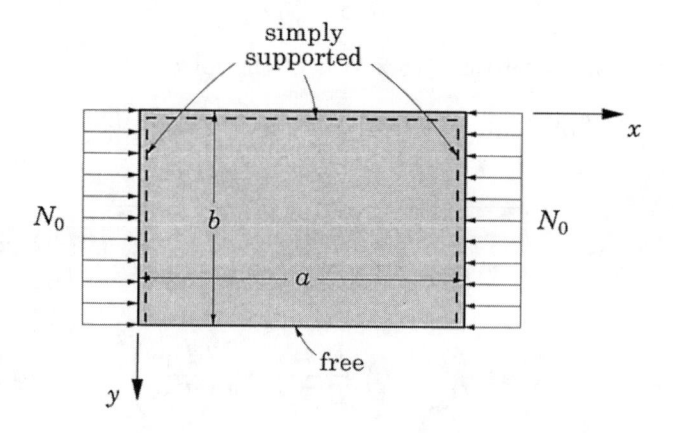

Figure 8.2.2. Uniformly compressed rectangular plate simply supported along two opposite sides ($x = 0, a$) perpendicular to the direction of compression, and simply supported on side $y = 0$ and free on side $y = b$ (SSSF).

One possible solution of these homogeneous equations is that $B = D = 0$. Then the deflection at each point of the plate is zero, which is the flat form of equilibrium of the plate. The buckled form of equilibrium of the plate is possible only if equations in (8.2.11) yield nonzero values of B and D. This requires the determinant of the two linear equations in (8.2.11) be zero. We obtain

$$\lambda_2 \Omega_1^2 \sinh \lambda_1 b \, \cos \lambda_2 b - \lambda_1 \Omega_2^2 \cosh \lambda_1 b \, \sin \lambda_2 b = 0 \quad (8.2.12)$$

where Ω_1 and Ω_2 are defined by

$$\Omega_1 = \left(\lambda_1^2 - \frac{D_{12}}{D_{22}} \alpha_m^2 \right), \quad \Omega_2 = \left(\lambda_2^2 + \frac{D_{12}}{D_{22}} \alpha_m^2 \right) \quad (8.2.13)$$

Since λ_1 and λ_2 contain N_0 [see Eq. (8.2.8a–c)], Eq. (8.2.12) can be solved for the smallest N_0 once the geometric and material parameters of the plate are known. The critical buckling load may be written as

$$N_{cr} = \kappa \frac{\pi^2 D_{22}}{b^2} \quad (8.2.14)$$

where κ is a numerical factor depending on the plate aspect ratio b/a and material properties. The magnitude of the corresponding critical compressive stress is given by $(\sigma_{xx})_{cr} = N_{cr}/h$. The general mode shape is given by

$$w(x,y) = (\Omega_1 \sinh \lambda_1 b \sin \lambda_2 y + \Omega_2 \sinh \lambda_1 y \sin \lambda_2 b) \sin \alpha_m x \quad (8.2.15)$$

Table 8.2.1 contains buckling loads of isotropic ($\nu = 0.25$) and orthotropic ($G_{12} = 0.5 E_2$, $\nu_{12} = 0.25$) plates (SSSF) for various values of a/b and modes $m = 1, 2$. The critical buckling load occurs in mode $m = 1$ for isotropic as well orthotropic plates with aspect ratio $0 < a/b \leq 6$. The mode shape associated with the critical buckling load is

$$w(x,y) = (\Omega_1 \sinh \lambda_1 b \sin \lambda_2 y + \Omega_2 \sinh \lambda_1 y \sin \lambda_2 b) \sin \frac{\pi x}{a}$$

8.2.3 Buckling of SSCF Plates

Here we consider buckling of uniformly compressed rectangular plates with side $y = 0$ clamped and side $y = b$ free (see Figure 8.2.3). The boundary conditions are

$$w = 0, \quad \frac{\partial w}{\partial y} = 0 \quad \text{at} \quad y = 0 \quad (8.2.16)$$

Table 8.2.1. Effect of plate aspect ratio and modulus ratio on the nondimensionalized buckling loads $\bar{N} = N_0 b^2/(\pi^2 D_{22})$ of rectangular plates (SSSF) under uniform compression $\hat{N}_{xx} = -N_0$.

	$\frac{E_1}{E_2} = 1$		$\frac{E_1}{E_2} = 3$		$\frac{E_1}{E_2} = 10$	
$\frac{a}{b}$	$m=1$	$m=2$	$m=1$	$m=2$	$m=1$	$m=2$
0.4	6.6367	25.2899	19.2688	75.4387	63.0270	250.448
0.6	3.1921	11.4675	8.8763	33.8266	28.3288	111.613
0.8	1.9894	6.3667	5.2425	19.2688	16.1880	63.027
1.0	1.4342	4.4036	3.5626	12.5334	10.5707	40.542
1.5	0.8880	2.2022	1.9075	5.8859	5.0268	18.338
2.0	0.6979	1.4342	1.3307	3.5626	3.0891	10.571
2.5	0.6104	1.0798	1.0648	2.4893	2.1932	6.977
3.0	0.5630	0.8879	0.9208	1.9075	1.7071	5.027
3.5	0.5345	0.7726	0.8341	1.5574	1.4142	3.851
4.0	0.5161	0.6979	0.7780	1.3307	1.2242	3.089
4.5	0.5034	0.6469	0.7396	1.1755	1.0939	2.567
5.0	0.4944	0.6104	0.7121	1.0648	1.0008	2.193
5.5	0.4877	0.5835	0.6918	0.9829	0.9319	1.917
6.0	0.4826	0.5630	0.6764	0.9208	0.8795	1.707

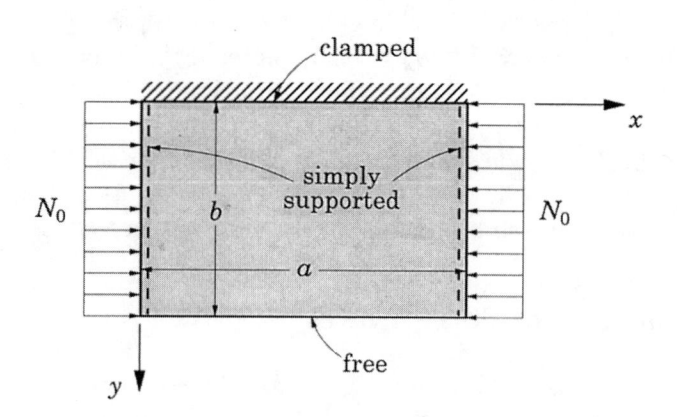

Figure 8.2.3. Uniformly compressed rectangular plate simply supported along two opposite sides ($x = 0, a$), clamped on side $y = 0$, and free on side $y = b$ (SSCF).

$$M_{yy} = 0, \quad V_y = -D_{22}\frac{\partial^3 w}{\partial y^3} - \bar{D}_{12}\frac{\partial^3 w}{\partial x^2 \partial y} = 0 \quad \text{at} \quad y = b \quad (8.2.17)$$

Substituting for w from Eq. (8.2.7) into the boundary conditions, we obtain $C = -A$, $D = -(\lambda_1/\lambda_2)B$, and

$$-\Omega_1 \left(A \cosh \lambda_1 b + B \sinh \lambda_1 b\right) + \Omega_2 \left(C \cos \lambda_2 b + D \sin \lambda_2 b\right) = 0$$
$$-\lambda_1 \Omega_1 \left(A \sinh \lambda_1 b + B \cosh \lambda_1 b\right) + \lambda_2 \Omega_2 \left(-C \sin \lambda_2 b + D \cos \lambda_2 b\right) = 0$$
$$(8.2.18)$$

The determinant of these equations, after substituting for $C = -A$ and $D = -(\lambda_1/\lambda_2)B$, is

$$2\Omega_1\Omega_2 + \left(\Omega_1^2 + \Omega_2^2\right) \cosh \lambda_1 b \, \cos \lambda_2 b$$
$$- \frac{1}{\lambda_1 \lambda_2}\left(\lambda_1^2 \Omega_2^2 - \lambda_2^2 \Omega_1^2\right) \sinh \lambda_1 b \, \sin \lambda_2 b = 0 \quad (8.2.19)$$

Nondimensional critical buckling loads $\bar{N} = N_0 b^2/(\pi^2 D_{22})$ of SSCF plates are presented in Table 8.2.2 for isotropic ($\nu = 0.25$) and orthotropic ($G_{12} = 0.5E_2$, $\nu_{12} = 0.25$) plates for various aspect ratios. Figures 8.2.4 and 8.2.5 contain plots of nondimensional buckling load versus plate aspect ratio for modes $m = 1, 2, 3$ of isotropic and orthotropic plates, respectively. It is seen that at the beginning, the buckling load of an isotropic plate decreases with an increase in the aspect ratio a/b. The minimum value of the buckling load $\bar{N}_{cr} = 1.329$ occurs at $a/b = 1.635$. There is a mode change at $a/b = 2.3149$ from $m = 1$ and $m = 2$, and the buckling load for this aspect ratio is $\bar{N}_{cr} = 1.503$. The minimum buckling load ($\bar{N}_{cr} = 1.329$) for mode $m = 2$ occurs at $a/b = 3.27$. For comparatively long isotropic plates, the critical buckling load can be taken with sufficient accuracy as $\bar{N}_{cr} = 1.329$. For orthotropic plates with $E_1/E_2 = 3$, the minimum buckling load ($\bar{N}_{cr} = 2.0606$) occurs at $a/b = 2.168$.

The actual boundary conditions in practical problems are often intermediate to simply supported and clamped type. For example, consider the T junction of the (vertical) web plate under compression and the (horizontal) flange plate at $y = 0$, which provides the elastic restraint to the web plate. Then the boundary condition at $y = 0$ can be expressed as (see Timoshenko and Gere [1])

Table 8.2.2. Nondimensionalized buckling loads $\bar{N}$ of SSCF rectangular plates under uniform axial compression $\hat{N}_{xx} = -N_0$.

$\frac{a}{b}$	$\frac{E_1}{E_2}=1$	$\frac{E_1}{E_2}=3$	$\frac{E_1}{E_2}=10$	$\frac{a}{b}$	$\frac{E_1}{E_2}=1$	$\frac{E_1}{E_2}=3$	$\frac{E_1}{E_2}=10$
0.5	4.518	12.665	40.674	3.5	1.336	2.174	3.239
1.0	1.698	3.861	10.871	4.0	1.386	2.076	3.529
1.5	1.339	2.407	5.529	4.5	1.339†	2.064	3.459†
2.0	1.386	2.076	3.838	5.0	1.329	2.110	3.242
2.5	1.432†	2.110	3.242	5.5	1.347	2.129†	3.138
3.0	1.339	2.407†	3.114	6.0	1.339†	2.076	3.114

† Denotes change to next higher mode.

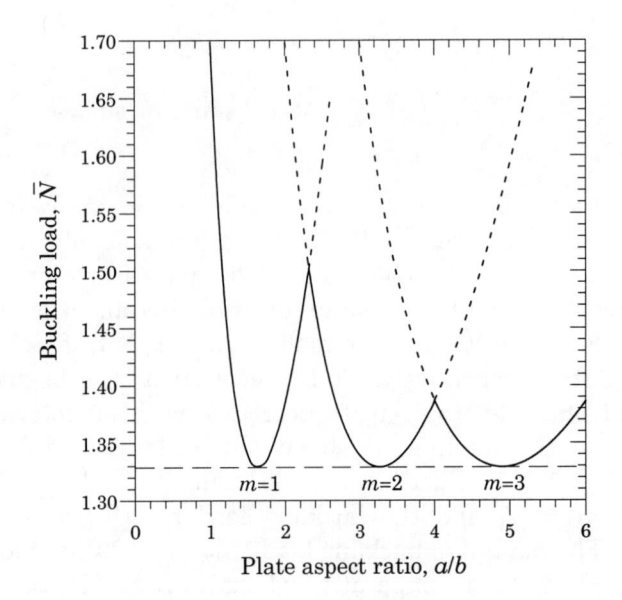

Figure 8.2.4. Nondimensionalized buckling load, $\bar{N} = N_0 b^2/(\pi^2 D_{22})$, versus plate aspect ratio a/b for isotropic SSCF plates.

$$-D_{22}\frac{\partial^2 w}{\partial y^2} - D_{12}\frac{\partial^2 w}{\partial x^2} = \mu\frac{\partial^3 w}{\partial x^2 \partial y} \qquad (8.2.20)$$

where μ is the torsional rigidity of the flange.

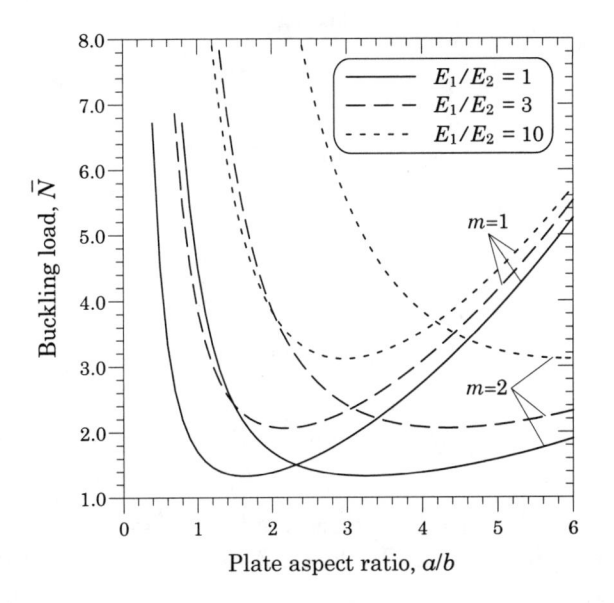

Figure 8.2.5 Nondimensionalized buckling load, $\bar{N} = N_0 b^2/(\pi^2 D_{22})$, versus plate aspect ratio a/b for orthotropic SSCF plates.

8.2.4 Buckling of SSCC Plates

Here we consider buckling of uniformly compressed rectangular plates with sides $y = 0, b$ clamped (see Figure 8.2.6). The boundary conditions are

$$w = 0, \quad \frac{\partial w}{\partial y} = 0 \quad \text{at} \quad y = 0, b \qquad (8.2.21)$$

The boundary conditions yield

$$A (\cosh \lambda_1 b - \cos \lambda_2 b) + B \left(\sinh \lambda_1 b - \frac{\lambda_1}{\lambda_2} \sin \lambda_2 b \right) = 0$$
$$A (\lambda_1 \sinh \lambda_1 b + \lambda_2 \sin \lambda_2 b) + B \lambda_1 (\cosh \lambda_1 b - \cos \lambda_2 b) = 0 \qquad (8.2.22)$$

The determinant of these equations is

$$2 (1 - \cosh \lambda_1 b \, \cos \lambda_2 b) + \left(\frac{\lambda_1}{\lambda_2} - \frac{\lambda_2}{\lambda_1} \right) \sinh \lambda_1 b \, \sin \lambda_2 b = 0 \qquad (8.2.23)$$

where λ_1 and λ_2 are defined by Eqs. (8.2.8a,b), respectively.

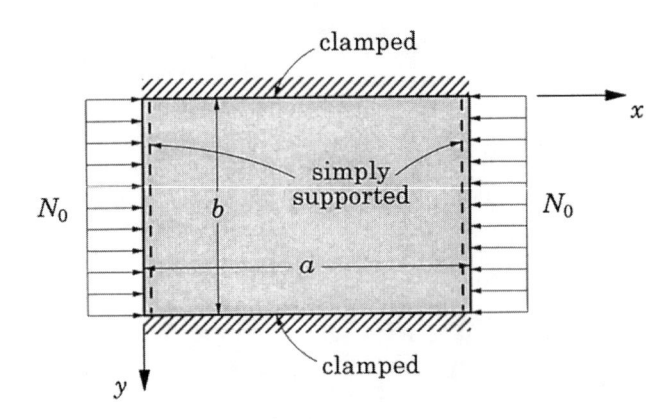

Figure 8.2.6. Uniformly compressed SSCC rectangular plate.

Nondimensional critical buckling loads $\bar{N}_{cr} = N_{cr}b^2/(\pi^2 D_{22})$ of SSCC plates are presented in Table 8.2.3 for isotropic ($\nu = 0.25$) and orthotropic ($G_{12} = 0.E_2$, $\nu_{12} = 0.25$) plates for various aspect ratios. Figures 8.2.7 and 8.2.8 contain plots of buckling load versus plate aspect ratio for modes $m = 1, 2, 3$ of isotropic and orthotropic plates. The minimum value of the buckling load $\bar{N}_{cr} = 0.697$ occurs at $a/b = 0.661$. There is a mode change at $a/b = 0.9349$ from $m = 1$ to $m = 2$, and the buckling load for this aspect ratio is $\bar{N}_{cr} = 8.097$. The minimum buckling load ($\bar{N}_{cr} = 0.697$) for mode $m = 2$ occurs at $a/b = 1.322$. For orthotropic plates with $E_1/E_2 = 3$, the minimum buckling load ($\bar{N}_{cr} = 10.866$) occurs at $a/b = 0.871$. The minimum buckling load ($\bar{N}_{cr} = 17.407$) for $E_1/E_2 = 10$ occurs at $a/b = 1.181$.

Table 8.2.3. Nondimensionalized buckling loads $\bar{N}_{cr}$ of SSCC rectangular plates under uniform axial compression $\hat{N}_{xx} = -N_0$.

$\frac{a}{b}$	$\frac{E_1}{E_2}=1$	$\frac{E_1}{E_2}=3$	$\frac{E_1}{E_2}=10$	$\frac{a}{b}$	$\frac{E_1}{E_2}=1$	$\frac{E_1}{E_2}=3$	$\frac{E_1}{E_2}=10$
0.4	9.448	22.473	66.256	1.4	7.001	11.631†	18.252
0.6	7.055	13.153	32.632	1.6	7.304	10.981	20.145
0.8	7.304	10.981	21.954	1.8	7.055†	10.882	19.559†
1.0	7.691†	11.163	18.198	2.0	6.972	11.163	18.198
1.2	7.055	12.518	17.415	2.2	7.069	11.337†	17.548

† Denotes change to next higher mode.

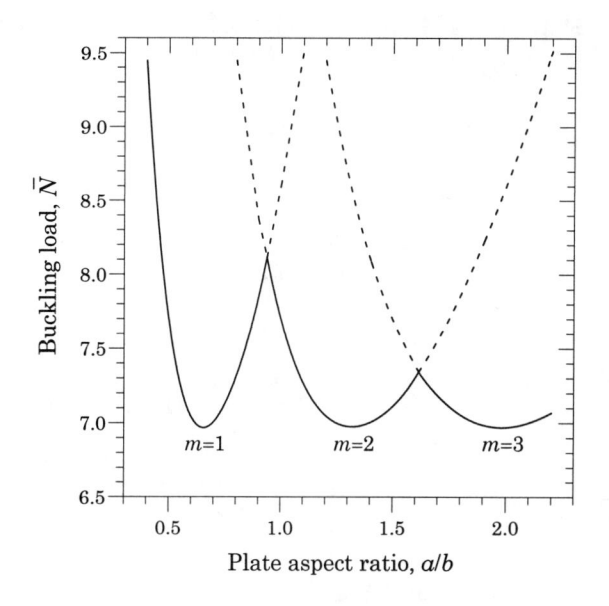

Figure 8.2.7. Nondimensionalized buckling load, $\bar{N} = N_0 b^2/(\pi^2 D_{22})$, versus plate aspect ratio a/b for isotropic SSCC plates.

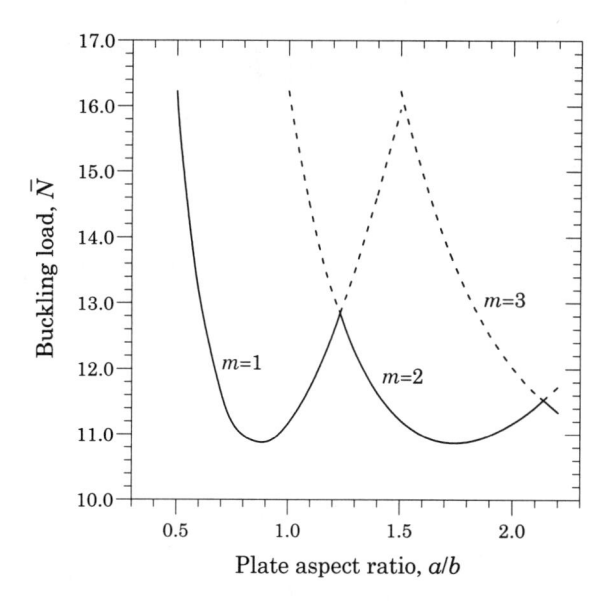

Figure 8.2.8 Nondimensionalized buckling load, $\bar{N} = N_0 b^2/(\pi^2 D_{22})$, versus plate aspect ratio a/b for orthotropic SSCC plates.

8.3 Buckling of Rectangular Plates Using the Rayleigh–Ritz Method

8.3.1 Analysis of the Lévy Plates

As was shown in section 7.3.1 for the bending case, the buckling equation obtained after the application of the Lévy method, namely Eq. (8.2.4), can also be solved using the Rayleigh–Ritz method. The weak form of Eq. (8.2.4) is

$$\int_0^b \left[D_{22} \frac{d^2W}{dy^2} \frac{d^2\delta W}{dy^2} + 2\alpha_m^2 \hat{D}_{12} \frac{dW}{dy} \frac{d\delta W}{dy} + \left(D_{11}\alpha_m^4 - N_0 \alpha_m^2 \right) W \delta W \right] dy$$
$$+ \left[\left(D_{22} \frac{d^3W}{dy^3} - 2\hat{D}_{12} \alpha_m^2 \frac{dW}{dy} \right) \delta W - D_{22} \frac{d^2W}{dy^2} \frac{d\delta W}{dy} \right]_0^b = 0 \qquad (8.3.1)$$

For a simply supported edge $[W = 0$ and $(d^2W/dy^2) = 0]$ or a clamped edge $[W = 0$ and $(dW/dy) = 0]$ the boundary terms in Eq. (8.3.1) vanish identically. However, for a free edge, the vanishing of bending moment and effective shear force require

$$D_{22} \frac{d^2W}{dy^2} - \alpha_m^2 D_{12} W = 0, \quad D_{22} \frac{d^3W}{dy^3} - \alpha_m^2 \bar{D}_{12} \frac{dW}{dy} = 0 \qquad (8.3.2)$$

where $\bar{D}_{12} = D_{12} + 4D_{66}$. Hence, on a free edge, the boundary expression of Eq. (8.3.1) can be simplified to

$$\left[\left(D_{22} \frac{d^3W}{dy^3} - 2\hat{D}_{12} \alpha_m^2 \frac{dW}{dy} \right) \delta W - D_{22} \frac{d^2W}{dy^2} \frac{d\delta W}{dy} \right]_0^b$$
$$= -\alpha_m^2 D_{12} \left[W \frac{d\delta W}{dy} + \frac{dW}{dy} \delta W \right]_0^b \qquad (8.3.3)$$

Thus, the weak form for SSSS, SSCC, and SSCS plates becomes

$$0 = \int_0^b \left[D_{22} \frac{d^2W}{dy^2} \frac{d^2\delta W}{dy^2} + 2\hat{D}_{12} \alpha_m^2 \frac{dW}{dy} \frac{d\delta W}{dy} \right.$$
$$\left. + \left(D_{11}\alpha_m^4 - N_0 \alpha_m^2 \right) W \delta W \right] dy \qquad (8.3.4)$$

and for SSSF and SSCF plates it is given by

$$0 = \int_0^b \left[D_{22} \frac{d^2W}{dy^2} \frac{d^2\delta W}{dy^2} + 2\hat{D}_{12}\alpha_m^2 \frac{dW}{dy}\frac{d\delta W}{dy} \right.$$
$$\left. + \left(D_{11}\alpha_m^4 - N_0 \alpha_m^2\right) W\delta W \right] dy$$
$$- \alpha_m^2 D_{12} \left[W \frac{d\delta W}{dy} + \frac{dW}{dy} \delta W \right]_0^b \quad (8.3.5)$$

Assume an approximation of the form

$$W(y) \approx \sum_{j=1}^{N} c_j \varphi_j(y) \quad (8.3.6)$$

where φ_j are the approximation functions [see Eqs. (7.3.10)–(7.3.20); replace x with y and a with b]. Substituting the approximation (8.3.6) into the weak form (8.3.5), we obtain

$$([R] - N_0[B])\{c\} = \{0\} \quad (8.3.7)$$

where

$$R_{ij} = \int_0^b \left(D_{22} \frac{d^2\varphi_i}{dy^2}\frac{d^2\varphi_j}{dy^2} + 2\alpha_m^2 \hat{D}_{12}\frac{d\varphi_i}{dy}\frac{d\varphi_j}{dy} + D_{11}\alpha_m^4 \varphi_i \varphi_j \right) dy$$
$$- \alpha_m^2 D_{12}\left[\varphi_j \frac{d\varphi_i}{dy} + \frac{d\varphi_j}{dy}\varphi_i \right]_0^b \quad (8.3.8a)$$

$$B_{ij} = \alpha_m^2 \int_0^b \varphi_i \varphi_j \, dy, \quad \alpha_m = \frac{m\pi}{a} \quad (8.3.8b)$$

The boundary term in R_{ij} is nonzero only for SSSF and SSCF plates. Equation (8.3.7) has a non-trivial solution, $c_i \neq 0$, only if the determinant of the coefficient matrix is zero

$$|[R] - N_0[B]| = 0 \quad (8.3.9)$$

Equation (8.3.9) yields an Nth order polynomial in N_0 that depends on m. Hence, for any given aspect ratio a/b, the critical buckling load is the smallest value of N_0 for all m.

Table 8.3.1 contains numerical values of nondimensional critical buckling loads $\bar{N}_{cr} = N_{cr}b^2/(\pi^2 D_{22})$ obtained using the algebraic approximation functions given in Eqs. (7.3.10)–(7.3.20) (replace x with y and a with b) for various boundary conditions. The accuracy of the buckling loads predicted by the three-paramater Rayleigh–Ritz approximation are in very good agreement with the analytical solutions derived in section 8.2.

Table 8.3.1. Nondimensionalized buckling loads $\bar{N}$ of rectangular isotropic ($\nu = 0.25$) plates under uniform compression $\hat{N}_{xx} = -N_0$ (Rayleigh–Ritz solutions).

$\frac{a}{b}$	N	SSSS	SSCC	SSCS	SSCF	SSSF	SSFF
0.5	1	6.334	7.725	7.915	4.896	4.456	4.000
	3	6.250	7.693	6.860	4.526	4.436	3.958
	E	6.250	7.691	6.853	4.518	4.404	—
1.0	1	4.258	8.606	8.149	2.050	1.456	1.000
	3	4.001	8.605	5.741	1.699	1.445	0.972
	E	4.000	8.604	5.740	1.698	1.434	—
1.5	1	4.497†	7.120†	7.040†	1.751	0.900	0.988
	3	4.341	7.116	5.432	1.339	0.894	0.426
	E	4.340	7.116	5.431	1.339	0.888	—
2.0	1	4.258	8.606	8.149	1.915	0.706	0.250
	3	4.001	8.605	5.741	1.386	0.702	0.238
	E	4.000	8.604	5.740	1.386	0.698	—

† Denotes change to the next higher mode; E denotes the 'exact' solution obtained in section 8.2.

8.3.2 General Formulation

In this section we consider buckling of orthotropic rectangular plates using the Rayleigh–Ritz method. Buckling problems of plates with combined bending and compression or under pure in-plane shear stress do not permit analytical solutions; therefore, we use the Rayleigh–Ritz method to solve such problems. Of course, the method can be used for any combination of boundary conditions and edges loads. The statement of the principle of virtual displacements or the minimum total potential energy for plates subjected to in-plane edge forces $\hat{N}_{xx}$, $\hat{N}_{yy}$, and $\hat{N}_{xy}$ can be expressed as

$$0 = \int_0^b \int_0^a \left[D_{11} \frac{\partial^2 w}{\partial x^2} \frac{\partial^2 \delta w}{\partial x^2} + D_{12} \left(\frac{\partial^2 w}{\partial y^2} \frac{\partial^2 \delta w}{\partial x^2} + \frac{\partial^2 w}{\partial x^2} \frac{\partial^2 \delta w}{\partial y^2} \right) \right.$$
$$+ 4 D_{66} \frac{\partial^2 w}{\partial x \partial y} \frac{\partial^2 \delta w}{\partial x \partial y} + D_{22} \frac{\partial^2 w}{\partial y^2} \frac{\partial^2 \delta w}{\partial y^2}$$
$$+ \hat{N}_{xx} \frac{\partial w}{\partial x} \frac{\partial \delta w}{\partial x} + \hat{N}_{yy} \frac{\partial w}{\partial y} \frac{\partial \delta w}{\partial y}$$
$$\left. + \hat{N}_{xy} \left(\frac{\partial w}{\partial y} \frac{\partial \delta w}{\partial x} + \frac{\partial w}{\partial x} \frac{\partial \delta w}{\partial y} \right) \right] dx dy \quad (8.3.10)$$

In general, the applied edge forces are functions of position and they are independent of each other. Let

$$\hat{N}_{xx} = -N_0, \quad \hat{N}_{yy} = -\gamma_1 N_0, \quad \hat{N}_{xy} = -\gamma_2 N_0 \quad (8.3.11)$$

where N_0 is a constant and γ_1 and γ_2 are possibly functions of position.

Using an N-parameter Rayleigh–Ritz solution of the form

$$w(x,y) \approx \sum_{j=1}^{N} c_j \varphi_j(x,y) \quad (8.3.12)$$

in Eq. (8.3.10), we obtain

$$([R] - N_0[B]) \{c\} = \{0\} \quad (8.3.13)$$

where

$$R_{ij} = \int_0^b \int_0^a \left[D_{11} \frac{\partial^2 \varphi_i}{\partial x^2} \frac{\partial^2 \varphi_j}{\partial x^2} + D_{12} \left(\frac{\partial^2 \varphi_i}{\partial y^2} \frac{\partial^2 \varphi_j}{\partial x^2} + \frac{\partial^2 \varphi_i}{\partial x^2} \frac{\partial^2 \varphi_j}{\partial y^2} \right) \right.$$
$$\left. + 4 D_{66} \frac{\partial^2 \varphi_i}{\partial x \partial y} \frac{\partial^2 \varphi_j}{\partial x \partial y} + D_{22} \frac{\partial^2 \varphi_i}{\partial y^2} \frac{\partial^2 \varphi_j}{\partial y^2} \right] dx dy \quad (8.3.14a)$$

$$B_{ij} = \int_0^b \int_0^a \left[\frac{\partial \varphi_i}{\partial x} \frac{\partial \varphi_j}{\partial x} + \gamma_1 \frac{\partial \varphi_i}{\partial y} \frac{\partial \varphi_j}{\partial y} + \gamma_2 \left(\frac{\partial \varphi_i}{\partial x} \frac{\partial \varphi_j}{\partial y} + \frac{\partial \varphi_i}{\partial y} \frac{\partial \varphi_j}{\partial x} \right) \right] dx dy$$
$$(8.3.14b)$$

As discussed in section 7.3.2, it is convenient to express the Rayleigh–Ritz approximation of rectangular plates in the form

$$w(x,y) \approx W_{mn}(x,y) = \sum_{i=1}^{M} \sum_{j=1}^{N} c_{ij} X_i(x) Y_j(y) \quad (8.3.15)$$

The functions X_i and Y_j for various boundary conditions are given in Eqs. (7.3.29)–(7.3.44). Substituting Eq. (8.3.15) into Eq. (8.3.10), we obtain Eq. (8.3.13) with the following definitions of the coefficients

$$R_{(ij)(k\ell)} = \int_0^b \int_0^a \left[D_{11} \frac{d^2 X_i}{dx^2} \frac{d^2 X_k}{dx^2} Y_j Y_\ell + D_{22} X_i X_k \frac{d^2 Y_j}{dy^2} \frac{d^2 Y_\ell}{dy^2} \right.$$
$$+ D_{12} \left(X_i \frac{d^2 X_k}{dx^2} \frac{d^2 Y_j}{dy^2} Y_\ell + \frac{d^2 X_i}{dx^2} X_k Y_j \frac{d^2 Y_\ell}{dy^2} \right)$$
$$\left. + 4 D_{66} \frac{dX_i}{dx} \frac{dX_k}{dx} \frac{dY_j}{dy} \frac{dY_\ell}{dy} \right] dx\, dy \quad (8.3.16a)$$

$$B_{(ij)(k\ell)} = \int_0^b \int_0^a \left[\frac{dX_i}{dx} \frac{dX_k}{dx} Y_j Y_\ell + \gamma_1 X_i X_k \frac{dY_j}{dy} \frac{dY_\ell}{dy} \right.$$
$$\left. + \gamma_2 \left(\frac{dX_i}{dx} X_k Y_j \frac{dY_\ell}{dy} + X_i \frac{dX_k}{dx} \frac{dY_j}{dy} Y_\ell \right) \right] dx\, dy \quad (8.3.16b)$$

As an example, consider a simply supported plate. For the choice of algebraic functions

$$X_i(x) = \left(\frac{x}{a}\right)^i - \left(\frac{x}{a}\right)^{i+1}, \quad Y_j(y) = \left(\frac{y}{b}\right)^i - \left(\frac{y}{b}\right)^{j+1} \quad (8.3.17)$$

and for $M = N = 1$, we obtain

$$R_{(11)(11)} = \frac{2b}{15a^3} D_{11} + \frac{2}{9ab} \hat{D}_{12} + \frac{2a}{15b^3} D_{22}$$
$$B_{(11)(11)} = \frac{1}{90} + \frac{\gamma_1}{90} + \gamma_2 \times 0 \quad (8.3.18)$$

Note that the buckling load under in-plane shear cannot be determined with one-parameter approximation. The buckling load under uniaxial compression along the x−axis is given by setting $\gamma_1 = 0$

$$N_{cr}^u = \frac{12b}{a^3} D_{11} + \frac{20}{ab} \hat{D}_{12} + \frac{12a}{b^3} D_{22} \quad (8.3.19a)$$

and the critical buckling load under biaxial compression is given by ($\gamma_1 = 1$)

$$N_{cr}^b = \frac{6b}{a^3} D_{11} + \frac{10}{ab} \hat{D}_{12} + \frac{6a}{b^3} D_{22} \quad (8.3.19b)$$

For square isotropic plates, the above expressions become

$$N_{cr}^u = 44 \frac{D}{a^2} = 4.458 \frac{D\pi^2}{a^2}, \quad N_{cr}^b = 22 \frac{D}{a^2} = 2.229 \frac{D\pi^2}{a^2}$$

The exact values for the two cases are $4D\pi^2/a^2$ and $2D\pi^2/a^2$, respectively (see Table 8.1.1). The results are in about -11.5% error, and the values do not change for $N = M = 2$.

For a square isotropic plate clamped on all sides and subjected to uniaxial compression or biaxial compression, the buckling loads obtained using the Rayleigh–Ritz method with algebraic polynomials given in Eqs. (7.3.43a,b) are (for $N = M = 1$ or 2)

$$N^u_{cr} = 10.943 \frac{D\pi^2}{a^2}, \quad N^b_{cr} = 5.471 \frac{D\pi^2}{a^2}$$

8.3.3 Buckling of a Simply Supported Plate Under Combined Bending and Compression

Let us consider a simply supported rectangular plate (see Figure 8.3.1) with distributed in-plane forces applied in the middle plane of the plate on sides $x = 0, a$. The distribution of the applied forces is assumed to be

$$\hat{N}_{xx} = -N_0 \gamma_1 = -N_0 \left(1 - c_0 \frac{y}{b}\right) \tag{8.3.20}$$

where N_0 is the magnitude of the compressive force at $y = 0$ and c_0 is a parameter that defines the relative bending and compression. For example, $c_0 = 0$ corresponds to the case of uniformly distributed compressive force, as discussed in section 8.2, and for $c_0 = 2$ we obtain the case of pure bending. All other values give a combination of bending and compression or tension.

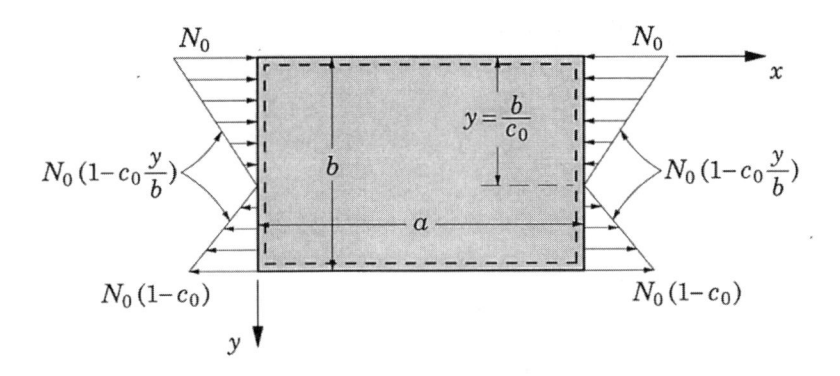

Figure 8.3.1. Buckling of simply supported plates under combined bending and compression.

380 THEORY AND ANALYSIS OF ELASTIC PLATES

We seek the deflection of the buckled plate, which is simply supported on all sides, in the form of a double sine series

$$w(x,y) \approx W_{MN} = \sum_{n=1}^{N}\sum_{m=1}^{M} c_{mn} \sin\frac{m\pi x}{a} \sin\frac{n\pi y}{b} \quad (8.3.21)$$

that satisfies the geometric as well as the force boundary conditions of the problem.

Substituting Eq. (8.3.21) for w and

$$\delta w = \sin\frac{p\pi x}{a} \sin\frac{q\pi y}{b}$$

into Eq. (8.3.10) with $\hat{N}_{yy} = \hat{N}_{xy} = 0$, we obtain

$$0 = \sum_{n=1}^{N}\sum_{m=1}^{M} c_{mn}\left\{D_{11}\left(\frac{m\pi}{a}\right)^2\left(\frac{p\pi}{a}\right)^2 + D_{22}\left(\frac{n\pi}{b}\right)^2\left(\frac{q\pi}{b}\right)^2\right.$$

$$\left. + D_{12}\left[\left(\frac{n\pi}{b}\right)^2\left(\frac{p\pi}{a}\right)^2 + \left(\frac{m\pi}{a}\right)^2\left(\frac{q\pi}{b}\right)^2\right]\right\}I_1$$

$$+ \sum_{n=1}^{N}\sum_{m=1}^{M} c_{mn}4D_{66}\left(\frac{m\pi}{a}\right)\left(\frac{n\pi}{b}\right)\left(\frac{p\pi}{a}\right)\left(\frac{q\pi}{b}\right)I_2$$

$$- \sum_{n=1}^{N}\sum_{m=1}^{M}\sum_{q=1}^{M} c_{mq}N_0\left(\frac{m\pi}{a}\right)\left(\frac{p\pi}{a}\right)I_{nq} \quad (8.3.22)$$

where I_1 and I_2 are nonzero only when $p = m$ and $q = n$

$$I_1 = \int_0^b\int_0^a \sin\frac{m\pi x}{a}\sin\frac{n\pi y}{b}\sin\frac{p\pi x}{a}\sin\frac{q\pi y}{b}\,dxdy = \frac{ab}{4}$$

$$I_2 = \int_0^b\int_0^a \cos\frac{m\pi x}{a}\cos\frac{n\pi y}{b}\cos\frac{p\pi x}{a}\cos\frac{q\pi y}{b}\,dxdy = \frac{ab}{4}$$

$$(8.3.23a)$$

and I_{nq} is defined as

$$I_{nq} = \int_0^b\int_0^a \left(1 - c_0\frac{y}{b}\right)\cos\frac{m\pi x}{a}\cos\frac{p\pi x}{a}\sin\frac{n\pi y}{b}\sin\frac{q\pi y}{b}\,dxdy$$

$$(8.3.23b)$$

GENERAL BUCKLING OF RECTANGULAR PLATES 381

which can be computed with the help of the following identities:

$$I_0 \equiv \int_0^b y \sin \frac{n\pi y}{b} \sin \frac{q\pi y}{b} \, dy$$

$$= \frac{b^2}{4} \quad \text{when } n = q$$

$$= 0 \quad \text{when } n \neq q \text{ and } n \pm q \text{ an even number}$$

$$= -\frac{b^2}{\pi^2} \frac{2nq}{(n^2 - q^2)^2} \quad \text{when } n \neq q \text{ and}$$

$$n \pm q \text{ an odd number} \qquad (8.3.24)$$

Thus $I_{nq} = 0$ if $p \neq m$ and

$$I_{nq} = \frac{a}{2}\left(\frac{b}{2} - \frac{c_0}{b}I_0\right) \qquad (8.3.25)$$

when $p = m$. For any m and n, Eq. (8.3.22) becomes

$$0 = \left[D_{11}\left(\frac{m\pi}{a}\right)^4 + D_{22}\left(\frac{n\pi}{b}\right)^4 + 2\hat{D}_{12}\left(\frac{n\pi}{b}\right)^2\left(\frac{m\pi}{a}\right)^2\right]c_{mn}$$

$$- N_0\left(\frac{m\pi}{a}\right)^2\left[c_{mn} - \frac{c_0}{2}\left(c_{mn} - \frac{8}{\pi^2}\sum_{q=1}^{M}\frac{2nqc_{mq}}{(n^2-q^2)^2}\right)\right] \qquad (8.3.26)$$

where the summation is taken over all numbers q such that $n \pm q$ is an odd number. Taking $m = 1$ in Eq. (8.3.26), we obtain

$$\left(D_{11}s^4 + D_{22}n^4 + 2\hat{D}_{12}n^2s^2\right)c_{1n}$$

$$= N_0 \frac{s^2 b^2}{\pi^2}\left[c_{1n}\left(1 - \frac{c_0}{2}\right) + \frac{8c_0}{\pi^2}\sum_{q=1}^{M}\frac{nqc_{1q}}{(n^2-q^2)^2}\right] \qquad (8.3.27)$$

where s denotes the aspect ratio $s = b/a$. A nontrivial solution, i.e., for nonzero c_{1i}, the determinant of the linear equations in (8.3.27) must be zero if the plate buckles.

For one-parameter approximation ($N = M = 1$), we obtain

$$N_0 = \frac{\pi^2}{s^2 b^2}\left(s^4 D_{11} + D_{22}n^4 + 2\hat{D}_{12}n^2s^2\right)\frac{1}{1 - 0.5c_0} \qquad (8.3.28)$$

which gives a satisfactory result only for small values of c_0, i.e., in cases where the bending stresses are small compared with the uniform compressive stress. In particular, when $c_0 = 0$, Eq. (8.3.28) gives the result in Eq. (8.1.22) for $m = 1$. Higher-order approximations yield sufficiently accurate results for the case of pure bending. Table 8.3.2 contains critical buckling loads $\bar{N} = N_0(b^2/\pi^2 D)$ obtained using two-parameter approximation, except for $c_0 = 2$, where the three-parameter approximation was used (see [1]).

Table 8.3.2. Nondimensionalized critical buckling loads $\bar{N}$ of simply supported rectangular isotropic plates under combined bending and compression $\hat{N}_{xy} = N_0(1 - c_0 y/b)$.

c_0	$\frac{a}{b} \rightarrow$	0.4	0.5	0.6	$\frac{2}{3}$	0.75	0.8	0.9	1.0	1.5
2		29.1	25.6	24.1	23.9	24.1	24.4	25.6	25.6	24.1
4/3		18.7	—	12.9	—	11.5	11.2	—	11.0	11.5
1		15.1	—	9.7	—	8.4	8.1	—	7.8	8.4
4/5		13.3	—	8.3	—	7.1	6.9	—	6.6	7.1
2/3		10.8	—	7.1	—	6.1	6.0	—	5.8	6.1

8.3.4 Buckling of a Simply Supported Plate Under In-plane Shear

When the plate is simply supported on all its edges and subjected to uniformly distributed in-plane shear force $\hat{N}_{xy} = -N_{xy}^0$ (see Figure 8.3.2), the Navier or Lévy solution procedure cannot be used to determine the critical buckling load. Hence, we will seek an approximate solution by a variational method.

Let us seek the solution in the form

$$w(x, y) \approx W_{MN} = \sum_{n=1}^{N} \sum_{m=1}^{M} c_{mn} \sin \frac{m\pi x}{a} \sin \frac{n\pi y}{b} \qquad (8.3.29)$$

The approximate solution satisfies the geometric ($w = 0$) as well as the natural ($M_{xx} = 0$ on sides $x = 0, a$ and $M_{yy} = 0$ on sides $y = 0, b$) boundary conditions of the problem. Therefore, both the Rayleigh–Ritz and Galerkin methods yield the same solutions. The Galerkin solution

GENERAL BUCKLING OF RECTANGULAR PLATES

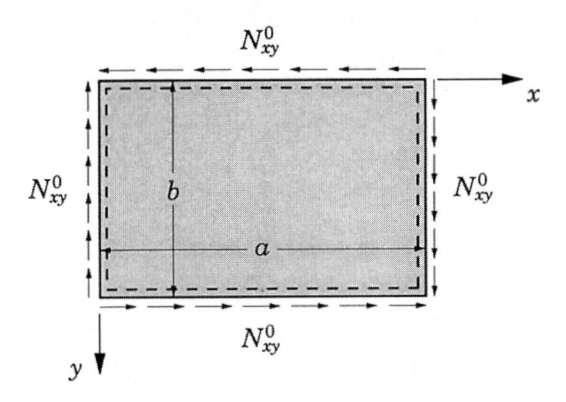

Figure 8.3.2. Buckling of rectangular plates under the action of shearing stresses.

is obtained by substituting Eq. (8.3.29) in the "weighted-residual" statement

$$0 = \int_0^b \int_0^a \left[D_{11} \frac{\partial^4 w}{\partial x^4} + 2\hat{D}_{12} \frac{\partial^4 w}{\partial x^2 \partial y^2} + D_{22} \frac{\partial^4 w}{\partial y^4} - 2N_{xy}^0 \frac{\partial^2 w}{\partial x \partial y} \right] \varphi_{pq} \, dx dy \tag{8.3.30}$$

We obtain

$$0 = \frac{ab}{4} \left[D_{11} \left(\frac{p\pi}{a}\right)^4 + (D_{12} + 2D_{66}) \left(\frac{p\pi}{a}\right)^2 \left(\frac{q\pi}{b}\right)^2 + D_{22} \left(\frac{q\pi}{b}\right)^4 \right] c_{pq}$$

$$- 2N_{xy}^0 \sum_{m=1}^N \sum_{n=1}^M \left(\frac{m\pi}{a}\right) \left(\frac{n\pi}{b}\right) I_{mp} I_{nq} \, c_{mn} \tag{8.3.31}$$

where

$$I_{mp} = \int_0^a \cos \frac{m\pi x}{a} \sin \frac{p\pi x}{a} \, dx$$

$$= \left(\frac{2a}{\pi^2}\right) \frac{p}{(p^2 - m^2)} \quad \text{for} \quad p^2 \neq m^2 \tag{8.3.32a}$$

$$I_{nq} = \int_0^b \cos \frac{n\pi y}{b} \sin \frac{q\pi y}{b} \, dy$$

$$= \left(\frac{2b}{\pi^2}\right) \frac{q}{(q^2 - n^2)} \quad \text{for} \quad q^2 \neq n^2 \tag{8.3.32b}$$

and the integral I_{mp} is zero when $p = m$ or $p \pm m$ is an even number, and I_{nq} is zero when $q = n$ or $q \pm n$ is an even number. The set of mn homogeneous equations (8.3.31) define an eigenvalue problem

$$\sum_{m=1}^{N}\sum_{n=1}^{M}\left(A_{(mn),(pq)} - N_{xy}^{0} G_{(mn)(pq)}\right) c_{mn} = 0$$

$$\left([A] - N_{xy}^{0}[G]\right)\{c\} = \{0\} \qquad (8.3.33a)$$

where

$$A_{(mn)(pq)} = \delta_{mp}\delta_{nq}\frac{ab}{4}\left[D_{11}\alpha_m^2\alpha_p^2 + 2\hat{D}_{12}\alpha_m\alpha_p\beta_n\beta_q + D_{22}\beta_n^2\beta_q^2\right]$$
$$G_{(mn)(pq)} = 2\alpha_m\beta_n\, I_{mp}I_{nq}$$
$$\hat{D}_{12} = D_{12} + 2D_{66}, \quad \alpha_m = \frac{m\pi}{a}, \quad \beta_n = \frac{n\pi}{b} \qquad (8.3.33b)$$

which has a nontrivial solution (i.e., $c_{mn} \neq 0$) when the determinant of the coefficient matrix is zero. Note that $[A]$ is a diagonal matrix while $[G]$ is a nonpositive-definite matrix; hence, the solution of (8.3.33a) requires an eigenvalue routine that is suitable for nonpositive-definite matrices. It is found that the solution of (8.3.33a) converges very slowly with increasing values of M and N. We note that $[G]$ does not exist for $M = N = 1$.

For $M = N = 2$, we find from Eq. (8.3.33a) the result ($g_{(11)(12)} = g_{(11)(21)} = 0$)

$$\left(\begin{bmatrix} a_{11} & 0 \\ 0 & a_{22} \end{bmatrix} - N_{xy}^{0}\begin{bmatrix} 0 & g_{(11)(22)} \\ g_{(22)(11)} & 0 \end{bmatrix}\right)\begin{Bmatrix} c_{11} \\ c_{22} \end{Bmatrix} = \begin{Bmatrix} 0 \\ 0 \end{Bmatrix} \qquad (8.3.34)$$

where the coefficients are given by

$$a_{(11)(11)} = \frac{\pi^4}{4sb^2}\left(D_{11}s^4 + 2\hat{D}_{12}s^2 + D_{22}\right), \quad a_{(22)(22)} = 16 a_{(11)(11)}$$

$$g_{(11)(11)} = g_{(22)(22)} = 0, \quad g_{(11)(22)} = g_{(22)(11)} = \frac{32}{9} \qquad (8.3.35)$$

where $s = b/a$ is the plate aspect ratio. Setting the determinant of the coefficient matrix in Eq. (8.3.34) to zero, we obtain

$$N_{xy}^{0} = \pm\frac{9\pi^4}{32sb^2}\left(D_{11}s^4 + 2\hat{D}_{12}s^2 + D_{22}\right) \qquad (8.3.36)$$

The two signs indicate that the value of the critical buckling load does not depend on sign.

Timoshenko and Gere [1] obtained the following equation for short isotropic plates ($a/b < 2$) using a five-term ($c_{11}, c_{22}, c_{13}, c_{31}, c_{33}$ and c_{42}) approximation:

$$\lambda^2 = \frac{s^4}{81(1+s^2)^4}\left[1 + \frac{81}{625} + \frac{81}{25}\left(\frac{1+s^2}{9+s^2}\right)^2 + \frac{81}{25}\left(\frac{1+s^2}{1+9s^2}\right)^2\right] \quad (8.3.37)$$

where

$$s = \frac{b}{a}, \quad \lambda = -\frac{\pi^4 D}{32ab N^0_{xy}} \quad (8.3.38)$$

For a square plate, Eq. (8.3.38) yields the critical buckling load

$$(N^0_{xy})_{cr} = 9.4\frac{\pi^2 D}{b^2} \quad (8.3.39)$$

whereas the value obtained with a larger (than 5) number of equations give 9.34 in place of 9.4. Table 8.3.3 contains the critical buckling loads $\bar{N} = (N^0_{xy})_{cr}(b^2/\pi^2 D)$ obtained using a large number of parameters (see [1]).

Table 8.3.3. Nondimensionalized critical buckling loads $\bar{N}$ of rectangular isotropic ($\nu = 0.25$) plates under uniform shear $\hat{N}_{xy} = N^0_{xy}$ (Rayleigh–Ritz solution).

$\frac{a}{b}$	1.0	1.2	1.4	1.5	1.6	1.8	2.0	2.5	3	4
$\bar{N}$	9.34	8.0	7.3	7.1	7.0	6.8	6.6	6.1	5.9	5.7

8.3.5 Buckling of Clamped Plates Under In-plane Shear

The total potential energy expression for the clamped rectangular plate under in-plane shear load N^0_{xy} is

$$\Pi(w) = \frac{1}{2}\int_0^b\int_0^a\left[D_{11}\left(\frac{\partial^2 w}{\partial x^2}\right)^2 + 2D_{12}\frac{\partial^2 w}{\partial x^2}\frac{\partial^2 w}{\partial y^2} + 4D_{66}\left(\frac{\partial^2 w}{\partial x \partial y}\right)^2 \right.$$
$$\left. + D_{22}\left(\frac{\partial^2 w}{\partial y^2}\right)^2 - 2N^0_{xy}\left(\frac{\partial w}{\partial x}\frac{\partial w}{\partial y}\right)\right]dxdy \quad (8.3.40)$$

386   THEORY AND ANALYSIS OF ELASTIC PLATES

The minimum total potential energy principle requires that $\delta\Pi = 0$. We have

$$0 = \int_0^b \int_0^a \left[D_{11} \frac{\partial^2 w}{\partial x^2} \frac{\partial^2 \delta w}{\partial x^2} + D_{12} \left(\frac{\partial^2 w}{\partial y^2} \frac{\partial^2 \delta w}{\partial x^2} + \frac{\partial^2 w}{\partial x^2} \frac{\partial^2 \delta w}{\partial y^2} \right) \right.$$

$$+ 4 D_{66} \frac{\partial^2 w}{\partial x \partial y} \frac{\partial^2 \delta w}{\partial x \partial y} + D_{22} \frac{\partial^2 w}{\partial y^2} \frac{\partial^2 \delta w}{\partial y^2}$$

$$\left. - N_{xy}^0 \left(\frac{\partial \delta w}{\partial x} \frac{\partial w}{\partial y} + \frac{\partial w}{\partial x} \frac{\partial \delta w}{\partial y} \right) \right] dx\,dy \quad (8.3.41)$$

We assume a Rayleigh–Ritz approximation of the form

$$w(x,y) \approx W_{MN}(x,y) = \sum_{j=1}^{N} \sum_{i=1}^{M} c_{ij}\, X_i(x)\, Y_j(y) \quad (8.3.42)$$

Substituting Eq. (8.3.42) into Eq. (8.3.41), we obtain

$$0 = \sum_{j=1}^{N} \sum_{i=1}^{M} c_{ij} \left\{ \int_0^b \int_0^a \left[D_{11} \frac{d^2 X_i}{dx^2} Y_j \frac{d^2 X_p}{dx^2} Y_q + D_{22} X_i \frac{d^2 Y_j}{dy^2} X_p \frac{d^2 Y_q}{dy^2} \right.\right.$$

$$+ 2\hat{D}_{12} \frac{dX_i}{dx} \frac{dY_j}{dy} \frac{dX_p}{dx} \frac{dY_q}{dy}$$

$$\left.\left. - N_{xy}^0 \left(\frac{dX_i}{dx} Y_j X_p \frac{dY_q}{dy} + X_i \frac{dY_j}{dy} \frac{dX_p}{dx} Y_q \right) \right] dx\,dy \right\} c_{ij} \quad (8.3.43a)$$

$$= \sum_{j=1}^{N} \sum_{i=1}^{M} c_{ij} \left(a_{(ij)(pq)} - N_{xy}^0 g_{(ij)(pq)} \right) c_{ij} \quad (8.3.43b)$$

Using the two-parameter approximation of the form

$$w(x,y) \approx c_{11} X_1(x) Y_1(y) + c_{22} X_2(x) Y_2(y) \quad (8.3.44)$$

with

$$X_1(x) = \sin \frac{4.73x}{a} - \sinh \frac{4.73x}{a} + 1.0178 \left(\cosh \frac{4.73x}{a} - \cos \frac{4.73x}{a} \right)$$

$$X_2(x) = \sin \frac{7.853x}{a} - \sinh \frac{7.853x}{a} + 0.9992 \left(\cosh \frac{7.853x}{a} - \cos \frac{7.853x}{a} \right)$$

$$Y_1(y) = \sin \frac{4.73y}{b} - \sinh \frac{4.73y}{b} + 1.0178 \left(\cosh \frac{4.73y}{b} - \cos \frac{4.73y}{b} \right)$$

$$Y_2(y) = \sin \frac{7.853y}{b} - \sinh \frac{7.853y}{b} + 0.9992 \left(\cosh \frac{7.853y}{b} - \cos \frac{7.853y}{b} \right)$$

$$(8.3.45)$$

we obtain

$$\begin{bmatrix} b_{11} & -N_{xy}^0 b_{12} \\ -N_{xy}^0 b_{12} & b_{22} \end{bmatrix} \begin{Bmatrix} c_{11} \\ c_{22} \end{Bmatrix} = -\begin{Bmatrix} 0 \\ 0 \end{Bmatrix} \qquad (8.3.46)$$

where

$$b_{11} = \frac{537.181}{a^4} D_{11} + \frac{324.829}{a^2 b^2} \hat{D}_{12} + \frac{537.181}{b^4} D_{22}, \quad b_{12} = \frac{23.107}{ab}$$

$$b_{22} = \frac{3791.532}{a^4} D_{11} + \frac{4227.255}{a^2 b^2} \hat{D}_{12} + \frac{3791.532}{b^4} D_{22} \qquad (8.3.47)$$

and $\hat{D}_{12} = D_{12} + 2D_{66}$. For a nontrivial solution, the determinant of the coefficient matrix in Eq. (8.3.46) should be zero, $b_{11} b_{22} - b_{12} b_{12} (N_{xy}^0)^2 = 0$. Solving for the buckling load N_{xy}^0, we obtain

$$N_{xy}^0 = \pm \frac{1}{a_{12}} \sqrt{a_{11} a_{22}} \qquad (8.3.48)$$

The ± sign indicates that the shear buckling load may be either positive or negative.

For an isotropic square plate, we have $a = b$ and $D_{11} = D_{22} = (D_{12} + 2D_{66}) = D$, and the shear buckling load predicted by Eq. (8.3.48) is

$$N_{xy}^0 = \pm 176 \frac{D}{a^2} \qquad (8.3.49)$$

whereas the "exact" critical buckling load is

$$N_{xy}^0 = \pm 145 \frac{D}{a^2} \qquad (8.3.50)$$

The two-term Rayleigh–Ritz solution (8.3.49) is over 21% in error.

This completes the discussion on the application of the Rayleigh–Ritz method to the buckling of rectangular plates. The Rayleigh–Ritz method proves to be very powerful in obtaining approximate critical buckling loads of rectangular plates for boundary conditions for which the Navier and Lévy solution procedures do not apply. In particular, buckling loads of rectangular plates clamped on all four edges can be determined using the Rayleigh–Ritz method.

Exercises

8.1 Consider buckling of uniformly compressed rectangular plates with edges $x = 0, a$ and $y = 0, b$ simply supported. The boundary conditions are

$$w = 0, \quad M_{yy} = -D_{12}\frac{\partial^2 w}{\partial x^2} - D_{22}\frac{\partial^2 w}{\partial y^2} = 0 \quad \text{at} \quad y = 0, b \quad (a)$$

Use the boundary conditions (a) to show that the constants $A - -C$ in Eq. (8.2.7) are given by $A = C = 0$ and

$$B \sinh \lambda_1 b + D \sin \lambda_2 b = 0, \quad \lambda_1^2 B \sinh \lambda_1 b - \lambda_2^2 D \sin \lambda_2 b = 0 \quad (b)$$

Show that the buckled form of equilibrium of the plate is possible only if

$$\lambda_2 = \frac{n\pi}{b} = \beta_n \quad (c)$$

Then, using Eqs. (8.2.8b), show that

$$N_0(m, n) = \frac{1}{\alpha_m^2} \left(D_{11}\alpha_m^4 + 2\alpha_m^2 \beta_n^2 \hat{D}_{12} + D_{22}\beta_n^4 \right) \quad (d)$$

which is the same as Eq. (8.1.21).

8.2 Use the *state-space approach* of Exercise 7.17 to find the general solution of Eq. (8.2.4) for the case of biaxial buckling. In particular, show that the characteristic equation is given by

$$\lambda^4 - C_2\lambda^2 - C_1 = 0 \quad (a)$$

Hint: The constants C_1 and C_2 of Exercise 7.17 for the present case are given by

$$C_1 = -\frac{1}{D_{11}}\left(D_{22}\beta_n^4 + \hat{N}_{yy}\beta_n^2\right), \quad C_2 = \frac{2\beta_n^2 \hat{D}_{12} + \hat{N}_{xx}}{D_{11}} \quad (b)$$

Then find the eigenvalues of the matrix $[T]$.

8.3 Consider the buckling of uniformly compressed rectangular SSCS plates with sides $x = 0, a$ simply supported, side $y = 0$ clamped, and side $y = b$ simply supported. The boundary conditions are

$$w = 0, \quad \frac{\partial w}{\partial y} = 0 \quad \text{at} \quad y = 0 \quad (a)$$

$$w = 0, \quad M_{yy} = -D_{22}\frac{\partial^2 w}{\partial y^2} - D_{12}\frac{\partial^2 w}{\partial x^2} = 0 \text{ at } y = 0 \quad \text{(b)}$$

Show that the characteristic equation is given by

$$\lambda_2 \tanh \lambda_1 b - \lambda_1 \tan \lambda_2 b = 0 \quad \text{(c)}$$

8.4 Establish the following identities with the help of Eqs. (8.2.8a,b):

$$\Omega_1 \bar{\Omega}_2 = -\Omega_1^2, \quad \Omega_2 \bar{\Omega}_1 = -\Omega_2^2$$

where ($\bar{D}_{12} = D_{12} + 4D_{66}$)

$$\Omega_1 = \lambda_1^2 - \frac{D_{12}}{D_{22}}\alpha_m^2, \quad \Omega_2 = \lambda_2^2 + \frac{D_{12}}{D_{22}}\alpha_m^2$$

$$\bar{\Omega}_1 = \lambda_1^2 - \frac{\bar{D}_{12}}{D_{22}}\alpha_m^2, \quad \bar{\Omega}_2 = \lambda_2^2 + \frac{\bar{D}_{12}}{D_{22}}\alpha_m^2$$

8.5 Establish the characteristic equation of SSSF plates, i.e., Eq. (8.2.12).

8.6 Derive the boundary condition on the *free edge* $y = b$ of a plate when it is biaxially compressed. *Hint:* Use Eqs. (3.5.11) and (3.5.15) to show that

$$V_y(b) = \left(M_{yy,y} + 2M_{xy,x} + \hat{N}_{yy}\frac{\partial w}{\partial y}\right)_{y=b}$$

8.7 Derive the characteristic equation associated with SSEF (i.e., elastically restrained at $y = 0$) plates using the boundary condition in Eq. (8.2.20) and $w = 0$ at $y = 0$.

8.8 Show that the one-parameter Rayleigh–Ritz approximation (8.3.15) with X_i and Y_j defined by Eq. (8.3.17) yields the following critical buckling load for a simply supported plate under combined bending and compression

$$N_{cr} = \left(\frac{12b}{a^3}D_{11} + \frac{20}{ab}\hat{D}_{12} + \frac{12a}{b^3}D_{22}\right)\frac{1}{(1 - 0.5c_0)}$$

8.9 Show that the two-parameter Rayleigh–Ritz approximation in Eq. (8.3.21) yields the equations [see Eq. (8.3.27)]

$$\left[\left(1 + \frac{a^2}{b^2}\right)^2 - N_0\frac{a^2}{\pi^2 D}(1 - 0.5c_0)\right]c_{11} - N_0 c_0\frac{16a^2}{9\pi^4 D}c_{12} = 0$$

$$-N_0 c_0 \frac{16a^2}{9\pi^4 D} c_{11} + \left[\left(1 + 4\frac{a^2}{b^2}\right)^2 - N_0 \frac{a^2}{\pi^2 D}(1 - 0.5c_0)\right] c_{12} = 0 \quad \text{(a)}$$

and show that the associated characteristic equation is

$$A\lambda^2 - B\lambda + C = 0 \tag{b}$$

where $\lambda = N_0(a^2/\pi^2 D)$ and

$$A = (1 - 0.5c_0)^2 - \left(\frac{16a^2}{9\pi^4 D}\right)^2$$

$$B = (1 - 0.5c_0)\left[\left(1 + \frac{a^2}{b^2}\right)^2 + \left(1 + 4\frac{a^2}{b^2}\right)^2\right]$$

$$C = \left(1 + \frac{a^2}{b^2}\right)^2 \left(1 + 4\frac{a^2}{b^2}\right)^2 \tag{c}$$

References for Additional Reading

1. Timoshenko, S. P. and Gere, J. M., *Theory of Elastic Stability*, Second Edition, McGraw–Hill, New York (1961).

2. Szilard, R., *Theory and Analysis of Plates, Classical and Numerical Methods*, Prentice–Hall, Englewood Cliffs, NJ (1974).

3. Reddy, J. N., *Energy and Variational Methods in Applied Mechanics*, John Wiley and Sons, New York (1984).

4. Reddy, J. N., *Mechanics of Laminated Composite Plates: Theory and Analysis*, CRC Press, Boca Raton, FL (1997).

5. Khdeir, A. A. and Reddy, J. N., "Buckling and Vibration of Laminated Composite Plates Using Various Plate Theories", *AIAA Journal*, **27**(12), 1808-1817 (1989).

Chapter Nine

Dynamic Analysis of Rectangular Plates

9.1 Introduction

9.1.1 Governing Equations

The equation of motion governing the linear bending of elastic plates is given by Eq. (3.8.9). In the absence of thermal loads and without elastic foundation (i.e., $k = 0$), the equation reduces to

$$D_{11}\frac{\partial^4 w_0}{\partial x^4} + 2\hat{D}_{12}\frac{\partial^4 w_0}{\partial x^2 \partial y^2} + D_{22}\frac{\partial^4 w_0}{\partial y^4}$$
$$+ I_0 \frac{\partial^2 w_0}{\partial t^2} - I_2 \left(\frac{\partial^4 w_0}{\partial t^2 \partial x^2} + \frac{\partial^4 w_0}{\partial t^2 \partial y^2} \right) = q \quad (9.1.1)$$

where $\hat{D}_{12} = 2(D_{12} + 2D_{66})$ and

$$I_0 = \rho_0 h, \quad I_2 = \frac{\rho_0 h^3}{12} \quad (9.1.2)$$

We wish to determine the natural frequencies of vibration and the transient response of a rectangular plate.

9.1.2 Natural Vibration

For natural vibration, the solution is assumed to be periodic

$$w_0(x, y, t) = w(x, y)e^{i\omega t} \quad (9.1.3)$$

where $i = \sqrt{-1}$ and ω is the frequency of natural vibration associated with mode shape w. Substituting Eq. (9.1.3) in Eq. (9.1.1) and setting

$q = 0$, we obtain

$$\left\{ D_{11} \frac{\partial^4 w}{\partial x^4} + 2\hat{D}_{12} \frac{\partial^4 w}{\partial x^2 \partial y^2} + D_{22} \frac{\partial^4 w}{\partial y^4} \right.$$
$$\left. - \omega^2 \left[I_0 w - I_2 \left(\frac{\partial^2 w}{\partial x^2} + \frac{\partial^2 w}{\partial y^2} \right) \right] \right\} e^{i\omega t} = 0 \quad (9.1.4)$$

which must hold for any time t. Hence, we have

$$D_{11} \frac{\partial^4 w}{\partial x^4} + 2\hat{D}_{12} \frac{\partial^4 w}{\partial x^2 \partial y^2} + D_{22} \frac{\partial^4 w}{\partial y^4}$$
$$- \omega^2 \left[I_0 w - I_2 \left(\frac{\partial^2 w}{\partial x^2} + \frac{\partial^2 w}{\partial y^2} \right) \right] = 0 \quad (9.1.5)$$

In this chapter, we will determine the solutions of Eq. (9.1.5) for various boundary conditions. In particular, we wish to find values of ω such that Eq. (9.1.5) has a non-trivial solution w.

9.1.3 Transient Analysis

The determination of the solution $w_0(x, y, t)$ of Eq. (9.1.1) for all times $t > 0$ under any applied load $q(x, y, t)$ is termed *transient response*. The transient response of rectangular plates can be determined by assuming that the solution is of the form

$$w_0(x, y, t) = w(x, y) T(t) \quad (9.1.6)$$

where $w(x, y)$ is only a function of the spatial coordinates x and y and T is a function of time t only. The Navier or Lévy methods can be used to determine the spatial part of the solution, and then the resulting ordinary differential equation in time is solved analytically or by numerical methods.

9.2 Natural Vibrations of Simply Supported Plates

Consider a rectangular plate with all four sides simply supported. The boundary conditions of a simply supported plate can be expressed in terms of w as

$$w = 0, \quad \frac{\partial^2 w}{\partial x^2} = \frac{\partial^2 w}{\partial y^2} = 0 \quad (9.2.1)$$

on all four edges. In the Navier solution procedure, we assume a solution of the form
$$w(x,y) = W_{mn} \sin\frac{m\pi x}{a} \sin\frac{n\pi y}{b} \qquad (9.2.2)$$
that satisfies the boundary conditions in Eq. (9.2.1). Equation (9.2.2) represents the mode shape of the plate for mode (m,n) and W_{mn} are the amplitudes of vibration. Substituting Eq. (9.2.2) into Eq. (9.1.5), we obtain the following relation for any m and n:
$$D_{11}\alpha_m^4 + 2\hat{D}_{12}\alpha_m^2\beta_n^2 + D_{22}\beta_n^4 - \omega^2\left[I_0 + \left(\alpha_m^2 + \beta_n^2\right)I_2\right] = 0 \qquad (9.2.3)$$
where $\alpha_m = m\pi/a$ and $\beta_n = n\pi/b$. Solving Eq. (9.2.3) for the natural frequency, we obtain
$$\omega^2 = \frac{\pi^4}{\tilde{I}_0 b^4}\left[D_{11}m^4\left(\frac{b}{a}\right)^4 + 2(D_{12}+2D_{66})m^2n^2\left(\frac{b}{a}\right)^2 + D_{22}n^4\right] \qquad (9.2.4)$$
where
$$\tilde{I}_0 = I_0 + I_2\left[\left(\frac{m\pi}{a}\right)^2 + \left(\frac{n\pi}{b}\right)^2\right] \qquad (9.2.5)$$

For different values of m and n, there correspond a unique frequency ω_{mn} and mode shape given by Eq. (9.2.2). The smallest value of ω_{mn} is called the *fundamental frequency*. When the rotatory inertia I_2 is not zero, it is not easy to find the lowest natural frequency. Equation (9.2.4) shows that the rotatory (or rotary) inertia has the effect of reducing the magnitude of the frequency of vibration and its relative effect depends on m and n, degree of orthotropy, and the plate thickness-to-side ratio h/b. For most plates with $h/b < 0.1$, the rotary inertia can be neglected.

When the rotatory inertia I_2 is neglected, the frequency of a rectangular orthotropic plate reduces to ($I_0 = \rho h$)
$$\omega_{mn}^2 = \frac{\pi^4}{\rho h b^4}\left[D_{11}m^4\left(\frac{b}{a}\right)^4 + 2(D_{12}+2D_{66})m^2n^2\left(\frac{b}{a}\right)^2 + D_{22}n^4\right] \qquad (9.2.6)$$

For isotropic rectangular plates, when the rotatory inertia I_2 is neglected, we have
$$\omega_{mn} = \frac{\pi^2}{b^2}\sqrt{\frac{D}{\rho h}}\left(m^2\frac{b^2}{a^2} + n^2\right) \qquad (9.2.7)$$

For a square isotropic plate, the expression becomes

$$\omega_{mn} = \frac{\pi^2}{a^2}\sqrt{\frac{D}{\rho h}}\left(m^2 + n^2\right) \quad (9.2.8)$$

and the fundamental frequency is given by

$$\omega_{11} = \frac{2\pi^2}{a^2}\sqrt{\frac{D}{\rho h}} \quad (9.2.9)$$

Nondimensionalized frequencies, $\bar{\omega}_{mn} = \omega_{mn}(b^2/\pi^2)\sqrt{\rho h/D_{22}}$, of isotropic ($\nu = 0.25$) and orthotropic ($G_{12}/E_2 = 0.5, /\nu_{12} = 0.25$) plates are presented in Table 9.2.1 for modulus ratios $E_1/E_2 = 3$ and 10. The fundamental frequency increases with modular ratio. The effect of including rotary inertia is to decrease the frequency of vibration, and the effect is negligible for plates with $h/b < 0.1$. Figure 9.2.1 shows a plot of nondimensionalized fundamental frequency $\bar{\omega}_{11}$ as a function of plate aspect ratio a/b for isotropic and orthotropic plates. The natural frequency increases with the degree of orthotropy E_1/E_2. For long plates, the frequency approaches that of a plate strip ($\bar{\omega}_{11} = 1.0$).

Table 9.2.1. Nondimensionalized fundamental frequencies $\bar{\omega}_{11}$ of simply supported (SSSS) isotropic and orthotropic plates[†].

$$\bar{\omega}_{11} = \omega_{11}(b^2/\pi^2)\sqrt{\rho h/D_{22}}$$

a/b	Isotropic			$E_1/E_2 = 3$			$E_1/E_2 = 10$		
	w/o	0.01	0.1	w/o	0.01	0.1	w/o	0.01	0.1
0.5	5.000	4.999	4.900	7.670	7.669	7.517	13.075	13.072	12.814
1.0	2.000	2.000	1.984	2.541	2.541	2.521	3.672	3.672	3.643
1.5	1.444	1.444	1.436	1.639	1.639	1.629	2.020	2.020	2.008
2.0	1.250	1.250	1.244	1.342	1.342	1.336	1.499	1.499	1.491
2.5	1.160	1.160	1.154	1.213	1.212	1.207	1.286	1.286	1.280
3.0	1.111	1.111	1.106	1.145	1.145	1.139	1.183	1.183	1.178

[†] w/o = without rotary inertia; the second and third columns contain frequencies when the rotary inertia is included for $h/b = 0.01$ and 0.1, respectively.

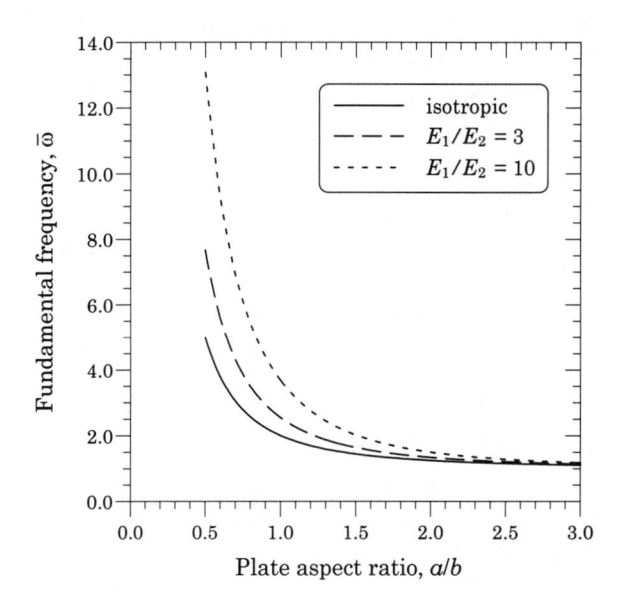

Figure 9.2.1 Nondimensionalized fundamental frequency versus aspect ratio for simply supported plates.

9.3 Natural Vibration of Plates with Two Parallel Sides Simply Supported

9.3.1 The Lévy Solution

The Lévy method introduced in the previous chapters can also be used to determine natural frequencies of rectangular plates for which two opposite edges are simply supported and the other two edges have any combination of fixed, hinged, and free boundary conditions.

Recall from section 8.2 that in the Lévy method, the partial differential equation (9.1.5) is reduced to an ordinary differential equation in y by assuming solution in the form of a single Fourier series

$$w(x,y) = W_m(y) \sin \alpha_m x, \quad \alpha_m = \frac{m\pi}{a} \quad (9.3.1)$$

that satisfies the simply supported boundary conditions

$$w = 0, \quad M_{xx} = -\left(D_{11}\frac{\partial^2 w}{\partial x^2} + D_{12}\frac{\partial^2 w}{\partial y^2}\right) = 0 \quad (9.3.2)$$

on edges $x = 0, a$. Substituting Eq. (9.3.1) into Eq. (9.1.5), we obtain

$$\left[D_{11}\alpha_m^4 - \left(I_0 + \alpha_m^2 I_2\right)\omega^2\right] W - \left(2\alpha_m^2 \hat{D}_{12} - I_2\omega^2\right)\frac{d^2 W}{dy^2} + D_{22}\frac{d^4 W}{dy^4} = 0 \quad (9.3.3)$$

The ordinary differential equation (9.3.3) obtained in the Lévy method can be solved for the natural frequencies and mode shapes either analytically or by means of the Rayleigh–Ritz method. We will discuss both procedures in the following sections. Since the effect of rotary inertia I_2 is negligible on the fundamental frequency in majority of cases, we will neglect it.

9.3.2 Analytical Solution

The form of the solution to Eq. (9.3.3) depends on the nature of the roots λ of the equation

$$D_{22}\lambda^4 - 2\alpha_m^2 \hat{D}_{12}\lambda^2 + \left(\alpha_m^4 D_{11} - I_0\omega^2\right) = 0 \quad (9.3.4)$$

There are two cases: (1) the case in which $\omega^2 \geq \alpha_m^4 D_{11}/I_0$, and (2) the case in which $\omega^2 < \alpha_m^4 D_{11}/I_0$. The general solution of Eq. (9.3.3) when $\omega^2 \geq \alpha_m^4 D_{11}/I_0$ is given by [see Eq. (8.2.7)]

$$W_n(x) = A \cosh \lambda_1 y + B \sinh \lambda_1 y + C \cos \lambda_2 y + D \sin \lambda_2 y \quad (9.3.5)$$

and the solution for the case $\omega^2 < \alpha_m^4 D_{11}/I_0$ is given by

$$W_n(x) = A \cosh \lambda_1 y + B \sinh \lambda_1 y + C \cosh \lambda_2 y + D \sinh \lambda_2 y \quad (9.3.6)$$

where [see Eqs. (8.2.8a,b); replace $\alpha_m^2 N_0$ with $I_0 \omega^2$]

$$(\lambda_1)^2 = \sqrt{\omega^2 \frac{I_0}{D_{22}} - \alpha_m^4 \frac{D_{11}}{D_{22}} + \alpha_m^4 \left(\frac{\hat{D}_{12}}{D_{22}}\right)^2} + \frac{\hat{D}_{12}}{D_{22}}\alpha_m^2 \quad (9.3.7a)$$

$$(\lambda_2)^2 = \sqrt{\omega^2 \frac{I_0}{D_{22}} - \alpha_m^4 \frac{D_{11}}{D_{22}} + \alpha_m^4 \left(\frac{\hat{D}_{12}}{D_{22}}\right)^2} - \frac{\hat{D}_{12}}{D_{22}}\alpha_m^2 \quad (9.3.7b)$$

and A, B, C, and D are integration constants, which are determined using the boundary conditions. However, we do not actually determine these constants. Instead, the values of ω are determined by setting the determinant of the coefficient matrix of the four equations among the constants A, B, C and D to zero. The procedure is exactly parallel to that used in determining the buckling loads. Indeed, the characteristic equations derived in connection with the Lévy solutions of plates for various boundary conditions hold for natural vibrations as well for the correspondence

$$N_0 = \frac{I_0}{\alpha_m^2}\omega^2 \qquad (9.3.8)$$

In the sections that follow, we present natural frequencies for various plates.

9.3.3 Vibration of SSSF Plates

Consider natural vibrations of rectangular plates with sides $x = 0, a$ and $y = 0$ simply supported and side $y = b$ free (see Figure 8.2.2). The boundary conditions edges $y = 0, b$ are

$$w = 0, \quad M_{yy} = -\left(D_{12}\frac{\partial^2 w}{\partial x^2} + D_{22}\frac{\partial^2 w}{\partial y^2}\right) = 0 \quad \text{at} \quad y = 0 \qquad (9.3.9)$$

$$M_{yy} = 0, \quad V_y = -\left(D_{22}\frac{\partial^3 w}{\partial y^3} + \bar{D}_{12}\frac{\partial^3 w}{\partial x^2 \partial y}\right) = 0 \quad \text{at} \quad y = b \qquad (9.3.10)$$

where $\bar{D}_{12} = D_{12} + 4D_{66}$. It should be noted that when rotary inertia is included, the shear force V_y will include an additional term $-I_2\omega^2(dw/dy)$.

For the solution in Eq. (9.3.5), the boundary conditions yield $A=C=0$ and

$$-\left(\Omega_1 \sinh \lambda_1 b\right) B + \left(\Omega_2 \sin \lambda_2 b\right) D = 0$$
$$\left(\lambda_1 \bar{\Omega}_1 \cosh \lambda_1 b\right) B + \left(\lambda_2 \bar{\Omega}_2 \cos \lambda_2 b\right) D = 0 \qquad (9.3.11)$$

where

$$\Omega_1 = \left(\lambda_1^2 - \frac{D_{12}}{D_{22}}\alpha_m^2\right), \quad \Omega_2 = \left(\lambda_2^2 + \frac{D_{12}}{D_{22}}\alpha_m^2\right) \qquad (9.3.12a)$$

$$\bar{\Omega}_1 = \left(\lambda_1^2 - \frac{\bar{D}_{12}}{D_{22}}\alpha_m^2\right), \quad \bar{\Omega}_2 = \left(\lambda_2^2 + \frac{\bar{D}_{12}}{D_{22}}\alpha_m^2\right) \qquad (9.3.12b)$$

Setting the determinant of the coefficient matrix of the two equations in (9.3.11) to zero, we obtain the following characteristic equation for the natural vibration of SSSF plates:

$$\lambda_2 \Omega_1 \bar{\Omega}_2 \sinh \lambda_1 b \, \cos \lambda_2 b - \lambda_1 \Omega_2 \bar{\Omega}_1 \cosh \lambda_1 b \, \sin \lambda_2 b = 0 \qquad (9.3.13)$$

For the solution in Eq. (9.3.6), the boundary conditions yield $A = C = 0$ and

$$\begin{aligned}(\Omega_1 \sinh \lambda_1 b) \, B + (\Omega_3 \sin \lambda_2 b) \, D &= 0 \\ (\lambda_1 \bar{\Omega}_1 \cosh \lambda_1 b) \, B + (\lambda_2 \bar{\Omega}_3 \cos \lambda_2 b) \, D &= 0\end{aligned} \qquad (9.3.14)$$

which results in the characteristic equation

$$\lambda_2 \Omega_1 \bar{\Omega}_3 \sinh \lambda_1 b \, \cosh \lambda_2 b - \lambda_1 \Omega_3 \bar{\Omega}_1 \cosh \lambda_1 b \, \sinh \lambda_2 b = 0 \qquad (9.3.15)$$

$$\Omega_3 = \left(\lambda_2^2 - \frac{D_{12}}{D_{22}} \alpha_m^2\right), \quad \bar{\Omega}_3 = \left(\lambda_2^2 - \frac{\bar{D}_{12}}{D_{22}} \alpha_m^2\right) \qquad (9.3.16)$$

Equations (9.3.13) and (9.3.15) can be solved iteratively, for example, using the bisection method for the roots ω_{mn} for a given m once the geometric and material parameters of the plate are known. Here, m gives the number of half-sine waves in the x direction, and n is the nth lowest frequency for a given value of m. Care must be taken in the iterative procedure not to miss any roots, from the least to a desired number. The mode shape for the case $\omega^2 > \alpha_m^4 D_{11}/I_0$ is given by

$$w(x, y) = (\Omega_1 \sinh \lambda_1 b \sin \lambda_2 y + \Omega_2 \sinh \lambda_1 y \sin \lambda_2 b) \sin \alpha_m x \qquad (9.3.17)$$

and for the case $\omega^2 < \alpha_m^4 D_{11}/I_0$ it is given by

$$w(x, y) = (\Omega_1 \sinh \lambda_1 b \sinh \lambda_2 y + \Omega_2 \sinh \lambda_1 y \sinh \lambda_2 b) \sin \alpha_m x \qquad (9.3.18)$$

For the SSSF plates there exist no real roots of Eq. (9.3.15).

Table 9.3.1 contains the first seven natural frequencies $\bar{\omega}$ of isotropic ($\nu = 0.25$) and orthotropic plates for various aspect ratios. The following material properties were used for orthotropic plates:

$$E_1 = 3E_2 \text{ and } 10E_2, \quad G_{12} = 0.5E_2, \quad \nu_{12} = 0.25 \qquad (9.3.19)$$

Table 9.3.1. Effect of plate aspect ratio and modulus ratio on the nondimensionalized frequencies $\bar{\omega}_{mn} = \omega_{mn} a^2 (\sqrt{\rho h / D_{22}})$ of rectangular plates (SSSF).

$\frac{E_1}{E_2}$	Mode (m,n)	\multicolumn{6}{c}{Plate aspect ratio, $\frac{a}{b}$}					
		0.5	1.0	1.5	2.0	2.5	3.0
1	(1,1)	10.356	11.820	13.950	16.491	19.278	22.217
	(1,2)	14.860	27.995	48.195	75.678	110.653	153.226
	(1,3)	23.742	62.076	124.811	212.363	324.839	462.273
	(2,1)	39.884	41.422	43.939	47.278	51.280	55.801
	(2,2)	44.650	59.437	82.156	111.981	148.811	192.779
	(2,3)	54.074	94.969	159.641	248.304	361.421	499.245
	(3,1)	89.131	90.628	93.200	96.752	101.181	106.376
3	(1,1)	17.471	18.629	20.447	22.770	25.461	28.412
	(1,2)	21.038	33.066	52.768	80.024	114.882	157.387
	(1,3)	28.884	66.067	128.423	215.816	328.213	465.602
	(2,1)	68.714	69.882	71.834	74.515	77.859	81.786
	(2,2)	72.220	84.150	104.267	132.266	167.896	211.074
	(2,3)	79.569	115.536	177.233	264.268	376.446	513.691
	(3,1)	154.139	155.280	157.234	159.973	163.462	167.660
10	(1,1)	31.421	32.089	33.192	34.693	36.541	38.686
	(1,2)	33.558	42.220	59.019	84.358	118.016	159.757
	(1,3)	38.999	71.181	131.225	217.577	365.412	466.531
	(2,1)	125.029	125.684	126.795	128.355	130.352	132.769
	(2,2)	127.012	134.232	147.800	168.878	198.202	236.077
	(2,3)	131.375	155.997	206.153	284.726	391.315	524.901
	(3,1)	281.055	281.692	282.789	284.342	286.347	288.798

9.3.4 Vibration of SSCF Plates

Next, consider vibration of rectangular plates with sides $x = 0, a$ simply supported, side $y = 0$ clamped, and side $y = b$ free (see Figure 8.2.3). The boundary conditions on edges $y = 0, b$ are

$$w = 0, \quad \frac{\partial w}{\partial y} = 0 \quad \text{at} \quad y = 0 \tag{9.3.20}$$

$$M_{yy} = 0, \quad V_y = -D_{22}\frac{\partial^3 w}{\partial y^3} - \bar{D}_{12}\frac{\partial^3 w}{\partial x^2 \partial y} = 0 \quad \text{at} \quad y = b \tag{9.3.21}$$

Substitution of Eq. (9.3.5) into the boundary conditions (9.3.20) and (9.3.21) results in the following characteristic equation:

$$2\Omega_1\Omega_2 + \left(\Omega_1^2 + \Omega_2^2\right)\cosh\lambda_1 b \cos\lambda_2 b$$
$$- \frac{1}{\lambda_1\lambda_2}\left(\lambda_1^2\Omega_2^2 - \lambda_2^2\Omega_1^2\right)\sinh\lambda_1 b \sin\lambda_2 b = 0 \quad (9.3.22)$$

Tables 9.3.2 and 9.3.3 contain natural frequencies of isotropic and orthotropic plates, respectively. Table 9.3.2 contains the first five roots (i.e., $n = 1, 2, \cdots, 5$) of Eq. (9.3.22) for $m = 1, 2, \cdots, 6$ and for aspect

Table 9.3.2. Effect of plate aspect ratio on the nondimensionalized frequencies $\bar{\omega}_{mn} = \omega_{mn} b^2 (\sqrt{\rho h/D})$ of isotropic ($\nu = 0.25$) rectangular plates (SSCF).

| $\frac{a}{b}$ | m | \multicolumn{6}{c}{$\omega_{mn} b^2 \sqrt{\rho h/D}$ for values of n} |
		1	2	3	4	5	6
0.5	1	41.958	63.434	103.670	162.867	241.577	340.009
	2	159.782	180.971	222.271	282.892	362.616	461.579
	3	356.671	377.544	418.698	479.750	560.319	660.219
	4	632.519	653.260	694.146	755.122	835.918	936.313
	5	987.303	1008.03	1048.70	1109.49	1190.24	1290.78
1.0	1	12.862	33.317	72.605	131.592	210.391	308.977
	2	41.958	63.434	103.670	162.867	241.577	340.009
	3	90.971	112.381	153.435	213.415	292.520	391.068
	4	159.782	180.971	222.271	282.892	362.616	461.579
	5	248.354	269.358	310.630	371.576	451.818	551.286
1.5	1	7.614	27.266	66.594	125.672	204.549	303.192
	2	20.335	41.357	80.848	139.805	218.533	317.055
	3	41.958	63.434	103.670	162.867	241.577	340.009
	4	72.430	93.902	134.759	194.467	273.394	371.864
	5	111.711	133.048	174.232	234.461	313.769	412.431
2.0	1	5.810	25.036	64.462	123.589	202.499	301.164
	2	12.862	33.317	72.605	131.592	210.391	308.977
	3	24.907	46.116	85.774	144.752	223.456	321.948
	4	41.958	63.434	103.670	162.867	241.577	340.009
	5	63.985	85.478	126.208	185.779	264.631	363.077

Table 9.3.3. Effect of the plate aspect ratio and the modulus ratio on the nondimensionalized frequencies $\bar{\omega}_{mn} = \omega_{mn}b^2(\sqrt{\rho h/D_{22}})$ of orthotropic rectangular plates (SSCF).

		\multicolumn{6}{c}{$\omega_{mn}b^2\sqrt{\rho h/D_{22}}$ for values of n}
$\frac{a}{b}$	m	1	2	3	4	5	6

Modulus ratio, $E_1/E_2 = 3$

$\frac{a}{b}$	m	1	2	3	4	5	6
0.5	1	69.901	84.921	118.563	173.237	249.143	345.828
	2	274.662	287.897	316.217	362.353	428.564	516.092
	3	616.269	628.968	655.256	696.923	755.968	834.208
1.0	1	19.042	36.395	74.275	132.684	211.209	309.646
	2	69.901	84.921	118.563	173.237	249.143	345.828
	3	155.162	168.997	199.340	249.263	320.417	413.114
1.5	1	9.880	28.126	67.051	125.991	204.806	303.416
	2	32.143	48.577	85.092	142.573	220.542	318.636
	3	69.901	84.921	118.563	173.237	249.143	345.828
2.0	1	6.828	25.381	64.654	123.733	202.622	301.276
	2	19.042	36.395	74.275	132.684	211.209	309.646
	3	40.145	56.166	91.948	148.825	226.397	324.236

Modulus ratio, $E_1/E_2 = 10$

$\frac{a}{b}$	m	1	2	3	4	5	6
0.5	1	125.694	134.707	158.222	202.607	270.537	361.680
	2	500.009	507.477	524.214	553.485	599.190	664.909
	3	1124.06	1131.15	1146.13	1170.67	1207.04	1257.87
1.0	1	32.331	44.871	78.843	135.351	212.940	310.868
	2	125.694	134.707	158.222	202.607	270.537	361.680
	3	281.625	289.557	308.403	342.994	397.943	475.999
1.5	1	15.254	30.479	68.106	126.582	205.192	303.694
	2	56.477	67.279	97.091	150.134	225.581	322.211
	3	125.694	134.707	158.222	202.607	270.537	361.680
2.0	1	94.590	26.242	65.017	123.939	202.759	301.378
	2	32.331	44.871	78.843	135.351	212.940	310.868
	3	71.171	81.370	109.294	160.232	234.148	329.778

402 THEORY AND ANALYSIS OF ELASTIC PLATES

ratios $a/b = 0.5, 1.0, 1.5$, and 2.0 of isotropic plates. The first six roots for $m = 1, 2, 3$ are presented in Table 9.3.3 for orthotropic plates. The orthotropic material properties used are

$$E_1 = 3E_2 \text{ and } 10E_2, \quad G_{12} = 0.4E_2, \quad \nu_{12} = 0.25 \quad (9.3.23)$$

Note that frequencies of certain mode and aspect ratios are identical to those of another mode and aspect ratio. For example, the frequencies ω_{1n} for aspect ratio $a/b = 0.5$ are identical to ω_{2n} for aspect ratio $a/b = 1.0$, ω_{3n} for aspect ratio $a/b = 1.5$, and ω_{4n} for aspect ratio $a/b = 2.0$. Similar correspondence can be found among other frequencies of Tables 9.3.2 and 9.3.3. The effect of orthotropy on the frequencies is to increase their magnitude. The increase is more in the fundamental and lower modes, and it is almost negligible in higher modes.

9.3.5 Vibration of SSCC Plates

The boundary conditions on edges $y = 0, b$ of SSCC rectangular plates (with sides $x = 0, a$ simply supported and sides $y = 0, b$ clamped; see Figure 8.2.6) are

$$w = 0, \quad \frac{\partial w}{\partial y} = 0 \quad (9.3.24)$$

Substitution of Eq. (9.3.5) into Eq. (9.3.24) yields the following characteristic equation:

$$2\left(1 - \cosh \lambda_1 b \, \cos \lambda_2 b\right) + \left(\frac{\lambda_1}{\lambda_2} - \frac{\lambda_2}{\lambda_1}\right) \sinh \lambda_1 b \, \sin \lambda_2 b = 0 \quad (9.3.25)$$

The frequency equation obtained with Eq. (9.3.6) is

$$2\left(1 - \cosh \lambda_1 b \, \cosh \lambda_2 b\right) + \left(\frac{\lambda_1}{\lambda_2} + \frac{\lambda_2}{\lambda_1}\right) \sinh \lambda_1 b \, \sinh \lambda_2 b = 0 \quad (9.3.26)$$

and it does not have any roots when $\rho h \omega^2 < \alpha_m^4 (D_{11}/D_{22})$.

The first six roots of Eq. (9.3.25) for $m = 1, 2, \cdots, 6$ and for $a/b = 0.5, 1.0, 1.5$, and 2.0 of isotropic ($\nu = 0.25$) rectangular plates are presented in Table 9.3.4. The first five frequencies for modulus ratios $E_1/E_2 = 1, 3$, and 10 ($\nu_{12} = 0.25$ and $G_{12} = 0.4$) are listed in Table 9.3.5.

Table 9.3.4. Effect of plate aspect ratio on the frequencies ($\bar{\omega}_{mn}$) of isotropic ($\nu = 0.25$) rectangular plates (SSCC).

$\frac{a}{b}$	m	\multicolumn{6}{c}{$\bar{\omega}_{mn} = \omega_{mn} b^2 \sqrt{\rho h / D}$ for values of n}
		1	2	3	4	5	6
0.5	1	54.743	94.585	154.775	234.585	333.952	452.911
	2	170.346	206.696	265.195	344.536	444.034	563.379
	3	366.816	401.078	457.437	535.161	633.640	752.449
	4	642.726	730.684	806.895	904.072	1021.87	1160.01
	5	1084.00	1255.06	1371.79	1509.03	1666.58	1844.29
	6	1431.87	1463.80	1516.91	1591.07	1686.13	1801.92
1.0	1	28.951	69.327	129.095	208.391	307.316	425.914
	2	54.743	94.585	154.775	234.585	333.952	452.911
	3	102.216	140.204	199.809	279.650	379.267	498.532
	4	170.346	206.696	265.195	344.536	444.034	563.379
	5	258.613	293.755	351.112	429.676	528.750	647.924
	6	366.816	401.078	457.437	535.161	633.640	752.449
1.5	1	25.043	65.008	124.516	203.637	302.441	420.952
	2	35.104	75.605	135.612	215.103	314.173	432.882
	3	54.743	94.585	154.775	234.585	333.952	452.911
	4	84.054	122.672	182.564	262.462	362.034	481.217
	5	122.667	160.060	219.323	299.045	398.663	517.984
	6	170.346	206.696	265.195	344.536	444.034	563.379
2.0	1	23.815	63.534	122.930	201.982	300.739	419.218
	2	28.951	69.327	129.095	208.391	307.316	425.914
	3	39.089	79.525	139.622	219.207	318.354	437.124
	4	54.743	94.585	154.775	234.585	333.952	452.911
	5	75.841	114.779	174.785	254.686	354.221	473.354
	6	102.216	140.204	199.809	279.650	379.267	498.532

Table 9.3.5. Effect of modulus ratio on the first five nondimensionalized frequencies ($\bar{\omega}_{mn}$) of square plates (SSCC).

$\frac{E_1}{E_2}$	\multicolumn{5}{c}{$\bar{\omega}_{mn} = \omega_{mn} a^2 (\sqrt{\rho h / D_{22}})$ for (m, n)}
1	$28.951^{(1,1)}$	$54.743^{(2,1)}$	$69.327^{(1,2)}$	$94.585^{(2,2)}$	$102.216^{(3,1)}$
3	$32.263^{(1,1)}$	$70.930^{(1,2)}$	$78.386^{(2,1)}$	$110.370^{(2,2)}$	$130.098^{(1,3)}$
10	$41.539^{(1,1)}$	$75.654^{(1,2)}$	$130.632^{(2,1)}$	$132.778^{(1,3)}$	$152.094^{(2,2)}$

9.3.6 Vibration of SSCS Plates

The boundary conditions on edges $y = 0, b$ of SSCS rectangular plates (with sides $x = 0, a$ and $y = b$ simply supported and side $y = 0$ clamped) are

$$w = 0, \quad \frac{\partial w}{\partial y} = 0 \quad \text{at } y = 0 \quad (9.3.27a)$$

$$w = 0, \quad M_{yy} = -\left(D_{12}\frac{\partial^2 w}{\partial x^2} + D_{22}\frac{\partial^2 w}{\partial y^2}\right) = 0 \quad \text{at } y = b \quad (9.3.27b)$$

These boundary conditions in conjunction with the solution in Eq. (9.3.5) yield the characteristic equation

$$\lambda_1 \cosh \lambda_1 b \, \sin \lambda_2 b - \lambda_2 \sinh \lambda_1 b \, \cos \lambda_2 b = 0 \quad (9.3.28)$$

The frequency equation obtained with Eq. (9.3.6) is

$$\lambda_1 \cosh \lambda_1 b \, \sinh \lambda_2 b - \lambda_2 \sinh \lambda_1 b \, \cosh \lambda_2 b = 0 \quad (9.3.29)$$

and it does not have any roots when $\rho h \omega^2 < \alpha_m^4 (D_{11}/D_{22})$. The mode shapes associated with the frequencies obtained from Eq. (9.3.28) are

$$w(x, y) = (\sin \lambda_2 b \sinh \lambda_1 y - \sinh \lambda_1 b \sin \lambda_2 y) \sin \alpha_m x \quad (9.3.30)$$

The first six frequencies for modulus ratios $E_1/E_2 = 1, 3$, and 10 ($\nu_{12} = 0.25$ and $G_{12} = 0.5$) are listed in Table 9.3.6, and the first six roots of Eq. (9.3.28) for $m = 1, 2, \cdots, 6$ and for $a/b = 0.5, 1.0, 1.5$, and 2.0 of isotropic ($\nu = 0.25$) rectangular plates are presented in Table 9.3.7.

Table 9.3.6. Effect of the modulus ratio on the first six nondimensionalized frequencies ($\bar{\omega}_{mn}$) of square plates (SSCS).

$\frac{E_1}{E_2}$	$\bar{\omega}_{mn} = \omega_{mn} a^2 (\sqrt{\rho h/D_{22}})$ for (m,n)					
1	23.646$^{(1,1)}$	51.674$^{(2,1)}$	58.646$^{(1,2)}$	86.134$^{(2,2)}$	100.270$^{(3,1)}$	113.228$^{(1,3)}$
3	28.365$^{(1,1)}$	61.867$^{(1,2)}$	77.345$^{(2,1)}$	106.306$^{(2,2)}$	115.933$^{(1,3)}$	156.998$^{(2,3)}$
10	38.596$^{(1,1)}$	67.244$^{(1,2)}$	118.951$^{(1,3)}$	130.017$^{(2,1)}$	149.195$^{(2,2)}$	188.854$^{(2,3)}$

Table 9.3.7. Effect of the plate aspect ratio on the nondimensionalized frequencies of isotropic ($\nu = 0.25$) rectangular plates (SSCS).

		\multicolumn{6}{c}{$\bar{\omega}_{mn} = \omega_{mn} b^2 \sqrt{\rho h/D}$ for values of n}				
$\frac{a}{b}$	m	1	2	3	4	5	6
0.5	1	86.134	140.846	215.294	309.403	423.178	556.634
	2	168.959	201.725	255.469	329.574	423.682	537.614
	3	365.950	450.482	523.760	617.315	730.935	864.474
	4	642.100	673.388	725.392	797.937	890.842	1003.94
	5	997.288	1028.25	1079.77	1151.75	1244.09	1356.66
	6	1431.47	1462.21	1513.39	1584.95	1676.83	1788.94
1.0	1	23.646	58.646	113.228	187.437	281.322	394.910
	2	51.674	86.134	140.846	215.294	309.403	423.178
	3	100.270	133.791	188.113	262.517	356.723	470.639
	4	168.959	201.725	255.469	329.574	423.682	537.614
	5	257.544	289.762	342.950	416.648	510.515	624.335
	6	365.950	397.768	450.482	523.760	617.315	730.935
1.5	1	18.901	53.776	108.221	182.335	276.156	389.697
	2	30.668	65.616	120.305	194.612	288.574	402.221
	3	51.674	86.134	140.846	215.294	309.403	423.178
	4	81.822	115.645	170.135	244.589	338.784	452.665
	5	120.947	154.191	208.327	282.652	376.845	490.782
	6	168.959	201.725	255.469	329.574	423.682	537.614
2.0	1	17.332	52.098	106.479	180.555	274.350	387.874
	2	23.646	58.646	113.228	187.437	281.322	394.910
	3	35.051	69.912	124.633	198.986	292.988	406.666
	4	51.674	86.134	140.846	215.294	309.403	423.178
	5	73.439	107.420	161.982	236.448	330.630	444.488
	6	100.270	133.791	188.113	262.517	356.723	470.639
3.0	1	16.252	50.909	105.238	179.285	273.061	386.572
	2	18.901	53.776	108.221	182.335	276.156	389.687
	3	23.646	58.646	113.228	187.437	281.322	394.910
	4	30.668	65.616	120.305	194.612	288.574	402.221
	5	40.015	74.762	129.497	203.889	297.929	411.639
	6	51.674	86.134	140.846	215.294	309.403	423.178

9.3.7 Vibration of SSFF Plates

The boundary conditions on edges $y = 0, b$ of SSFF rectangular plates (with sides $x = 0, a$ simply supported and sides $y = 0, b$ free) are

$$M_{yy} = -\left(D_{12}\frac{\partial^2 w}{\partial x^2} + D_{22}\frac{\partial^2 w}{\partial y^2}\right) = 0 \quad \text{at } y = 0, b \qquad (9.3.31\text{a})$$

$$V_y = -\left(D_{22}\frac{\partial^3 w}{\partial y^3} + \bar{D}_{12}\frac{\partial^3 w}{\partial x^2 \partial y}\right) = 0 \quad \text{at } y = 0, b \qquad (9.3.31\text{b})$$

These boundary conditions in conjunction with the solution in Eq. (9.3.5) yield the characteristic equation

$$2\left(1 - \cosh \lambda_1 b \, \cos \lambda_2 b\right) + \left(\mu_0 - \frac{1}{\mu_0}\right) \sinh \lambda_1 b \, \sin \lambda_2 b = 0 \qquad (9.3.32)$$

where

$$\mu_0 = \frac{\lambda_1}{\lambda_2}\frac{\Omega_2}{\Omega_1}\frac{\bar{\Omega}_1}{\bar{\Omega}_2} \qquad (9.3.33)$$

and Ω_i and $\bar{\Omega}_i$ are defined by Eqs. (9.3.12a,b). The frequency equation obtained with Eq. (9.3.6) is

$$2\left(1 - \cosh \lambda_1 b \, \cosh \lambda_2 b\right) + \left(\mu_0 + \frac{1}{\mu_0}\right) \sinh \lambda_1 b \, \sinh \lambda_2 b = 0 \qquad (9.3.34)$$

which *does* have some roots when $\rho h \omega^2 < \alpha_m^4 (D_{11}/D_{22})$.

The first eight frequencies for modulus ratios $E_1/E_2 = 1, 3$, and 10 ($\nu_{12} = 0.25$ and $G_{12} = 0.5$) are listed in Table 9.3.8. For $E_1/E_2 = 1$ and 3, the frequencies ω_{11}, ω_{21}, and ω_{31} are the only frequencies among the first nine frequencies of SSFF plates with $\rho h \omega^2 < \alpha_m^4(D_{11}/D_{22})$ [i.e., roots of Eq. (9.3.34)]. The remaining frequencies are the roots of Eq. (9.3.32). For $E_1/E_2 = 10$ there are no roots of Eq. (9.3.34). Table 9.3.9 contains the first five roots of Eq. (9.3.32) or (9.3.34), whichever is the minimum, for $m = 1, 2, \cdots, 4$ and for $a/b = 0.5, 0.6, 0.8, 1.0$, and 2.0 of isotropic ($\nu = 0.3$) rectangular plates. Finally, Figure 9.3.1 contains plots of frequency parameters $\bar{\omega} = \omega_{mn} a^2 (\sqrt{\rho h/D})$ for various a/b ratios of a rectangular isotropic ($\nu = 0.3$) plate. Each set of three curves corresponds to $n = 1, 2$, and 3, and the three sets corresponds to $m = 1, 2$, and 3. The intersections indicate that two modes can exist simultaneously. For example, the third root for $m = 1$ and the second root for $m = 2$ can exist simultaneously for a plate having an a/b ratio of approximately 1.25.

Table 9.3.8. Effect of the modulus ratio on the first eight natural frequencies ($\bar{\omega}_{mn}$) of square plates (SSFF).

$\frac{E_1}{E_2}$	$\bar{\omega}_{mn} = \omega_{mn} a^2 (\sqrt{\rho h / D_{22}})$ for (m, n)							
	ω_{11}	ω_{12}	ω_{13}	ω_{21}	ω_{22}	ω_{23}	ω_{31}	ω_{32}
1	9.710*	16.491	37.176	39.128*	47.278	71.559	88.282*	96.752
3	17.013*	22.770	42.319	68.204*	74.515	95.091	153.583*	159.973
10	34.693	49.863	84.358	128.355	141.431	168.878	284.342	296.666

* Frequencies computed using Eq. (9.3.34); all other frequencies are computed using Eq. (9.3.32).

Table 9.3.9. Effect of the plate aspect ratio on the nondimensionalized frequencies of isotropic ($\nu = 0.3$) rectangular plates (SSFF).

$\frac{a}{b}$	m	$\bar{\omega}_{mn} = \omega_{mn} a^2 \sqrt{\rho h / D}$ for values of n				
		1	2	3	4	5
0.5	1	9.736	11.684	17.685	27.756	42.384
	2	39.188	41.196	47.966	59.065	74.525
	3	88.363	90.294	97.342	108.918	125.016
	4	157.260	159.080	166.291	178.095	194.542
0.6	1	9.713	12.434	20.715	34.935	55.909
	2	39.138	42.042	51.556	67.149	88.991
	3	88.287	91.133	101.118	117.513	140.301
	4	157.159	159.888	170.125	186.928	210.311
0.8	1	9.670	14.170	27.943	52.748	90.033
	2	39.040	44.147	60.249	86.824	124.707
	3	88.135	93.289	110.481	138.656	177.963
	4	156.955	162.014	179.766	208.911	249.424
1.0	1	9.631	16.135	36.725	75.283	133.704
	2	38.945	46.738	70.739	111.024	169.537
	3	87.987	96.040	122.039	164.694	224.756
	4	156.752	164.786	191.866	236.260	298.100
2.0	1	9.512	27.522	105.489	260.870	496.633
	2	38.525	64.538	146.901	301.132	534.817
	3	87.285	116.831	206.579	363.202	596.068
	4	155.780	186.951	282.958	444.097	678.146

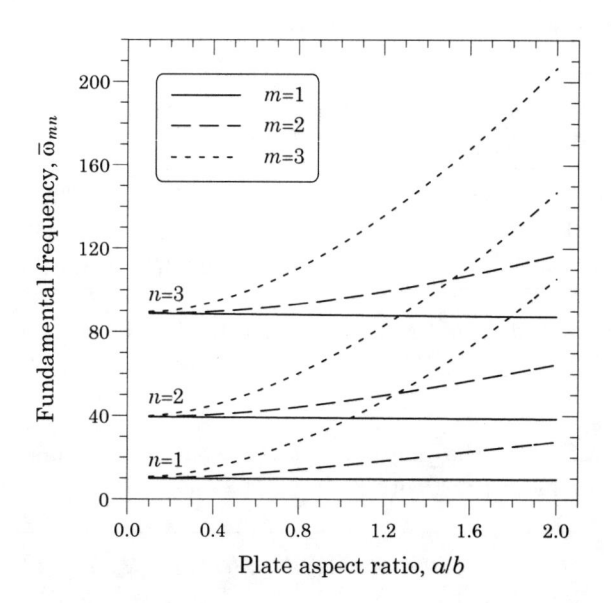

Figure 9.3.1 Plots of the frequency parameters $\omega_{mn} a^2 (\sqrt{\rho h/D})$ for various a/b ratios of isotropic ($\nu = 0.3$) rectangular plates (SSFF).

9.3.8 The Rayleigh–Ritz Solutions

As discussed in section 8.3.1, Eq. (9.3.3) can also be solved using the Rayleigh–Ritz method. The weak form of Eq. (9.3.3) is given by

$$\int_0^b \left[D_{22} \frac{d^2W}{dy^2} \frac{d^2\delta W}{dy^2} + k_1 \frac{dW}{dy} \frac{d\delta W}{dy} + k_2 W \delta W \right] dy$$

$$+ \left[\left(D_{22} \frac{d^3W}{dy^3} - k_1 \frac{dW}{dy} \right) \delta W - D_{22} \frac{d^2W}{dy^2} \frac{d\delta W}{dy} \right]_0^b = 0 \quad (9.3.35)$$

where

$$k_1 = \left(2\alpha_m^2 \hat{D}_{12} - I_2 \omega^2 \right), \quad k_2 = \left[D_{11} \alpha_m^4 - \left(I_0 + \alpha_m^2 I_2 \right) \omega^2 \right] \quad (9.3.36)$$

For a simply supported edge [$W = 0$ and $(d^2W/dy^2) = 0$] or a clamped edge [$W = 0$ and $(dW/dy) = 0$], the boundary terms in Eq. (9.3.35) vanish identically. However, for a free edge the vanishing of bending moment and effective shear force require

$$D_{22}\frac{d^2W}{dy^2} - \alpha_m^2 D_{12} W = 0, \quad D_{22}\frac{d^3W}{dy^3} - \alpha_m^2 \bar{D}_{12}\frac{dW}{dy} = 0 \quad (9.3.37)$$

where $\bar{D}_{12} = D_{12} + 4D_{66}$. These conditions can be used, as shown in Eqs. (8.3.3)–(8.3.5), to simplify the weak form (9.3.35) for various boundary conditions on sides $y = 0, b$.

Assume an N-parameter Rayleigh–Ritz approximation of the form

$$W(y) \approx \sum_{j=1}^{N} c_j \varphi_j(y) \quad (9.3.38)$$

where φ_j are the functions given in Eqs. (7.3.10)–(7.3.20). Substituting the approximation (9.3.38) into the simplified weak form (9.3.35), we obtain

$$\left([R] - \omega^2 [B]\right)\{c\} = \{0\} \quad (9.3.39)$$

where [c.f. Eq. (8.3.8a,b)]

$$R_{ij} = \int_0^b \left(D_{22}\frac{d^2\varphi_i}{dy^2}\frac{d^2\varphi_j}{dy^2} + 2\alpha_m^2 \hat{D}_{12}\frac{d\varphi_i}{dy}\frac{d\varphi_j}{dy} + D_{11}\alpha_m^4 \varphi_i \varphi_j \right) dy$$

$$- \alpha_m^2 D_{12} \left[\varphi_j \frac{d\varphi_i}{dy} + \frac{d\varphi_j}{dy}\varphi_i \right]_0^b \quad (9.3.40a)$$

$$B_{ij} = \int_0^b \left[I_2 \alpha_m^2 \frac{d\varphi_i}{dy}\frac{d\varphi_j}{dy} + \left(I_0 + \alpha_m^2 I_2\right)\varphi_i \varphi_j \right] dy \quad (9.3.40b)$$

where the boundary term in the expression for R_{ij} is zero for SSSS, SSCC, and SSCS plates.

Equation (9.3.39) is an eigenvalue problem much the same way as Eq. (8.3.9), and it can be solved for the first n frequencies ω_{mn}^2 for any choice of m. Due to the similarity of the buckling problem and natural vibration problem, we will not discuss any examples of application of Eq. (8.3.39).

9.4 Natural Vibration of Plates with General Boundary Conditions

9.4.1 The Rayleigh–Ritz Solution

In the preceding sections, we discussed analytical solutions based on the Navier solution method for plates with all sides simply supported and solutions based on the Lévy method for plates with two opposite edges simply supported while the other two had any combination of simple support, clamped, and free edge conditions. For rectangular plates with boundary conditions that do not fall into the above categories (e.g., a plate with all four edges clamped), the natural freqencies may be computed using approximate methods. Here we consider the Rayleigh–Ritz method of solution.

The weak form of Eq. (9.1.5) governing plates undergoing natural vibration can be expressed as [c.f. Eq. (8.3.10)]

$$0 = \int_0^b \int_0^a \left\{ D_{11} \frac{\partial^2 w}{\partial x^2} \frac{\partial^2 \delta w}{\partial x^2} + D_{12} \left(\frac{\partial^2 w}{\partial y^2} \frac{\partial^2 \delta w}{\partial x^2} + \frac{\partial^2 w}{\partial x^2} \frac{\partial^2 \delta w}{\partial y^2} \right) \right.$$
$$+ 4 D_{66} \frac{\partial^2 w}{\partial x \partial y} \frac{\partial^2 \delta w}{\partial x \partial y} + D_{22} \frac{\partial^2 w}{\partial y^2} \frac{\partial^2 \delta w}{\partial y^2}$$
$$\left. - \omega^2 \left[I_0 w \delta w + I_2 \left(\frac{\partial w}{\partial x} \frac{\partial \delta w}{\partial x} + \frac{\partial w}{\partial y} \frac{\partial \delta w}{\partial y} \right) \right] \right\} dx dy \quad (9.4.1)$$

Using an N–parameter Rayleigh–Ritz solution of the form

$$w(x, y) \approx \sum_{j=1}^{N} c_j \varphi_j(x, y) \quad (9.4.2)$$

in Eq. (9.4.1), we obtain

$$\left([R] - \omega^2 [B] \right) \{c\} = \{0\} \quad (9.4.3)$$

where

$$R_{ij} = \int_0^b \int_0^a \left[D_{11} \frac{\partial^2 \varphi_i}{\partial x^2} \frac{\partial^2 \varphi_j}{\partial x^2} + D_{12} \left(\frac{\partial^2 \varphi_i}{\partial y^2} \frac{\partial^2 \varphi_j}{\partial x^2} + \frac{\partial^2 \varphi_i}{\partial x^2} \frac{\partial^2 \varphi_j}{\partial y^2} \right) \right.$$
$$\left. + 4 D_{66} \frac{\partial^2 \varphi_i}{\partial x \partial y} \frac{\partial^2 \varphi_j}{\partial x \partial y} + D_{22} \frac{\partial^2 \varphi_i}{\partial y^2} \frac{\partial^2 \varphi_j}{\partial y^2} \right] dx dy \quad (9.4.4a)$$

$$B_{ij} = \int_0^b \int_0^a \left[I_0 \varphi_i \varphi_j + I_2 \left(\frac{\partial \varphi_i}{\partial x} \frac{\partial \varphi_j}{\partial x} + \frac{\partial \varphi_i}{\partial y} \frac{\partial \varphi_j}{\partial y} \right) \right] dx dy \quad (9.4.4b)$$

As discussed in sections 7.3.2 and 8.3.2, it is convenient to express the Rayleigh–Ritz approximation of rectangular plates in the form

$$w(x,y) \approx W_{mn}(x,y) = \sum_{i=1}^{M}\sum_{j=1}^{N} c_{ij}\, X_i(x)Y_j(y) \qquad (9.4.5)$$

The functions X_i and Y_j for various boundary conditions are given in Eqs. (7.3.29)–(7.3.44). These functions satisfy only the geometric boundary conditions of the problem. Substituting Eq. (9.4.5) into Eq. (9.4.1), we obtain Eq. (9.4.3) in which $R_{(ij)(k\ell)}$ of $[R]$ are defined by Eq. (8.3.16a) and coefficients $B_{(ij)(k\ell)}$ of $[B]$ are defined by

$$B_{ij,k\ell} = \int_0^b \int_0^a \left[I_0 X_i X_k Y_j Y_\ell + I_2 \left(\frac{dX_i}{dx}\frac{dX_k}{dx} Y_j Y_\ell + X_i X_k \frac{dY_j}{dy}\frac{dY_\ell}{dy} \right) \right] dx\,dy \qquad (9.4.6)$$

The size of the matrix $[R]$ is $MN \times MN$, and Eq. (9.4.3) can be used to calculate first MN frequencies of the infinite set.

9.4.2 Simply Supported Plates (SSSS)

For a rectangular plate simply supported on all sides (see Figure 9.4.1), we may select the following approximation functions

$$\varphi_{ij} = X_i(x)Y_j(y) \qquad (9.4.7a)$$

$$X_i(x) = \left(\frac{x}{a}\right)^i - \left(\frac{x}{a}\right)^{i+1}, \quad Y_j(x) = \left(\frac{y}{b}\right)^j - \left(\frac{y}{b}\right)^{j+1} \qquad (9.4.7b)$$

For isotropic ($\nu = 0.25$) square plates with $M = N = 2$ [see Eq. (9.4.5)], the matrices $[R]$ and $[B]$ are given by

$$[R] = \begin{bmatrix} 0.4346 & 0.2173 & 0.2173 & 0.1086 \\ 0.2173 & 0.2314 & 0.1086 & 0.1157 \\ 0.2173 & 0.1086 & 0.2314 & 0.1157 \\ 0.1086 & 0.1157 & 0.1157 & 0.0993 \end{bmatrix} \times 10^{-1}$$

$$[B] = \begin{bmatrix} 1.1111 & 0.5556 & 0.5556 & 0.2778 \\ 0.5556 & 0.3175 & 0.2778 & 0.1587 \\ 0.5556 & 0.2778 & 0.3175 & 0.1587 \\ 0.2778 & 0.1587 & 0.1587 & 0.0907 \end{bmatrix} \times 10^{-5} \qquad (9.4.8)$$

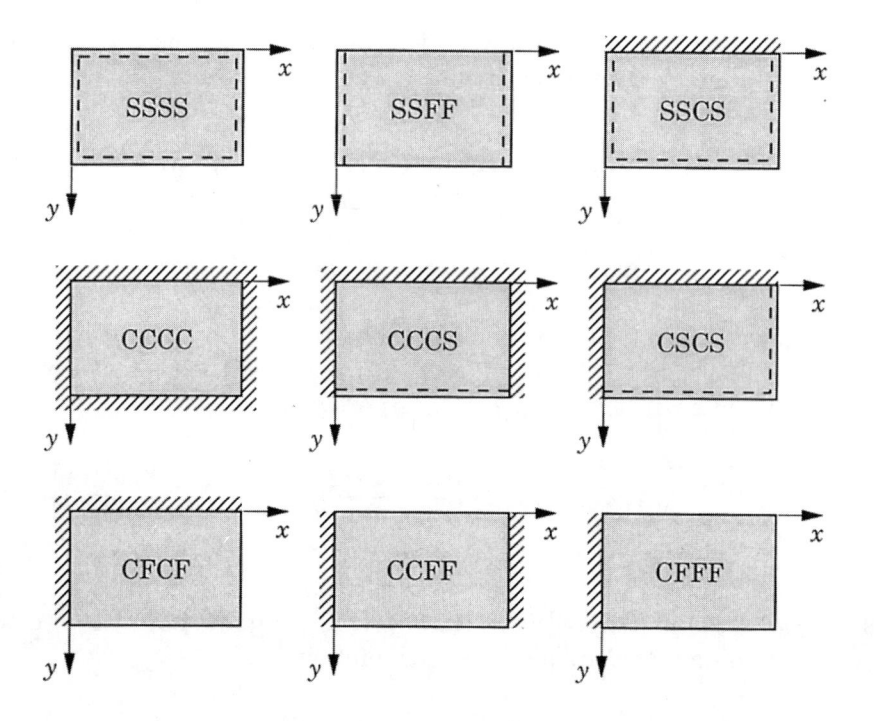

Figure 9.4.1 Rectangular plates with various boundary conditions along their edges.

where the following notation is used to store the coefficients

$$R_{11} = R_{11,11}, \quad R_{12} = R_{11,12}, \quad R_{13} = R_{11,21}, \quad R_{14} = R_{11,22}$$
$$R_{21} = R_{12,11}, \quad R_{22} = R_{12,12}, \quad R_{23} = R_{12,21}, \quad R_{24} = R_{12,22}$$
$$R_{31} = R_{21,11}, \quad R_{32} = R_{21,12}, \quad R_{33} = R_{21,21}, \quad R_{34} = R_{21,22}$$
$$R_{41} = R_{22,11}, \quad R_{42} = R_{22,12}, \quad R_{43} = R_{22,21}, \quad R_{44} = R_{22,22}$$

The first four frequencies ($\bar{\omega}_{mn} = \omega_{mn} a^2 \sqrt{\rho h/D}$) are

$$\bar{\omega}_{11} = 20.973, \quad \bar{\omega}_{12} = 58.992, \quad \bar{\omega}_{21} = 59.007, \quad \bar{\omega}_{22} = 92.529 \quad (9.4.9)$$

The exact frequencies [see Eq. (9.2.8)] are

$$\bar{\omega}_{11} = 19.739, \quad \bar{\omega}_{12} = \bar{\omega}_{21} = 49.348, \quad \bar{\omega}_{22} = 88.826 \quad (9.4.10)$$

For larger values of N and M, the Rayleigh–Ritz method will yield increasingly accurate and more frequencies.

9.4.3 Clamped Plates (CCCC)

Both the Rayleigh–Ritz and Lévy methods cannot be used to determine the frequencies of clamped plates. Therefore, here we use the Rayleigh–Ritz method to determine the natural frequencies. For a rectangular plate clamped on all sides, we may select the following approximation functions

$$\varphi_{ij} = X_i(x)Y_j(y) \tag{9.4.11}$$

$$X_i(x) = \left(\frac{x}{a}\right)^{i+1} - 2\left(\frac{x}{a}\right)^{i+2} + \left(\frac{x}{a}\right)^{i+3} \tag{9.4.12a}$$

$$Y_j(y) = \left(\frac{y}{b}\right)^{j+1} - 2\left(\frac{y}{b}\right)^{j+2} + \left(\frac{y}{b}\right)^{j+3} \tag{9.4.12b}$$

For isotropic ($\nu = 0.25$) square plates with $M = N = 2$, the matrices $[R]$ and $[B]$ are given by

$$[R] = \begin{bmatrix} 0.2902 & 0.1451 & 0.1451 & 0.0725 \\ 0.1451 & 0.1007 & 0.0725 & 0.0503 \\ 0.1451 & 0.0725 & 0.1007 & 0.0503 \\ 0.0725 & 0.0503 & 0.0503 & 0.0335 \end{bmatrix} \times 10^{-3}$$

$$[B] = \begin{bmatrix} 0.2520 & 0.1260 & 0.1260 & 0.0630 \\ 0.1260 & 0.0687 & 0.0630 & 0.0344 \\ 0.1260 & 0.0630 & 0.0687 & 0.0344 \\ 0.0630 & 0.0344 & 0.0344 & 0.0187 \end{bmatrix} \times 10^{-7} \tag{9.4.13}$$

The first four nondimensionalized frequencies ($\bar{\omega}_{mn} = \omega_{mn} a^2 \sqrt{\rho h/D}$) are

$$\bar{\omega}_{11} = 36.000, \quad \bar{\omega}_{12} = 74.296, \quad \bar{\omega}_{21} = 74.297, \quad \bar{\omega}_{22} = 108.592 \tag{9.4.14a}$$

The approximate frequencies obtained by Ödman [9] (see Leissa [8]) are

$$\bar{\omega}_{11} = 35.999, \quad \bar{\omega}_{12} = 73.405, \quad \bar{\omega}_{21} = 73.405, \quad \bar{\omega}_{22} = 108.237 \tag{9.4.14b}$$

Table 9.4.1 contains the first four natural frequencies of clamped rectangular plates. The results are obtained with $M = N = 2$ in Eq. (9.4.5) (c.f., Tables 4.28 and 4.29 on pages 63 and 64, respectively, of Leissa [8]; no mention is made of Poisson's ratio used). The frequencies

Table 9.4.1. Effect of the plate aspect ratio on the nondimensionalized frequencies of isotropic ($\nu = 0.25$) and orthotropic clamped rectangular plates (CCCC).

		\multicolumn{6}{c}{$\omega_{mn}a^2\sqrt{\rho h/D_{22}}$ for values of a/b}
m	n	0.25	0.5	0.667	1.0	1.5	2.0

Isotropic Plates ($\nu = 0.25$)

m	n	0.25	0.5	0.667	1.0	1.5	2.0
1	1	22.890	24.647	27.047	36.000	60.856	98.590
	2	24.196	31.867	41.899	74.296	94.273	127.466
2	1	63.466	65.234	69.297	74.297	151.419	260.938
	2	64.943	71.941	80.394	108.592	180.887	287.766

Orthotropic Plates ($E_1 = 10E_2$, $G_{12} = 0.5E_2$, $\nu = 0.25$)

m	n	0.25	0.5	0.667	1.0	1.5	2.0
1	1	71.164	71.840	72.792	76.826	91.637	120.569
	2	71.677	74.939	80.036	101.555	167.460	230.036
2	1	199.206	199.898	200.708	203.511	212.381	271.393
	2	199.793	202.613	206.212	219.947	265.488	349.607

listed for the orthotropic plate are the first four frequencies. The mode shape is of the form

$$w(x,y) = X_1(c_1 Y_1 + c_2 Y_2) + X_2(c_3 Y_1 + c_4 Y_2) \qquad (9.4.15)$$

The vectors $\{c\}^i$ for the four modes in the case of $a/b = 0.25$, for example, are given by

$$\{c\}^1 = \begin{Bmatrix} 1.0 \\ 0.0 \\ 0.0 \\ 0.0 \end{Bmatrix}, \{c\}^2 = \begin{Bmatrix} -0.5 \\ 1.0 \\ 0.0 \\ 0.0 \end{Bmatrix}, \{c\}^3 = \begin{Bmatrix} -0.50 \\ 0.08 \\ 1.00 \\ 0.16 \end{Bmatrix}, \{c\}^4 = \begin{Bmatrix} 0.25 \\ -0.50 \\ -0.50 \\ 1.00 \end{Bmatrix}$$

The vectors $\{c\}^i$ for the four modes in the case of $a/b = 1$ are given by

$$\{c\}^1 = \begin{Bmatrix} 1.0 \\ 0.0 \\ 0.0 \\ 0.0 \end{Bmatrix}, \{c\}^2 = \begin{Bmatrix} -0.5 \\ 1.0 \\ 0.0 \\ 0.0 \end{Bmatrix}, \{c\}^3 = \begin{Bmatrix} -0.5 \\ 0.0 \\ 1.0 \\ 0.0 \end{Bmatrix}, \{c\}^4 = \begin{Bmatrix} 0.25 \\ -0.50 \\ -0.50 \\ 1.00 \end{Bmatrix}$$

9.4.4 CCCS Plates

Consider natural vibrations of rectangular plates with sides $x = 0, a$ and $y = 0$ clamped and side $y = b$ simply supported (see Figure 9.4.1). For this case, the approximation functions are given by

$$X_i = \left(\frac{x}{a}\right)^{i+1} - 2\left(\frac{x}{a}\right)^{i+2} + \left(\frac{x}{a}\right)^{i+3}, \quad Y_j = \left(\frac{y}{b}\right)^{j+1} - \left(1 - \frac{y}{b}\right)^{j+2} \tag{9.4.16}$$

The first four natural frequencies of rectangular plates are presented in Table 9.4.2 for various aspect ratios.

Table 9.4.2. Effect of the plate aspect ratio on the nondimensionalized frequencies of isotropic and orthotropic rectangular plates (CCCS).

Mode (m, n)	\multicolumn{6}{c}{$\omega_{mn} a^2 \sqrt{\rho h / D_{22}}$ for values of a/b}
	0.25	0.5	0.667	1.0	1.5	2.0

Isotropic Plates ($\nu = 0.25$)

(1,1)	22.840	24.215	25.913	31.849	48.217	73.573
(1,2)	24.654	34.421	46.951	72.025	86.045	108.452
(2,1)	63.411	64.957	66.663	86.497	179.342	310.663
(2,2)	65.444	74.254	84.914	120.087	208.448	337.474

Orthotropic Plates ($E_1 = 10 E_2, G_{12} = 0.5 E_2, \nu = 0.25$)

(1,1)	71.146	71.685	72.365	74.942	83.725	101.066
(1,2)	71.859	76.161	82.957	111.066	193.427	219.912
(2,1)	199.162	199.773	200.452	202.629	208.753	319.871
(2,2)	200.015	203.605	208.276	226.356	285.872	392.681

9.4.5 CSCS Plates

Here we consider natural vibrations of rectangular plates with sides $x = 0$ and $y = 0$ clamped and sides $x = a$ and $y = b$ simply supported (see Figure 9.4.1). For this case, the approximation functions are

$$X_i = \left(\frac{x}{a}\right)^{i+1} - \left(1 - \frac{x}{a}\right)^{i+2}, \quad Y_j = \left(\frac{y}{b}\right)^{j+1} - \left(1 - \frac{y}{b}\right)^{j+2} \tag{9.4.17}$$

416   THEORY AND ANALYSIS OF ELASTIC PLATES

The first four natural frequencies of rectangular plates are presented in Table 9.4.3 for various aspect ratios.

Table 9.4.3. Effect of the plate aspect ratio on the nondimensionalized frequencies of isotropic and orthotropic rectangular plates (CSCS).

Mode (m,n)	\multicolumn{6}{c}{$\omega_{mn} a^2 \sqrt{\rho h / D_{22}}$ for values of a/b}
	0.25	0.5	0.667	1.0	1.5	2.0
Isotropic Plates ($\nu = 0.25$)						
(1,1)	15.985	17.818	19.991	27.087	44.979	71.272
(1,2)	18.386	30.012	43.638	84.399	98.194	120.056
(2,1)	75.839	77.413	79.132	84.517	178.042	309.642
(2,2)	77.892	86.649	97.076	131.386	218.411	346.580
Orthotropic Plates ($E_1 = 10 E_2$, $G_{12} = 0.5 E_2$, $\nu = 0.25$)						
(1,1)	49.070	49.815	50.754	54.272	65.694	86.527
(1,2)	50.052	55.904	64.724	97.935	185.878	258.208
(2,1)	238.427	239.047	239.728	241.879	247.752	315.079
(2,2)	239.240	242.788	247.265	264.141	319.351	420.586

9.4.6 CFCF, CCFF, and CFFF Plates

First we consider natural vibrations of rectangular plates with sides $x = 0$ and $y = 0$ clamped and sides $x = a$ and $y = b$ free (see Figure 9.4.1). The first choice of approximation functions is

$$X_i = \left(\frac{x}{a}\right)^{i+1}, \quad Y_j = \left(\frac{y}{b}\right)^{j+1} \tag{9.4.18}$$

The first four natural frequencies of rectangular plates are presented in Table 9.4.4 for various aspect ratios (first row of numbers for each mode). These results differ considerably from those presented by Young [6] (see Leissa [8], p. 73). The reason for this is that the functions in Eq. (9.4.18) do not satisfy the force boundary conditions on the free edges when $M = N = 2$. The following functions:

$$X_i(x) = \left(\frac{x}{a}\right)^{i+1} - \frac{2i}{i+2}\left(\frac{x}{a}\right)^{i+2} + \frac{i(i+1)}{(i+2)(i+3)}\left(\frac{x}{a}\right)^{i+3} \tag{9.4.19a}$$

Table 9.4.4. Effect of the plate aspect ratio on the nondimensionalized frequencies of isotropic ($\nu = 0.3$) rectangular plates (CFCF).

Mode (m,n)	$\omega_{mn}a^2\sqrt{\rho h/D}$ for values of a/b					
	0.25	0.5	0.667	1.0	1.5	2.0
(1, 1)[†]	3.719	4.408	5.162	7.229	11.614	17.634
	3.685	4.317	5.020	6.996	11.296	17.269
(1, 2)	6.347	12.739	18.982	33.888	42.710	50.955
	5.095	9.257	13.478	24.696	30.327	37.030
(2,1)	35.041	35.832	36.828	43.044	82.863	143.328
	22.888	23.447	24.079	27.056	54.177	93.789
(2,2)	39.111	50.516	60.882	86.665	136.984	202.065
	24.552	29.848	35.066	49.113	78.898	119.391

[†] The first row corresponds to the frequencies obtained using the functions in Eq. (9.4.18), while the second row corresponds to those obtained using the functions in Eqs. (9.4.19a,b).

$$Y_j(y) = \left(\frac{y}{b}\right)^{j+1} - \frac{2j}{j+2}\left(\frac{y}{b}\right)^{j+2} + \frac{j(j+1)}{(j+2)(j+3)}\left(\frac{y}{b}\right)^{j+3} \quad (9.4.19b)$$

satisfy the force boundary conditions

$$M_{xx} = V_x = 0 \text{ on side } x = a \text{ and } M_{yy} = V_y = 0 \text{ on side } y = b \quad (9.4.20)$$

The first four natural frequencies obtained with the functions in Eqs. (9.4.19a,b) are listed in the second row of each mode. The values for the square plate are: 6.996, 24.696, 27.056, and 49.113, which are close to those of Young: 6.958, 24.80, 26.80, and 48.05.

Next, consider the natural vibrations of CCFF and CFFF plates. For CCFF plates (sides $x = 0, a$ clamped and sides $y = 0, b$ free) the approximation functions are

$$X_i = \left(\frac{x}{a}\right)^{i+1} - 2\left(\frac{x}{a}\right)^{i+2} + \left(\frac{x}{a}\right)^{i+3}, \quad Y_j = \left(\frac{y}{b}\right)^{j-1} \quad (9.4.21)$$

For CFFF plates, we have the choice of

$$X_i = \left(\frac{x}{a}\right)^{i+1}, \quad Y_j = \left(\frac{y}{b}\right)^{j-1} \quad (9.4.22)$$

418 THEORY AND ANALYSIS OF ELASTIC PLATES

or
$$X_i(x) = \left(\frac{x}{a}\right)^{i+1} - \frac{2i}{i+2}\left(\frac{x}{a}\right)^{i+2} + \frac{i(i+1)}{(i+2)(i+3)}\left(\frac{x}{a}\right)^{i+3}, \quad Y_j = \left(\frac{y}{b}\right)^{j-1} \quad (9.4.23)$$

The first four natural frequencies of CCFF and CFFF plates are presented in Table 9.4.5 for various aspect ratios. The mode shapes associated with these plates are of the form given in Eq. (9.4.15). The amplitudes $\{c\}^i$ associated with the four modes of CCFF plates are

$$\{c\}^1 = \begin{Bmatrix} 1.0 \\ 0.0 \\ 0.0 \\ 0.0 \end{Bmatrix}, \quad \{c\}^2 = \begin{Bmatrix} 0.5 \\ -1.0 \\ 0.0 \\ 0.0 \end{Bmatrix}, \quad \{c\}^3 = \begin{Bmatrix} 0.5 \\ 0.0 \\ -1.0 \\ -0.0 \end{Bmatrix}, \quad \{c\}^4 = \begin{Bmatrix} 0.25 \\ -0.5 \\ -0.5 \\ 1.0 \end{Bmatrix}$$

Table 9.4.5. Effect of plate aspect ratio on the nondimensionalized frequencies of isotropic ($\nu = 0.3$) CCFF and CFFF plates.

Mode	\multicolumn{6}{c}{$\omega_{mn}a^2\sqrt{\rho h/D}$ for values of a/b}
(m,n)	0.25	0.5	0.667	1.0	1.5	2.0
CCFF Plates						
(1,1)	22.450	22.450	22.450	22.450	22.450	22.450
(1,2)	22.729	23.546	24.364	26.563	30.945	36.199
(2,1)	62.928	62.928	62.928	62.928	62.928	62.928
(2,2)	63.294	64.380	65.487	68.550	74.987	83.167
CFFF Plates[†]						
(1,1)	3.533	3.533	3.533	3.533	3.533	3.533
	3.516	3.516	3.516	3.516	3.516	3.516
(1,2)	4.176	5.612	6.715	8.984	12.383	15.792
	4.139	5.514	6.555	8.664	11.794	14.937
(2,1)	34.807	34.807	34.807	34.807	34.807	34.807
	22.712	22.712	22.712	22.712	22.712	22.712
(2,2)	35.684	38.205	40.666	47.029	58.946	72.446
	23.378	25.289	27.151	31.938	40.800	50.721

[†] The first row corresponds to the frequencies obtained using the functions in Eq. (9.4.22), while the second row corresponds to those obtained using the functions in Eq. (9.4.23).

The amplitudes of CFFF plates based on Eq. (9.4.22) are

$$\{c\}^1 = \begin{Bmatrix} 1.0 \\ 0.0 \\ -0.4 \\ 0.0 \end{Bmatrix}, \quad \{c\}^2 = \begin{Bmatrix} 0.5 \\ -1.0 \\ -0.3 \\ 0.6 \end{Bmatrix}, \quad \{c\}^3 = \begin{Bmatrix} 0.8 \\ 0.0 \\ -1.0 \\ 0.0 \end{Bmatrix}, \quad \{c\}^4 = \begin{Bmatrix} 0.4 \\ -0.8 \\ -0.5 \\ 1.0 \end{Bmatrix}$$

and those based on Eq. (9.4.23) are

$$\{c\}^1 = \begin{Bmatrix} 1.0 \\ 0.0 \\ 0.2 \\ 0.0 \end{Bmatrix}, \quad \{c\}^2 = \begin{Bmatrix} 0.5 \\ -1.0 \\ -0.4 \\ 0.9 \end{Bmatrix}, \quad \{c\}^3 = \begin{Bmatrix} 0.5 \\ 0.0 \\ -1.0 \\ 0.0 \end{Bmatrix}, \quad \{c\}^4 = \begin{Bmatrix} 0.3 \\ -0.5 \\ -0.5 \\ 1.0 \end{Bmatrix}$$

It is interesting to note that the predicted frequencies ω_{11} and ω_{21} are independent of the plate aspect ratio, a/b. This can be attributed to the low order of approximation used in the y coordinate direction.

9.5 Transient Analysis

9.5.1 Spatial Variation of the Solution

The equation of motion governing bending deflection w_0 of an orthotropic plate, assuming no applied in-plane and thermal forces, is

$$D_{11}\frac{\partial^4 w_0}{\partial x^4} + 2(D_{12} + 2D_{66})\frac{\partial^4 w_0}{\partial x^2 \partial y^2} + D_{22}\frac{\partial^4 w_0}{\partial y^4} - q(x,y,t)$$
$$+ I_0 \ddot{w}_0 - I_2\left(\frac{\partial^2 \ddot{w}_0}{\partial x^2} + \frac{\partial^2 \ddot{w}_0}{\partial y^2}\right) = 0 \quad (9.5.1)$$

Suppose that the plate is simply supported with the boundary conditions

$$w_0(x,0,t) = 0, \quad w_0(x,b,t) = 0, \quad w_0(0,y,t) = 0, \quad w_0(a,y,t) = 0$$
$$M_{xx}(0,y,t) = 0, \quad M_{xx}(a,y,t) = 0, \quad M_{yy}(x,0,t) = 0, \quad M_{yy}(x,b,t) = 0 \quad (9.5.2)$$

for $t > 0$, and assume that the initial conditions are

$$w_0(x,y,0) = d_0(x,y), \quad \frac{\partial w_0}{\partial t}(x,y,0) = v_0(x,y) \quad \text{for all } x \text{ and } y \quad (9.5.3)$$

where d_0 and v_0 are the initial displacement and velocity, respectively.

We assume the following expansion of the transverse deflection to satisfy the boundary conditions (9.5.2) for any time $t \geq 0$

$$w_0(x,y,t) = \sum_{n=1}^{\infty} \sum_{m=1}^{\infty} W_{mn}(t) \sin\alpha_m x \sin\beta_n y \qquad (9.5.4)$$

Similarly, we assume that the transverse load, initial displacement, and initial velocity can be expanded as

$$q(x,y,t) = \sum_{n=1}^{\infty} \sum_{m=1}^{\infty} Q_{mn}(t) \sin\alpha_m x \sin\beta_n y \qquad (9.5.5)$$

$$d_0(x,y) = \sum_{n=1}^{\infty} \sum_{m=1}^{\infty} D_{mn} \sin\alpha_m x \sin\beta_n y \qquad (9.5.6)$$

$$v_0(x,y) = \sum_{n=1}^{\infty} \sum_{m=1}^{\infty} V_{mn} \sin\alpha_m x \sin\beta_n y \qquad (9.5.7)$$

where $\alpha_m = m\pi/a$, $\beta_n = n\pi/b$, and Q_{mn}, for example, is given by

$$Q_{mn}(t) = \frac{4}{ab} \int_0^b \int_0^a q(x,y,t) \sin\alpha_m x \sin\beta_n y \, dx dy \qquad (9.5.8)$$

Substituting the expansions (9.5.4) and (9.5.5) into Eq. (9.5.1), we obtain

$$\sum_{n=1}^{\infty} \sum_{m=1}^{\infty} \left\{ W_{mn} \left[D_{11}\alpha_m^4 + 2(D_{12} + 2D_{66})\alpha_m^2\beta_n^2 + D_{22}\beta_n^4 \right] \right.$$
$$\left. + \left[I_0 + I_2 \left(\alpha_m^2 + \beta_n^2 \right) \right] \ddot{W}_{mn} - Q_{mn} \right\} \sin\alpha_m x \sin\beta_n y = 0 \quad (9.5.9)$$

Since the above expression must hold for all x and y, it follows that

$$W_{mn} \left[D_{11}\alpha_m^4 + 2(D_{12} + 2D_{66})\alpha_m^2\beta_n^2 + D_{22}\beta_n^4 \right]$$
$$+ \left[I_0 + I_2 \left(\alpha_m^2 + \beta_n^2 \right) \right] \ddot{W}_{mn} - Q_{mn} = 0 \qquad (9.5.10a)$$

or

$$K_{mn} W_{mn}(t) + M_{mn} \ddot{W}_{mn} = Q_{mn}(t) \qquad (9.5.10b)$$

where

$$K_{mn} = D_{11}\alpha_m^4 + 2(D_{12} + 2D_{66})\alpha_m^2\beta_n^2 + D_{22}\beta_n^4$$
$$M_{mn} = I_0 + I_2\left(\alpha_m^2 + \beta_n^2\right) \qquad (9.5.10c)$$

9.5.2 Time Integration

The second-order differential equation (9.5.10a) can be solved either exactly or numerically. The numerical time integration methods will be discussed in the subsequent chapters. To solve it exactly, we first write Eq. (9.5.10a) in the form

$$\frac{d^2W_{mn}}{dt^2} + \left(\frac{K_{mn}}{M_{mn}}\right)W_{mn} = \frac{1}{M_{mn}}Q_{mn}(t) \equiv \hat{Q}_{mn}(t) \qquad (9.5.11)$$

The solution of Eq. (9.5.11) is given by

$$W_{mn}(t) = C_1 e^{\lambda_1 t} + C_2 e^{\lambda_2 t} + W_{mn}^p(t) \qquad (9.5.12)$$

where C_1 and C_2 are constants to be determined using the initial conditions, $W_{mn}^p(t)$ is the particular solution

$$W_{mn}^p(t) = \int^t \frac{r_1(\tau)r_2(t) - r_1(t)r_2(\tau)}{r_1(\tau)\dot{r}_2(\tau) - \dot{r}_1(\tau)r_2(\tau)} \hat{Q}_{mn}(\tau)\,d\tau \qquad (9.5.13a)$$

with $r_1(t) = e^{\lambda_1 t}$ and $r_2(t) = e^{\lambda_2 t}$, and λ_1 and λ_2 are the roots of the equation

$$\lambda^2 + \frac{K_{mn}}{M_{mn}} = 0; \quad \lambda_1 = -i\mu, \ \lambda_2 = i\mu, \ i = \sqrt{-1}, \ \mu = \sqrt{\frac{K_{mn}}{M_{mn}}} \qquad (9.5.13b)$$

The solution becomes

$$W_{mn}(t) = A\cos\mu t + B\sin\mu t + W_{mn}^p(t) \qquad (9.5.14a)$$

$$W_{mn}^p(t) = \frac{1}{2i\mu}\left(e^{i\mu t}\int^t e^{-i\mu\tau}\hat{Q}_{mn}(\tau)d\tau - e^{-i\mu t}\int^t e^{i\mu\tau}\hat{Q}_{mn}(\tau)d\tau\right) \qquad (9.5.14b)$$

Once the load distribution, both spatially and with time, is known, the solution can be determined from Eq. (9.5.14a).

For a step loading, $Q_{mn}(t) = Q^0_{mn} H(t)$, where $H(t)$ denotes the Heaviside step function, Eq. (9.5.14a) takes the form

$$W_{mn}(t) = A_{mn} \cos \mu t + B_{mn} \sin \mu t + \frac{1}{K_{mn}} Q^0_{mn} \qquad (9.5.15a)$$

Using the initial conditions (9.5.3), we obtain

$$A_{mn} = D_{mn} - \frac{1}{K_{mn}} Q^0_{mn}, \quad B_{mn} = \frac{V_{mn}}{\mu} \qquad (9.5.15b)$$

Thus the final solution (9.5.4) is given by

$$w_0(x,y,t) = \sum_{n=1}^{\infty} \sum_{m=1}^{\infty} \left[D_{mn} \cos \mu t + \frac{V_{mn}}{\mu} \sin \mu t + \frac{Q^0_{mn}}{K_{mn}} (1 - \cos \mu t) \right]$$

$$\times \sin \alpha_m x \sin \beta_n y \qquad (9.5.16)$$

The coefficients Q^0_{mn} were given in Table 6.3.1 for various types of distributions. The same holds for D_{mn} and V_{mn}.

It should be noted that the procedure outlined above is valid irrespective of how one arrives at Eq. (9.5.10b), e.g., Eq. (9.5.10b) could have been obtained using the Rayleigh–Ritz method or other methods. The exact solution of the differential equation (9.5.10b) can also be obtained using the Laplace transform method (see Table 4.6.1 for the Laplace transform pairs). Once the solution w_0 is known, stresses can be computed using the constitutive equations.

Exercises

9.1 Consider the natural vibration of a rectangular plate with all edges simply supported. Use the Lévy solution method and establish the result in Eq. (9.2.6).

9.2 Assuming that $(I_0 \omega^2 / D_{22}) < \alpha_m^2 (D_{11}/D_{22})$, show that the frequency equation of SSSF plates is

$$\Omega_1 \Lambda_2 \cosh \lambda_2 b \, \sinh \lambda_1 b - \Lambda_1 \Omega_2 \cosh \lambda_1 b \, \sinh \lambda_2 b = 0$$

where

$$\Omega_1 = \lambda_1^2 - \frac{D_{12}}{D_{22}} \alpha_m^2, \quad \Omega_2 = \lambda_2^2 - \frac{D_{12}}{D_{22}} \alpha_m^2$$

$$\Lambda_1 = \lambda_1^2 - \frac{\bar{D}_{12}}{D_{22}}\alpha_m^2, \quad \Lambda_2 = \lambda_2^2 - \frac{\bar{D}_{12}}{D_{22}}\alpha_m^2$$

and $\bar{D}_{12} = D_{12} + 4D_{66}$.

9.3 Assuming that $(I_0\omega^2/D_{22}) < \alpha_m^2(D_{11}/D_{22})$, show that the frequency equation of SSCC plates is

$$2\lambda_1\lambda_2\left(1 - \cosh\lambda_1 b\cosh\lambda_2 b\right) + \left(\lambda_1^2 + \lambda_2^2\right)\sinh\lambda_1 b\sinh\lambda_2 b = 0$$

9.4 Assuming that $(I_0\omega^2/D_{22}) < \alpha_m^2(D_{11}/D_{22})$, show that the frequency equation of SSCS plates is

$$\lambda_2\cosh\lambda_2 b\sinh\lambda_1 b - \lambda_1\cosh\lambda_1 b\sinh\lambda_2 b = 0$$

9.5 Assuming that $(I_0\omega^2/D_{22}) < \alpha_m^2(D_{11}/D_{22})$, show that the frequency equation of SSCF plates is

$$\lambda_1\lambda_2\left[\Omega_1\Lambda_1 + \Omega_2\Lambda_2 - (\Omega_1\Lambda_2 + \Omega_2\Lambda_1)\cosh\lambda_1 b\cosh\lambda_2 b\right]$$
$$+ \left(\lambda_1^2\Omega_2\Lambda_1 + \lambda_2^2\Omega_1\Lambda_2\right)\sinh\lambda_1 b\sinh\lambda_2 b = 0$$

9.6 Assuming that $(I_0\omega^2/D_{22}) < \alpha_m^2(D_{11}/D_{22})$, show that the frequency equation of SSCC plates is given by Eq. (9.3.26).

9.7 Assuming that $(I_0\omega^2/D_{22}) > \alpha_m^2(D_{11}/D_{22})$, show that the frequency equation of SSCS plates is given by Eq. (9.3.28).

9.8 Assuming that $(I_0\omega^2/D_{22}) < \alpha_m^2(D_{11}/D_{22})$, show that the frequency equation of SSCS plates is given by Eq. (9.3.29).

9.9 Derive the approximation functions X_i and Y_j in Eqs. (9.4.19a,b) for CFCF plates so that they satisfy the geometric and force boundary conditions.

9.10 Derive the explicit form of the coefficients $R_{(ij)(k\ell)}$ and $B_{(ij)(k\ell)}$ for CSCS plates using the functions in Eq. (9.4.17).

References for Additional Reading

1. Timoshenko, S. P. and Young, D. H., *Vibration Problems in Engineering,* Third Edition, D. Van Nostrand, New York (1955).
2. Szilard, R., *Theory and Analysis of Plates, Classical and Numerical Methods,* Prentice–Hall, Englewood Cliffs, NJ (1974).

3. Reddy, J. N., *Energy and Variational Methods in Applied Mechanics*, John Wiley and Sons, New York (1984).

4. Reddy, J. N., *Mechanics of Laminated Composite Plates: Theory and Analysis*, CRC Press, Boca Raton, FL (1997).

5. Hearman, R. F. S., "The Frequency of Flexural Vibration of Rectangular Orthotropic Plates with Clamped or Supported Edges," *Journal of Applied Mechanics*, **26**(4), 537–540 (1959).

6. Young, D., "Vibrations of Rectangular Plates by the Ritz Method", *Journal of Applied Mechanics*, **17**, 448–453 (1950).

7. Warburton, G. B., "The Vibration of Rectangular Plates", *Proceedings of the Institution of Mechanical Engineering*, Series A, **168**(12), 371–384 (1954).

8. Leissa, A. W., *Vibration of Plates*, NASA SP–160, NASA, Washington, D.C., 1969.

9. Ödman, S. T. A., "Studies of Boundary Value Problems. Part II. Characteristic Functions of Rectangular Plates", *Proc. NR 24*, Swedish Cement and Concrete Research Institute, Royal Institute of Technology, Stockholm, 7–62 (1955).

10. Khdeir, A. A., "Free Vibration and Buckling of Symmetric Cross-Ply Laminated Plates by an Exact Method", *Journal of Sound and Vibration*, **126**(3), 447–461 (1988).

11. Reddy, J. N. and Khdeir, A. A., "Buckling and Vibration of Laminated Composite Plates Using Various Plate Theories", *AIAA Journal*, **27**(12), 1808–1817 (1989).

12. Young, D. and Felgar, F. P., *Tables of Characteristic Functions Representing the Normal Modes of Vibration of a Beam*, University of Texas, Austin, TX, Publication No. 4913 (1949).

Chapter Ten

Analysis of Rectangular Plates Using Shear Deformation Plate Theories

10.1 First-Order Shear Deformation Plate Theory

10.1.1 Preliminary Comments

The preceding chapters of the book were devoted to the study of bending, buckling, and natural vibrations of rectangular plates using the classical plate theory (CPT), in which transverse normal and shear stresses are neglected. The first-order shear deformation plate theory (FSDT) extends the kinematics of the CPT by relaxing the normality restriction (see section 3.2) and allowing for arbitrary but constant rotation of transverse normals (see Figure 10.1.1).

The more significant difference between the classical and first-order theories is the effect of including transverse shear deformation on the predicted deflections, frequencies, and buckling loads. As will be seen in the sequel, the classical plate theory underpredicts deflections and overpredicts frequencies as well as buckling loads of plates with side-to-thickness ratios of the order of 20 or less (i.e., thick plates). For this reason alone it is necessary to use the first-order shear deformation plate theory in the analysis of relatively thick plates. In this chapter, we develop analytical solutions of rectangular plates using the first-order shear deformation theory. The primary objective is to bring out the effect of shear deformation on deflections, stresses, frequencies, and buckling loads.

10.1.2 Kinematics

Under the same assumptions and restrictions as in the classical laminate theory but relaxing the normality condition, the displacement field of the first-order theory can be expressed in the form

$$u(x,y,z,t) = u_0(x,y,t) + z\phi_x(x,y,t)$$
$$v(x,y,z,t) = v_0(x,y,t) + z\phi_y(x,y,t)$$
$$w(x,y,z,t) = w_0(x,y,t) \tag{10.1.1}$$

where $(u_0, v_0, w_0, \phi_x, \phi_y)$ are unknown functions to be determined. As before, (u_0, v_0, w_0) denote the displacements of a point on the plane $z = 0$ and ϕ_x and ϕ_y are the rotations of a transverse normal about the $y-$ and $x-$axes, respectively (see Figure 10.1.1). The quantities $(u_0, v_0, w_0, \phi_x, \phi_y)$ are called the *generalized displacements*.

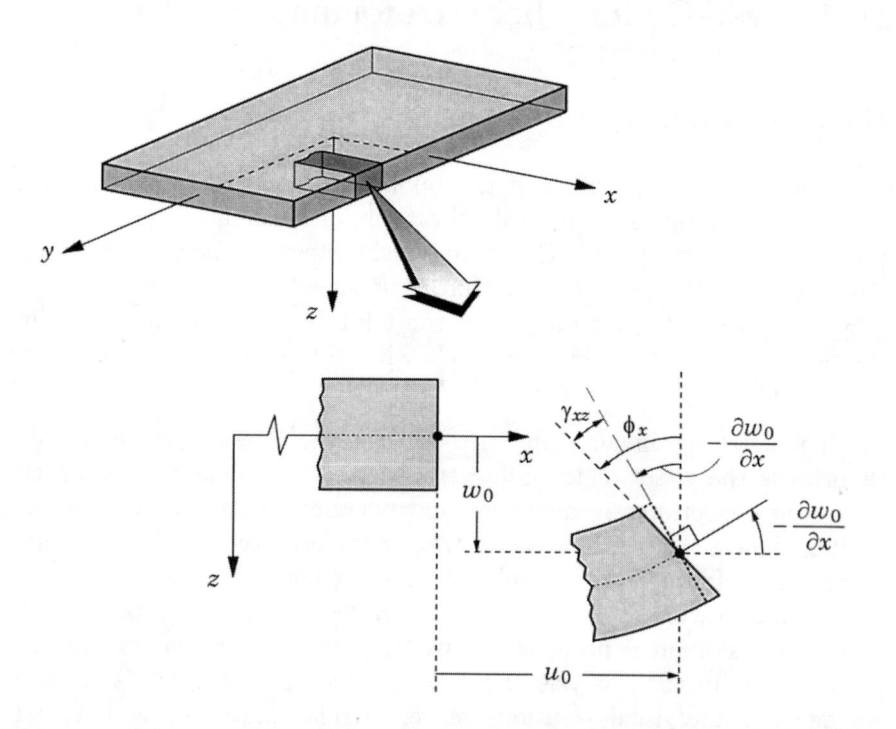

Figure 10.1.1. Undeformed and deformed geometries of an edge of a plate under the assumptions of the first-order shear deformation plate theory (FSDT).

The notation that ϕ_x denotes the rotation of a transverse normal about the y-axis and ϕ_y denotes the rotation about the x-axis may be confusing to some because they do not follow the right-hand rule. However, the notation has been used extensively in the literature, and we will not depart from it. If (β_x, β_y) denote the rotations about the x and y axes, respectively, that follow the right-hand rule, then

$$\beta_x = \phi_y, \quad \beta_y = -\phi_x \qquad (10.1.2)$$

For thin plates, i.e., when the plate in-plane characteristic dimension to thickness ratio is on the order of 50 or greater, the rotation functions ϕ_x and ϕ_y should approach the respective slopes of the transverse deflection:

$$\phi_x = -\frac{\partial w_0}{\partial x}, \quad \phi_y = -\frac{\partial w_0}{\partial y} \qquad (10.1.3)$$

Indeed, the first-order shear deformation theory can be derived from the classical plate theory using the conditions (10.1.3) as constraints (see Reddy [1], pp. 552–554).

The von Kármán nonlinear strains associated with the displacement field (10.1.1) are

$$\varepsilon_{xx} = \frac{\partial u_0}{\partial x} + \frac{1}{2}\left(\frac{\partial w_0}{\partial x}\right)^2 + z\frac{\partial \phi_x}{\partial x}$$

$$\gamma_{xy} = \left(\frac{\partial u_0}{\partial y} + \frac{\partial v_0}{\partial x} + \frac{\partial w_0}{\partial x}\frac{\partial w_0}{\partial y}\right) + z\left(\frac{\partial \phi_x}{\partial y} + \frac{\partial \phi_y}{\partial x}\right)$$

$$\varepsilon_{yy} = \frac{\partial v_0}{\partial y} + \frac{1}{2}\left(\frac{\partial w_0}{\partial y}\right)^2 + z\frac{\partial \phi_y}{\partial y}$$

$$\gamma_{xz} = \frac{\partial w_0}{\partial x} + \phi_x, \quad \gamma_{yz} = \frac{\partial w_0}{\partial y} + \phi_y, \quad \varepsilon_{zz} = 0 \qquad (10.1.4)$$

Note that the strains $(\varepsilon_{xx}, \varepsilon_{yy}, \gamma_{xy})$ are linear through the plate thickness, while the transverse shear strains $(\gamma_{xz}, \gamma_{yz})$ are constant. The strains in Eq. (10.1.4) can be expressed in the vector form as

$$\begin{Bmatrix} \varepsilon_{xx} \\ \varepsilon_{yy} \\ \gamma_{yz} \\ \gamma_{xz} \\ \gamma_{xy} \end{Bmatrix} = \begin{Bmatrix} \varepsilon_{xx}^0 \\ \varepsilon_{yy}^0 \\ \gamma_{yz}^0 \\ \gamma_{xz}^0 \\ \gamma_{xy}^0 \end{Bmatrix} + z \begin{Bmatrix} \varepsilon_{xx}^1 \\ \varepsilon_{yy}^1 \\ 0 \\ 0 \\ \gamma_{xy}^1 \end{Bmatrix}, \quad \text{or} \quad \{\varepsilon\} = \{\varepsilon^0\} + z\{\varepsilon^1\} \qquad (10.1.5a)$$

$$\{\varepsilon^0\} = \begin{Bmatrix} \frac{\partial u_0}{\partial x} + \frac{1}{2}\left(\frac{\partial w_0}{\partial x}\right)^2 \\ \frac{\partial v_0}{\partial y} + \frac{1}{2}\left(\frac{\partial w_0}{\partial y}\right)^2 \\ \frac{\partial w_0}{\partial y} + \phi_y \\ \frac{\partial w_0}{\partial x} + \phi_x \\ \frac{\partial u_0}{\partial y} + \frac{\partial v_0}{\partial x} + \frac{\partial w_0}{\partial x}\frac{\partial w_0}{\partial y} \end{Bmatrix}, \quad \{\varepsilon^1\} = \begin{Bmatrix} \frac{\partial \phi_x}{\partial x} \\ \frac{\partial \phi_y}{\partial y} \\ 0 \\ 0 \\ \frac{\partial \phi_x}{\partial y} + \frac{\partial \phi_y}{\partial x} \end{Bmatrix} \quad (10.1.5b)$$

10.1.3 Equations of Motion

The governing equations of the first-order plate theory can be derived using the dynamic version of the principle of virtual displacements (or Hamilton's principle of section 2.2.3)

$$0 = \int_0^T (\delta U + \delta V - \delta K)\, dt \quad (10.1.6)$$

where the virtual strain energy δU, virtual work done by applied forces δV, and the virtual kinetic energy δK are given by

$$\delta U = \int_{\Omega_0} \left\{ \int_{-\frac{h}{2}}^{\frac{h}{2}} \left[\sigma_{xx}\left(\delta\varepsilon_{xx}^0 + z\delta\varepsilon_{xx}^1\right) + \sigma_{yy}\left(\delta\varepsilon_{yy}^0 + z\delta\varepsilon_{yy}^1\right) \right.\right.$$
$$\left.\left. + \sigma_{xy}\left(\delta\gamma_{xy}^0 + z\delta\gamma_{xy}^1\right) + \sigma_{xz}\delta\gamma_{xz}^0 + \sigma_{yz}\delta\gamma_{yz}^0 \right] dz \right\} dxdy \quad (10.1.7a)$$

$$\delta V = -\int_{\Gamma_\sigma} \int_{-\frac{h}{2}}^{\frac{h}{2}} [\hat{\sigma}_{nn}(\delta u_n + z\delta\phi_n) + \hat{\sigma}_{ns}(\delta u_s + z\delta\phi_s) + \hat{\sigma}_{nz}\delta w_0]\, dzds$$
$$- \int_{\Omega_0} (q_b + q_t - kw_0)\,\delta w_0\, dxdy \quad (10.1.7b)$$

$$\delta K = \int_{\Omega_0} \int_{-\frac{h}{2}}^{\frac{h}{2}} \rho_0 \left[(\dot{u}_0 + z\dot{\phi}_x)(\delta\dot{u}_0 + z\delta\dot{\phi}_x) + (\dot{v}_0 + z\dot{\phi}_y)(\delta\dot{v}_0 + z\delta\dot{\phi}_y) \right.$$
$$\left. + \dot{w}_0\delta\dot{w}_0 \right] dz\, dxdy \quad (10.1.8)$$

where Ω_0 denotes the undeformed mid-plane of the plate, h the total thickness, ρ_0 the density of the plate, and k the modulus of the elastic foundation.

Substituting for $\delta U, \delta V$, and δK from Eqs. (10.1.7a,b) and (10.1.8) into the virtual work statement in Eq. (10.1.6) and integrating through the thickness of the laminate, we obtain

$$0 = \int_0^T \left\{ \int_{\Omega_0} \left[N_{xx}\delta\varepsilon_{xx}^0 + M_{xx}\delta\varepsilon_{xx}^1 + N_{yy}\delta\varepsilon_{yy}^0 + M_{yy}\delta\varepsilon_{yy}^1 + N_{xy}\delta\gamma_{xy}^0 \right. \right.$$
$$+ M_{xy}\delta\gamma_{xy}^1 + Q_x\delta\gamma_{xz}^0 + Q_y\delta\gamma_{yz}^0 + kw_0\delta w_0 - q\delta w_0$$
$$- I_1 \left(\dot\phi_x\delta\dot u_0 + \dot\phi_y\delta\dot v_0 + \delta\dot\phi_x\dot u_0 + \delta\dot\phi_y\dot v_0 \right)$$
$$- I_0 \left(\dot u_0\delta\dot u_0 + \dot v_0\delta\dot v_0 + \dot w_0\delta\dot w_0 \right) - I_2 \left(\dot\phi_x\delta\dot\phi_x + \dot\phi_y\delta\dot\phi_y \right) \bigg] dx dy$$
$$\left. - \int_{\Gamma_\sigma} \left(\hat N_{nn}\delta u_n + \hat N_{ns}\delta u_s + \hat M_{nn}\delta\phi_n + \hat M_{ns}\delta\phi_s + \hat Q_n\delta w_0 \right) ds \right\} dt$$
$$(10.1.9)$$

where $q = q_b + q_t$, the stress resultants $(N_{xx}, N_{yy}, N_{xy}, M_{xx}, M_{yy}, M_{xy})$ and the inertias (I_0, I_1, I_2) were defined in Eqs. (3.4.4) and (3.4.9), respectively, $(N_{nn}, N_{ns}, M_{nn}, M_{ns})$ in Eq. (3.4.7) [also, see Eqs. (3.5.7) and (3.5.8)], and the *transverse force resultants* (Q_x, Q_y) are defined by

$$\left\{ \begin{matrix} Q_x \\ Q_y \end{matrix} \right\} = \int_{-\frac{h}{2}}^{\frac{h}{2}} \left\{ \begin{matrix} \sigma_{xz} \\ \sigma_{yz} \end{matrix} \right\} dz \qquad (10.1.10)$$

Shear Correction Factors

Since the transverse shear strains are represented as constant through the laminate thickness, it follows that the transverse shear stresses will also be constant. It is well known from elementary theory of homogeneous beams that the transverse shear stress variation is parabolic through the beam thickness. This discrepancy between the actual stress state and the constant stress state predicted by the first-order theory is often corrected in computing the transverse shear force resultants (Q_x, Q_y) by multiplying the integrals in Eq. (10.1.10) with a parameter K, called *shear correction coefficient*:

$$\left\{ \begin{matrix} Q_x \\ Q_y \end{matrix} \right\} = K \int_{-\frac{h}{2}}^{\frac{h}{2}} \left\{ \begin{matrix} \sigma_{xz} \\ \sigma_{yz} \end{matrix} \right\} dz \qquad (10.1.11)$$

This amounts to modifying the plate transverse shear stiffnesses. The factor K is computed such that the strain energy due to transverse shear stresses in Eq. (10.1.11) equals the strain energy due to the true transverse stresses predicted by the three-dimensional elasticity theory.

For example, consider a homogeneous beam with rectangular cross section, with width b and height h. The actual shear stress distribution through the thickness of the beam is given by

$$\sigma_{xz}^c = \frac{3Q_0}{2bh}\left[1 - \left(\frac{2z}{h}\right)^2\right], \quad -\frac{h}{2} \leq z \leq \frac{h}{2} \quad (10.1.12)$$

where Q_0 is the transverse load. The transverse shear stress in the first-order theory is a constant, $\sigma_{xz}^f = Q_0/bh$. The strain energies due to transverse shear stresses in the two theories are

$$U_s^c = \frac{1}{2G_{13}} \int_A (\sigma_{xz}^c)^2 \, dA = \frac{3Q_0^2}{5G_{13}bh}$$
$$U_s^f = \frac{1}{2G_{13}} \int_A (\sigma_{xz}^f)^2 \, dA = \frac{Q_0^2}{2G_{13}bh} \quad (10.1.13)$$

The shear correction factor is the ratio of U_s^f to U_s^c, which gives $K = 5/6$. The shear correction factor, in general, depends on the geometry and material properties of the plate.

Returning to the virtual work statement in Eq. (10.1.9), we substitute for the virtual strains into Eq. (10.1.9) and integrate by parts to relieve the virtual generalized displacements $(\delta u_0, \delta v_0, \delta w_0, \delta\phi_x, \delta\phi_y)$ in Ω_0 of any differentiation, so that we can use the fundamental lemma of variational calculus. We obtain

$$\begin{aligned}
0 = \int_0^T \int_{\Omega_0} &\bigg[-\left(N_{xx,x} + N_{xy,y} - I_0\ddot{u}_0 - I_1\ddot{\phi}_x\right)\delta u_0 \\
&- \left(N_{xy,x} + N_{yy,y} - I_0\ddot{v}_0 - I_1\ddot{\phi}_y\right)\delta v_0 \\
&- \left(M_{xx,x} + M_{xy,y} - Q_x - I_2\ddot{\phi}_x - I_1\ddot{u}_0\right)\delta\phi_x \\
&- \left(M_{xy,x} + M_{yy,y} - Q_y - I_2\ddot{\phi}_y - I_1\ddot{v}_0\right)\delta\phi_y \\
&- (Q_{x,x} + Q_{y,y} - kw_0 + \mathcal{N} + q - I_0\ddot{w}_0)\delta w_0 \bigg] dxdy \\
+ \int_0^T \int_\Gamma &\bigg[\left(N_{nn} - \hat{N}_{nn}\right)\delta u_n + \left(N_{ns} - \hat{N}_{ns}\right)\delta u_s + \left(Q_n - \hat{Q}_n\right)\delta w_0 \\
&+ \left(M_{nn} - \hat{M}_{nn}\right)\delta\phi_n + \left(M_{ns} - \hat{M}_{ns}\right)\delta\phi_s \bigg] dsdt \quad (10.1.14)
\end{aligned}$$

where $\mathcal{N}$ and $\mathcal{P}$ were defined in Eq. (3.4.14)

$$\mathcal{N} = \frac{\partial}{\partial x}\left(N_{xx}\frac{\partial w_0}{\partial x} + N_{xy}\frac{\partial w_0}{\partial y}\right) + \frac{\partial}{\partial y}\left(N_{xy}\frac{\partial w_0}{\partial x} + N_{yy}\frac{\partial w_0}{\partial y}\right)$$
$$\mathcal{P} = \left(N_{xx}\frac{\partial w_0}{\partial x} + N_{xy}\frac{\partial w_0}{\partial y}\right)n_x + \left(N_{xy}\frac{\partial w_0}{\partial x} + N_{yy}\frac{\partial w_0}{\partial y}\right)n_y$$
(10.1.15)

and the boundary expressions were arrived by expressing ϕ_x and ϕ_y in terms of the normal and tangential rotations, (ϕ_n, ϕ_s):

$$\phi_x = n_x\phi_n - n_y\phi_s , \quad \phi_y = n_y\delta\phi_n + n_x\delta\phi_s \qquad (10.1.16)$$

The Euler–Lagrange equations are

$$\delta u_0: \quad \frac{\partial N_{xx}}{\partial x} + \frac{\partial N_{xy}}{\partial y} = I_0\frac{\partial^2 u_0}{\partial t^2} + I_1\frac{\partial^2 \phi_x}{\partial t^2} \qquad (10.1.17)$$

$$\delta v_0: \quad \frac{\partial N_{xy}}{\partial x} + \frac{\partial N_{yy}}{\partial y} = I_0\frac{\partial^2 v_0}{\partial t^2} + I_1\frac{\partial^2 \phi_y}{\partial t^2} \qquad (10.1.18)$$

$$\delta w_0: \quad \frac{\partial Q_x}{\partial x} + \frac{\partial Q_y}{\partial y} - kw_0 + \mathcal{N} + q = I_0\frac{\partial^2 w_0}{\partial t^2} \qquad (10.1.19)$$

$$\delta\phi_x: \quad \frac{\partial M_{xx}}{\partial x} + \frac{\partial M_{xy}}{\partial y} - Q_x = I_2\frac{\partial^2 \phi_x}{\partial t^2} + I_1\frac{\partial^2 u_0}{\partial t^2} \qquad (10.1.20)$$

$$\delta\phi_y: \quad \frac{\partial M_{xy}}{\partial x} + \frac{\partial M_{yy}}{\partial y} - Q_y = I_2\frac{\partial^2 \phi_y}{\partial t^2} + I_1\frac{\partial^2 v_0}{\partial t^2} \qquad (10.1.21)$$

The natural boundary conditions are

$$N_{nn} - \hat{N}_{nn} = 0 , \quad N_{ns} - \hat{N}_{ns} = 0 , \quad Q_n - \hat{Q}_n = 0$$
$$M_{nn} - \hat{M}_{nn} = 0 , \quad M_{ns} - \hat{M}_{ns} = 0 \qquad (10.1.22)$$

where
$$Q_n \equiv Q_x n_x + Q_y n_y + \mathcal{P} \qquad (10.1.23)$$

Thus, the primary and secondary variables of the theory are

primary variables: $u_n, u_s, w_0, \phi_n, \phi_s$
secondary variables: $N_{nn}, N_{ns}, Q_n, M_{nn}, M_{ns}$ (10.1.24)

The initial conditions of the theory involve specifying the values of the displacements and their first derivatives with respect to time at $t = 0$:

$$u_n = u_n^0, \quad u_s = u_s^0, \quad w_0 = w_0^0, \quad \phi_n = \phi_n^0, \quad \phi_s = \phi_s^0$$
$$\dot{u}_n = \dot{u}_n^0, \quad \dot{u}_s = \dot{u}_s^0, \quad \dot{w}_0 = \dot{w}_0^0, \quad \dot{\phi}_n = \dot{\phi}_n^0, \quad \dot{\phi}_s = \dot{\phi}_s^0 \qquad (10.1.25)$$

for all points in Ω_0.

10.1.4 Plate Constitutive Equations

The plate constitutive equations in Eqs. (3.6.6) and (3.6.7) are valid also for the first-order plate theory. In addition, we have the following constitutive equations:

$$\begin{Bmatrix} Q_y \\ Q_x \end{Bmatrix} = K \int_{-\frac{h}{2}}^{\frac{h}{2}} \begin{Bmatrix} \sigma_{yz} \\ \sigma_{xz} \end{Bmatrix} dz$$

$$= K \begin{bmatrix} A_{44} & A_{45} \\ A_{45} & A_{55} \end{bmatrix} \begin{Bmatrix} \gamma_{yz} \\ \gamma_{xz} \end{Bmatrix} \qquad (10.1.26)$$

where the extensional stiffnesses A_{44}, A_{45}, and A_{55} are defined by

$$(A_{44}, A_{45}, A_{55}) = \int_{-\frac{h}{2}}^{\frac{h}{2}} (Q_{44}, Q_{45}, Q_{55}) \, dz$$

$$= \sum_{k=1}^{N} (Q_{44}^k, Q_{45}^k, Q_{55}^k) h_k \qquad (10.1.27)$$

where N denotes the number of layers in the plate, Q_{ij}^k are the elastic stiffnesses [see Eqs. (1.4.37) and (1.4.38)], and h_k is the thickness of the kth layer. When thermal effects are not present, the stress resultants in an *orthotropic* plate are related to the generalized displacements $(u_0, v_0, w_0, \phi_x, \phi_y)$ by

$$\begin{Bmatrix} N_{xx} \\ N_{yy} \\ N_{xy} \end{Bmatrix} = \begin{bmatrix} A_{11} & A_{12} & 0 \\ A_{12} & A_{22} & 0 \\ 0 & 0 & A_{66} \end{bmatrix} \begin{Bmatrix} \frac{\partial u_0}{\partial x} + \frac{1}{2}(\frac{\partial w_0}{\partial x})^2 \\ \frac{\partial v_0}{\partial y} + \frac{1}{2}(\frac{\partial w_0}{\partial y})^2 \\ \frac{\partial u_0}{\partial y} + \frac{\partial v_0}{\partial x} + \frac{\partial w_0}{\partial x}\frac{\partial w_0}{\partial y} \end{Bmatrix} \qquad (10.1.28)$$

$$\begin{Bmatrix} M_{xx} \\ M_{yy} \\ M_{xy} \end{Bmatrix} = \begin{bmatrix} D_{11} & D_{12} & 0 \\ D_{12} & D_{22} & 0 \\ 0 & 0 & D_{66} \end{bmatrix} \begin{Bmatrix} \frac{\partial \phi_x}{\partial x} \\ \frac{\partial \phi_y}{\partial y} \\ \frac{\partial \phi_x}{\partial y} + \frac{\partial \phi_y}{\partial x} \end{Bmatrix} \qquad (10.1.29)$$

$$\begin{Bmatrix} Q_y \\ Q_x \end{Bmatrix} = K \begin{bmatrix} A_{44} & 0 \\ 0 & A_{55} \end{bmatrix} \begin{Bmatrix} \frac{\partial w_0}{\partial y} + \phi_y \\ \frac{\partial w_0}{\partial x} + \phi_x \end{Bmatrix} \qquad (10.1.30)$$

10.1.5 Equations of Motion in Terms of Displacements

The equations of motion (10.1.17)–10.1.21) can be expressed in terms of displacements $(u_0, v_0, w_0, \phi_x, \phi_y)$ by substituting for the force and moment resultants from Eqs. (10.1.28)–(10.1.30). For homogeneous plates, the equations of motion take the form

$$A_{11}\left(\frac{\partial^2 u_0}{\partial x^2} + \frac{\partial w_0}{\partial x}\frac{\partial^2 w_0}{\partial x^2}\right) + A_{12}\left(\frac{\partial^2 v_0}{\partial y \partial x} + \frac{\partial w_0}{\partial y}\frac{\partial^2 w_0}{\partial y \partial x}\right) +$$

$$A_{66}\left(\frac{\partial^2 u_0}{\partial y^2} + \frac{\partial^2 v_0}{\partial x \partial y} + \frac{\partial^2 w_0}{\partial x \partial y}\frac{\partial w_0}{\partial y} + \frac{\partial w_0}{\partial x}\frac{\partial^2 w_0}{\partial y^2}\right) -$$

$$\left(\frac{\partial N_{xx}^T}{\partial x} + \frac{\partial N_{xy}^T}{\partial y}\right) = I_0\frac{\partial^2 u_0}{\partial t^2} \qquad (10.1.31)$$

$$A_{66}\left(\frac{\partial^2 u_0}{\partial y \partial x} + \frac{\partial^2 v_0}{\partial x^2} + \frac{\partial^2 w_0}{\partial x^2}\frac{\partial w_0}{\partial y} + \frac{\partial w_0}{\partial x}\frac{\partial^2 w_0}{\partial y \partial x}\right) +$$

$$A_{12}\left(\frac{\partial^2 u_0}{\partial x \partial y} + \frac{\partial w_0}{\partial x}\frac{\partial^2 w_0}{\partial x \partial y}\right) + A_{22}\left(\frac{\partial^2 v_0}{\partial y^2} + \frac{\partial w_0}{\partial y}\frac{\partial^2 w_0}{\partial y^2}\right) -$$

$$\left(\frac{\partial N_{xy}^T}{\partial x} + \frac{\partial N_{yy}^T}{\partial y}\right) = I_0\frac{\partial^2 v_0}{\partial t^2} \qquad (10.1.32)$$

$$KA_{55}\left(\frac{\partial^2 w_0}{\partial x^2} + \frac{\partial \phi_x}{\partial x}\right) + KA_{44}\left(\frac{\partial^2 w_0}{\partial y^2} + \frac{\partial \phi_y}{\partial y}\right) - kw_0 +$$

$$\mathcal{N}(u_0, v_0, w_0) + q(x,y) = I_0\frac{\partial^2 w_0}{\partial t^2} \qquad (10.1.33)$$

$$D_{11}\frac{\partial^2 \phi_x}{\partial x^2} + D_{12}\frac{\partial^2 \phi_y}{\partial y \partial x} + D_{66}\left(\frac{\partial^2 \phi_x}{\partial y^2} + \frac{\partial^2 \phi_y}{\partial y \partial x}\right) -$$

$$KA_{55}\left(\frac{\partial w_0}{\partial x} + \phi_x\right) - \left(\frac{\partial M_{xx}^T}{\partial x} + \frac{\partial M_{xy}^T}{\partial y}\right) = I_2\frac{\partial^2 \phi_x}{\partial t^2} \qquad (10.1.34)$$

$$D_{66}\left(\frac{\partial^2 \phi_x}{\partial x \partial y} + \frac{\partial^2 \phi_y}{\partial x^2}\right) + D_{12}\frac{\partial^2 \phi_x}{\partial x \partial y} + D_{22}\frac{\partial^2 \phi_y}{\partial y^2} -$$

$$KA_{44}\left(\frac{\partial w_0}{\partial y} + \phi_y\right) - \left(\frac{\partial M_{xy}^T}{\partial x} + \frac{\partial M_{yy}^T}{\partial y}\right) = I_2\frac{\partial^2 \phi_y}{\partial t^2} \qquad (10.1.35)$$

Equations (10.1.31)–(10.1.35) describe five second-order, nonlinear, partial differential equations in terms of the five generalized displacements. When the strains and rotations are assumed to be small, Eqs. (10.1.31) and (10.1.32) are uncoupled from Eqs. (10.1.33)–(10.1.35). In the following sections, we consider bending, buckling, and natural vibrations using these linear equations of bending.

10.2 The Navier Solutions of FSDT

10.2.1 General Solution

As in the case of the classical plate theory, exact analytical solutions of the first-order shear deformation plate theory can be obtained for simply supported plates using Navier's method: the simply supported boundary conditions for the first-order shear deformation plate theory (FSDT) can be expressed as (Figure 10.2.1)

$$w_0(x,0,t) = 0, \quad w_0(x,b,t) = 0, \quad w_0(0,y,t) = 0, \quad w_0(a,y,t) = 0$$

$$\phi_x(x,0,t) = 0, \quad \phi_x(x,b,t) = 0, \quad \phi_y(0,y,t) = 0, \quad \phi_y(a,y,t) = 0$$

$$M_{yy}(x,0,t) = 0, \quad M_{yy}(x,b,t) = 0, \quad M_{xx}(0,y,t) = 0, \quad M_{xx}(a,y,t) = 0$$
$$(10.2.1)$$

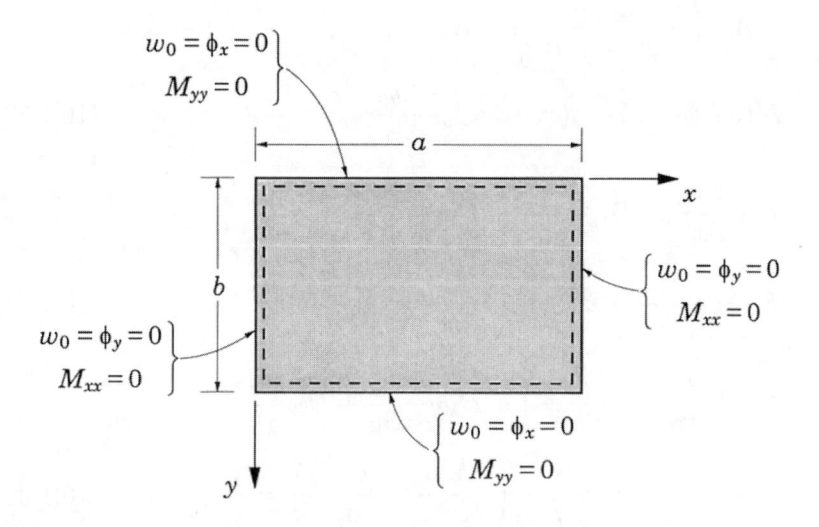

Figure 10.2.1. The simply supported boundary conditions of the first-order shear deformation theory.

The boundary conditions in Eq. (10.2.1) are satisfied by the following expansions of the displacements:

$$w_0(x,y,t) = \sum_{n=1}^{\infty}\sum_{m=1}^{\infty} W_{mn}(t) \sin\frac{m\pi x}{a} \sin\frac{n\pi y}{b} \qquad (10.2.2)$$

$$\phi_x(x,y,t) = \sum_{n=1}^{\infty}\sum_{m=1}^{\infty} X_{mn}(t) \cos\frac{m\pi x}{a} \sin\frac{n\pi y}{b} \qquad (10.2.3)$$

$$\phi_y(x,y,t) = \sum_{n=1}^{\infty}\sum_{m=1}^{\infty} Y_{mn}(t) \sin\frac{m\pi x}{a} \cos\frac{n\pi y}{b} \qquad (10.2.4)$$

where a and b denote the dimensions of the rectangular plate. The mechanical and thermal loads are also expanded in double Fourier sine series

$$q(x,y,t) = \sum_{n=1}^{\infty}\sum_{m=1}^{\infty} Q_{mn}(t) \sin\frac{m\pi x}{a} \sin\frac{n\pi y}{b} \qquad (10.2.5)$$

$$\Delta T(x,y,z,t) = \sum_{n=1}^{\infty}\sum_{m=1}^{\infty} T_{mn}(z,t) \sin\frac{m\pi x}{a} \sin\frac{n\pi y}{b} \qquad (10.2.6)$$

where

$$Q_{mn}(t) = \frac{4}{ab}\int_0^a\int_0^b q(x,y,t) \sin\frac{m\pi x}{a} \sin\frac{n\pi y}{b} dx dy \qquad (10.2.7)$$

$$T_{mn}(z,t) = \frac{4}{ab}\int_0^a\int_0^b \Delta T(x,y,z,t) \sin\frac{m\pi x}{a} \sin\frac{n\pi y}{b} dx dy \qquad (10.2.8)$$

Substitution of Eqs. (10.2.2)–(10.2.6) into Eqs. (10.1.33)–(10.1.35) yields the following equations for the coefficients (W_{mn}, X_{mn}, Y_{mn}):

$$\begin{bmatrix} \hat{s}_{11}+\tilde{s}_{11} & \hat{s}_{12} & \hat{s}_{13} \\ \hat{s}_{12} & \hat{s}_{22} & \hat{s}_{23} \\ \hat{s}_{13} & \hat{s}_{23} & \hat{s}_{33} \end{bmatrix} \begin{Bmatrix} W_{mn} \\ X_{mn} \\ Y_{mn} \end{Bmatrix} + \begin{bmatrix} \hat{m}_{11} & 0 & 0 \\ 0 & \hat{m}_{22} & 0 \\ 0 & 0 & \hat{m}_{33} \end{bmatrix} \begin{Bmatrix} \ddot{W}_{mn} \\ \ddot{X}_{mn} \\ \ddot{Y}_{mn} \end{Bmatrix}$$

$$= \begin{Bmatrix} Q_{mn} \\ 0 \\ 0 \end{Bmatrix} - \begin{Bmatrix} 0 \\ \alpha_m M^1_{mn} \\ \beta_n M^2_{mn} \end{Bmatrix} \qquad (10.2.9)$$

where $\hat{s}_{ij}$ and $\hat{m}_{ij}$ are defined by

$$\hat{s}_{11} = K(A_{55}\alpha_m^2 + A_{44}\beta_n^2) + k, \quad \tilde{s}_{11} = \hat{N}_{xx}\alpha_m^2 + \hat{N}_{yy}\beta_n^2$$
$$\hat{s}_{12} = KA_{55}\alpha_m, \quad \hat{s}_{13} = KA_{44}\beta_n, \quad \hat{s}_{22} = (D_{11}\alpha_m^2 + D_{66}\beta_n^2 + KA_{55})$$
$$\hat{s}_{23} = (D_{12}+D_{66})\alpha_m\beta_n, \quad \hat{s}_{33} = (D_{66}\alpha_m^2 + D_{22}\beta_n^2 + KA_{44})$$
$$\hat{m}_{11} = I_0, \quad \hat{m}_{22} = I_2, \quad \hat{m}_{33} = I_2 \qquad (10.2.10)$$

where the thermal coefficients M_{mn}^1 and M_{mn}^2 are defined by Eqs. (6.2.4a,b) on page 156, and $\alpha_m = m\pi/a$ and $\beta_n = n\pi/b$.

10.2.2 Bending Analysis

The static solution can be obtained from Eq. (10.2.9) by setting the time derivative terms and edge forces ($\hat{N}_{xx}, \hat{N}_{yy}$) to zero:

$$\begin{bmatrix} \hat{s}_{11} & \hat{s}_{12} & \hat{s}_{13} \\ \hat{s}_{12} & \hat{s}_{22} & \hat{s}_{23} \\ \hat{s}_{13} & \hat{s}_{23} & \hat{s}_{33} \end{bmatrix} \begin{Bmatrix} W_{mn} \\ X_{mn} \\ Y_{mn} \end{Bmatrix} = \begin{Bmatrix} Q_{mn} \\ 0 \\ 0 \end{Bmatrix} - \begin{Bmatrix} 0 \\ \alpha_m M_{mn}^1 \\ \beta_n M_{mn}^2 \end{Bmatrix} \qquad (10.2.11)$$

Solution of Eq. (10.2.11) for each $m, n = 1, 2, \ldots$ gives (W_{mn}, X_{mn}, Y_{mn}), which can then be used to compute the solution (w_0, ϕ_x, ϕ_y) from Eqs. (10.2.2)–(10.2.4). We obtain

$$W_{mn} = \frac{1}{b_{mn}} \left[b_0 Q_{mn} + \hat{s}_{12} \left(\alpha_m M_{mn}^1 \hat{s}_{33} - \beta_n M_{mn}^2 \hat{s}_{23} \right) \right.$$
$$\left. - \hat{s}_{13} \left(\alpha_m M_{mn}^1 \hat{s}_{23} - \beta_n M_{mn}^2 \hat{s}_{22} \right) \right]$$

$$X_{mn} = \frac{1}{b_0} \left[b_1 W_{mn} - \left(\alpha_m M_{mn}^1 \hat{s}_{33} - \beta_n M_{mn}^2 \hat{s}_{23} \right) \right]$$

$$Y_{mn} = \frac{1}{b_0} \left[b_2 W_{mn} + \left(\alpha_m M_{mn}^1 \hat{s}_{23} - \beta_n M_{mn}^2 \hat{s}_{22} \right) \right] \qquad (10.2.12)$$

where

$$b_{mn} = \hat{s}_{11} b_0 + \hat{s}_{12} b_1 + \hat{s}_{13} b_2 \ , \ b_0 = \hat{s}_{22} \hat{s}_{33} - \hat{s}_{23} \hat{s}_{23}$$
$$b_1 = \hat{s}_{23} \hat{s}_{13} - \hat{s}_{12} \hat{s}_{33}, \ b_2 = \hat{s}_{12} \hat{s}_{23} - \hat{s}_{22} \hat{s}_{13} \qquad (10.2.13)$$

When thermal effects are absent, we have

$$W_{mn} = \frac{b_0}{b_{mn}} Q_{mn}, \ X_{mn} = \frac{b_1}{b_{mn}} Q_{mn}, \ Y_{mn} = \frac{b_2}{b_{mn}} Q_{mn} \qquad (10.2.14)$$

The bending moments are given by

$$M_{xx} = -\sum_{n=1}^{\infty} \sum_{m=1}^{\infty} (D_{11} \alpha_m X_{mn} + D_{12} \beta_n Y_{mn}) \sin \alpha_m x \ \sin \beta_n y$$

$$M_{yy} = -\sum_{n=1}^{\infty} \sum_{m=1}^{\infty} (D_{12} \alpha_m X_{mn} + D_{22} \beta Y_{mn}) \sin \alpha_m x \sin \beta_n y$$

$$M_{xy} = D_{66} \sum_{n=1}^{\infty} \sum_{m=1}^{\infty} (\beta_n X_{mn} + \alpha_m Y_{mn}) \ \cos \alpha_m x \ \cos \beta_n y \qquad (10.2.15)$$

The in-plane stresses and transverse shear stresses can be computed using the constitutive equations:

$$\begin{Bmatrix} \sigma_{xx} \\ \sigma_{yy} \\ \sigma_{xy} \end{Bmatrix} = -z \sum_{n=1}^{\infty} \sum_{m=1}^{\infty} \begin{Bmatrix} (Q_{11}\alpha_m X_{mn} + Q_{12}\beta_n Y_{mn}) \sin\alpha_m x \sin\beta_n y \\ (Q_{12}\alpha_m X_{mn} + Q_{22}\beta_n Y_{mn}) \sin\alpha_m x \sin\beta_n y \\ -Q_{66}(\beta_n X_{mn} + \alpha_m Y_{mn}) \cos\alpha_m x \cos\beta_n y \end{Bmatrix}$$
(10.2.16)

$$\begin{Bmatrix} \sigma_{yz} \\ \sigma_{xz} \end{Bmatrix} = \sum_{n=1}^{\infty} \sum_{m=1}^{\infty} \begin{Bmatrix} Q_{44}(Y_{mn} + \beta_n W_{mn}) \sin\alpha_m x \cos\beta_n y \\ Q_{55}(X_{mn} + \alpha_m W_{mn}) \cos\alpha_m x \sin\beta_n y \end{Bmatrix} \quad (10.2.17)$$

The transverse shear stresses can also be computed using 3-D equilibrium equations in terms of stresses [see Chapter 6, Eqs. (6.2.25)–(6.2.28)]. They are given by

$$\sigma_{xz} = -\frac{h^2}{8}\left[1 - \left(\frac{2z}{h}\right)^2\right](T_{11}X_{mn} + T_{12}Y_{mn})\cos\alpha_m x \sin\beta_n y$$

$$\sigma_{yz} = -\frac{h^2}{8}\left[1 - \left(\frac{2z}{h}\right)^2\right](T_{12}X_{mn} + T_{22}Y_{mn})\sin\alpha_m x \cos\beta_n y$$

$$\sigma_{zz} = \frac{h^3}{48}\left\{\left[1 + \left(\frac{2z}{h}\right)^3\right] - 3\left[1 + \left(\frac{2z}{h}\right)\right]\right\}(T_{31}X_{mn} + T_{32}Y_{mn})$$
$$\times \sin\alpha_m x \sin\beta_n y \qquad (10.2.18a)$$

where

$$T_{11} = \alpha_m^2 Q_{11} + \beta_n^2 Q_{66}, \quad T_{12} = \alpha_m\beta_n(Q_{12} + Q_{66})$$
$$T_{22} = \alpha_m^2 Q_{66} + \beta_n^2 Q_{22}, \quad T_{31} = \alpha_m^3 Q_{11} + \alpha_m\beta_n^2(2Q_{66} + Q_{12})$$
$$T_{32} = \alpha_m^2\beta_n(Q_{12} + 2Q_{66}) + \beta_n^3 Q_{22} \qquad (10.2.18b)$$

Numerical results for the maximum transverse deflection and stresses of simply supported plates are discussed next. The following nondimensionalizations are used to present results in graphical and tabular forms:

$$\bar{w} = w_0\left(\frac{E_2 h^3}{b^4 q_0}\right), \quad \hat{w} = w_0\left(\frac{D_{22}}{b^4 q_0}\right) \times 10^2$$

$$\bar{\sigma}_{xx} = \sigma_{xx}\left(\frac{h^2}{b^2 q_0}\right), \quad \bar{\sigma}_{yy} = \sigma_{yy}\left(\frac{h^2}{b^2 q_0}\right), \quad \bar{\sigma}_{xy} = \sigma_{xy}\left(\frac{h^2}{b^2 q_0}\right)$$

$$\bar{\sigma}_{xz} = \sigma_{xz}\left(\frac{h}{bq_0}\right), \quad \bar{\sigma}_{yz} = \sigma_{yz}\left(\frac{h}{bq_0}\right) \qquad (10.2.19)$$

Table 10.2.1 contains the maximum nondimensionalized deflections ($\hat{w}$) and stresses of simply supported square plates under sinusoidally distributed load (SSL) and uniformly distributed load (UDL), and for different side-to-thickness ratios ($E_1 = 25E_2$, $G_{12} = G_{13} = 0.5E_2$, $G_{23} = 0.2E_2$, $\nu_{12} = 0.25$, $K = 5/6$, $k = 0$). The stresses were evaluated at the locations indicated below:

$$\bar{\sigma}_{xx}(a/2, b/2, \tfrac{h}{2}), \quad \bar{\sigma}_{yy}(a/2, b/2, \tfrac{h}{2}), \quad \bar{\sigma}_{xy}(a, b, -\tfrac{h}{2})$$

The transverse shear stresses are calculated using the constitutive equations as well as equilibrium equations. They are the maximum at the locations indicated below:

$$\bar{\sigma}_{xz}(0, b/2, \tfrac{h}{2}), \quad \bar{\sigma}_{yz}(a/2, 0, \tfrac{h}{2})$$

Of course, the constitutively-derived transverse stresses are independent of the z coordinate. The nondimensionalized quantities in the classical plate theory are independent of the side-to-thickness ratio. The influence of transverse shear deformation is to increase the transverse deflection. The difference between the deflections predicted by the first-order shear deformation theory and classical plate theory decreases with the increase in the ratio a/h (see Figure 10.2.2).

Table 10.2.1. Effect of the transverse shear deformation on deflections and stresses in isotropic and orthotropic square plates subjected to distributed loads* (table is continued on the next page).

Load	$\frac{b}{h}$	$\hat{w}$	$\bar{\sigma}_{xx}$	$\bar{\sigma}_{yy}$	$\bar{\sigma}_{xy}$	$\bar{\sigma}_{xz}$	$\bar{\sigma}_{yz}$
Isotropic plates ($\nu = 0.25$)							
SL	10	0.2702	0.1900	0.1900	0.1140	0.1910	0.1910
						0.2387	0.2387[†]
	20	0.2600	0.1900	0.1900	0.1140	0.1910	0.1910
						0.2387	0.2387
	50	0.2572	0.1900	0.1900	0.1140	0.1910	0.1910
						0.2387	0.2387
	100	0.2568	0.1900	0.1900	0.1140	0.1910	0.1910
						0.2387	0.2387
	CPT	0.2566	0.1900	0.1900	0.1140	—	—
						0.2387	0.2387

ANALYSIS USING SHEAR DEFORMATION PLATE THEORIES

Load	$\frac{b}{h}$	$\hat{w}$	$\bar{\sigma}_{xx}$	$\bar{\sigma}_{yy}$	$\bar{\sigma}_{xy}$	$\bar{\sigma}_{xz}$	$\bar{\sigma}_{yz}$
UL (19)[‡]	10	0.4259	0.2762	0.2762	0.2085	0.3927	0.3927
						0.4909	0.4909
	20	0.4111	0.2762	0.2762	0.2085	0.3927	0.3927
						0.4909	0.4909
	50	0.4070	0.2762	0.2762	0.2085	0.3927	0.3927
						0.4909	0.4909
	100	0.4060	0.2762	0.2762	0.2085	0.3927	0.3927
						0.4909	0.4909
	CPT	0.4062	0.2762	0.2762	0.2085	—	—
						0.4909	0.4909

Orthotropic plates

SL	10	0.0533	0.5248	0.0338	0.0246	0.3452	0.0367
						0.4315	0.0459
	20	0.0404	0.5350	0.0286	0.0222	0.3501	0.0319
						0.4376	0.0399
	50	0.0367	0.5380	0.0270	0.0214	0.3515	0.0304
						0.4394	0.0380
	100	0.0362	0.5385	0.0267	0.0213	0.3517	0.0302
						0.4397	0.0377
	CPT	0.0360	0.5387	0.0267	0.0213	—	—
						0.4398	0.0376
UL (19)	10	0.0795	0.7706	0.0352	0.0539	0.6147	0.1529
						0.7684	0.1911
	20	0.0607	0.7828	0.0272	0.0487	0.6194	0.1466
						0.7742	0.1833
	50	0.0553	0.7860	0.0249	0.0468	0.6207	0.1452
						0.7756	0.1814
	100	0.0545	0.7865	0.0245	0.0464	0.6206	0.1449
						0.7757	0.1812
	CPT	0.0543	0.7866	0.0244	0.0463	—	—
						0.7758	0.1811

[*] See Figure 10.2.1 for the plate geometry and coordinate system.

[†] Stresses computed using the stress equilibrium equations (10.2.18a,b).

[‡] The numbers in parenthesis denotes the maximum values of m and n used to evaluate the series.

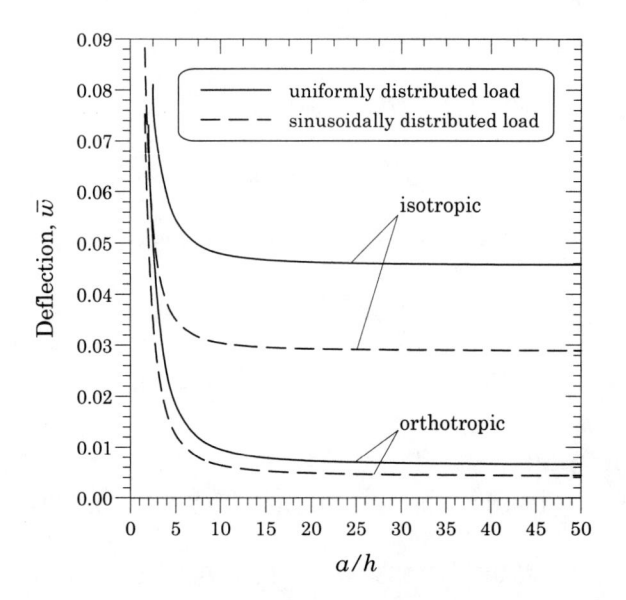

Figure 10.2.2. Nondimensionalized center transverse deflection ($\bar{w}$) versus side-to-thickness ratio (a/h) for simply supported square plates.

10.2.3 Buckling Analysis

For buckling analysis, we assume that the only applied loads are the in-plane compressive forces

$$\hat{N}_{xx} = -N_0, \quad \hat{N}_{yy} = -\gamma N_0, \quad \gamma = \frac{\hat{N}_{yy}}{\hat{N}_{xx}} \tag{10.2.20}$$

and all other mechanical and thermal loads are zero (and $k = 0$). From Eq. (10.2.9) we have

$$\begin{bmatrix} \hat{s}_{11} - N_0\left(\alpha_m^2 + \gamma\beta_n^2\right) & \hat{s}_{12} & \hat{s}_{13} \\ \hat{s}_{12} & \hat{s}_{22} & \hat{s}_{23} \\ \hat{s}_{13} & \hat{s}_{23} & \hat{s}_{33} \end{bmatrix} \begin{Bmatrix} W_{mn} \\ X_{mn} \\ Y_{mn} \end{Bmatrix} = \begin{Bmatrix} 0 \\ 0 \\ 0 \end{Bmatrix} \tag{10.2.21}$$

For a non-trivial solution the determinant of the coefficient matrix in Eq. (10.2.21) must be zero. This gives the following expression for the buckling load:

$$N_0 = \left(\frac{1}{\alpha_m^2 + \gamma\beta_n^2}\right)\left[\frac{K^2 A_{44} A_{55} \hat{c}_{33} + (KA_{55}\alpha_m^2 + KA_{44}\beta_n^2)c_1}{c_1 + KA_{44}c_2 + KA_{55}c_3 + K^2 A_{44} A_{55}}\right]$$

$$= \left(\frac{1}{\alpha_m^2 + \gamma\beta_n^2}\right)\left[\frac{c_0 + \left(\frac{\alpha_m^2}{KA_{44}} + \frac{\beta_n^2}{KA_{55}}\right)c_1}{1 + \frac{c_1}{K^2 A_{44} A_{55}} + \frac{c_2}{KA_{55}} + \frac{c_3}{KA_{44}}}\right] \quad (10.2.22)$$

$$c_0 = D_{11}\alpha_m^4 + 2(D_{12} + 2D_{66})\alpha_m^2\beta_n^2 + D_{22}\beta_n^4, \quad c_1 = c_2 c_3 - (c_4)^2 > 0$$
$$c_2 = D_{11}\alpha_m^2 + D_{66}\beta_n^2, \quad c_3 = D_{66}\alpha_m^2 + D_{22}\beta_n^2, \quad c_4 = (D_{12} + D_{66})\alpha_m\beta_n \quad (10.2.23)$$

When the effect of transverse shear deformation is neglected (i.e., set K to a large value), Eq. (10.2.22) yields the result (8.1.7) obtained using the classical plate theory. The expression in (10.2.22) is of the form

$$\frac{c_0 + k_1}{1 + k_2} \quad \text{with} \quad k_1 < k_2 \quad \text{which implies} \quad c_0 \geq \frac{c_0 + k_1}{1 + k_2}$$

indicating that transverse shear deformation has the effect of *reducing* the buckling load (as long as $c_0 > 1$).

In general, no conclusions can be drawn from the complicated expression of the buckling load concerning its minimum. For an isotropic plate ($D_{11} = D_{22} = D_{12} + 2D_{66} = D$), the critical buckling load occurs for $m = n = 1$ and it is given by

$$N_{cr} = 4D\left(\frac{\pi}{a}\right)^2 \frac{\left[1 + \frac{3(1-\nu^2)\pi^2(h/a)^2}{K}\right]}{\left[1 + \frac{72(1+\nu)(1-\nu^2)\pi^4(h/a)^4}{K^2} + \frac{6(1+\nu)(3-\nu)\pi^2(h/a)^2}{K}\right]} \quad (10.2.24)$$

Table 10.2.2 contains the critical buckling loads $\bar{N} = N_{cr}b^2/(\pi^2 D_{22})$ as a function of the plate aspect ratio a/b, side-to-thickness ratio b/h, and modulus ratio E_1/E_2 for uniaxial ($\gamma = 0$) and biaxial ($\gamma = 1$) compression. The classical plate theory (CPT) results are also included for comparison (see Table 8.1.1). The effect of transverse shear deformation is significant for lower aspect ratios, thick plates, and larger modular ratios. For thin plates, irrespective of the aspect ratio and modular ratio, the buckling loads predicted by the shear deformation plate theory are very close to those of the classical plate theory (see Figure 10.2.3).

Table 10.2.2. Nondimensionalized buckling loads $\bar{N}$ of simply supported plates under in-plane uniform compression ($\gamma = 0$) and biaxial compression ($\gamma = 1$).

γ	$\frac{a}{b}$	$\frac{h}{b}$	$\frac{E_1}{E_2}=1$	$\frac{E_1}{E_2}=3$	$\frac{E_1}{E_2}=10$	$\frac{E_1}{E_2}=25$
0	0.5	10	5.523	11.583	23.781	34.701
		20	6.051	13.779	35.615	68.798
		100	6.242	14.669	42.398	100.750
		CPT	6.250	14.708	42.737	102.750
	1.0	10	3.800	5.901	11.205	19.252
		20	3.948	6.309	12.832	25.412
		100	3.998	6.452	13.460	28.357
		CPT	4.000	6.458	13.488	28.495
	1.5	10	$4.045^{(2,1)\dagger}$	5.664	8.354	13.166
		20	$4.262^{(2,1)}$	5.942	8.959	15.077
		100	$4.337^{(2,1)}$	6.037	9.173	15.823
		CPT	$4.340^{(2,1)}$	6.042	9.182	15.856
	3.0	10	$3.800^{(3,1)}$	$5.664^{(2,1)}$	$8.354^{(2,1)}$	$13.166^{(2,1)}$
		20	$3.948^{(3,1)}$	$5.942^{(2,1)}$	$8.959^{(2,1)}$	14.052
		100	$3.998^{(3,1)}$	$6.037^{(2,1)}$	$9.173^{(2,1)}$	14.264
		CPT	$4.000^{(3,1)}$	$6.042^{(2,1)}$	$9.182^{(2,1)}$	14.273
1	0.5	10	4.418	9.405	$15.191^{(1,3)}$	$17.773^{(1,3)}$
		20	4.841	11.070	$21.565^{(1,3)}$	$30.073^{(1,4)}$
		100	4.993	11.737	$25.241^{(1,3)}$	$40.157^{(1,4)}$
		CPT	5.000	11.767	$25.427^{(1,3)}$	$40.784^{(1,4)}$
	1.0	10	1.900	3.015	5.662	$7.518^{(1,2)}$
		20	1.974	3.173	6.433	$9.308^{(1,2)}$
		100	1.999	3.227	6.731	$10.156^{(1,2)}$
		CPT	2.000	3.229	6.744	$10.196^{(1,2)}$
	1.5	10	1.391	1.788	2.614	4.093
		20	1.431	1.841	2.769	4.651
		100	1.444	1.858	2.823	4.869
		CPT	1.444	1.859	2.825	4.879
	3.0	10	1.079	1.151	1.227	1.375
		20	1.103	1.172	1.251	1.414
		100	1.111	1.179	1.259	1.426
		CPT	1.111	1.179	1.260	1.427

$\dagger$ Denotes mode numbers (m,n) at which the critical buckling load occured; $(m,n) = (1,1)$ for all other cases.

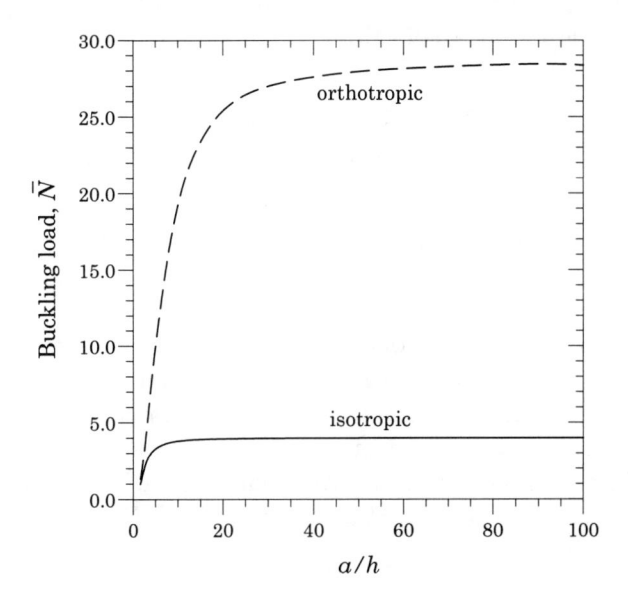

Figure 10.2.3. Nondimensionalized critical buckling load ($\bar{N}$) versus side-to-thickness ratio (a/h) for simply supported square plates.

10.2.4 Natural Vibration

For free vibration, we set the thermal and mechanical loads to zero and seek periodic solution to Eq. (10.2.9) in the form

$$W_{mn}(t) = W_{mn}^0 e^{i\omega t}, \quad X_{mn}(t) = X_{mn}^0 e^{i\omega t}, \quad Y_{mn}(t) = Y_{mn}^0 e^{i\omega t} \quad (10.2.25)$$

and obtain the following 3×3 system of eigenvalue problem:

$$\left(\begin{bmatrix} \hat{s}_{11} & \hat{s}_{12} & \hat{s}_{13} \\ \hat{s}_{12} & \hat{s}_{22} & \hat{s}_{23} \\ \hat{s}_{13} & \hat{s}_{23} & \hat{s}_{33} \end{bmatrix} - \omega^2 \begin{bmatrix} \hat{m}_{11} & 0 & 0 \\ 0 & \hat{m}_{22} & 0 \\ 0 & 0 & \hat{m}_{33} \end{bmatrix} \right) \begin{Bmatrix} W_{mn}^0 \\ X_{mn}^0 \\ Y_{mn}^0 \end{Bmatrix} = \begin{Bmatrix} 0 \\ 0 \\ 0 \end{Bmatrix} \quad (10.2.26)$$

where $\hat{s}_{ij}$ and $\hat{m}_{ij}$ are defined in Eq. (10.2.10). Setting the determinant of the coefficient matrix in Eq. (10.2.26) yields the frequency equation.

If the rotatory inertia I_2 is omitted (i.e., $\hat{m}_{22} = \hat{m}_{33} = 0$), the frequency equation can be solved for ω^2

$$\omega^2 = \frac{1}{\hat{m}_{11}}\left(\hat{s}_{11} - \frac{\hat{s}_{13}\hat{s}_{23} - \hat{s}_{12}\hat{s}_{33}}{\hat{s}_{22}\hat{s}_{33} - \hat{s}_{23}\hat{s}_{23}}\hat{s}_{12} - \frac{\hat{s}_{12}\hat{s}_{23} - \hat{s}_{13}\hat{s}_{22}}{\hat{s}_{22}\hat{s}_{33} - \hat{s}_{23}\hat{s}_{23}}\hat{s}_{13}\right) \quad (10.2.27)$$

Table 10.2.3 contains the first four frequencies of isotropic plates. The effect of the shear correction factor is to decrease the frequencies, i.e., the smaller the K, the smaller are the frequencies. The rotary inertia (RI) also has the effect of decreasing the frequencies. Table 10.2.4 contains fundamental natural frequencies of square plates for various values of side-to-thickness ratio a/b and modulus ratios $E_1/E_2 = 1$ and 10. Figure 10.2.4 shows the effect of transverse shear deformation and rotary inertia on fundamental natural frequencies of orthotropic square plates. The effect of rotary inertia is negligible in FSDT and therefore not shown in the figure.

Table 10.2.3. Effect of the shear deformation, rotatory inertia, and shear correction coefficient on nondimensionalized natural frequencies of simply supported isotropic square plates ($\bar{\omega} = \omega(a^2/h)\sqrt{\rho/E}$; $\nu = 0.3, a/h = 10$).

m	n	CPT[†] w/o RI	CPT with RI	K	FSDT w/o RI	FSDT with RI
1	1	5.973	5.925	5/6	5.812	5.769
				2/3	5.773	5.732
				1.0	5.838	5.794
2	1	14.933	14.635	5/6	13.980	13.764
				2/3	13.769	13.568
				1.0	14.127	13.899
2	2	23.893	23.144	5/6	21.582	21.121
				2/3	21.103	20.688
				1.0	21.922	21.424
3	1	29.867	28.709	5/6	26.378	25.734
				2/3	25.682	25.115
				1.0	26.875	26.171

[†] w/o RI means without rotary inertia.

Table 10.2.4. Effect of shear deformation, material orthotropy, rotatory inertia on dimensionless fundamental frequencies of simply supported square plates [$\bar{\omega} = \omega(a^2/h)\sqrt{\rho/E_2}$, $K = 5/6$].

Theory	$a/h \to$	5	10	20	25	50	100
$E_1/E_2 = 1$, $\nu = 0.25$							
FSDT	w-RI[†]	5.232	5.694	5.835	5.853	5.877	5.883
	w/o-RI	5.349	5.736	5.847	5.860	5.879	5.883
CPT	(5.885)[‡]	5.700	5.837	5.873	5.877	5.883	5.885
$E_1 = 10E_2$, $G_{12} = G_{13} = 0.5E_2$, $G_{23} = 0.2E_2$, $\nu_{12} = 0.25$							
FSDT	w-RI	7.709	9.509	10.218	10.316	10.450	10.484
	w/o-RI	7.810	9.567	10.238	10.329	10.454	10.486
CPT	(10.496)	10.167	10.411	10.475	10.483	10.493	10.495

[†] w-RI = with rotatory inertia; w/o-RI = without rotatory inertia.

[‡] Value in the parenthesis is the frequency without rotatory inertia.

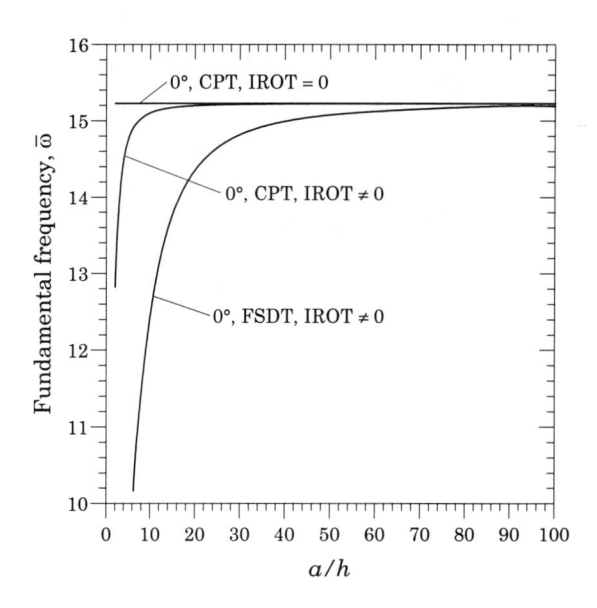

Figure 10.2.4. Fundamental frequency ($\bar{\omega}$) versus side-to-thickness ratio (a/b) for simply supported orthotropic plates.

10.3 Third-Order Plate Theories

10.3.1. General Comments

In principle, it is possible to expand the displacement field of a plate in terms of the thickness coordinate up to any desired degree. However, due to the algebraic complexity and computational effort involved with higher-order theories in return for a marginal gain in accuracy, theories higher than third order have not been attempted. The reason for expanding the displacements up to the cubic term in the thickness coordinate is to have quadratic variation of the transverse shear strains and transverse shear stresses through the plate thickness. This avoids the need for shear correction coefficients used in the first-order shear deformation plate theory (FSDT).

There are many papers on third-order theories (see [17–32]). Although many of them seem to differ from each other on the surface, the displacement fields of these theories are related. Here we present the original third-order shear deformation plate theory (TSDT) of Reddy [25,26] that contains other lower-order plate theories, including the classical plate theory and first-order shear deformation plate theory as special cases. For additional details, see Chapter 11 of Reddy [1].

10.3.2. Displacement Field

The third-order plate theory to be presented here is also based on the same assumptions as the classical and first-order plate theories, except that the assumption on the straightness and normality of a transverse normal after deformation is relaxed by expanding the displacements (u, v, w) as cubic functions of the thickness coordinate. Figure 10.3.1 shows the kinematics of deformation of a transverse normal on edge $y = 0$ in the classical, first-order, and third-order plate theories.

Consider the displacement field

$$u(x,y,z,t) = u_0(x,y,t) + z\phi_x(x,y,t) - \frac{4z^3}{3h^2}\left(\phi_x + \frac{\partial w_0}{\partial x}\right)$$

$$v(x,y,z,t) = v_0(x,y,t) + z\phi_y(x,y,t) - \frac{4z^3}{3h^2}\left(\phi_y + \frac{\partial w_0}{\partial y}\right)$$

$$w(x,y,z,t) = w_0(x,y,t) \qquad (10.3.1)$$

where (u_0, v_0, w_0) have the same meaning as in the first-order shear deformation theory except that (ϕ_x, ϕ_y) now denote slopes of a

ANALYSIS USING SHEAR DEFORMATION PLATE THEORIES   447

transverse normal at $z = 0$. Note that the displacement field in Eq. (10.3.1) suggests that a straight line perpendicular to the undeformed mid-plane becomes a cubic curve after the plate deforms (see Figure 10.3.1). As will be seen shortly, this choice of the displacement field is arrived to satisfy the following stress-free boundary conditions on the top and bottom faces of the plate:

$$\sigma_{xz}(x, y, \pm\frac{h}{2}) = \sigma_{yz}(x, y, \pm\frac{h}{2}) = 0 \qquad (10.3.2)$$

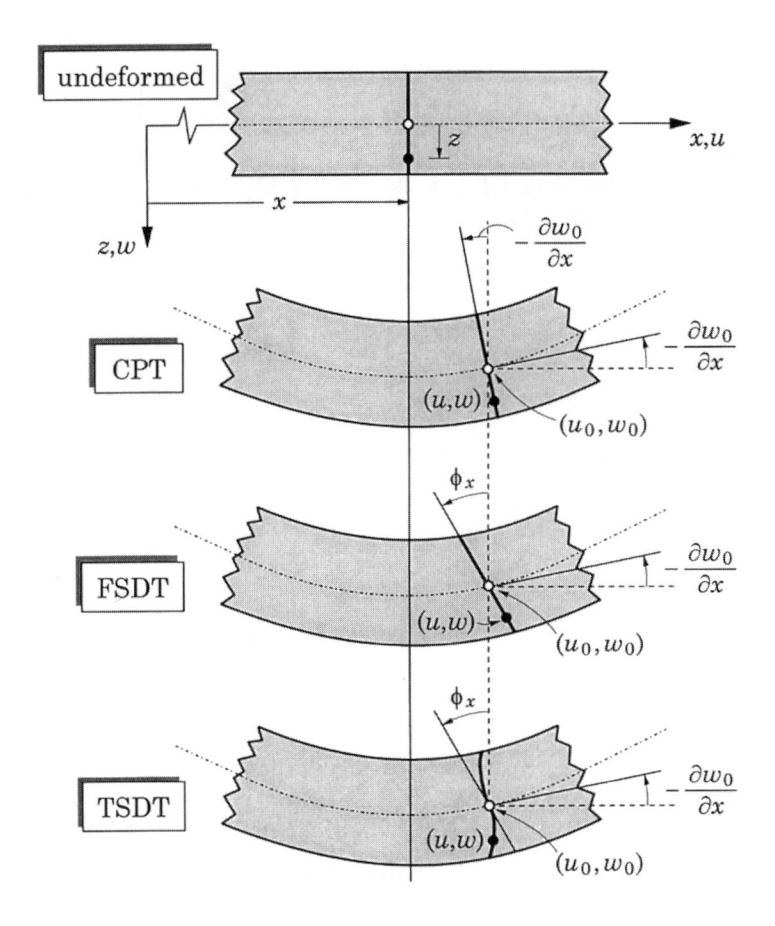

Figure 10.3.1. Deformation of a transverse normal according to the classical, first-order, and third-order plate theories.

10.3.3. Strains and Stresses

Substitution of the displacement field (10.3.1) into the linear strain-displacement relations of Eq. (1.4.8) yields the strains

$$\begin{Bmatrix} \varepsilon_{xx} \\ \varepsilon_{yy} \\ \gamma_{xy} \end{Bmatrix} = \begin{Bmatrix} \varepsilon_{xx}^{(0)} \\ \varepsilon_{yy}^{(0)} \\ \gamma_{xy}^{(0)} \end{Bmatrix} + z \begin{Bmatrix} \varepsilon_{xx}^{(1)} \\ \varepsilon_{yy}^{(1)} \\ \gamma_{xy}^{(1)} \end{Bmatrix} + z^3 \begin{Bmatrix} \varepsilon_{xx}^{(3)} \\ \varepsilon_{yy}^{(3)} \\ \gamma_{xy}^{(3)} \end{Bmatrix} \quad (10.3.3a)$$

$$\begin{Bmatrix} \gamma_{yz} \\ \gamma_{xz} \end{Bmatrix} = \begin{Bmatrix} \gamma_{yz}^{(0)} \\ \gamma_{xz}^{(0)} \end{Bmatrix} + z^2 \begin{Bmatrix} \gamma_{yz}^{(2)} \\ \gamma_{xz}^{(2)} \end{Bmatrix} \quad (10.3.3b)$$

where

$$\begin{Bmatrix} \varepsilon_{xx}^{(0)} \\ \varepsilon_{yy}^{(0)} \\ \gamma_{xy}^{(0)} \end{Bmatrix} = \begin{Bmatrix} \frac{\partial u_0}{\partial x} \\ \frac{\partial v_0}{\partial y} \\ \frac{\partial u_0}{\partial y} + \frac{\partial v_0}{\partial x} \end{Bmatrix}, \quad \begin{Bmatrix} \varepsilon_{xx}^{(1)} \\ \varepsilon_{yy}^{(1)} \\ \gamma_{xy}^{(1)} \end{Bmatrix} = \begin{Bmatrix} \frac{\partial \phi_x}{\partial x} \\ \frac{\partial \phi_y}{\partial y} \\ \frac{\partial \phi_x}{\partial y} + \frac{\partial \phi_y}{\partial x} \end{Bmatrix} \quad (10.3.4a)$$

$$\begin{Bmatrix} \varepsilon_{xx}^{(3)} \\ \varepsilon_{yy}^{(3)} \\ \gamma_{xy}^{(3)} \end{Bmatrix} = -c_1 \begin{Bmatrix} \left(\frac{\partial \phi_x}{\partial x} + \frac{\partial^2 w_0}{\partial x^2}\right) \\ \left(\frac{\partial \phi_y}{\partial y} + \frac{\partial^2 w_0}{\partial y^2}\right) \\ \left(\frac{\partial \phi_x}{\partial y} + \frac{\partial \phi_y}{\partial x} + 2\frac{\partial^2 w_0}{\partial x \partial y}\right) \end{Bmatrix} \quad (10.3.4b)$$

$$\begin{Bmatrix} \gamma_{yz}^{(0)} \\ \gamma_{xz}^{(0)} \end{Bmatrix} = \begin{Bmatrix} \phi_y + \frac{\partial w_0}{\partial y} \\ \phi_x + \frac{\partial w_0}{\partial x} \end{Bmatrix}, \quad \begin{Bmatrix} \gamma_{yz}^{(2)} \\ \gamma_{xz}^{(2)} \end{Bmatrix} = -c_2 \begin{Bmatrix} \left(\phi_y + \frac{\partial w_0}{\partial y}\right) \\ \left(\phi_x + \frac{\partial w_0}{\partial x}\right) \end{Bmatrix} \quad (10.3.4c)$$

and

$$c_1 = \frac{4}{3h^2}, \quad c_2 = 3c_1 = \frac{4}{h^2} \quad (10.3.5)$$

The bending (or membrane) stresses ($\sigma_{xx}, \sigma_{yy}, \sigma_{xy}$) and transverse shear stresses (σ_{xz}, σ_{yz}) are given by

$$\begin{Bmatrix} \sigma_{xx} \\ \sigma_{yy} \\ \sigma_{xy} \end{Bmatrix} = \begin{bmatrix} Q_{11} & Q_{12} & 0 \\ Q_{12} & Q_{22} & 0 \\ 0 & 0 & Q_{66} \end{bmatrix} \begin{Bmatrix} \varepsilon_{xx} \\ \varepsilon_{yy} \\ \gamma_{xy} \end{Bmatrix} \quad (10.3.6a)$$

$$\begin{Bmatrix} \sigma_{yz} \\ \sigma_{xz} \end{Bmatrix} = \begin{bmatrix} Q_{44} & 0 \\ 0 & Q_{55} \end{bmatrix} \begin{Bmatrix} \gamma_{yz} \\ \gamma_{xz} \end{Bmatrix} \quad (10.3.6b)$$

Thus, the membrane stresses vary as cubic functions of the thickness coordinate (z) while the transverse shear stresses are quadratic through the thickness. Note that the transverse normal stress σ_{zz} is assumed to be zero, like in CPT and FSDT, and this assumption was used in deriving the plane-stress-reduced stiffnessess Q_{ij} in section 1.4 [see Eqs. (1.4.35)–(1.4.37)], which are related to the engineering constants E_1, E_2, ν_{12}, G_{12}, G_{13}, and G_{23} by

$$Q_{11} = \frac{E_1}{1-\nu_{12}\nu_{21}}, \quad Q_{12} = \frac{\nu_{12}E_2}{1-\nu_{12}\nu_{21}}, \quad Q_{22} = \frac{E_2}{1-\nu_{12}\nu_{21}}$$
$$Q_{66} = G_{12}, \quad Q_{44} = G_{23}, \quad Q_{55} = G_{13} \qquad (10.3.7)$$

It can be readily seen that the transverse shear stresses are zero at $z = \pm(h/2)$

$$\sigma_{yz} = Q_{44}\left(\gamma_{yz}^{(0)} + z^2\gamma_{yz}^{(2)}\right) = Q_{44}\left[1 - 4\left(\frac{z}{h}\right)^2\right]\left(\phi_y + \frac{\partial w_0}{\partial y}\right) \quad (10.3.8a)$$

$$\sigma_{xz} = Q_{44}\left(\gamma_{xz}^{(0)} + z^2\gamma_{xz}^{(2)}\right) = Q_{55}\left[1 - 4\left(\frac{z}{h}\right)^2\right]\left(\phi_x + \frac{\partial w_0}{\partial x}\right) \quad (10.3.8b)$$

10.3.4 Equations of Motion

The equations of motion of the third-order plate theory can be derived using the dynamic version of the principle of virtual displacements. The virtual strain energy δU, virtual work done by applied forces δV, and the virtual kinetic energy δK are given by

$$\begin{aligned}\delta U = \int_{\Omega_0}\Bigg\{&\int_{-\frac{h}{2}}^{\frac{h}{2}}\bigg[\sigma_{xx}\left(\delta\varepsilon_{xx}^{(0)} + z\delta\varepsilon_{xx}^{(1)} - c_1 z^3 \delta\varepsilon_{xx}^{(3)}\right)\\
&+ \sigma_{yy}\left(\delta\varepsilon_{yy}^{(0)} + z\delta\varepsilon_{yy}^{(1)} - c_1 z^3 \delta\varepsilon_{yy}^{(3)}\right)\\
&+ \sigma_{xy}\left(\delta\gamma_{xy}^{(0)} + z\delta\gamma_{xy}^{(1)} - c_1 z^3 \delta\gamma_{xy}^{(3)}\right)\\
&+ \sigma_{xz}\left(\delta\gamma_{xz}^{(0)} + z^2\delta\gamma_{xz}^{(2)}\right) + \sigma_{yz}\left(\delta\gamma_{yz}^{(0)} + z^2\delta\gamma_{yz}^{(2)}\right)\bigg]dz\Bigg\}dxdy\\
= \int_{\Omega_0}\bigg(&N_{xx}\delta\varepsilon_{xx}^{(0)} + M_{xx}\delta\varepsilon_{xx}^{(1)} - c_1 P_{xx}\delta\varepsilon_{xx}^{(3)} + N_{yy}\delta\varepsilon_{yy}^{(0)} + M_{yy}\delta\varepsilon_{yy}^{(1)}\\
&- c_1 P_{yy}\delta\varepsilon_{yy}^{(3)} + N_{xy}\delta\gamma_{xy}^{(0)} + M_{xy}\delta\gamma_{xy}^{(1)} - c_1 P_{xy}\delta\gamma_{xy}^{(3)}\\
&+ Q_x\delta\gamma_{xz}^{(0)} - c_2 R_x\delta\gamma_{xz}^{(2)} + Q_y\delta\gamma_{yz}^{(0)} - c_2 R_y\delta\gamma_{yz}^{(2)}\bigg)dxdy \quad (10.3.9)\end{aligned}$$

450 THEORY AND ANALYSIS OF ELASTIC PLATES

$$\delta V = -\int_{\Omega_0} \left[(q_b - kw_0)\,\delta w_0 + q_t \delta w_0\right] dx dy$$

$$-\int_\Gamma \int_{-\frac{h}{2}}^{\frac{h}{2}} \left[\hat{\sigma}_{nn}\left(\delta u_n + z\delta\phi_n - c_1 z^3 \delta\varphi_n\right)\right.$$

$$\left. + \hat{\sigma}_{ns}\left(\delta u_s + z\delta\phi_s - c_1 z^3 \delta\varphi_{ns}\right) + \hat{\sigma}_{nr}\delta w_0\right] dz d\Gamma$$

$$= -\int_{\Omega_0} (q - kw_0)\,\delta w_0 dx dy - \int_\Gamma (\,\hat{N}_{nn}\delta u_n + \hat{M}_{nn}\delta\phi_n - c_1 \hat{P}_{nn}\delta\varphi_n$$

$$+ \hat{N}_{ns}\delta u_s + \hat{M}_{ns}\delta\phi_s - c_1 \hat{P}_{ns}\delta\varphi_{ns} + \hat{Q}_n \delta w_0\,)\,d\Gamma \qquad (10.3.10)$$

$$\delta K = \int_{\Omega_0} \int_{-\frac{h}{2}}^{\frac{h}{2}} \rho_0 \left[\left(\dot{u}_0 + z\dot{\phi}_x - c_1 z^3 \dot{\varphi}_x\right)\left(\delta\dot{u}_0 + z\delta\dot{\phi}_x - c_1 z^3 \delta\dot{\varphi}_x\right)\right.$$

$$+ \left(\dot{v}_0 + z\dot{\phi}_y - c_1 z^3 \dot{\varphi}_y\right)\left(\delta\dot{v}_0 + z\delta\dot{\phi}_y + z^3 \delta\dot{\varphi}_y\right)$$

$$\left. + \dot{w}_0 \delta\dot{w}_0 \right] dv$$

$$= \int_{\Omega_0} \left\{ I_0 \dot{u}_0 \delta\dot{u}_0 + I_0 \dot{v}_0 \delta\dot{v}_0 + \left[I_2 \dot{\phi}_x - c_1 I_4 \left(\dot{\phi}_x + \frac{\partial \dot{w}_0}{\partial x}\right)\right] \delta\dot{\phi}_x \right.$$

$$- c_1 \left[I_4 \dot{\phi}_x - c_1 I_6 \left(\dot{\phi}_x + \frac{\partial \dot{w}_0}{\partial x}\right)\right] \left(\delta\dot{\phi}_x + \frac{\partial \delta\dot{w}_0}{\partial x}\right)$$

$$+ \left[I_2 \dot{\phi}_y - c_1 I_4 \left(\dot{\phi}_y + \frac{\partial \dot{w}_0}{\partial y}\right)\right] \delta\dot{\phi}_y$$

$$\left. - c_1 \left[I_4 \dot{\phi}_y - c_1 I_6 \left(\dot{\phi}_y + \frac{\partial \dot{w}_0}{\partial y}\right)\right] \left(\delta\dot{\phi}_y + \frac{\partial \delta\dot{w}_0}{\partial y}\right) \right\} dx dy$$

$$(10.3.11)$$

where Ω_0 denotes the midplane of the laminate, and

$$\begin{Bmatrix} N_{\alpha\beta} \\ M_{\alpha\beta} \\ P_{\alpha\beta} \end{Bmatrix} = \int_{-\frac{h}{2}}^{\frac{h}{2}} \sigma_{\alpha\beta} \begin{Bmatrix} 1 \\ z \\ z^3 \end{Bmatrix} dz, \quad \begin{Bmatrix} Q_\alpha \\ R_\alpha \end{Bmatrix} = \int_{-\frac{h}{2}}^{\frac{h}{2}} \sigma_{\alpha z} \begin{Bmatrix} 1 \\ z^2 \end{Bmatrix} dz \quad (10.3.12)$$

$$I_i = \int_{-\frac{h}{2}}^{\frac{h}{2}} \rho_0 (z)^i \, dz \quad (i = 0, 1, 2, \cdots, 6) \qquad (10.3.13)$$

In Eq. (10.3.12), α and β take the symbols x and y. The same definitions hold for the stress resultants with a hat, which are specified. The inertias I_1 and I_3 are zero for homogeneous plates.

The Euler–Lagrange equations of the third-order shear deformation theory (TSDT) can be derived by substituting for $\delta U, \delta V$, and δK from Eqs. (10.3.9)–(10.3.11) into the virtual work statement, noting that the virtual strains can be written in terms of the generalized displacements using Eqs. (10.3.3a,b) and (10.3.4a–c), integrating by parts to relieve the virtual generalized displacements $\delta u_0, \delta v_0, \delta w_0, \delta\phi_x$, and $\delta\phi_y$ in Ω_0 of any differentiation, and using the fundamental lemma of calculus of variations. The equations involve higher-order moments P's and R's [see Eq. (10.3.12)], which are mathematically similar to the traditional moments M's and shear forces Q's. However, they do present a difficulty when one is required to specify the higher-order resultants on the boundary.

We obtain the following Euler–Lagrange equations:

$$\frac{\partial N_{xx}}{\partial x} + \frac{\partial N_{xy}}{\partial y} = I_0 \ddot{u}_0 \tag{10.3.14}$$

$$\frac{\partial N_{xy}}{\partial x} + \frac{\partial N_{yy}}{\partial y} = I_0 \ddot{v}_0 \tag{10.3.15}$$

$$\frac{\partial \bar{Q}_x}{\partial x} + \frac{\partial \bar{Q}_y}{\partial y} + c_1 \left(\frac{\partial^2 P_{xx}}{\partial x^2} + 2\frac{\partial^2 P_{xy}}{\partial x \partial y} + \frac{\partial^2 P_{yy}}{\partial y^2} \right) + q - kw_0$$
$$= I_0 \ddot{w}_0 - c_1^2 I_6 \left(\frac{\partial^2 \ddot{w}_0}{\partial x^2} + \frac{\partial^2 \ddot{w}_0}{\partial y^2} \right) + c_1 J_4 \left(\frac{\partial \ddot{\phi}_x}{\partial x} + \frac{\partial \ddot{\phi}_y}{\partial y} \right) \tag{10.3.16}$$

$$\frac{\partial \bar{M}_{xx}}{\partial x} + \frac{\partial \bar{M}_{xy}}{\partial y} - \bar{Q}_x = K_2 \ddot{\phi}_x - c_1 J_4 \frac{\partial \ddot{w}_0}{\partial x} \tag{10.3.17}$$

$$\frac{\partial \bar{M}_{xy}}{\partial x} + \frac{\partial \bar{M}_{yy}}{\partial y} - \bar{Q}_y = K_2 \ddot{\phi}_y - c_1 J_4 \frac{\partial \ddot{w}_0}{\partial y} \tag{10.3.18}$$

where

$$\bar{M}_{\alpha\beta} = M_{\alpha\beta} - c_1 P_{\alpha\beta}, \quad \bar{Q}_\alpha = Q_\alpha - c_2 R_\alpha \tag{10.3.19}$$

$$I_i = \int_{-\frac{h}{2}}^{\frac{h}{2}} \rho_0 \, (z)^i \, dz \quad (i = 0, 2, \cdots, 6) \tag{10.3.20a}$$

$$J_i = I_i - c_1 I_{i+2}, \quad K_2 = I_2 - 2c_1 I_4 + c_1^2 I_6 \tag{10.3.20b}$$

and (P_{xx}, P_{yy}, P_{xy}) and (R_x, R_y) denote the higher-order stress resultants

$$\begin{Bmatrix} P_{xx} \\ P_{yy} \\ P_{xy} \end{Bmatrix} = \int_{-\frac{h}{2}}^{\frac{h}{2}} \begin{Bmatrix} \sigma_{xx} \\ \sigma_{yy} \\ \sigma_{xy} \end{Bmatrix} z^3 dz, \quad \begin{Bmatrix} R_x \\ R_y \end{Bmatrix} = \int_{-\frac{h}{2}}^{\frac{h}{2}} \begin{Bmatrix} \sigma_{yz} \\ \sigma_{xz} \end{Bmatrix} z^2 dz \tag{10.3.21}$$

The primary and secondary variables of the theory are

$$\text{Primary Variables}: \quad u_n, u_s, w_0, \frac{\partial w_0}{\partial n}, \phi_n, \phi_s \qquad (10.3.22)$$

$$\text{Secondary Variables}: \quad N_{nn}, N_{ns}, \bar{V}_n, P_{nn}, \bar{M}_{nn}, \bar{M}_{ns} \qquad (10.3.23)$$

where u_n, u_s, N_{nn}, N_{ns}, and so on have the same meaning as in the classical and first-order plate theory [see Eqs. (3.5.4)–(3.5.8)],

$$\begin{aligned}\bar{V}_n \equiv c_1 &\left[\left(\frac{\partial P_{xx}}{\partial x} + \frac{\partial P_{xy}}{\partial y}\right)n_x + \left(\frac{\partial P_{xy}}{\partial x} + \frac{\partial P_{yy}}{\partial y}\right)n_y\right] \\ &- c_1\left[\left(J_4\ddot{\phi}_x - c_1 I_6 \frac{\partial \ddot{w}_0}{\partial x}\right)n_x + \left(J_4\ddot{\phi}_y - c_1 I_6 \frac{\partial \ddot{w}_0}{\partial y}\right)n_y\right] \\ &+ (\bar{Q}_x n_x + \bar{Q}_y n_y) + c_1\frac{\partial P_{ns}}{\partial s} \end{aligned} \qquad (10.3.24)$$

P_{nn} and P_{ns} are defined by [see Eq. (3.5.8)]

$$\left\{\begin{array}{c} P_{nn} \\ P_{ns} \end{array}\right\} = \begin{bmatrix} n_x^2 & n_y^2 & 2n_x n_y \\ -n_x n_y & n_x n_y & n_x^2 - n_y^2 \end{bmatrix} \left\{\begin{array}{c} P_{xx} \\ P_{yy} \\ P_{xy} \end{array}\right\} \qquad (10.3.25)$$

and (n_x, n_y) are the direction cosines of the unit normal on the boundary.

The stress resultants are related to the strains by the relations

$$\left\{\begin{array}{c} \{N\} \\ \{M\} \\ \{P\} \end{array}\right\} = \begin{bmatrix} [A] & [0] & [0] \\ [0] & [D] & [F] \\ [0] & [F] & [H] \end{bmatrix} \left\{\begin{array}{c} \{\varepsilon^{(0)}\} \\ \{\varepsilon^{(1)}\} \\ \{\varepsilon^{(3)}\} \end{array}\right\} \qquad (10.3.26)$$

$$\left\{\begin{array}{c} \{Q\} \\ \{R\} \end{array}\right\} = \begin{bmatrix} [A] & [D] \\ [D] & [F] \end{bmatrix} \left\{\begin{array}{c} \{\gamma^{(0)}\} \\ \{\gamma^{(2)}\} \end{array}\right\} \qquad (10.3.27)$$

$$(A_{ij}, D_{ij}, F_{ij}, H_{ij}) = \int_{-\frac{h}{2}}^{\frac{h}{2}} Q_{ij}\left(1, z^2, z^4, z^6\right) dz \qquad (10.3.28a)$$

$$(A_{ij}, D_{ij}, F_{ij}) = \int_{-\frac{h}{2}}^{\frac{h}{2}} Q_{ij}\left(1, z^2, z^6\right) dz \qquad (10.3.28b)$$

The stiffnesses in Eq. (10.3.28a) are defined for $i, j = 1, 2, 6$ and those in Eq. (10.3.28b) are defined for $i, j = 4, 5$. The coefficients A_{ij} and D_{ij} were given in terms of the material constants and plate thickness h in Eq. (3.6.11) for single-layer orthotropic plates, and in Eq. (3.6.14) for symmetrically laminated plates with principal material axes coinciding with those of the plate. The higher-order stiffness coefficients introduced in the third-order theory are given by

$$F_{ij} = \frac{1}{5}\sum_{k=1}^{N} Q_{ij}^{(k)} \left[(z_{k+1})^5 - (z_k)^5\right]$$

$$H_{ij} = \frac{1}{7}\sum_{k=1}^{N} Q_{ij}^{(k)} \left[(z_{k+1})^7 - (z_k)^7\right] \qquad (10.3.29)$$

This completes the development of the third-order shear deformation theory (TSDT). Note that the equations of motion of the first-order theory are obtained from the present third-order theory by setting $c_1 = 0$. The displacement field in Eq. (10.3.1) contains, as special cases, the displacement fields used by other researchers to derive a third-order plate theory, as shown in Table 10.3.1. Therefore, the third-order plate theories reported in the literature, despite the different looks of the assumed displacement fields, are indeed equivalent.

Table 10.3.1. Relationship of the displacements of other third-order theories to the one in Eq. (10.3.1).

References	Displacement Field[†]	Correspondence with Eq. (10.3.1)[‡]
Vlasov [19], Jemielita[20]	$u_\alpha = u_\alpha^0 + f(z)\psi_\alpha - c_1 z^3 u_{3,\alpha}^0$	$\varphi_\alpha = \psi_\alpha + u_{3,\alpha}^0$
Schmidt [21]	$u_\alpha = u_\alpha^0 - z u_{3,\alpha}^0 + \frac{3}{2} f(z)\varepsilon_\alpha$	$\varphi_\alpha = \frac{3}{2}\varepsilon_\alpha$
Krishna Murty [23], Levinson [24], Reddy [25]	$u_\alpha = u_\alpha^0 - z u_{3,\alpha}^0 - c_3 f(z)\theta_\alpha$	$\varphi_\alpha = -c_3 \theta_\alpha$
Reddy [26]	$u_\alpha = u_\alpha^0 + f(z)\hat{\phi}_\alpha - z u_{3,\alpha}^b$ $-c_1 h z^3 u_{3,\alpha}^0$	$\varphi_\alpha = \hat{\phi}_\alpha + u_{3,\alpha}^s$ $w_0 = u_3^s + u_3^b$
Reddy [32]	$u_\alpha = u_\alpha^0 - z u_{3,\alpha}^0 + f(z)\phi_\alpha$	$\varphi_\alpha = \phi_\alpha$

[†] $c_1 = \frac{4}{3h^2}$, $c_3 = \frac{3}{4h}$, $c_4 = \frac{5}{3h^2}$, $f(z) = z(1 - c_1 z^2)$. [‡] $\varphi_\alpha \equiv \phi_\alpha + u_{3,\alpha}^0$, $u_3^0 = w_0$.

10.4 The Navier Solutions of TSDT

10.4.1 Preliminary Comments

Except for additional higher-order terms, the third-order plate theory equations are very similar to those of the first-order plate theory. They too admit exact solutions for the case of simply supported rectangular plates. Following the same steps as in the case of the first-order plate theory, here we present the Navier solutions of the third-order plate theory for bending, buckling, and vibrations of rectangular plates. Since the stretching deformation is uncoupled from the bending, we can analyze three equations, Eqs. (10.3.16)–(10.3.18), independent of Eqs. (10.3.14) and (10.3.15).

The simply supported boundary conditions on the edges of a rectangular plate for the third-order shear deformation plate theory are

$$w_0(x,0,t) = 0, \quad w_0(x,b,t) = 0, \quad w_0(0,y,t) = 0, \quad w_0(a,y,t) = 0 \tag{10.4.1a}$$

$$\phi_x(x,0,t) = 0, \quad \phi_x(x,b,t) = 0, \quad \phi_y(0,y,t) = 0, \quad \phi_y(a,y,t) = 0 \tag{10.4.1b}$$

$$\bar{M}_{xx}(0,y,t) = 0, \quad \bar{M}_{xx}(a,y,t) = 0, \quad \bar{M}_{yy}(x,0,t) = 0, \quad \bar{M}_{yy}(x,b,t) = 0 \tag{10.4.1c}$$

where the origin of the coordinate system is taken at the upper left corner (see Figure 10.4.1).

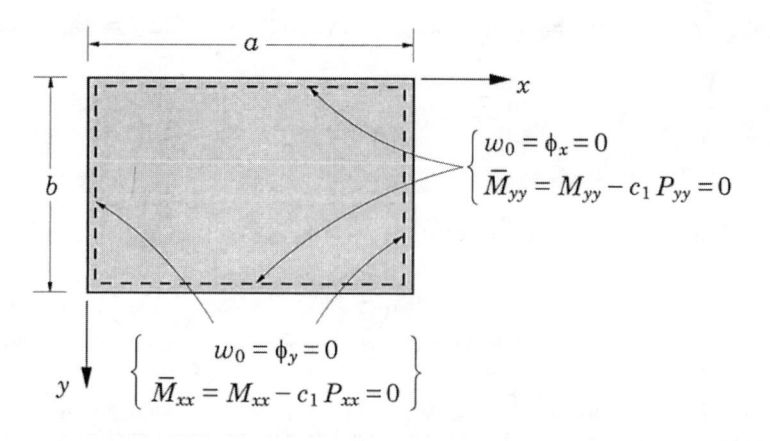

Figure 10.4.1. Simply supported boundary conditions for Reddy's third-order shear deformation plate theory.

10.4.2 General Solution

The boundary conditions in (10.4.1a–c) are satisfied by the following expansions of the generalized displacements:

$$w_0(x,y,t) = \sum_{n=1}^{\infty}\sum_{m=1}^{\infty} W_{mn}(t) \sin \alpha_m x \, \sin \beta_n y \quad (10.4.2a)$$

$$\phi_x(x,y,t) = \sum_{n=1}^{\infty}\sum_{m=1}^{\infty} X_{mn}(t) \cos \alpha_m x \, \sin \beta_n y \quad (10.4.2b)$$

$$\phi_y(x,y,t) = \sum_{n=1}^{\infty}\sum_{m=1}^{\infty} Y_{mn}(t) \sin \alpha_m x \, \cos \beta_n y \quad (10.4.2c)$$

where $\alpha_m = m\pi/a$ and $\beta_n = n\pi/b$. The transverse load q is also expanded in double Fourier sine series

$$q(x,y,t) = \sum_{n=1}^{\infty}\sum_{m=1}^{\infty} Q_{mn}(t) \sin \alpha_m x \, \sin \beta_n y \quad (10.4.3a)$$

$$Q_{mn}(z,t) = \frac{4}{ab}\int_0^a \int_0^b q(x,y,t) \, \sin \alpha_m x \, \sin \beta_n y \, dxdy \quad (10.4.3b)$$

Substitution of Eqs. (10.4.2) and (10.4.3) into Eqs. (10.3.16)–(10.3.18) results in the following equations for the undetermined coefficients (W_{mn}, X_{mn}, Y_{mn})

$$\begin{bmatrix} \hat{s}_{11}+\tilde{s}_{11} & \hat{s}_{12} & \hat{s}_{13} \\ \hat{s}_{12} & \hat{s}_{22} & \hat{s}_{23} \\ \hat{s}_{13} & \hat{s}_{23} & \hat{s}_{33} \end{bmatrix} \begin{Bmatrix} W_{mn} \\ X_{mn} \\ Y_{mn} \end{Bmatrix} + \begin{bmatrix} \hat{m}_{11} & \hat{m}_{12} & \hat{m}_{13} \\ \hat{m}_{12} & \hat{m}_{22} & 0 \\ \hat{m}_{23} & 0 & \hat{m}_{33} \end{bmatrix} \begin{Bmatrix} \ddot{W}_{mn} \\ \ddot{X}_{mn} \\ \ddot{Y}_{mn} \end{Bmatrix}$$

$$= \begin{Bmatrix} Q_{mn} \\ 0 \\ 0 \end{Bmatrix} - \begin{Bmatrix} 0 \\ \alpha_m \hat{M}_{mn}^1 \\ \beta_n \hat{M}_{mn}^2 \end{Bmatrix} \quad (10.4.4)$$

where $\hat{s}_{ij}$ and $\hat{m}_{ij}$ are defined by

$$\hat{s}_{11} = \bar{A}_{55}\alpha_m^2 + \bar{A}_{44}\beta_n^2 + k$$
$$\quad + c_1^2\left[H_{11}\alpha_m^4 + 2(H_{12}+2H_{66})\alpha_m^2\beta_n^2 + H_{22}\beta_n^4\right]$$
$$\hat{s}_{12} = \bar{A}_{55}\alpha - c_1\left[\hat{F}_{11}\alpha_m^3 + (\hat{F}_{12}+2\hat{F}_{66})\alpha_m\beta_n^2\right]$$
$$\hat{s}_{13} = \bar{A}_{44}\beta_n - c_1\left[\hat{F}_{22}\beta_n^3 + (\hat{F}_{12}+2\hat{F}_{66})\alpha_m^2\beta_n\right]$$

$$\hat{s}_{22} = \bar{A}_{55} + \bar{D}_{11}\alpha_m^2 + \bar{D}_{66}\beta_n^2, \quad \hat{s}_{23} = (\bar{D}_{12} + \bar{D}_{66})\alpha_m\beta_n$$
$$\hat{s}_{33} = \bar{A}_{44} + \bar{D}_{66}\alpha_m^2 + \bar{D}_{22}\beta_n^2, \quad \tilde{s}_{11} = \hat{N}_{xx}\alpha_m^2 + \hat{N}_{yy}\beta_n^2 \quad (10.4.5a)$$
$$\hat{m}_{11} = I_0 + c_1^2 I_6 \left(\alpha_m^2 + \beta_n^2\right), \quad \hat{m}_{12} = -c_1 J_4 \alpha_m$$
$$\hat{m}_{13} = -c_1 J_4 \beta_n, \quad \hat{m}_{22} = K_2, \quad \hat{m}_{33} = K_2 \quad (10.4.5b)$$
$$\hat{D}_{ij} = D_{ij} - c_1 F_{ij}, \quad \hat{F}_{ij} = F_{ij} - c_1 H_{ij} \quad \bar{D}_{ij} = \hat{D}_{ij} - c_1 \hat{F}_{ij} \quad (10.4.6a)$$

for $i, j = 1, 2, 6$, and

$$\hat{A}_{ij} = A_{ij} - c_2 D_{ij}, \quad \hat{D}_{ij} = D_{ij} - c_2 \hat{F}_{ij}, \quad \bar{A}_{ij} = \hat{A}_{ij} - c_2 \hat{D}_{ij} \quad (10.4.6b)$$

for $i, j = 4, 5$. The thermal resultants are defined by

$$\begin{Bmatrix} N_{xx}^T \\ N_{yy}^T \\ N_{xy}^T \end{Bmatrix} = \sum_{n=1}^{\infty} \sum_{m=1}^{\infty} \begin{Bmatrix} N_{mn}^1(t) \\ N_{mn}^2(t) \\ N_{mn}^6(t) \end{Bmatrix} \sin\alpha_m x \; \sin\beta_n y \quad (10.4.7a)$$

$$\begin{Bmatrix} M_{xx}^T \\ M_{yy}^T \\ M_{xy}^T \end{Bmatrix} = \sum_{n=1}^{\infty} \sum_{m=1}^{\infty} \begin{Bmatrix} M_{mn}^1(t) \\ M_{mn}^2(t) \\ M_{mn}^6(t) \end{Bmatrix} \sin\alpha_m x \; \sin\beta_n y \quad (10.4.7b)$$

$$\begin{Bmatrix} P_{xx}^T \\ P_{yy}^T \\ P_{xy}^T \end{Bmatrix} = \sum_{n=1}^{\infty} \sum_{m=1}^{\infty} \begin{Bmatrix} P_{mn}^1(t) \\ P_{mn}^2(t) \\ P_{mn}^6(t) \end{Bmatrix} \sin\alpha_m x \; \sin\beta_n y \quad (10.4.7c)$$

where

$$\{N_{mn}(t)\} = \int_{-\frac{h}{2}}^{\frac{h}{2}} [Q]\{\alpha\} \, T_{mn}(z,t) \, dz \quad (10.4.8a)$$

$$\{M_{mn}(t)\} = \int_{-\frac{h}{2}}^{\frac{h}{2}} [Q]\{\alpha\} \, T_{mn}(z,t) \, z \, dz \quad (10.4.8b)$$

$$\{P_{mn}(t)\} = \int_{-\frac{h}{2}}^{\frac{h}{2}} [Q]\{\alpha\} \, T_{mn}(z,t) \, z^3 \, dz \quad (10.4.8c)$$

and

$$\Delta T(x,y,z,t) = \sum_{m=1}^{\infty} \sum_{n=1}^{\infty} \left(T_{mn}^0 + z T_{mn}^1\right) \sin\alpha_m x \; \sin\beta_n y \quad (10.4.9a)$$

$$\left(T_{mn}^0, T_{mn}^1\right) = \frac{4}{ab} \int_0^a \int_0^b (T_0, T_1) \, \sin\alpha_m x \; \sin\beta_n y \, dx dy \quad (10.4.9b)$$

$$\{\hat{M}_{mn}\} = \{M_{mn}\} - c_1\{P_{mn}\} \tag{10.4.10}$$

Equation (10.4.4) can be specialized to static bending analysis, buckling, and natural vibration.

10.4.3 Bending Analysis

The static solution can be obtained from Eqs. (10.4.4) by setting the time derivative terms to zero. The resulting equations are identical to those in Eq. (10.2.11) but with $\hat{s}_{ij}$ defined by Eq. (10.4.5a).

The in-plane stresses can be computed from the equations

$$\begin{Bmatrix} \sigma_{xx} \\ \sigma_{yy} \\ \sigma_{xy} \end{Bmatrix} = \begin{bmatrix} Q_{11} & Q_{12} & 0 \\ Q_{12} & Q_{22} & 0 \\ 0 & 0 & Q_{66} \end{bmatrix} \left(\begin{Bmatrix} \varepsilon_{xx} \\ \varepsilon_{yy} \\ \gamma_{xy} \end{Bmatrix} - \begin{Bmatrix} \alpha_{xx} \\ \alpha_{yy} \\ 2\alpha_{xy} \end{Bmatrix} \Delta T \right) \tag{10.4.11}$$

$$\begin{Bmatrix} \varepsilon_{xx} \\ \varepsilon_{yy} \\ \gamma_{xy} \end{Bmatrix} = \begin{Bmatrix} \varepsilon_{xx}^{(0)} \\ \varepsilon_{yy}^{(0)} \\ \gamma_{xy}^{(0)} \end{Bmatrix} + z \begin{Bmatrix} \varepsilon_{xx}^{(1)} \\ \varepsilon_{yy}^{(1)} \\ \gamma_{xy}^{(1)} \end{Bmatrix} + z^3 \begin{Bmatrix} \varepsilon_{xx}^{(3)} \\ \varepsilon_{yy}^{(3)} \\ \gamma_{xy}^{(3)} \end{Bmatrix}$$

$$= \sum_{m=1}^{\infty} \sum_{n=1}^{\infty} \begin{Bmatrix} (zS_{mn}^{xx} + c_1 z^3 T_{mn}^{xx}) \sin \alpha_m x \sin \beta_n y \\ (zS_{mn}^{yy} + c_1 z^3 T_{mn}^{yy}) \sin \alpha_m x \sin \beta_n y \\ (zS_{mn}^{xy} + c_1 z^3 T_{mn}^{xy}) \cos \alpha_m x \cos \beta_n y \end{Bmatrix} \tag{10.4.12}$$

$$\begin{Bmatrix} S_{mn}^{xx} \\ S_{mn}^{yy} \\ S_{mn}^{xy} \end{Bmatrix} = \begin{Bmatrix} -\alpha_m X_{mn} \\ -\beta_n Y_{mn} \\ \beta_n X_{mn} + \alpha_m Y_{mn} \end{Bmatrix} \tag{10.4.13a}$$

$$\begin{Bmatrix} T_{mn}^{xx} \\ T_{mn}^{yy} \\ T_{mn}^{xy} \end{Bmatrix} = \begin{Bmatrix} \alpha_m X_{mn} + \alpha_m^2 W_{mn} \\ \beta_n Y_{mn} + \beta_n^2 W_{mn} \\ -(\beta_n X_{mn} + \alpha_m Y_{mn} + 2\alpha_m \beta_n W_{mn}) \end{Bmatrix} \tag{10.4.13b}$$

The transverse shear stresses from the constitutive equations are given by

$$\sigma_{yz} = (1 - c_2 z^2) \sum_{m=1}^{\infty} \sum_{n=1}^{\infty} Q_{44} (Y_{mn} + \beta_n W_{mn}) \sin \alpha_m x \cos \beta_n y$$

$$\sigma_{xz} = (1 - c_2 z^2) \sum_{m=1}^{\infty} \sum_{n=1}^{\infty} Q_{55} (X_{mn} + \alpha_m W_{mn}) \cos \alpha_m x \sin \beta_n y$$

$$\tag{10.4.14}$$

where $c_2 = 4/h^2$. Note that the transverse shear stresses are quadratic through the thickness.

The transverse shear stresses can also be determined using the equilibrium equations of 3-D elasticity. In the absence of thermal effects, they are given by

$$\sigma_{xz} = \sum_{m=1}^{\infty} \sum_{n=1}^{\infty} \left[\frac{1}{2}(z^2 - z_b^2)\mathcal{B}_{mn} + \frac{c_1}{4}(z^4 - z_b^4)\mathcal{E}_{mn} \right] \cos\alpha_m x \; \sin\beta_n y$$

$$\sigma_{yz} = \sum_{m=1}^{\infty} \sum_{n=1}^{\infty} \left[\frac{1}{2}(z^2 - z_b^2)\mathcal{D}_{mn} + \frac{c_1}{4}(z^4 - z_b^4)\mathcal{F}_{mn} \right] \sin\alpha_m x \; \cos\beta_n y$$

(10.4.15)

where $z_b = -h/2$ and

$$\mathcal{B}_{mn} = \left[\left(\alpha_m^2 Q_{11} + \beta_n^2 Q_{66} \right) X_{mn} + \alpha_m \beta_n (Q_{12} + Q_{66}) Y_{mn} \right]$$
$$\mathcal{D}_{mn} = \left[\alpha_m \beta_n (Q_{12} + Q_{66}) X_{mn} + \left(\alpha_m^2 Q_{66} + \beta_n^2 Q_{22} \right) Y_{mn} \right]$$
$$\mathcal{F}_{mn} = - \left[\beta^3 Q_{22} + \alpha_m^2 \beta_n (Q_{12} + 2Q_{66}) \right] W_{mn} - \mathcal{D}_{mn} \quad (10.4.16)$$

Table 10.4.1 contains the maximum nondimensionalized deflections ($\bar{w}$) and stresses ($\bar{\sigma}$) of simply supported, isotropic ($\nu = 0.3$), square plates under sinusoidally distributed load (SSL) and uniformly distributed load (UDL), and for different side-to-thickness ratios. The nondimensionalizations used are the same as in Eq. (10.2.19), and the locations of the maximum stresses are the same as given before:

$$\bar{\sigma}_{xx}(\frac{a}{2}, \frac{b}{2}, \frac{h}{2}), \quad \bar{\sigma}_{yy}(\frac{a}{2}, \frac{b}{2}, \frac{h}{2}), \quad \bar{\sigma}_{xy}(a, b, -\frac{h}{2})$$

$$\bar{\sigma}_{xz}(0, \frac{b}{2}, \frac{h}{2}), \quad \bar{\sigma}_{yz}(\frac{a}{2}, 0, \frac{h}{2}) \quad (10.4.17)$$

Table 10.4.2 contains the results for orthotropic plates. The following material properties (see [2,3]) are used:

$$E_1 = 20.83 \text{ msi}, \quad E_2 = 10.94 \text{ msi}, \quad \nu_{12} = 0.44, \quad k = 0$$
$$G_{12} = 6.10 \text{ msi}, \quad G_{13} = 3.71 \text{ msi}, \quad G_{23} = 6.19 \text{ msi} \quad (10.4.18)$$

The difference between the deflections predicted by the first-order theory ($K = 5/6$) and the third-order theory is not significant. This should be

Table 10.4.1. Effect of transverse shear deformation on deflections and stresses in isotropic ($\nu = 0.3$) square plates subjected to distributed loads.

Theory	$\frac{b}{h}$	$10\bar{w}$	$\bar{\sigma}_{xx}$	$\bar{\sigma}_{xy}$	$(\bar{\sigma}_{xz})^{\dagger}$	$\bar{\sigma}_{xz}$
Sinusoidal Load						
CPT*		0.2803	0.1976	0.1064	—	0.2387
FSDT	5	0.3435	0.1976	0.1064	0.1910	0.2387
TSDT	5	0.3433	0.2050	0.1104	0.2381	0.2365
FSDT	10	0.2961	0.1976	0.1064	0.1910	0.2387
TSDT	10	0.2961	0.1994	0.1074	0.2386	0.2382
FSDT	100	0.2804	0.1976	0.1064	0.1910	0.2387
TSDT	100	0.2804	0.1976	0.1064	0.2387	0.2387
Uniform Load$(m, n = 1, 3, \cdots, 19)$						
CPT*		0.4436	0.2873	0.1946	—	0.4909
FSDT	5	0.5355	0.2873	0.1946	0.3927	0.4909
TSDT	5	0.5354	0.2944	0.2112	0.4840	0.4668
FSDT	10	0.4666	0.2873	0.1946	0.3927	0.4909
TSDT	10	0.4666	0.2890	0.1990	0.4890	0.4843
FSDT	100	0.4438	0.2873	0.1946	0.3927	0.4909
TSDT	100	0.4438	0.2873	0.1946	0.4909	0.4909

† Transverse shear stress computed using the constitutive equation (10.4.14); the next column shows the transverse shear stress computed using the stress equilibrium equation (10.4.15).

* CPT solutions are independent of the ratio a/h.

clear from an examination of the higher-order stiffnesses F_{ij} and H_{ij}, which contain the fifth and seventh powers, respectively, of the plate thickness h, and therefore are expected to contribute little to the displacements. The main advantage of the third-order theory is that the transverse shear strains and stresses are represented quadratically, a state of stress that is close to the 3-D elasticity solution, and consequently *no shear correction coefficients are needed*. Note that the constitutively-derived transverse shear stresses of TSDT are close to the equilibrium-derived values, and they are zero at $z = \pm(h/2)$.

Table 10.4.2. Effect of transverse shear deformation on deflections and stresses in orthotropic plates subjected to uniformly distributed load ($m, n = 1, 3, \cdots, 29$).

Theory	$\frac{b}{h}$	$10\bar{w}$	$\bar{\sigma}_{xx}$	$\bar{\sigma}_{yy}$	$\bar{\sigma}_{xy}$	$\bar{\sigma}_{xz}$	$\bar{\sigma}_{yz}$
Aspect Ratio, $\frac{a}{b} = 1.0$							
CPT		0.3022	0.3609	0.2162	0.1939	—	—
						0.5499	0.4393[†]
FSDT	100	0.3024	0.3609	0.2163	0.1939	0.4399	0.3514
						0.5499	0.4393
TSDT	100	0.3024	0.3609	0.2163	0.1939	0.5498	0.4393
						0.5496	0.4392
FSDT	20	0.3079	0.3597	0.2171	0.1941	0.4394	0.3520
						0.5493	0.4401
TSDT	20	0.3079	0.3608	0.2175	0.1957	0.5477	0.4395
						0.5437	0.4381
FSDT	10	0.3252	0.3562	0.2196	0.1947	0.4381	0.3537
						0.5476	0.4422
TSDT	10	0.3252	0.3604	0.2213	0.2010	0.5420	0.4400
						0.5279	0.4348
Aspect Ratio, $\frac{a}{b} = 2.0$							
CPT		0.5862	0.1021	0.1346	0.0804	—	—
						0.3159	0.3264
FSDT	100	0.5864	0.1021	0.1346	0.0804	0.2527	0.2611
						0.3159	0.3264
TSDT	100	0.5864	0.1021	0.1346	0.0804	0.3159	0.3264
						0.3159	0.3264
FSDT	20	0.5908	0.1019	0.1348	0.0803	0.2524	0.2613
						0.3155	0.3266
TSDT	20	0.5908	0.1020	0.1349	0.0807	0.3151	0.3263
						0.3141	0.3256
FSDT	10	0.6046	0.1013	0.1352	0.0800	0.2515	0.2615
						0.3143	0.3269
TSDT	10	0.6046	0.1017	0.1357	0.0816	0.3128	0.3259
						0.3089	0.3233

[†] Stresses computed using the stress equilibrium equations.

10.4.4 Buckling Analysis

The buckling problem associated with the third-order shear deformation theory is again given by Eq. (10.2.21), with $\hat{S}_{ij}$ given by Eq. (10.4.5a); see Exercise 10.3. Table 10.4.3 contains the critical buckling loads $\bar{N} = N_{cr}b^2/(\pi^2 D_{22})$ as a function of the plate aspect ratio a/b, side-to-thickness ratio b/h, and modulus ratio E_1/E_2 for uniaxial ($\gamma = 0$) and biaxial ($\gamma = 1$) compression. It is clear that the difference between the buckling loads predicted by the third-order plate theory and the first-order plate theory is negligible. For the biaxial case, there is a small difference between the two solutions, with the buckling loads predicted by the third-order theory being slightly lower. Thus, there is no real advantage of using the third-order plate theory for the prediction of buckling loads.

Table 10.4.3. Nondimensionalized buckling loads $\bar{N}$ of simply supported plates under in-plane uniform compression ($\gamma = 0$) and biaxial compression ($\gamma = 1$).

γ	$\frac{a}{b}$	$\frac{h}{b}$	$\frac{E_1}{E_2} = 1$ FSDT	TSDT	$\frac{E_1}{E_2} = 10$ FSDT	TSDT
0	1.0	10	3.800	3.800	11.205	11.209
		20	3.948	3.948	12.832	12.832
		100	3.998	3.998	13.460	13.460
		CPT	4.000	4.000	13.488	13.488
	3.0	10	$3.800^{(3,1)}$	$3.800^{(3,1)}$	$8.354^{(2,1)}$	$8.355^{(2,1)\dagger}$
		20	$3.948^{(3,1)}$	$3.948^{(3,1)}$	$8.959^{(2,1)}$	$8.959^{(2,1)}$
		100	$3.998^{(3,1)}$	$3.998^{(3,1)}$	$9.173^{(2,1)}$	$9.173^{(2,1)}$
		CPT	$4.000^{(3,1)}$	$4.000^{(3,1)}$	$9.182^{(2,1)}$	$9.182^{(2,1)}$
1	1.0	10	1.900	1.900	5.662	5.605
		20	1.974	1.974	6.433	6.416
		100	1.999	1.999	6.731	6.730
		CPT	2.000	2.000	6.744	6.744
	3.0	10	1.079	1.079	1.227	1.195
		20	1.103	1.103	1.251	1.243
		100	1.111	1.111	1.259	1.259
		CPT	1.111	1.111	1.260	1.260

$\dagger$ Denotes mode numbers (m, n) at which the critical buckling load occured; $(m, n) = (1, 1)$ for all other cases.

10.4.5 Natural Vibration

For natural vibration, Eq. (10.4.4) becomes

$$\left(\begin{bmatrix} \hat{s}_{11} & \hat{s}_{12} & \hat{s}_{13} \\ \hat{s}_{12} & \hat{s}_{22} & \hat{s}_{23} \\ \hat{s}_{13} & \hat{s}_{23} & \hat{s}_{33} \end{bmatrix} - \omega^2 \begin{bmatrix} \hat{m}_{11} & \hat{m}_{12} & \hat{m}_{13} \\ \hat{m}_{12} & \hat{m}_{22} & 0 \\ \hat{m}_{13} & 0 & \hat{m}_{33} \end{bmatrix}\right) \left\{\begin{array}{c} W^0_{mn} \\ X^0_{mn} \\ Y^0_{mn} \end{array}\right\} = \left\{\begin{array}{c} 0 \\ 0 \\ 0 \end{array}\right\}$$
(10.4.19)

where $\hat{s}_{ij}$ and $\hat{m}_{ij}$ are defined in Eqs. (10.4.5a,b).

Table 10.4.4 contains natural frequencies $\hat{\omega} = \omega h(\sqrt{\rho/G})$ of isotropic ($\nu = 0.3$) rectangular plates as predicted by the first-order (FSDT), third-order (TSDT), and classical (CPT) plate theories. Once again, we find that the difference between the frequencies predicted by the first- and third-order theories is negligible, and that both are in close agreement with the 3-D elasticity results.

Table 10.4.4. Nondimensionalized frequencies $\hat{\omega}$ of simply supported isotropic ($\nu = 0.3$) rectangular plates ($a/h = 10$, $k = 0$).

$\frac{b}{a}$	m	n	Exact	TSDT	FSDT	CPT[a]	CPT
1.0	1	1	0.0932[b]	0.0930	0.0930	0.0955	0.0963
	2	1	0.2260	0.2219	0.2219	0.2360	0.2408
	2	2	0.3421	0.3406	0.3406	0.3732	0.3853
	1	3	0.4171	0.4106	0.4149	0.4629	0.4816
	2	3	0.5239	0.5208	0.5206	0.5951	0.6261
$\sqrt{2}$	1	1	0.0704[c]	0.0704	0.0704	0.0718	0.0722
	1	2	0.1376	0.1374	0.1373	0.1427	0.1445
	2	1	0.2018	0.2013	0.2012	0.2128	0.2167
	1	3	0.2431	0.2423	0.2423	0.2591	0.2649
	2	2	0.2634	0.2625	0.2625	0.2821	0.2889
	2	3	0.3612	0.3596	0.3595	0.3957	0.4093
	1	4	0.3800	0.3783	0.3782	0.4182	0.4334
	3	1	0.3987	0.3968	0.3967	0.4406	0.4575
	3	2	0.4535	0.4511	0.4509	0.5073	0.5297
	2	4	0.4890	0.4863	0.4861	0.5513	0.5779
	3	3	0.5411	0.5378	0.5375	0.6168	0.6501

[a] CPT solution without rotatory inertia and all other solutions include rotatory inertia; [b] 3-D elasticity solution of Srinivas and Rao [3]; [c] 3-D elasticity solution of Reismann and Lee [4]; also see Reddy [2], pp. 381–383.

Exercises

10.1 Show that the solution of Eq. (10.2.11) is given by

$$W_{mn} = \frac{1}{b_{mn}}\left(a_1 Q_{mn} - a_2\alpha_m M^1_{mn} - a_3\beta_n M^2_{mn}\right)$$

$$X_{mn} = \frac{1}{b_{mn}}\left(a_2 Q_{mn} - a_4\alpha_m M^1_{mn} - a_5\beta_n M^2_{mn}\right)$$

$$Y_{mn} = \frac{1}{b_{mn}}\left(a_3 Q_{mn} - a_5\alpha_m M^1_{mn} - a_6\beta_n M^2_{mn}\right) \quad (a)$$

where

$$b_{mn} = \begin{vmatrix} \hat{s}_{11} & \hat{s}_{12} & \hat{s}_{13} \\ \hat{s}_{12} & \hat{s}_{22} & \hat{s}_{23} \\ \hat{s}_{13} & \hat{s}_{23} & \hat{s}_{33} \end{vmatrix} = \hat{s}_{11}a_1 + \hat{s}_{12}a_2 + \hat{s}_{13}a_3$$

$$a_1 = \hat{s}_{22}\hat{s}_{33} - \hat{s}_{23}\hat{s}_{23}, \quad a_2 = \hat{s}_{23}\hat{s}_{13} - \hat{s}_{12}\hat{s}_{33}, \quad a_3 = \hat{s}_{12}\hat{s}_{23} - \hat{s}_{22}\hat{s}_{13}$$

$$a_4 = \hat{s}_{11}\hat{s}_{33} - \hat{s}_{13}\hat{s}_{13}, \quad a_5 = \hat{s}_{12}\hat{s}_{13} - \hat{s}_{11}\hat{s}_{23}, \quad a_6 = \hat{s}_{11}\hat{s}_{22} - \hat{s}_{12}\hat{s}_{12}$$

$$(b)$$

10.2 Derive the expressions for transverse stresses in FSDT presented in Eqs. (10.2.18a,b).

10.3 Show that the expression for the critical buckling load, using Eq. (10.2.21), is

$$N_0 = \frac{1}{\alpha_m^2 + \gamma\beta_n^2}\left(\hat{s}_{11} + \frac{\hat{s}_{12}\hat{s}_{33} - \hat{s}_{13}\hat{s}_{23}}{\hat{s}_{22}\hat{s}_{33} - \hat{s}_{23}\hat{s}_{23}}\hat{s}_{12} + \frac{\hat{s}_{22}\hat{s}_{13} - \hat{s}_{23}\hat{s}_{12}}{\hat{s}_{22}\hat{s}_{33} - \hat{s}_{23}\hat{s}_{23}}\hat{s}_{13}\right)$$

10.4 Show that the equations of equilibrium [see Eqs. (10.1.33)–(10.1.35)] for bending of the first-order shear deformation plate theory can be expressed in matrix form as

$$\begin{bmatrix} L_{11} & L_{12} & L_{13} \\ L_{12} & L_{22} & L_{23} \\ L_{13} & L_{23} & L_{33} \end{bmatrix} \begin{Bmatrix} w_0 \\ \phi_x \\ \phi_y \end{Bmatrix} = \begin{Bmatrix} q \\ 0 \\ 0 \end{Bmatrix} \quad (a)$$

where the coefficients $L_{ij} = L_{ji}$ are

$$L_{11} = -KA_{55}\frac{\partial^2}{\partial x^2} - KA_{44}\frac{\partial^2}{\partial y^2}$$

$$L_{12} = -KA_{55}\frac{\partial}{\partial x}, \quad L_{13} = -KA_{44}\frac{\partial}{\partial y}$$

464   THEORY AND ANALYSIS OF ELASTIC PLATES

$$L_{22} = -KA_{55} + D_{11}\frac{\partial^2}{\partial x^2} + D_{66}\frac{\partial^2}{\partial y^2}$$

$$L_{23} = (D_{12} + D_{66})\frac{\partial^2}{\partial x \partial y}$$

$$L_{33} = -KA_{44} + D_{66}\frac{\partial^2}{\partial x^2} + D_{22}\frac{\partial^2}{\partial y^2} \quad \text{(b)}$$

10.5 In the Lévy method, the generalized displacements are expressed as products of undetermined functions and known trigonometric functions so as to satisfy the simply supported boundary conditions. Suppose that the edges at $y = 0, b$ of a rectangular plate are simply supported

$$w_0 = \phi_x = N_{yy} = M_{yy} = 0 \quad \text{(a)}$$

Then the displacement field (w_0, ϕ_x, ϕ_y) can be represented as

$$\begin{Bmatrix} w_0(x,y) \\ \phi_x(x,y) \\ \phi_y(x,y) \end{Bmatrix} = \sum_{n=1}^{\infty} \begin{Bmatrix} W_n(x) \\ X_n(x) \\ Y_n(x) \end{Bmatrix} \cos \beta_n y \quad \text{(b)}$$

where $\beta_n = n\pi/b$. The transverse load is also expanded as

$$q(x,y) = \sum_{n=1}^{\infty} q_n \sin \beta_n y \quad \text{(c)}$$

$$q_n(x) = \frac{2}{b}\int_0^b q(x,y)\sin \beta_n y \, dy \quad \text{(d)}$$

Show that the substitution of the displacement field (b) into governing equations (10.1.33)–(10.1.35) [or Eq. (a) of Exercise 10.4] results in the following three ordinary differential equations

$$W_n'' = C_{11}W_n + C_{12}X_n' + C_{13}Y_n + C_0 q_n$$
$$X_n'' = C_{21}W_n' + C_{22}X_n + C_{23}Y_n'$$
$$Y_n'' = C_{31}W_n + C_{32}X_n' + C_{33}Y_n \quad \text{(e)}$$

where a prime denotes the derivative with respect to x. The coefficients in Eq. (e) are given by

$$C_0 = -\frac{1}{KA_{55}}, \quad C_{11} = \frac{A_{44}}{A_{55}}\beta_n^2, \quad C_{12} = -1, \quad C_{13} = \frac{A_{44}}{A_{55}}\beta_n$$

$$C_{21} = \frac{KA_{55}}{D_{11}}, \quad C_{22} = \frac{(KA_{55} + D_{66}\beta_n^2)}{D_{11}}, \quad C_{23} = \frac{(D_{12} + D_{66})}{D_{11}}\beta_n$$

$$C_{31} = -\frac{KA_{44}}{D_{66}}\beta_n, \quad C_{32} = \frac{(D_{12} + D_{66})}{D_{66}}\beta_n, \quad C_{33} = \frac{(KA_{44} + D_{22}\beta_n^2)}{D_{66}}$$

(f)

10.6 Formulate the state-space solution procedure (see [10–16]) for the bending of simply supported plates using FSDT. In particular, show that Eq. (e) of Exercise 10.5 can be expressed as

$$\mathbf{Z}' = \mathbf{TZ} + \mathbf{F} \tag{a}$$

$$Z_1 = W_n, \quad Z_2 = W'_n$$
$$Z_3 = X_n, \quad Z_4 = X'_n, \quad Z_5 = Y_n, \quad Z_6 = Y'_n \tag{b}$$

where the matrix $\mathbf{T}$ is the 6×6 matrix

$$[T] = \begin{bmatrix} 0 & 1 & 0 & 0 & 0 & 0 \\ C_{11} & 0 & 0 & C_{12} & C_{13} & 0 \\ 0 & 0 & 0 & 1 & 0 & 0 \\ 0 & C_{21} & C_{22} & 0 & 0 & C_{23} \\ 0 & 0 & 0 & 0 & 0 & 1 \\ C_{31} & 0 & 0 & C_{32} & C_{33} & 0 \end{bmatrix} \tag{c}$$

the load vector $\mathbf{F}$ is given by

$$\{F\} = \{0, C_0 q_n, 0, 0, 0, 0\}^T \tag{d}$$

and various quantities appearing in Eqs. (c) and (d) are defined in Exercise 10.5.

10.7 Formulate the Lévy type solution procedure for the natural vibration of simply supported plates using FSDT. In particular, show that the operator $[T]$ in Eq. (c) of Exercise 10.6 holds with

$$C_{11} = \frac{\beta_n^2 K A_{44} - I_0 \omega_n^2}{K A_{55}}$$

$$C_{22} = \frac{K A_{55} + \beta_n^2 D_{66} - I_2 \omega_n^2}{D_{11}}$$

$$C_{33} = \frac{\beta_n^2 D_{22} + K A_{44} - I_2 \omega_n^2}{D_{66}} \tag{a}$$

10.8 Formulate the Lévy type solution procedure for the buckling of simply supported plates under in-plane compressive loads using FSDT. In particular, show that the operator $[T]$ in Eq. (c) of Exercise 10.6 holds with

$$C_0 = -\frac{1}{K A_{55} + \hat{N}_{xx}}, \quad C_{11} = \left(\frac{K A_{44} + \hat{N}_{yy}}{K A_{55} + \hat{N}_{xx}}\right) \beta_n^2$$

$$C_{12} = -\frac{K A_{55}}{K A_{55} + \hat{N}_{xx}}, \quad C_{13} = \left(\frac{K A_{44}}{K A_{55} + \hat{N}_{xx}}\right) \beta_n \tag{a}$$

10.9 Show that the von Kármán nonlinear strains of TSDT are

$$\left\{ \begin{array}{c} \varepsilon_{xx}^{(0)} \\ \varepsilon_{yy}^{(0)} \\ \gamma_{xy}^{(0)} \end{array} \right\} = \left\{ \begin{array}{c} \frac{\partial u_0}{\partial x} + \frac{1}{2}\left(\frac{\partial w_0}{\partial x}\right)^2 \\ \frac{\partial v_0}{\partial y} + \frac{1}{2}\left(\frac{\partial w_0}{\partial x}\right)^2 \\ \frac{\partial u_0}{\partial y} + \frac{\partial v_0}{\partial x} + \frac{\partial w_0}{\partial x}\frac{\partial w_0}{\partial y} \end{array} \right\}$$

10.10 Show that the nonlinear equations of motion of TSDT are given by Eqs. (10.3.14), (10.3.15), (10.3.17), (10.3.18), and

$$\frac{\partial \bar{Q}_x}{\partial x} + \frac{\partial \bar{Q}_y}{\partial y} + \frac{\partial}{\partial x}(N_{xx}\frac{\partial w_0}{\partial x} + N_{xy}\frac{\partial w_0}{\partial y}) + \frac{\partial}{\partial y}(N_{xy}\frac{\partial w_0}{\partial x} + N_{yy}\frac{\partial w_0}{\partial y})$$
$$+ c_1\left(\frac{\partial^2 P_{xx}}{\partial x^2} + 2\frac{\partial^2 P_{xy}}{\partial x \partial y} + \frac{\partial^2 P_{yy}}{\partial y^2}\right) + q = I_0\ddot{w}_0 - c_1^2 I_6\left(\frac{\partial^2 \ddot{w}_0}{\partial x^2} + \frac{\partial^2 \ddot{w}_0}{\partial y^2}\right)$$
$$+ c_1\left[I_3\left(\frac{\partial \ddot{u}_0}{\partial x} + \frac{\partial \ddot{v}_0}{\partial y}\right) + J_4\left(\frac{\partial \ddot{\phi}_x}{\partial x} + \frac{\partial \ddot{\phi}_y}{\partial y}\right)\right]$$

References for Additional Reading

1. Reddy, J. N., *Mechanics of Laminated Composite Plates: Theory and Analysis*, CRC Press, Boca Raton, Florida (1997).
2. Reddy, J. N., *Energy and Variational Methods in Applied Mechanics*, John Wiley, New York (1984).
3. Srinivas, S. and Rao, A. K., "Bending, Vibration, and Buckling of Simply Supported Thick Orthotropic Rectangular Plates and Laminates," *International Journal of Solids and Structures*, **6**, 1463–1481 (1970).
4. Reismann, H. and Lee, Y.-C., "Forced Motions of Rectangular Plates", *Developments in Theoretical and Applied Mechanics*, **4**, Pergamon, New York, p. 3 (1969).
5. Pagano, N. J., "Exact Solutions for Rectangular Bidirectional Composites and Sandwich Plates", *Journal of Composite Materials*, 4(1), 20–34 (1970).
6. Pagano, N. J., and Hatfield, S. J., "Elastic Behavior of Multilayered Bidirectional Composites", *AIAA Journal*, **10**(7), 931–933 (1972).
7. Reddy, J. N. and Hsu, Y. S., "Effects of Shear Deformation and Anisotropy on the Thermal Bending of Layered Composite Plates", *Journal of Thermal Stresses*, **3**, 475–493 (1980).

8. Reddy, J. N. and Chao, W. C., "A Comparison of Closed-Form and Finite Element Solutions of Thick Laminated Anisotropic Rectangular Plates", *Nuclear Engineering and Design*, **64**, 153-167 (1981).

9. Reddy, J. N., "On the Solutions to Forced Motions of Rectangular Composite Plates", *Journal of Applied Mechanics*, **49**, 403–408 (1982).

10. Brogan, W. L., *Modern Control Theory*, Prentice–Hall, Englewood Cliffs, NJ (1985).

11. Franklin, J. N., *Matrix Theory*, Prentice–Hall, Englewood Cliffs, NJ (1968).

12. Khdeir, A. A., Librescu, L., and Reddy, J. N., "Analytical Solution of a Refined Shear Deformation Theory for Rectangular Composite Plates", *International Journal of Solids and Structures*, **23**(10), 1447–1463 (1987).

13. Khdeir, A. A. and Librescu, L., "Analysis of Symmetric Cross-Ply Laminated Elastic Plates Using a Higher-Order Theory: Part I-Stress and Displacement", *Composite Structures*, **9**, 189–213 (1988).

14. Khdeir, A. A. and Librescu, L., "Analysis of Symmetric Cross-Ply Laminated Elastic Plates Using a Higher-Order Theory: Part II-Buckling and Free Vibration", *Composite Structures*, **9**, 259–277 (1988).

15. Reddy, J. N. and Khdeir, A. A., "Buckling and Vibration of Laminated Composite Plates Using Various Plate Theories", *AIAA Journal*, **27**(12), 1808–1817 (1989).

16. Nosier, A. and Reddy, J. N., "On Vibration and Buckling of Symmetric Laminated Plates According to Shear Deformation Theories", *Acta Mechanica*, **94** (3-4), 123–170 (1992).

17. Basset, A. B., "On the Extension and Flexure of Cylindrical and Spherical Thin Elastic Shells", *Philosophical Transactions of the Royal Society*, (London) Series A, **181**(6), 433–480 (1890).

18. Hildebrand, F. B., Reissner, E., and Thomas, G. B., "Notes on the Foundations of the Theory of Small Displacements of Orthotropic Shells", NACA TN-1833, Washington, D.C. (1949).

19. Vlasov, B. F., "Ob uravnieniakh izgiba plastinok (On equations of bending of plates)", (in Russian), *Doklady Akademii Nauk Azerbeijanskoi SSR*, **3**, 955–959 (1957).

20. Jemielita, G., "Techniczna Teoria Plyt Srednieej Grubbosci", (Technical Theory of Plates with Moderate Thickness), *Rozprawy Inzynierskie, Polska Akademia Nauk*, **23**(3), 483–499 (1975).
21. Schmidt, R., "A Refined Nonlinear Theory for Plates with Transverse Shear Deformation", *Journal of the Industrial Mathematics Society*, **27**(1), 23–38 (1977).
22. Lo, K. H., Christensen, R. M., and Wu, E. M., "A High-Order Theory of Plate Deformation: Part 1: Homogeneous Plates", *Journal of Applied Mechanics*, **44**(4), 663–668 (1977).
23. Krishna Murty, A. V., "Higher Order Theory for Vibration of Thick Plates", *AIAA Journal*, **15**(2), 1823–1824 (1977).
24. Levinson, M., "An Accurate, Simple Theory of the Static and Dynamics of Elastic Plates", *Mechanics Research Communications*, **7**(6), 343–350 (1980).
25. Reddy, J. N., "A Simple Higher-Order Theory for Laminated Composite Plates", *Journal of Applied Mechanics*, **51**, 745–752 (1984).
26. Reddy, J. N., "A Small Strain and Moderate Rotation Theory of Laminated Anisotropic Plates", *Journal of Applied Mechanics*, **54**, 623–626 (1987).
27. Bhimaraddi, A. and Stevens, L. K., "A Higher Order Theory for Free Vibration of Orthotropic, Homogeneous and Laminated Rectangular Plates", *Journal of Applied Mechanics*, **51**, 195–198 (1984).
28. Di Sciuva, M., "A Refined Transverse Shear Deformation Theory for Multilayered Anisotropic Plates", *Atti della Academia delle Scienze di Torino*, **118**, 269–295 (1984).
29. Reddy, J. N. and Phan, N. D., "Stability and Vibration of Isotropic Orthotropic and Laminated Plates According to a Higher-Order Shear Deformation Theory", *Journal of Sound & Vibration*, **98**, 157–170 (1985).
30. Phan, N. D., and Reddy, J. N., "Analysis of Laminated Composite Plates Using a Higher-Order Shear Deformation Theory", *International Journal for Numerical Methods in Engineering*, **21**, 2201–2219 (1985).
31. Krishna Murty, A. V., "Flexure of Composite Plates", *Composite Structures*, **7**(3), 161–177 (1987).
32. Reddy, J. N., "A General Non-Linear Third-Order Theory of Plates with Moderate Thickness", *International Journal of Non-linear Mechanics*, **25**(6), 677–686 (1990).

Chapter Eleven

Finite Element Models of Beams and Plates

11.1 Introduction

In Chapters 4 through 10, the Navier, Lévy, and Rayleigh–Ritz solutions to the equations of beams and circular and rectangular plates were presented. Analytical or variational solutions cannot be readily developed when the geometry of the plate is not circular or rectangular, different portions of the plate boundary are subjected to different boundary conditions, or non-linearities are involved. In such cases, one must resort to approximate methods of analysis that are capable of predicting accurate solutions.

The *finite element method* is a powerful numerical method for the solution of differential equations that arise in various fields of engineering and applied science. The basic idea of the finite element method is to view a given domain as an assemblage of simple geometric shapes, called *finite elements*, for which it is possible to *systematically* generate the approximation functions $\varphi_i(x,y)$ needed in the solution of differential equations over a typical element. The approximation functions $\varphi_i(x,y)$ are often constructed using ideas from interpolation theory, and hence they are also called *interpolation functions*. Thus, the finite element method is a piecewise (or elementwise) application of the Rayleigh–Ritz method, Galerkin's method, sub-domain method, least-squares method, and so on. For a given differential equation, it is possible to develop different *finite element models*, depending on the choice of the method used to generate algebraic equations among the undetermined coefficients of the approximate solution. The ability to represent geometrically complicated domains and ease of of application of physical boundary conditions made the finite element method a practical tool of engineering analysis and design. For a detailed introduction to the finite element method, the reader is advised to consult [1–7].

In this chapter, we develop finite element models of classical and first-order shear deformation theories of beams and plates. The objective is to introduce the reader to the finite element method in the context of the material covered in this book. While the coverage is not exhaustive in terms of solving complicated problems, it is hoped that the reader gains a basic understanding of the finite elements used in the analysis of beam and plate problems.

It is important to bear in mind that any approximate method is a means to analyze a practical engineering problem and that analysis is not an end in itself but rather an aid to design. *The value of the theory and analytical solutions presented in the preceding chapters to gain insight into the behavior of simple plate problems is immense in the numerical modeling of complicated problems by any numerical method, including the finite element method.* Those who are quick to use a computer rather than *think* about the problem to be analyzed may find it difficult to interpret or explain the computer-generated results.

First, we present the finite element models of beams, which are treated as one-dimensional problems. Those readers who have a good background in the finite element method may skip this part or browse through it. Next, finite element models of the classical plate theory are presented. Finally, finite element models of the first-order theory are developed. The developments presented herein are fairly complete for basic understanding of the finite element models of beams and plates, and one may consult other books on the finite element method for additional details or advanced topics.

11.2 Finite Element Models of Beams

11.2.1 Euler–Bernoulli Beam Elements

The equation of motion governing the bending of a beams according to the Euler–Bernoulli beam theory is given by

$$\frac{\partial^2}{\partial x^2}\left(EI\frac{\partial^2 w_0}{\partial x^2}\right) - \hat{N}_{xx}\frac{\partial^2 w_0}{\partial x^2} - q + I_0\frac{\partial^2 w_0}{\partial t^2} - I_2\frac{\partial^4 w_0}{\partial x^2 \partial t^2} = 0 \quad (11.2.1)$$

where $\hat{N}_{xx}$ is the axial (tensile) load, $q(x,t)$ is the distributed transverse load, ρ the material density, E the modulus of elasticity, I is the moment of inertia about the axis of bending (y), and I_0 and I_2 are mass inertias

$$I_0 = \int_{-\frac{h}{2}}^{\frac{h}{2}} \rho \, dz, \quad I_2 = \int_{-\frac{h}{2}}^{\frac{h}{2}} \rho z^2 \, dz \qquad (11.2.2)$$

First, the domain of the beam, $0 < x < L$ (see Figure 11.2.1a), is divided into a set of subintervals, called *beam finite elements*. The division of a domain into a collection of sub-domains facilitates accurate representation of the geometry (e.g., variable cross section), material properties (e.g., beams made of different material along the span), and the solution itself. Discontinuities in the geometry, loading, or material properties automatically make the deflection to be only piecewise continuous and thus a piecewise representation of the deflection in such cases is necessary. The set of elements representing the total domain is called the *finite element mesh*. A typical element occupies the spatial domain between $x = x_A$ and $x = x_B$. The end points, $x = x_A$ and $x = x_B$, of an element are called *element nodes*. The end points $x = 0$ and $x = L$ as well as the connecting points of all elements of the mesh are called *global nodes* (see Figure 11.2.1a). Note that x_A and x_B denote the global x-coordinates of the ends of the typical element. Each element can have its own geometric properties: length, $h_e \equiv x_B - x_A$, cross-sectional area, A_e, and moment of inertia, I_e; material properties: E_e, I_0^e, and I_2^e; and applied load: q_e.

Next, a typical element is isolated from the mesh and its structural behavior is *modeled* using a variational method. In the present study, we use the Rayleigh–Ritz method with algebraic polynomials for the approximation functions $\varphi_i(x)$. Recall that the Rayleigh–Ritz method uses either the total potential energy expression or the virtual work statement of the governing equation of the beam element. Here we use the virtual work statement of the governing equation to develop the *finite element model*, which is a set of relations among the displacements and forces (i.e., primary and secondary variables) of the element nodes.

When a portion of the body is isolated, we must show its *free-body diagram* (i.e., show all its applied and reactive forces). For a typical beam element, the free-body diagram is shown in Figure 11.2.1b. The ends $x = x_A$ and $x = x_B$ are denoted as *element nodes* 1 and 2, respectively. The generalized forces Q_1^e and Q_3^e denote the transverse shear forces, whereas Q_2^e and Q_4^e denote the bending moments at nodes 1 and 2. Figure 11.2.1b defines the notation used for the generalized displacements and generalized forces.

472 THEORY AND ANALYSIS OF ELASTIC PLATES

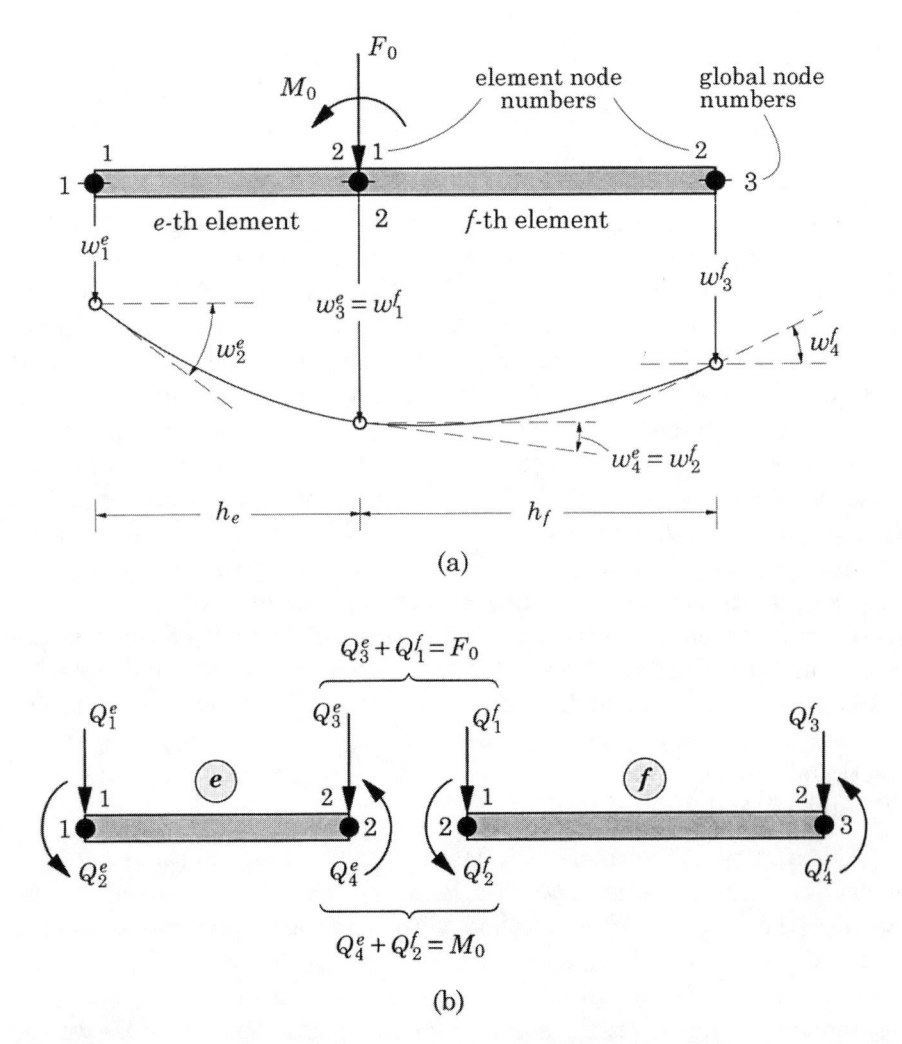

Figure 11.2.1. Domain, finite element subdivision, and typical finite element: (a) beam structure; (b) a typical beam finite element with displacement and force degrees of freedom.

As will be shown shortly, the generalized displacements must be continuous throughout the beam. In particular, they must be unique at the nodes common to elements. Therefore, the generalized displacements must be carried as nodal values in order to impose their continuity across the elements.

Spatial Discretization

The virtual work statement associated with Eq. (11.2.1a) over the typical element $\Omega^e = (x_A, x_B)$ is given by (see Reddy [1-3])

$$0 = \int_0^T \int_{x_A}^{x_B} \left(EI \frac{\partial^2 \delta w_0}{\partial x^2} \frac{\partial^2 w_0}{\partial x^2} + \hat{N}_{xx} \frac{\partial \delta w_0}{\partial x} \frac{\partial w_0}{\partial x} - \delta w_0 q \right.$$

$$\left. + I_0 \delta w_0 \frac{\partial^2 w_0}{\partial t^2} + I_2 \frac{\partial \delta w_0}{\partial x} \frac{\partial^3 w_0}{\partial x \partial t^2} \right) dx\, dt$$

$$- \int_0^T \left[Q_1^e\, \delta w_0(x_A) - Q_3^e\, \delta w_0(x_B) \right.$$

$$\left. - Q_2^e \left(-\frac{\partial \delta w_0}{\partial x} \right)_{x_A} - Q_4^e \left(-\frac{\partial \delta w_0}{\partial x} \right)_{x_B} \right] dt \quad (11.2.3a)$$

where δw_0 denotes the virtual deflection and Q_i^e are related to the transverse deflection w_0 by the relations

$$Q_1^e(t) = \left[\frac{\partial}{\partial x}\left(EI \frac{\partial^2 w_0}{\partial x^2} - I_2 \frac{\partial^2 w_0}{\partial t^2} \right) - \hat{N}_{xx} \frac{\partial w_0}{\partial x} \right]_{x=x_A}$$

$$Q_3^e(t) = -\left[\frac{\partial}{\partial x}\left(EI \frac{\partial^2 w_0}{\partial x^2} - I_2 \frac{\partial^2 w_0}{\partial t^2} \right) - \hat{N}_{xx} \frac{\partial w_0}{\partial x} \right]_{x=x_B}$$

$$Q_2^e(t) = \left[EI \frac{\partial^2 w_0}{\partial x^2} \right]_{x=x_A}, \quad Q_4^e(t) = -\left[EI \frac{\partial^2 w_0}{\partial x^2} \right]_{x=x_B} \quad (11.2.3b)$$

We note from the boundary terms that deflection and rotation, w_0 and $\theta \equiv -\partial w_0/\partial x$, are the *primary variables* of the problem, and they must be continuous at every point of the domain $0 < x < L$. The primary variables are the generalized displacements

$$w_0(x_A, t) = w_1^e(t), \quad \theta(x_A) \equiv -\frac{\partial w_0}{\partial x}\Big|_{x=x_A} = w_2^e(t)$$

$$w_0(x_B, t) = w_3^e(t), \quad \theta(x_B) \equiv -\frac{\partial w_0}{\partial x}\Big|_{x=x_B} = w_4^e(t) \quad (11.2.4)$$

and the secondary variables are the generalized forces defined in Eq. (11.2.3b). Here w_1^e and w_3^e denote transverse deflections and w_2^e and w_4^e denote the slopes at nodes 1 and 2 (see Figure 11.2.1b); w_i^e and Q_i^e ($i = 1, 2, 3, 4$) are called the element nodal displacement and force degrees of freedom.

474   THEORY AND ANALYSIS OF ELASTIC PLATES

The next step is to develop the finite element model based on the virtual work statement (11.2.3a) using the Rayleigh–Ritz method. To this end, we seek an approximation of w_0 over the element in the form

$$w_0(x,t) \approx \sum_{i=1}^{n} c_i^e(t)\varphi_i^e(x)$$

Since both w_0 and θ must be continuous everywhere in the domain, they should be continuous at the interface of any two elements. Therefore, the continuity of the generalized displacements w_i^e at element interfaces must be enforced. This in turn requires carrying w_i^e as the undetermined parameters $c_i = w_i^e$:

$$w_0(x,t) \approx w_1^e(t)\varphi_1^e(x) + w_2^e(t)\varphi_2^e(x) + w_3^e(t)\varphi_3^e(x) + w_4^e(t)\varphi_4^e(x) \quad (11.2.5)$$

so that the displacement continuity can be enforced by simply equating the corresponding nodal values of elements connected at a node. In view of the definitions (11.2.4), the approximation functions $\varphi_i^e(x)$ of Eq. (11.2.5) must satisfy the properties

$$\varphi_1^e(x_A) = 1, \quad \varphi_2^e(x_A) = 0, \quad \varphi_3^e(x_A) = 0, \quad \varphi_4^e(x_A) = 0,$$
$$\varphi_1^e(x_B) = 0, \quad \varphi_2^e(x_B) = 0, \quad \varphi_3^e(x_B) = 1, \quad \varphi_4^e(x_B) = 0,$$
$$\frac{\partial \varphi_1^e}{\partial x}\bigg|_{x_A} = 0, \quad \frac{\partial \varphi_2^e}{\partial x}\bigg|_{x_A} = -1, \quad \frac{\partial \varphi_3^e}{\partial x}\bigg|_{x_A} = 0, \quad \frac{\partial \varphi_4^e}{\partial x}\bigg|_{x_A} = 0,$$
$$\frac{\partial \varphi_1^e}{\partial x}\bigg|_{x_B} = 0, \quad \frac{\partial \varphi_2^e}{\partial x}\bigg|_{x_B} = 0, \quad \frac{\partial \varphi_3^e}{\partial x}\bigg|_{x_A} = 0, \quad \frac{\partial \varphi_4^e}{\partial x}\bigg|_{x_A} = -1 \quad (11.2.6)$$

Thus there are four conditions on each of the four functions. Hence, we must select a four-parameter expansion, i.e., cubic polynomial for each of the functions

$$\varphi_i^e(x) = a_0^{(i)} + a_1^{(i)}x + a_2^{(i)}x^2 + a_3^{(i)}x^3 \quad (11.2.7)$$

and determine the four constants $a_j^{(i)}$ ($j = 0, 1, 2, 3$) using the four conditions. We obtain

$$\varphi_1^e(x) = 1 - 3\left(\frac{x - x_A}{h_e}\right)^2 + 2\left(\frac{x - x_A}{h_e}\right)^3$$

$$\varphi_2^e(x) = -(x - x_A)\left(1 - \frac{x - x_A}{h_e}\right)^2$$

$$\varphi_3^e(x) = 3\left(\frac{x - x_A}{h_e}\right)^2 - 2\left(\frac{x - x_A}{h_e}\right)^3$$

$$\varphi_4^e(x) = -(x - x_A)\left[\left(\frac{x - x_A}{h_e}\right)^2 - \frac{x - x_A}{h_e}\right] \quad (11.2.8)$$

The approximation functions φ_i^e ($i = 1, 2, 3, 4$) are called *Hermite cubic interpolation functions*, which were derived by interpolating the function and its derivatives at the nodes. In the interest of brevity, the element label 'e' will be omitted in the following discussion.

Finite Element Model

Substituting for w_0 from Eq. (11.2.5) and $\delta w_0 = \varphi_i(x)$ into Eq. (11.2.3a), we obtain the following ith equation of the four equations for any time $t > 0$:

$$
\begin{aligned}
0 = & \int_{x_A}^{x_B} \left[EI \frac{d^2\varphi_i}{dx^2} \left(\sum_{j=1}^{4} w_j^e(t) \frac{d^2\varphi_j}{dx^2} \right) + \hat{N}_{xx} \frac{d\varphi_i}{dx} \left(\sum_{j=1}^{4} w_j^e(t) \frac{d\varphi_j}{dx} \right) \right. \\
& \left. - \varphi_i q + I_0 \varphi_i \left(\sum_{j=1}^{4} \varphi_j \frac{d^2 w_j^e}{dt^2} \right) + I_2 \frac{d\varphi_i}{dx} \left(\sum_{j=1}^{4} \frac{d\varphi_j}{dx} \frac{d^2 w_j^e}{dt^2} \right) \right] dx \\
& - Q_1^e \varphi_i(x_A) - Q_3^e \varphi_i(x_B) - Q_2^e \left(-\frac{d\varphi_i}{dx} \right)_{x_A} - Q_4^e \left(-\frac{d\varphi_i}{dx} \right)_{x_B} \\
= & \sum_{j=1}^{4} \int_{x_A}^{x_B} \left[\left(EI \frac{d^2\varphi_i}{dx^2} \frac{d^2\varphi_j}{dx^2} + \hat{N}_{xx} \frac{d\varphi_i}{dx} \frac{d\varphi_j}{dx} \right) w_j^e - \varphi_i q \right. \\
& \left. + \left(I_0 \varphi_i \varphi_j + I_2 \frac{d\varphi_i}{dx} \frac{d\varphi_j}{dx} \right) \frac{d^2 w_j^e}{dt^2} \right] dx - Q_1^e \varphi_i(x_A) - Q_3^e \varphi_i(x_B) \\
& - Q_2^e \left(-\frac{d\varphi_i}{dx} \right)_{x_A} - Q_4^e \left(-\frac{d\varphi_i}{dx} \right)_{x_B} \\
\equiv & \sum_{j=1}^{4} \left[\left(K_{ij}^e + G_{ij}^e \right) w_j^e + M_{ij}^e \frac{d^2 w_j^e}{dt^2} \right] - F_i^e \quad (11.2.9a)
\end{aligned}
$$

where

$$
K_{ij}^e = \int_{x_A}^{x_B} EI \frac{d^2\varphi_i}{dx^2} \frac{d^2\varphi_j}{dx^2} \, dx, \quad G_{ij}^e = \int_{x_A}^{x_B} \hat{N}_{xx} \frac{d\varphi_i}{dx} \frac{d\varphi_j}{dx} \, dx
$$

$$
M_{ij}^e = \int_{x_A}^{x_B} \left(I_0 \varphi_i \varphi_j + I_2 \frac{d\varphi_i}{dx} \frac{d\varphi_j}{dx} \right) dx
$$

$$
F_i^e = q_i^e + Q_1^e \varphi_i(x_A) + Q_3^e \varphi_i(x_B) + Q_2^e \left(-\frac{d\varphi_i}{dx} \right)_{x_A} + Q_4^e \left(-\frac{d\varphi_i}{dx} \right)_{x_B}
$$

$$
q_i^e = \int_{x_A}^{x_B} \varphi_i q \, dx \quad (11.2.9b)
$$

Equation (11.2.9a) can be written in matrix form as

$$([K^e] + [G^e])\{w^e\} + [M^e]\{\ddot{w}^e\} = \{F^e\}, \quad t > 0 \tag{11.2.10}$$

where $[K^e]$ is known as the element stiffness matrix, $[G^e]$ the element geometric stiffness matrix, $[M^e]$ the element mass matrix, and $\{F^e\}$ the element force vector. All of the matrices are symmetric: $[K^e] = [K^e]^T$ or $K_{ij}^e = K_{ji}^e$, etc. In view of the properties (11.2.6) of the approximation functions $\varphi_j^e(x)$, the expression for F_i^e can be simplified to

$$F_i^e = q_i^e + Q_i^e \tag{11.2.11}$$

which are composed of nodal forces q_i^e due to the distributed load q and the nodal reactions Q_i^e, which, in general, are not known a priori.

Equation (11.2.10) is called the *semidiscrete finite element model* because it represents only the spatial approximation of the solution. The set of ordinary differential equations (11.2.10) in time must be further reduced to algebraic equations by means of a time-approximation scheme to complete the discretization.

The coefficient matrices $[K^e], [G^e]$, and $[M^e]$ and force components q_i^e can be computed by evaluating the integrals in Eq. (11.2.9b). For example, when the geometry and material properties are elementwise constant, we have (the label 'e' on the variables is omitted for brevity)

$$[K^e] = \frac{2EI}{h^3} \begin{bmatrix} 6 & -3h & -6 & -3h \\ -3h & 2h^2 & 3h & h^2 \\ -6 & 3h & 6 & 3h \\ -3h & h^2 & 3h & 2h^2 \end{bmatrix}, \quad \{q^e\} = \frac{q_0 h}{12} \begin{Bmatrix} 6 \\ -h \\ 6 \\ h \end{Bmatrix} \tag{11.2.12}$$

$$[M^e] = \frac{I_0 h}{420} \begin{bmatrix} 156 & -22h & 54 & 13h \\ -22h & 4h^2 & -13h & -3h^2 \\ 54 & -13h & 156 & 22h \\ -13h & -3h^2 & 22h & 4h^2 \end{bmatrix} + \frac{I_2}{30h} [H^e] \tag{11.2.13}$$

$$[H^e] = \begin{bmatrix} 36 & -3h & -36 & -3h \\ -3h & 4h^2 & 3h & -h^2 \\ -36 & 3h & 36 & 3h \\ -3h & -h^2 & 3h & 4h^2 \end{bmatrix}, \quad [G^e] = \frac{\hat{N}_{xx}}{30h} [H^e] \tag{11.2.14}$$

If the problem is one of static bending under applied transverse load q and axial load $\hat{N}_{xx}$, then all variables are only functions of position and the time-derivative term is omitted in Eq. (11.2.10). The element equation (11.2.10) takes the form

$$([K^e] + [G^e])\{w^e\} = \{F^e\} \qquad (11.2.15)$$

If the problem is one of studying buckling under the axial compressive load, $\hat{N}_{xx} = -N^0_{xx}$, Eq. (11.2.10) reduces to

$$\left([K^e] - N^0_{xx}[\bar{G}^e]\right)\{w^e\} = \{Q^e\} \qquad (11.2.16)$$

where $[\bar{G}^e]$ is the same as $[G^e]$ but without $\hat{N}_{xx}$ [see Eq. (11.2.9b)].

In the case of natural vibrations of a beam, we assume $\hat{N}_{xx} = 0$ and that the nodal values $w^e_i(t)$ and reaction forces are periodic in time

$$w^e_j(t) = W^e_j e^{-i\omega t}, \quad Q^e_j(t) = \bar{Q}^e_j e^{-i\omega t}, \quad i = \sqrt{-1} \qquad (11.2.17)$$

Then Eq. (11.2.10) reduces to

$$\left([K^e] - \omega^2[M^e]\right)\{W^e\} = \{\bar{Q}^e\} \qquad (11.2.18)$$

We note that both the buckling and natural vibration problems are eigenvalue problems in that we must determine the eigenvalues λ such that

$$([A] - \lambda[B])\{W\} = \{0\} \qquad (11.2.19)$$

for any nontrivial (i.e., nonzero) $\{W\}$. We will return to these topics later in this chapter.

Time Discretization

In the case of time dependent response of beams, Eq. (11.2.10) governs the motion, and the ordinary differential equations in time must be integrated to determine the nodal values $w^e_j(t)$ as functions of time. Although the matrix equation (11.2.10) can be solved, in principle, using the Laplace transforms for certain load cases, it is a tedious exercise and involves more computational effort than using a numerical time-approximation scheme. Here we consider the Newmark family of time-integration schemes that are used widely in structural dynamics.

In the Newmark method, the first and second time derivatives are approximated as

$$\{\dot{w}^e\}_{s+1} = \{\dot{w}^e\}_s + a_1\{\ddot{w}^e\}_s + a_2\{\ddot{w}^e\}_{s+1}$$
$$\{\ddot{w}^e\}_{s+1} = a_3\left(\{w^e\}_{s+1} - \{w^e\}_s\right) - a_4\{\dot{w}^e\}_s - a_5\{\ddot{w}^e\}_s \quad (11.2.20a)$$

where the superposed dot denotes differentiation with respect to time,

$$a_1 = (1-\alpha)\Delta t_s, \quad a_2 = \alpha\Delta t_s, \quad a_3 = \frac{2}{\gamma(\Delta t_s)^2}$$

$$a_4 = \Delta t_s a_3, \quad a_5 = \frac{(1-\gamma)}{\gamma} \quad (11.2.20b)$$

Δt is the time increment, $\Delta t_s = t_{s+1} - t_s$, and $\{\cdot\}_s$, for example, denotes the value of the enclosed vector at time t_s.

The parameters α and γ in the Newmark scheme are selected such that the scheme is either stable or conditionally stable; i.e., the error introduced through the time approximation (11.2.20a) does not grow unboundedly as we march in time. All schemes for which $\gamma \geq \alpha \geq 1/2$ are unconditionally stable. Schemes for which $\gamma < \alpha$ and $\alpha \geq 0.5$ are conditionally stable, and the stability condition is

$$\Delta t \leq \Delta t_{cr} \equiv \left[\frac{1}{2}\omega_{max}(\alpha - \gamma)\right]^{-\frac{1}{2}} \quad (11.2.21)$$

where ω_{max} denotes the maximum frequency associated with the finite element equations (11.2.18).

The Newmark family contains the following widely used schemes:

$\alpha = \frac{1}{2}, \quad \gamma = \frac{1}{2},$ the constant average acceleration method (stable)

$\alpha = \frac{1}{2}, \quad \gamma = \frac{1}{3},$ the linear acceleration method (conditionally stable)

$\alpha = \frac{1}{2}, \quad \gamma = \frac{1}{6},$ the Fox–Goodwin scheme (conditionally stable)

$\alpha = \frac{1}{2}, \quad \gamma = 0,$ the centered difference method (conditionally stable)

$\alpha = \frac{3}{2}, \quad \gamma = \frac{8}{5},$ the Galerkin method (stable)

$\alpha = \frac{3}{2}, \quad \gamma = 2,$ the backward difference method (stable) $\quad (11.2.22)$

Premultiplying the second equation in (11.2.20a) with $[M^e]_{s+1}$ and using Eq. (11.2.10) at $t = t_{s+1}$ to replace $[M^e]_{s+1}\{w^e\}_{s+1}$, we obtain

$$[\hat{K}^e]\{w^e\}_{s+1} = \{\hat{F}^e\} \quad (11.2.23)$$

where

$$[\hat{K}^e] = ([K^e]_{s+1} + [G^e]_{s+1}) + a_3[M^e]_{s+1}$$
$$\{\hat{F}^e\} = \{F^e\}_{s+1} + [M^e]_{s+1}(a_3\{w^e\}_s + a_4\{\dot{w}^e\}_s + a_5\{\ddot{w}^e\}_s) \quad (11.2.24)$$

Note that for the centered difference scheme ($\gamma = 0$), it is necessary to use an alternative form of Eq. (11.2.23).

Equation (11.2.23) represents a system of algebraic equations among the (discrete) values of $\{w^e(t)\}$ at time $t = t_{s+1}$ in terms of known values at time $t = t_s$. At the first time step (i.e., $s = 0$), the values $\{w^e\}_0 = \{w^e(0)\}$ and $\{\dot{w}^e\}_0 = \{\dot{w}^e(0)\}$ are known from the initial conditions of the problem, and Eq. (11.2.10) is used to determine $\{\ddot{w}^e\}_0$ at $t = 0$:

$$\{\ddot{w}^e\}_0 = [M^e]^{-1}\{\{F^e\} - ([K^e] + [G^e])\{w^e\}_0\} \quad (11.2.25)$$

Assembly of Element Equations

The element equations (11.2.15), (11.2.16), (11.2.18), or (11.2.23) cannot be solved at the element level because we have eight unknowns (w_j^e and Q_j^e) but only four equations. This is consistent with the fact one cannot solve the element equations independent of the total problem. To obtain exactly the same number of equations as there are unknowns in a problem, it is necessary to combine the equations of various elements in a meaningful way. First, we note that the primary variables, namely, the deflection and rotation, must be continuous everywhere, including at the element interfaces. For example, if the node 2 of element e is connected to node 1 of element f, then the continuity requirement can be expressed as (see Figure 11.2.2a)

$$w_3^e = w_1^f, \quad w_4^e = w_2^f \quad (11.2.26)$$

Since elements e and f are connected at a node, the reaction forces from the two elements at the common node should be in equilibrium (see Figure 11.2.2b)

$$Q_3^e + Q_1^f = F_0, \quad Q_4^e + Q_2^f = M_0 \quad (11.2.27)$$

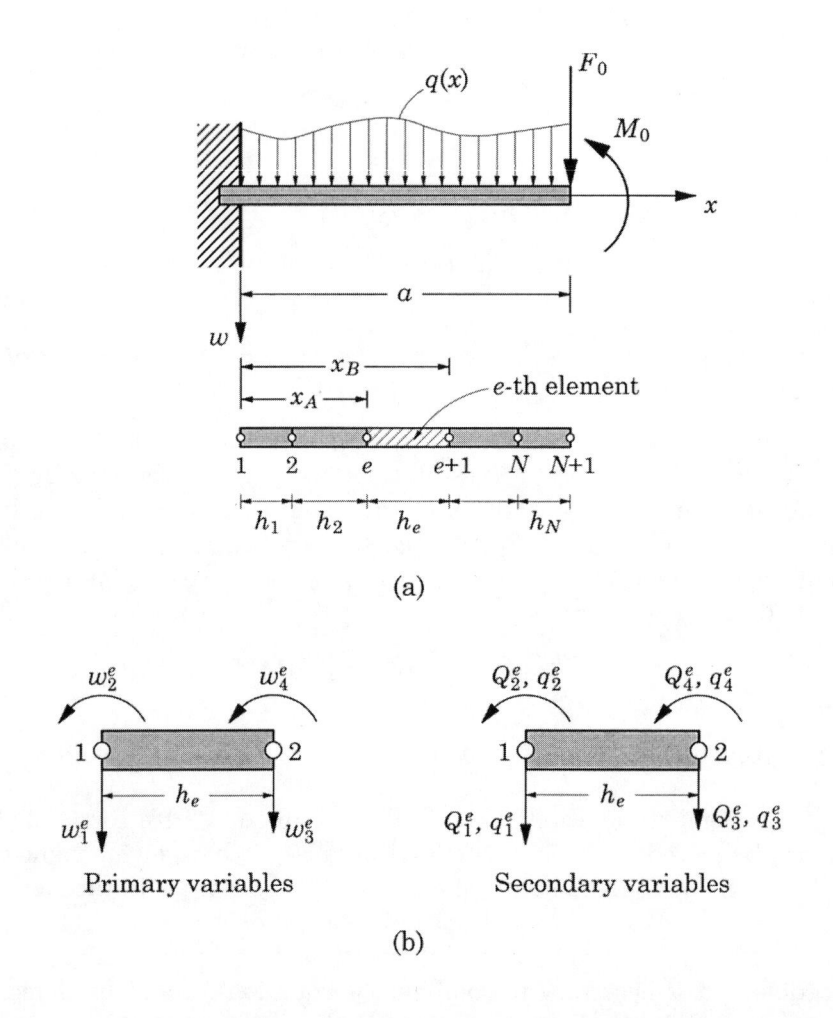

Figure 11.2.2. Assembly of two Euler–Bernoulli beam finite elements, Ω^e and Ω^f: (a) continuity of the generalized displacements, w_i^e; (b) balance of the generalized internal forces, Q_i^e.

where F_0 and M_0 are the applied transverse concentrated load (positive up) and bending moment (clockwise positive), respectively, at the node where the two elements are connected. If there is no applied load F_0 or bending moment M_0 at the node, their values in Eq. (11.2.27) should be set to zero.

Thus, the assembly of all elements in a finite element mesh is based on the Inter-element continuity of primary variables and balance of secondary variables. Note that the external nodal forces q_i^e due to the distributed load do not enter Eq. (11.2.27). The balance equations (11.2.27) require that we must add the third equation of element e with the first equation of element f, and add the fourth equation of element e with the second equation of element f. Therefore, there will be only two equations for the node common to the two elements. If a beam is divided into N elements (see Figure 11.2.1a), there will be a total $N+1$ nodes, including $N-1$ internal nodes. Consequently, there are $2(N-1)$ equations from the internal nodes and two equations from each of the boundary nodes (node 1 and $N+1$) — a total of $2(N+1)$ equations. In view of the balance conditions (11.2.27), generalized forces at the internal nodes are replaced by known applied forces. The generalized forces at node 1 and node $N+1$ will remain a part of the $2(N+1)$ equations. In addition, there are $2(N+1)$ generalized displacements. Thus the total number of unknowns is equal to $2(N+1)+4$, whereas the number of equations is $2(N+1)$. The additional four equations are provided by the four boundary conditions of the problem. The boundary conditions involve specifying either a generalized displacement or the corresponding generalized force, but never both from the same pair.

Imposition of Boundary Conditions

To illustrate the imposition of boundary conditions on the assembled equations, we consider the static bending of a beam of length $L = 3h$, clamped at $x = 0$, and on a roller support at $x = 2h$. Suppose that a uniformly distributed load of intensity q_0 (per unit area) is distributed over the span $h \leq x \leq 2h$, and the free end $x = 3h$ is subjected to a point load F_0 and bending moment M_0 (see Figure 11.2.3). The beam must be modeled, because of the loading and boundary conditions, using at least three elements.

The geometric boundary conditions of the problem are

$$w_0(0) = 0, \quad \frac{dw_0}{dx}(0) = 0, \quad w_0(2h) = 0 \qquad (11.2.28a)$$

Expressing the geometric conditions in terms of the global nodal variables, we have

$$U_1 = 0, \quad U_2 = 0, \quad U_5 = 0 \qquad (11.2.28b)$$

482 THEORY AND ANALYSIS OF ELASTIC PLATES

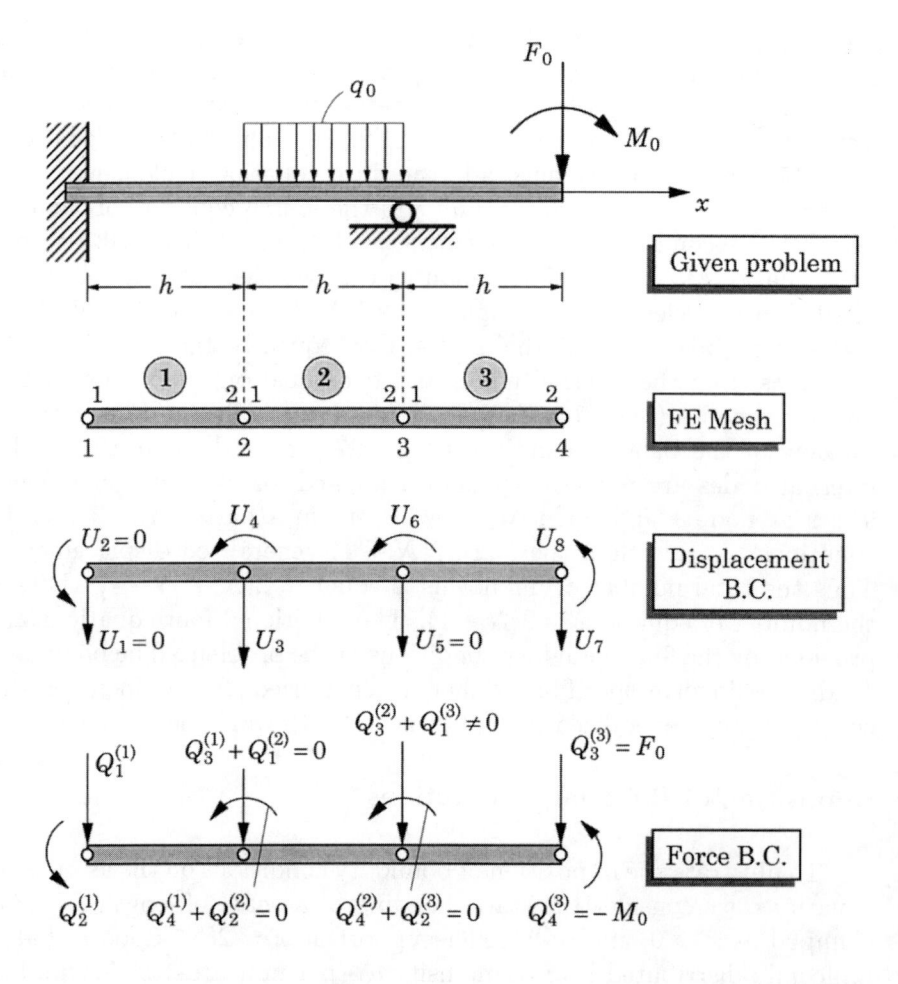

Figure 11.2.3. Bending of an indeterminate beam. Illustration of continuity of displacements, equilibrium of internal forces, and boundary conditions.

The force boundary conditions are

$$-\left[\frac{\partial}{\partial x}\left(EI\frac{\partial^2 w_0}{\partial x^2}\right)\right]_{x=3h} = F_0, \quad \left[EI\frac{\partial^2 w_0}{\partial x^2}\right]_{x=3h} = -M_0 \quad (11.2.29a)$$

which can be expressed in terms of the nodal forces [see Eq. (11.2.3b)]

$$Q_3^{(3)} = F_0, \quad Q_4^{(3)} = -M_0 \quad (11.2.29b)$$

The equilibrium of internal forces requires

$$Q_3^{(1)} + Q_1^{(2)} = 0, \quad Q_4^{(1)} + Q_2^{(2)} = 0, \quad Q_4^{(2)} + Q_2^{(3)} = 0 \qquad (11.2.29c)$$

The reaction force $Q_1^{(1)}$ and bending moment $Q_2^{(1)}$ at $x = 0$ and reaction force $Q_3^{(2)}$ at $x = 2h$ are unknown and can be determined in the post-computation.

Using the boundary conditions (11.2.28b) and (11.2.29b) and equilibrium conditions (11.2.29c) in the assembled equations, we obtain

$$\{U\} = \begin{Bmatrix} 0 \\ 0 \\ U_3 \\ U_4 \\ 0 \\ U_6 \\ U_7 \\ U_8 \end{Bmatrix}, \quad \{F\} = \frac{q_0 h}{12} \begin{Bmatrix} 0 \\ 0 \\ 6 \\ -h \\ 6 \\ h \\ 0 \\ 0 \end{Bmatrix} + \begin{Bmatrix} Q_1^{(1)} \\ Q_2^{(1)} \\ 0 \\ 0 \\ Q_3^{(2)} + Q_1^{(3)} \\ 0 \\ F_0 \\ -M_0 \end{Bmatrix} \qquad (11.2.30)$$

There are five unknown generalized displacements and three unknown generalized forces to be determined from the eight equations. Often, the unknown generalized displacements are determined first, and the generalized forces are determined subsequently. To determine the unknown generalized displacements $U_3, U_4, U_6, U_7,$ and U_8, we use the corresponding equations, equations 3, 4, 6, 7, and 8, which do not contain the unknown generalized forces. Since the specified displacements are zero, one can omit the rows and columns (1, 2, and 5) corresponding to the specified displacements. When the specified values are nonzero, the products of stiffnesses with the known displacements are moved to the right side of the equation. We have

$$\begin{bmatrix} K_{33}^{(1)} + K_{11}^{(2)} & K_{34}^{(1)} + K_{12}^{(2)} & K_{14}^{(2)} & 0 & 0 \\ K_{43}^{(1)} + K_{21}^{(2)} & K_{44}^{(1)} + K_{22}^{(2)} & K_{24}^{(2)} & 0 & 0 \\ K_{41}^{(2)} & K_{42}^{(2)} & K_{44}^{(2)} + K_{22}^{(3)} & K_{23}^{(3)} & K_{24}^{(3)} \\ 0 & 0 & K_{32}^{(3)} & K_{33}^{(3)} & K_{34}^{(3)} \\ 0 & 0 & K_{42}^{(3)} & K_{43}^{(3)} & K_{44}^{(3)} \end{bmatrix} \begin{Bmatrix} U_3 \\ U_4 \\ U_6 \\ U_7 \\ U_8 \end{Bmatrix}$$

$$= \frac{q_0 h}{12} \begin{Bmatrix} 6 \\ -h \\ h \\ 0 \\ 0 \end{Bmatrix} + \begin{Bmatrix} 0 \\ 0 \\ 0 \\ F_0 \\ -M_0 \end{Bmatrix} \qquad (11.2.31a)$$

The unknown generalized forces can be computed from equations 1, 2, and 5 of Eq. (11.2.27a):

$$\left\{ \begin{array}{c} Q_1^{(1)} \\ Q_2^{(1)} \\ Q_3^{(2)} + Q_1^{(3)} \end{array} \right\} = \begin{bmatrix} K_{13}^{(1)} & K_{14}^{(1)} & 0 & 0 & 0 \\ K_{23}^{(1)} & K_{24}^{(1)} & 0 & 0 & 0 \\ K_{31}^{(2)} & K_{32}^{(2)} & K_{34}^{(2)} + K_{12}^{(3)} & K_{13}^{(3)} & K_{14}^{(3)} \end{bmatrix} \left\{ \begin{array}{c} U_3 \\ U_4 \\ U_6 \\ U_7 \\ U_8 \end{array} \right\}$$

$$- \frac{q_0 h}{2} \left\{ \begin{array}{c} 0 \\ 0 \\ 1 \end{array} \right\} \qquad (11.2.31b)$$

This completes the finite element model development of Euler-Bernoulli beams. The Hermite cubic beam finite element gives exact values of the deflection and slope *at the nodes* whenever the flexural stiffness is elementwise constant. Once the nodal values of the generalized displacements are known, the transverse deflection and its derivatives up to and including order three can be determined from Eq. (11.2.8a) at any point within the element. Of course, only the deflection and its first derivatives will be continuous at the nodes, and the second and third derivatives of w_0 will be discontinuous. If shear force and bending moments at any interior points are computed using Eqs. (11.2.3b), they will be discontinuous. Alternatively, they can be computed at the nodes using Eq. (11.2.31b). These are unique and the most accurate, if not exact, values of the shear forces and bending moments at the nodes.

11.2.2 Timoshenko Beam Elements

The equations of motion of the Timoshenko beam theory (see Example 2.2.2) are

$$\frac{\partial}{\partial x}\left[KGA\left(\frac{\partial w_0}{\partial x} + \phi_x\right)\right] + \hat{N}_{xx}\frac{\partial^2 w_0}{\partial x^2} + q = I_0 \frac{\partial^2 w_0}{\partial t^2} \quad (11.2.32a)$$

$$\frac{\partial}{\partial x}\left(EI\frac{\partial \phi_x}{\partial x}\right) - KGA\left(\frac{\partial w_0}{\partial x} + \phi_x\right) = I_2 \frac{\partial^2 \phi_x}{\partial t^2} \quad (11.2.32b)$$

where ϕ_x denotes the rotation of a transverse normal line about the axis of bending, K the shear correction coefficient, G the shear modulus, and

A the area of cross section. All other variables have the same meaning as before.

Since the terminology and steps involved in the finite element analysis of beams are already presented, attention is focused on the spatial approximation of Eqs. (11.2.32a,b) to arrive at their finite element model. A number of Timoshenko beam finite elements have appeared in the literature [8–12]. They differ from each other in the choice of approximation functions used for the transverse deflection w_0 and rotation ϕ_x or in the variational form used to develop the finite element model. Some are based on equal interpolation and others on unequal interpolation of w_0 and ϕ_x.

The Timoshenko beam finite element with linear interpolation of both w_0 and ϕ_x is the simplest element. However, it is very stiff in the thin beam limit, i.e., as the length-to-thickness ratio becomes large (say, 100). Such behavior is known as *shear locking*. The locking is due to the inconsistency of the interpolation used for w_0 and ϕ_x. To overcome the locking, one may use equal interpolation for both w_0 and ϕ_x but use a lower-order polynomial for the shear strain, $\gamma_{xz} = (dw_0/dx) + \phi_x$. This is often realized by using selective integration, in which reduced-order integration is used to evaluate the stiffness coefficients associated with the transverse shear strain, and all other coefficients of the stiffness matrix are evaluated using full integration. This selective integration Timoshenko beam element is known to exhibit spurious energy modes. An approximation consistent with Eqs. (11.2.32a,b) provides a Timoshenko beam element that is the best alternative to selective integration Timoshenko beam element. Here we review the derivations of these elements briefly (see Reddy [12]).

General Finite Element Model

The finite element model of the Timoshenko beam equations can be constructed using the virtual work statement in Eq. (g) of Example 2.2.2, where the shear force Q_x and bending moment M_{xx} are known in terms of the generalized displacements (w_0, ϕ_x) by the relations

$$Q_x = KGA\left(\frac{\partial w_0}{\partial x} + \phi_x\right), \quad M_{xx} = EI\frac{\partial \phi_x}{\partial x} \qquad (11.2.33)$$

The virtual work statement is equivalent to the following two statements (the third statement corresponds to the axial displacement u_0 and not required here)

$$0 = \int_{x_A}^{x_B} \left[KGA \frac{\partial \delta w_0}{\partial x} \left(\frac{\partial w_0}{\partial x} + \phi_x \right) + \hat{N}_{xx} \frac{\partial \delta w_0}{\partial x} \frac{\partial w_0}{\partial x} \right.$$
$$\left. + \delta w_0 I_0 \frac{\partial^2 w_0}{\partial t^2} - \delta w_0 q \right] dx - Q_1^e \delta w_0(x_A) - Q_3^e \delta w_0(x_B)$$

$$0 = \int_{x_A}^{x_B} \left[EI \frac{\partial \delta \phi_x}{\partial x} \frac{\partial \phi_x}{\partial x} + KGA \, \delta \phi_x \left(\frac{\partial w_0}{\partial x} + \phi_x \right) + I_2 \delta \phi_x \frac{\partial^2 \phi_x}{\partial t^2} \right] dx$$
$$- Q_2^e \delta \phi_x(x_A) - Q_4^e \delta \phi_x(x_B) \tag{11.2.34}$$

where δw_0 and $\delta \phi_x$ are the virtual displacements. The Q_i^e have the same physical meaning as in the Euler–Bernoulli beam element [see Eq. (11.2.3b)], but their relationship to the transverse deflection and rotation, w_0 and ϕ_x, is different:

$$Q_1^e(t) = -\left[KGA \left(\frac{\partial w_0}{\partial x} + \phi_x \right) + \hat{N}_{xx} \frac{\partial w_0}{\partial x} \right]_{x=x_A}$$

$$Q_3^e(t) = \left[KGA \left(\frac{\partial w_0}{\partial x} + \phi_x \right) + \hat{N}_{xx} \frac{\partial w_0}{\partial x} \right]_{x=x_B}$$

$$Q_2^e(t) = -\left[EI \frac{\partial \phi_x}{\partial x} \right]_{x=x_A}, \quad Q_4^e(t) = \left[EI \frac{\partial \phi_x}{\partial x} \right]_{x=x_B} \tag{11.2.35}$$

An examination of the virtual work statements (11.2.34) suggests that $w_0(x,t)$ and $\phi_x(x,t)$ are the primary variables and therefore must be carried as nodal variables. In general w_0 and ϕ_x need not be approximated by the same degree polynomials, although w_0 should be approximated using the same or greater degree polynomials than those used for ϕ_x. Suppose that

$$w_0(x,t) = \sum_{j=1}^{m} w_j^e(t) \psi_j^e(x), \quad \phi_x(x,t) = \sum_{j=1}^{n} \Phi_j^e(t) \Theta_j^e(x), \quad m \geq n \tag{11.2.36}$$

where $\psi_j^e(x)$ and $\Theta_j^e(x)$ are the Lagrange interpolation functions of degree $(m-1)$ and $(n-1)$, respectively. Substitution of (11.2.36) for w_0 and ϕ_x, and $\delta w_0 = \psi_i^e$ and $\delta \phi_x = \Theta_i^e$, into Eq. (11.2.34) yields the semidiscrete finite element model

$$0 = \sum_{j=1}^{m} \left(K_{ij}^{11} + G_{ij}^e \right) w_j^e + \sum_{k=1}^{n} K_{ik}^{12} \Phi_k^e + \sum_{j=1}^{m} M_{ij}^{11} \frac{d^2 w_j^e}{dt^2} - F_i^1$$

$$0 = \sum_{j=1}^{m} K_{jk}^{12} w_j^e + \sum_{j=1}^{n} K_{kj}^{22} \Phi_j^e + \sum_{j=1}^{n} M_{ij}^{22} \frac{d^2 \Phi_j^e}{dt^2} - F_k^2 \tag{11.2.37a}$$

where

$$K_{ij}^{11} = \int_{x_A}^{x_B} KGA\frac{d\psi_i}{dx}\frac{d\psi_j}{dx}\,dx, \quad G_{ij}^e = \int_{x_A}^{x_B} \hat{N}_{xx}\frac{d\psi_i}{dx}\frac{d\psi_j}{dx}\,dx$$

$$K_{ij}^{12} = \int_{x_A}^{x_B} KGA\frac{d\psi_i}{dx}\Theta_j\,dx$$

$$K_{ij}^{22} = \int_{x_A}^{x_B} \left(EI\frac{d\Theta_i}{dx}\frac{d\Theta_j}{dx} + KGA\Theta_i\Theta_j\right)dx$$

$$M_{ij}^{11} = \int_{x_A}^{x_B} I_0\psi_i\psi_j\,dx, \quad M_{ij}^{22} = \int_{x_A}^{x_B} I_2\Theta_i\Theta_j\,dx$$

$$F_i^1 = \int_{x_A}^{x_B} \psi_i q\,dx + Q_1^e\psi_i(x_A) + Q_3^e\psi_i(x_B)$$

$$F_i^2 = Q_2^e\Theta_i(x_A) + Q_4^e\Theta_i(x_B) \tag{11.2.37b}$$

The element equations (11.2.37a) can be expressed in matrix form as

$$\begin{bmatrix}[K^{11}] & [K^{12}]\\ [K^{12}]^T & [K^{22}]\end{bmatrix}\begin{Bmatrix}\{w\}\\ \{\Phi\}\end{Bmatrix} + \begin{bmatrix}[G] & [0]\\ [0] & [0]\end{bmatrix}\begin{Bmatrix}\{w\}\\ \{\Phi\}\end{Bmatrix}$$

$$+ \begin{bmatrix}[M^{11}] & [0]\\ [0] & [M^{22}]\end{bmatrix}\begin{Bmatrix}\{\ddot{w}\}\\ \{\ddot{\Phi}\}\end{Bmatrix} = \begin{Bmatrix}\{F^1\}\\ \{F^2\}\end{Bmatrix} \tag{11.2.38}$$

Next we discuss the choice of the approximation functions ψ_i and Θ_i, which dictate different finite element models.

Reduced Integration Linear Timoshenko Beam Element (RIE)

When equal interpolation (especially linear) of w_0 and ϕ_x is used (i.e., $\psi_i = \Theta_i$ and $m = n$) and full (or exact) integration is used to evaluate the coefficients K_{ij}^{11}, K_{ij}^{12} and K_{ij}^{22}, the element is known to be very stiff. This behavior is due to the inability of the element to represent a constant state of transverse shear strain $\gamma_{xz} = \phi_x + (\partial w_0/\partial x)$, and the phenomenon is known as shear locking. To avoid locking, the shear strain $\phi_x + (\partial w_0/\partial x)$ is approximated, while using equal interpolation of w_0 and ϕ, as a constant by using reduced integration of the shear stiffnesses (i.e., the second term in the definition of K_{ij}^{22} is evaluated using one-point integration; the mass coefficients M_{ij}^{11} and M_{ij}^{22} are always evaluated exactly).

For linear interpolation of both w_0 and ϕ_x and for constant values of EI and KGA, the element stiffness matrix (with reduced integration) is given by

$$[K^e] = \frac{2EI}{\mu_0 h^3} \begin{bmatrix} 6 & -3h & -6 & -3h \\ -3h & h^2(1.5+6\Omega) & 3h & h^2(1.5-6\Omega) \\ -6 & 3h & 6 & 3h \\ -3h & h^2(1.5-6\Omega) & 3h & h^2(1.5+6\Omega) \end{bmatrix} \quad (11.2.39)$$

where $\Omega = EI/KGAh^2$, $\mu_0 = 12\Omega$, and h is the element length. We note that the force vector $\{f\}$ associated with this element does not contain any components corresponding to the moment degrees of freedom. The element is not completely free of locking. Unless two or more elements are used in each span of a beam structure, one does not obtain good results with this element.

Consistent Interpolation Timoshenko Beam Element (CIE)

When $(m-1)$st degree polynomials for w_0 and $(m-2)$nd degree polynomials for ϕ_x are used and the coefficients K_{ij}^{11}, K_{ij}^{12}, and K_{ij}^{22} are evaluated using full integration, the element is termed a *consistent interpolation element*. For example, consider the case in which quadratic interpolation of w_0 and linear interpolation of ϕ_x are used. The element stiffness matrix for this case is 5×5, with the end nodes having two degrees of freedom and the middle node having only the deflection as the degree of freedom. By eliminating the degree of freedom of the middle node (i.e., condense it out), the matrix order can be reduced to 4×4. For constant values of EI and KGA, the 4×4 element stiffness matrix and 4×1 force vector (after the condensation) are given by

$$[K^e] = \frac{2EI}{\mu_0 h^3} \begin{bmatrix} 6 & -3h & -6 & -3h \\ -3h & h^2(1.5+6\Omega) & 3h & h^2(1.5-6\Omega) \\ -6 & 3h & 6 & 3h \\ -3h & h^2(1.5-6\Omega) & 3h & h^2(1.5+6\Omega) \end{bmatrix} \quad (11.2.40)$$

$$\{f^e\} = \begin{Bmatrix} f_1 + 0.5f_3 \\ -0.125f_3 h \\ f_2 + 0.5f_3 \\ 0.125f_3 h \end{Bmatrix}, \quad f_i = \int_{x_A}^{x_B} q(x)\varphi_i \, dx \quad (11.2.41)$$

where φ_i are the quadratic interpolation functions and node 3 is the center node. The nodal degrees of freedom associated with this element are the same as those in the RIE element. It is interesting to see that the stiffness matrix for CIE is the same as that for RIE. However, the load vector is not the same as that for the RIE. For uniform loading, the force vector of CIE becomes

$$\{f\} = \frac{q_0 h}{12} \begin{Bmatrix} 6 \\ -h \\ 6 \\ h \end{Bmatrix} \qquad (11.2.42)$$

which is the same as that of the Euler–Bernoulli element.

It is also possible to extend the idea of static condensation to higher-order elements. For example, cubic interpolation of w_0 and quadratic interpolation of ϕ_x will yield a 7×7 stiffness matrix. By eliminating the degrees of freedom associated with the interior nodes, one may reduce the size of the stiffness matrix to 4×4. An alternate formulation of this element is discussed next. *The elimination of the displacement degree of freedom associated with interior nodes is also valid for all static problems, linear or nonlinear. However, it is not valid for dynamic problems.*

Interdependent Interpolation Element (IIE)

Recall that the Euler–Bernoulli element gives exact nodal values because the element is based on the exact polynomial solution of the homogeneous problem (i.e., problem with $q = 0$). Analogously, we can develop a Timoshenko beam element based on the exact solution of the homogeneous form of the static equilibrium equations associated with Eqs. (11.2.32a,b). Here we consider such an element.

Consider the following equilibrium equations associated with Eqs. (11.2.32a,b)

$$\frac{d}{dx}\left[KGA\left(\frac{dw_0}{dx} + \phi_x\right)\right] = 0 \qquad (11.2.43)$$

$$\frac{d}{dx}\left(EI\frac{d\phi}{dx}\right) - KGA\left(\frac{dw_0}{dx} + \phi\right) = 0 \qquad (11.2.44)$$

Equation (11.2.43) requires that the shear strain be constant over each element. Assuming that EI and KGA are constant and integrating Eqs. (11.2.43) and (11.2.44) with respect to x, we obtain (see Reddy [12])

$$w_0(x) = -\left(c_1 \frac{x^3}{6} + c_2 \frac{x^2}{2} + c_3 x + c_4\right) + h^2 \Omega c_1 x \qquad (11.2.45a)$$

$$\phi(x) = c_1 \frac{x^2}{2} + c_2 x + c_3 \qquad (11.2.45b)$$

where c_1 through c_4 are the constants of integration and $\Omega = (EI/KGAh^2)$.

Equations (11.2.45a,b) indicate that the same four constants define the solution for w_0 and ϕ_x. This in turn suggests that the finite element interpolation used for w_0 and ϕ_x are interdependent. If we use independent approximations of w_0 and ϕ_x, they may lead to locked elements for low order polynomials and to redundant degrees of freedom for high order polynomials. The optimal choice would be one that is suggested by Eqs. (11.2.45a,b).

To develop the interdependent interpolation element (IIE), we first define the nodal degrees of freedom and then express them in terms of the constants c_1, c_2, c_3, and c_4 using Eqs. (11.2.45a,b). For the sake of convenience, we use the local coordinate $\bar{x}$ such that $0 < \bar{x} < h$ to formulate the element. Let

$$w_1^e \equiv w_0(0), \quad w_2^e \equiv \phi(0), \quad w_3^e \equiv w_0(h), \quad w_4^e \equiv \phi(h) \qquad (11.2.46)$$

Using (11.2.45a,b) we can write

$$\begin{Bmatrix} w_1^e \\ w_2^e \\ w_3^e \\ w_4^e \end{Bmatrix} = \frac{1}{6} \begin{bmatrix} 0 & 0 & 0 & -6 \\ 0 & 0 & 6 & 0 \\ -h^3\gamma & -3h^2 & -6h & -6 \\ 3h^2 & 6h & 6 & 0 \end{bmatrix} \begin{Bmatrix} c_1 \\ c_2 \\ c_3 \\ c_4 \end{Bmatrix} \qquad (11.2.47)$$

and the inverse relations are given by

$$\begin{Bmatrix} c_1 \\ c_2 \\ c_3 \\ c_4 \end{Bmatrix} = \frac{1}{\mu h^3} \begin{bmatrix} -12 & 6h & 12 & 6h \\ 6h & -4h^2\lambda & -6h & -2h^2\gamma \\ 0 & h^3\mu & 0 & 0 \\ -h^3\mu & 0 & 0 & 0 \end{bmatrix} \begin{Bmatrix} w_1^e \\ w_2^e \\ w_3^e \\ w_4^e \end{Bmatrix}$$

$$\equiv [S^e]\{w^e\} \qquad (11.2.48)$$

where

$$\gamma = 1 - 6\Omega, \quad \lambda = 1 + 3\Omega, \quad \mu = 1 + 12\Omega, \quad \Omega = \frac{EI}{KGAh^2} \qquad (11.2.49)$$

Next we obtain the finite element interpolation functions for this element using Eqs. (11.2.45a,b) and (11.2.48). We have

$$\begin{Bmatrix} w_0 \\ \phi_x \end{Bmatrix} = \frac{1}{h^3\mu} \begin{bmatrix} -\frac{x^3}{6} + \Omega x h^2 & -\frac{x^2}{2} & -x & -1 \\ \frac{x^2}{2} & x & 1 & 0 \end{bmatrix} [S^e]\{w^e\}$$

$$= \begin{bmatrix} \varphi_1^{(1)} & \varphi_2^{(1)} & \varphi_3^{(1)} & \varphi_4^{(1)} \\ \varphi_1^{(2)} & \varphi_2^{(2)} & \varphi_3^{(2)} & \varphi_4^{(2)} \end{bmatrix} \begin{Bmatrix} w_1^e \\ w_2^e \\ w_3^e \\ w_4^e \end{Bmatrix}$$

$$= \begin{Bmatrix} \sum_{j=1}^{4} w_j^e \varphi_j^{(1)}(x) \\ \sum_{j=1}^{4} w_j^e \varphi_j^{(2)}(x) \end{Bmatrix} \quad (11.2.50)$$

where

$$\varphi_1^{(1)} = \frac{1}{\mu}\left[\mu - 12\Omega\left(\frac{x}{h}\right) - 3\left(\frac{x}{h}\right)^2 + 2\left(\frac{x}{h}\right)^3\right]$$

$$\varphi_2^{(1)} = -\frac{h}{\mu}\left(\frac{x}{h}\right)\left[\left(1 - \frac{x}{h}\right)^2 + 6\Omega\left(1 - \frac{x}{h}\right)\right]$$

$$\varphi_3^{(1)} = \frac{1}{\mu}\left[3\left(\frac{x}{h}\right)^2 - 2\left(\frac{x}{h}\right)^3 + 12\Omega\frac{x}{h}\right]$$

$$\varphi_4^{(1)} = \frac{h}{\mu}\left(\frac{x}{h}\right)\left[\left(\frac{x}{h}\right) - \left(\frac{x}{h}\right)^2 + 6\Omega\left(1 - \frac{x}{h}\right)\right] \quad (11.2.51)$$

$$\varphi_1^{(2)} = \frac{6}{h\mu}\left(\frac{x}{h}\right)\left(1 - \frac{x}{h}\right)$$

$$\varphi_2^{(2)} = \frac{1}{\mu}\left[\mu - 4\left(\frac{x}{h}\right) + 3\left(\frac{x}{h}\right)^2 + 12\Omega\left(1 - \frac{x}{h}\right)\right]$$

$$\varphi_3^{(2)} = -\frac{6}{h\mu}\left(\frac{x}{h}\right)\left(1 - \frac{x}{h}\right)$$

$$\varphi_4^{(2)} = \frac{1}{\mu}\left[-2\left(\frac{x}{h}\right) + 3\left(\frac{x}{h}\right)^2 + 12\Omega\left(\frac{x}{h}\right)\right] \quad (11.2.52)$$

Note that the interpolation functions depend on the material properties through Ω. For dynamic problems (i.e., vibration and transient analysis), one may set $\Omega = 0$ in Eqs. (11.2.51) and (11.2.52) and use the resulting functions to develop the finite element model.

Now the element stiffness matrix of the IIE element can be derived directly using the approximation (11.2.50) in Eq. (11.2.34). For static problems, we obtain

$$[K]\{u\} = \{q\} + \{Q\} \quad (11.2.53)$$

where

$$K_{ij} = \int_{x_a}^{x_b} \left[EI \frac{d\varphi_i^{(2)}}{dx} \frac{d\varphi_j^{(2)}}{dx} + KGA \left(\varphi_i^{(2)} + \frac{d\varphi_i^{(1)}}{dx} \right) \left(\varphi_j^{(2)} + \frac{d\varphi_j^{(1)}}{dx} \right) \right] dx$$

$$q_i = \int_{x_a}^{x_b} q(x)\varphi_i^{(1)}(x)\, dx \qquad (11.2.54)$$

For constant values of EI and KGA and uniformly distributed transverse load $q = q_0$, we obtain the following element equations:

$$\frac{2EI}{\mu h^3} \begin{bmatrix} 6 & -3h & -6 & -3h \\ -3h & 2h^2\lambda & 3h & h^2\gamma \\ -6 & 3h & 6 & 3h \\ -3h & h^2\gamma & 3h & 2h^2\lambda \end{bmatrix} \begin{Bmatrix} w_1^e \\ w_2^e \\ w_3^e \\ w_4^e \end{Bmatrix} = \frac{q_0 h}{12} \begin{Bmatrix} 6 \\ -h \\ 6 \\ h \end{Bmatrix} + \begin{Bmatrix} Q_1^e \\ Q_2^e \\ Q_3^e \\ Q_4^e \end{Bmatrix} \qquad (11.2.55)$$

Note that although the interpolation functions depend on Ω, the load vector does not depend on it (for uniform loading). Although not shown here, the consistent interpolation element based on independent cubic interpolation of w_0 and quadratic interpolation of ϕ_x, and condensing out the degrees of freedom associated with the interior nodes should yield the result in Eq. (11.2.55).

The element stiffness matrix in Eq. (11.2.55) degenerates to that for the Euler–Bernoulli beam element (EBE) when $\Omega = 0$, or the shear rigidity KGA is infinite. The IIE gives the exact values of w_0 and ϕ_x at the nodes *independent of the load distribution*. Since IIE contains EBE, accounts for shear deformation, and gives exact nodal values, the element can be used in lieu of EBE in computer programs that are designed for linear frame structural analysis, since both elements have the same number and type of degrees of freedom and IIE includes transverse shear deformation.

Table 11.2.1 shows a comparison of the finite element solutions obtained with one, two, and four elements in half beam with the exact beam solutions of a simply supported beam under uniformly distributed transverse load. The following data is used in obtaining the numerical results:

$$E = 10^6, \ \nu = 0.25, \ K = \frac{5}{6}, \ q_0 = 1 \qquad (11.2.56)$$

Two different length-to-height ratios, $L/h = 10$, and 100 are considered. Table 11.2.2 contains the results for sinusoidally distributed load. Clearly, more than two elements of CIE and RIE are required to obtain

Table 11.2.1. Comparison of the finite element solutions with the exact maximum deflection and rotation of a simply supported isotropic beam under uniformly distributed transverse load (N = number of elements used in half beam).

Element	$N=1$	$N=2$	$N=4$	$N=1$	$N=2$	$N=4$
$L/h=10$		$w_0 \times 10^2$			$-\phi_x \times 10^3$	
RIE	0.09750	0.14438	0.15609	0.37500	0.46875	0.49219
CIE	0.12875	0.15219	0.15805	0.50000	0.50000	0.50000
IIE[†]	0.16000	0.16000	0.16000	0.50000	0.50000	0.50000
EBE[†]	0.15265	0.15265	0.15265	0.50000	0.50000	0.50000
$L/h=100$		$w_0 \times 10^{-2}$			$-\phi_x$	
RIE	0.09379	0.14066	0.15238	0.37500	0.46875	0.49219
CIE	0.12504	0.14847	0.15433	0.50000	0.50000	0.50000
IIE[†]	0.15629	0.15629	0.15629	0.50000	0.50000	0.50000
EBE[†]	0.15265	0.15265	0.15265	0.50000	0.50000	0.50000

[†] Exact values compared to the respective beam theories.

Table 11.2.2. Comparison of the finite element solutions with the exact maximum deflection and rotation of a simply supported isotropic beam under sinusoidally distributed transverse load (N = number of elements used in half beam).

Element	$N=1$	$N=2$	$N=4$	$N=1$	$N=2$	$N=4$
$L/h=100$		$w_0 \times 10^{-2}$			$-\phi_x$	
RIE	0.07639	0.11079	0.12007	0.30543	0.36702	0.38204
CIE	0.09679	0.11682	0.12163	0.38705	0.38702	0.38702
IIE[†]	0.12322	0.12322	0.12322	0.38702	0.38702	0.38702
EBE[†]	0.12319	0.12319	0.12319	0.38702	0.38702	0.38702

[†] Exact values compared to the respective beam theories.

acceptable solutions. On the other hand, IIE yields exact nodal values with one element. Elements that give exact nodal values are called *superconvergent elements*.

11.3 Finite Element Models of CPT

11.3.1 Introduction

In this section, the displacement finite element model of Eq. (3.8.9) governing the motion of plates according to the classical plate theory is developed. While the basic idea of the finite element method carries over from one-dimensional beam problems to two-dimensional plate problems, finite element models of plates are considerably complicated by the fact that two-dimensional problems are described by partial differential equations over geometrically complex regions. The boundary Γ of a two-dimensional domain Ω is, in general, a curve. Therefore, the two-dimensional finite elements must be simple geometric shapes that can be used to approximate a given two-dimensional domain as well as the solution over it. Consequently, the finite element solution will have errors due to the approximation of the solution as well as the domain. In two dimensions, there is more than one geometric shape that can be used as a finite element (see Figure 11.3.1). As we shall see shortly, the interpolation functions depend not only on the number of nodes in the element, but also on the shape of the element. The shape of the element must be such that its geometry is uniquely defined by a set of points, which serves as the element nodes in the development of the interpolation functions. As will be discussed later in this section, a triangle is the simplest geometric shape, followed by a rectangle. As in the case of beams, the finite element model of a plate theory is developed using its virtual work statement and an interpolation of the displacement field over an element. In the following sections, several standard plate bending elements are discussed.

11.3.2 General Formulation

The virtual work statement of the classical plate theory over typical finite element Ω^e is given by [from Eq. (3.4.10)]

$$0 = \int_0^T \int_{\Omega^e} \left[-\frac{\partial^2 \delta w_0}{\partial x^2} M_{xx} - 2\frac{\partial^2 \delta w_0}{\partial x \partial y} M_{xy} - \frac{\partial^2 \delta w_0}{\partial y^2} M_{yy} + k\delta w_0 w_0 \right.$$
$$+ \frac{\partial \delta w_0}{\partial x}\left(\hat{N}_{xx}\frac{\partial w_0}{\partial x} + \hat{N}_{xy}\frac{\partial w_0}{\partial y}\right) + \frac{\partial \delta w_0}{\partial y}\left(\hat{N}_{xy}\frac{\partial w_0}{\partial x} + \hat{N}_{yy}\frac{\partial w_0}{\partial y}\right)$$
$$\left. + I_0 \delta w_0 \ddot{w}_0 + I_2\left(\frac{\partial \delta w_0}{\partial x}\frac{\partial \ddot{w}_0}{\partial x} + \frac{\partial \delta w_0}{\partial y}\frac{\partial \ddot{w}_0}{\partial y}\right) - \delta w_0 q \right] dxdydt$$

FINITE ELEMENT MODELS OF BEAMS AND PLATES

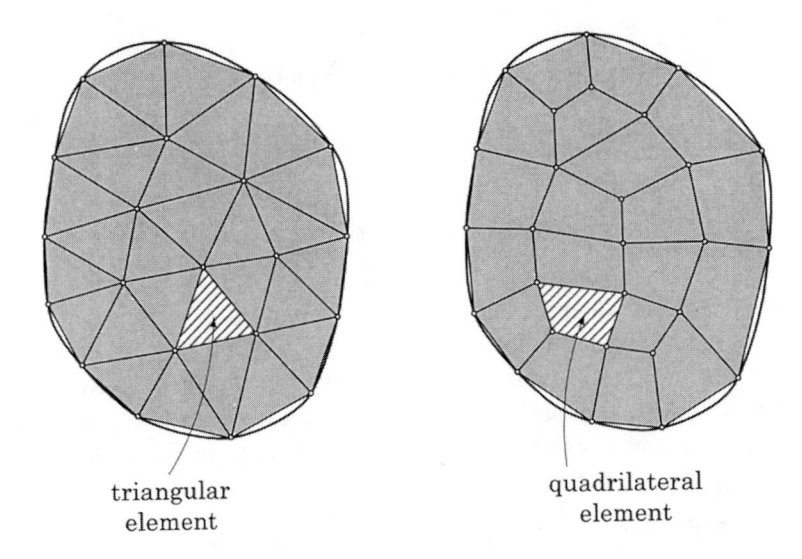

Figure 11.3.1. Finite element discretization of a two-dimensional domain using triangular and quadrilateral elements.

$$+ \int_0^T \oint_{\Gamma^e} \left[-\delta w_0 \left(\frac{\partial M_{xx}}{\partial x} + \frac{\partial M_{xy}}{\partial y} + \hat{N}_{xx} \frac{\partial w_0}{\partial x} + \hat{N}_{xy} \frac{\partial w_0}{\partial y} \right) n_x \right.$$
$$- \delta w_0 \left(\frac{\partial M_{xy}}{\partial x} + \frac{\partial M_{yy}}{\partial y} + \hat{N}_{xy} \frac{\partial w_0}{\partial x} + \hat{N}_{yy} \frac{\partial w_0}{\partial y} \right) n_y$$
$$+ I_2 \delta w_0 \left(\frac{\partial \ddot{w}_0}{\partial x} n_x + \frac{\partial \ddot{w}_0}{\partial y} n_y \right) + \frac{\partial \delta w_0}{\partial x} (M_{xx} n_x + M_{xy} n_y)$$
$$\left. + \frac{\partial \delta w_0}{\partial y} (M_{xy} n_x + M_{yy} n_y) \right] ds \, dt \qquad (11.3.1)$$

where (n_x, n_y) denote the direction cosines of the unit normal on the element boundary Γ^e and (M_{xx}, M_{xy}, M_{yy}) are defined by Eq. (6.1.9a–c). Integration by parts of the inertia terms in the last equation is necessitated by the symmetry considerations of the mass matrix in the finite element model.

We note from the boundary terms in Eq. (11.3.1) that w_0, $\partial w_0/\partial x$, and $\partial w_0/\partial y$ are the primary variables (or generalized displacements),

and

$$T_x \equiv M_{xx}n_x + M_{xy}n_y, \quad T_y \equiv M_{xy}n_x + M_{yy}n_y \tag{11.3.2}$$

$$Q_n \equiv \left(\frac{\partial M_{xx}}{\partial x} + \frac{\partial M_{xy}}{\partial y} + \hat{N}_{xx}\frac{\partial w_0}{\partial x} + \hat{N}_{xy}\frac{\partial w_0}{\partial y} - I_2\frac{\partial^3 w_0}{\partial x \partial t^2}\right)n_x$$

$$+ \left(\frac{\partial M_{xy}}{\partial x} + \frac{\partial M_{yy}}{\partial y} + \hat{N}_{xy}\frac{\partial w_0}{\partial x} + \hat{N}_{yy}\frac{\partial w_0}{\partial y} - I_2\frac{\partial^3 w_0}{\partial y \partial t^2}\right)n_y \tag{11.3.3}$$

are the secondary degrees of freedom (or generalized forces). Thus, finite elements based on the classical plate theory require continuity of the transverse deflection and its normal derivative across element boundaries. Also, to satisfy the constant displacement (rigid body mode) and constant strain requirements, the polynomial expansion for w_0 should be a complete quadratic.

Assume finite element approximation of the form

$$w_0(x,y,t) = \sum_{j=1}^{n} \Delta_j^e(t)\varphi_j^e(x,y) \tag{11.3.4}$$

where Δ_j^e are the values of w_0 and its derivatives at the nodes, and φ_j^e are the interpolation functions, the specific form of which will depend on the geometry of the element and the nodal degrees of freedom interpolated. Substituting approximations (11.3.4) for w_0 and φ_i^e for the virtual displacement δw_0 into Eq. (11.3.1), we obtain the ith equation

$$\sum_{j=1}^{n}\left[\left(K_{ij}^e + G_{ij}^e\right)\Delta_j^e + M_{ij}^e\ddot{\Delta}_j^e\right] = F_i^e + F_i^{eT} \tag{11.3.5}$$

where $i, j = 1, 2, \cdots, n$. The coefficients of the stiffness matrix $K_{ij}^e = K_{ji}^e$, mass matrix $M_{ij}^e = M_{ji}^e$, geometric stiffness (or stability) matrix $G_{ij}^e = G_{ji}^e$, and force vectors F_i^e and F_i^{eT} are defined as follows:

$$K_{ij}^e = \int_{\Omega^e}\left[D_{11}T_{ij}^{xxxx} + D_{12}\left(T_{ij}^{xxyy} + T_{ij}^{yyxx}\right) + 4D_{66}T_{ij}^{xyxy}\right.$$
$$\left. + D_{22}T_{ij}^{yyyy} + kS_{ij}^{00}\right]dxdy$$

$$G_{ij}^e = \int_{\Omega^e}\left[\hat{N}_{xx}S_{ij}^{xx} + \hat{N}_{xy}\left(S_{ij}^{xy} + S_{ij}^{yx}\right) + \hat{N}_{yy}S_{ij}^{yy}\right]dxdy$$

$$M_{ij}^e = \int_{\Omega^e}\left[I_0 S_{ij}^{00} + I_2\left(S_{ij}^{xx} + S_{ij}^{yy}\right)\right]dxdy$$

$$F_i^e = \int_{\Omega^e} q\varphi_i^e \, dxdy + \oint_{\Gamma^e} \left(Q_n\varphi_i^e - T_x\frac{\partial \varphi_i^e}{\partial x} - T_y\frac{\partial \varphi_i^e}{\partial y} \right) ds$$

$$F_i^{eT} = \int_{\Omega^e} \left(\frac{\partial^2 \varphi_i^e}{\partial x^2} M_{xx}^T + 2\frac{\partial^2 \varphi_i^e}{\partial x \partial y} M_{xy}^T + \frac{\partial^2 \varphi_i^e}{\partial y^2} M_{yy}^T \right) dxdy \quad (11.3.6a)$$

where N_{xx}^T, M_{xx}^T, etc. are the thermal force and moment resultants, and

$$T_{ij}^{\xi\eta\zeta\mu} = \frac{\partial^2 \varphi_i^e}{\partial \xi \partial \eta} \frac{\partial^2 \varphi_j^e}{\partial \zeta \partial \mu} \, dxdy, \quad S_{ij}^{\xi\eta} = \frac{\partial \varphi_i^e}{\partial \xi} \frac{\partial \varphi_j^e}{\partial \eta} \, dxdy \quad (11.3.6b)$$

and ξ, η, ζ, and μ take on the symbols x and y. In matrix notation, Eq. (11.3.5) can be expressed as

$$([K^e] + [G^e])\{\Delta^e\} + [M^e]\{\ddot{\Delta}^e\} = \{F^e\} + \{F^{eT}\} \quad (11.3.7)$$

This completes the finite element model development of the classical laminate theory. The finite element model in Eq. (11.3.5) or (11.3.7) is called a *displacement finite element model* because it is based on equations of motion expressed in terms of the displacements, and the generalized displacements are the primary nodal degrees of freedom.

11.3.3 Plate Bending Elements

There exists a large body of literature on triangular and rectangular plate bending finite elements of isotropic or orthotropic plates based on the classical plate theory (e.g., see [13–20]). There are two kinds of plate bending elements of the classical plate theory. A *conforming element* is one in which the inter-element continuity of w_0, $\theta_x \equiv \partial w_0/\partial x$, and $\theta_y \equiv \partial w_0/\partial y$ (or $\partial w_0/\partial n$) are satisfied, and a *nonconforming element* is one in which the continuity of the normal slope, $\partial w_0/\partial n$, is not satisfied.

An effective nonconforming triangular element (the BCIZ triangle) was developed by Bazeley, Cheung, Irons, and Zienkiewicz [14], and it consists of three degrees of freedom $(w_0, \theta_x, \theta_y)$ at the three vertex nodes (see Figure 11.3.2). Thus, a nine-term polynomial approximation of w_0 is required

$$w_0(x,y,t) = \sum_{k=1}^{9} \Delta_k^e(t)\, \varphi_k^e(x,y) \quad (11.3.8)$$

where Δ_k^e the values of w_0, $\partial w_0/\partial x$, and $\partial w_0/\partial y$ at the three nodes:

$$\{\Delta^e\}^T = \{w_1, \theta_{x1}, \theta_{y1}, w_2, \theta_{x2}, \theta_{y2}, w_3, \theta_{x3}, \theta_{y3}\}^e \quad (11.3.9)$$

The interpolation functions φ_k^e for this element can be expressed in terms of the area coordinates (see Reddy [1]) as

498   THEORY AND ANALYSIS OF ELASTIC PLATES

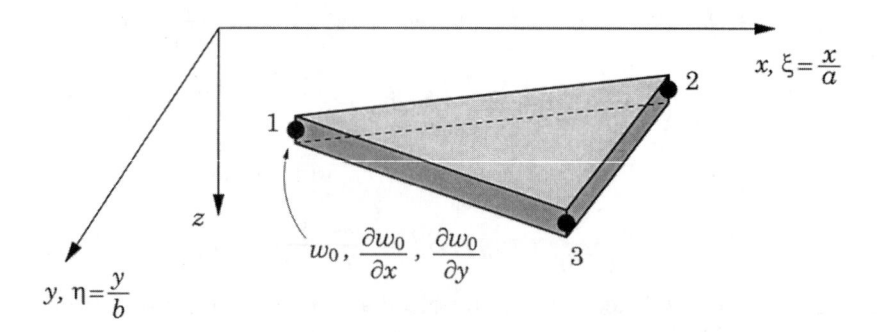

Figure 11.3.2. A nonconforming triangular element with three degrees of freedom ($w_0, \partial w_0/\partial x, \partial w_0/\partial y$) per node.

$$\begin{Bmatrix} \varphi_1^e \\ \varphi_2^e \\ \varphi_3^e \\ \varphi_4^e \\ \varphi_5^e \\ \varphi_6^e \\ \varphi_7^e \\ \varphi_8^e \\ \varphi_9^e \end{Bmatrix} = \begin{Bmatrix} L_1 + L_1^2 L_2 + L_1^2 L_3 - L_1 L_2^2 - L_1 L_3^2 \\ x_{31}(L_3 L_1^2 - f) - x_{12}(L_1^2 L_2 + f) \\ y_{31}(L_3 L_1^2 + f) - y_{12}(L_1^2 L_2 + f) \\ L_2 + L_2^2 L_3 + L_2^2 L_1 - L_2 L_3^2 - L_2 L_1^2 \\ x_{12}(L_1 L_2^2 - f) - x_{23}(L_2^2 L_3 + f) \\ y_{12}(L_1 L_2^2 + f) - y_{23}(L_2^2 L_3 + f) \\ L_3 + L_3^2 L_1 + L_3^2 L_2 - L_3 L_1^2 - L_3 L_2^2 \\ x_{23}(L_2 L_3^2 - f) - x_{31}(L_3^2 L_1 + f) \\ y_{23}(L_2 L_3^2 + f) - y_{31}(L_3^2 L_1 + f) \end{Bmatrix} \quad (11.3.10a)$$

where $f = 0.5 L_1 L_2 L_3$, $x_{ij} = x_i - x_j$, $y_{ij} = y_i - y_j$, (x_i, y_i) being the global coordinates of the ith node, and L_i being the area coordinates defined within an element (see page 510)

$$L_i = \frac{A_i}{A}, \quad A = \sum_{i=1}^{3} A_i \quad (11.3.10b)$$

A conforming triangular element due to Clough and Tocher [15] is an assemblage of three triangles as shown in Figure 11.3.3. The normal slope continuity is enforced at the mid-side nodes between the sub-triangles. In each sub-triangle, the transverse deflection is represented by the polynomial ($i = 1, 2, 3$)

$$w_0^i(x,y) = a_i + b_i \xi + c_i \eta + d_i \xi \eta + e_i \xi^2 + f_i \eta^2 + g_i \xi^3 + h_i \xi^2 \eta + k_i \xi \eta^2 + \ell_i \eta^3 \quad (11.3.11)$$

FINITE ELEMENT MODELS OF BEAMS AND PLATES

where (ξ, η) are the local coordinates, as shown in the Figure 11.3.3. The thirty coefficients are reduced to nine, three $(w_0, \partial w_0/\partial x, \partial w_0/\partial y)$ at each vertex of the triangle, by equating the variables from the vertices of each sub-triangle at the common points and normal slope between the mid-side points of sub-triangles.

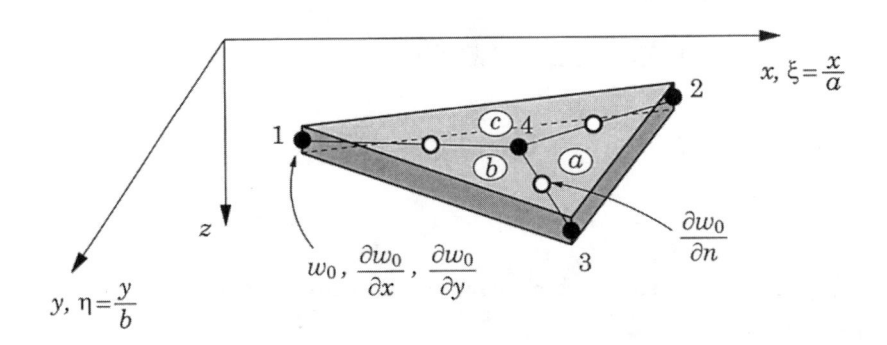

Figure 11.3.3. A conforming triangular element.

A nonconforming rectangular element has w_0, θ_x, and θ_y as the nodal variables (see Figure 11.3.4). The element was developed by Melosh [16] and Zienkiewicz and Cheung [17]. The normal slope variation is cubic along an edge, whereas there are only two values of $\partial w_0/\partial n$ available on the edge. Therefore, the cubic polynomial for the normal derivative of w_0 is not the same on the edge common to two elements. The interpolation functions for this element can be expressed compactly as

$$\varphi_i^e = g_{i1} \ (i = 1, 4, 7, 10); \quad \varphi_i^e = g_{i2} \ (i = 2, 5, 8, 11)$$
$$\varphi_i^e = g_{i3} \ (i = 3, 6, 9, 12) \tag{11.3.12a}$$
$$g_{i1} = \frac{1}{8}(1 + \xi_0)(1 + \eta_0)(2 + \xi_0 + \eta_0 - \xi^2 - \eta^2)$$
$$g_{i2} = \frac{1}{8}\xi_i(\xi_0 - 1)(1 + \eta_0)(1 + \xi_0)^2$$
$$g_{i3} = \frac{1}{8}\eta_i(\eta_0 - 1)(1 + \xi_0)(1 + \eta_0)^2$$
$$\xi = (x - x_c)/2, \ \eta = (y - y_c)/b, \ \xi_0 = \xi\xi_i, \ \eta_0 = \eta\eta_i \tag{11.3.12b}$$

where $2a$ and $2b$ are the sides of the rectangle, and (x_c, y_c) are the global coordinates of the center of the rectangle.

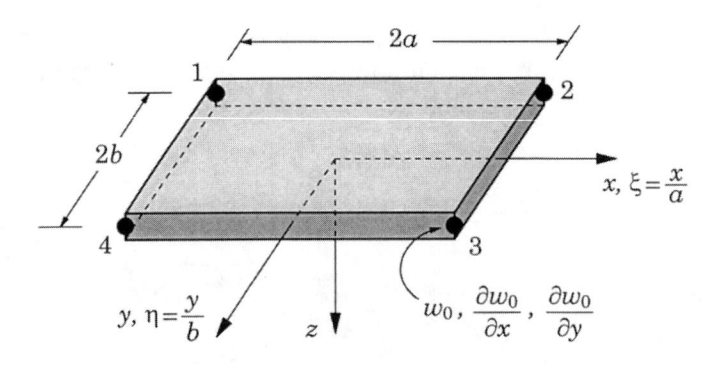

Figure 11.3.4. A nonconforming rectangular element.

A conforming rectangular element with w_0, $\partial w_0/\partial x$, $\partial w_0/\partial y$, and $\partial^2 w_0/\partial x \partial y$ as the nodal variables was developed by Bogner, Fox, and Schmidt [18]. The interpolation functions for this element (see Figure 11.3.5) are

$$\varphi_i^e = g_{i1} \ (i = 1, 5, 9, 13); \quad \varphi_i^e = g_{i2} \ (i = 2, 6, 10, 14)$$
$$\varphi_i^e = g_{i3} \ (i = 3, 7, 11, 15); \quad \varphi_i^e = g_{i4} \ (i = 4, 8, 12, 16) \quad (11.3.13a)$$

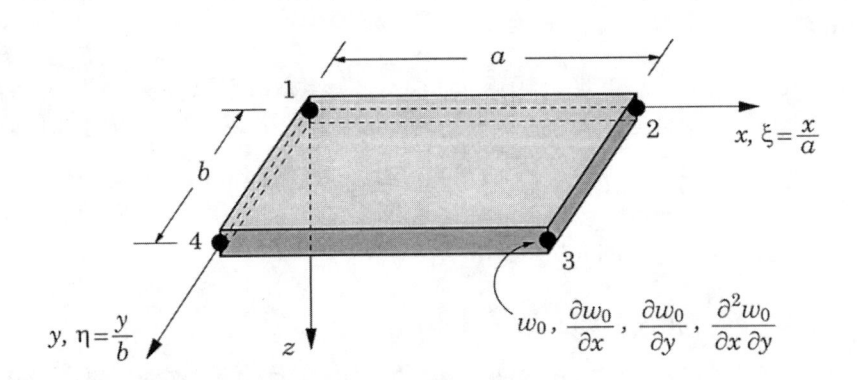

Figure 11.3.5. A conforming rectangular element.

$$g_{i1} = \frac{1}{16}(\xi + \xi_i)^2(\xi_0 - 2)(\eta + \eta_i)^2(\eta_0 - 2)$$
$$g_{i2} = \frac{1}{16}\xi_i(\xi + \xi_i)^2(1 - \xi_0)(\eta + \eta_i)^2(\eta_0 - 2)$$
$$g_{i3} = \frac{1}{16}\eta_i(\xi + \xi_i)^2(\xi_0 - 2)(\eta + \eta_i)^2(1 - \eta_0)$$
$$g_{i4} = \frac{1}{16}\xi_i\eta_i(\xi + \xi_i)^2(1 - \xi_0)(\eta + \eta_i)^2(1 - \eta_0) \quad (11.3.13b)$$

The combined conforming element has four degrees of freedom per node, whereas the nonconforming element has three degrees of freedom per node. For the conforming rectangular element the total number of bending nodal degrees of freedom per element is 16, and for the nonconforming element, the total number is 12.

11.3.4 Fully Discretized Finite Element Models

Static Bending

In the case of static bending under applied mechanical and thermal loads, Eq. (11.3.7) reduces to

$$([K^e] + [G^e])\{\Delta^e\} = \{F^e\} + \{F^{eT}\} \quad (11.3.14)$$

where it is understood that all time-derivative terms are zero.

Once the nodal values of generalized displacements (w_0, $\partial w_0/\partial x$, $\partial w_0/\partial y$) have been obtained by solving the assembled equations of a problem, the strains are evaluated in each element by differentiating the displacements. The strains and stresses are the most accurate if they are computed at the center of the element [1,3].

Buckling

In the case of buckling under applied in-plane compressive and shear edge loads, Eq. (11.3.7) reduces to

$$\left([K^e] - \lambda[\hat{G}^e]\right)\{\Delta^e\} = \{\bar{F}\} \quad (11.3.15)$$

where

$$\lambda = -\hat{N}_{xx}/\hat{N}^0_{xx} = -\hat{N}_{yy}/\hat{N}^0_{yy} = -\hat{N}_{xy}/\hat{N}^0_{xy} \quad (11.3.16a)$$

$$\hat{G}^e_{ij} = \int_{\Omega^e} \left[\hat{N}^0_{xx}S^{xx}_{ij} + \hat{N}^0_{xy}\left(S^{xy}_{ij} + S^{yx}_{ij}\right) + \hat{N}^0_{yy}S^{yy}_{ij}\right] dxdy$$

$$\bar{F}_k = \oint_{\Gamma^e} \left(Q_n\varphi^e_k - T_x\frac{\partial \varphi^e_k}{\partial x} - T_y\frac{\partial \varphi^e_k}{\partial y}\right) ds \quad (11.3.16b)$$

Natural Vibration

In the case of natural vibration, the response of the plate is assumed to be periodic. Equation (11.3.7) becomes

$$\left([K^e] - \omega^2[M^e]\right)\{\Delta^e\} = \{\bar{F}^e\} \tag{11.3.17}$$

where ω denotes the frequency of natural vibration.

Transient Response

Since Eq. (11.3.7) is identical in form to Eq. (11.2.10), the time approximations discussed for Eq. (11.2.10) in Eqs. (11.2.20)–(11.2.25) are also valid for Eq. (11.3.17). We have

$$[\hat{K}^e]_{s+1}\{\Delta^e\}_{s+1} = \{\hat{F}^e\} \tag{11.3.18a}$$

where

$$\begin{aligned}[\hat{K}^e]_{s+1} &= ([K^e]_{s+1} + [G^e]_{s+1}) + a_3[M^e]_{s+1} \\ \{\hat{F}^e\} &= \{F^e\}_{s+1} + [M^e]_{s+1}\{\tilde{\ddot{\Delta}}^e\} \end{aligned} \tag{11.3.18b}$$

$$\{\tilde{\ddot{\Delta}}^e\} = \left(a_3\{\Delta^e\}_s + a_4\{\dot{\Delta}^e\}_s + a_5\{\ddot{\Delta}^e\}_s\right) \tag{11.3.18c}$$

and parameters a_1, a_2, and so on are defined in Eq. (11.2.20b) and $\delta t_s = t_{s+1} - t_s$ is the time step; the notation δt in place of Δt for time step should be obvious in the context.

11.3.5 Numerical Results

Here we use the conforming (C) and nonconforming (NC) rectangular finite elements to analyze plates for bending and natural vibration. A note should be made of the fact that the finite element model developed herein is not restricted to any particular geometry, boundary conditions, or loading. Solution symmetries available in a problem should be taken advantage of to identify the computational domain because they reduce computational effort. For example, a 2×2 mesh in a quadrant of the plate is the same as 4×4 mesh in the total plate, and the results obtained with the two meshes would be identical, within the round-off errors of the computation, if the solution exhibits biaxial symmetry. The notation "$m \times n$ mesh" denotes m subdivisions along the x–axis and n subdivisions along the y–axis with the same type of elements. A solution is symmetric about a line only if (a) the geometry, including boundary conditions, (b) the material properties, and (c) the loading are symmetric about the line. The boundary conditions along a line of symmetry should be correctly identified and imposed in the finite element model. When one is not sure of the solution symmetry, it is advised that the whole plate be modeled.

Bending

For a rectangular orthotropic plate with simply supported edges, a quadrant of the plate may be used as the computational domain. The boundary conditions along the symmetry lines are shown in Figure 11.3.6. In the case of conforming element, it is necessary that the cross-derivative $\partial^2 w_0/\partial x \partial y$ be also set to zero at the center of the plate when a quarter-plate model is used. Otherwise, the results will be less accurate. The stresses in the finite element analysis were computed at the center of the elements. The foundation modulus k is set to zero.

$$\text{CPT:} \quad \frac{\partial w_0}{\partial x} = 0 \text{ at } x=0; \quad \frac{\partial w_0}{\partial y} = 0 \text{ at } y=0$$

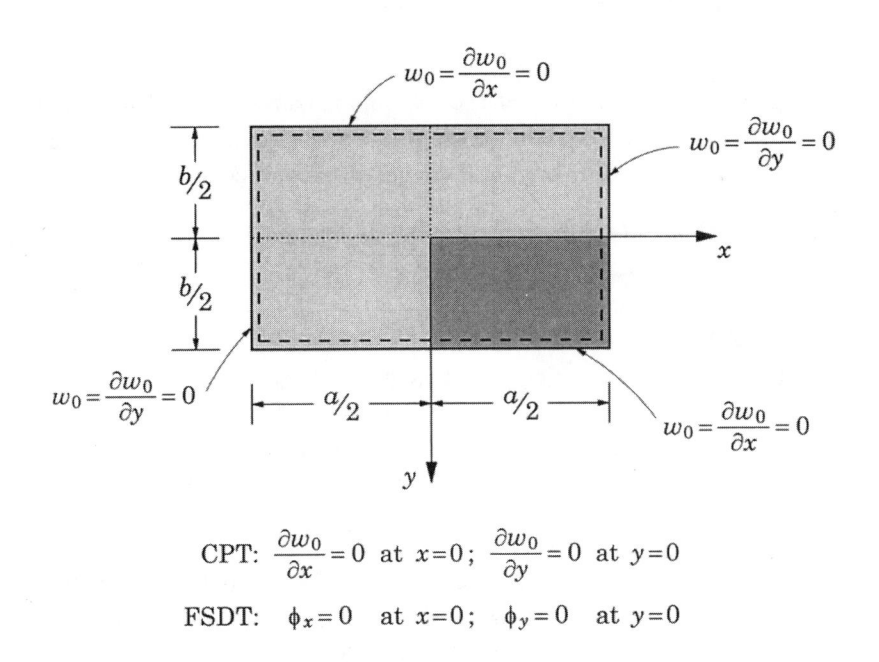

$$\text{FSDT:} \quad \phi_x = 0 \text{ at } x=0; \quad \phi_y = 0 \text{ at } y=0$$

Figure 11.3.6. Symmetry boundary conditions for rectangular plates.

Table 11.3.1 shows a comparison of finite element solutions with the analytical solutions of simply supported isotropic and orthotropic square plates under uniformly distributed transverse load. The analytical solutions were evaluated using $m, n = 1, 3, \cdots, 19$. The exact maximum deflection occurs at $x = y = 0$, maximum stresses σ_{xx} and σ_{yy} occur at $(0, 0, h/2)$, and the maximum shear stress σ_{xy} occurs at $(a/2, b/2, -h/2)$. The geometric boundary conditions of the computational domain (see the shaded quadrant in Figure 11.3.6) are

$$\frac{\partial w_0}{\partial x} = 0 \text{ at } x = 0; \quad \frac{\partial w_0}{\partial y} = 0 \text{ at } y = 0 \qquad (11.3.19)$$

$$w_0 = \frac{\partial w_0}{\partial x} = 0 \text{ at } x = \frac{a}{2}; \quad w_0 = \frac{\partial w_0}{\partial y} = 0 \text{ at } y = \frac{b}{2} \qquad (11.3.20)$$

In addition, $\partial^2 w_0/\partial x \partial y = 0$ was used at $x = y = 0$ for the conforming element. Therefore, the locations of the maximum normal stresses are $(a/8, b/8)$, $(a/16, b/16)$, and $(a/32, b/32)$ for uniform meshes 2×2, 4×4, and 8×8, respectively, while those of σ_{xy} are $(3a/8, 3b/8)$, $(7a/16, 7b/16)$, and $(15a/32, 15b/32)$ for the three meshes.

Table 11.3.1. A comparison of the maximum transverse deflections and stresses[†] of simply supported (SSSS) square $(a = b)$ plates under a uniformly distributed transverse load.

Variable	Nonconforming			Conforming			Analytical
	2×2	4×4	8×8	2×2	4×4	8×8	solution
Isotropic Plate ($\nu = 0.25$)							
$\bar{w}$	4.8571	4.6425	4.5883	4.7619	4.5989	4.5739	4.5701
$\bar{\sigma}_{xx}$	0.2405	0.2673	0.2740	0.2239	0.2631	0.2731	0.2762
$\bar{\sigma}_{yy}$	0.2405	0.2673	0.2740	0.2239	0.2631	0.2731	0.2762
$\bar{\sigma}_{xy}$	0.1713	0.1964	0.2050	0.1688	0.1936	0.2040	0.2085
Orthotropic Plate ($E_1/E_2 = 25$, $G_{12} = G_{13} = 0.5 E_2$, $\nu_{12} = 0.25$)							
$\bar{w}$	0.7082	0.6635	0.6531	0.7532	0.6651	0.6517	0.6497
$\bar{\sigma}_{xx}$	0.7148	0.7709	0.7828	0.5772	0.7388	0.7759	0.7866
$\bar{\sigma}_{yy}$	0.0296	0.0253	0.0246	0.0283	0.0249	0.0246	0.0244
$\bar{\sigma}_{xy}$	0.0337	0.0421	0.0444	0.0369	0.0416	0.0448	0.0463

[†] $\bar{w} = 10^2 [w_0 E_2 h^3/(q_0 a^4)]$, $\bar{\sigma} = \sigma h^2/(q_0 a^2)$.

Similar results are presented in Table 11.3.2 for a clamped square plate. The locations of the normal stresses reported for the three meshes are:

$$2 \times 2: \ (a/8, b/8); \quad 4 \times 4: \ (a/16, b/16); \quad 8 \times 8: \ (a/32, b/32)$$

and the locations of the shear stress for the three meshes are:

$$2 \times 2: \ (3a/8, 3b/8); \quad 4 \times 4: \ (7a/16, 7b/16); \quad 8 \times 8: \ (15a/32, 15b/32)$$

These stresses are not necessarily the maximum ones in the plate. For example, for an 8×8 mesh, the maximum normal stress in the isotropic plate is found to be 0.2316 at $(0.46875a, 0.03125b, -h/2)$ and the maximum shear stress is 0.0225 at $(0.28125a, 0.09375b, -h/2)$ for the non-conforming element.

Table 11.3.2. Maximum transverse deflections and stresses[†] of clamped (CCCC), isotropic and orthotropic, square plates ($a = b$) under a uniformly distributed transverse load.

Variable	Nonconforming			Conforming		
	2×2	4×4	8×8	2×2	4×4	8×8
Isotropic Plate ($\nu = 0.25$)						
$\bar{w}$	1.5731	1.4653	1.4342	1.7245	1.5327	1.4539
$\bar{\sigma}_{xx}$	0.0987	0.1238	0.1301	0.1115	0.1247	0.1305
$\bar{\sigma}_{yy}$	0.0987	0.1238	0.1301	0.1115	0.1247	0.1305
$\bar{\sigma}_{xy}$	0.0497	0.0222	0.0067	0.0700	0.0297	0.0086
Orthotropic Plate ($E_1/E_2 = 25, G_{12} = G_{13} = 0.5E_2, \nu_{12} = 0.25$)						
$\bar{w}$	0.1434	0.1332	0.1314	0.3120	0.1889	0.1467
$\bar{\sigma}_{xx}$	0.1962	0.2491	0.2598	0.2287	0.2700	0.2650
$\bar{\sigma}_{yy}$	0.0085	0.0046	0.0042	0.0171	0.0126	0.0074
$\bar{\sigma}_{xy}$	0.0076	0.0046	0.0019	0.0104	0.0049	0.0019

[†] $\bar{w} = 10^2 [w_0 E_2 h^3/(q_0 a^4)]$, $\bar{\sigma} = \sigma h^2/(q_0 a^2)$.

The conforming element yields slightly better solutions than the nonconforming element, and both elements show good convergence. However, convergence of the displacements is always faster than stresses for the displacement-based finite elements, and the rate of convergence of stresses is two orders less than that of displacements for the CPT-based element. Since the stresses in the finite element analysis are computed at locations different from the analytical solutions, they are expected to be different. Mesh refinement not only improves the accuracy of the solution, but the Gauss point locations also get closer to the true locations of the maximum values.

Vibration

For natural vibration, the use of solution symmetry yields symmetric modes and associated frequencies. Since, fundamental (i.e., minimum) frequency does not always correspond to a symmetric mode, one should exercise caution in finite element modeling of vibration response.

Table 11.3.3 contains fundamental frequencies for isotropic and orthotropic simply supported and clamped plates. The results were obtained using the conforming plate bending element. Only a quadrant was modeled and rotary inertia was included ($b/h = 0.1$).

Table 11.3.3. A comparison of the fundamental frequencies[†] of simply supported (SSSS) and clamped (CCCC) rectangular plates.

$\frac{a}{b}$	SSSS Plates			CCCC Plates		
	2×2	4×4	Exact	2×2	4×4	R–Ritz[‡]
Isotropic Plate ($\nu = 0.25$)						
0.5	4.8301	4.9752	4.9003	7.8213	9.1185	9.9891
1.0	1.9736	1.9959	1.9838	3.3103	3.5203	3.6476
1.5	1.4144	1.4395	1.4359	2.3435	2.5861	2.7404
2.0	1.2074	1.2439	1.2436	1.9551	2.2781	2.4973
Orthotropic Plate ($E_1/E_2 = 25$, $G_{12} = G_{13} = 0.5E_2$, $\nu_{12} = 0.25$)						
0.5	17.6756	19.8085	19.8682	24.4927	33.1670	40.5568
1.0	4.9917	5.2801	5.2947	8.1276	10.1619	11.6406
1.5	2.5558	2.6388	2.6390	4.6561	5.2791	5.6483
2.0	1.7387	1.7785	1.7759	3.3358	3.6045	3.7424

[†] $\bar{\omega} = \omega \frac{b^2}{\pi^2} \sqrt{\frac{\rho h}{D_{22}}}$; foundation modulus $k = 0$.

[‡] R–Ritz= Rayleigh–Ritz solution discussed in section 9.4.3.

11.4 Finite Element Models of FSDT
11.4.1 Virtual Work Statements

Following the procedure described in section 11.3, we can develop the finite element models of the equations governing the first-order shear deformation plate theory. We consider the linear equations of motion of FSDT from Eqs. (10.1.19)–(10.1.21), which are in terms of the stress resultants but equivalent to Eqs. (10.1.33) through (10.1.35). The generalized displacements of FSDT are (w_0, ϕ_x, ϕ_y). The virtual work statements required for the finite element model development can be identified from Eq. (10.1.9), and they are

$$\begin{aligned}
0 = \int_{\Omega^e} &\left[\frac{\partial \delta w_0}{\partial x} Q_x + \frac{\partial \delta w_0}{\partial y} Q_y - \delta w_0 q + k w_0 \delta w_0 + I_0 \delta w_0 \frac{\partial^2 w_0}{\partial t^2} \right. \\
&+ \frac{\partial \delta w_0}{\partial x} \left(\hat{N}_{xx} \frac{\partial w_0}{\partial x} + \hat{N}_{xy} \frac{\partial w_0}{\partial y} \right) \\
&+ \left. \frac{\partial \delta w_0}{\partial y} \left(\hat{N}_{xy} \frac{\partial w_0}{\partial x} + \hat{N}_{yy} \frac{\partial w_0}{\partial y} \right) \right] dxdy \\
&- \oint_{\Gamma^e} \left[\left(Q_x + \hat{N}_{xx} \frac{\partial w_0}{\partial x} + \hat{N}_{xy} \frac{\partial w_0}{\partial y} \right) n_x \right. \\
&+ \left. \left(Q_y + \hat{N}_{xy} \frac{\partial w_0}{\partial x} + \hat{N}_{yy} \frac{\partial w_0}{\partial y} \right) n_y \right] \delta w_0 \, ds \quad (11.4.1a)
\end{aligned}$$

$$0 = \int_{\Omega^e} \left(\frac{\partial \delta \phi_x}{\partial x} M_{xx} + \frac{\partial \delta \phi_x}{\partial y} M_{xy} + \delta \phi_x Q_x + I_2 \delta \phi_x \frac{\partial^2 \phi_x}{\partial t^2} \right) dxdy$$
$$- \oint_{\Gamma^e} T_x \delta \phi_x \, ds \quad (11.4.1b)$$

$$0 = \int_{\Omega^e} \left(\frac{\partial \delta \phi_y}{\partial x} M_{xy} + \frac{\partial \delta \phi_y}{\partial y} M_{yy} + \delta \phi_y Q_y + I_2 \delta \phi_y \frac{\partial^2 \phi_y}{\partial t^2} \right) dxdy$$
$$- \oint_{\Gamma^e} T_y \delta \phi_y \, ds \quad (11.4.1c)$$

We note from the boundary terms in Eqs. (11.4.1a–c) that u_0, v_0, w_0, ϕ_x, ϕ_y) are the primary variables (or generalized displacements). Unlike in the classical plate theory, the rotations (ϕ_x, ϕ_y) are independent of w_0. Note also that no derivatives of w_0 are in the list of the primary variables and therefore the element is called C^0 element.

508   THEORY AND ANALYSIS OF ELASTIC PLATES

The secondary variables are

$$T_x \equiv M_{xx}n_x + M_{xy}n_y, \quad T_y \equiv M_{xy}n_x + M_{yy}n_y$$

$$Q_n \equiv \left(Q_x + \hat{N}_{xx}\frac{\partial w_0}{\partial x} + \hat{N}_{xy}\frac{\partial w_0}{\partial y}\right)n_x$$
$$+ \left(Q_y + \hat{N}_{xy}\frac{\partial w_0}{\partial x} + \hat{N}_{yy}\frac{\partial w_0}{\partial y}\right)n_y \qquad (11.4.2)$$

11.4.2 Lagrange Interpolation Functions

The virtual work statements in Eqs. (11.4.1a–c) of the first-order shear deformation plate theory contain at the most only the first derivatives of the dependent variables (w_0, ϕ_x, ϕ_y). Therefore, they can all be approximated using the Lagrange interpolation functions. In principle, w_0 and (ϕ_x, ϕ_y) can be approximated with differing degrees of functions. For simplicity, we use the same order of interpolation for all variables. Let

$$w_0(x,y,t) = \sum_{j=1}^{n} w_j(t)\psi_j^e(x,y) \qquad (11.4.3)$$

$$\phi_x(x,y,t) = \sum_{j=1}^{n} S_j^1(t)\psi_j^e(x,y) \qquad (11.4.4a)$$

$$\phi_y(x,y,t) = \sum_{j=1}^{n} S_j^2(t)\psi_j^e(x,y) \qquad (11.4.4b)$$

where ψ_j^e are Lagrange interpolation functions. One can use linear, quadratic, or higher-order interpolations of these sets. A discussion of the Lagrange interpolation functions is presented next.

The simplest Lagrange element in two dimensions is the triangular element with nodes at its vertices (see Figure 11.4.1), and its interpolation functions have the form

$$\psi_i^e(x,y) = a_i + b_i x + c_i y \qquad (11.4.5)$$

The functions are continuous in x and y and *complete* (i.e., contain all lower order terms of the polynomial). An element with quadratic variation of the dependent variables requires six nodes, because a complete quadratic polynomial in two dimensions has six coefficients:

$$\psi_i^e(x,y) = a_i + b_i x + c_i y + d_i xy + e_i x^2 + f_i y^2 \qquad (11.4.6)$$

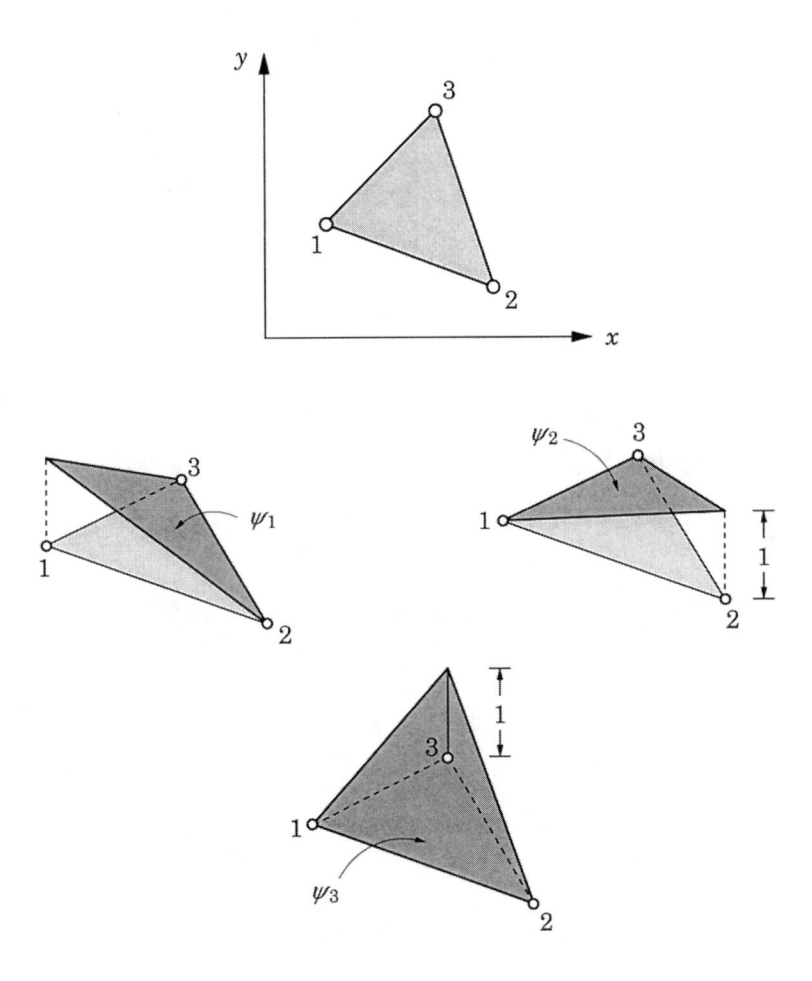

Figure 11.4.1. Linear Lagrange triangular element and its interpolation functions.

The three vertex nodes uniquely describe the geometry of the element (as in the linear element), and the other three nodes are placed at the midpoints of the sides (see Figure 11.4.2). The interpolation functions for linear and quadratic triangular elements are presented below in terms of the area coordinates, L_i (see Figure 11.4.2):

$$L_i = \frac{A_i}{A}, \quad A = \sum_{i=1}^{3} A_i \tag{11.4.7a}$$

510   THEORY AND ANALYSIS OF ELASTIC PLATES

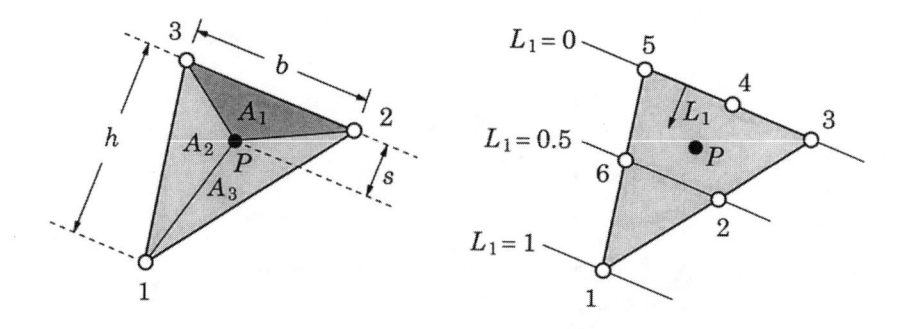

Figure 11.4.2. Quadratic Lagrange triangular element.

$$\left\{ \begin{array}{c} \psi_1^e \\ \psi_2^e \\ \psi_3^e \end{array} \right\} = \left\{ \begin{array}{c} L_1 \\ L_2 \\ L_3 \end{array} \right\}, \quad \left\{ \begin{array}{c} \psi_1^e \\ \psi_2^e \\ \psi_3^e \\ \psi_4^e \\ \psi_5^e \\ \psi_6^e \end{array} \right\} = \left\{ \begin{array}{c} L_1(2L_1 - 1) \\ L_2(2L_2 - 1) \\ L_3(2L_3 - 1) \\ 4L_1L_2 \\ 4L_2L_3 \\ 4L_3L_1 \end{array} \right\} \quad (11.4.7b)$$

The simplest rectangular element has four nodes at its vertices (see Figure 11.4.3) that define the geometry. The interpolation functions for this element have the form

$$\psi_i^e(x,y) = a_i + b_i x + c_i y + d_i xy \quad (11.4.8)$$

A complete quadratic polynomial representation

$$\psi_i^e(x,y) = a_i + b_i x + c_i y + d_i xy + e_i x^2 + f_i y^2 + g_i x^2 y \\ + h_i xy^2 + k_i x^2 y^2 \quad (11.4.9)$$

for a rectangular element contains nine parameters and hence nine nodes (see Figure 11.4.4a).

The linear and quadratic Lagrange interpolation functions of rectangular elements are given below in terms of the element coordinates (ξ, η), called the *natural coordinates*:

$$\left\{ \begin{array}{c} \psi_1^e \\ \psi_2^e \\ \psi_3^e \\ \psi_4^e \end{array} \right\} = \frac{1}{4} \left\{ \begin{array}{c} (1-\xi)(1-\eta) \\ (1+\xi)(1-\eta) \\ (1+\xi)(1+\eta) \\ (1-\xi)(1+\eta) \end{array} \right\} \quad (11.4.10)$$

FINITE ELEMENT MODELS OF BEAMS AND PLATES   511

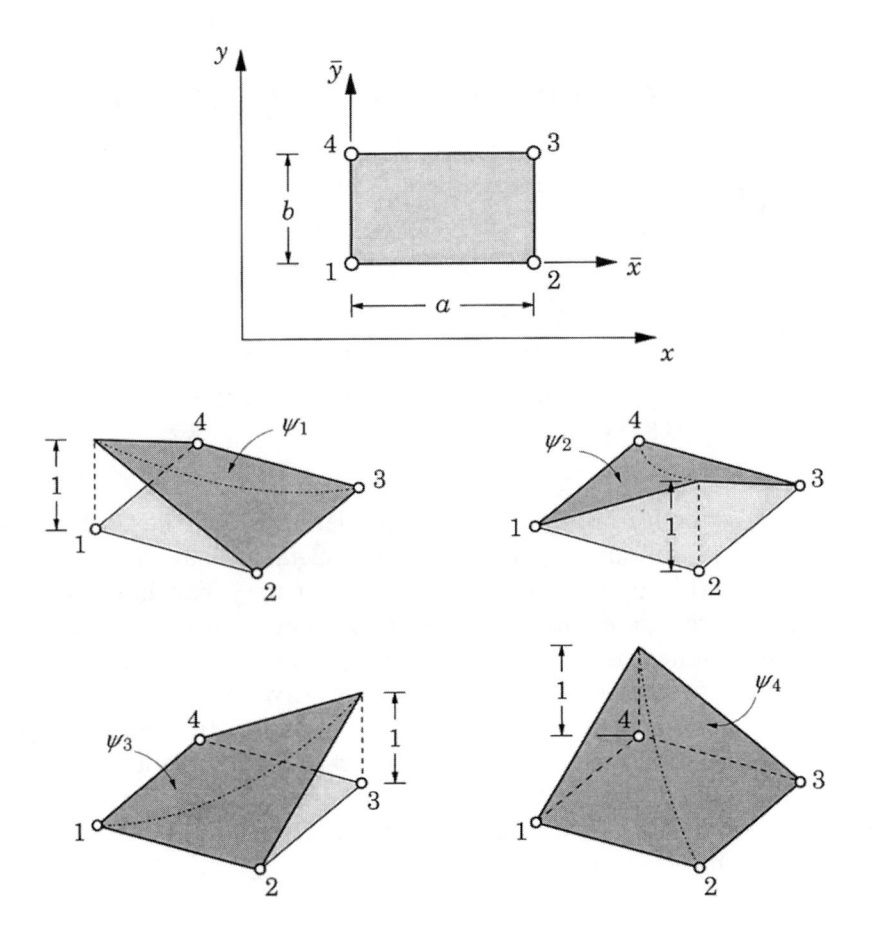

Figure 11.4.3. Linear rectangular element and its interpolation functions.

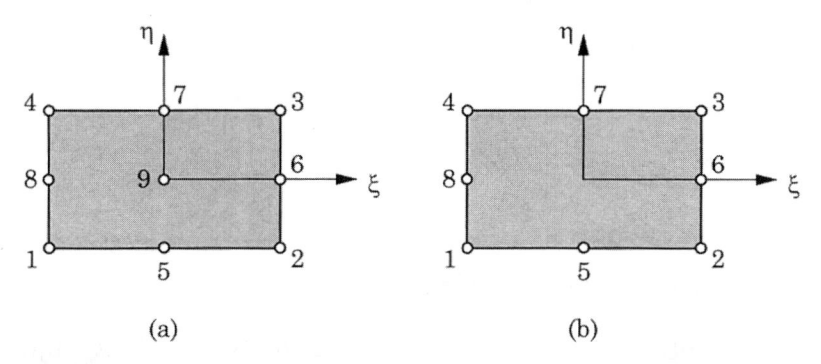

(a)   (b)

Figure 11.4.4. Nine- and eight-node quadratic rectangular elements.

and

$$\begin{Bmatrix} \psi_1^e \\ \psi_2^e \\ \psi_3^e \\ \psi_4^e \\ \psi_5^e \\ \psi_6^e \\ \psi_7^e \\ \psi_8^e \\ \psi_9^e \end{Bmatrix} = \frac{1}{4} \begin{Bmatrix} (1-\xi)(1-\eta)(-\xi-\eta-1) + (1-\xi^2)(1-\eta^2) \\ (1+\xi)(1-\eta)(\xi-\eta-1) + (1-\xi^2)(1-\eta^2) \\ (1+\xi)(1+\eta)(\xi+\eta-1) + (1-\xi^2)(1-\eta^2) \\ (1-\xi)(1+\eta)(-\xi+\eta-1) + (1-\xi^2)(1-\eta^2) \\ 2(1-\xi^2)(1-\eta) - (1-\xi^2)(1-\eta^2) \\ 2(1+\xi)(1-\eta^2) - (1-\xi^2)(1-\eta^2) \\ 2(1-\xi^2)(1+\eta) - (1-\xi^2)(1-\eta^2) \\ 2(1-\xi)(1-\eta^2) - (1-\xi^2)(1-\eta^2) \\ 4(1-\xi^2)(1-\eta^2) \end{Bmatrix}$$

(11.4.11)

The *serendipity* family of Lagrange elements are those elements which have no interior nodes. Serendipity elements have fewer nodes compared to the higher-order Lagrange elements. The interpolation functions of the serendipity elements are not complete, and they cannot be obtained using tensor products of one-dimensional Lagrange interpolation functions. Instead, an alternative procedure must be employed, as discussed in [1]. The interpolation functions for the quadratic serendipity element are given in Eq. (11.4.12) below (also see Figure 11.4.4b). Although the interpolation functions are not complete because the last term in Eq. (11.4.9) is omitted, the serendipity elements have proven to be very effective in most practical applications (*serendipity* !).

$$\begin{Bmatrix} \psi_1^e \\ \psi_2^e \\ \psi_3^e \\ \psi_4^e \\ \psi_5^e \\ \psi_6^e \\ \psi_7^e \\ \psi_8^e \end{Bmatrix} = \frac{1}{4} \begin{Bmatrix} (1-\xi)(1-\eta)(-\xi-\eta-1) \\ (1+\xi)(1-\eta)(\xi-\eta-1) \\ (1+\xi)(1+\eta)(\xi+\eta-1) \\ (1-\xi)(1+\eta)(-\xi+\eta-1) \\ 2(1-\xi^2)(1-\eta) \\ 2(1+\xi)(1-\eta^2) \\ 2(1-\xi^2)(1+\eta) \\ 2(1-\xi)(1-\eta^2) \end{Bmatrix} \qquad (11.4.12)$$

11.4.3 Finite Element Model

Substituting Eqs. (11.4.3) and (11.4.4a,b) for (w_0, ϕ_x, ϕ_y) into Eqs. (11.4.1a–c), we obtain the following semidiscrete finite element model of the first-order plate theory:

$$\left(\begin{bmatrix}[K^{11}] & [K^{12}] & [K^{13}]\\ [K^{12}]^T & [K^{22}] & [K^{23}]\\ [K^{13}]^T & [K^{23}]^T & [K^{33}]\end{bmatrix} + \begin{bmatrix}[G] & [0] & [0]\\ [0] & [0] & [0]\\ [0] & [0] & [0]\end{bmatrix}\right)\begin{Bmatrix}\{w^e\}\\ \{S^1\}\\ \{S^2\}\end{Bmatrix}$$
$$+ \begin{bmatrix}[M^{11}] & [0] & [0]\\ [0] & [M^{22}] & [0]\\ [0] & [0] & [M^{33}]\end{bmatrix}\begin{Bmatrix}\{\ddot{w}^e\}\\ \{\ddot{S}^1\}\\ \{\ddot{S}^2\}\end{Bmatrix} = \begin{Bmatrix}\{F^1\}\\ \{F^2\}-\{F^{T2}\}\\ \{F^3\}-\{F^{T3}\}\end{Bmatrix}$$
(11.4.13a)

or

$$[K^e]\{\Delta^e\} + [M^e]\{\ddot{\Delta}^e\} = \{F^e\} \tag{11.4.13b}$$

where the coefficients of the submatrices $[K^{\alpha\beta}]$ and $[M^{\alpha\beta}]$ and vectors $\{F^\alpha\}$ are defined for $(\alpha,\beta = 1,2,3)$ by the expressions

$$K_{ij}^{11} = \int_{\Omega^e}\left(KA_{55}\frac{\partial\psi_i^e}{\partial x}\frac{\partial\psi_j^e}{\partial x} + KA_{44}\frac{\partial\psi_i^e}{\partial y}\frac{\partial\psi_j^e}{\partial y} + k\psi_i^e\psi_j^e\right)dxdy$$

$$K_{ij}^{12} = \int_{\Omega^e}KA_{55}\frac{\partial\psi_i^e}{\partial x}\psi_j^e\,dxdy, \quad K_{ij}^{13} = \int_{\Omega^e}KA_{44}\frac{\partial\psi_i^e}{\partial y}\psi_j^e\,dxdy$$

$$K_{ij}^{22} = \int_{\Omega^e}\left(D_{11}\frac{\partial\psi_i^e}{\partial x}\frac{\partial\psi_j^e}{\partial x} + D_{66}\frac{\partial\psi_i^e}{\partial y}\frac{\partial\psi_j^e}{\partial y} + KA_{55}\psi_i^e\psi_j^e\right)dxdy$$

$$K_{ij}^{23} = \int_{\Omega^e}\left(D_{12}\frac{\partial\psi_i^e}{\partial x}\frac{\partial\psi_j^e}{\partial y} + D_{66}\frac{\partial\psi_i^e}{\partial y}\frac{\partial\psi_j^e}{\partial x}\right)dxdy$$

$$K_{ij}^{33} = \int_{\Omega^e}\left(D_{66}\frac{\partial\psi_i^e}{\partial x}\frac{\partial\psi_j^e}{\partial x} + D_{22}\frac{\partial\psi_i^e}{\partial y}\frac{\partial\psi_j^e}{\partial y} + KA_{44}\psi_i^e\psi_j^e\right)dxdy$$

$$M_{ij}^{11} = \int_{\Omega^e}I_0\psi_i^e\psi_j^e\,dxdy, \quad M_{ij}^{22} = \int_{\Omega^e}I_2\psi_i^e\psi_j^e\,dxdy = M_{ij}^{33}$$

$$F_i^1 = \int_{\Omega^e}q\psi_i^e\,dxdy + \oint_{\Gamma^e}Q_n\psi_i^e\,ds$$

$$F_i^2 = \oint_{\Gamma_e}T_x\psi_i^e\,dxdy, \quad F_i^3 = \int_{\Gamma_e}T_y\psi_i^e\,dxdy$$

$$F_i^{T2} = \oint_{\Gamma_e}\left(\frac{\partial\psi_i^e}{\partial x}M_{xx}^T + \frac{\partial\psi_i^e}{\partial y}M_{xy}^T\right)dxdy$$

$$F_i^{T3} = \int_{\Gamma_e}\left(\frac{\partial\psi_i^e}{\partial x}M_{xy}^T + \frac{\partial\psi_i^e}{\partial y}M_{yy}^T\right)dxdy \tag{11.4.14}$$

where M_{xx}^T, M_{yy}^T and M_{xy}^T are the thermal moments.

514 THEORY AND ANALYSIS OF ELASTIC PLATES

The displacement-based C^0 plate bending element of Eq. (11.4.13b) is often referred to in the finite element literature as the *Mindlin plate element* due to the fact that it is based on the so-called Mindlin plate theory, which is labeled in this book as the first-order shear deformation plate theory (FSDT). When the bilinear rectangular element is used for all generalized displacements (w_0, ϕ_x, ϕ_y), the element stiffness matrices are of the order 12×12; and for the nine-node quadratic element they are 27×27 (see Figure 11.4.5).

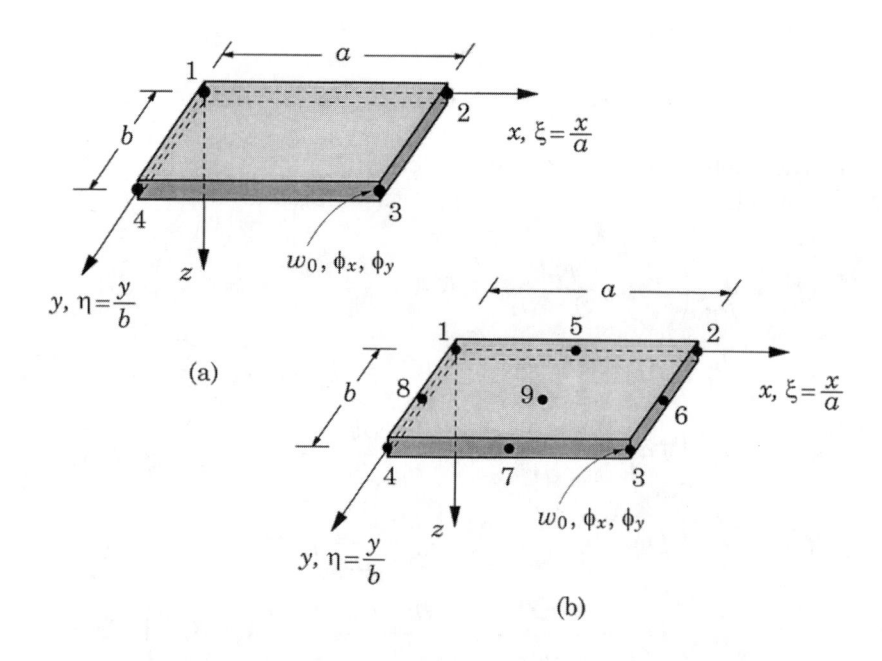

Figure 11.4.5. Linear and nine-node quadratic rectangular elements for the first-order shear deformation theory.

Equation (11.4.13b) can be simplified for static bending, buckling, natural vibration, and transient analyses, as described for the classical plate element in section 11.3. The simplifications are obvious and therefore are not repeated for the FSDT element. Numerical results for bending, buckling, and natural vibration will be discussed next.

The C^0-plate bending elements based on the first-order shear deformation plate theory are among the simplest available in the literature. Unfortunately, when lower order (quadratic or less) equal interpolation of the transverse deflection and rotations is used, the elements become excessively stiff in the thin plate limit, yielding displacements that are too small compared to the true solution. As discussed earlier, this type of behavior is known as shear locking. There are a number of papers on the subject of shear locking and elements developed to alleviate the problem (see [21–28]). A commonly used technique is to under-integrate the transverse shear stiffnesses (i.e., all coefficients in $K_{ij}^{\alpha\beta}$ that contain A_{44} and A_{55}). Higher-order elements or refined meshes of lower-order elements experience relatively less locking, but sometimes at the expense of rate of convergence. With the suggested Gauss rule, highly distorted elements tend to have slower rates of convergence but they give sufficiently accurate results.

11.4.4 Numerical Results

The effect of the integration rule and the convergence characteristics of the C^0 finite element model based on equal interpolation is illustrated using a simply supported, orthotropic, square plate constructed of four orthotropic layers of equal thickness $h/4$, where h is the total thickness of the plate; the principal material axis 1 of the top and bottom layer coincide with plate x axis, while the middle two layers have their principal axis 1 coinciding with the y−axis of the plate. Such a laminate is often denoted with (0/90/90/0). The material properties used for each orthotropic lamina are those typical of a graphite–epoxy material

$$E_1 = 25E_2, \ G_{12} = G_{13} = 0.5E_2, \ G_{23} = 0.2E_2, \ \nu_{12} = 0.25, \ K = \frac{5}{6} \quad (11.4.15)$$

The lamina stiffnesses in the principal material coordinates are (in msi)

$$Q_{11} = 25.063, \quad Q_{12} = 0.25063, \quad Q_{22} = 1.0025$$

$$Q_{44} = 0.2, \quad Q_{55} = Q_{66} = 0.5 \quad (11.4.16)$$

and the plate stiffnesses are [see Eqs. (3.6.11) and (10.1.27)]

$$D_{11} = 1.8379 \times 10^6 h^3, \quad D_{12} = 2.0886 \times 10^4 h^3, \quad D_{22} = 3.3417 \times 10^5 h^3$$

$$D_{66} = 4.1667 \times 10^4 h^3, \quad A_{44} = A_{55} = 3.5 \times 10^5 h \quad (11.4.17)$$

The plate is subjected to a sinusoidally distributed transverse load

$$q(x,y) = q_0 \cos \frac{\pi x}{a} \sin \frac{\pi y}{b} \quad (11.4.18)$$

As noted earlier, the stresses in the finite element analysis are computed at the reduced Gauss points, irrespective of the Gauss rule used for the evaluation of the element stiffness coefficients [29,30]. The Gauss point locations differ for each mesh used. The Gauss point coordinates A and B are shown in Table 11.4.1 for various meshes of linear and quadratic rectangular elements.

Table 11.4.1. The Gauss point locations at which the stresses are computed in the finite element analysis.

Coordinate	2L	4L	8L	2Q8/2Q9	4Q8/4Q9
A	$0.125a$	$0.0625a$	$0.03125a$	$0.05283a$	$0.02642a$
B	$0.375a$	$0.4375a$	$0.46875a$	$0.44717a$	$0.47358a$

The finite element solutions (FES) (see Reddy and Chao [31]) are compared with the 3-D elasticity solution (ELS) (see Pagano [32]) and the closed-form solutions (CFS) (see [1]) in Table 11.4.2 for three side-to-thickness ratios $a/h = 10, 20,$ and 100. The following nondimensionalizations are used $[\sigma(\cdot, z_0) = -\sigma(\cdot, -z_0)]$ in Table 11.4.2:

$$\bar{w} = w_0(0,0)\frac{E_2 h^3}{b^4 q_0}, \quad \bar{\sigma}_{xx} = \sigma_{xx}(0,0,\frac{h}{2})\frac{h^2}{b^2 q_0}$$

$$\bar{\sigma}_{yy} = \sigma_{yy}(0,0,\frac{h}{4})\frac{h^2}{b^2 q_0}, \quad \bar{\sigma}_{xy} = \sigma_{xy}(a/2, b/2, -\frac{h}{2})\frac{h^2}{b^2 q_0}$$

$$\bar{\sigma}_{xz} = \sigma_{xz}(a/2, 0, 0)\frac{h}{bq_0}, \quad \bar{\sigma}_{yz} = \sigma_{yz}(0, b/2, 0)\frac{h}{bq_0} \qquad (11.4.19)$$

where the origin of the coordinate system is taken at the center of the plate, $-a/2 \leq x \leq a/2, -a/2 \leq y \leq a/2$, and $-h/2 \leq z \leq h/2$. The notation nL stands for $n \times n$ uniform mesh of linear rectangular elements, nQ8 for $n \times n$ uniform mesh of eight-node quadratic elements, and nQ9 for $n \times n$ uniform mesh of nine-node quadratic elements in a quarter plate. Various stresses in the finite element analysis are evaluated at the Gauss points as indicated below:

$$\sigma_{xx}(A, A, \frac{h}{2}), \quad \sigma_{yy}(A, A, \frac{h}{4}), \quad \sigma_{xy}(B, B, -\frac{h}{2})$$

$$\sigma_{xz}(B, A) \text{ in layers 1 and 3}, \quad \sigma_{yz}(A, B) \text{ in layer 2} \qquad (11.4.20)$$

where the values of (A, B) are given in Table 11.4.1.

Table 11.4.2. Effect of reduced integration on the nondimensionalized maximum deflections $\bar{w}$ and stresses $\bar{\sigma}$ of simply supported laminate (0/90/90/0) square plates subjected to sinusoidal loading.

a/h	Source	$\bar{w} \times 10^2$	$\bar{\sigma}_{xx}$	$\bar{\sigma}_{yy}$	$\bar{\sigma}_{xy}$	$\bar{\sigma}_{xz}$	$\bar{\sigma}_{yz}$
Finite Element Solutions[†]							
10	2L–F	0.5901	0.3339	0.2454	0.0163	0.316	0.125
	2L–R	0.6508	0.3799	0.2838	0.0187	0.335	0.107
	2L–S	0.6655	0.3796	0.2882	0.0189	0.353	0.114
	4L–F	0.6427	0.4512	0.3280	0.0219	0.389	0.129
	4L–R	0.6599	0.4668	0.3406	0.0227	0.395	0.123
	4L–S	0.6632	0.4667	0.3419	0.0227	0.400	0.125
	2Q8–F	0.6605	0.4831	0.3492	0.0234	0.404	0.126
	2Q8–R	0.6615	0.4842	0.3509	0.0234	0.404	0.126
	2Q8–S	0.6613	0.4844	0.3509	0.0233	0.405	0.126
	2Q9–F	0.6551	0.4790	0.3400	0.0231	0.399	0.126
	2Q9–R	0.6633	0.4841	0.3508	0.0234	0.404	0.125
	2Q9–S	0.6631	0.4844	0.3509	0.0233	0.404	0.126
	8L–S	0.6628	0.4907	0.3565	0.0238	0.412	0.128
	4Q8–S	0.6626	0.4954	0.3589	0.0240	0.414	0.128
	4Q8–S	0.6627	0.4954	0.3589	0.0240	0.414	0.128
Analytical Solutions							
10	CFS	0.6627	0.4989	0.3614	0.0241	0.416	0.129
	ELS	0.7370	0.5590	0.4010	0.0276	0.301	0.196
Finite Element Solutions							
20	2L–F	0.3236	0.2645	0.1491	0.0111	0.303	0.139
	2L–R	0.4712	0.4036	0.2289	0.0170	0.353	0.089
	2L–S	0.4760	0.4043	0.2308	0.0171	0.373	0.094
	4L–F	0.4346	0.4365	0.2451	0.0183	0.395	0.123
	4L–R	0.4863	0.4940	0.2777	0.0207	0.415	0.103
	4L–S	0.4874	0.4942	0.2782	0.0207	0.420	0.105
	2Q8–F	0.4876	0.5082	0.2828	0.0214	0.424	0.106
	2Q8–R	0.4901	0.5117	0.2870	0.0214	0.424	0.106
	2Q8–S	0.4901	0.5120	0.2870	0.0214	0.424	0.106

(Table continued on the next page)

a/h	Source	$\bar{w} \times 10^2$	$\bar{\sigma}_{xx}$	$\bar{\sigma}_{yy}$	$\bar{\sigma}_{xy}$	$\bar{\sigma}_{xz}$	$\bar{\sigma}_{yz}$
	2Q9-F	0.4891	0.5083	0.2829	0.0214	0.424	0.106
	2Q9-R	0.4915	0.5118	0.2870	0.0213	0.424	0.106
	2Q9-S	0.4915	0.5120	0.2870	0.0213	0.424	0.105
	8L–S	0.4902	0.5189	0.2912	0.0218	0.433	0.108
	4Q8-S	0.4911	0.5236	0.2936	0.0219	0.434	0.108
	4Q9-S	0.4912	0.5236	0.2936	0.0219	0.434	0.108
Analytical Solutions							
20	CFS	0.4912	0.5273	0.2956	0.0221	0.437	0.109
	ELS	0.5128	0.5430	0.3080	0.0230	0.328	0.156
Finite Element Solutions							
100	2L–F	0.0315	0.0299	0.0151	0.0012	0.230	0.211
	2L–R	0.4107	0.4129	0.2076	0.0164	0.360	0.082
	2L–S	0.4120	0.4140	0.2082	0.0164	0.381	0.086
	4L–F	0.1034	0.1203	0.0604	0.0048	0.298	0.221
	4L–R	0.4281	0.5045	0.2535	0.0200	0.422	0.098
	4L–S	0.4284	0.5048	0.2537	0.0207	0.428	0.097
	2Q8-F	0.4143	0.4900	0.2435	0.0199	0.430	0.100
	2Q8-R	0.4319	0.5214	0.2621	0.0206	0.435	0.102
	2Q8-S	0.4319	0.5214	0.2620	0.0206	0.433	0.102
	2Q9-F	0.4193	0.4946	0.2420	0.0201	0.431	0.098
	2Q9-R	0.4339	0.5224	0.2625	0.0207	0.432	0.098
	2Q9-S	0.4339	0.5224	0.2550	0.0206	0.430	0.097
	8L–S	0.4324	0.5297	0.2662	0.0210	0.441	0.100
	4Q8-S	0.4336	0.5344	0.2685	0.0212	0.441	0.100
	4Q9-S	0.4337	0.5344	0.2685	0.0212	0.441	0.100
Analytical Solutions							
100	CFS	0.4337	0.5382	0.2704	0.0213	0.445	0.101
	ELS	0.4347	0.5390	0.2710	0.0214	0.339	0.139
	CPT*	0.4313	0.5387	0.2667	0.0213	(0.339)	(0.138)

† F = full integration; R = reduced integration; S = selective integration: full integration of all except the transverse shear coefficients, which are evaluated using reduced integration rule.

* The CPT solution is independent of side-to-thickness ratio, a/h.

An examination of the numerical results presented in Table 11.4.2 shows that the FSDT finite element with equal interpolation of all generalized displacements does not experience shear locking for thick plates even when full integration rule is used. Shear locking is evident when the element is used to model thin plates ($a/h \geq 100$) with full integration rule (F). Also, higher-order elements show less locking but with slower convergence. The element behaves uniformly well for thin and thick plates when the reduced (R) or selectively reduced integration (S) rule is used.

The finite element results are in excellent agreement with the closed-form solutions of the first-order shear deformation theory (FSDT). The displacements converge faster than stresses. This is expected with all displacement finite element models because the rate of convergence of gradients of the solution is one order less than the rate of convergence of the solution. Of course, the finite element solutions based on FSDT *should not* be expected to agree well with the 3-D elasticity solution (ELS); the finite element solutions should only converge to the closed-form solutions (CFS) of FSDT.

Next we consider a sandwich plate subjected to sinusoidally distributed transverse loading. The sandwich plate is treated as a three layer plate with different thickness of the face sheets (layers 1 and 3) and the core (layer 2). The face sheets are assumed to be orthotropic with the following material properties

$$E_1 = 25E_2, \ E_2 = 10^6 \text{ psi}, \ G_{12} = G_{13} = 0.5E_2, \ G_{23} = 0.2E_2, \ \nu_{12} = 0.25 \tag{11.4.21}$$

and the core material is transversely isotropic and is characterized by the following material properties:

$$E_1 = E_2 = 10^6 \text{ psi}, \ G_{13} = G_{23} = 0.06 \times 10^6 \text{ psi}, \ \nu_{12} = 0.25$$

$$G_{12} = \frac{E_1}{2(1+\nu_{12})} = 0.016 \times 10^6 \text{ psi} \tag{11.4.22}$$

Each face sheet is assumed to be one-tenth of the total thickness of the sandwich plate ($a = b$). The finite element results obtained with 4×4 mesh of eight-node quadratic elements with reduced integration (4Q8-R) are compared with the closed-form solution and elasticity solution of Pagano [33] in Table 11.4.3. The stresses are nondimensionalized as before, and their locations with respect to a coordinate system whose origin is at the center of the plate are as follows:

Table 11.4.3. Comparison of nondimensionalized maximum deflections and stresses in a simply supported sandwich plate subjected to sinusoidally varying transverse load ($h_1 = h_3 = 0.1h$, $h_2 = 0.8h$, $K = 5/6$).

a/h	Source	$\bar{w} \times 10^2$	$\bar{\sigma}_{xx}$	$\bar{\sigma}_{yy}$	$\bar{\sigma}_{xy}$	$\bar{\sigma}_{xz}$	$\bar{\sigma}_{yz}$
2	ELS	—	3.278	0.4517	0.2338	0.185	0.1399
	CFS	14.928	0.7378	0.2328	0.1263	0.1084	0.0781
						(0.2481)	(1085)[†]
	FEM	14.927	0.7326	0.2311	0.1254	0.1077	0.0776
4	ELS	—	1.556	0.2595	0.1481	0.239	0.1072
	CFS	4.7666	0.8918	0.1562	0.0907	0.1229	0.0537
						(0.2808)	(0.0746)
	FEM	4.7663	0.8856	0.1551	0.0901	0.1221	0.0534
10	ELS	—	1.153	0.1104	0.0717	0.300	0.0527
	CFS	1.5604	1.0457	0.0798	0.0552	0.1374	0.0293
						(0.3134)	(0.0408)
	FEM	1.5603	1.0384	0.0792	0.0548	0.1365	0.0278
20	ELS	—	1.110	0.0700	0.0511	0.317	0.0361
	CFS	1.0524	1.0831	0.0612	0.0466	0.1409	0.0234
						(0.3213)	(0.0325)
	FEM	1.0523	1.0755	0.0608	0.0462	0.1399	0.0233
50	ELS	—	1.099	0.0569	0.0446	0.323	0.0306
	CFS	0.9063	1.0947	0.0554	0.0439	0.1420	0.0216
						(0.3238)	(0.0300)
	FEM	0.9061	1.087	0.0551	0.0436	0.1410	0.0214
100	ELS	—	1.098	0.0550	0.0437	0.324	0.0297
	CFS	0.8852	1.0964	0.0546	0.0435	0.1422	0.0213
						(0.3242)	(0.0296)
	FEM	0.8851	1.0887	0.0542	0.0432	0.1412	0.0161
	CPT	0.8782	1.0970	0.0543	0.0433	(0.3243)	(0.0295)

[†] Values computed from equilibrium equations.

$$\sigma_{xx}(0,0,\frac{h}{2}),\ \sigma_{yy}(0,0,\frac{h}{2}),\ \sigma_{xy}(a/2,b/2,-\frac{h}{2})$$
$$\sigma_{xz}(0,b/2,0),\ \sigma_{yz}(a/2,0,0) \qquad (11.4.23)$$

The results indicate that the effect of shear deformation on deflections is significant in sandwich plates even at large values of a/h. The equilibrium-derived transverse shear stresses are surprisingly close to those predicted by the elasticity theory for $a/h \geq 10$, while those computed from constitutive equations are considerably underestimated for small side-to-thickness ratios. The transverse shear stress component σ_{yz} is significantly overestimated by CPT. Figures 11.4.6 and 11.4.7 show the variation of the transverse shear stresses through the thickness of the sandwich plates for side-to-thickness ratios $a/h = 2, 10$, and 100.

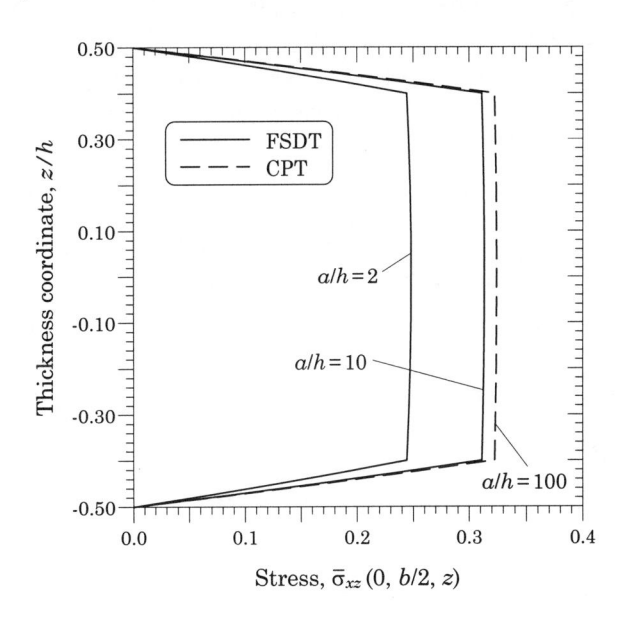

Figure 11.4.6. Distribution of transverse shear stress σ_{xz} through the thickness of a simply supported sandwich plate under sinusoidally distributed transverse load.

The same sandwich plate discussed above is analyzed for simply supported and clamped boundary conditions when uniformly distributed load is used. Once again a quarter plate model is used with 4×4 mesh of quadratic FSDT elements and 8×8 mesh of CPT conforming cubic

522   THEORY AND ANALYSIS OF ELASTIC PLATES

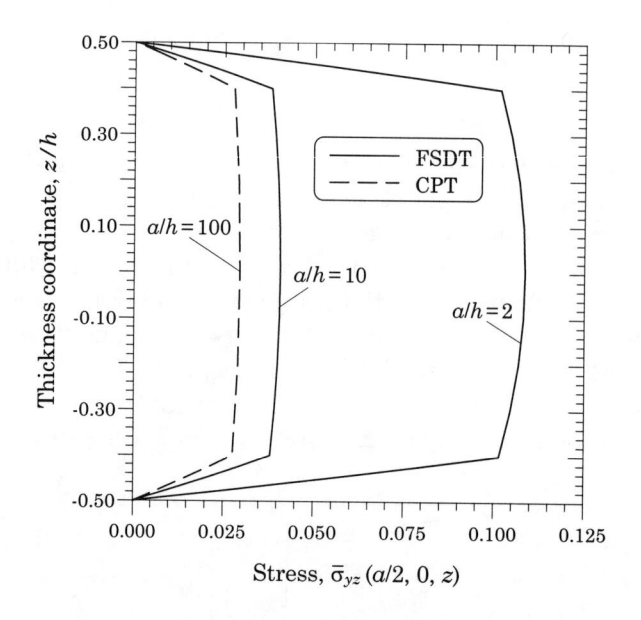

Figure 11.4.7. Distribution of transverse shear stress σ_{yz} through the thickness of a simply supported sandwich plate under sinusoidally distributed transverse load.

elements. The results are presented in Table 11.4.4. The effect of shear deformation on the deflections is even more significant in clamped plates than in simply supported plates.

Effects of side-to-thickness ratio, integration, and type of element on the nondimensionalized fundamental frequency $\bar{\omega}$ of simply supported square laminates (0/90/90/0) on the accuracy of the natural frequencies can be seen from the results presented in Table 11.4.5. The lamina material properties are taken to be

$$E_1 = 40E_2, \ G_{12} = G_{13} = 0.6E_2, \ G_{23} = 0.5E_2, \ \nu_{12} = 0.25 \quad (11.4.24)$$

A 2×2 mesh in a quarter plate is used to obtain the results (rotary inertia included). From the results obtained, it is clear that both full (F) and selective (S) integrations give good results for thick plates ($a/h \leq 10$), whereas reduced integration (R) gives the best results for thin plates ($a/h \geq 100$). However, the reduced integration and selective integration rules both give good results for a wide range of side-to-thickness ratios.

Table 11.4.4. Nondimensionalized maximum deflections and stresses in a square sandwich plate with simply supported and clamped boundary conditions ($h_1 = h_3 = 0.1h$, $h_2 = 0.8h$, $K = 5/6$).

a/h	Source	$\bar{w} \times 10^2$	$\bar{\sigma}_{xx}$	$\bar{\sigma}_{yy}$	$\bar{\sigma}_{xy}$	$\bar{\sigma}_{xz}$	$\bar{\sigma}_{yz}$
Simply supported plate under uniformly distributed load							
10	4Q8-R	2.3370	1.5430	0.0883	0.1136	0.2396	0.0991
50	4Q8-R	1.3671	1.5964	0.0526	0.0916	0.2433	0.0881
100	4Q8-R	1.3359	1.5978	0.0514	0.0906	0.2394	0.0880
CLPT	8CC-F[‡]	1.3296	1.5830	0.0509	0.0906	–	–
Clamped plate under uniformly distributed load							
10	4Q9-R[†]	1.2654	0.5018	0.0550	0.0120	0.2318	0.1445
50	4Q9-R	0.3111	0.5356	0.0108	0.0039	0.2406	0.1160
100	4Q9-R	0.2785	0.5347	0.0094	0.0030	0.2400	0.1148
CPT	8CC-F[‡]	0.2951	0.5401	0.0145	0.0605	–	–

[†] The 4Q9-S element gives the same results as 4Q9-R.

[‡] 8 × 8 mesh of conforming cubic elements with full integration for stiffness coefficient evaluation and one-point Gauss rule for stresses.

Table 11.4.5. Effects of side-to-thickness ratio, integration, and type of element on the nondimensionalized fundamental frequency $\bar{\omega} = \omega(a^2/h)\sqrt{\rho/E_2}$ of simply supported square laminates (0/90/90/0).

a/h	Serendipity Element			Lagrange Element			Exact
	F	R	S	F	R	S	
2	5.502	5.503	5.501	5.502	5.503	5.501	5.500
10	15.174	15.179	15.159	15.182	15.193	15.172	15.143
100	19.171	18.841	18.808	19.225	18.883	18.933	18.836

Exercises

11.1 (a) Derive finite element interpolation functions using w_0, $\theta = -dw_0/dx$, and d^2w_0/dx^2 as the nodal variables of an element with two (end) nodes with a total of six degrees of freedom per element. Note that you must select a complete polynomial containing six parameters

$$w_0(x) = c_1 + c_2 x + c_3 x^2 + c_4 x^3 + c_5 x^4 + c_6 x^5$$

and derive the Hermite interpolation functions.

(b) Use the finite element approximation to compute the stiffness matrix $[K^e]$.

11.2 Derive finite element Hermite interpolation functions using w_0 and $\theta = -dw_0/dx$ as the nodal variables of an element with three nodes (two end nodes and the middle node), with a total of six degrees of freedom per element. As in Exercise 11.1, you must select a complete polynomial containing six parameters and derive the Hermite interpolation functions.

11.3 Consider the spring-supported cantilever beam shown in Figure P11.3. Determine the deflection w_0 in the spring under uniformly distributed load of intensity q_0. Use the Hermite cubic beam finite element to solve the problem.

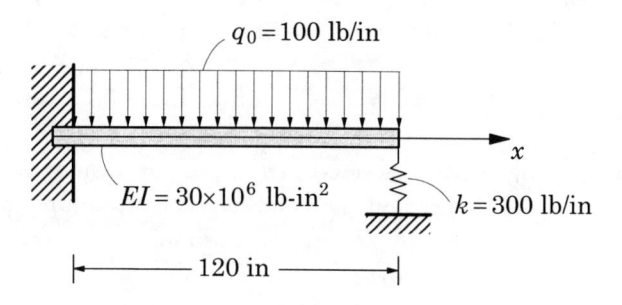

$q_0 = 100$ lb/in

$EI = 30 \times 10^6$ lb-in^2

$k = 300$ lb/in

120 in

Figure P11.3

11.4 Determine the natural frequencies of vibration of a cantilever beam of length L using one Euler–Bernoulli beam element. Assume that the flexural stiffness EI and mass per unit length m are constant and $I_2 = 0$. The first two exact natural frequencies are $\omega_1 = (3.516/L^2)\sqrt{EI/m}$ and $\omega_2 = (22.03/L^2)\sqrt{EI/m}$. (Ans: $\omega_1 = (3.533/L^2)\sqrt{EI/m}$ and $\omega_2 = (34.81/L^2)\sqrt{EI/m}$.)

11.5 Determine the natural frequencies of vibration of a clamped–clamped beam of length L using one Euler–Bernoulli beam element. Assume that the flexural stiffness EI and mass per unit length m are constant.

11.6 Determine the natural frequencies of vibration of a cantilever beam of length L using one linear Timoshenko beam element. Assume that the flexural stiffness EI, shear stiffness KGA, and mass per unit length m are constant. Compare your answer with exact frequencies.

11.7 Show that the weak form of the modified Timoshenko beam theory (see Exercise 2.9) over a typical element is

$$0 = \int_{x_a}^{x_b} \left[EI \left(\frac{d\delta\phi}{dx} - \frac{d^2\delta w^b}{dx^2} \right) \left(\frac{d\phi}{dx} - \frac{d^2 w^b}{dx^2} \right) \right.$$
$$+ KGA \left(\delta\phi + \frac{d\delta w^s}{dx} \right) \left(\phi + \frac{dw^s}{dx} \right) - q \left(\delta w^b + \delta w^s \right) \bigg] dx$$
$$- M_1 \delta\phi(x_a) - M_2 \delta\phi(x_b) - V_1 \delta w^s(x_a) - V_2 \delta w^s(x_b)$$
$$- Q_1 \delta w^b(x_a) - Q_2 \left(-\frac{d\delta w^b}{dx} \right)_{x_a} - Q_3 \delta w^b(x_b) - Q_4 \left(-\frac{d\delta w^b}{dx} \right)_{x_b}$$

where M_i, V_i, and Q_i are the generalized nodal forces

$$M_1 \equiv -M(x_a) = \left[EI \left(\frac{d\phi}{dx} - \frac{d^2 w^b}{dx^2} \right) \right]_{x_a}$$

$$M_2 \equiv M(x_b) = \left[-EI \left(\frac{d\phi}{dx} - \frac{d^2 w^b}{dx^2} \right) \right]_{x_b}$$

$$V_1 \equiv -Q_x(x_a) = \left[KGA \left(\phi + \frac{dw^s}{dx} \right) \right]_{x_a}$$

$$V_2 \equiv Q_x(x_b) = \left[-KGA \left(\phi + \frac{dw^s}{dx} \right) \right]_{x_b}$$

$$Q_1 \equiv -\left(\frac{dM_x}{dx} \right)_{x_a} = -Q_x(x_a) = V_1$$

$$Q_3 \equiv \left(\frac{dM_x}{dx} \right)_{x_b} = Q_x(x_b) = V_2$$

$$Q_2 \equiv -M_x(x_a) = M_1, \quad Q_4 \equiv M_x(x_b) = M_2$$

11.8 (*Continuation of Exercise 11.7*) From the weak form of Exercise 11.7 it is clear that ϕ and w^s can be interpolated using the Lagrange interpolation, and the lowest admissible functions are linear. However, the condition that the shear force be element-wise constant for element-wise constant values of EI

$$KGA\left(\phi + \frac{dw^s}{dx}\right) = \text{constant} \quad (a)$$

in turn requires that w^s be quadratic. On the other hand, w^b requires Hermite cubic interpolation. Let (ϕ, w^b, w^s) be interpolated as

$$\phi(x) \approx \sum_{i}^{m} \Phi_i \psi_i^{(1)}, \quad w^s(x) \approx \sum_{i}^{n} W_i^s \psi_i^{(2)}, \quad w^b(x) \approx \sum_{i}^{p} W_i^b \varphi_i \quad (b)$$

where Φ_i, W_i^s, and W_i^b denote the nodal values of ϕ, w^s, and w^b, respectively, $\psi_i^{(1)}$ and $\psi_i^{(2)}$ are linear and quadratic interpolation functions, and φ_i are the Hermite cubic interpolation functions. Substitute the interpolations (b) into the weak form of Exercise 11.7 and obtain the following finite element equations:

$$\begin{bmatrix} [A] & [B] & [C] \\ [B]^T & [D] & [0] \\ [C]^T & [0] & [G] \end{bmatrix} \begin{Bmatrix} \{\Phi\} \\ \{W^b\} \\ \{W^s\} \end{Bmatrix} = \begin{Bmatrix} \{0\} \\ \{F^b\} \\ \{F^s\} \end{Bmatrix} + \begin{Bmatrix} \{M\} \\ \{Q\} \\ \{V\} \end{Bmatrix} \quad (c)$$

Define the cofficients A_{ij}, B_{ij}, and so on in Eq. (c) and evaluate them numerically.

11.9 Use the conditions

$$\Delta_1 \equiv w(x_a) = W_1^b + W_1^s, \quad \Delta_2 \equiv \left(-\frac{dw^b}{dx} + \phi\right)_{x_a} = W_2^b + \Phi_1$$

$$\Delta_3 \equiv w(x_b) = W_3^b + W_2^s, \quad \Delta_4 \equiv \left(-\frac{dw^b}{dx} + \phi\right)_{x_b} = W_4^b + \Phi_2 \quad (a)$$

to reduce the 9×9 system of equations in (c) of Exercise 11.8 to the 4×4 system of equations given in Eq. (11.2.55).

11.10 The principle of total potential energy for axisymmetric bending of annular plates is given by

$$0 = -\int_b^a \left(M_{rr}\frac{d^2\delta w_0}{dr^2} + M_{\theta\theta}\frac{1}{r}\frac{d\delta w_0}{dr}\right) r\,dr - \int_b^a q\delta w_0\, r\,dr \quad (a)$$

where

$$M_{rr} = \int_{-\frac{h}{2}}^{\frac{h}{2}} \sigma_{rr} z \, dz = -\left(D_{11} \frac{d^2 w_0}{dr^2} + D_{12} \frac{1}{r} \frac{dw_0}{dr} \right)$$

$$M_{\theta\theta} = \int_{-\frac{h}{2}}^{\frac{h}{2}} \sigma_{\theta\theta} z \, dz = -\left(D_{12} \frac{d^2 w_0}{dr^2} + D_{22} \frac{1}{r} \frac{dw_0}{dr} \right) \quad (b)$$

For a typical element located between radial distances $r = r_a$ and $r = r_b$, Eq. (a) can be written as

$$0 = \int_{r_a}^{r_b} \left[D_{11} r \frac{d^2 \delta w_0}{dr^2} \frac{d^2 w_0}{dr^2} + D_{12} \left(\frac{d^2 \delta w_0}{dr^2} \frac{dw_0}{dr} + \frac{d\delta w_0}{dr} \frac{d^2 w_0}{dr^2} \right) \right.$$
$$\left. + D_{22} \frac{1}{r} \frac{d\delta w_0}{dr} \frac{dw_0}{dr} - q\delta w_0 \right] dr - Q_1 \delta w_0(r_a) - Q_3 \delta w_0(r_b)$$
$$- Q_2 \left(-\frac{d\delta w_0}{dr} \right)_{r_a} - Q_4 \left(-\frac{d\delta w_0}{dr} \right)_{r_b} \quad (c)$$

Here, (Q_1, Q_3) denote the shear forces and (Q_2, Q_4) denote the bending moments at the nodes:

$$Q_1 = -\left[\frac{d}{dr}(rM_{rr}) - M_{\theta\theta} \right]_{r=r_a}, \quad Q_3 = \left[\frac{d}{dr}(rM_{rr}) - M_{\theta\theta} \right]_{r=r_b}$$
$$Q_2 = -[rM_{rr}]_{r=r_a}, \quad Q_4 = [rM_{rr}]_{r=r_b} \quad (d)$$

The primary variables of the formulation are w_0 and $-dw_0/dr$, which are required to be continuous from element to element. This requires an Hermite cubic interpolation of $w_0(r)$ over an element:

$$w_0(r) = \sum_{j=1}^{4} \Delta_j \phi_j(r) \quad (e)$$

where Δ_1 and Δ_3 are the transverse displacements at $r = r_a$, $r = r_b$, Δ_2 and Δ_4 are the rotations $-dw_0/dr$ at $r = r_a$ and $r = r_b$, and ϕ_i are the Hermite interpolation functions

$$\phi_1(r) = 1 - 3\xi^2 + 2\xi^3$$
$$\phi_2(r) = -h\xi(1-\xi)^2$$
$$\phi_3(r) = 3\xi^2 - 2\xi^3$$
$$\phi_4(r) = -h\xi(\xi^2 - \xi) \quad (f)$$

528   THEORY AND ANALYSIS OF ELASTIC PLATES

and $\xi = \bar{r}/h$ and $\bar{r}$ denotes the element coordinate such that $r = \bar{r} + r_a$. Show that the finite element model is of the form

$$[K^e]\{\Delta^e\} = \{f^e\} + \{Q^e\} \qquad (g)$$

$$K_{ij}^e = \int_{r_a}^{r_B} \left[D_{11} r \frac{d^2\phi_i}{dr^2} \frac{d^2\phi_j}{dr^2} + D_{22} \frac{1}{r} \frac{d\phi_i}{dr} \frac{d\phi_j}{dr} \right.$$
$$\left. + D_{12} \left(\frac{d^2\phi_i}{dr^2} \frac{d\phi_j}{dr} + \frac{d\phi_i}{dr} \frac{d^2\phi_j}{dr^2} \right) \right] dr$$

$$f_i^e = \int_{r_a}^{r_b} \phi_i q \, r dr \qquad (h)$$

11.11 The governing equations of equilibrium of the axisymmetric bending of a circular plate are

$$-\frac{d}{dr}(rM_{rr}) + M_{\theta\theta} = Dr\frac{d}{dr}\left[\frac{1}{r}\frac{d}{dr}\left(r\frac{dw_0}{dr}\right)\right] \qquad (a)$$

The general solution of Eq. (a) in the absence of the distributed load can be obtained as

$$w_0^c(r) = -\frac{1}{4D}\left[c_1 r^2 (\log r - 1) + c_2 r^2 + 4c_3 \log r + 4c_4\right]$$
$$= \hat{c}_1 + \hat{c}_2 r^2 + \hat{c}_3 \log r + \hat{c}_4 r^2 \log r \qquad (b)$$
$$\frac{dw_0^c}{dr} = 2\hat{c}_2 r + \hat{c}_3 \frac{1}{r} + \hat{c}_4 r (1 + 2\log r) \qquad (c)$$

Define the generalized displacements and forces by the definitions

$$u_1 \equiv w_0(r_a), \quad u_2 \equiv \left(-\frac{dw_0}{dr}\right)_{r_a}, \quad u_3 \equiv w_0(r_b), \quad u_4 \equiv \left(-\frac{dw_0}{dr}\right)_{r_b}$$

$$Q_1 \equiv -2\pi(rQ_r)_{r_a}, \quad Q_2 \equiv -2\pi(rM_{rr})_{r_a}$$
$$Q_3 \equiv 2\pi(rQ_r)_{r_b}, \quad Q_4 \equiv 2\pi(rM_{rr})_{r_b} \qquad (d)$$

and show that the element stiffness matrix for the Euler–Bernoulli theory of circular plates is $[K^e]$ with the following coefficients:

$$K_{11}^c = \frac{16\pi D}{\Delta}(r_a^2 - r_b^2), \quad K_{12} = \frac{8r_a \pi D}{\Delta}\left[-r_a^2 + r_b^2(1 - 2\log\beta)\right]$$

FINITE ELEMENT MODELS OF BEAMS AND PLATES   529

$$K_{13} = -\frac{16\pi D}{\Delta}(r_a^2 - r_b^2), \quad K_{14} = \frac{8r_b\pi D}{\Delta}\left[-r_b^2 + r_a^2(1 + 2\log\beta)\right]$$

$$K_{22} = \frac{2\pi D}{\Delta}\left\{2r_a^2 r_b^2\left[-1 + 2\log\beta(2 - \log\beta) + \nu\right.\right.$$
$$\left.\left. + 2\nu\left((\log r_b)^2 + 2\log r_a \log\beta\right) + r_a^4(3 + \nu) + r_b^4(\nu - 1)\right]\right\}$$

$$K_{23} = -\frac{8r_a\pi D}{\Delta}\left[-r_a^2 + r_b^2(1 - 2\log\beta)\right]$$

$$K_{24} = \frac{8r_a r_b \pi D}{\Delta}\left[-r_a^2(1 + \log\beta) + r_b^2(1 - \log\beta)\right]$$

$$K_{33} = \frac{16\pi D}{\Delta}(r_a^2 - r_b^2), \quad K_{34} = -\frac{8r_b\pi D}{\Delta}\left[-r_b^2 + r_a^2(1 + 2\log\beta)\right]$$

$$K_{44} = \frac{8\pi D}{\Delta}\left[r_b^4 - r_a^4 + r_a^2 r_b^2(\nu + 4\log\beta)\right] - K_{22}$$

$$\Delta = \left[4r_a^2 r_b^2(\log\beta)^2 - (r_a^2 - r_b^2)^2\right] \tag{e}$$

11.12 The governing equations of the first-order shear deformation plate theory for the axisymmetric bending of isotropic circular plates are given by

$$-\frac{d}{dr}(rM_{rr}) + M_{\theta\theta} + rQ_r = 0, \quad -\frac{d}{dr}(rQ_r) = rq \tag{a, b}$$

$$M_{rr} = D\left(\frac{d\phi}{dr} + \frac{\nu}{r}\phi\right), \quad M_{\theta\theta} = D\left(\nu\frac{d\phi}{dr} + \frac{1}{r}\phi\right) \tag{c}$$

$$Q_r = GAK\left(\phi + \frac{dw_0}{dr}\right) \tag{d}$$

The general solution of Eqs. (a,b) is

$$w_0(r) = \frac{1}{4D}\left\{\left[\Gamma\log r - r^2(\log r - 1)\right]c_1 - c_2 r^2 - 4c_3 \log r - 4c_4\right\}$$
$$= \left[\hat{c}_1 + \hat{c}_2 r^2 + \hat{c}_3 \log r + \hat{c}_4 r^2 \log r\right] \tag{e}$$

$$\phi(r) = \frac{1}{4D}\left[c_1 r(2\log r - 1) + 2c_2 r + 4\frac{c_3}{r}\right]$$
$$= -2\hat{c}_2 r - \frac{\hat{c}_3}{r} - \hat{c}_4\left[r(1 + 2\log r) + \frac{1}{r}\Gamma\right] \tag{f}$$

where $D = Eh^3/12(1 - \nu^2)$ and $\Gamma = (4D/GAK)$. Then show that the element stiffness matrix is given by

$$[K] = [G][H]^{-1} \tag{g}$$

530   THEORY AND ANALYSIS OF ELASTIC PLATES

$$[H] = \begin{bmatrix} 1 & r_a^2 & \log r_a & r_a^2 \log r_a \\ 0 & -2r_a & -\frac{1}{r_a} & -r_a(1+2\log r_a) - \frac{1}{r_a}\Gamma \\ 1 & r_b^2 & \log r_b & r_b^2 \log r_b \\ 0 & -2r_b & -\frac{1}{r_b} & -r_b(1+2\log r_b) - \frac{1}{r_b}\Gamma \end{bmatrix} \begin{Bmatrix} \hat{c}_1 \\ \hat{c}_2 \\ \hat{c}_3 \\ \hat{c}_4 \end{Bmatrix}$$

$$[G] = 2\pi D \begin{bmatrix} 0 & 0 & 0 & 4 \\ 0 & 2(1+\nu)r_a & -\frac{(1-\nu)}{r_a} & \Theta_a \\ 0 & 0 & 0 & -4 \\ 0 & -2(1+\nu)r_b & \frac{(1-\nu)}{r_b} & -\Theta_b \end{bmatrix}$$

$$\Lambda_\alpha = [2(1+\nu)\log r_\alpha + (3+\nu)]\, r_\alpha, \quad \Theta_\alpha = \left[\Lambda_\alpha - \frac{(1-\nu)}{r_\alpha}\Gamma\right]$$

The generalized displacements and forces are the same as those defined in Eq. (d) of Exercise 11.11.

11.13 Consider a rectangular classical plate theory element of dimension $2a \times 2b$. The displacement w_0 and slopes $\theta_x \equiv \partial w_0/\partial x$ and $\theta_y \equiv \partial w_0/\partial y$ are taken as the nodal degrees of freedom. Suppose that $w_0(x,y)$ is represented by the polynomial

$$\begin{aligned} w_0(x,y) =& c_1 + c_2 x + c_3 y + c_4 x^2 + c_5 xy + c_6 y^2 + c_7 x^3 + c_8 x^2 y \\ & c_9 xy^2 + c_{10} y^3 + c_{11} x^3 y + c_{12} xy^3 \end{aligned}$$

Show that the inter-element continuity of the displacement is guaranteed. Check if the continuity of the normal derivative (for example, θ_x on side $x = 0$) is guaranteed.

11.14 For the plate element in Exercise 11.13, the deflection can be rewritten as

$$w_0(x,y) = \varphi_1(x,y)w_1 + \varphi_2(x,y)\theta_{x1} + \varphi_3(x,y)\theta_{y1} + \cdots + \varphi_{12}(x,y)\theta_{x4}$$

Take the coordinate system at the center of the element such that $-a < x < a$ and $-b < y < b$, and derive the first three interpolation functions using the following interpolation properties:

(i) $\varphi_1 = 1$ at node 1 and zero at all other nodes; $\partial \varphi_1/\partial x = 0$ and $\partial \varphi_1/\partial y = 0$ at all nodes.

(ii) $\varphi_2 = 0$ and $\partial \varphi_2/\partial x = 0$ at all nodes; and $\partial \varphi_2/\partial y = 1$ at node 1, and zero at all other nodes.

(iii) $\varphi_3 = 0$ and $\partial \varphi_3/\partial y = 0$ at all nodes; and $\partial \varphi_2/\partial x = -1$ at node 1, and zero at all other nodes.

References for Additional Reading

1. Reddy, J. N., *An Introduction to the Finite Element Method*, Second Edition, McGraw-Hill, New York (1993).
2. Reddy, J. N., *Energy and Variational Methods in Applied Mechanics*, John Wiley, New York (1984).
3. Reddy, J. N., *Mechanics of Laminated Composite Plates, Theory and Analysis*, CRC Press, Boca Raton, FL (1997).
4. Burnett, D. S., *Finite Element Analysis*, Addison-Wesley, Reading, MA (1987).
5. Zienkiewicz, O. C. and Taylor, R. L., *The Finite Element Method, Vol. 1: Linear Problems*, McGraw-Hill, New York (1989).
6. Hughes, T. J. R., *The Finite Element Method, Linear Static and Dynamic Finite Element Analysis*, Prentice-Hall, Englewood Cliffs, NJ (1987).
7. Bathe, K. J., *Finite Element Procedures*, Prentice-Hall, Englewood Cliffs, NJ (1996).
8. Gere, J. M. and Weaver, Jr., W., *Analysis of Framed Structures*, D. von Nostrand, New York (1965).
9. Przemieniecki, J. S., *Theory of Matrix Structural Analysis*, McGraw-Hill, New York (1968).
10. Severn, R. T., "Inclusion of Shear Deflection in the Stiffness Matrix for a Beam Element", *Journal of Strain Analysis*, **5**, 239–241 (1970).
11. Nickell, R. E. and Secor, G.A., "Convergence of Consistently Derived Timoshenko Beam Finite Elements", *International Journal for Numerical Methods in Engineering*, **5**, 243–253 (1972).
12. Reddy, J. N., "On Locking-Free Shear Deformable Beam Finite Elements", *Computer Methods in Appl. Mech. and Engineering*, **149**, 113–132 (1997).
13. Hrabok, M. M. and Hrudey, T. M., "A Review and Catalog of Plate Bending Finite Elements", *Computers and Structures* **19**(3), 479–495 (1984).
14. Bazeley, G. P., Cheung, Y. K., Irons, B. M., and Zienkiewicz, O. C., "Triangular Elements in Bending – Conforming and Non-Conforming Solutions", *Proceedings of the Conference on Matrix Methods in Structural Mechanics*, Air Force Institute of Technology, Wright-Patterson Air Force Base, Ohio, 547–576 (1965).

15. Clough, R. W. and Tocher, J. L., "Finite Element Stiffness Matrices for Analysis of Plates in Bending", *Proceedings of the Conference on Matrix Methods in Structural Mechanics,* Air Force Institute of Technology, Wright–Patterson Air Force Base, Ohio, 515–545 (1965).

16. Melosh, R. J., "Basis of Derivation of Matrices for the Direct Stiffness Method," *AIAA Journal,* **1**, 1631–1637 (1963).

17. Zienkiewicz, O. C. and Cheung, Y. K., "The Finite Element Method for Analysis of Elastic Isotropic and Orthotropic Slabs," *Proceeding of the Institute of Civil Engineers,* London, **28**, 471–488 (1964).

18. Bogner, F. K., Fox, R. L., and Schmidt, L. A., Jr., "The Generation of Interelement-Compatible Stiffness and Mass Matrices by the Use of Interpolation Formulas", *Proceedings of the Conference on Matrix Methods in Structural Mechanics,* Air Force Institute of Technology, Wright–Patterson Air Force Base, Ohio, 397–443 (1965).

19. Fraeijis de Veubeke, B., "A Conforming Finite Element for Plate Bending", *International Journal of Solids and Structures,* **4**(1), 95–108 (1968).

20. Irons, B. M., "A Conforming Quartic Triangular Element for Plate Bending", *Int. J. Numer. Meth. Engng.,* **1**, 29–45 (1969).

21. Reddy, J. N., "Simple Finite Elements with Relaxed Continuity for Non-Linear Analysis of Plates", *Proceedings of the Third International Conference in Australia on Finite Element Methods,* University of New South Wales, Sydney, July 2–6, 1979.

22. Reddy, J. N., "A Penalty Plate-Bending Element for the Analysis of Laminated Anisotropic Composite Plates", *International Journal for Numerical Methods in Engineering,* **15**(8), 1187–1206 (1980).

23. Huang, H. C. and Hinton, E., "A Nine-Node Lagrangian Plate Element with Enhanced Shear Interpolation", *Engineering Computations,* **1**, 369–379 (1984).

24. Averill, R. C. and Reddy, J. N., "Behavior of Plate Elements Based on the First-Order Shear Deformation Theory", *Engineering Computations,* **7**, 57–74 (1990).

25. Zienkiewicz, O. C., Too, J. J. M., and Taylor, R. L., "Reduced Integration Technique in General Analysis of Plates and Shells", *International Journal for Numerical Methods in Engineering,* **3**, 275–290 (1971).

26. Hughes, T. J. R., Cohen, M., and Haroun, M., "Reduced and Selective Integration Techniques in the Finite Element Analysis of Plates", *Nuclear Engineering and Design,* **46**, 203–222 (1981).

27. Belytschko, T., Tsay, C. S., and Liu, W. K., "Stabilization Matrix for the Bilinear Mindlin Plate Element", *Computer Methods in Applied Mechanics and Engineering*, **29**, 313–327 (1981).
28. Bathe, K. J. and Dvorkin, E. N., "A Four-Node Plate Bending Element Based on Mindlin/Reissner Plate Theory and Mixed Interpolation", *International Journal for Numerical Methods in Engineering*, **21**, 367–383 (1985).
29. Barlow, J., "Optimal Stress Location in Finite Element Models", *International Journal for Numerical Methods in Engineering*, **10**, 243–251 (1976).
30. Barlow, J., "More on Optimal Stress Points — Reduced Integration Element Distortions and Error Estimation", *International Journal for Numerical Methods in Engineering*, **28**, 1486–1504 (1989).
31. Reddy, J. N. and Chao, W. C., "A Comparison of Closed-Form and Finite Element Solutions of Thick Laminated Anisotropic Rectangular Plates", *Nuclear Engineering and Design*, **64**, 153–167 (1981).
32. Pagano, N. J., "Exact Solutions for Composite Laminates in Cylindrical Bending", *Journal of Composite Materials*, **3**, 398–411 (1967).
33. Pagano, N. J., "Exact Solutions for Rectangular Bidirectional Composites and Sandwich Plates", *Journal of Composite Materials*, **4**, 20–34 (1970).

SUBJECT INDEX

Adjoint, 14
Adjunct, see Adjoint
Admissible configurations, 48
Admissible variations, 49
Alternating symbol, 5
Analytical (exact) solutions:
   of beams, 82,90,96,489
   of plate strips, 145,148,154,162
   see Navier's and Lévy solutions
Anisotropic body, 34
Anisotropic plates, 127
Annular plates, 182
   boundary conditions, 195
   clamped, 215, 224
   simply supported, 207
Approximation functions, 74
   for bars, 80,85
   for beams, 88,91,95
   for circular plates, 199,202
   221–223,236
   for rectangular plates, 301,304,
   336,340–344,413–418
Area coordinates, 498,509
Asymmetrical bending, 226,229,230
Axisymmetric bending, 192,194,526

Basis vectors, 3
   Cartesian, 3
   cylindrical, 10
   orthonormal, 3,12
Beam finite elements:
   Euler–Bernoulli, 470
   Timoshenko, 484
Beam theory:
   Euler–Bernoulli, 25
   Third-order, 99
   Timoshenko, 67

Bending rigidity, 135
Bending stiffnesses, 128,135
Bessel functions, 239
Boundary conditions:
   annular plates, 195
   circular plates, 188,195
   clamped (CLPT), 125,140,147,
   160
   CLPT, 121–124,140
   elastically supported, 160
   essential, 48,59,69,124
   force, see natural
   free, 140,147,159
   geometric, see essential
   homogeneous form of, 48,75
   natural, 59,69,124
   simply supported(CLPT), 125,
   140,147,159
Buckling load:
   biaxial compression, 355
   critical, 104,152,155,355
   in-plane shear, 382, 385
   plate strips, 155
   rectangular plates, 353
   uniaxial compression, 358

Cartesian basis, 3,10,12
Castigliano's theorems, 102,103
Cauchy stress formula, 30
Cauchy stress tensor, 30
Characteristic polynomials, 160,
   163,166
Circular inclusion, 218
Circular plates:
   asymmetric bending of, 226
   axisymmetric bending of, 192
   classical theory of, 184–191

equations of motion of, 186
first-order theory of, 529
on elastic foundation, 220
with clamped edges, 212–220
under thermal loads, 223
Classical plate theory (CPT):
circular plates, 184–191
displacement field, 110
equations of motion, 120
finite element models, 494
strains, 112
Collocation method, 94
Compatibility conditions, 26
Completeness, 75
Compliance coefficients, 36
Concentrated forces at corners, 261
Conforming element, 497
Consistent interpolation, 488
Constitutive equations:
for anisotropic materials, 34
for orthotropic materials, 35
thermoelastic, 38,42,99,126,189
Contracted notation, 35
Convergence, 74
Coordinate functions:
see: approximation functions
Coordinates:
Cartesian, 3
cylindrical, 10
material, 21
polar, 180
rectangular, 3
spherical, 12
transformation of, 179
Cramer's rule, 15
Critical buckling load, 104,152,155
Curl operator, 8,44
Cylindrical bending, 138, 143
Cylindrical coordinates, 10

Deformation, 21
Del operator, 7,11,13,44
Dirac delta function, 94,104,257,306
Directional derivative, 7
Displacement field of:
circular plates, 184
CLPT, 110
Euler–Bernoulli beam, 25
FSDT, 426
Third-order beam, 99
Timoshenko beam, 67,100
TSTD, 446
Divergence, 8,11
Divergence theorem, 61
Dummy index, 4
Dyads:
see: second-order tensors
Dynamic analysis, 65,167,239,391
also see: transient response

Effective shear force, 124,187,261,
299,324,334,397,399,409
Elastic coefficients, 35
for plane stress, 37
Elastic foundation, 55,64,153,116,
182,194,220,289
Engineering constants:
see: material constants
Epsilon–delta identity, 5
Equations of equilibrium:
3-D elasticity, 31–33
circular plates, 182,189
cylindrical coordinates, 45
Euler–Bernoulli beam theory, 65
Timoshenko beam theory, 101
Equations of motion:
3-D elasticity, 31–33,141
circular plates, 186–188
CLPT, 120,137,139

SUBJECT INDEX 537

cylindrical bending, 138,144
third-order beam theory, 100
Timoshenko beam theory, 69
Euler–Bernoulli hypothesis, 25
Euler–Lagrange equations, 58–60, 83,120,188

Fiber, 134
Finite element method, 90,469
Finite element model:
    displacement, 497
    of beams, 475–492
    of plates (CPT), 494
    of plates (FSDT), 513
First-order shear deformation theory (FSDT):
    bending solutions of, 436
    buckling analysis of, 440
    displacement field of, 426
    equations of motion of, 431
    finite element models, 513
    Navier's solution, 434
    plate constitutive equation, 432
    shear correction factors, 429
    transverse force resultants, 429
    vibration analysis of, 443
Flexural rigidity:
    see: bending rigidity
Force resultants, 115,186,189
    thermal, 129,190
Four point bending, 174
Fourier coefficients, 74,168
Free edge, 125
Free vibration:
    see: natural vibration
Frequency equation:
    see: characteristic equation
    of circular plates, 241
    of plate strips, 166
    of rectangular plates, 398,400–404

Fully discretized model, 501
Functional, 57,60
Fundamental frequency, 164,393
Fundamental lemma, 58

Galerkin method, 92–94
Gauss quadrature, 515
Gradient operator:
Gradient vector, 7
    see: del operator
Graphite-epoxy material, 131
Green–Lagrange strain tensor, 22

Hamilton's principle, 65–69,83
Heaviside step function, 104
Hermite cubic interpolation functions, 475
Hooke's law, generalized, 34
Hyperelastic materials, 35

Infinitesimal strain tensor, 24
Interdependent interpolation, 489
Internal virtual work, 51
Internal work, 52
Interpolation functions:
    Hermite, 469,475,491,497–501
    Lagrange, 486,508,510–512
Invariant, 2
Isotropic materials, 36
Isotropic plates, 290

Kinematics, 21
Kinetic energy, 65,83
Kinetics, 21
Kirchhoff assumptions, 109
Kirchhoff free-edge condition, 124, 189
Kirchhoff hypothesis, 108
Kronecker delta, 5

Lagrange element:
    interpolation functions of, 486, 508,510–512
    linear, 511
    quadratic, 511
    rectangular, 511
    triangular, 509
Lagrangian, 66
Laminate stiffnesses, 129
Laminated plates, 129
Laplace transform, 172
Laplacian, 11,13,44,180
Least squares method, 94
Lévy solution, 278,294,364
Linear functional, 57
Linear independence, 75
Logarithmic singularity, 220

Marcus moment, 191
Material constants, 36,39,131
Material stiffnesses, 126
Matrices, 13
    determinant of, 14,43
    orthogonal, 18
Matrix material, 134
Method of superposition, 330
Micromechanics, 134
Moment resultants, 115,186,189
    thermal, 129,139,190

Natural coordinates, 510
Natural vibration of:
    bars, 84
    beams, 161
    circular plates, 239
    CPT, 391
    FSDT, 443
    TSDT, 462
Navier's solution, 168
    bending, 256,436,457
    buckling, 354,440,461
    vibration, 393,443,462
Newmark family of
    approximations, 477–479
Nodal circles, 241
Nodal diameter, 240
Nodal line, 240
Nonconforming element, 497,499
Norm, 74
Normal derivative, 61

Orthonormal basis, 3
Orthotropic material, 35,133

Particular solution, 75,157,226, 283,295,328
Permutation symbol, 5
Petrov–Galerkin method, 94
Physical vector, 2
Plane strain, 143
Plane stress-reduced stiffnesses, 37,126
Plane stress state, 37
Plate stiffnesses, 129,135, 189,432,452
Plate strip, 138
    buckling, 157
    on elastic foundation, 155
Plates:
    annular, 195,207,215,526
    circular, 179,528
    classical theory of, 107,494
    first-order theory of, 425,507
    third-order theory of, 446
    with distributed loads, 269
    with distributed moments, 291
Poisson's equation, 180
Polar coordinates, 180
Potential energy functional, 65,72, 79,98,104,175,235

SUBJECT INDEX

Primary variables, 59,69,120,473
Principle of the minimum
   total potential energy, 71,170

Quadratic functional, 57

Rayleigh–Ritz approximation:
   see: Rayleigh–Ritz solution
Rayleigh–Ritz method, 73–78,151
   3-D elasticity, 31–33
Rayleigh–Ritz solution, 82,86,169,
   175,199,202–204,206,211–215,
   220–223,236,245,300,334,377,408
Reactions at corners of plates, 268
Reactions at edges of plates, 261
Rectangular plates, 253
   under hydrostatic load, 258,273
   under line load, 265
   under point load, 265,273
   under sinusoidal load, 265,273
   under uniform load, 258,273
   with clamped edges, 313,340,372,
     413
   with distributed edge moments, 291
   with patch loading, 274
   with simply supported edges, 253,
     354,392,411,434,454
   with thermal loads, 275
   with various boundary conditions,
     309,340,412
Reduced integration, 487
Rotary inertia, 120,166
Rotatory inertia, 165
   also see: rotary inertia

Secondary variable, 59,69,120
Semidiscrete model, 476
Serendipity family, 512
Shear correction coefficient, 68,429
Shear correction factor:
   see: shear correction coefficient
Shear deformation, 426
Shear force resultants, 187,429
Shear locking, 485, 487,515
Spatial discretization, 473
Stability:
   see: buckling
State-space approach, 281,351,
   388,465
Strain components:
   in cylindrical coordinates, 23,45
   in Cartesian coordinates, 23
   infinitesimal, 24
   nonlinear, 23–25,111,185
   of CLPT, 113
   of circular plates, 185
   of Euler–Bernoulli beams, 25
Strain energy density, 35,70,83
Stress components:
   Cartesian, 29
   cylindrical, 32
Stress tensor, 29
Summation convention, 4
Symmetric laminates, 129

Tensor:
   alternating, 5
   second order, 4,20
   transformation of, 18
Thermal coefficients of expansion
   42,127,134
Thermal stresses, 262,276
Thermoelastic constitutive relations,
   38,126
Third-order beam theory, 99
Third-order plate theory:
   bending, 457
   buckling, 461
   displacement field of, 446
   equations of motion of, 451

natural vibration, 462
resultants of, 451
Time approximations, 477,478
Timoshenko beam theory, 484
Total potential energy:
see: potential energy functional
Transformation of:
material stiffness, 42,127
strain components, 41
stress components, 40
tensors, 18–20
thermal coefficients, 42,127
vectors, 16
Transient response, 86,95,167,171,
392,419,477,502
Transverse shear deformation, 438
True stress, 30

Unit dummy force method, 100

Variational operator, 56
Vector differential, 8
Vector gradient, 7
Vector transformations, 16
Vectors:
cross product of, 5
curl of, 8
divergence of, 8
dot product of, 5
stress, 28

Vibration:
of beams, 415
of circular plates, 238
of rectangular plates, 416
Virtual complementary
strain energy, 54
Virtual displacements, 47–50
principle of, 47,62,88,107,
170,185,198
Virtual forces, 49–52
Virtual strains, 52,63,67,118,186,
428, 449
Virtual strain energy, 53
Virtual work, 51–55
external, 51
internal, 52
Voigt-Kelvin notation, 35
von Kármán strains, 112,186,466

Weak form, 78
classical plate theory, 117,186,
198,233,300,334,374,377,386,
410,494
Euler–Bernoulli beam theory, 88,
151,170
first-order plate theory, 429,507
third-order plate theory, 449
Timoshenko beam theory, 69,71
Weight functions, 93
Weighted-residual method, 93,95